Molecular Electronics

An Introduction to Theory and Experiment

World Scientific Series in Nanoscience and Nanotechnology

Series Editor: Mark Reed *(Yale University)*

Vol. 1 Molecular Electronics: An Introduciton to Theory and Experiment
*Juan Carlos Cuevas (Universidad Autónoma de Madrid, Spain) &
Elke Scheer (Universität Konstanz, Germany)*

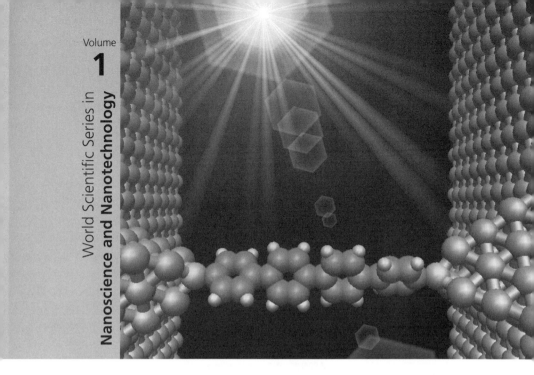

Volume
1

World Scientific Series in
Nanoscience and Nanotechnology

Molecular Electronics
An Introduction to Theory and Experiment

Juan Carlos Cuevas
Universidad Autónoma de Madrid, Spain
Elke Scheer
Universität Konstanz, Germany

 **World Scientific**

NEW JERSEY · LONDON · SINGAPORE · BEIJING · SHANGHAI · HONG KONG · TAIPEI · CHENNAI

Published by

World Scientific Publishing Co. Pte. Ltd.

5 Toh Tuck Link, Singapore 596224

USA office: 27 Warren Street, Suite 401-402, Hackensack, NJ 07601

UK office: 57 Shelton Street, Covent Garden, London WC2H 9HE

British Library Cataloguing-in-Publication Data
A catalogue record for this book is available from the British Library.

World Scientific Series in Nanoscience and Nanotechnology — Vol. 1
MOLECULAR ELECTRONICS
An Introduction to Theory and Experiment

Copyright © 2010 by World Scientific Publishing Co. Pte. Ltd.

ISBN-13 978-981-4282-58-1
ISBN-10 981-4282-58-8

Printed in Singapore by World Scientific Printers.

To our families

Preface

The trend in the miniaturization of electronic devices has naturally led to the question of whether or not it is possible to use single molecules as active elements in nanocircuits for a variety of applications. The recent developments in nanofabrication techniques have made possible the old dream of contacting individual molecules and exploring their electronic transport properties. Moreover, it has been shown that molecules can indeed mimic the behavior of some of today's microelectronic components, and even strategies to interconnect molecular devices have already been developed. These achievements have given rise to what is nowadays known as *Molecular Electronics*. There are still many problems and challenges to be faced to make this novel electronics a viable technology, but the exploration of molecular-scale circuits has already led to the discovery of many fundamental effects. In this sense, molecular electronics has become a new interdisciplinary field of science, in which knowledge from traditional disciplines like physics, chemistry, engineering and biology is combined to understand the electrical and thermal conduction at the molecular scale.

This book provides a comprehensive overview of the rapidly developing field of molecular electronics. It focuses on our present understanding of the electrical conduction in single-molecule circuits and presents a thorough introduction to the experimental techniques and the theoretical concepts. To be precise, our goal in this monograph is two-fold. On the one hand, we want to provide a true textbook for advanced undergraduate and graduate students both in physics and chemistry who are interested in the field of molecular electronics or nanoelectronics in general. Our idea is to take a student with a good background in quantum mechanics all the way to be able to follow the specialized literature in molecular electronics or to start working in this field. On the other hand, we also want provide a thorough review of the recent activities in molecular electronics from which newcomers and specialists in the field can benefit.

Bearing these goals in mind, this book has been written in a self-contained and unified way. It contains four parts that can be read independently. In the first two ones we review the basic experimental techniques and the main theoretical concepts concerning the electronic transport in atomic-scale junctions. These two parts are meant to be textbook material for an advanced course in molecular electronics. In particular, we have included a collection of exercises at the end of most chapters, which in many cases are motivated by recent experiments in the field. On the other hand, Part 3 contains two chapters in which we describe at an introductory level the physics of metallic atomic-size contacts and we also point out some of the remaining challenges and open problems in this context. Finally, Part 4 is devoted to the electrical and thermal transport in molecular circuits, with special emphasis on single-molecule junctions. Here, we do not only review the recent activities in the field of molecular electronics, but we also introduce the addressed topics at a basic level. In this sense, we have often included unpublished material and additional exercises to help the reader to gain a deeper insight into the fundamental concepts involved in the field of molecular electronics.[1]

We have tried to cover in this monograph as many aspects of molecular electronics as possible, but obviously the selection is limited for space reasons and it reflects unavoidably our own research interests. We also want to apologize with those authors that feel that their contribution was not properly highlighted in the review part of this monograph, but it is by now impossible to include all the huge amount of work done in this field. Finally, we just hope to have achieved, at least partially, the goal that truly motivated the writing of this book, namely the sincere will to provide a useful book for the new generation of researchers that should consolidate molecular electronics as a solid pillar of the emerging nanoscience.

---

[1]See section 1.3 for a more detailed description of the structure and scope of the book.

Acknowledgments

It would not have been possible to write the book without the help of many coworkers and colleagues. First of all, we want to thank Edith Goldberg for encouraging one of us (JCC) to give a postgraduate course on molecular electronics in the fall of 2008 in Santa Fe (Argentina). The excellent students who attended that course demonstrated that, after a 50-hours course and without any previous knowledge about this field, one can master the basic concepts and techniques that now form the body of this monograph. This fact provided the final boost that we needed to collect all our notes and turn them into this book.

Similarly, for the experimental point of view of this book, the students in the graduate course at Konstanz served as test candidates. Some of them even got contaminated by this exciting field and went on asking questions what finally resulted in contributions to this book. Very valuable input came from my colleague Artur Erbe who was the real expert in molecular electronics in our Department until he left to Dresden.

We also want to express our gratitude to Alvaro Martín Rodero, who not only introduced one of us (JCC) to the exciting field of nanoelectronics, but also contributed decisively to this manuscript with his personal notes, which are the basis of several chapters of the theoretical background. The same holds for Hilbert von Löhneysen and Cristián Urbina who sent the other one of us (ES) to perform experiments with nanoelectronic circuits.

We would especially like to thank our coworkers Fabian Pauly, Janne K. Viljas, Michael Häfner, Sören Wohlthat, Stefan Bilan, Linda A. Zotti, Cécile Bacca, Stefan Bächle, Tobias Böhler, Uta Eberlein, Stefan Egle, Daniel Guhr, Ning Kang, Thomas Kirchner, Christian Kreuter, Shou-Peng Liu, Youngsang Kim, Hans-Fridtjof Pernau, Olivier Schecker, Christian Schirm, Dima Sysoiev, Simon Verleger, and Reimar Waitz. They have contributed to this manuscript with many results, special figures and very important suggestions and critical comments about the text.

Thanks go also sincerely to our colleagues who have read different parts of the manuscript and have provided helpful comments: Douglas Natelson, Abraham Nitzan, Wilson Ho, Latha Venkataraman, and Arunava Majumdar.

This monograph reflects our view of this field, which has emerged thanks to the collaboration and exchange of ideas with many colleagues over the years. So in this respect, we want to thank Alfredo Levy Yeyati, Gerd Schön, Jan Heurich, Wolfgang Wenzel, Jan M. van Ruitenbeek, Nicolás Agraït, Gabino Rubio, Roel Smit, Oren Tal, Markus Dreher, Peter Nielaba, Christoph Sürgers, Maya Lukas, Christoph Strunk, Sophie Guéron, Richard Berndt, Paul Leiderer, Wolfgang Belzig, Marcel Mayor, Thomas Huhn, Andreas Marx, Ulrich Steiner, and Ulrich Groth.

We also want acknowledge the contribution of all the authors who have kindly granted us the permission to reprint their work in this monograph.

Finally, I (JCC) want to thank my parents and brothers for being always by my side. I also want to thank Ana for being so patient and share my time with this book for too many nights and weekends. ES thanks her family for continuous support and reminding me steadily of what is really important in life.

Contents

PART 1

Brief history of the field and experimental techniques

Chapter 1

The birth of molecular electronics

How does the electrical current flow through a single molecule? Can a molecule mimic the behavior of an ordinary microelectronics component or maybe provide a new electronic functionality? How can a single molecule be addressed and incorporated into an electrical circuit? How to interconnect molecular devices and integrate them into complex architectures? These questions and related ones are by no means new and, as we shall see later in this chapter, they were already posed many decades ago. The difference is that we are now in position to at least address them in the usual scientific manner, i.e. by providing quantitative experimental and theoretical results. The advances in the last two or three decades, both in nanofabrication techniques and in the quantum theory of electronic transport, allow us now to explore and to understand the basic properties of rudimentary electrical circuits in which molecules are used as basic building blocks. It is worth stressing right from the start that we do not yet have definitive answers for the questions posed above. However, a tremendous progress has been made in recent years and some concepts and techniques have already been firmly established. In this sense, one of main goals of this book is to review such progress, but more importantly, this monograph is intended to provide a solid basis for the new generation of researchers that should take the field of molecular electronics to the next level.

Molecular electronics, as used in this book, is defined as the field of science that investigates the electronic and thermal transport properties of circuits in which individual molecules (or an assembly of them) are used as basic building blocks.[1] Obviously, some of the feature dimensions of such

---

[1] Molecular electronics, in the sense used here, should not be confused with organic electronics, the field in which molecular materials are investigated as possible constituents of a variety of macroscopic electronic devices.

3

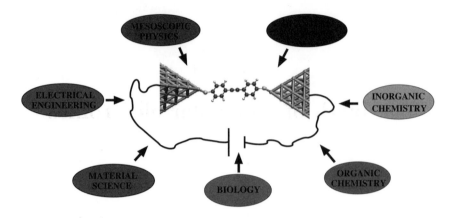

Fig. 1.1   Molecular electronics: An interdisciplinary field.

molecular circuits are of the order of nanometers (or even less) and therefore, molecular electronics should be viewed as a subfield of nanoscience or nanotechnology in which traditional disciplines like physics, chemistry, material science, electrical engineering and biology play a fundamental role (see Fig. 1.1). Molecular electronics, in the sense of a potential technology, is based on the bottom-up approach where the idea is to assemble elementary pieces to form more complex structures, as opposed to the top-down approach where the idea is to shrink macroscopic systems and components. Molecular electronics has emerged from the constant quest for new technologies that could complement the silicon-based electronics, which in the meantime it has become a true nanotechnology. It seems very unlikely that molecular electronics will ever replace the silicon-based electronics, but there are good reasons to believe that it can complement it by providing, for instance, novel functionalities out of the scope of traditional solid state devices. More importantly, molecular electronics has become in recent years a true field of science where many basic questions and quantum phenomena are being investigated. In this sense, the importance of molecular electronics is unquestionable and we are convinced that different traditional disciplines will benefit from advances in this new field.

In the rest of this introductory chapter, we shall first try to answer the questions of why it is worth pursuing molecular electronics research and why it is interesting to work in a field like this. Then, in section 1.2 we shall briefly review the complex history of this field to set the stage for this book. Finally, in section 1.3 we shall clearly define the scope of this monograph and explain its structure.

1.1 Why molecular electronics?

Every researcher is sooner or later confronted with natural questions like "why do you work in your field?" or "what is your research good for?" Of course, the answers are always personal, but in the case of molecular electronics they also depend on whether one's interests are closer to fundamental science or to technological applications. From the point of view of basic science, molecular electronics offers, for instance, the possibility to investigate electronic and thermal conduction at the smallest imaginable scale, where the physics is completely dominated by quantum mechanical effects. The small feature dimensions of molecular circuits together with the great variety of electrical, mechanical and optical properties of molecules can give rise to countless new physical phenomena. Molecular junctions are also ideal systems where to investigate and shed new light into the fundamental electron transfer mechanisms that play a key role both in chemistry and biology. These reasons and many others make molecular electronics a very attractive field of basic research. Moreover, we should never forget that the history of science proves that the exploration of new territories and the subsequent discovery of novel phenomena often lead to unexpected technological applications. History also teaches us that there is no technology without basic understanding and thus, the future of molecular electronics as an emerging technology depends on our ability to understand the fundamental mechanisms that govern the electronic conduction at the molecular scale.

From a technological point of view, there are also good reasons to investigate the use of molecules as electronically active elements for a variety of applications. In comparison with the silicon-based technology, which is already a nanotechnology in the sense that the structure sizes are in the range of nanometers,[2] molecular electronics could in principle offer the following major advantages [2]:

- Size. The reduce size of small molecules (between 1 and 10 nm) could lead to a higher packing density of devices with the subsequent advantages in cost, efficiency, and power dissipation.

---

[2]The next generation of transistors for advanced microprocessors will have gate lengths of 22 nm and a SiO_2 gate oxide thickness of less than 1.2 nm [1].

- Speed. Although most molecules are poorly conductive, good molecular wires could reduce the transit time of typical transistors ($\sim 10^{-14}$ s), reducing so the time needed for an operation.
- Assembly and recognition. One can exploit specific intermolecular interactions to form structures by nanoscale self-assembly. Molecular recognition can be used to modify electronic behavior, providing both switching and sensing capabilities on the single-molecule scale.
- New functionalities. Special properties of molecules, like the existence of distinct stable geometric structures or isomers, could lead to new electronic functions that are not possible to implement in conventional solid state devices.
- Synthetic tailorability. By choice of composition and geometry, one can extensively vary a molecule's transport, binding, optical, and structural properties. The tools of molecular synthesis are highly developed.

Molecules have also obvious disadvantages such as instabilities at high temperatures. Moreover, the fabrication of reliable molecular junctions requires sometimes to control matter at an unprecedented level, which can be not only difficult, but also slow and costly. Anyway, the advantages described above are sufficient to motivate the exploration of a molecule-based electronics.

1.2 A brief history of molecular electronics

It is always difficult to trace back the history of an emerging field and to summarize it in a few pages. Anyway, even at the risk of being unfair leaving out some important contributors, we find necessary to say a few words about the history of molecular electronics as a tribute to those visionary scientists that made possible that we are now working in this fascinating field. Our brief account here is partially based on a delightful (non-scientific) article by Choi and Mody [3], which reviews the history of molecular electronics paying special attention to its social aspects.

We start this historical review in 1950's, after the revolution in electronics due to the invention of the transistor and the subsequent introduction of integrated circuits. In that context and in view of the difficulties to radically miniaturize the existent electronic components, Arthur von Hippel, a German physicist working at the MIT, formulated in 1956 the basis of a bottom-up approach that he called *molecular engineering* [4]. He argued:

> Instead of taking prefabricated materials and trying to devise engineering applications consistent with their macroscopic properties, one builds materials from their atoms and molecules for the purpose at hand ...

The concept of molecular engineering introduced by von Hippel [5] led to the first notion of "molecular electronics", which crystallized in a collaboration between the company Westinghouse and the US Air Force at the end of the 1950's. Westinghouse had begun a program to implement von Hippel's ideas and it applied for the financial support of the US Air Force, which at that time was receptive to new ideas and alternatives to the recently introduced integrated circuits. The Air Force organized a conference on "Molecular Electronics" and invited scientists and engineers from military and private research labs. In this conference, colonel C.H. Lewis, director of Electronics at the Air Research and Development Command, expressed the need for a *breakthrough* in electronics in the following way:

> Instead of taking known materials which will perform explicit electronic functions, and reducing them in size, we should build materials which due to their inherent molecular structure will exhibit certain electronic property phenomena. We should synthesize, that is, tailor materials with predetermined electronic characteristic. Once we can correlate electronic property phenomena with the chemical, physical, structural, and molecular properties of matter, we should be able to tailor materials with predetermined characteristics. We could design and create materials to perform desired functions. Inherent dependability might eventually result. We call this more exact process of constructing materials with predetermined electrical characteristics MOLECULAR ELECTRONICS.

This is probably the first time that the term molecular electronics was used publicly, although it originally referred to a new strategy for the fabrication of electronic components, and it had yet little to do with the vision of using individual molecules as electronically active elements. Fig. 1.2 summarizes the vision of colonel Lewis, where molecular electronics should constitute be the next breakthrough in electronics, although it was not yet clear what molecular electronics was supposed to mean.

The collaboration between Westinghouse and the US Air Force, which started after the mentioned conference, lasted a few years and certain

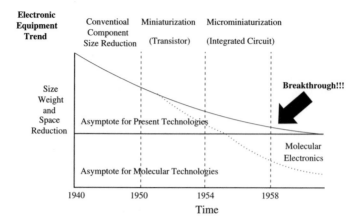

Fig. 1.2   Graph presented by colonel Lewis of the US Air Force in the first conference on molecular electronics held in November 1958. Here, one can see the trend in the miniaturization of the electronic components during the 1940's and 1950's. According to Lewis, molecular electronics should have constituted the next breakthrough in electronics by the end of the 1950's. Adapted from [3].

progress was indeed made in the development of new fabrication strategies. However, these initiatives were not able to compete with the steady miniaturization of the semiconductor-based electronic devices and they were soon abandoned.

From a more scientific point of view, one can consider that molecular electronics, as we understand it today, started at the end of the 1960's and the beginning of 1970's. At that time, different groups started to investigate experimentally the electronic transport through molecular monolayers. For instance, Hans Kuhn, a Swiss chemist working at the University of Göttingen, and his coworkers studied at that time new ways of fabricating the so-called Langmuir-Blodgett films.[3] They were able to not only master the fabrication of these molecular films, but also to sandwich them between metal electrodes and to measure the electrical conductivity of the resulting junctions. In Fig. 1.3 we reproduce the experimental results of Ref. [6] for the low-bias conductivity of $Al/S(n)/Hg$ junctions, where $S(n)$ stands for a monolayer of Cd salt of fatty acid $CH_3(CH_2)_{n-2}COOH$ of different chain lengths. There one can see the exponential decay of the conductivity with the length of the molecules, which is still a very important issue in today's

---

[3] A Langmuir-Blodgett film contains one or more monolayers of an organic material, deposited from the surface of a liquid onto a solid by immersing the solid substrate into the liquid. A monolayer is adsorbed homogeneously with each immersion or emersion step, thus films with very accurate thickness can be formed.

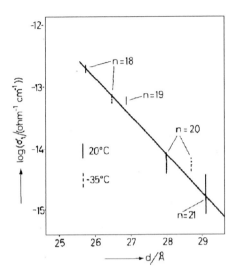

Fig. 1.3 Measurements of the low-bias tunneling conductivity (σ_t) vs. the distance (d) between the electrodes in $Al/S(n)/Hg$ junctions. Here, $S(n)$ stands for monolayers of Cd salt of fatty acid $CH_3(CH_2)_{n-2}COOH$ with different lengths (n ranges between 18 and 21). The solid line is a linear fit to the experiment data. The measurements were performed at two different temperatures: 20 and -35 °C. Reprinted with permission from [6]. Copyright 1971, American Institute of Physics.

molecular electronics (see Chapter 13). This type of experimental results can be considered as the starting point of molecular electronics as a modern field of science.

The idea of molecular electronics reappeared in the States at the beginning of the 1970's at IBM and thanks to the enthusiasm of Ari Aviram, a synthetic chemist. Aviram was working at that time on charge-transfer salts, which had recently been discovered to be reasonably good conductors in their solid form. Although Aviram's task at IBM was to synthesize new types of charge-transfer salts, he started working on the theory of electron transfer through single organic molecules in collaboration with Mark Ratner,[4] at that time at New York University. In the course of their investigations, Aviram and Ratner saw a clear analogy between charge-transfer salts like TTF-TCNQ (tetrathiafulvalene-tetracyanoquinodimethane), with a functional unit (TTF) rich in electrons and another unit (TCNQ) poor in electrons, and traditional semiconductor diodes. In 1974 they published a now-famous paper on "molecular rectifiers" [8] in which they described

---

[4]Indeed Ratner was officially Aviram's thesis advisor during that time.

how a modified charge-transfer salt could operate as a traditional diode in an electrical circuit. This is probably the first proposal to use a single molecule as an electronic component, which is something that lies at the heart of the modern molecular electronics. Aviram and Ratner's idea was considered during a long time a theoretical curiosity that could not be tested experimentally and in this sense, it did not have much impact in the scientific community at that time.

In the late 1970's and early 1980's other scientists started to work on ideas similar to Aviram-Ratner's unimolecular concept. Let us mention for instance the name of Forrest Carter, a chemist at the Naval Research Laboratory, who was certainly influenced by Feynman's (1960) famous "Room at the Bottom" speech [9]. Carter introduced concepts such as molecular computing or cellular automata, where the essence was to use individual molecules as the ultimate electronic components or as elementary units where to store bits of information in a hypothetical molecular computer. These ideas were to a large extend purely theoretical and they were no supported by real experiments. However, Carter was able to nucleate a first molecular electronics community around him and, in particular, the organization of a series of conferences on molecular electronics in the 1980's played an important role in the history of this field. People like Robert Metzger, Mark Reed and others, who played later an important role in molecular electronics, attended those conferences and they were inspired by the discussions held there.

As for many other fields in nanoscience, the invention of the scanning tunneling microscope (STM) by Gerd Binnig and Heinrich Rohrer (at IBM Zurich) in 1981 [10, 11] changed the panorama for molecular electronics. The STM was the first tool that provided a practical way to "see", "touch", and manipulate matter at the atomic scale (see Fig. 1.4). Soon after its invention, it became clear to the STM could provide a realistic way to address single molecules and to study their electronic transport properties.

Since the original experiments of Kuhn and coworkers [7], many different groups studied the electrical conductivity through Langmuir-Blodgett (LB) multilayers and even monolayers. For instance, Fujihira and co-workers demonstrated an LB monolayer photodiode already back in 1985 [13], which is probably the first unimolecular electronic device. In the 1990's one of the main goals in this context was to confirm the ideas of Aviram and Ratner about unimolecular rectification. The Aviram-Ratner mechanism, slightly modified, was confirmed by Robert Metzger's group in both macroscopic and nanoscopic conductivity measurements through a monolayer of

Fig. 1.4 Principle of a local probe like the scanning tunneling microscope: The gentle touch of a nanofinger. If the interaction between tip and sample decays sufficiently rapidly on the atomic scale, only the two atoms that are closest to each other are able to "feel" each other. Reprinted with permission from [12]. Copyright 1999 by the American Physical Society.

γ-hexadecyl-quinolinium tricyanoquinomethanide in 1997 [14].

At the end of the 1980's and the beginning of the 1990's the appearance of the metallic atomic-sized contacts had an important impact in the nanoscience community. Different groups showed that the STM and the recently introduced mechanically controllable break-junction (MCBJ) technique[5] could be used to fabricate metallic wires of atomic dimensions (for a review, see Ref. [15]). Since then these nanowires have become an endless source of new physical phenomena and have played a crucial role in the fields of mesoscopic physics and nanoelectronics. The relevance of these systems for molecular electronics is two-fold. On the one hand, they provide the basis to contact individual molecules with dimensions on the range of a few nanometers, which is out of the scope of conventional lithographies. On the other hand, the atomic contacts (or atomic-size contacts) have allowed establishing the connection between the quantum properties of single atoms and the macroscopic electrical properties of the circuits in which they are embedded, which is an important lesson for molecular electronics.[6]

In 1997 the collaboration between the groups of Mark Reed (a physicist at Yale University) and James Tour (a synthetic chemist at the University of South Carolina) led to the publication of the results of what is often considered as the first transport experiment in single-molecule junctions [16].[7] These authors used the MCBJ technique to contact benzenedithiol

---

[5]This technique will be described in the next chapter.

[6]The physics of these metallic nanowires will be described in the third part of this monograph.

[7]Let us clarify that the first transport measurements involving single molecules were indeed performed with the STM, but the experiment of Reed *et al.* is the first one realized

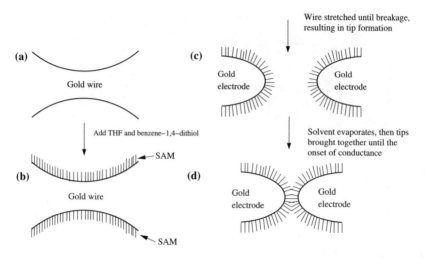

Fig. 1.5   Schematics of the first transport measurements through single-molecule junctions performed with the MCBJ technique [16]. (a) The gold wire of the break-junction before breaking and tip formation.   (b) After addition of benzene-1,4-dithiol, self-assembled monolayers (SAMs) form on the gold wire surfaces. (c) Mechanical breakage of the wire in solution produces two opposing gold contacts that are SAM-covered. (d) After the solvent is evaporated, the gold contacts are slowly moved together until the onset of conductance is achieved.

molecules with gold electrodes (the principle of this experiment is schematically illustrated in Fig. 1.5).[8] The importance of this experiment is that it triggered off the realization of many others in the same spirit. Indeed, our review on single-molecule conduction in the last part of this book will cover the activities from the appearance of this experiment on.

At the end of the 1990's new experimental techniques were introduced and additional results were reported showing that molecules can indeed mimic the behavior of ordinary microelectronics components. Thus for instance, Reed's group adapted the so-called nanopore technique (see Chapter 3) to form metal-self-assembled monolayer-metal heterojunctions. With this technique it was shown that junctions based on certain organic molecules can exhibit, for instance, rectifying behavior [17] or a very pronounced negative differential resistance [18]. On the other hand, James Heath and Fraser Stoddart groups joined efforts to show that junctions based on rotaxanes and catenanes could act as reconfigurable switches [19, 20].

---

in a symmetric structure that could in principle be integrated in more complex circuits.
[8]This experiment will be described in detail in section 14.1.1.

Techniques like electromigration [21], which were specially designed to contact single molecules, were developed at the turn of the century. These methods made possible to incorporate a gate electrode in single-molecule junctions and thus, to mimic the measurements performed in solid state devices like transistors or in nanostructures like quantum dots. With the use of these techniques it was possible to show that single-molecule junctions can behave as a new kind of single-electron transistors [22] or that they can exhibit basic physical phenomena like Coulomb blockade or the Kondo effect [23, 24], which are well-known in the context of other nanoscopic structures.

These results obtained in academic institutions and research laboratories attracted the attention of global players in information technology like HP, IBM and others that decided to set up small molecular electronics research groups. This gave a new impulse to the field by providing very important missing ingredients like, for instance, strategies to link molecular devices with each other and with external systems. As an example we can mention the nanoscale circuits based on a configurable crossbar architecture introduced by Stanley Williams and coworkers at the HP Laboratories in Palo Alto [25], see Fig. 1.6(a-d). This strategy was used, for instance, to show that molecular crossbar circuits fabricated from a molecular monolayer of [2]rotaxanes can function as an ultra-high-density memory [26], see Fig. 1.6(e-f). The working principle of these molecular memories is supposed to be based on the ability of molecules like rotaxanes to switch between two metastable states upon the application of an external bias voltage. The actual origin of the switching behavior in these molecular junctions has been heavily debated and, in some cases, it has shown that the metal electrodes or the metal-molecule interface are responsible for the switching mechanism rather than the molecules themselves (see e.g. Ref. [27]). The controversy about these results, and also about some of the original experiments mentioned above, led to the extended belief that molecular electronics was going through a midlife crisis [28], although it was no more than a teenager. In the meantime, the situation concerning the molecular memories has been clarified to a large extend and more recently the densest memory circuit ever made (10^{11} bits cm^{-2}) was fabricated using a monolayer of bistable [2]rotaxane molecules as the data storage elements [29]. Although many scientific and engineering challenges, such as device robustness, remain to be addressed before these devices can be practical, these results show clearly the potential of a molecule-based electronics.

On the other hand, the efforts in recent years of numerous research

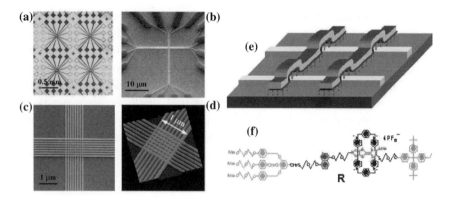

Fig. 1.6   Nanoscale molecular-switch crossbar circuits. (a) An optical microscope image of an array of four test circuits, showing that each has 16 contact pads with micron-scale connections leading to nanoscale circuits in the center. (b) An image taken with a scanning electron microscope (SEM) showing two mutually perpendicular arrays of nanowires connected to their micron-scale connections. (c) A SEM image showing that the two sets of nanowires cross each other in the central area. (d) A 3D image of the crossbar taken with an atomic force microscope. (e) Schematic representation of the crossbar circuit structure in which monolayer of the [2]rotaxane is sandwiched between an array of Pt/Ti nanowires on the bottom and an array of Pt/Ti nanowires on the top. (f) Molecular structure of the bistable [2]rotaxane **R**. Reprinted with permission from [26]. Copyright 2003 IOP Publishing Ltd.

groups world-wide have established molecular electronics as a true field of science, where there is a lot of new physics and chemistry to be learned. Although it is still difficult to fabricate reliable molecular junctions, in particular at the single-molecule level, and there are other basic problems to be solved, many concepts and techniques are by now well established and they are precisely the subject of this book. For us, it is clear that molecular electronics has reappeared this time to stay forever with us. In the next years we shall surely contemplate many basic discoveries in this field and some of them will hopefully lead to new and unforeseen technological applications.

1.3   Scope and structure of the book

By now molecular electronics is a very broad field with many different interesting aspects and special topics. These topics can be divided in a natural way into those related to the development and potential applications of molecular devices and those concerning the novel physical phenomena that take place in molecular-scale junctions. In this monograph we are interested in the latter type of topics and, in particular, we shall focus our attention on the understanding of the basic mechanisms that dominate the electronic transport at the molecular scale. To be precise, we shall concentrate on the analysis of the properties of single-molecule junctions, although some examples of junctions based on molecular assemblies will also be presented and discussed.

Our main goal in this monograph is two-fold. On the one hand, we want to provide a true textbook on molecular electronic for advanced undergraduate and graduate students both in physics and chemistry. The book has been designed so that, by the end of it, a student with a background in quantum mechanics and some elementary notions of solid state physics[9] and organic chemistry[10] should be able to start doing research in the field of molecular electronics. On the other hand, we also want to provide a thorough review of the activities on single-molecule conduction over the last ten years, from which both newcomers and researches working in the field can profit.

With this double goal in mind, we have divided this monograph into four parts that can be read independently.[11] The first two are meant as textbook material that can be used for a regular course, while the last two ones are closer to a topical review. Part 1 includes, apart from this introductory chapter, a detailed description of the experimental techniques that are currently being used to fabricate both atomic-scale wires and molecular junctions as well as the basic principles of transport measurements. Here, we have tried to explain both the basis of the different techniques as well as their advantages and disadvantages. Moreover, we have included in section 3.2 a brief discussion about the main molecules used in molecular electronics and their basic properties, which can be viewed as an accelerated course in organic chemistry.

Part 2 contains an extensive theoretical background that provides a basic introduction both to the transport mechanisms in nanoscale systems and to the standard theoretical techniques that are used to describe the transport in molecular systems. We want to stress that this theory part is not just meant for theoreticians and theory-inclined students, but for every-

---

[9]For the students in chemistry we recommend the brief introduction to solid state physics provided in Chapter 4 of Ref. [30] or in Chapter 3 of Ref. [31].

[10]For the students in physics we recommend the brief introduction to organic chemistry provided in Chapter 5 of Ref. [31].

[11]There is indeed a fifth part that contains an appendix about the second quantization formalism of quantum mechanics.

body. All the topics are discussed in a didactic and self-contained manner so that students without a previous knowledge on these topics should be able, after reading this part, to follow the theory papers in this field. To be precise, this part starts in Chapter 4 with an introduction to the scattering (or Landauer) approach, which provides an appealing framework to describe coherent transport in nanostructures. Then, we go on with several chapters devoted to Green's function techniques (Chapters 5-8), which provide powerful tools to compute equilibrium and nonequilibrium properties of atomic-scale junctions beyond the capabilities of the scattering approach. Finally, Chapters 9 and 10 deal with the two most widely used electronic structure methods in molecular electronics, namely the tight-binding approach and density functional theory. These methods in combination with the Green's function techniques provide the starting point for the realistic description of the transport properties of atomic and molecular junctions. Let us emphasize that at the end of every chapter one can find several exercises that have been chosen to illustrate the main concepts.

Part 3 presents a basic description of the physics of atomic-sized contacts. Although this is not the main topic of the book, it is crucial to have a basic knowledge about the transport properties of the metallic wires that are then used as electrodes in molecular junctions. We have divided this part into two chapters where we describe the physics of non-magnetic atomic contacts (Chapter 11) and magnetic ones (Chapter 12).

Finally, Part 4 presents a detailed review on the transport through molecular junctions. We have organized the material according to the physical mechanism which dominates the transport properties. Thus, we start this part with two chapters devoted to the coherent transport in molecular junctions (Chapters 13 and 14). Then, we discuss in Chapter 15 the physics of the so-called molecular transistors, which are nothing but weakly coupled molecular junctions where the transport is dominated by electronic correlations that lead to phenomena like Coulomb blockade or the Kondo effect. We then proceed to discuss in Chapters 16 and 17 the role of molecular vibrations in the electrical current through molecular junctions. Chapter 19 is devoted to other transport properties beyond conductance and we discuss there, in particular, shot noise and thermal transport in molecular conductors. The optical properties of current-currying molecular junctions are the subject of Chapter 20. Chapter 18 deals with the electronic transport in long molecules where the hopping (or incoherent) transport regime is realized. Finally, we conclude this part in Chapter 21 with a list of topics that have not been addressed in this monograph and we indicate where to find information about them. It is worth remarking that these chapters have been written so that they can be read almost independently. This way a reader can concentrate on those topics or chapters that are of special interest for him/her.

Parts 3 and 4 are meant for both students and researchers working in the field. We do not only review what has recently been done in the field, but we also introduce the different topics at a elementary level. In this sense, whenever it was possible, we have provided simple arguments and suggested additional exercises. These two parts are intended for both experimentalists and theoreticians and, most of the time, we have intentionally avoided the typical separation between experiment and theory, which we find particularly harmful in this field.

Let us close this chapter with some recommendations about the existent literature. For those who want a quick overview about molecular electronics, we recommend the short reviews of Refs. [2, 32–37]. A nice general overview of the field can be found in Chapter 20 of Ref. [31]. For more extensive introductions, we recommend Ref. [38] for the theory in molecular systems and Refs. [39–41] for a discussion of the experimental techniques used in molecular electronics. There already exist several books that deal with different aspects of molecular electronics, see e.g. Refs. [42–49]. Most of them consist of a collection of articles written by different authors, but they are very useful if one wants a more detailed discussion of certain topics. Concerning the theory of quantum transport or transport in nanoscale systems, which is one of the central subjects of this manuscript, we recommend the monographs of Refs. [50–53].

Chapter 2

Fabrication of metallic atomic-size contacts

2.1 Introduction

In this chapter we shall present the most common methods which have been developed during the last years for the fabrication of metallic atomic-size contacts. Both the contacting methods and the physical properties of atomic contacts found the basis for contacting single molecules. On the other hand, these techniques have been further refined for contacting molecules. These refinements are now also used for studying atomic contacts. Therefore, the decision in which chapter one or the other method is described is somewhat arbitrary. Manifold variations of the techniques exist and are permanently improved further. The aim of this chapter is to introduce into the most important principles and to compare the techniques regarding their advantages and drawbacks.

As important as the sample preparation is the quality of the electronic transport measurements. When dealing with tiny contacts, care has to be taken to reduce the influence of the measurement onto the contact itself. We will therefore end this chapter with a few brief remarks about the most common measurement setups and possible artifacts.

2.2 Techniques involving the scanning electron microscope (STM)

One of the most versatile tools for the fabrication of atomic-size contacts and atomic chains is the scanning tunneling microscope (STM) (for a review, see Ref. [54]). It has been used for that purpose from the very beginning of its invention [55]. While in the standard application of an STM a fine metallic tip is held at distance from a counter electrode (in general

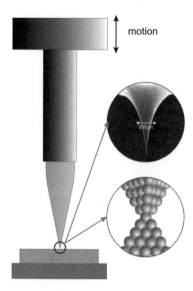

Fig. 2.1   Working principle of the fabrication of atomic contacts with an scanning tunneling microscope (STM). The electron micrograph shows a STM tip. The width at half
length is in the order of 100 to 200 μm. The lower inset gives an artist's view of the
atomic arrangement of an atomic contact. Courtesy of C. Bacca.

a metallic surface) by making use of the exponential distance dependence
of the tunneling current, the tip can also be indented into the surface and
carefully withdrawn until an atomic size contact or short atomic wire forms.
An artist's view of the STM geometry and the atomic configuration of a
contact is shown in Fig. 2.1. For many metals it has been shown that the
tip will be covered by several atomic layers of the metal of the counter electrode upon repeated indentation such that clean contacts may be formed
consisting of the same metal for both electrodes.

The main advantages of the STM in this application are its speed and
versatility. When the electrodes forming the contacts are prepared in ultra
high vacuum conditions, the STM furthermore allows to gather information
about the topography of the two electrodes on a somewhat larger than the
atomic scale before or after the formation of the contact. Since however,
the tip is usually pressed into the substrate and the atomic-size contact is
formed when withdrawing, the exact atomic configuration of the atomic-
contact cannot be measured directly.

This problem is partially solved when the contact is formed upon approaching [56]. For good metals the distance dependence of the conduc-

tance follows an exponential increase until a sudden "jump to contact" occurs which is marked by a step-like increase of the conductance. The jump indicates the formation of a chemical bond between the tip and the electrode and thus the formation of a single-atom contact. The geometry of the substrate side of the contact can be well controlled by first preparing and characterizing a clean terrace of a single crystalline substrate and subsequently evaporating a sub-monolayer small amount of metal atoms onto it. The surface can then be scanned and the tip can be approached right on top of one of the extra atoms. This technique enables to form heterojunctions, i.e. contacts between two different metals. The determination of the atomic configuration on the tip-side of the contact remains unsolved, though.

Spectroscopic measurements on the scale of electron volts allow one to deduce information about the cleanliness and the electronic structure of the metal [57].

The main drawbacks are its limited stability with respect to the change of external parameters such as the temperature or magnetic fields and the short lifetime of the contacts in general because of the sensitivity of the STM to vibrations. In the early years of STM-based atomic contact studies they were furthermore limited to rather high temperatures in the range of 10 K or higher. This drawback has been overcome in the last years. Nowadays ultra high vacuum (UHV) STMs, which work with sufficient stability at temperatures below 1 K and in strong magnetic fields are even commercially available.

2.3 Methods using atomic force microscopes (AFM)

Another scanning probe technique which complements STM in many aspects is the atomic force microscope (AFM). Instead of the tunnel current an AFM uses the distance dependence of the force between a fine tip and a surface. Depending on the chemical nature of both the tip and the surface this force consists of several contributions and its distance dependence may be complex and even nonmonotonic. The working principle of the AFM is based on measuring the force by recording the deflection of a cantilever that carries the tip. The deflection can be detected by optical means or by the detuning of an oscillator circuit due to the deflection. The AFM has become a very versatile tool in surface science which works in various environments and temperature ranges. In surface science the main advan-

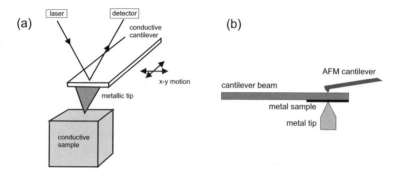

Fig. 2.2   Fabrication and characterization of atomic contacts with an atomic force microscope (AFM). (a) The conductive AFM uses a conductive cantilever and metallic tip for recording the electrical signal. The deflection of the cantilever beam is detected optically and used for recording the topographic information of sample. After Ref. [59]. (b) In the combined AFM-STM the sample is clamped to a cantilever. The metallic contact is formed between the sample and the metal tip. The metal tip is part of an STM and records the electrical signal. The deflection of the cantilever is recorded with a separate AFM. This signal is used for measuring the force acting on the cantilever when the atomic contact rearranges. After Ref. [58].

tage of AFM as compared to STM is its possibility to work on insulating substrates. For the fabrication and characterization of atomic contacts the AFM is in use in two different variations. The first one is the combination with an STM which records the current while the AFM measures the force that is necessary to form or break the contacts [58]. The second one is the so-called conductive AFM which uses a metal-covered tip on a metallic surface and both quantities, the current and the force, are available simultaneously, Fig. 2.2 [54]. The force signal can be used to determine the topography.

2.4   Contacts between macroscopic wires

Transient atomic chains and contacts with lifetimes in the millisecond range can also be fabricated in a table-top experiment first demonstrated by N. Garcia and coworkers [60], which we call here "dangling-wire contacts". Two metal wires in loose contact to each other are excited to mechanical vibrations, such that the contact opens and closes repeatedly. One end of each wire is connected to the poles of a voltage source and the current is recorded with a fast oscilloscope. This method is in principle particularly versatile because it enables the formation of heterojunctions between

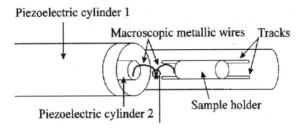

Fig. 2.3 Experimental setup used to visualize contacts between macroscopic metallic electrodes inside a scanning electron microscope (SEM). Adapted with permission from [61]. Copyright 1997 by the American Physical Society.

various metals. However, in order to provide clean metallic contacts a thorough cleaning of the wires would be required, similar to the tip and surface preparation in a STM. Another drawback is the lack of control of the distance of the electrodes. It is thus mostly used as demonstration experiment in schools with Au-Au contacts. The method has later been improved by attaching the wires to piezo tubes. This realization thus resembles contacts fabricated in the STM and have also been used within the chamber of an scanning electron microscope for simultaneous imaging and conductance measurements, see Fig. 2.3.

2.5   Transmission electron microscope

Another interesting method for preparing and imaging atomic contacts are transient structures forming in a transmission electron microscope (TEM) when irradiating thin metal films onto dewetting substrates [62, 63]. The high energy impact caused by the intensive electron beam locally melts the metal film causing the formation of constrictions which eventually shrink down to the atomic size and finally pinch-off building a vacuum tunnel gap. A typical system for these studies is Au on glassy carbon substrates. Several variations of this principle have been developed that allows one to contact both electrodes forming the contact, see Fig. 2.4. The high electron current density necessary for imaging causes also high local temperatures resulting in short lifetimes of these contacts. However, they offer the unique possibility to simultaneously perform conductance measurements and imaging with atomic precision. Similar results have been obtained with variations of the STM inside a TEM [64]. This method enabled to directly prove the existence of single-atom contacts, single-atom wide and several atom

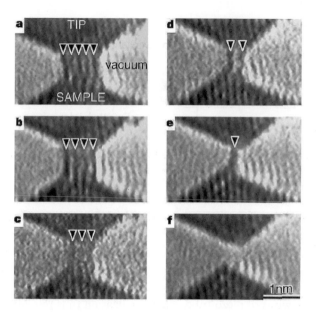

Fig. 2.4   High resolution TEM images of short atomic wires fabricated with an STM inside the vacuum chamber of the TEM. The arrows indicate the number of atomic rows. In panel f the contact is broken and forms a tunnel contact. Reprinted by permission from Macmillan Publishers Ltd: Nature [63], copyright 1998.

long chains as well as to establish a correlation between contact size and conductance [63, 62, 65]. For Au and Ag contacts it has been shown that preferably well ordered contacts with growing directions corresponding to the symmetry axes of the crystal structure are formed.

2.6   Mechanically controllable break-junctions (MCBJ)

Already before the development of the first STM another technique enabling the fabrication of atomic-size contacts and tunable tunnel contacts has been put forward. The first realizations include the needle-anvil or wedge-wedge point contact technique pioneered by Yanson and co-workers (for a review see [66]) and the squeezable tunnel junction method described by Moreland and Hansma [67] and Moreland and Ekin [68] who used metal electrodes on two separate substrates which are then carefully adjusted with respect to each other. The needle-anvil technique was mainly used to form contacts with diameters of typically several nanometers and thus having

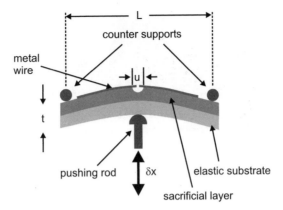

Fig. 2.5 Working principle of the MCBJ (not to scale) with the metal wire, the elastic substrate, the insulating sacrificial layer, the pushing rod, the counter supports and the dimensions used for calculating the reduction ratio (see text).

hundreds or thousands of atoms in the narrowest cross section. These two techniques formed the starting point for the development of the mechanically controllable break-junctions (MCBJ) by C. Muller and coworkers [69], which nowadays is applied for the fabrication of atomic contacts in various subforms, the most common of which are the so-called notched-wire [70] and thin-film MCBJs [71]. The working principle which is depicted in Fig. 2.5 is the same for both variations: A suspended metallic bridge is fixed on a flexible substrate, which itself is mounted in a three-point bending mechanism consisting of a pushing rod and two counter-supports. The position of the pushing rod relative to the counter supports is controlled by a motor or piezo drive or combinations of both. The electrodes on top of the substrate are elongated by increasing the bending of the substrate. The elongation can be reduced again by pulling back the pushing rod and thus reducing the curvature of the substrate. In order to break a junction to the tunneling regime, considerable displacements of the pushing rod and thus important bending of the substrate is required. Therefore the most common substrates are metals with a relatively high elastic limit like spring steel or bronze. The substrates are covered by an electrically isolating material such as polyimide before the junction can be fixed on it.

The notched-wire MCBJ, an example of which is shown in Fig. 2.6, uses a thin metallic wire (diameter 50 μm to 200 μm) with a short, knife-cut constriction to a diameter of 20 μm to 50 μm. The wire is glued at both sides of the notch to the substrate and connected electrically to the

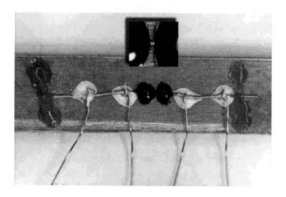

Fig. 2.6   The 100 nm wide gold wire is glued with epoxy resin (black) onto the substrate. The electrical contact is made by thin copper wires glued with silver paint. The inset shows a zoom into the notch region between the two black drops of epoxy resin. Reprinted from [15]. Copyright 2003, with permission from Elsevier.

measurement circuit at both ends. The distance between the glue drops is of the order of 50 μm to 200 μm.

Variations of this method have been put forward which enable contacting of reactive or brittle materials out of which no wires can be formed [72]. For this purpose the sample preparation is performed in protective environment. A beam-shaped piece of the material is cut in a non-reactive liquid such as dodecanol or other slowly evaporating alcohols, or glycerine. Four holes are drilled into the metal and a wedge is cut in the middle between the holes. An example is shown in Fig. 2.7. The beam is screwed with the help of two electrically isolating bolts to the substrate, one on each side of the wedge. The remaining two holes serve for screwing metallic wires to the beam for the conductance measurements.

For a version which enables scanning the two electrodes with respect to each other, at first two piezo tubes are glued to the substrate. The metal wire is then glued on top of the piezos. After mechanically breaking the wire, the piezos are polarized such that they are bent and the two parts of the wire are sliding along each other [73]. This realization corresponds to a high-stability STM, but with very restricted scan possibility. It is therefore used only sparsely. Finally, simultaneous force and conductance measurements are possible when adding a tuning fork like in AFMs. Details of this very sophisticated method are given in Ref. [75].

Fig. 2.8 shows two examples of thin-film MCBJs, which were fabricated using the usual techniques of nanofabrication, i.e. electron beam lithography and metal deposition by evaporation. There are mainly two differences to

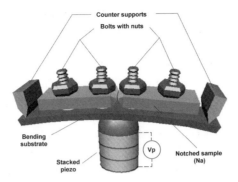

Fig. 2.7 Principle of the MCBJ technique adapted for reactive metals. Reprinted from [15]. Copyright 2003, with permission from Elsevier.

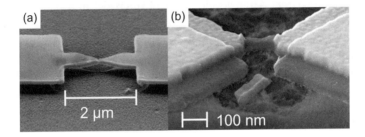

Fig. 2.8 Lithographic MCBJ. (a) Electron micrograph of a thin-film MCBJ made of cobalt on polyimide taken under an inclination angle of 60° with respect to the normal. The distance between the rectangular shaped electrodes is 2 μm, the thickness of the thin film is 100 nm and the width of the constriction at its narrowest part is approximately 100 nm. (b) Electron micrograph of a thin-film MCBJ made of cobalt (medium grey) with leads made of gold (light grey) taken under an inclination angle of 50° with respect to the normal. The distance between the rectangular shaped electrodes is 2 μm, the thickness of the Co film is 80 nm, of the Au film is 100 nm and the width of the constriction at its narrowest part is approximately 100 nm. The sample has been fabricated using shadow evaporation through a suspended mask such that two images of the mask exist. The Au shadow of the bridge is broken off.

standard nanostructuring. The first one is the substrate, which in case of MCBJs has to provide sufficient elastic flexibility without breaking or irreversible bending. The second difference is the final etching step which is needed to suspend the nanobridge (with typical dimensions of 2 μm in length and 100 nm × 100 nm at the narrowest part of the constriction) above the substrate by partial removal of a sacrificial layer underneath the metal film. Fig. 2.9 summarizes the fabrication procedure. A piece of metal with a typical thickness of a few hundred micrometers serves as substrate.

The metal should have a high elastic deformation limit. Typical metals are bronze or spring steel. For particular purposes, in particular when capacitive effects have to be minimized, the metal is replaced by a plastic substrate. Both metal or plastic are thoroughly polished to reduce the roughness to less than a micrometer. The remaining corrugations are then filled with a thin layer of polyimide (thickness 1-2 μm), which is spin-coated and hardbaked in vacuum. The polyimide also serves as electrical insulator between the nanostructure and the substrate. Subsequently the electron resist is spin-coated and thermally treated as required for electron beam structuring. Fig. 2.9(c) shows an example in which a double-layer resist is used. The double-layer is necessary for, e.g. evaporation of the metal under arbitrary angle. The next step is electron-beam writing in a scanning electron microscope equipped with a pattern generator or in a commercial electron-beam writer. After development of the resist in a selective solvent the resist mask remains on top of the polyimide layer. The mask itself may be partially suspended when using a double-layer resist. Subsequently the metal will be deposited either by evaporation, sputtering, chemical vapor deposition or other means. Shadow evaporation, i.e. evaporation of several materials under different angles can be used for forming contacts between different metals or for supplying nanobridges of one metal with electrodes made of another metal. The advantage of the shadow-evaporation technique lies at first in its self-alignment property because the same mask is used for all metal depositions. The second advantage is given by the fact that all depositions can be made in a single vacuum step, which enables one to fabricate clean interfaces between the metals. After the metal deposition the mask is stripped in a more aggressive solvent. Finally the structure is exposed to an isotropic oxygen plasma which attacks the polyimide layer. Consequently its thickness is reduced and all narrow metal parts, like the nanobridge become suspended like a bridge.

Both versions of the technique - the notched-wire MCBJs and the lithographic (or thin-film) MCBJs - share the idea of enhanced stability due to the formation of the contact by breaking the very same piece of metal on a single substrate and by transformation of the motion of the actuator into a much reduced motion of the electrodes perpendicular to it. The small dimensions of the freestanding bridge-arms give rise to high mechanical eigenfrequencies, much higher than the ones of the setup. As a result the system is less sensitive to mechanical perturbations by vibrations.

Assuming homogeneous beam-bending of the substrate we can calculate the reduction ratio r between the length change of the bridge u and the

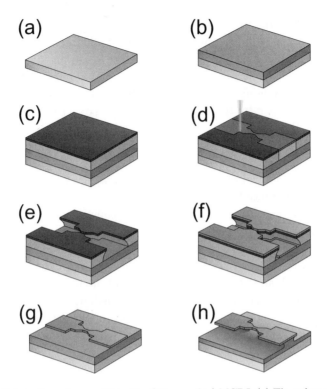

Fig. 2.9 Fabrication scheme of thin-film (lithographic) MCBJ. (a) The substrate (metal, plastic) is polished mechanically. (b) the sacrificial wafer (polyimide) is spin-coated and baked. (c) The resin (typically a bi-layer electron sensitive organic material) is spin-coated and baked. (d) The resin is exposed in an electron beam writer or a scanning electron microscope equipped with a pattern generator in the desired pattern. (e) The chip is developed in a solvent which selectively removes the exposed parts of the resin. The result is a mask, which resides on the sacrificial layer, in the shape of the exposed pattern. (f) The metal is deposited by evaporation or sputtering. (g) The mask with the metal on top of it is lifted-off in a more aggressive solvent which attacks the unexposed parts of the resin. The result is a metal layer in the shape of exposed pattern. (h) Finally the thickness of the sacrificial layer is reduced in an isotropic plasma. The narrow parts of the metal pattern are suspended and form the bridge which will be broken in the MCBJ mechanism.

motion of the pushing rod x (see Fig. 2.5).

$$r = \frac{6tu}{L^2}, \qquad (2.1)$$

where t is the thickness of the substrate, u the length of the free-standing bridge arms and L the distance of the counter supports. This quantity denotes the factor with which any motion of the pushing rod is reduced when it is transferred to the point contact. In a real MCBJ setup, however, the beam-bending is in general non-uniform. Furthermore, also the sacrificial layer has a finite elasticity and is deformed when bending the MCBJ. These effects can be accounted for by a correction factor, which enhances r by a factor of roughly 4 [76]. The effective reduction ratio has a typical value of 10^{-3} to 10^{-2} for the notched-wire MCBJs and 10^{-6} to 10^{-4} for the thin-film MCBJs. The relatively weak reduction ratio of the notched-wire MCBJs usually requires the use of a piezo drive for controlling and stabilizing single-atom contacts, while the lithographic MCBJs can be controlled with purely mechanical drives, i.e. a dc-motor with a combination of gear boxes, and a differential screw.

A common realization of a bending mechanism suitable for thin-film MCBJs and use at low temperatures $T < 1$ K is shown in Fig. 2.10. A rotary axis is connected to a differential screw which consists of a thread, the two sections of which have a slightly different pitch. The typical values for the pitches A and B are 0.7 to 0.8 mm and pitch differences 50 μm to 150 μm. Each full turn of the axis changes the distance between the sample holder and the ground plate by the difference of the pitches. The shape of the end of the pushing rod can be semi-cylindrical or wedge shaped, depending on the desired deformation of the substrate. Because of the off-line axis arrangement of rotary axis and pushing rod several guiding rods are needed to reduce torque and ensure linear motion of the sample holder with respect to the ground plate. The pushing rod can be designed such that it hosts a piezo tube. The MCBJ is electrically contacted via spring contacts or by gluing the wiring to it via silver paint. The thermal contact of the sample to the thermal bath can additionally be provided by thick wires and copper braid. Care has to be taken when choosing the materials combination of the thread and its counterpart to avoid friction because lubrification at low temperature and in vacuum is difficult.

Typical motion speeds of the piezo drive lie between 10 nm/s and 10 μm/s corresponding to results in 10 pm/s to 100 nm/s for the electrodes forming the atomic contacts. For purely mechanical drive these values are 10 nm/s to 1 μm/s for the pushing rod and 10 fm/s to 10 nm/s for the contact. Due to the in-built reduction also the piezo-driven setups are in general slower than STM systems. The high stability enables comprehensive studies on the very same atomic contact at various values of control parameters such as fields and temperature.

On the other hand the small r values require considerable absolute motion of the pushing rod and deformation of the substrate in order to achieve

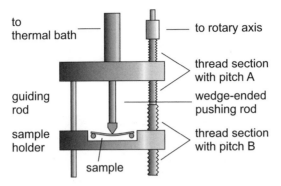

to
thermal bath — to rotary axis

thread section
with pitch A

guiding
rod

wedge-ended
pushing rod

sample
holder

thread section
with pitch B

sample

Fig. 2.10   Sample holder with differential screw for thin-film MCBJ. A motor drives a rotary axis which ends in a thread with two different pitches. Rotating the axis results in varying distance between ground plate and sample holder. The sample resides on two counter supports connected to the sample holder. It is bent by the pushing rod which is attached to the ground plate. Three guiding rods (only one of which is shown) ensure smooth and linear motion.

sufficient displacements of the electrodes. This reduces the possible choices of the substrate material considerably.

MCBJ mechanisms have been developed for various environments including ambient conditions, vacuum, very low temperatures [77] or liquid solutions [78]. The latter one is of particular interest for the study of single-molecule junctions and will be explained in detail in the following Chapter 3. The disadvantages of MCBJs as compared to STM techniques are the small speed and the fact that the surrounding area of the contact cannot easily be scanned. As for STM setups clean contacts can only be guaranteed when working in good vacuum conditions. The sample preparation itself, however, does not require clean conditions because the atomic contacts are only formed during the measurement by breaking the bulk of the electrodes.

2.7   Electromigration technique

A third method for the formation of atomic-size contacts is controlled burning of a wire by electromigration (see Fig. 2.11). This technique has been optimized for the formation of nanometer sized gaps for trapping individual molecules or other nanoobjects [79, 80]. Before the wire finally fails and the current drops drastically, atomic size contacts are formed for a rather short time span [81–83]. During the electromigration process the trans-

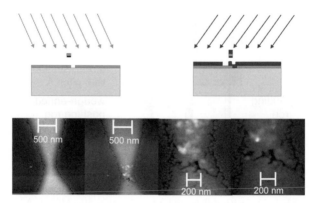

Fig. 2.11   Electromigration technique.   *Top*: Fabrication via shadow evaporation through a suspended mask of an electrode structure to be used for producing atomic contacts by electromigration. The arrows indicate the directions from which the metal is deposited. The electromigration will nucleate at the thinnest part of the electrode structure. *Bottom*: Series of atomic force microscope images taken in the tapping mode of an electromigrated contact made of Au on Si in different phases of the electromigration process. From left to right: before electromigration ($R = 40\ \Omega$), $R = 105\ \Omega$, $R = 630\ \Omega$, $R = 30.000\ \Omega$. Courtesy of D. Stöffler and R. Hoffmann.

port changes from ohmic behavior, i.e. limited by scattering events of the electrons to wave-like electronic transport, which can be described by the Landauer picture (see Chapter 4).

The term electromigration denotes a process in which ions are moved due to high electrical current densities. We concentrate here on the electromigration behavior of metals. It has been understood that several effects contribute to the total force acting on a metal atom which forms the conductor, the two most important being the so-called direct force due to the electric field. It causes the electrical current and thus points into the direction of the field. The second one is caused by momentum transfer of the conduction electrons onto the ions. It has opposite sign and is called the wind force. When the total force overcomes the binding force of the ions, they start to diffuse but can be pinned again at defects or positions where the current density and driving force falls below this threshold value. Depending on the material, the temperature, the crystallinity, the surface roughness, and many other parameters either the direct force may exceed the wind force or vice versa [84]. Therefore the exact direction of the material transport depends on the microscopic structure of the wires. In many cases the motion of the material is such that the cross section of the conductor is locally reduced and its electrical resistance increases. The higher resistance causes higher losses, enhanced dissipation, increasing temperature in the wire which further enhances the dissipation of ions. An important role plays the temperature of the lattice because the diffusion and the threshold current strongly depend on temperature. Electromigration has become one of the most important origins of failures in integrated circuits, due to the miniaturization of the metallic interconnects without reducing the current by the same factor. Consequently, electromigration has widely been studied in electrical engineering with the aim to achieve the highest possible threshold current density for it to set in and the smallest diffusion speed [85].

For the formation of atomic contacts a high threshold current is not important but the possibility for controlling speed, shape and size of the final structure. One of the most important preconditions is to define the position at which the electromigration starts, and the contact forms. For this purpose a short and thin metallic wire is fabricated by lithographic methods as described in the previous section. Typical dimensions are a length and width of 50 to 100 nm and a thickness of 10 to 20 nm. The thin wire is connected to wider and thicker electrodes which consequently have smaller resistivity. A convenient method to fabricate these structures is shadow evaporation through a suspended mask as shown in Fig. 2.11. First, thin layers of the metal (typical thickness 10 nm) are evaporated under the angles Θ and $-\Theta$. The angle is chosen such that both layers slightly overlap underneath the suspended part of the mask. Afterwards a thick layer of the electrode metal is deposited perpendicular to the substrate plane. The ideal structure would consist of a single-crystalline wire in the thin part of the wire, the boundaries of which are covered by the thick electrodes in order to avoid electromigration of possible contaminants from the grain boundaries. It is advantageous to work on a substrate with high thermal conductivity in order to control the temperature.

The electromigration process itself is performed such that an electrical current is continuously ramped up while the resistivity is monitored. As soon as the resistance starts to increase a computer-controlled feedback loop controls the current such that the rate of the resistance increase is kept constant or slowed down. The resistance increase is partially due to the temperature increase caused by the Joule heating of the driving current. Although it has been shown that in the ohmic regime the current density is the quantity which determines the diffusion of the ions, it is advantageous to control the voltage in order to produce atomic size contacts. When the resistance increases the current becomes smaller, which helps to limit the migration speed. The low-resistive electrodes ensure that the voltage drops locally making the driving force acting only locally as well. Consequently, the dissipation and Joule heat generation are local as well. The procedure should be stopped when the desired resistance is achieved. For the study of atomic contacts the interesting regime is reached when the resistance exceeds roughly one kiloohm. For usual metals this corresponds to contacts with a narrowest cross section of roughly 10 atoms. An important finding is that the behavior changes markedly when the size of the smallest cross section corresponds to a few atoms. However, the exact position of the position at which the wire finally breaks is difficult to predict. As will be explained in Chapter 11 the electrical transport of contacts of this size is determined by the wave properties of the electrons rather than by collisions with defects. If this happens the resistance may start to decrease again before the wire finally is burned through. This non-monotonous behavior complicates the control scheme further. Several control schemes have been put forward which are optimized for various sample geometries, metals and working conditions such as vacuum or low temperature [21, 81–83, 86]. So far only a few studies exist in which the electromigration process has been imaged in detail, although these kind of studies are very insightful. One example is shown in Fig. 2.11, where AFM images have been taken after discrete electromigration steps. A particularly nice series of TEM images showing that the most dramatic shape changes occur during the final phase can be found in Ref. [82].

An important difference to STM techniques and MCBJs is the fact that the wire forming the contact is in solid contact with a substrate. The advantages are at first ultimate stability which will become important when studying atomic or molecular junctions as a function of external fields (see Chapters 12 and 20). The second advantage lies in the fact that no particular requirements exist for the properties of the substrate, besides the fact that it should be sufficiently insulating. Often silicon - the standard substrate in microelectronics - is used. With suitable doping it can be used as back-gate for inducing an electric potential and building a three-terminal device. This technique is important for studying effects like Coulomb blockade, which will be explained in Chapters 11 and 15.

The main drawback of the electromigration technique is the fact that it is a single-shot experiment: Once an atomic contact has been established there is only limited possibility to fine tune its atomic configuration, in particular coming back to a larger contact is almost impossible. After burning

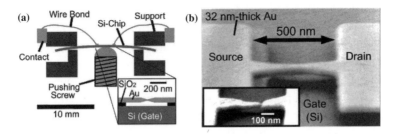

Fig. 2.12   Electromigrated MCBJ with gate on silicon substrate. (a) Working principle and (b) electron micrograph of an electromigrated MCBJ. The substrate is doped silicon and can be used as back-gate. Reprinted with permission from [86]. Copyright 2005 American Chemical Society.

through the wire it cannot be closed again. As described before, the control of the final part of the electromigration process is tricky because the character of the transport changes from ohmic to wave-like. A combination of electromigration with the lithographic MCBJ technique overcomes this problem: a thin-film MCBJ is thinned-out by electromigration to a narrow constriction with a cross section of less than 10 nm (see Fig. 2.12). The substrate is then bent carefully for completely breaking the wire or arranging single-atom contacts. This last step is reversible and repeatable for studying small contacts [87] or trapped nanoobjects [86]. Because only the very last part of the breaking requires mechanical deformation of the substrate it is rather fast and enables the use of more brittle substrates such as silicon.

2.8   Electrochemical methods

A completely distinct method for the formation of atomic-size contacts uses electrochemical deposition and removal of metal atoms. Electrochemical deposition of metals is a standard technique for surface treatment and in micromachining. For the purpose of forming atomic contacts basically the same principles are used. The main difference to the macroscopic techniques is the shape of the starting electrodes and the feedback which controls the deposition speed. Nanocontact formation by electrochemical methods starts from metal electrodes with a gap or with a continuous wire that is first broken either mechanically or by electromigration. The working principle is depicted in Fig. 2.13. The electrode structure is then immersed into an electrolyte containing metal ions. The electrochemical setup is adapted from the three-electrode cyclic voltammetry principle [88]. The deposition and dissolution of metal is controlled by applying an electrical potential difference between a so-called counter electrode and the electrodes forming the nanocontact, which serve as "working electrodes". A fourth electrode defines the reference potential. The conductance is monitored and used as control signal for the potentiostat which controls the deposition rate. The typical control voltages are in the range of 20 mV to 1 V and can be adjusted to optimize the electrochemical process. It should exceed the bias voltage if one aims at symmetric deposition on both electrodes forming the contact. Obviously the place at which the fastest deposition and dissolution takes place can further be controlled by the size and the polarity of the bias voltage. A typical metal combination is gold as electrode material because of its weak chemical reactivity and silver for the formation of the atomic contacts [89, 90]. Silver is easily dissolved in acids, like e.g. in nitric acid, and simultaneously silver atomic contacts have well understood transport properties, as will be further detailed in Chapter 11. One main advantage of this technique is its versatility, since electrochemical deposition methods on the macroscale have been developed for almost all metals. A further advantage is the simplicity of the working principle, in particular the simplicity with which the starting electrodes can be produced: macroscopic wires as well as deposited thin films [91, 92] or STM setups [93] are possible. Furthermore the contacts are mechanically stable because no suspended parts are required.

Electrochemical contacts are often regarded to be three-terminal devices: The two electrodes forming the contact correspond to source and drain, the control electrode to the gate electrode in the language of semiconductor transistors. Since the electrochemical control involves diffusion of ions, it is slower than the usual electrostatic gating in semiconductor technology. It is however much faster than the purely mechanical control used in the lithographic MCBJ technique. One obvious drawback is the fact that the control mechanism requires liquid environment. It is not obvious how one can bring the contacts into dry environment, vacuum or low temperatures. Anyhow, after removal of the electrochemical environment the contacts cannot be varied anymore (or one of the other techniques, e.g. MCBJ or electromigration, have to be applied for this purpose).

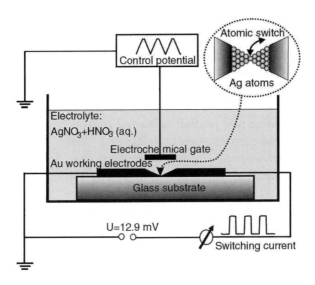

Fig. 2.13 Setup for the electrochemical fabrication and control of atomic contacts. For particular choices of the control potential the atomic contact can be switched between defined conductance values and thus a "switching current" is recorded. Reprinted with permission from [89]. Copyright 2003 by the American Physical Society.

2.9 Recent developments

As mentioned in the introduction of this chapter, many variations of the standard methods described above have been developed. In particular, combinations of the archetypical methods have been described. As an example we present here two new versions of the MCBJ technique. The first one has been introduced by Waitz *et al.* [94]. It uses thin-film-wires on silicon membranes with a thickness of a few hundred nanometers. The membrane is deformed by a fine tip on the rear side. At variance to the MCBJ techniques on bulk substrates the elasticity of the membrane rather than the bending determines the stretching of the metal wire, see Fig. 2.14. The deformation of the substrate is applied locally and it is thus possible to address particular positions while the rest of the circuit on the substrate remains mainly unaffected. This is important when the MCBJ is embedded in a more complex electronic circuit close to the atomic contact, which should not be affected when changing the atomic contact. Such complex circuits are required e.g. for studying Coulomb blockade, which we will describe in Chapter 15. Another advantage of this method as compared to bulk substrates is that the membranes are electrically insulating or only poorly conducting. This reduces the capacitance of the circuit to ground and is advantageous when fast measurements are required. A further difference to standard MCBJ techniques is that smaller suspended length of the metal wire can be used. This enhances stability and reduces often undesired effects such as magnetostriction when investigating magnetotransport as explained in Chapter 12. Finally, by combining this membrane MCBJs with electromigration it is possible to control atomic-size contacts at room temperature without suspension at all [95].

The second recent improvement, which we want to describe here, is the successful incorporation of a gate electrode into the lithographic MCBJ techniques without combination with electromigration [96]. It is based on the lithographic MCBJ technique on metallic substrates using two lithography steps. In the first step a thin and rather narrow metallic gate strip is patterned. The gate is then covered by an approximately 50 nm thick insulating sacrificial layer and the resist system for the second lithography step in which the nanobridge is patterned. After evaporation of the nanobridge metal the sacrificial layer removed by dry etching as in the conventional process for lithographic MCBJs. The result is shown in Fig. 2.15: a nanobridge that is suspended approximately 50 nm above the gate electrode. With this technique three-terminal devices with controllable source-drain coupling are now possible.

2.10   Electronic transport measurements

Usually the first electrical characterization of nanoscale contacts is the measurement of the linear conductance as a function of an outer parameter such as temperature, magnetic field or size of the junction. The next more complex quantity is the nonlinear conductance, i.e. measurements of the current-voltage (I-V) characteristics or the differential conductance. Since these quantities belong to the most common properties of any material characterization their correct measurement is supposed to be trivial, and manifold sophisticated equipment is on the market. In fact, several suppliers of electronic measurement units offer information material or seminars about low-level, high-resolution electronic measurements, and we encourage our readers to access this literature. Therefore textbooks about nanoscience only rarely address this issue. However, when dealing with nanoobjects it is not easy how to perform a good conductance measurement. In this section we will not give a complete overview over the various techniques. But since

Fig. 2.14   MCBJ on silicon membranes. *Top*: Working principle of the membrane MCBJ (not to scale). One or several lithographic MCBJs are defined on the front side of the membrane. A glass or graphite tip is scanned along the rear side of the membrane with the help of micromechanically controlled scan tables. The vertical motion of the tip controls the deformation of the membrane. The close ups at the right side illustrate the deformation of the membrane with a graphite tip, the rupture of the nanobridge, and give an artist's view of the atomic arrangement of a single-atom contact. The thickness of the membrane is in the order of 300 nm, the lateral dimension of the membrane is typically 1 mm × 1 mm. The length of the suspended bridge is smaller than the one for lithographic MCBJs on massive substrates. The thickness of the sacrificial layer is in the order of 100 nm only. When reducing the lateral size of the constriction first by electromigration, non-suspended metal bridges can be used. *Bottom*: optical micrograph of a membrane carrying two MCBJs made of gold. The tip is positioned underneath the lower bridge where the membrane is deformed. The size of the membrane is 0.6 mm × 0.6 mm.

the scope of this book is to serve as textbook for beginners in the field of molecular electronics, we want to sensitize the reader to this issue. The particular facts which have to be taken into account in molecular conductance measurements are the following:

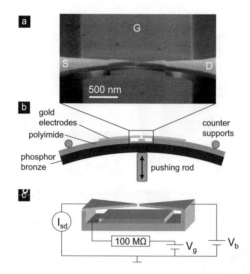

Fig. 2.15   MCBJ with gate electrode on bulk substrate. (a) Scanning electron micrograph of a lithographic MCBJ with gate electrode, (b) working principle of the MCBJ, (c) and electronic circuit for the gated MCBJ. Reprinted with permission from [96]. Copyright 2009 American Chemical Society.

- Wide range of conductances from nanosiemens (corresponding to $10^{-5} G_0$ ($G_0 = 2e^2/h$ is the conductance quantum with e the elementary charge and h Planck's constant) to siemens.[1]
- Correct choice of bias voltage to assure working in the linear regime is difficult because the effects giving rise to nonlinearities happen on varying voltage scales ranging from microvolt to volt.
- Self-heating of the contacts due to Joule dissipation is not always easy to detect and to discriminate from the intrinsic properties of the sample.
- Sudden voltage spikes and jumps may destroy the sample. Therefore abrupt switching actions in the electrical measurement circuit have to be minimized, often hampering optimum range adjustment.
- Extreme variation of the differential conductance within small changes of the bias.
- Limited lifetime of the junctions to study.

The typical signal sizes which have to be resolved are of the order of a few nanovolts for the voltage and picoamperes for the current. For par-

---

[1]1 Siemens is the inverse of $1 \, \Omega = 1$ Volt/Ampere and thus the unit of the conductance in the international system of units (SI).

ticular experiments the requirements might even be stronger. The relative measurement accuracy which is required for most investigations is 10^{-4} or better corresponding to a resolution of typically 14 bits when expressed in digital units. These requirements mean that one often works at the resolution limit of commercial electronic equipment. When enhancing the size of the excitation signal to obtain response signals well above the noise floor one risks to at least smear out the electronic characteristics of the sample by warming it up. In the worst case the sample is destroyed by the heat dissipation.

When designing a measurement circuit the first choice that one has to take is whether one feeds the current and measures the voltage or vice versa. For measurements of the linear conductance, or when the I-Vs are mainly linear, the most important criterion is to optimize the signal-to-noise-ratio. The general rule is that measuring voltage is the better solution for small conductances whereas measuring current is good for high conductance values. When, however, a well-defined energy difference between source and drain is required, e.g. for investigating Coulomb blockade[2], a voltage bias is obviously the best choice. For other purposes the transport current is the decisive quantity and has to be defined. When dealing with hysteretic I-Vs or junctions revealing negative differential resistance (NDR) (see section 13.7) the measurement strategy is crucial for reaching all interesting parts of the I-Vs. Similar choices have to be made concerning the position of the electric ground level of the circuit and whether one pole of the sample will be directly connected to it.

Small nonlinearities in the I-Vs may easily disappear in the noise floor of the electronic circuit. They are much easier to detect with a low-noise lock-in amplifier working at a small but finite frequency. When the electric circuit under study is biased with a harmonic voltage signal, the lock-in detector measures directly the first derivative of the I-V when locking it on the bias frequency. The second derivative (which is an important quantity for detecting vibrational excitations (see Chapter 16) can then be determined by numerical differentiation of the dI/dV. Alternatively it can be directly measured when recording the response at twice the excitation frequency.[3]

In any case the energy scale given by the excitation voltage has to be kept smaller than the width of the vibrational resonances under study.

---

[2]Coulomb blockade and related effects shall be explained in Chapter 15.

[3]Practically all companies producing lock-in amplifiers offer tutorial material available online.

Furthermore the excitation energy has to be smaller than the temperature, otherwise the spectra will be smeared out.

Abrupt changes of the conductance as a function of the bias or another parameter, e.g. the conformation of the junction, result in abrupt changes of the dissipated power as well. On the one hand this is a difficult task for the measurement electronics to cope with. On the other hand this forces one to take precautions, i.e. introduce measures for current limitation, which themselves hamper a perfect voltage bias.

The limited lifetime of the junctions forces one to perform fast measurements, a fact resulting in limited signal to noise ratios and limited statistical information. Atomic and molecular junctions at room temperature reveal intrinsic noise caused by atomic motion. Therefore low-temperature experiments are very appealing. In standard cryostats the wires are rather long and thermalization requires higher cable resistances. Additional measures for high-frequency filtering are required. All these facts reduce the bandwidth of the measurement circuit. As a result it is not trivial to perform fast measurements at low temperatures.

As will be explained in Chapters 13 and 19, many important properties of quantum transport cannot be revealed from conductance measurements alone, but more complex transport properties such as shot noise or thermoelectric voltage have to be studied.

Obviously, for a meaningful noise measurement one has to discriminate the shot noise signal from the undesired but unavoidable noise of the measurement circuit. A fruitful method to do so is a correlation measurement using two identical sets of cables [97, 98]. All noise signals which originate from the wiring are uncorrelated to each other. Signals from the sample are fed into both wires. They are correlated and are recorded in a spectrum analyzer. Only those parts are processed further. An example of such a wiring is shown in Fig. 2.16. It is particularly demanding to measure shot noise at high frequency. A successful solution based on coupled quantum dots has been reported in Ref. [99] and a version using superconducting tunnel contacts in Ref. [100].

For measuring the thermopower a small voltage signal has to be detected which is created by a small temperature gradient across the sample. This means that this temperature difference has to be applied and detected with high precision. One example where this has been successfully achieved is given in Fig. 2.17. It is designed for detecting the conductance and the thermopower of molecular junctions at room temperature [101, 102]. Another setup used for measuring the thermopower in atomic contacts at

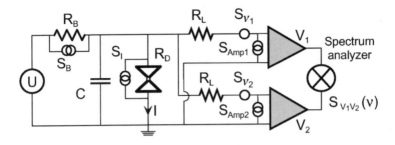

Fig. 2.16 Schematic experimental setup for measuring the voltage dependence of the shot noise of an atomic contact. An atomic contact (double triangle symbol), of dynamic resistance R_D, is current biased through a resistance R_B. The voltage V across the contact is measured by two low noise preamplifiers through two nominally identical lossy lines with total resistance R_L in each line and the total capacitance C introduced by the setup across the contact. The spectrum analyzer measures the cross-correlation spectrum of the two voltage lines. The S_i ($i = B$, Amp1,Amp2) are the known current noise sources associated with the bias resistor and the two amplifiers. S_I represents the signal of interest, i.e. the shot noise associated with the current through the contact. S_{v_1} and S_{v_2} represent the voltage noise sources of each line (amplifier 1 connecting leads). Reprinted with permission from [98]. Copyright 2001 by the American Physical Society.

low temperature is presented shown in Fig. 19.7 and explained there.

With these examples we will finish our short and incomplete list of electronic measurement setups. Our aim was to make clear that although the fabrication of atomic and molecular junctions is not simple, the correct measurement of their electronic transport properties might be even more demanding.

2.11 Exercises

2.1 Vacuum: Estimate the number of gas atoms per area impinging on a surface at normal pressure, in high vacuum ($p = 10^{-6}$ mbar), and in ultra high vacuum ($p = 10^{-10}$ mbar) during one minute. Let us assume that all incoming gas atoms stick to the surface. How thick is the gas layer after 10 minutes?

2.2 Nanowires and atomic contacts: Let us consider a cylindrical nanowire made of Au. Au has a lattice constant of $a = 0.41$ nm.

(a) Estimate the number of atoms in the cross section for a wire with diameter 10 nm, 5 nm, and 1 nm.

(b) Estimate the number of surface atoms for these wires with a length of 5 nm.

(c) Calculate the ratio between surface atoms and bulk atoms in these nanowires.

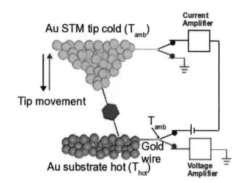

Fig. 2.17   Schematic description of the experimental setup for measuring thermoelectric voltage based on an STM break junction. Individual molecules (symbolized by a hexagon) are trapped between the Au STM tip kept at ambient temperature and a heated Au substrate kept at temperature ΔT above the ambient. When the tip approaches the substrate, a voltage bias is applied and the current is monitored to estimate the conductance. When the conductance reaches a threshold of 0.1 G_0, the voltage bias and the current amplifier are disconnected. A voltage amplifier is then used to measure the induced thermoelectric voltage, while the tip is gradually pulled away from the substrate. Reprinted with permission from [102]. Copyright 2008 American Chemical Society.

2.3 Mechanically controllable break-junctions: Let us consider a MCBJ setup with a separation of the counter supports of $L = 10$ mm, a substrate thickness of $t = 0.5$ mm and a suspended length of $u = 2$ μm. For simplicity let us neglect the insulating sacrificial layer between substrate and metal wire. Calculate the required displacement of the pushing rod for elongating the junction by 10 nm assuming homogeneous bending, when the MCBJ is installed into a differential screw with a pitch difference of 100 μm.

2.4 Joule heating: (a) Calculate the power dissipated in an atomic contact (initially at room temperature) with a resistance of 10 kΩ when a voltage of 10 mV is applied.

(b) Assume that the dissipated power heats up a spherical volume containing 1000 atoms of a material with a specific heat of 130 J/(kg·K). Assume that the sample is only possible to dissipate energy into the environment by radiation. What is the temperature increase?

(c) Perform the same estimation when the sample is surrounded by a material with heat conductivity of 300 W/(K·m).

(d) Repeat the set of estimations for a molecular contact with a resistance of 10 MΩ.

Chapter 3

Contacting single molecules: Experimental techniques

3.1 Introduction

In this chapter we shall present the most common methods for contacting molecules. Although we are mainly interested in single molecule devices, we shall also introduce the most basic methods which are in use for contacting molecular ensembles, since many interesting effects in molecular electronics have first been observed in devices containing these assemblies. Of course, this list can never be complete because new methods and variations of existing ones are constantly being developed. Let us remark that we shall focus here on methods to contact molecules with metal electrodes. Devices including at least one semiconductor electrode have also been realized and examples will be briefly described in section 13.7. Finally, as in the previous chapter, we shall compare the performance of the various techniques and indicate their most common applications.

In the fabrication of molecular junctions not only the kind of the electrodes used is crucial, but also the deposition method of the molecules. Thus, any report about electric current through molecular junctions has to address the "protocol", i.e. the precise contacting scheme including the way how, the moment when, and the conditions under which the molecules are brought into electric contact with the electrodes. For this reason, we shall introduce in this chapter the most common deposition methods, then we shall turn to single-molecule contacting schemes and we shall end by addressing the ensemble techniques.

Particularly interesting are techniques which enable the fabrication of three-terminal devices. In these systems, two of the terminals serve to inject the current and measure the voltage, while the third one acts as a gate that controls the electrostatic potential in the molecule. The incorporation of this third electrode is crucial for revealing the transport mechanism and it allows us to tune the current through a molecular junction, very much like in the transistors fabricated with the standard semiconductor technology.

For the sake of completeness, the first part of this chapter will be devoted to introduce the standard molecules in use in molecular electronics as well as to describe their basic properties.

3.2   Molecules for molecular electronics

Part of the fascination of molecular electronics lies in the fact that the molecular toolbox is almost infinite, which makes us believe that it is possible to find an appropriate molecule for any imaginable application. So far, however, only a few classes of molecules have been explored in molecular electronics. In this section we shall introduce some of these molecules and discuss their basic properties. But before doing that, it is convenient to recall the most common functional elements in digital electronic circuits that molecules are supposed to mimic. The main elements and their requirements are the following:

- Conducting wires: low resistance, high ampacity.
- Insulators: high resistivity, high breakdown voltage.
- Switches: high on/off resistance ratio, reliable switching, small leak current in off position.
- Storage elements: long storage time, low loss.

When extending the scope to cover also logic circuits one additionally has to consider:

- Diodes: high forward/backward current ratio.
- Amplifiers: high gain.

Finally, since most of the existing devices containing molecules are composite devices in which the molecules are connected to either metal or semiconductor electrodes yet another function has to be realized:

- Anchoring groups: reliable contact between functional molecular unit and electrode.

In order to be able to compete with standard semiconductor technology, the time constants of all devices have to be small, i.e. capacitances and/or

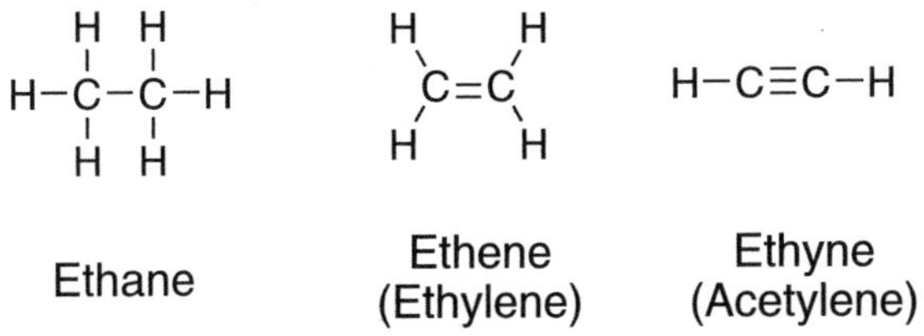

Fig. 3.1 Examples of hydrocarbons. *Left:* Ethane with C-C single bond. *Middle:* Ethene with one C-C double bond. *Right:* Ethyne with one C-C triple bond.

resistances have to be small. Since dissipation is already one of the most severe problems in nowadays semiconductor devices, signal sizes, i.e. the level of the current should be considerably smaller than in those devices. Since our main interest lies in exploring the fundamental properties of molecular electronic devices, we shall not pay attention to those requirements for the rest of this book.

From the very beginning of molecular electronics, it has been become clear that carbon-based molecules offer the required versatility to realize most of these desired functionalities. Carbon is the basis of a great variety of solid structures including graphite, diamond, graphene, and molecules like the cage-shaped fullerenes and - last but not least - the quasi one-dimensional nanotubes.

3.2.1 *Hydrocarbons*

Another very rich class of carbon-based molecules is the hydrocarbons with the possibility to tune their degree of conjugation. The electronic richness of both classes stems from the fact that the degree of hybridization of the molecular orbitals depends on the conformation and the environment. The carbon atom has four valence electrons which in the case of diamond are sp^3 hybridized corresponding to a tetrahedral arrangement of the bonds in space. This conformation is realized in the saturated hydrocarbons with the sum formula C_nH_{2n+2} which are called *alkanes*.[1] Each carbon atom has four direct neighbors, either C or H atoms and all bonds are σ-bonds, see Fig. 3.1. Bigger alkanes with $n \geq 4$ exist in several isomers, some of which are ring-shaped (*cycloalkanes*). Since all electrons are used for forming chemical bonds they are basically localized and the alkanes are insulating.

---

[1]The transport through alkane-based molecular junctions will be discussed in section 14.1.2.

In graphite the valence electrons are sp^2 hybridized in the graphite plane with an angle of 120° between the bonds. The fourth electronic orbital has p character with its lobes pointing perpendicular to the graphite plane. The wave functions of neighboring carbon atoms overlap and form the electronic π-system, which in case of graphite is responsible for the in plane and the finite plane-to-plane conductance. The same situation takes place in the *alkene* hydrocarbons containing one carbon-carbon double bond, see Fig. 3.1. Interesting for molecular electronics are *polyenes* with the sum formula C_nH_{n+2}, which contain more than one double bond. When these double bonds are alternating with single bonds, the wave function of π-system is extended over the whole molecule. These molecules are called conjugated or aromatic molecules. The criterion of aromaticity is $4n + 2$ π-electrons.

The carbons in hydrocarbons may furthermore be triply bond in sp-hybrids forming *alkynes*. When alternated with single-bonds these linear bonds are very stable and give also rise to delocalized wave functions as in the conjugated species with double bonds.

The delocalization of the wave function is broken when the double or triple bonds do not alternate with single bonds. Furthermore, the conjugation can be tuned by introducing an angle between the planes of the individual cyclic parts. The consequences of breaking the conjugation for the conductance of a molecular junction will be discussed in section 13.5.

In a very common representation only the bonds are shown: single bonds as single lines, double bonds as double lines, triple bonds as triple lines. The carbon atoms themselves are not displayed. The positions of the carbon atoms are at the kinks between these lines. Neither the hydrogen atoms nor the bonds to them are drawn. The number and positions of them can be deduced by fulfilling the valence four at each carbon. As an example we show in Fig. 3.2(a) the polyene hexatriene (consisting of six carbons and with three double bonds) in various representations.

As for the alkanes larger species of alkenes and alkynes arrive in several isomers. When two doubly-bond carbon atoms are surrounded by different groups one has to distinguish between the *cis* conformation, in which the neighboring groups are on the same side of the double bond, and the *trans* conformation with the neighbor groups being located on opposite sides of the double bond. A *cis-trans* conformation change sets the basis for a class of molecules with in-built switching functionality.[2]

---

[2]The most popular species of molecular switches are those which can be addressed optically. Many realizations are based on two ground types of switching (cis/trans conforma-

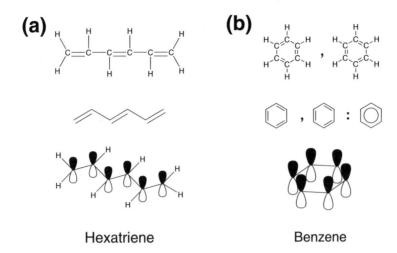

Hexatriene   Benzene

Fig. 3.2 Various representations of the hexatriene and the benzene molecule. (a) The polyene hexatriene is chosen as an example for a conjugated linear hydrocarbon molecule. (b) The benzene molecule. Top and center panel: Because of the delocalization of the π-electrons the positions of the double bonds are not defined. Therefore, they are often symbolized by an inner ring.

The typical conformations of polyenes are zigzag-shaped lines reflecting the preferred 120° orientation of the sp^2 hybrid. When building the angle to the same side cyclic molecules are formed. The ideal cyclic polyene geometry is the benzene molecule consisting of six carbons forming planar ring with perfect conjugation, see Fig. 3.2(b). Since the π-electrons are delocalized over the whole ring, it is not obvious between which carbons the double bonds and where the single bonds have to be drawn. Therefore, one often uses a notation in which the π-electrons are symbolized by an inner ring.

Molecules consisting of several benzene rings merged along one bond are called *polycyclic aromates*. The most prominent examples are naphtalene, consisting of two benzene rings, anthracene consisting of three rings in a linear arrangement, tetracene with four and pentacene with five rings in series. Also angular arrangements of the rings or combinations with rings containing five carbon are used. Examples are shown in Fig. 3.3. Also five-rings (cyclopentadiene) and less often seven-rings (cycloheptatriene) are possible. They are aromatic if six π-electrons per ring exist. In the case

tion switching and ring opening/ring closure). These types of molecules are introduced in section 20.7.

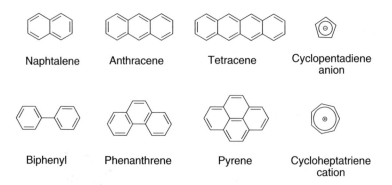

Fig. 3.3   Examples of polycyclic molecules.

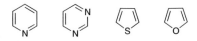

Fig. 3.4   Examples of the most common heterocyclic aromates.

of cyclopentadiene this means that an extra electron has to be added to the ring to provide a stable π-electron sextet (anion), while in cycloheptatriene one electron charge has to be withdrawn (cation), see Fig. 3.3.

In *heterocyclic molecules* one or more carbon atoms are replaced by an atom of another species. Some heterocycles in use in molecular electronics are depicted in Fig. 3.4. The most common substituents are sulfur, nitrogen and oxygen. Because of their chemical valence they posses more electrons than the carbons. In hexagonal rings the additional electrons do not contribute to the π-system, but may be used for forming bonds to other atoms, e.g. to the metal electrodes. In five-rings they help stabilizing the conjugation.

3.2.2   *All carbon materials*

As mentioned in the beginning, also pure carbon molecules are promising for molecular electronics. Carbon nanotubes are sheets of graphite which are rolled together. They have diameters ranging from 1 nm to several tens of nanometers and length of up to millimeters. Depending on the orientation of the long axis with respect to the hexagons various nanotubes with

C_{60}

Fig. 3.5 Line representation of the bonds of the fullerene molecule C_{60}.

varying electronic properties are possible.[3] Since defect-free carbon nanotubes are ballistic conductors they may serve as interconnects for bridging long distances.

Finally, the combination of pure carbon hexagons with pentagons, but without hydrogen sets the basis for the fullerenes. Since the bond length in pentagons is smaller than in hexagons, these molecules are not planar but have a curvature. The most famous fullerene is C_{60} (see Fig. 3.5) consisting of 20 hexagons and 12 pentagons in the same conformation as in a soccer ball. It has a completely delocalized π system, making it also a good candidate for molecular electronics applications.

3.2.3 *DNA and DNA derivatives*

A completely different class of molecules is based on our genetic information carrying molecule DNA. It is very tempting to use DNA because of the rich versatility, the possibility to tune the length from short to very long, and its self-reproduction properties. After almost two decades of research on DNA-based electronics it seems now to be clear that DNA by itself is too poorly conducting for real electronic applications. However, it may serve as template for assembling better conducting molecules or metal-molecule combinations. Furthermore, DNA derivatives are under study which seem to have more fortunate electronic properties. In section 18.3 we shall discuss the transport properties of DNA-based molecular junctions.

---

[3] An excellent review about the conformation and resulting electronic properties of nanotubes is given in Ref. [103].

3.2.4   *Metal-molecule contacts: anchoring groups*

A common problem in molecular electronics is the difficulty to form stable and electronically transparent chemical bonds of the molecules to the metal electrodes. Among the manifold possibilities one particular solution has been chosen as standard system. This is the combination of a sulfur atom to gold electrodes. The reason to choose gold lies in the fact that it is inert to chemical reactions, which allows to prepare clean surfaces and tips. The drawback of this inertia is the fact that it hardly undergoes chemical reactions with other species. One of the rare exceptions is sulfur in its thiol (sulfur-hydrogen) form. This bond is mechanically stable with a force in the order of 1.5 nN [104]. The thiol-gold binding scheme has successfully been tested in self-assembled monolayers (SAM) (see below) on flat surfaces as well as in single-molecule contacts on tips. It provides sufficient electronic transparency for most applications. This is the reason why alkanedithiols (i.e. alkanes with thiol endgroups at both ends) and benzenedithiols (a benzene ring with thiols usually at opposite ends) represent the testbeds for molecular electronic circuits. The alkanedithiols are the archetypical insulators, while benzenedithiol is the most simple aromatic molecule which can be coupled to metal electrodes. However, alternatives to the thiol bonding scheme are also under study, as it will be described in section 14.2.

3.2.5   *Conclusions: molecular functionalities*

We want to close this section by pointing out which molecules can be considered as possible candidates for various electronic components in molecular circuits:

- Conducting wires: polyenes and alkynes.
- Insulators: alkanes.
- Switches: cis/trans conformation changes of manifold molecules, the prototype being azobenzene, consisting of two benzene rings connected via a C=C double bond. In many examples the conjugation is reduced in the trans isomer because the π-systems of both parts are not coplanar. The second prototype of switches are ring-opening-ring-closure transformations which can be triggered optically, see section 20.7. In these switches one of the hydrocarbon rings or heterocycles is opened thereby affecting the conjugation of the π-system.
- Storage elements: all kinds of molecules with at least two states

may serve as storage elements, including among others conformations, redox states, spin states, and vibrational states. Examples will be discussed in Chapters 13, 15, and 16.

- Diodes: molecules which consist of two different, and electronically decoupled parts. An example is the famous suggestion by Aviram and Ratner [8] mentioned in the first chapter.
- Amplifiers: in principle all molecules the electronic levels of which can be tuned by a gate electrode might act as amplifiers. Although electronic three-terminal devices following this principle of bipolar transistors have been demonstrated, they do not provide current amplification yet.
- Anchoring groups: thiols, amines, nitros, cyanos or heterocycles with the substituent atoms serving as linkers to the metal electrodes (see 14.2).

3.3 Deposition of molecules

Molecular deposition methods are manifold because of the rich variety of molecules in use. In most experiments the molecules are deposited from solution onto the metal films forming the electrodes. Various solvents and a wide range of concentrations are used. The molecules are allowed to chemisorb to the metal electrodes. After an incubation time the molecular solution is rinsed away with pure solvent. For low-temperature measurements the devices are then dried in a gas (nitrogen) flow. In some cases the electronic measurements are performed without drying, in solution - either in presence of the pure solvent or with the molecular solution. A variation of this deposition from solution is spin-coating. A drop of the molecular solution is given on the substrate which is mounted on the chuck of spin-coater. Upon rotation of the substrate the solution is wide-spread over the wafer such that a very small concentration of molecules on the substrate is achieved. As an example we mention individual carbon nanotubes, which after spin-coating can be localized by atomic force microscopy or other techniques. A particular nanotube can subsequently be contacted via lithographically defined metal electrodes.

Many molecules, in particular rod-like molecules form self-assembled monolayers (SAM) on metal surfaces. For that purpose the substrate covered with the metal layer is dipped into the molecular solution. The mostly amphiphilic molecules are equipped with one anchoring group that facilitates the chemical adsorption on the surface. The most common combination for molecular electronics devices is thiol-terminated molecules for adsorption on gold surfaces. The molecules organize such that they form ordered monolayers (see Fig. 3.6). This procedure sounds simple, but in practice many parameters have to be well controlled for obtaining reproducible SAM quality. A recent review of this technique is given in Ref. [105].

Another highly developed technique is monolayer formation via the Langmuir-Blodgett (LB) technique [106, 107]. A LB film consists of one or more monolayers of an organic material, deposited from the surface of a liquid onto a metal surface by immersing the solid substrate into the liquid. The molecules form a monolayer on the surface of the solution. The monolayer is transferred to the substrate when dipping it into the solution. Upon repetition of the immersion a multilayer consisting of several monolayers and, thus films with very accurate thickness can be formed (see Fig. 3.6). The film formation relies on the fact that amphiphilic molecules with a hydrophilic head and a hydrophobic tail are used. These molecules assemble vertically onto the substrate. For other molecules a horizontal adsorption may be favored, yielding low-density films. The density and ordering can be enhanced by concentrating the molecular layer on the surface of the solution with a spatula before the substrate is dipped into it.

In particular, for the preparation of samples for low-temperature measurements, remainders of the solvent may hamper the formation of clean metal-molecule-metal junctions. Therefore, alternative "dry" deposition methods have been developed. Gaseous molecules (like e.g. hydrogen, oxygen, nitrogen, carbonmonoxide, methane) can be deposited directly from the gas phase by condensation on the cold metal electrodes. Very stable molecules, like the fullerenes or DNA bases may be evaporated thermally from various sources including Knudsen cells or tungsten boats, which are Joule heated by driving a current through them. More sensitive molecules can be deposited using electrospray ionization (ESI). The method starts with a solution in which the molecules to be ionized are dissolved. An electrospray of this solution is created by a strong electric field, which originates from a voltage applied between the spray needle and the end of a capillary. Due to the strong field at the tip apex, charged droplets are created, which are directed towards the capillary, which forms the connection to a vacuum chamber where the already prepared metal electrodes are located [108]. With this method well-controlled submonolayer molecular films may be deposited onto substrates in ultra-high vacuum (UHV).

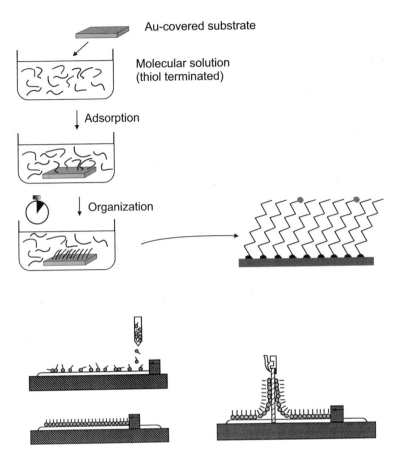

Fig. 3.6   *Top:* Formation of a self-assembled monolayer (SAM) shown for two species of alkanethiols on a gold-covered substrate. The substrate is immersed into the molecular solution. The molecules adsorb assemble with the thiol-terminated end on the substrate. After an incubation time a self-assembled monolayer is formed. *Bottom:* Fabrication of a Langmuir-Blodgett (LB) film. The left panel shows a droplet of an amphiphilic molecule dissolved in a volatile solvent. It is spread on the water-air interface of the trough. The solvent evaporates and leaves a diluted and disordered monolayer behind which is then compressed with the help of a moving barrier. The right panel shows how the monolayer is transferred onto the substrate. Reprinted with permission from Ref. [106].

3.4   Contacting single molecules

The fabrication of single molecule electronic devices is a difficult task. The main problem lies in the size of the molecules, which is usually smaller than the resolution of lithographic methods. Thus, sophisticated techniques have to be applied for forming nanometer-size metal gaps. Most of the longer molecules are not conductive enough to be studied in single-molecule devices, but are rather investigated in ensembles.[4] Furthermore, the coupling between the molecule and the electrodes plays an important role. Another consequence of the small size is the difficulty to image the geometry of the junction and to prove that one deals indeed with a single-molecule device. So far, no method exists which allows one to perform systematic measurements of the electronic transport and to characterize the geometry of a given junction with atomic precision. Therefore, several methods are used and are permanently improved. This enables to distinguish between the properties of the metal-molecule combination and the influence of the contacting scheme. The methods may be divided into two main classes. The first one produces rather stable devices, however, the geometry of it cannot be varied and contamination cannot be excluded. Besides the stability, the possibility to add a third electrode is an important advantage. For that purpose, metal electrodes with small volume are desirable for reducing the shielding of the electric field. The second class enables clean contacts and modification of the junction geometry, but offers only limited stability.

The majority of methods in use for contacting individual molecules are based on one of the techniques described in the previous chapter, since contacting single molecules requires at least one atomically fine metal electrode.

3.4.1 *Electromigration technique*

The electromigration technique described in the previous chapter is successfully used for the fabrication of pairs of metal electrodes for contacting single molecules [21, 109, 110]. For this purpose, the electromigration has to be stopped when the contact is broken and the electrodes form a nanometer-size gap. In vacuum this would be signaled by a sudden increase of the resistance above the typical resistance of a single-atom contact. However, because clean interfaces are needed for achieving well-shaped single-molecule junctions, the molecules are usually deposited - by one of the methods mentioned above - *before* the electromigration process. This complicates the control sequence needed for stopping the electromigration at the right moment, because molecules short-cut the gap resistance. Therefore, many junctions are prepared in parallel and the statistical behavior is determined. Since the metal wire is at ambient conditions before the

---

[4]The transport properties of long molecules will be addressed in Chapter 18.

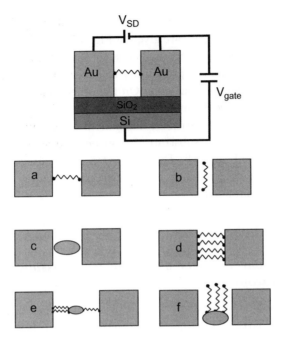

Fig. 3.7   Three-terminal devices and possible artifacts in molecular contacts. *Top panel:* Schematic diagram of electromigration gap and measurement configuration. *Bottom panel:* Six models describing possible geometries formed within the electromigration gap by molecule(s) and contaminant metal particles. (a) Single-molecule contact as desired. The molecule is chemisorbed with both ends at the metal electrodes. (b) Single-molecule in the vacuum gap between the electrodes. The molecule is not chemisorbed. (c) Metal-nanoparticle bridging the gap between the electrodes. (d) Multi-molecule contact. (e) The molecules are coupled indirectly via a metal nanoparticle to the electrodes. (f) The molecules are not chemisorbed to the electrodes but to a metal nanoparticle. After Ref. [109].

deposition of the molecules, all kinds of contaminants might be present and have to be carefully removed before deposition of the molecules.

With this technique, all kinds of current-voltage characteristics have been measured ranging from ohmic behavior to Coulomb-blockade behavior.[5] The tunnel contacts may be formed by vacuum gaps (without molecules), single-molecule or multi-molecule contacts. One particular problem of the method is the risk to form small metal grains, the transport

---

[5]The various possible transport mechanisms will be described in Chapters 13 and 15.

properties of which resemble molecular contacts [109, 110]. Some examples of possible contact geometries are given in Fig. 3.7. Finally, the metal grain may be contacted to both electrodes via one or several molecules. Thus, the yield of this method, i.e. the probability to have a single-molecule contact is in the order of a few percent only. On the other hand, the junctions are extremely stable and well suited for systematic studies of their transport behavior at varying temperature or magnetic field. Because the electrodes are in direct contact to a substrate, it can be used as a back-gate forming a three-terminal device.[6] By applying a gate voltage the transport mechanism can be detected and at least partial information of the contact geometry can be obtained.

3.4.2   *Molecular contacts using the transmission electron microscope*

In order to obtain very strong coupling between the metal electrodes and the molecule, a particular method has been put forward. It includes furthermore the possibility to image the contact geometry, because the molecules form suspended junctions over slits in thin membranes and can thus be inspected by transmission electron microscopy (TEM). Several variations have been reported, which are optimized for the various molecules. The common point is that the metal electrodes, which have been pre-patterned on a thin membrane or a TEM inspection grid, are rapidly heated up by an intensive electron or laser beam above their melting temperature. Contamination atoms are distilled out of the electrodes and defects are driven out as well. The molecules are brought into contact while the metal is liquid. During recrystallization parts of the molecule are soldered into the electrodes resulting in small contact resistances.

This method has been demonstrated to work for long molecules like DNA and carbon nanotubes [112] as well as for chains of clusters [113] (see Fig. 3.8). Possible risks are, of course, destruction of the molecule by the high-energy impact of the laser or electron beam or the hot metal electrodes as well as formation of metal whiskers shorting the molecular junction.

---

[6]The physical results obtained with these devices are discussed in particular in Chapters 15 and 16.

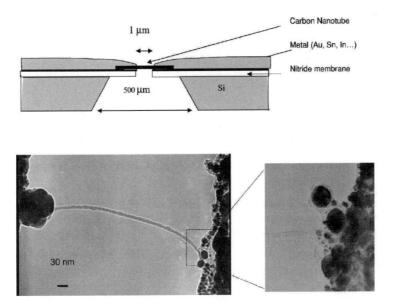

Fig. 3.8   Contacting individual molecules in a transmission electron microscope. *Top:* Schematics of the sample geometry. A long molecule is suspended over a slit in a thin membrane and soldered at its ends to two metal electrodes. *Bottom:* Transmission electronic micrograph (TEM) of nanotubes, suspended across a slit between two metallic pads, and detailed view of the contact region showing the metal molten by the laser beam. Reprinted with permission from [112]. Copyright 2003 by the American Physical Society.

3.4.3   *Gold nanoparticle dumbbells*

A very elegant method for overcoming the size mismatch between the resolution of lithographic methods in use for the definition of the electrodes and the molecules has been described by T. Dadosh *et al.* [114]. The authors use gold nanoparticles (GNPs) with a typical diameter of 10 nm. The molecules to be contacted are functionalized at both ends with thiol anchoring groups, which have a high affinity to gold. By these thiol bonds the molecules are attached to the GNPs such that two of them are combined to form a dumbbell. Those dumbbells now have a suitable size for bridging lithographically defined nanogaps and can be deposited onto them straightforwardly. A further advantage of the method is that the statistical behavior of the molecules in contact with the GNPs can be studied by various non-contact methods such as optical spectroscopic measurements before deposition onto the electrodes (see Fig. 3.9).

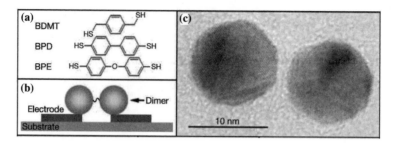

Fig. 3.9 (a) The structures of three molecules studied with the dumbbell technique: 1,4-benzenedimethanethiol (BDMT), 4,4′-biphenyldithiol (BPD) and bis-(4-mercaptophenyl)-ether (BPE). (b) The dimer contacting scheme. (c) TEM image of a BDMT dimer made of 10-nm colloidal gold particles. The separation between the two particles corresponds approximately to the BDMT length (0.9 nm). Adapted with permission from MacMillan Publishers Ltd: Nature [114], copyright 2005.

3.4.4   *Scanning probe techniques*

Conceptually, the most straightforward method for contacting a singe molecule with a fine tip is to deposit the molecule on a metallic substrate and to approach the molecule with the tip until one or several atoms of the molecule are chemisorbed to the tip. However, this is not as simple as it sounds and this method is only suitable for certain molecules. Even if the process is successful, the interpretation of the subsequent conductance measurements is not simple because in STM the electronic signal is convoluted with the topographic information. Furthermore, the presence of the tip may disturb or even destroy the molecule.[7] Therefore, various variations of the STM technique have been developed. They all have in common the difficulty to add a third electrode for gating. A certain but nonlocal gate effect can be achieved via electrochemical gates (see below). STM-based techniques are particularly suitable for gathering statistical information because many contacts can be studied in relatively short time. As already explained in the previous chapter, the price for the high flexibility is the low stability and the, in general, short lifetimes of the junctions.

3.4.4.1   *Direct contact*

The direct contacting scheme mentioned above requires first a careful preparation and characterization of the surface. Subsequently a sub-monolayer of the molecules is deposited. For stable molecules such as the fullerene

---

[7]This can be checked by comparing topographic and spectroscopic results, though.

C_{60} this can be performed via evaporation [115, 116]. The surface is then scanned and a suitable molecule is selected. Depending on the physical question to study, an isolated molecule or a member of a larger aggregate can be chosen. As described in the previous chapter, for single-atom contacts, the formation of a single-molecule chemical junction is signaled by a sudden increase of the conductance. When this is achieved, the approach can be stopped and spectroscopic investigations can be performed. From the electronic point of view this contacting method usually results in asymmetric contacts, meaning that the molecule is electronically better coupled to the substrate than to the tip. This is important for the interpretation of the transport properties, which will be discussed in Part 4. Often the coupling to the substrate is in the "strong" regime while the electrons have to tunnel from the molecule onto the tip and vice versa, i.e. it is in the "weak coupling" regime. Therefore, this method is most suitable for molecules which are only loosely bound to the substrate, e.g. by a single atom or a few atoms, like for C_{60}, where the binding is given through one pentagon or one hexagon of carbon atoms.

3.4.4.2 *Contacting rod-like molecules*

Rod-like or planar molecules have the tendency to lay flat on the surface. In that case the current will not flow along the molecule, but most probably transverse it perpendicularly finding the path of smallest resistance. For those molecules several variations of the scanning probe technique have been put forward. The first method is particularly suitable for imaging and spectroscopy on the molecular orbitals [117]. After preparation of the clean metallic surface, a monomolecular layer of an insulator, e.g. a salt is deposited. The molecules are then evaporated on top of this thin layer which acts as a tunnel barrier between substrate and molecule.

Another possibility is to directly deposit the molecules onto the metal surface, but to design the molecules such that they have edge atoms with high chemical affinity to the tip metal. The tip is then approached to one of these atoms until a chemical bond is formed. Upon carefully withdrawing the tip the molecule is peeled off the substrate, as illustrated in Fig. 3.10. During the peel-off process spectroscopic measurements can be performed which enables to identify the varying charge-transport mechanisms and to quantify the coupling strength [118, 119]. This will be explained in more detail in section 14.4.

The spatial resolution of the STM imaging can be enhanced by suit-

Fig. 3.10 Schematics of the contact formation process of a molecular junction with the STM. Four stages of the contact formation during approach (1,2) and retraction (3,4) are shown. At (3) the chemical bond between the contact atom and the substrate is broken and the molecular wire is formed. Reprinted with permission from [119]. Copyright 2008 IOP Publishing Ltd.

able functionalization of the STM tip, e.g. with hydrogen molecules [122]. Recently, it has been demonstrated that molecular orbitals can be even better resolved by atomic force microscopy when the tips are terminated with carbon monoxide (CO) molecules [120, 121].

Finally, an elegant way to contact rod-like molecules is to embed the molecules into a matrix of less conducting molecules, such that the long axis of the molecules is almost perpendicular to the substrate, see Fig. 3.11(a). With the techniques described in section 3.3 a self-assembled layer of weakly conducting molecules is prepared. A standard combination would be alkanes with one thiol anchor group on a gold substrate. The thiol binds chemically to the gold releasing the non-thiolated ends to the top of the SAM. The properties of the SAM are chosen such that free places or defects exist at which the study molecule can be incorporated. When scanning the sample with an STM tip the positions of the better conducting molecules can be located and spectroscopic measurements can be performed [123].

In a variation of this technique the study molecules are equipped with two highly reacting anchoring groups, e.g. thiols. One end attaches to the gold surface, the other one pointing to the top of the SAM. These thiols can the be used as binding places for gold nanoparticles (GNPs) , see Fig. 3.11(b). Depending on the density of the study molecule and the size of the GNPs, one or several molecules are contacted with the same GNP. In this way a very stable molecular junction consisting of substrate, molecule and GNP is fabricated. The prepared sample is then investigated with an STM [124] or a conductive AFM [125]. The tip is either brought into strong contact with the GNP, such that the tip-GNP contact has negligible resistance. Or the transport properties due to the presence of the GNP have to be incorporated in modeling the transport for deducing the properties of the molecular junction. The obvious advantage of this latter method is the high stability of the device. Both variations share the in-built possibility

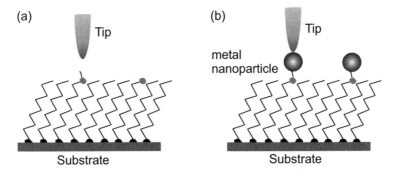

Fig. 3.11 (a) Scanning tunneling microscopy (STM) study of electron transport through a target molecule inserted into an ordered array of reference molecules. (b) STM or conducting atomic force microscopy (AFM) measurement of conductance of a molecule with one end attached to a substrate and the other end bound to a metal nanoparticle. After Ref. [40].

to perform statistical investigations because hundreds of junctions can be prepared on the same chip. The main drawbacks are the complex sample preparation and the limited versatility because successful embedding into the matrix is not obvious.[8]

3.4.4.3 *STM in liquid environment*

A very powerful tool is the use of an STM in liquid environment. The surface and the tip are prepared as usual for forming atomic contacts, but immersed into a solvent, in which the molecules under study can be dissolved. The tip can be sharpened and covered with substrate atoms by repeated indentation into the substrate. Then molecules are added to the solvent. After an incubation time needed for chemical binding to the substrate, the tip is repeatedly approached to the surface and withdrawn while the conductance is recorded. Upon closing the gap a metal-molecule-metal junction consisting of several molecules is formed. When withdrawing the tip, the molecules loose the contact to the electrodes not all at once but in an irregular series. The result is a step-like decrease of the conductance as a function of the distance which varies from repetition to repetition. After breaking the contact to the last molecule, a new junction can be formed. The molecules which get stuck to either the substrate or the tip are replaced by fresh molecules diffusing in from the solution. After a while a

---

[8]Examples of transport measurements performed with this technique will be described in Chapter 13.

new position on the substrate can be chosen. The method is very suitable for gathering statistical information about the preferred conductance values of molecular junctions. However, the stability of the molecular junctions is usually not sufficient for spectroscopic measurements [36, 126].[9]

3.4.5   *Mechanically controllable break-junctions (MCBJs)*

Mechanically controllable break-junctions (MCBJs) (see section 2.6) are used for contacting molecules in various environments. For measurements at room temperature under ambient conditions, the molecules are usually deposited from solution. After a reaction period, the remainder of the molecular solution is rinsed in pure solvent and blew dry with nitrogen. At variance to the electromigrated junctions, the molecules are usually deposited *after* forming the electrode gap by breaking the MCBJ. For polarizable molecules it might be helpful to apply a voltage in order to pull one or several molecules into the junction. The junction is then carefully closed until a measurable current flows. Depending on the molecule, the closing traces show plateaus which signal the formation of a molecular junction containing one or several molecules. At room temperature the electrode atoms are rather mobile and the molecular junctions have only limited lifetime of a few minutes. This is, however, a much longer time span than usually achieved with STM setups and is sufficient for measuring I-V characteristics. On the other hand, only limited statistical information can be acquired because of aging effects of the junctions. After several opening or closing cycles no molecular junctions form any more. For recording conductance histograms it is advantageous to perform the measurements in liquid environment, as it was first proposed by Grüter *et al.* [78]. Fig. 3.12 shows a slightly different setup. A pipette is pressed onto the inner part of MCBJ electrodes and sealed with gasket made of a flexible and solvent resistant material (polydimethylsiloxane (PDMS)). The molecular solution is continuously pulled through the pipette, while the MCBJ is opened and closed and the conductance is recorded. Molecules which leave the junctions are replaced by fresh ones from the solution as discussed earlier for STM setups.

Much longer lifetimes of molecular junctions can be achieved at low temperature. Furthermore, the thermal smearing of the electronic properties is considerably reduced. For that purpose several protocols have been

---

[9]This technique has been applied for transport measurements through DNA. Examples will be discussed in section 18.3.

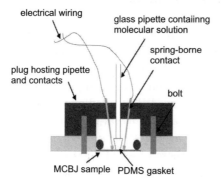

electrical wiring

glass pipette contaiinng
molecular solution

spring-borne
contact

plug hosting pipette
and contacts

bolt

MCBJ sample   PDMS gasket

Fig. 3.12   A PDMS-sealed glass pipette, in which the molecular solution circulates, is pressed onto the central part of MCBJ chip with the help of a plug screwed to the sample holder. The electrical contacts are realized in this case via spring-borne contacts outside the gasket.

developed. When starting with a deposition of the molecules from solution, the solvent and any humidity has to be carefully removed in order to obtain clean molecular junctions without tunnel barriers due to ice formation. For this purpose it is helpful to make use of strong metal molecule binding: A molecular junction is formed at room temperature. When breaking it again it may happen that the breaking does not occur between the molecule and the metal electrode, but that one or several gold atoms remain attached to the molecule leaving a gap between two metal atoms. The junction is then cooled down and the metal-metal gap is closed again. Of course, water films or other kinds of contamination may form on the metal surfaces as well, but they can be pushed out of the contact such that a good electrical contact can be established.

The problem of ice formation can be solved when forming the electrode gap at low temperatures under cryogenic vacuum conditions. Even though the surface of the native break-junction might be covered with water or other contaminants fresh and clean metal tips are formed. Small molecules, which at ambient conditions are in the gaseous phase (like e.g. hydrogen, oxygen, carbonmonoxide, methane), may be condensed directly onto the cold MCBJ electrodes with a nanometer-size separation [127]. Other small molecules with low evaporation temperature (e.g. water) are first vaporized and then condensed. Similarly, stable molecules like the fullerenes can be evaporated on an opened MCBJ at low temperature [128].

3.5   Contacting molecular ensembles

One main problem in single-molecule studies lies in the fact that the electronic transport depends crucially on the exact coupling between the molecule and the metal electrodes, i.e. on the precise atomic arrangement of the contacts.[10] As a result pronounced sample-to-sample and junction-to-junction variations are observed. Repeated measurements are needed to deduce the typical behavior of a given metal-molecule system. The influence of varying contact geometry averages out in devices containing ensembles of molecules. Furthermore, these ensembles are contacted with rigid and robust electrodes. These devices usually provide better mechanical stability and longer life-times allowing long-time systematic measurements and the variation of outer control parameters like temperature or magnetic field.

However, when interpreting data recorded on ensemble devices one has to bear in mind possible interaction effects between the molecules themselves which might affect their electronic properties. Furthermore, also without interaction effects it is not straightforward to infer the single-molecule junction behavior from the ensemble because the number of molecules which contribute to the transport may be smaller than the total number of molecules in the ensemble, if not all are contacted equally. For instance, some of the molecules forming the ensemble might be in strong coupling to the electrodes while others are only weakly coupled. As a result the transport characteristics may show superpositions of various transport mechanisms. Furthermore, ensemble structures are necessarily larger in space than single-molecule devices giving limits to their maximum integration density. From the point of view of fundamental research the most promising strategy is to compare the results from single-molecule contact schemes with ensemble measurements for revealing the robust properties of the given molecule-metal system. We shall restrict ourselves to methods suitable for small ensembles ranging from roughly a few hundred molecules to several thousand molecules. Very efficient methods have been developed for contacting large area molecular films, which are however, out of scope of this monograph.

3.5.1   *Nanopores*

One technique which produces rather small ensembles of molecules uses pores in thin freestanding membranes. The method has been used in

---

[10]This issue will be addressed in Chapters 13 and 14.

the 1980's and 1990's for fabricating nanometer-sized metallic contacts for point contact spectroscopy [129]. However, no single-atom contacts can be achieved. A single crystalline silicon wafer is covered from both sides with a thin layer of silicon nitride with a typical thickness of 50 nm to 100 nm. The rear side of the wafer is patterned by optical lithography with squares of typical lateral size of 100 μm. Using first plasma etching for attacking through the nitride, then wet-etching in hydrofluoric acid the squares are etched through the bulk of the silicon wafer. The wet etching process is anisotropic. It attacks particular crystal orientations of the silicon much faster than others. As a result inclined etch walls are formed thereby reducing the size of the squares. The inclined walls become covered with a native silicon oxide layer during the following process steps. Furthermore, the acid attacks silicon much faster than silicon nitride. The process can thus be stopped controllably when a suspended silicon nitride membrane is obtained. Now the membranes are patterned from the front side via electron beam lithography with a small dot in each membrane. Using plasma etching a small pore is drilled into the membrane with a typical diameter of 10 to 50 nm.

The formation of molecular junctions requires three further steps [130]. First, a metal electrode - usually gold - is evaporated from the top side. The device is then immersed into the molecular solution until a SAM has formed. After a suitable reaction time which depends on the molecule-metal combination the sample is rinsed and dried and the second metal electrode is deposited by evaporation onto the rear side, see Fig. 3.13. Care has to be taken that the SAM is not destroyed by thermal impact coming from the metal atoms. With this technique thermally stable molecular ensemble junctions are obtained which are particularly suitable for studies of the temperature dependence of the transport properties. A difficulty of the method lies in the fact that the quality of the first deposited electrode cannot be characterized; it might be covered with water or other contaminants which could hamper the formation of a high-quality SAM. A similar objection was made concerning the second molecule-metal interface: The molecular layer is exposed to ambient conditions before the deposition of the second electrode.[11]

---

[11] We will discuss data recorded with this sample species in Chapter 13.

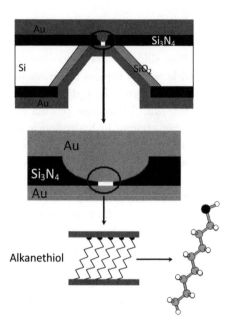

Fig. 3.13   Molecular junctions in nanopores. A small molecular ensemble is contacted with metal electrodes in a nanometer-sized pore in a silicon-nitride membrane. Top schematic is the cross section of a silicon wafer with a nanometer-scale pore etched through a suspended silicon nitride membrane. Middle and bottom schematics show a Au-SAM-Au junction formed in the pore area. The structure of octanethiol is shown as an example. Reprinted with permission from [130]. Copyright 2003 by the American Physical Society.

3.5.2   *Shadow masks*

Another method to fabricate small ensemble devices uses the self-alignment property of shadow masks. The sample fabrication scheme is shown in Fig. 3.14. Via e-beam lithography a suspended mask is produced with a geometry of a wire that is interrupted by a small gap. A first metal layer is evaporated perpendicularly through this mask. The next step is the deposition of a SAM of the molecules. Alternatively molecules can be evaporated on top of the metal under the same angle. Subsequently a second metal layer is evaporated under an inclined angle such that the edge of the metal film covers the molecular layer. The resolution limits of the lithography used for the preparation of the mask restrict the contact size to roughly 50 nm in width. The overlay length is given by the evaporation angle and is usually chosen in the range of 20 to 50 nm. It has been

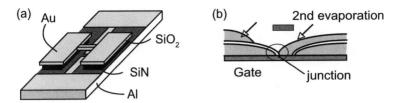

Fig. 3.14 Production of shadow mask on silicon substrate. (a) The shadow mask is defined via electron-beam lithography in a Si_3N_4/SiO_2 double layer using two dry etching steps. (b) The bridge in the center of the structure is used to separate two metal contacts, which are evaporated vertically onto the substrate. A SAM is deposited on both electrodes. In a second step metal is evaporated under an angle that allows a small overlap between this top electrode and one of the bottom electrodes. If this overlap is small enough, transport through single or a few molecules can be possibly measured. Reprinted with permission from [111]. Copyright 2005, American Institute of Physics.

shown that the smoothness of the first metal layer is mandatory for avoiding shortcuts between both electrodes. A second problem of this method is the risk of destroying the SAM by the heat impact during the evaporation of the top electrode or of creating metal grains [109].

3.5.3 *Conductive polymer electrodes*

These problems are partially overcome by a technique described by Akkerman *et al.* [131]. The fabrication method is shown in Fig. 3.15. In a first lithography step metal lines are fabricated and then a second resist is spread over the sample. In the next step this resist is patterned with holes via electron-beam lithography. The molecular ensemble is deposited into these holes. Next, the whole substrate is overcast with a highly conductive polymer which provides the second electrode. The polymer is finally capped by a planar top metal electrode. The result is a very robust molecular junction because the SAM remains embedded into the resist. Furthermore, the deposition of the conductive polymer is less aggressive to the SAM than standard metal deposition techniques. At variance to most of the previously described methods the contact scheme intrinsically gives rise to asymmetric contacts.[12] The fact that at least one of the metal electrodes is not in direct contact with the molecular ensemble can be helpful when exciting the molecular system optically, as described in Chapter 20.

---

[12]The importance of the metal-molecule contact shall be discussed in detail in section 14.2.

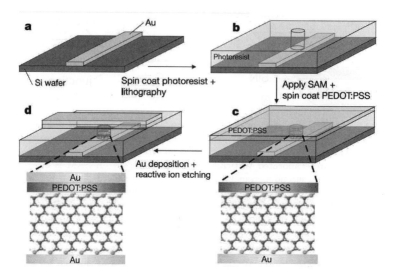

Fig. 3.15   Processing steps of a large-area molecular junction. (a) Gold electrodes are vapor-deposited on a silicon wafer and a photoresist is spin-coated. (b) Holes are photolithographically defined in the photoresist. (c) An alkane dithiolSAM is sandwiched between a gold bottom electrode and the highly conductive polymer PEDOT:PSS as a top electrode. (d) The junction is completed by vapor-deposition of gold through a shadow mask, which acts as a self-aligned etching mask during reactive ion etching of the PEDOT:PSS. The dimensions for these large-area molecular diodes range from 10 to 100 mm in diameter. Reprinted with permission from MacMillan Publishers Ltd: Nature [131], copyright 2006.

3.5.4   *Microtransfer printing*

A method which combines gentle deposition of the top electrode with the ability to fabricate arrays of molecular junctions with similar contact properties is given by the micro- or nanotransfer printing technique. It produces stable contacts on a substrate and involves also the formation of a SAM (Fig. 3.16). At first an array of bottom electrodes is fabricated using lithographic methods or evaporation through a mechanical mask. Subsequently a SAM of the molecules to study is formed on the substrate. The molecules are functionalized at their top ends with an anchoring group suitable for binding to the metal of the top electrode. In a separate fabrication line a stamp made of a flexible material such as PDMS is fabricated. The stamp is topographically patterned in the geometry of the top electrodes. The metal of the top electrode is evaporated onto it. This stamp is then pressed onto the substrate. During this step the metal is transferred from the stamp to

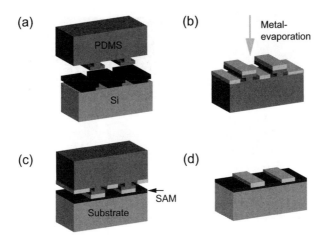

Fig. 3.16   Production of nanoscale features by nano transfer printing (nTP). (a) The features are defined by electron beam lithography in a polymethylmethacrylate (PMMA) double layer on a silicon substrate. The elastomeric polydimethylsiloxane (PDMS) is cast into the structures and cured at 60° C. Fluorination of the substrate before this step ensures easy separation of PDMS and substrate after the curing. (b) Layers of 10-30 nm metal gold are evaporated onto the PDMS stamp. (c) Alkanedithiols form a monolayer on a GaAs substrate. The gold on the PDMS stamp binds to this monolayer and is transferred to the substrate. (d) The patterned gold film that forms is transferred on top of the GaAs substrate. Good binding to the monolayer is proved by the scotch tape test. Reprinted with permission from [111]. Copyright 2005, American Institute of Physics.

the substrate, thus forming an array of molecular junctions. This technique enables junctions with areas ranging from less than a micrometer squared - and thus named nanotransfer printing (nTP) - up to several hundred square micrometers - microtransfer printing (μTP) [111, 132]. Besides the in-built statistical information of molecular ensembles the quality of the SAM and the contacts can be investigated by comparing contacts with varying area. Furthermore, the contacts may be gated by applying voltages to the substrate.

3.5.5   Gold nanoparticle arrays

Finally, it is possible to form networks of single-molecule junctions combining the robustness and statistical richness of ensemble studies with the fact that each junction is formed by a single molecule or a very small number of molecules only [133, 134]. The fabrication scheme is shown in Fig. 3.17. At first gold nanoparticles (GNP) with a diameter of roughly 10 nanometers

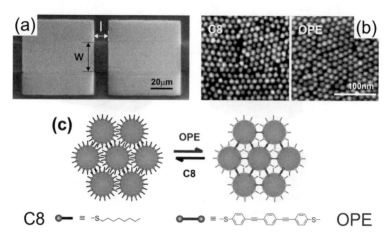

Fig. 3.17 Contacting molecular networks with gold nanoparticles. (a) Electron microscopy image of a device: two square-shaped gold contacts were evaporated on top of a nanoparticle array line of width w. (b) Electron micrographs of the array structure before and after OPE (oligo-phenylene-ethynylene) exchange. (c) Schematic of the molecular-exchange process. *Left:* self-assembled alkanethiol-capped nanoparticles before exchange. *Right:* During the exchange process. The OPE molecules displace part of the alkane chains and interlink neighboring nanoparticles to form a network of molecular junctions. Adapted with permission from [133]. Copyright Wiley-VCH Verlag GmbH & Co. KGaA.

are covered with a spherical ligand shell. The thickness of the ligand shell corresponds to half the length of the molecules which shall be assembled between the GNPs later. A dense-packed, well-ordered, two-dimensional array with an approximate size of 10 μm $\times$ 20 μm of these dressed GNPs is deposited onto a substrate which is subsequently patterned with metallic electrodes for performing the contacts to the measurement circuit. The array contains approximately a million nanoparticles. The molecules forming the ligand shell can be replaced with an exchange reaction by the molecules to be studied electrically. By using network analysis methods the typical properties of an individual molecular junction can be at least partially deduced from the behavior of the network. Besides the particular stability and in-built ensemble averaging, this method is suitable for the investigation of very small signals, such as electrical response to optical activation of photochromic molecules [135], see section 20.7.

3.6   Exercises

3.1 Molecular ensembles: Estimate the number of alkanedithiol molecules in the cross section of a nanopore with a diameter of 50 nm (see section 3.5.1). For estimating the diameter of a molecule assume a C-C bond length of 0.15 nm, a C-H bond length of 0.11 nm and a bonding angle of $110°$ between adjacent C-C bonds. Furthermore, assume a densely packed SAM in a triangular arrangement.

3.2 Molecular arrays: Let us consider the technique shown in Fig. 3.17. Assume that the exchange reaction was perfect. Furthermore, assume that each pair of nanoparticles is connected via a single molecule. How many molecules will contribute to the transport if the array has a size of 20 μm $\times$ 10 μm. What is the effective circuit diagram of this network? What happens when the exchange reaction has a yield of 50%? What is the minimum rate for the exchange reaction in order to obtain at least one conducting path between the ends of the array (percolation threshold)?

3.3 Optical activation of molecules: In Chapter 20 we will present experiments in which molecular contacts were excited by light irradiation. Therefore, we want to estimate here the probability that a molecule in contact with metal electrodes will be hit by a photon of the light source. Assume a single decanedithiol molecule which spans the gap between two gold electrodes. The electrodes have been fabricated with the MCBJ technique and have a cross section of 100 nm times 100 nm. The break forms a slit with perfectly flat walls perpendicular to the direction of light irradiation. The width of the slit is given by the length of the molecule. Typical light intensities of the experiments are $P = 1$ mW focused on an area of $s = 100$ μm^2 with a light wavelength of $\lambda = 400$ nm. Consider different positions of the molecule in the slit: (a) Top of the slit. (b) Center of the slit. (c) Bottom of the slit.

PART 2
Theoretical background

Chapter 4

The scattering approach to phase-coherent transport in nanocontacts

4.1 Introduction

The electrical conduction in macroscopic metallic wires is described by Ohm's law, which establishes that the current is proportional to the applied voltage. The constant of proportionality is simply the conductance, G, which for a given sample grows linearly with the transverse area S and it is inversely proportional to its length L, i.e.

$$G = \sigma \frac{S}{L}, \tag{4.1}$$

where σ is the conductivity of the sample, which is a material specific property. The conductance will be a key quantity in our analysis of the transport properties of atomic and molecular junctions. However, concepts like Ohm's law are not applicable at the atomic scale. Atomic-size conductors are a limiting case of mesoscopic systems in which quantum coherence plays a central role in the transport properties.

In mesoscopic systems one can identify different transport regimes according to the relative size of various length scales. These scales are, in turn, determined by different scattering mechanisms. A fundamental length scale is the *phase-coherence length*, L_φ, which measures the distance over which the information about the phase of the electron wave function is preserved. Phase coherence can be destroyed by inelastic scattering mechanisms such as electron-electron and electron-phonon interactions. Scattering of electrons by magnetic impurities, with internal degrees of freedom, also degrades the phase but elastic scattering by (static) non-magnetic impurities does not affect the coherence length. Information on the coherence length can be obtained experimentally, for instance, by studying the so-called weak localization [50]. A typical value for Au at $T = 1\,\mathrm{K}$ is around $1\,\mu\mathrm{m}$ [136],

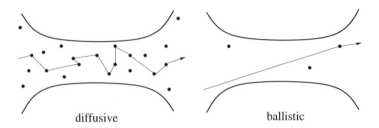

diffusive   ballistic

Fig. 4.1   Schematic illustration of a diffusive (left) and ballistic (right) conductor.

while at room temperature it becomes of the order of a few tens of nm. The mesoscopic regime is determined by the condition $L < L_\varphi$, where L is a typical length scale of our sample.

Another important length scale is the elastic mean free path ℓ, which roughly measures the distance between elastic collisions with static impurities. The regime $\ell \ll L$ is called *diffusive*. In a semi-classical picture the electron motion in this regime can be viewed as a random walk of step size ℓ among the impurities. On the other hand, when $\ell > L$ we reach the *ballistic* regime in which the electron momentum can be assumed to be constant and only limited by scattering with the boundaries of the sample. These two regimes are illustrated in Fig. 4.1.

In the previous discussion we have implicitly assumed that the typical dimensions of the sample are much larger than the Fermi wavelength λ_F. However, when dealing with atomic-scale junctions the contact width W is of the order of a few nanometers or even less and thus we have $W \sim \lambda_F$. We thus enter into the *full quantum* limit which cannot be described by semi-classical arguments. A main challenge for the theory is to derive the conductance of an atomic-scale conductor from microscopic principles.

In this chapter we shall introduce the scattering (or Landauer) approach, which is presently the most popular theoretical formalism to describe the coherent transport in nanodevices. The central idea of this approach, already put forward by Rolf Landauer in the late 1950's [137], is that if one can ignore inelastic interactions, a transport problem can always be viewed as a scattering problem. This means in practice that transport properties like the electrical conductance are intimately related to the transmission probability for an electron to cross the system. Our introduction to the scattering approach will be divided into two main parts. First, using heuristic arguments we shall show the relation between conductance and transmission, which is summarized in the so-called Landauer formula. This formula will then be used to discuss basic concepts such as the tunnel effect or resonant tunneling. Second, we shall present a more rigorous formulation of this approach that will be used to compute various transport properties such as shot noise and thermoelectric coefficients. Finally, we shall conclude this chapter with a discussion of the limitations of the scattering formalism.

4.2   From mesoscopic conductors to atomic-scale junctions

On the basis of Ohm's law one would expect the conductance of a metallic wire to scale as R^2, where R is its radius. Deviations from such a scaling law were already discussed by Maxwell [138], who studied with classical arguments the conductance of a diffusive constriction, where the contact radius is large compared to the mean free path. He found that the conductance scales linearly with the contact radius, i.e.

$$G = 2R\sigma. \tag{4.2}$$

where σ is the conductivity.

As we shrink a conductor to well below the mean free path, the conductance departs from the value expected from the previous expression. In 1965 Sharvin [139] considered the propagation of electrical current through a ballistic contact by approximating it with a classical problem of dilute gas flow through an orifice. He reasoned that if the potential difference between the two half-spaces is eV, the conduction electrons passing through the orifice should change their velocity by the amount $\Delta v = \pm eV/p_{\mathrm{F}}$, where p_{F} is the Fermi momentum.[1] The net current will be $I = ne\Delta vS$, where $S = \pi R^2$ is the contact area and taking into account the Fermi-Dirac statistics for electrons, $n = 4\pi p_F^3/(3h^3)$, one gets the conductance for a circular ballistic point-contact

$$G = \frac{2e^2}{h}\left(\frac{\pi R}{\lambda_{\mathrm{F}}}\right)^2 = \frac{2e^2}{h}\left(\frac{k_{\mathrm{F}}R}{2}\right)^2, \tag{4.3}$$

where e is the electron charge and h is the Planck's constant. Notice that for ballistic contacts the conductance is proportional to the contact area, like in Ohm's law, but the proportionality constant $2e^2/h$ has a quantum nature. An important difference between the two lies in the fact that G is

---

[1]This is just an approximation and the exact treatment includes an integration of the projection of Δv along the orifice axis over the solid angle of 2π. Anyway, the phenomenological result is only a factor $8/3$ different from the exact one [140].

independent of the length of the conductor and is determined only by its cross-section radius R. It is remarkable that the Sharvin formula, being based on semiclassical arguments, holds well for all ballistic contacts with diameters down to a few nanometers. In the context of atomic contacts, it is customary to use a slightly modified version of this equation in which the so-called Weyl correction is introduced [141, 142]. This correction comes from the fact that the Heisenberg uncertainty principle for Fermi electrons in a narrow contact, $2p_F R \geq \hbar$, gives a small correction to the conductance and the resulting semiclassical formula takes the form

$$G = \frac{2e^2}{h} \left(\frac{k_F R}{2} \right)^2 \left(1 - \frac{2}{k_F R} + \cdots \right), \qquad (4.4)$$

where k_F is the wave vector. This equation is valid for a contact in the form of a wire. For an orifice the numerator of the last fraction should be 1 instead of 2. Eq. (4.4), valid for contacts down to a few nanometers in diameter [143], is often used to establish the relationship between the conductance and the radius of a contact.

Due to limitations of the semiclassical approach, Eq. (4.4) does not account for purely quantum effects which dominate when the size of the contact becomes so small that the wave nature of an electron can no longer be ignored. Rolf Landauer [137] showed, already back in the 1950's, that in the latter case "conductance is transmission", i.e. in order to determine the total conductance one has to solve the Schrödinger equation, find the current-carrying eigenmodes, calculate their transmission values and sum up their contributions. Mathematically, this is summarized by in the Landauer formula

$$G = \frac{2e^2}{h} \sum_{n=1}^{N} T_n, \qquad (4.5)$$

where the summation is performed over all available conduction modes and T_n are their individual transmissions. If the transmission of a mode is perfect, it contributes exactly one quantum unit of conductance, $G_0 = 2e^2/h \sim (12.9 \text{ k}\Omega)^{-1}$. This formula shows that by changing the size of the contact, one can change the number of modes contributing to the conductance and thus the conductance itself in a step-like manner (see discussion below). This is clearly at variance with the situations described above. The derivation of the Landauer formula and the discussion of its physical implications is the subject of the rest of the next sections.

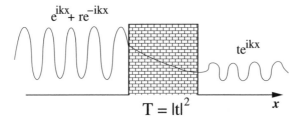

$$e^{ikx} + re^{-ikx}$$

$$te^{ikx}$$

$$T = |t|^2$$

x

Fig. 4.2   Wave function (plane wave) impinging on a potential barrier. The wave is partially reflected with a probability amplitude r and partially transmitted with a probability $T = |t|^2$.

4.3   Conductance is transmission: Heuristic derivation of the Landauer formula

In a typical transport experiment on a nanoscale device, the sample is connected to macroscopic electrodes by a set of *leads* (or electrodes) which allow us to inject currents and fix voltages. The electrodes act as ideal electron reservoirs in thermal equilibrium with a well-defined temperature and chemical potential. The basic idea of the scattering approach is to relate the transport properties with the transmission and reflection probabilities for carriers incident on the sample. In this one-electron approach phase-coherence is assumed to be preserved on the entire sample and inelastic scattering is restricted to the electron reservoirs only. Instead of dealing with complex processes taking place inside the reservoirs, they enter into the description as a set of boundary conditions. In spite of its simplicity, this approach has been very successful in explaining many experiments on nanodevices.

Before turning to the description of the general scattering formalism, it is instructive to understand the relation between current and transmission with a simple heuristic argument. Let us consider a one-dimensional situation, like the one depicted in Fig. 4.3. Here, the potential simulates the central part of a junction, where electrons are elastically scattered before reaching one of the electrodes. We assume that when the electrons are inside the reservoirs, they are in thermal equilibrium at the temperature of the corresponding electrode. Let us now consider a plane wave, $(1/\sqrt{L})e^{ikx}$, that is impinging on the potential barrier from the left (L represents the length of the system). This wave is partially reflected with a probability amplitude r and partially transmitted with a probability $T = |t|^2$. We can now compute the electrical current density, J_k, carried by an electron described by this wave function. It is given by the quantum-mechanical expression

$$J_k = \frac{\hbar}{2mi}\left[\psi^*(x)\frac{d\psi}{dx} - \psi(x)\frac{d\psi^*}{dx}\right] = \frac{e}{L}v(k)T(k), \qquad (4.6)$$

where $v(k) = \hbar k/m$ is the group velocity and we have computed the current on the right hand side of the scattering potential (remember that the current is conserved and thus its value is independent of where it is evaluated).

In a solid state device there are many electrons contributing to the current. Therefore, we have to introduce a sum over k (strictly speaking over the positive values). Moreover, we have to take into account the Pauli principle, which means in practice that we have to introduce a factor $f_L(k)[1 - f_R(k)]$, where $f_{L,R}$ is the Fermi function of the electron reservoir on the left (L) or on the right (R) of the potential barrier. These Fermi functions take also into account the fact that the corresponding chemical potential can be shifted by an applied bias voltage, V. The blocking factor above ensures that only those states that were initially occupied on the left and empty on the right contribute to the current flowing from left to right, $J_{L \to R}$, which adopts the form

$$J_{L \to R} = \frac{e}{L}\sum_k v(k)T(k)f_L(k)[1 - f_R(k)]. \qquad (4.7)$$

Now, we can convert the sum into an integral with the usual replacement: $(1/L)\sum_k g(k) \to 1/(2\pi)\int g(k)dk$. Thus,

$$J_{L \to R} = \frac{e}{2\pi}\int dk\ v(k)T(k)f_L(k)[1 - f_R(k)]. \qquad (4.8)$$

We now change from the variable k to energy, E, introducing the density of states $dk/dE = (dE/dk)^{-1} = m/(\hbar^2 k)$, since $E = \hbar^2 k^2/(2m)$.[2] Due to the cancellation between the group velocity and the density of states, the left-to-right current can be written as

$$J_{L \to R} = \frac{e}{h}\int dE\ T(E)f_L(E)[1 - f_R(E)]. \qquad (4.9)$$

Analogously, we can show that the current from right to left can be written as

$$J_{R \to L} = \frac{e}{h}\int dE\ T(E)f_R(E)[1 - f_L(E)], \qquad (4.10)$$

---

[2]Here, we are assuming that the conduction electrons can be described by a non-interacting electron (or Fermi) gas.

where we have used the fact that the transmission probability is the same, no matter in which direction the barrier is crossed.

Now, the total current[3] $I(V) = J_{L \to R} - J_{R \to L}$ can be simply expressed as

$$I(V) = \frac{2e}{h} \int_{-\infty}^{\infty} dE\, T(E)[f_L(E) - f_R(E)]. \qquad (4.11)$$

Here, we have introduced an extra factor 2 to account for the spin degeneracy that usually exists in the systems that we shall analyze. This expression is the simplest version of the so-called Landauer formula and it illustrates the close relation between current and transmission. At zero temperature $f_L(E)$ and $f_R(E)$ are step functions, equal to 1 below $E_F + eV/2$ and $E_F - eV/2$, respectively, and 0 above this energy. If we moreover assume low voltages (linear regime), this expression reduces to $I = GV$, where the conductance is $G = (2e^2/h)T$, where the transmission is evaluated at the Fermi energy.

This simple calculation demonstrates that a perfect single mode conductor between two electrodes has a finite resistance, given by the universal quantity $h/2e^2 \approx 12.9$ kΩ. This is an important difference with respect to macroscopic leads, where one expects to have zero resistance for the perfectly conducting case. The proper interpretation of this result was first pointed out by Imry [144], who associated the finite resistance with the resistance arising at the interfaces between the leads and the sample.

4.4   Penetration of a potential barrier: Tunnel effect

As it is clear from Eq. (4.11), the transmission probability plays a central role in Landauer approach. For this reason, it is worth reminding how this quantum mechanical quantity can be computed in some simple situations of special interest. For the sake of concreteness, we shall focus our discussion in this section on the analysis of the transmission through a single potential barrier. This simple problem not only illustrates some fundamental issues, but it also provides a basic model widely used for the understanding of tunneling currents in a great variety of situations such as tunnel junctions based on insulating barriers, STM and even single-molecule junctions, as we shall show later in this book.

---

[3]Since we are in a 1D situation, there is no difference between total current and current density.

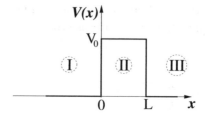

Fig. 4.3   Rectangular potential barrier of height V_0 and width L.

Let us consider the potential barrier of height V_0 depicted in Fig. 4.3. Our goal is to compute the probability to cross such a barrier as a function of the energy, E, of an incoming electron. Classical mechanics tell us that an incident particle will always be reflected when $E < V_0$, and it will always be transmitted when $E > V_0$. We all know that in quantum mechanics a particle can pass through a barrier, even when its energy is lower than the barrier height. This phenomenon is known as *quantum tunneling* or simply *tunnel effect* and it lies at the heart of the whole physics discussed in this book.

In order to compute the transmission we proceed in the standard way. We first determine the wave functions in the three different regions defined in Fig. 4.3, and then we match these functions and their first spatial derivatives at the boundaries ($x = 0$ and $x = L$). Let us first consider the case of $E < V_0$. In this case, the solutions of the Schrödinger equation in the three regions are of the form

$$\psi_{\mathrm{I}} = a_1 e^{ik_1 x} + b_1 e^{-ik_1 x}, \ \psi_{\mathrm{II}} = a_2 e^{k_2 x} + b_2 e^{-k_2 x}, \ \psi_{\mathrm{III}} = a_3 e^{ik_3 x}, \quad (4.12)$$

where

$$k_1 = k_3 = \frac{\sqrt{2mE}}{\hbar} \ \text{and} \ k_2 = \frac{\sqrt{2m(V_0 - E)}}{\hbar}. \quad (4.13)$$

Note that we have assumed that the effective mass is the same everywhere and we have discarded the incoming term ($b_3 e^{-ik_3 x}$) in ψ_{III} because we are considering here the problem of an a wave function impinging on the barrier from the left.

Using now the continuity of the wave function and its first derivative at $x = 0$ and $x = L$, we arrive at the following relationships

$$a_1 + b_1 = a_2 + b_2 \ ; \ ik_1 a_1 - ik_1 b_1 = k_2 a_2 - k_2 b_2 \quad (4.14)$$
$$a_2 e^{k_2 L} + b_2 e^{-k_2 L} = a_3 e^{ik_1 L} \ ; \ k_2 a_2 e^{k_2 L} - k_2 b_2 e^{-k_2 L} = ik_1 a_3 e^{ik_1 L}.$$

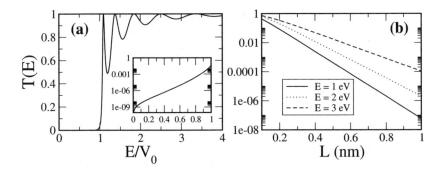

Fig. 4.4 (a) Transmission probability vs. energy for a symmetric potential barrier of height $V_0 = 4$ eV and width $L = 1$ nm. The inset shows a blow-up of the region $E < V_0$. (b) Transmission as a function of the width of the potential barrier ($V_0 = 4$ eV) for different values of the energy. In both cases the mass is assumed to be the electron mass.

Solving these equations, we obtain the following expression for the energy dependence of the transmission coefficient

$$T = \left|\frac{a_3}{a_1}\right|^2 = \frac{1}{1 + \left(\frac{k_1^2 + k_2^2}{2k_1 k_2}\right)^2 \sinh^2(k_2 L)} = \frac{4E(V_0 - E)}{4E(V_0 - E) + V_0^2 \sinh^2(k_2 L)}.$$

(4.15)

Proceeding in a similar way, one can compute the transmission for $E > V_0$ and the result is (see Exercise 4.2)

$$T = \frac{1}{1 + \left(\frac{k_1^2 - k_2^2}{2k_1 k_2}\right)^2 \sin^2(k_2 L)} = \frac{4E(E - V_0)}{4E(E - V_0) + V_0^2 \sin^2(k_2 L)}. \qquad (4.16)$$

The energy and length dependence of the transmission of this potential barrier are illustrated in Fig. 4.4. The most prominent feature is maybe the exponential dependence of the transmission on the barrier width for energies $E < V_0$, see Fig. 4.4(b). According to Eq. (4.15), this decay is given by $T \propto \exp(-2k_2 L) = \exp(-2L\sqrt{2m(V_0 - E)}/\hbar)$, i.e. the slopes in Fig. 4.4(b) are mainly determined by the square root of the difference between the electron energy and the barrier height. Since the transmission determines the conductance, this model provides a natural explanation for the exponential decay of the low-bias conductance as a function of the distance between the electrodes in all kind of tunnel barriers. It also tells us that such decay is simply governed by the work function of the metals involved.

Landauer formula shows that the linear conductance at low temperatures is determined by the transmission at the Fermi energy. However, the analysis of the current-voltage (I-V) characteristics requires the knowledge of the energy dependence and, strictly speaking, also of the voltage dependence of the transmission probability, see Eq. (4.11). In the case of a rectangular barrier, the voltage can be introduced in an approximate way as shown in Fig. 4.5(a). The computation of the transmission and in turn of the I-V curves is then a simple problem, see Exercise 4.3. A more appropriate way of describing the effect of the voltage is shown in Fig. 4.5(b), where a linear drop in the potential with the barrier region has been assumed.

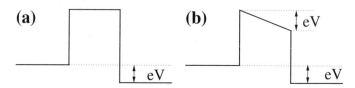

Fig. 4.5   Rectangular potential barrier under the application of a voltage: (a) approximation and (b) actual potential profile.

The analysis of the transmission through a potential like the one of Fig. 4.5(b), or any other smooth barrier, can be tackled with the help of the WKB approximation [145, 146] (see Exercise 4.4). This is precisely what Simmons did in 1963 [147] in his celebrated model. He considered the problem of the tunnel effect between metallic electrodes separated by a thin insulating film. He derived a general formula for the I-V curves for a barrier of arbitrary shape, and we reproduce here his result for the particular case of a rectangular barrier. Simmons showed that zero-temperature net current density in this case can be written as [147]

$$J = J_0 \left\{ \varphi_B \exp(-A\sqrt{\varphi_B}) - (\varphi_B + eV) \exp(-A\sqrt{\varphi_B + eV}) \right\}, \quad (4.17)$$

where φ_B is the average barrier height relative to the negative electrode and s_B is the barrier width s_B, see Fig. 4.6. Moreover,

$$A = \frac{2\alpha s_B}{\hbar}\sqrt{2m} \quad \text{and} \quad J_0 = \frac{e}{2\pi h \alpha^2 s_B^2}, \quad (4.18)$$

where α is a dimensionless correction factor of order unity. Eq. (4.17) can be simplified in three distinct cases depending on the applied voltage:

Low-voltage range. For very small voltages ($eV \sim 0$), see Fig. 4.6(a), the average barrier height φ_B is independent of the applied voltage and equals the zero voltage barrier height $\varphi_0 = (\varphi_1 + \varphi_2)/2$. Then, Eq. (4.17) can be simplified into

$$J = J_L V \quad \text{with} \quad J_L = \frac{e^2 \sqrt{2m\varphi_B}}{4\pi^2 \alpha \hbar^2 s_B} \exp(-A\sqrt{\varphi_B}). \quad (4.19)$$

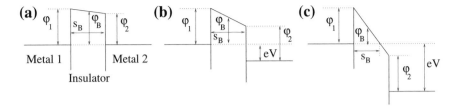

Fig. 4.6   Tunneling through a junction in which two metallic electrodes are separated by a thin insulating film, which is modeled as a rectangular potential barrier. The three panels show the three distinct voltage ranges discussed in the text.

Here, $\alpha = 1$. As it can be seen in Eq. (4.19), the current density is a linear function of the applied voltage V (Ohmic regime).

Intermediate-voltage range. For a medium applied voltage $eV < \varphi_0$, see Fig. 4.6(b), the average barrier height φ_B is given by $(\varphi_1 + \varphi_2 - eV)/2$. The current density can then be simplified to (assuming that $\alpha = 1$)

$$J = J_L(V + \gamma V^3) \quad \text{with} \quad \gamma = \frac{(Ae)^2}{96\varphi_0} - \frac{Ae^2}{32\varphi_0^{3/2}}. \tag{4.20}$$

This expression can be used to determine both the height and the barrier width in terms of the coefficients γ and J_L.

High-voltage range. For voltages $eV > \varphi_0$, see Fig. 4.6(c), the average barrier height is reduced to $\varphi_1/2$ and even the barrier width is reduced. Eventually, the voltage is high enough so that the Fermi level of electrode 2 is lower than the conduction band of electrode 1. In this case, tunneling from electrode 2 in electrode 1 is not possible since there are no empty states in electrode 1 to tunnel to. As for electrons tunneling from electrode 1 into electrode 2, all states in electrode 2 are empty. This is analog to field emission from a metal into vacuum. Then, the current density can be simplified to

$$J = \frac{2.2e^3}{8\pi h} \frac{F^2}{\varphi_1} \exp\left(-\frac{8\pi\sqrt{2m}\varphi_1^{3/2}}{2.96ehF}\right), \tag{4.21}$$

with the electric field strength in the insulator $F = V/s$, where s is the thickness of the insulating field.

In the case of vacuum tunneling (or tunneling through an insulator), we should be aware of the fact that whilst the electron is in the tunnel gap, it will induce image charges in the two electrodes. This serves to modify the barrier potential. The net effect of this is to reduce the average barrier height and hence increase the transmission probability. For an analysis of

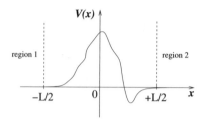

Fig. 4.7   The potential $V(x)$ under consideration varies in an arbitrary way within the interval $-L/2 \leq x \leq +L/2$ and goes to zero outside this interval.

these "image forces" for the case of the rectangular barrier discussed here, see Ref. [147].

It is worth mentioning that the problem of the rectangular barrier under an applied voltage, see Fig. 4.5(b), can be solved exactly using the full Airy functions. This was done by Grundlach [148], who showed that the current exhibits oscillations as a function of voltage that are superimposed in the WKB result discussed above.

4.5   The scattering matrix

In the next section we shall present a more rigorous discussion of the scattering formalism, where the concept of scattering matrix plays a central role. The definition and properties of this matrix are described in many quantum mechanics textbooks, but for the sake of completeness, we have included here a brief discussion of this subject.

4.5.1   *Definition and properties of the scattering matrix*

In order to keep our discussion at a simple level, we study here a one-dimensional situation. Let us consider a potential $V(x)$ which is zero outside the region defined by $|x| > L/2$, but which varies in an arbitrary way inside this interval, see Fig. 4.7. The equation satisfied by every wave function $\psi(x)$ associated with a stationary state of energy E is

$$\left\{ \frac{d^2}{dx^2} + \frac{2m}{\hbar^2}[E - V(x)] \right\} \psi(x) = 0. \tag{4.22}$$

The most general solution $\psi(x)$ of Eq. (4.22) in the region $x < -L/2$ (region 1) for a given value of E can be written as

$$\psi_k(x) = a_1 e^{ikx} + b_1 e^{-ikx}, \tag{4.23}$$

where $k = \sqrt{2mE/\hbar^2}$, while in the region $x > +L/2$ (region 2) it has the form

$$\psi_k(x) = a_2 e^{-ikx} + b_2 e^{ikx}, \qquad (4.24)$$

Here, the different coefficients depend on k, as well as on the shape of the potential under study. Notice that with our notation, the amplitudes a_i ($i = 1, 2$) correspond to the incoming waves impinging on the potential region, whereas the amplitudes b_i correspond to the outgoing waves.

The scattering matrix is defined as the 2×2 matrix that relates the incoming and outgoing amplitudes as follows

$$\begin{pmatrix} b_1 \\ b_2 \end{pmatrix} = \hat{S} \begin{pmatrix} a_1 \\ a_2 \end{pmatrix}, \qquad (4.25)$$

where $\hat{S}$ is usually written as

$$\hat{S} = \begin{pmatrix} r & t' \\ t & r' \end{pmatrix}. \qquad (4.26)$$

Here, r and r' are reflection amplitudes and t and t' are the transmission amplitudes associated to this potential.

Are all these four elements independent? What are the properties of the scattering matrix? A first property of the S-matrix can be deduced from the conservation of the current. Let us remind that in quantum mechanics, the current associated with a wave function $\psi(x)$ is given by

$$J(x) = \frac{\hbar}{2mi} \left[\psi^*(x) \frac{d\psi}{dx} - \psi(x) \frac{d\psi^*}{dx} \right]. \qquad (4.27)$$

Differentiating, we find

$$\frac{d}{dx} J(x) = \frac{\hbar}{2mi} \left[\psi^*(x) \frac{d^2\psi}{dx^2} - \psi(x) \frac{d^2\psi^*}{dx^2} \right]. \qquad (4.28)$$

Taking into account Eq. (4.22), we obtain

$$\frac{d}{dx} J(x) = 0. \qquad (4.29)$$

Therefore, the current $J(x)$ associated with a stationary state is the same at all points of the x-axis. Note, moreover, that Eq. (4.29) is simply the one-dimensional analog of the relation (continuity equation)

$$\nabla \cdot \mathbf{J}(\mathbf{r}) = 0, \qquad (4.30)$$

which is valid for any stationary state of a particle moving in three-dimensional space. According to Eq. (4.29), the current $J(x)$ has the same value, no matter in which region it is evaluated. Then, computing the current in regions 1 and 2 we have

$$J(x) = \frac{\hbar k}{m} \left[|a_1|^2 - |b_1|^2 \right] = \frac{\hbar k}{m} \left[|b_2|^2 - |a_2|^2 \right], \qquad (4.31)$$

which implies that

$$|a_1|^2 + |a_2|^2 = |b_1|^2 + |b_2|^2. \qquad (4.32)$$

This relation can be used to establish the first property of the scattering matrix in the following way

$$|b_1|^2 + |b_2|^2 = (b_1^*, b_2^*) \begin{pmatrix} b_1 \\ b_2 \end{pmatrix} = (a_1^*, a_2^*) \, \hat{S}^\dagger \hat{S} \begin{pmatrix} a_1 \\ a_2 \end{pmatrix} =$$

$$(a_1^*, a_2^*) \begin{pmatrix} a_1 \\ a_2 \end{pmatrix} = |a_1|^2 + |a_2|^2, \qquad (4.33)$$

which simply implies that $\hat{S}$ is a unitary matrix, i.e.

$$\hat{S}^\dagger = \hat{S}^{-1}. \qquad (4.34)$$

In terms of the matrix elements, this relation reads

$$|r|^2 + |t|^2 = 1 \; ; \; r^*t' + t^*r' = 0$$
$$(t')^*r + (r')^*t = 0 \; ; \; |r'|^2 + |t'|^2 = 1. \qquad (4.35)$$

Notice that the second and third relations are indeed the same.

If the potential $V(x)$ is real, which means in particular that there is no magnetic field applied, an additional property can be derived as follows. If $\psi(x)$ is a solution of Eq. (4.22), then $\psi^*(x)$ is also a solution. This new solution can be written as

$$\psi^*(x) = a_1^* e^{-ikx} + b_1^* e^{ikx} \qquad \text{if } x < -L/2$$
$$\psi^*(x) = a_2^* e^{ikx} + b_2^* e^{-ikx} \qquad \text{if } x > +L/2.$$

Notice that in this solution the coefficients a_i^* correspond to the outgoing amplitudes, while b_i^* represent the incoming amplitudes. Therefore, by definition they are related via the scattering matrix as follows

$$\begin{pmatrix} a_1^* \\ a_2^* \end{pmatrix} = \hat{S} \begin{pmatrix} b_1^* \\ b_2^* \end{pmatrix}, \qquad (4.36)$$

which can be rewritten as

$$\begin{pmatrix} b_1 \\ b_2 \end{pmatrix} = (\hat{S}^*)^{-1} \begin{pmatrix} a_1 \\ a_2 \end{pmatrix}, \qquad (4.37)$$

If we now compare this relation with Eq. (4.25), we arrive at

$$(\hat{S})^{-1} = \hat{S}^*. \tag{4.38}$$

If we now combine this with the fact that the scattering matrix is unitary, we have that $\hat{S}$ is symmetric

$$(\hat{S})^T = \hat{S} \Rightarrow t' = t. \tag{4.39}$$

In the presence of a magnetic field, this latter relation changes and one can show that reversing the magnetic field B transposes the S-matrix

$$\hat{S}(B) = \hat{S}^T(-B) \Rightarrow t'(B) = t(-B). \tag{4.40}$$

The demonstration is left to the reader as an exercise (see Exercise 4.7).

4.5.2 *Combining scattering matrices*

It is interesting to discuss how one can combine different scattering matrices in a problem in which there are several scattering potentials. Let us for instance consider the case of two potential barriers of arbitrary shape. This situation is schematically represented in Fig. 4.8. We shall include in the scattering matrix a superindex indicating to which potential barrier it corresponds, $\hat{S}^{(i)}$ ($i = 1, 2$). These matrices $\hat{S}^{(i)}$ relate the incoming and outgoing amplitudes across the corresponding potential barrier as follows (see Fig. 4.8)

$$\begin{pmatrix} b_1 \\ b_2 \end{pmatrix} = \hat{S}^{(1)} \begin{pmatrix} a_1 \\ a_2 \end{pmatrix}; \qquad \begin{pmatrix} a_2 \\ b_3 \end{pmatrix} = \hat{S}^{(2)} \begin{pmatrix} b_2 \\ a_3 \end{pmatrix}. \tag{4.41}$$

Notice that we have already used the fact that a_2 is at the same time the incoming amplitude for the potential 1 and the outgoing amplitude for potential 2. Something similar happens with b_2.

Our problem is to find in terms of the matrix elements of $\hat{S}^{(i)}$ the total scattering matrix $\hat{S}_{\text{Tot}}$ that relates the incoming and outgoing amplitudes of the two scatterers, i.e.

$$\begin{pmatrix} b_1 \\ b_3 \end{pmatrix} = \hat{S}_{\text{Tot}} \begin{pmatrix} a_1 \\ a_3 \end{pmatrix}; \qquad \hat{S}_{\text{Tot}} = \begin{pmatrix} r & t' \\ t & r' \end{pmatrix}. \tag{4.42}$$

This can be easily done eliminating a_2 and b_2 from Eq. (4.41) and the final result can be written as

$$r = r^{(1)} + t'^{(1)} r^{(2)} \left[1 - r'^{(1)} r^{(2)} \right]^{-1} t^{(1)} \; ; \; t = t^{(2)} \left[1 - r'^{(1)} r^{(2)} \right]^{-1} t^{(1)}$$

$$r' = r'^{(2)} + t^{(2)} \left[1 - r'^{(1)} r^{(2)} \right]^{-1} r'^{(1)} t'^{(2)} \; ; \; t' = t'^{(1)} \left[1 - r^{(2)} r'^{(1)} \right]^{-1} t'^{(2)}. \tag{4.43}$$

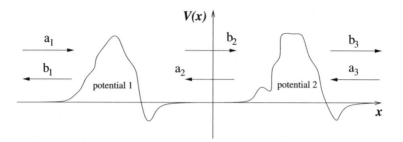

Fig. 4.8   Combination of two potential barriers of arbitrary shape. The coefficients a_i and b_i represent the different incoming and outgoing amplitudes with respect to the potential barrier i.

This result allows us to compute now very easily, for instance, the total transmission through the combined structure. According to the previous equations

$$T = |t|^2 = \frac{T_1 T_2}{1 - 2\sqrt{R_1 R_2}\cos\theta + R_1 R_2}, \qquad (4.44)$$

where $T_i = |t^{(i)}|^2 = |t'^{(i)}|^2$, $R_i = |r^{(i)}|^2 = |r'^{(i)}|^2$ and $\theta = \text{phase}(r'^{(1)}) + \text{phase}(r^{(2)})$ is the phase shift acquired in one round-trip between the scatterers.

This result can be used to study a very important phenomenon for us, namely the *resonant tunneling*. In a double barrier system (or in a potential well) one can have bound states in the region between the two scattering centers. Then, the transmission probability in this system exhibits resonances at energies close to the position of those bound states. The width of the transmission peaks depends upon the transmissivity of the barriers, while the distance between peaks is mainly determined by the distance between the barriers. These facts can be shown with the help of Eq. (4.44), as it is illustrated in Exercise 4.8.

4.6   Multichannel Landauer formula

We present in this section a more rigorous derivation of Landauer formula, where the important concept of conduction channel will arise. This formulation will also be the starting point for the extension of the scattering formalism to the description of other transport properties such as shot noise or thermoelectric coefficients. This section is based on Refs. [149, 150] and we refer the reader to them for more details.

We consider a mesoscopic sample connected to two reservoirs (terminals, probes), to be referred to as "left" (L) and "right" (R). It is assumed that the reservoirs are so large that they can be characterized by a temperature $T_{L,R}$ and a chemical potential $\mu_{L,R}$; the distribution functions of electrons in the reservoirs, defined via these parameters, are then Fermi distribution functions

$$f_\alpha(E) = [\exp[(E - \mu_\alpha)/k_B T_\alpha] + 1]^{-1}, \quad \alpha = L, R \qquad (4.45)$$

(see Fig. 4.9). Far from the sample, we can assume that transverse (across the leads) and longitudinal (along the leads) motion of electrons are separable. In the longitudinal (from left to right) direction the system is open, and is characterized by the continuous wave vector k_l. It is advantageous to separate incoming (to the sample) and outgoing states, and to introduce the longitudinal energy $E_l = \hbar^2 k_l^2/2m$ as a quantum number. Transverse motion is quantized and described by the discrete index n (corresponding to transverse energies $E_{L,R;n}$, which can be different for the left and right leads). These states are in the following referred to as *transverse (quantum) channels*. We write thus $E = E_n + E_l$. Since E_l needs to be positive, for a given total energy E only a finite number of channels exists. The number of incoming channels is denoted $N_{L,R}(E)$ in the left and right lead, respectively.

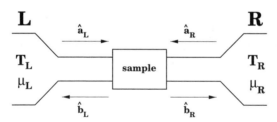

Fig. 4.9   Two-terminal scattering problem for the case of one transverse channel.

We now introduce *creation* and *annihilation operators* of electrons in the scattering states.[4] In principle, we could have used the operators which refer to particles in the states described by the quantum numbers n, k_l. However, the scattering matrix relates current amplitudes and not wave function amplitudes. Thus, we introduce operators $\hat{a}_{Ln}^\dagger(E)$ and $\hat{a}_{Ln}(E)$ which create and annihilate electrons with total energy E in the transverse

---

[4]The second quantization language will be used here at a very simple level. A discussion of this formalism is included in Appendix A and it will be widely used in the following chapters.

channel n in the left lead, which are incident upon the sample.[5] In the same way, the creation $\hat{b}_{Ln}^\dagger(E)$ and annihilation $\hat{b}_{Ln}(E)$ operators describe electrons in the outgoing states. They obey anti-commutation relations

$$\hat{a}_{Ln}^\dagger(E)\hat{a}_{Ln'}(E') + \hat{a}_{Ln'}(E')\hat{a}_{Ln}^\dagger(E) = \delta_{nn'}\delta(E - E')$$
$$\hat{a}_{Ln}(E)\hat{a}_{Ln'}(E') + \hat{a}_{Ln'}(E')\hat{a}_{Ln}(E) = 0$$
$$\hat{a}_{Ln}^\dagger(E)\hat{a}_{Ln'}^\dagger(E') + \hat{a}_{Ln'}^\dagger(E')\hat{a}_{Ln}^\dagger(E) = 0. \tag{4.46}$$

Similarly, we introduce creation and annihilation operators $\hat{a}_{Rn}^\dagger(E)$ and $\hat{a}_{Rn}(E)$ for incoming states and $\hat{b}_{Rn}^\dagger(E)$ and $\hat{b}_{Rn}(E)$ for outgoing states in the right lead (Fig. 4.9).

The operators $\hat{a}$ and $\hat{b}$ are related via the scattering matrix $\hat{S}$,

$$\begin{pmatrix} \hat{b}_{L1} \\ \vdots \\ \hat{b}_{LN_L} \\ \hat{b}_{R1} \\ \vdots \\ \hat{b}_{RN_R} \end{pmatrix} = \hat{S} \begin{pmatrix} \hat{a}_{L1} \\ \vdots \\ \hat{a}_{LN_L} \\ \hat{a}_{R1} \\ \vdots \\ \hat{a}_{RN_R} \end{pmatrix}. \tag{4.47}$$

The creation operators $\hat{a}^\dagger$ and $\hat{b}^\dagger$ obey a similar relation with the Hermitian conjugated matrix $\hat{S}^\dagger$.

The matrix $\hat{S}$ has dimensions $(N_L + N_R) \times (N_L + N_R)$. Its size, as well as the matrix elements, depends on the total energy E. It has the block structure

$$\hat{S} = \begin{pmatrix} \hat{r} & \hat{t}' \\ \hat{t} & \hat{r}' \end{pmatrix}. \tag{4.48}$$

Here the square diagonal blocks $\hat{r}$ (size $N_L \times N_L$) and $\hat{r}'$ (size $N_R \times N_R$) describe electron reflection back to the left and right reservoirs, respectively. The off-diagonal, rectangular blocks $\hat{t}$ (size $N_R \times N_L$) and $\hat{t}'$ (size $N_L \times N_R$) are responsible for the electron transmission through the sample. The properties of the matrix $\hat{S}$ are a straightforward generalization to a multi-mode case of those discussed in the previous section. Thus for instance, the flux conservation in the scattering process implies that $\hat{S}$ is quite generally unitary. In the presence of time-reversal symmetry the scattering matrix is also symmetric.

---

[5]We shall denote here the operators with a "hat" to distinguish them from the amplitudes of the previous section.

The current operator in the left lead (far from the sample) is expressed in a standard way,

$$\hat{I}_L(z,t) = \frac{\hbar e}{2im} \int d\mathbf{r}_\perp \left[\hat{\Psi}_L^\dagger(\mathbf{r},t) \frac{\partial}{\partial z} \hat{\Psi}_L(\mathbf{r},t) - \left(\frac{\partial}{\partial z} \hat{\Psi}_L^\dagger(\mathbf{r},t) \right) \hat{\Psi}_L(\mathbf{r},t) \right],$$

(4.49)

where the field operators $\hat{\Psi}$ and $\hat{\Psi}^\dagger$ are defined as

$$\hat{\Psi}_L(\mathbf{r},t) = \int dE e^{-iEt/\hbar} \sum_{n=1}^{N_L(E)} \frac{\chi_{Ln}(\mathbf{r}_\perp)}{(2\pi\hbar v_{Ln}(E))^{1/2}} \left[\hat{a}_{Ln} e^{ik_{Ln}z} + \hat{b}_{Ln} e^{-ik_{Ln}z} \right]$$

(4.50)

and

$$\hat{\Psi}_L^\dagger(\mathbf{r},t) = \int dE e^{iEt/\hbar} \sum_{n=1}^{N_L(E)} \frac{\chi_{Ln}^*(\mathbf{r}_\perp)}{(2\pi\hbar v_{Ln}(E))^{1/2}} \left[\hat{a}_{Ln}^\dagger e^{-ik_{Ln}z} + \hat{b}_{Ln}^\dagger e^{ik_{Ln}z} \right].$$

(4.51)

Here $\mathbf{r}_\perp$ is the transverse coordinate(s) and z is the coordinate along the leads (measured from left to right), χ_n^L are the transverse wave functions, and we have introduced the wave vector, $k_{Ln} = \hbar^{-1}[2m(E - E_{Ln})]^{1/2}$ (the summation only includes channels with real k_{Ln}), and the velocity of carriers $v_n(E) = \hbar k_{Ln}/m$ in the n-th transverse channel.

After some algebra, the expression for the current can be cast into the form[6]

$$\hat{I}_L(t) = \frac{e}{h} \sum_n \int dE dE' e^{i(E-E')t/\hbar} \left[\hat{a}_{Ln}^\dagger(E)\hat{a}_{Ln}(E') - \hat{b}_{Ln}^\dagger(E)\hat{b}_{Ln}(E') \right].$$

(4.52)

Using Eq. (4.47) we can now express the current in terms of the $\hat{a}$ and $\hat{a}^\dagger$ operators alone,

$$\hat{I}_L(t) = \frac{e}{h} \sum_{\alpha\beta} \sum_{mn} \int dE dE' e^{i(E-E')t/\hbar} \hat{a}_{\alpha m}^\dagger(E) A_{\alpha\beta}^{mn}(L; E, E') \hat{a}_{\beta n}(E').$$

(4.53)

Here the indices α and β label the reservoirs and may assume values L or R. The matrix A is defined as

$$A_{\alpha\beta}^{mn}(L; E, E') = \delta_{mn}\delta_{\alpha L}\delta_{\beta L} - \sum_k S_{L\alpha;mk}^\dagger(E)S_{L\beta;kn}(E'), \qquad (4.54)$$

and $S_{L\alpha;mk}(E)$ is the element of the scattering matrix relating $\hat{b}_{Lm}(E)$ to $\hat{a}_{\alpha k}(E)$. Note that Eq. (4.53) is independent of the coordinate z along the lead.

---

[6]Here, we have used the fact that the velocities $v_n(E)$ vary with energy quite slowly, typically on the scale of the Fermi energy, and neglected their energy dependence.

Let us now derive the average current from Eq. (4.53). For a system at thermal equilibrium the quantum statistical average of the product of an electron creation operator and annihilation operator of a Fermi gas is

$$\langle \hat{a}^{\dagger}_{\alpha m}(E) \hat{a}_{\beta n}(E') \rangle = \delta_{\alpha\beta} \delta_{mn} \delta(E - E') f_{\alpha}(E). \qquad (4.55)$$

Using Eq. (4.53) and Eq. (4.55) and taking into account the unitarity of the scattering matrix $\hat{S}$, we obtain

$$I \equiv \langle I_L \rangle = \frac{e}{h} \int_{-\infty}^{\infty} dE \, \text{Tr} \left[\hat{t}^{\dagger}(E) \hat{t}(E) \right] \left[f_L(E) - f_R(E) \right]. \qquad (4.56)$$

Here the matrix t is the off-diagonal block of the scattering matrix, $t_{mn} = S_{RL;mn}$. In the zero-temperature limit and for a small applied voltage Eq. (4.56) gives a conductance

$$G = \frac{e^2}{h} \text{Tr} \left[\hat{t}^{\dagger}(E_{\text{F}}) \hat{t}(E_{\text{F}}) \right], \qquad (4.57)$$

where E_{F} is the Fermi energy. Eq. (4.57) establishes the relation between the scattering matrix evaluated at the Fermi energy and the conductance. It is a basis invariant expression. The matrix $\hat{t}^{\dagger} \hat{t}$ can be diagonalized; it has a real set of eigenvalues (*transmission coefficients*) $T_n(E)$ (not to be confused with temperature), each of them assumes a value between zero and one. The corresponding eigenfunctions will be referred to as *eigenchannels or conduction channels*. In this natural basis we have instead of Eq. (4.56)

$$I = \frac{e}{h} \sum_{n} \int_{-\infty}^{\infty} dE \, T_n(E) \left[f_L(E) - f_R(E) \right]. \qquad (4.58)$$

and thus for the conductance

$$G = \frac{e^2}{h} \sum_{n} T_n. \qquad (4.59)$$

Eq. (4.59) is known as a multi-channel generalization of Landauer formula. Notice also that in the last formulas there is a difference of a factor 2 with respect to Eq. (4.11). The reason is that in the discussion above we have not assumed spin degeneracy.

For a constriction of only one atom in cross section one can estimate the number of conductance channels as $N \simeq (k_{\text{F}} R/2)^2$, which is between 1 and 3 for most metals. We shall see that the actual number of channels is determined by the valence orbital structure of the atoms. In the case of molecular junctions, it turns out that, apart from a few notable exceptions, the conductance is dominated by a single conduction channel.

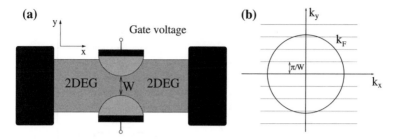

Fig. 4.10 (a) Schematic representation of a point contact defined in a two-dimensional electron gas (2DEG) by means of a split gate on top of the heterostructure. (b) Allowed states in the point contact constriction, which correspond to quantized values for $k_y = \pm n\pi/W$, and continuous values for k_x. The formation of these 1D subbands gives rise of a quantized conductance.

Let us emphasize that we have focused our discussion on a two-terminal configuration. The scattering approach was extended by Büttiker to describe the electronic transport in multi-terminal situations and this formalism (generally referred to as Landauer-Büttiker's formalism) has been widely used in the interpretation of mesoscopic experiments. We shall not discuss this generalization here and we refer the reader to Refs. [50, 149, 150, 171] for more details about this formalism.

4.6.1   *Conductance quantization in 2DEG: Landauer formula at work*

As a simple illustration of the use of Landauer formula, we shall now briefly discuss the conductance quantization in quantum point contacts defined in semiconductor hetero-structures (for a detailed discussion of this topic, see Refs. [151, 152]). It is well-known that in a semiconductor heterostructure like GaAs-AlGaAs one can confine the electrons in the two-dimensional interface between the two materials. Additionally, one can define electro-statically a point contact by means of a split gate on top of the heterostructure. This is schematically represented in Fig. 4.10(a). In this way one can define short and narrows constrictions in the two-dimensional electron gas (2DEG), of variable width $0 < W < 250$ nm comparable to the Fermi wavelength $\lambda_F \approx 40$ nm and much shorter than the mean free path $l \approx 10$ μm.

Van Wees *et al.* [153] and Wharam *et al.* [154] independently discovered a sequence of steps in the conductance of such a point contact as its width was varied by means of the voltage on the split gate (see Fig. 4.11).

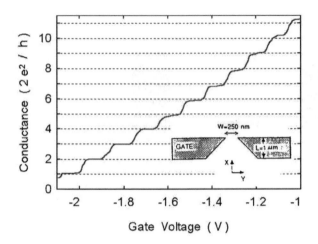

Fig. 4.11   Point contact conductance as a function of gate voltage at 0.6 K, demonstrating the conductance quantization in units of $2e^2/h$. The constriction width increases with increasing voltage on the gate (see inset). Reprinted with permission from [153]. Copyright 1988 by the American Physical Society.

The steps are near integer multiples of $2e^2/h$, after correction for a gate-voltage-independent series resistance from the wide 2DEG regions. This phenomenon is referred to as *conductance quantization*.

An elementary explanation of this effect relies on two facts: (i) the 2DEGs are ballistic systems (at least along the constriction) and the only scattering takes place against the potential walls defined by the split gates and (ii) the momentum of the electron is quantized in the transverse direction giving rise to 1D subbands. Since every subband that contributes to the transport (or conduction channel) has a perfect transparency and the number of them is obviously an integer, it follows from the two-terminal Landauer formula that the low temperature conductance G is quantized,

$$G = (2e^2/h)N, \qquad (4.60)$$

as observed experimentally. Here, N is the total number of open conduction channels and the prefactor 2 accounts for the spin degeneracy. This number can be simply calculated assuming a square-well lateral confining potential of width W. In the constriction, the electron momentum along the transport direction (x-direction) can take any value, while the transverse momentum k_y is quantized and can only take the following values: $k_y = \pm n\pi/W$ with $n = 1, 2, ..., N$, see Fig. 4.10(b). Since the current is only carried by those electrons at the Fermi energy (or with momentum equal to the Fermi momentum k_F), the number of subbands is simply given by $N = \text{Int}[k_F W/\pi]$. Therefore, a new subband is made available for transport every time the width of the gate is increased by approximately half of the Fermi wavelength. This explains the stair-like behavior seen in Fig. 4.11.

A detailed explanation of the necessary conditions to observe the conductance quantization requires a more rigorous treatment of the confinement potential and the corresponding analysis of the mode coupling at the entrance and exit of the constriction. A more realistic model is discussed in Exercise 4.9.

4.7   Shot noise

Shot noise is another important quantity for characterizing the transport properties of nanoscale systems [150, 155]. It refers to the time-dependent current fluctuations due to the discreteness of the electron charge. In a mesoscopic conductor these fluctuations have a quantum origin, arising from the quantum mechanical probability of electrons being transmitted or reflected from the sample. In contrast to thermal noise, shot noise only appears in the presence of transport, i.e. in a non-equilibrium situation.

Shot noise measurements provide information on temporal correlations between the electrons. In a tunnel junction, where the electrons are transmitted randomly and correlation effects can be neglected, the transfer of carriers of charge q is described by Poisson statistics and the amplitude of the current fluctuations is $2qI$. In nanoscale conductors correlations may suppress the shot noise below this value. Even when electron-electron interactions can be neglected the Pauli principle provides a source for electron correlations.

The relation between shot noise and the transmitted charge unit q has allowed the detection of the carrier charge in exotic situations such as the fractional quantum Hall regime [156, 157], where the charge can be fractional and depends on the filling factor. It has also allowed to show that the sub-gap transport in superconducting atomic contacts takes place in big shots of multiple ne charges associated with multiple Andreev reflection processes [158, 159, 98].

The interest in shot noise in molecular electronics lies in the fact that this quantity depends on the transmission coefficients in a nonlinear manner. Thus, the shot noise can provide valuable information, not contained in the conductance, about the number of conduction channels and their

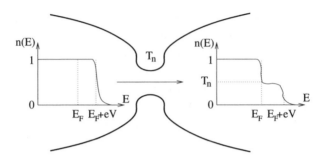

Fig. 4.12   In a quantum point contact with bias voltage, V, the transmission probability, T_n, determines the distribution function, $n(E)$, of a transmitted state as a function of its energy, E. In the right reservoir, states with energy lower than the Fermi energy are all occupied, while right-moving states with higher energy can only be coming from the left reservoir, and therefore their average occupation is equal to the transmission probability, T_n. This argument applies to every individual conduction channel.

transmission coefficients. This will be discussed in detail in Chapter 19.

Qualitatively, the shot noise in nanocontacts can be understood from the diagram in Fig. 4.12. Let us consider the right moving states in this contact, which have been transmitted through the junction with an excess energy between 0 and eV. Their average occupation number, $\bar{n}$, is given by the transmission probability T_n. For the fluctuations in this number we find

$$\overline{\Delta n^2} = \overline{n^2} - \bar{n}^2 = T_n(1 - T_n), \qquad (4.61)$$

where in the last step we used the fact that $\overline{n^2} = \bar{n}$, since for fermions n is either zero or one. Hence, the fluctuations in the current are suppressed both for $T_n = 1$ and for $T_n = 0$. According to Eq. (4.61) the fluctuations will be maximal when the electrons have a probability of one half to be transmitted. The shot noise is thus a non-linear function of the transmission coefficients, as we anticipated above.

We shall now derive in a rigorous manner the main results concerning shot noise in a two-terminal device within the scattering formalism. For this purpose, we follow again Ref. [150]. Since are concerned with the fluctuations of the current away from its average value, we then introduce the operators $\Delta \hat{I}(t) \equiv \hat{I}(t) - \langle I \rangle$, where $\hat{I}$ is the current operator evaluated in a given reservoir, let us say, the left one. We define the correlation function $P(t - t')$ of the current in a given contact as

$$P(t - t') \equiv \frac{1}{2} \left\langle \Delta \hat{I}(t)\Delta \hat{I}(t') + \Delta \hat{I}(t')\Delta \hat{I}(t) \right\rangle. \qquad (4.62)$$

Note that in the absence of time-dependent external fields, as we assume here, the correlation function must be function of only $t - t'$. Its Fourier transform,

$$2\pi\delta(\omega + \omega')P(\omega) \equiv \left\langle \Delta\hat{I}(\omega)\Delta\hat{I}(\omega') + \Delta\hat{I}(\omega')\Delta\hat{I}(\omega) \right\rangle, \qquad (4.63)$$

is sometimes referred to as *noise power*.

To find the noise power we need the quantum statistical expectation value of products of four operators $\hat{a}$. For a Fermi gas at equilibrium this expectation value is

$$\left\langle \hat{a}^\dagger_{\alpha k}(E_1)\hat{a}_{\beta l}(E_2)\hat{a}^\dagger_{\gamma m}(E_3)\hat{a}_{\delta n}(E_4) \right\rangle -$$
$$\left\langle \hat{a}^\dagger_{\alpha k}(E_1)\hat{a}_{\beta l}(E_2) \right\rangle \left\langle \hat{a}^\dagger_{\gamma m}(E_3)\hat{a}_{\delta n}(E_4) \right\rangle$$
$$= \delta_{\alpha\delta}\delta_{\beta\gamma}\delta_{kn}\delta_{ml}\delta(E_1 - E_4)\delta(E_2 - E_3)f_\alpha(E_1)\left[1 - f_\beta(E_2)\right]. \qquad (4.64)$$

Here $f_\alpha(E)$ is the corresponding Fermi distribution. Now, making use of the current operator of Eq. (4.53) and of the expectation value of Eq. (4.64), we arrive at the following expression for the noise power

$$P(\omega) = \frac{e^2}{h} \sum_{\gamma\delta} \sum_{mn} \int dE A^{mn}_{\gamma\delta}(L; E, E + \hbar\omega)A^{nm}_{\delta\gamma}(L; E + \hbar\omega, E)$$
$$\times \left\{ f_\gamma(E)\left[1 - f_\delta(E + \hbar\omega)\right] + \left[1 - f_\gamma(E)\right]f_\delta(E + \hbar\omega) \right\}. \qquad (4.65)$$

Note that with respect to frequency, it has the symmetry properties $P(\omega) = P(-\omega)$. In the rest of this discussion, we shall only be interested in the *zero-frequency* noise.[7] For the noise power at $\omega = 0$ we obtain

$$P \equiv P(0) = \frac{e^2}{h} \sum_{\gamma\delta} \sum_{mn} \int dE A^{mn}_{\gamma\delta}(L; E, E)A^{nm}_{\delta\gamma}(L; E, E) \qquad (4.66)$$
$$\times \left\{ f_\gamma(E)\left[1 - f_\delta(E)\right] + \left[1 - f_\gamma(E)\right]f_\delta(E) \right\}.$$

Eq. (4.66) can now be used to predict the low frequency noise properties of arbitrary multi-channel phase-coherent conductors. But before presenting the general result, let us first discuss two limiting cases of special interest:

Equilibrium noise. If the system is in thermal equilibrium at temperature T, the distribution functions in both reservoirs coincide and are equal to $f(E)$. Using the property $f(1 - f) = -k_{\rm B}T\partial f/\partial E$ and employing the unitarity of the scattering matrix, one can arrive at the following result

$$P = 4k_{\rm B}TG, \qquad (4.67)$$

---

[7]Zero-frequency noise actually means that the frequency is small in comparison with the relevant frequency scales of the problem, but large enough to neglect the $1/f$ noise that is present in almost any system.

where G is the linear conductance given by

$$G = \frac{e^2}{h} \int_{-\infty}^{\infty} dE \left(-\frac{\partial f}{\partial E}\right) \text{Tr}\left[\hat{t}^\dagger(E)\hat{t}(E)\right]. \qquad (4.68)$$

This is the thermal, or *Nyquist-Johnson* noise. In the approach discussed here it is a consequence of the thermal fluctuations of occupation numbers in the reservoirs. This is the manifestation of the fluctuation-dissipation theorem: equilibrium fluctuations are proportional to the corresponding generalized susceptibility, in this case to the conductance.

Zero-temperature shot noise. In the zero-temperature limit the Fermi distribution in each reservoir is a step function $f_\alpha(E) = \theta(\mu_\alpha - E)$. Utilizing the representation of the scattering matrix (4.48), and taking into account that the unitarity of the matrix $\hat{S}$ implies $\hat{r}^\dagger \hat{r} + \hat{t}^\dagger \hat{t} = 1$, after some algebra we can rewrite Eq. (4.66) as

$$P = \frac{2e^2}{h} \text{Tr}\left(\hat{r}^\dagger \hat{r} \hat{t}^\dagger \hat{t}\right) e|V|, \qquad (4.69)$$

where the scattering matrix elements are evaluated at the Fermi level. Like the expression of the conductance, Eq. (4.57), we can express this result in the basis of eigenchannels with the help of the transmission probabilities T_n and reflection probabilities $R_n = 1 - T_n$,

$$P = \frac{2e^3|V|}{h} \sum_n T_n (1 - T_n). \qquad (4.70)$$

We see that the non-equilibrium (shot) noise is not simply determined by the conductance of the sample. Instead, it is determined by a sum of products of transmission and reflection probabilities of the conduction channels. Only in the limit of low-transparency $T_n \ll 1$ in *all* conduction channels is the shot noise given by the *Poisson value*, discussed by Schottky,

$$P = \frac{2e^3|V|}{h} \sum_n T_n = 2e\langle I \rangle. \qquad (4.71)$$

It is clear that zero-temperature shot noise is always suppressed in comparison with the Poisson value. In particular, neither closed ($T_n = 0$) nor open ($T_n = 1$) channels contribute to shot noise; the maximal contribution comes from channels with $T_n = 1/2$. The suppression below the Poissonian limit given by Eq. (4.71) was one of the aspects of noise in mesoscopic systems which triggered many of the subsequent theoretical and experimental works. A convenient measure of *sub-Poissonian shot noise* is the *Fano factor F*, which is the ratio of the actual shot noise and the Poisson noise that would be measured if the system produced noise due to single independent electrons,

$$F = \frac{P}{2e\langle I \rangle}.$$ (4.72)

For energy-independent transmission and/or in the linear regime the Fano factor is

$$F = \frac{\sum_n T_n(1 - T_n)}{\sum_n T_n}.$$ (4.73)

The Fano factor assumes values between 0 (all channels are transparent) and 1 (Poissonian noise). In particular, for one channel it becomes $(1 - T)$.

The general result for arbitrary temperature and voltage for the noise power of the current fluctuations in a two-terminal conductor is

$$P = \frac{2e^2}{h} \sum_n \int_{-\infty}^{\infty} dE \, \{ T_n(E) \left[f_L(1 - f_L) + f_R(1 - f_R) \right] +$$

$$T_n(E) \left[1 - T_n(E) \right] (f_L - f_R)^2 \}.$$ (4.74)

Here the first two terms are the equilibrium noise contributions, and the third term is the non-equilibrium or shot noise contribution to the power spectrum. Note that this term is second order in the distribution function. At high energies, in the range where the Fermi distribution function is well approximated by a Maxwell-Boltzmann distribution, it is negligible compared to the equilibrium noise described by the first two terms. According to Eq. (4.74) the shot noise term enhances the noise power compared to the equilibrium noise.

In the practically important case, when the scale of the energy dependence of transmission coefficients $T_n(E)$ is much larger than both the temperature and applied voltage, these quantities in Eq. (4.74) may be replaced by their values taken at the Fermi energy. We obtain then

$$P = \frac{2e^2}{h} \left[2k_{\mathrm{B}}T \sum_n T_n^2 + eV \coth \left(\frac{eV}{2k_{\mathrm{B}}T} \right) \sum_n T_n (1 - T_n) \right],$$ (4.75)

where V is again the voltage applied between the left and right reservoirs. The full noise is a complicated function of temperature and applied voltage rather than a simple superposition of equilibrium and shot noise. For low voltages $eV \ll k_{\mathrm{B}}T$ one recovers the result of pure thermal noise, i.e. $P = 4k_{\mathrm{B}}TG$. Eq. (4.75) is the starting point for the analysis of experimental results on noise in atomic and molecular junctions, see section 19.1.

4.8 Thermal transport and thermoelectric phenomena

The scattering formalism is by no means restricted to the description of the electronic transport. It has also been extended to describe thermal transport and thermoelectric cross-effects [160–163] and in what follows we present a discussion of these transport properties within the scattering approach.[8]

Let us consider a generic two-terminal device like in the previous sections. In equilibrium, the electron reservoirs are at chemical potential μ and temperature T. In the regime of linear response, the current I and heat flow Q are related to the chemical potential difference $\Delta\mu$ and the temperature difference ΔT by the constitutive equations

$$\begin{pmatrix} I \\ Q \end{pmatrix} = \begin{pmatrix} G & L \\ M & K \end{pmatrix} \begin{pmatrix} \Delta\mu/e \\ \Delta T \end{pmatrix}. \tag{4.76}$$

The thermoelectric coefficients L and M are related by an Onsager relation, which in the absence of a magnetic field is

$$M = -LT. \tag{4.77}$$

Equation (4.76) is often re-expressed with the current I rather than the electrochemical potential $\Delta\mu$ as an independent variable,

$$\begin{pmatrix} \Delta\mu/e \\ Q \end{pmatrix} = \begin{pmatrix} R & S \\ \Pi & -\kappa \end{pmatrix} \begin{pmatrix} I \\ \Delta T \end{pmatrix}. \tag{4.78}$$

The resistance R is the reciprocal of the isothermal conductance G. The thermopower S is defined as

$$S \equiv \left(\frac{\Delta\mu/e}{\Delta T} \right)_{I=0} = -L/G. \tag{4.79}$$

The Peltier coefficient Π, defined as

$$\Pi \equiv \left(\frac{Q}{I} \right)_{\Delta T=0} = M/G = ST, \tag{4.80}$$

is proportional to the thermopower S in view of the Onsager relation (4.77). Finally, the thermal conductance κ is defined as

$$\kappa \equiv -\left(\frac{Q}{\Delta T} \right)_{I=0} = -K\left(1 + \frac{S^2 GT}{K} \right). \tag{4.81}$$

In order to compute all the thermoelectric coefficients, we still need to determine the heat current, which in the spirit of the scattering formalism

---

[8]It is worth stressing that we shall only consider the contribution of the electrons to the thermal transport properties. In general, phonons can also play an important role.

will be expressed in terms of the transmission and reflections coefficients of the system. Let us assume that the left electrode has a temperature T_1, while the right one has a temperature T_2. Following Ref. [160], the total entropy current moving to the right on the left lead will be given by[9]

$$J_{1S}^{\rightarrow} = -\frac{k_B}{h} \int [f_1 \ln f_1 + (1 - f_1) \ln(1 - f_1)] \, dE, \qquad (4.82)$$

where $f_1 = f(E, \mu_1, T_1)$ denotes the Fermi function on the left electrode. On the other hand, the entropy current going to the left on the same lead is given by

$$J_{1S}^{\leftarrow} = -\frac{k_B}{h} \int [(R_{11}f_1 + T_{12}f_2) \ln(R_{11}f_1 + T_{12}f_2) + $$
$$(1 - R_{11}f_1 - T_{12}f_2) \ln(1 - R_{11}f_1 - T_{12}f_2)] \, dE, \qquad (4.83)$$

where $T_{12} \equiv \text{Tr}\{\hat{t}^\dagger \hat{t}\}$ is the total transmission of the contact and $R_{11} \equiv \text{Tr}\{\hat{r}^\dagger \hat{r}\}$ is the corresponding reflection coefficient.

By subtracting (4.82) and (4.83) the following expression for the heat current is obtained [160]

$$Q_1 = T J_{1S} = \frac{1}{h} \int T_{12}(E)(E - \mu) [f_1 - f_2] \, dE, \qquad (4.84)$$

where T and μ are the average temperature and chemical potential.

Therefore, the thermoelectric coefficients are given in the scattering formalism by [160, 162]

$$G = -\frac{2e^2}{h} \int_{-\infty}^{\infty} dE \, \frac{\partial f}{\partial E} T_{12}(E), \qquad (4.85)$$

$$L = -\frac{2e^2}{h} \frac{k_B}{e} \int_{-\infty}^{\infty} dE \, \frac{\partial f}{\partial E} T_{12}(E) \frac{E - \mu}{k_B T}, \qquad (4.86)$$

$$\frac{K}{T} = \frac{2e^2}{h} \left(\frac{k_B}{e}\right)^2 \int_{-\infty}^{\infty} dE \, \frac{\partial f}{\partial E} T_{12}(E) \left[\frac{E - \mu}{k_B T}\right]^2. \qquad (4.87)$$

These integrals are convolutions of $T_{12}(E)$, which characterizes the conductor, and a kernel of the form $\epsilon^m df/d\epsilon$, $m = 0, 1, 2$, with $\epsilon \equiv (E - \mu)/k_B T$, and f the Fermi function $f(\epsilon) = [\exp(\epsilon) + 1]^{-1}$. Both $df/d\epsilon$ and $\epsilon^2 df/d\epsilon$ are symmetric functions of ϵ, which is why the conductance, G, and the thermal conductances K and κ are determined to first order by $T_{12}(\mu)$. (The term

---

[9]Notice that the expression in the square bracket is the entropy density of noninteracting electrons, distributed according to an arbitrary non-equilibrium distribution function f_1, see pag. 54 of Ref. [164].

within brackets in equation (4.81) is usually small.) In contrast, $\epsilon df/d\epsilon$ is an antisymmetric function of ϵ, so that the thermoelectric cross-coefficients L, S, M, and Π are determined mainly by the derivative $dT_{12}(E)/dE$ at $E = \mu$. This is substantiated by a Sommerfeld expansion of the integrals in Eqs. (4.85)-(4.87), valid for a smooth function $T_{12}(E)$ to lowest order in $k_{\mathrm{B}}T/\mu$ [162]

$$G \approx \frac{2e^2}{h} T_{12}(\mu) \tag{4.88}$$

$$L \approx \frac{2e^2}{h} L_0 eT \left(\frac{dT_{12}(E)}{dE} \right)_{E=\mu} \tag{4.89}$$

$$K \approx -\frac{2e^2}{h} L_0 TT_{12}(\mu), \tag{4.90}$$

with $L_0 \equiv (k_{\mathrm{B}}/e)^2 \pi^2/3$ the Lorentz number. In this approximation $K = -L_0 TG$, so that for $S^2 \ll L_0$ one finds from Eq. (4.81) the Wiedemann-Franz relation: $\kappa \approx L_0 TG$.

Thermoelectrical effects have been experimentally studied in detail in 2DEG quantum point contacts by van Houten *et al.* [163]. In the context of atomic and molecular junctions, special attention has been paid to the thermopower. As we shall discuss in section 19.3, the thermopower contains valuable information about these systems that is not contained in the electrical conductance.

4.9   Limitations of the scattering approach

The scattering formalism has been very successful explaining many basic transport phenomena in a great variety of nanostructures. It has also been extended to other situations of interest for the purpose of this book, such as e.g. photon-assisted transport [165]. For space reasons we have to end here our discussion of this formalism, and for more details we recommend the the reviews of Refs. [150, 151, 166–168], the didactic book of S. Datta [50] and the book on mesoscopic physics of Y. Imry [169].

In spite of its great success, the scattering approach is far from being a complete theory of quantum transport. In this sense, it is important to be aware of its limitations. Among them we want to emphasize two of special interest for the scope of this book:

(i) The scattering approach gives no hints on how to compute the transmission or, more generally, the scattering matrix. In particular, it does not tell us how to determine the actual transmission of an atomic contact or a molecular circuit. In this sense, one might think that this formalism has merely replaced a problem by another. This would be, of course, unfair. The scattering approach can be combined with simple models, as we showed in section 4.4, or with more sophisticated techniques like random matrix theory [170] to predict the transport properties of a great variety of systems such as diffusive wires, chaotic cavities, superconducting nanostructures, resonant tunneling systems, tunnel junctions, etc.

(ii) The scattering picture is an one-electron theory which is valid only as long as inelastic scattering processes can be neglected. In this formalism one assumes that the electron propagation is a fully quantum coherent process over the entire sample. According to normal Fermi-liquid theory, such a description would be strictly valid at zero temperature and only for electrons at the Fermi energy. At finite bias the coherent propagation may be limited by inelastic scattering processes due to electron-phonon and electron-electron collisions. The theoretical description of transport in situations where inelastic interactions play an important role requires more sophisticated methods like the Green's function techniques that will be described in the next chapters.

Let us mention that there is a phenomenological way of describing the effect of inelastic or phase-breaking mechanisms within the scattering approach, which is due to Büttiker [171]. In this description the inelastic scattering events are simulated by the addition of voltage probes distributed over the sample. The chemical potential on these probes is fixed by imposing the condition of no net current flow through them. Thus, although the presence of the probes does not change the total current through the sample, they introduce a randomization of the phase which tends to destroy phase coherence. The current in such a structure will contain a coherent component, corresponding to those electrons which go directly from one lead to the other, and an inelastic component, corresponding to those electrons which enter into at least one of the voltage probes in their travel between the leads.

4.10 Exercises

4.1 Transmission through a potential step: Show that the transmission probability as a function of energy, E, for the potential step shown in Fig. 4.13 is given by

$$T(E) = \begin{cases} 4k_1 k_2/(k_1 + k_2)^2 & \text{if } E > V_0 \\ 0 & \text{if } E < V_0 \end{cases}$$

where $k_1 = \sqrt{2mE/\hbar^2}$, $k_2 = \sqrt{2m(E - V_0)/\hbar^2}$ and m is the electron mass.

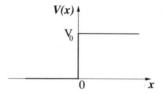

Fig. 4.13   Potential step of height V_0.

4.2 Penetration of a rectangular barrier: Show that the probability for an electron to cross the rectangular barrier shown in Fig. 4.3 for energies $E > V_0$ is given by Eq. (4.16).

4.3 A rectangular barrier under an applied voltage: Consider the rectangular barrier under an applied bias shown in Fig. 4.5(a). Show that the energy and voltage dependence of the transmission for $E < V_0$ is given by

$$T(E,V) = \left| \frac{4k_1 k_2}{2(k_1 k_3 - k_2^2)\sinh(k_2 L) + 2ik_2(k_1 + k_3)\cosh(k_2 L)} \right|^2 \frac{k_3}{k_1},$$

where $k_1 = \sqrt{2mE}/\hbar$, $k_2 = \sqrt{2m(V_0 - E)}/\hbar$ and $k_3 = \sqrt{2m(E + eV)}/\hbar$.

Use this result and the Landauer formula [Eq. (4.11)] to compute the zero-temperature current-voltage characteristics for a barrier of height $V_0 = 4$ eV and width $L = 1$ nm.

4.4 Penetration of an arbitrary potential barrier: Let us consider the potential barrier shown in Fig 4.14. Here, in a region $x < a$ (region I), $V(x) = V_0 = const.$; when $x > a$, $V(x)$ is a positive and smooth function decreasing monotonically from the positive value $V_a = V(a)$ to $V(\infty) = 0$.

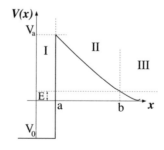

Fig. 4.14   Arbitrary potential barrier.

Use the WKB approximation to show that the transmission coefficient through that barrier is given by

$$T(E) = 4\frac{\sqrt{(V_a - E)(E - V_0)}}{V_a - V_0}e^{-2\tau}, \text{ where } \tau = \int_a^b \frac{\sqrt{2m[V(x) - E]}}{\hbar}dx.$$

Hint: The WKB approximation is nicely explained, e.g., in Ref. [146].

4.5 Resonant tunneling in a finite square well: Analyze the transmission coefficient in the case of the square well shown in Fig 4.15. In particular, show that in the energy range $E > V_3$ this coefficient is given by

$$T(E) = \frac{4k_1 k_3 k_2^2}{k_2^2(k_1 + k_3)^2 \cos^2(k_2 L) + (k_2^2 + k_1 k_3)^2 \sin^2(k_2 L)},$$

where $L = a - b$ and k_i is the electron momentum in the region $i =$I,II,III.

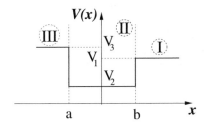

Fig. 4.15   Square well.

Show also that the transmission coefficient above exhibits resonances as a function of energy. In particular, calculate the position of those resonances and show that the transmission maxima are given by $4k_1 k_3/(k_1 + k_3)^2$.

4.6 Transmission through a delta function barrier: Let us model a one-dimensional conductor with the following Hamiltonian

$$\mathbf{H} = -\frac{\hbar^2}{2m}\frac{\partial^2}{\partial x^2} + V_0\delta(x),$$

where V_0 is the strength of the delta potential that acts at $x = 0$.

(a) Demonstrate that the boundary conditions for the scattering states $\psi_k(x)$, k being the electron momentum, are: (i) continuity at $x = 0$ and (ii) $\psi_k'(x = 0^+) - \psi_k'(x = 0^-) = (2mV_0/\hbar)\psi(x = 0)$, where the prime symbol indicates derivative with respect to x.

(b) Use the previous result to show that the transmission probability through this delta potential can be expressed as: $T = 1/(1 + Z^2)$, where $Z \equiv mV_0/(\hbar^2 k)$.

4.7 Scattering matrix:

(a) Show that in the presence of a magnetic field the scattering matrix fulfills the property of Eq. (4.40).

(b) Derive the relations of Eq. (4.43).

4.8 Resonant tunneling through a symmetric double barrier: Consider a symmetric double barrier system formed by combining two square barriers (see Exercise 4.2) of height V_0 and width L that are separated a distance d.

(a) Compute the total transmission through this system for energies smaller than V_0. Hint: Use the idea of the combination of scattering matrices, see Eq. (4.44) in section 4.5.2, and the results of Exercise 4.2.

(b) As in the case of the potential well of Exercise 4.5, the transmission in this double barrier system exhibits pronounced resonances. Find the position of those resonances and show that, in the limit in which they are well separated, the transmission around one of those resonances can be written as

$$T(E) = \frac{4\Gamma_L \Gamma_R}{(E - \epsilon_0)^2 + (\Gamma_L + \Gamma_R)^2},$$

where ϵ_0 is the position of the resonance and $\Gamma_{L,R}$ are the scattering rates associated to the left and right potential barriers. Find an expression for these rates in terms of the transmissions of the barriers. Hints: (i) The resonances are well separated when the transmissions T_1 and T_2 are small ($R_1, R_2 \approx 1$). (ii) The round-trip phase shift that appears in Eq. (4.44) is $\theta = 2kd$, where k is the electron momentum in the region between the two barriers.

4.9 Conductance quantization in a 2DEG: One of the most successful applications of the Landauer formula is the explanation of the conductance quantization that takes place in split-gate constrictions (or quantum-point contacts) in a two-dimensional electron gas (2DEG). A useful model to study the occurrence of conductance steps is the so-called saddle point model used by Büttiker in Ref. [172]. In this model it is assumed that near the bottleneck of the constriction the electrostatic potential can be expressed as

$$V(x,y) = V_0 - \frac{1}{2}m\omega_x^2 x^2 + \frac{1}{2}m\omega_y^2 y^2. \tag{4.91}$$

Here, V_0 is the electrostatic potential at the saddle, ω_x characterizes the curvature of the potential barrier in the constriction and ω_y the lateral confinement. Show that for this potential the transmission probabilities are given by

$$T_n(E) = \frac{1}{\exp[\pi(E - V_0 - (n + 1/2)\omega_x)/\omega_y] + 1}.$$

Using this expression in combination with the Landauer formula, find the criteria for the observation of well-defined conductance steps at low temperatures.

4.10 Shot noise and thermopower in a quantum point contact: Use the saddle point model of the previous exercise to study the shot noise [155] and the thermopower [161, 163] in a quantum point contact as a function of the Fermi energy (or gate voltage).

Chapter 5

Introduction to Green's function techniques for systems in equilibrium

The discussion of the scattering formalism in the previous chapter has left two basic questions open: (i) How to calculate the elastic transmission of real systems such as atomic and molecular junctions? and (ii) how to generalize Landauer formula to take into account correlation effects and inelastic mechanisms? Indeed, both questions can be answered, at least to a large extent, with the help of Green's function techniques. For this reason, we initiate here a series of three chapters devoted to this subject.

We are aware of the fact that at this point part of the readership will be certainly tempted to jump to the next part of the book. The words Green's functions cause in many people an immediate rejection because they associate them to some obscure theoretical techniques reserved to specialists. We believe that this judgment is a bit unfair. The degree of difficulty of the Green's function techniques depends primarily on the type of problems addressed. Thus for instance, we shall show that what is required to answer the first question posed above reduces to a standard problem of linear algebra that should be accessible to any student with a background in quantum mechanics. The answer to the second question requires however more elaborate methods, which will also be presented in this book. With this distinction in mind, we shall guide you through the next three chapters indicating the type of problems that we have in mind and we shall warn you about the possible difficulties.

In our discussion on the Green's function techniques we shall start in this chapter by introducing the subject concentrating ourselves on the case of electronic systems in equilibrium. This chapter is meant to give a first insight into what Green's functions in quantum mechanics are, what kind of physical information they contain and how they can be calculated in some simple situations. Having in mind the first question above, we shall focus on the analysis of noninteracting systems. Then, the next chapter will deal with the diagrammatic theory, which provides a systematic perturbative approach to compute the Green's functions of many-body systems where correlations and inelastic mechanisms in general play a fundamental role. Finally, since our final goal is the analysis of the transport properties of atomic-scale junctions, we shall present in Chapter 7 the Keldysh formalism that allows us to compute the Green's functions of nonequilibrium systems. Then, at the end of that chapter, we shall apply this formalism to the calculation of the transmission in some illustrative examples.

This chapter is organized as follows. First, we shall remind the reader of the basics of the Schrödinger and Heisenberg representations of quantum mechanics. Then, we shall introduce the retarded and advanced Green's functions in energy space for a noninteracting electron system and show how they can be computed in certain simple examples. We shall then define the general (valid also for interacting systems) time-dependent retarded, advanced and causal Green's functions and analyze their main analytical properties, their relation with the observables of interest and how they can be computed, in principle, with the so-called equation-of-motion method.

One last comment before we get started. We shall constantly make use of the second quantization formalism in our discussion of the Green's functions techniques. So, if you are not very familiar with this formalism, we strongly recommend you to read Appendix A.

5.1   The Schrödinger and Heisenberg pictures

Let us start by reviewing the two most standard pictures or representations in quantum mechanics. The usual way to introduce quantum mechanics makes use of the so-called *Schrödinger picture*, which is based on the time-dependent Schrödinger equation

$$i\hbar\frac{\partial}{\partial t}\Psi_{\rm S}(t) = {\bf H}\Psi_{\rm S}(t), \tag{5.1}$$

where $\bf H$ is the time-independent Hamiltonian of the system and $\Psi_{\rm S}(t)$ is the time-dependent wave function. Let us stress that in what follows, unless said otherwise, we shall set $\hbar = 1$ to simplify the different formulas and the operators will be written in boldface.

The previous equation has the formal solution

$$\Psi_{\rm S}(t) = e^{-i{\bf H}(t-t_0)}\Psi_{\rm S}(t_0), \tag{5.2}$$

where t_0 is an arbitrary initial time. Here, the exponential of any operator $\mathbf{A}$ is defined, as usual, by means of its Taylor series

$$\exp(\mathbf{A}) = \sum_{n=0}^{\infty} \frac{1}{n!} \mathbf{A}^n. \tag{5.3}$$

From this result, it is obvious that the operator $\exp[-i\mathbf{H}(t - t_0)]$ is the time-evolution operator in the Schrödinger picture, in the sense that by acting on the wave function at a initial time, t_0, this operator transforms it into the wave function at the time t. If we take $t_0 = 0$, we have

$$\Psi_{\mathrm{S}}(t) = e^{-i\mathbf{H}t} \Psi_{\mathrm{S}}(0). \tag{5.4}$$

For the moment, since we are only interested in equilibrium situations, we shall assume that the operators describing the observables in this representation, $\mathbf{O}_{\mathrm{S}}$, do not have any explicit time dependence.

Another typical representation in quantum mechanics is the so-called *Heisenberg picture*, which can be defined from the Schrödinger one by means of the following unitary transformation

$$\Psi_{\mathrm{H}}(t) = e^{i\mathbf{H}t} \Psi_{\mathrm{S}}(t) = \Psi_{\mathrm{S}}(0)$$
$$\mathbf{O}_{\mathrm{H}}(t) = e^{i\mathbf{H}t} \mathbf{O}_{\mathrm{S}} e^{-i\mathbf{H}t}. \tag{5.5}$$

Thus, in Heisenberg picture the time dependence has been transferred from the wave functions to the operators. The wave function in this representation is stationary and equal to the wave function in Schrödinger picture at time zero, i.e. $\Psi_{\mathrm{H}} = \Psi_{\mathrm{S}}(0)$, whereas the operators, $\mathbf{O}_{\mathrm{H}}(t)$, do depend explictly on time. Their time evolution can be obtained by taking the derivative with respect to time in the previous equation

$$i\frac{\partial}{\partial t}\mathbf{O}_{\mathrm{H}} = [\mathbf{O}_{\mathrm{H}}, \mathbf{H}], \tag{5.6}$$

which is the equation of motion of an operator in this representation (see Exercise 5.1).

Both representations are equivalent in the sense that the expectation values are the same, irrespective of the picture used. This is a simple consequence of the fact that both representations are related by means of a unitary transformation.

5.2   Green's functions of a noninteracting electron system

Green's functions are commonly used in traditional contexts such as classical mechanics and electromagnetism. In those cases, Green's functions are defined as the inverse of differential operators. One can indeed proceed in a similar way with the Schrödinger equation, which is a second order differential equation. As an illustration, let us consider the problem of an electron in an one-dimensional system, which is described by the Schrödinger equation

$$\mathbf{H}(x)\Psi(x) = E\Psi(x). \tag{5.7}$$

Now, we define the electron Green's function (or propagator) as

$$[E - \mathbf{H}(x)]\, G(x, x') = \delta(x - x'), \tag{5.8}$$

where

$$\mathbf{H}(x) = -\frac{1}{2m}\frac{\partial^2}{\partial x^2} + V(x), \tag{5.9}$$

$V(x)$ being an external potential acting on the electron. Notice that the Green's function is a complex function that depends both on the spatial coordinates and on the energy, E.

In the case of a free electron, $V(x) = V_0 = $ constant, the Green's function can be obtained exactly (see Exercise 5.2). Indeed, one can show that a solution is given by

$$G(x - x', E) = -\frac{i}{v}e^{ik|x-x'|}, \tag{5.10}$$

where $k = \sqrt{2m(E - V_0)}$, $v = k/m$ and we have included the energy, E, as an argument. As it will become clear later on, one can interpret the Green's function as the propagation amplitude of an electron. In this sense, the previous expression corresponds to the propagation of a free electron at energy E from the position x' to the right $(x - x' > 0)$ or to the left $(x - x' < 0)$.

It is important to notice that there is another solution that corresponds to the time-reserved solution as compared with the previous one:

$$G(x - x', E) = \frac{i}{v}e^{-ik|x-x'|}. \tag{5.11}$$

This simply reflects the fact that the Green's function is not completely determined until we specify the boundary conditions for its differential equation.

Eq. (5.10) corresponds to the so-called *retarded Green's function*, G^r, while Eq. (5.11) corresponds to the *advanced Green's function*, G^a. Although the time does not appear explicitly in these functions, we shall show later that one can relate G^r [Eq. (5.10)] with the propagation of an electron forwards in time, while G^a [Eq. (5.11)] is the corresponding time-reversed function (describing the electron propagation backwards in time).

An easy way to obtain the retarded/advanced function in the previous problem is by introducing an infinitesimal imaginary part in the energy in the expression defining $G(x-x')$. Thus, the substitution $E \to E \pm i\eta$ selects the retarded Green's function for the plus sign and the advanced one for the minus sign. A rigorous definition of the retarded Green's function for this one-dimensional problem would then be

$$\lim_{\eta \to 0} \left[E + i\eta - \mathbf{H}(x) \right] G^r(x, x') = \delta(x - x'), \tag{5.12}$$

and a similar one for the advanced function.

This definition for the one-dimensional problem can be generalized to any single-particle problem. If $\mathbf{H}$ is the Hamilton operator of the system, we can define the retarded and advanced Green's functions as

$$\mathbf{G}^{r,a}(E) = \lim_{\eta \to 0} \left[(E \pm i\eta)\mathbf{1} - \mathbf{H} \right]^{-1}, \tag{5.13}$$

where we have written the equation as an operator identity in order to have an expression that is independent of the representation. Here, $\mathbf{1}$ is the identity operator. It is possible to write the previous equation in an alternative form in terms of the eigenfunctions and eigenvalues of $\mathbf{H}$ ($\mathbf{H}|\psi_n\rangle = \epsilon_n|\psi_n\rangle$):

$$\mathbf{G}^{r,a}(E) = \sum_n \frac{|\psi_n\rangle\langle\psi_n|}{E - \epsilon_n \pm i\eta}, \tag{5.14}$$

where from now on the limit $\lim_{\eta \to 0}$ is implicitly assumed in all the expressions in which the parameter η appears. Are you able to show the equivalence of Eqs. (5.13) and (5.14)? If not, see hints in Exercise 5.3. Eq. (5.14) shows that the Green's functions (for a noninteracting case) have poles precisely at the eigenenergies, ϵ_n, of the system. This is the first important piece of information contained in these functions.

From the previous equations, one can deduce a number of important properties of the functions $G^{r,a}$. Let us discuss the most useful ones for our purposes:

Property 1. The imaginary part of the Green's functions is related to the density of states of the system. To demonstrate this, let us remind that the local density of states in a given position $\mathbf{r}$ can be written in terms of the eigenstates of $\mathbf{H}$ as follows

$$\rho(\mathbf{r}, E) = \sum_n |\langle \mathbf{r}|\psi_n\rangle|^2 \delta(E - \epsilon_n). \tag{5.15}$$

From Eq. (5.14) we can write

$$G^{r,a}(\mathbf{r}, E) = \sum_n \frac{\langle \mathbf{r} | \psi_n \rangle \langle \psi_n | \mathbf{r} \rangle}{E - \epsilon_n \pm i\eta}, \tag{5.16}$$

and comparing these last two equations, one obtains

$$\rho(\mathbf{r}, E) = \mp \frac{1}{\pi} \text{Im}\left\{ G^{r,a}(\mathbf{r}, E) \right\}. \tag{5.17}$$

Here, we have used the relation

$$\frac{1}{E \pm i\eta} = \mathcal{P}\left(\frac{1}{E}\right) \mp i\pi\delta(E), \tag{5.18}$$

where $\mathcal{P}$ denotes a Cauchy principal value.

If we use a discrete basis of atomic orbitals, we would have

$$\rho_i(E) = \mp \frac{1}{\pi} \text{Im}\left\{ G_{ii}^{r,a}(E) \right\}, \tag{5.19}$$

where i indicates that the density of states has been projected onto the atom (or site) i.

Property 2. The diagonal Green's functions satisfy in any basis that $\text{Im}\{G_{ii}^r(E)\} \leq 0$ and $\text{Im}\{G_{ii}^a(E)\} \geq 0$. This is obvious from Eq. (5.14).

Property 3. The real and imaginary parts of $G^{r,a}$ are related through a Hilbert transformation:

$$\text{Re}\left\{ \mathbf{G}^{r,a}(E) \right\} = \mp \mathcal{P} \int_{-\infty}^{\infty} \frac{dE'}{\pi} \frac{\text{Im}\left\{ \mathbf{G}^{r,a}(E') \right\}}{E - E'}. \tag{5.20}$$

This is a consequence of the pole structure of Eq. (5.14) and it can be easily shown with the help of Eq. (5.18). As a result of this relation, $\mathbf{G}^{r,a}(E)$ can be written as

$$\mathbf{G}^{r,a}(E) = \int_{-\infty}^{\infty} dE' \frac{\rho(E')}{E - E' \pm i\eta}, \tag{5.21}$$

where we have defined the density operator $\rho(E) \equiv \mp \text{Im}\{\mathbf{G}^{r,a}(E)\}/\pi$. This way of writing the Green's function in terms of the density of states is known as *spectral representation* and, as we shall show below, it is also valid in the case of interacting systems.

Property 4. An important consequence of the spectral representation is the asymptotic form of the diagonal Green's functions for $E \to \infty$. As $\rho_i(E)$ is a bounded function, one has

$$\lim_{E \to \infty} G_{ii}^{r,a}(E) = \frac{1}{E}. \tag{5.22}$$

This is a consequence of the fact that the energy integral of $\rho_i(E)$ is equal to 1, i.e.

$$\int_{-\infty}^{\infty} dE \, \rho_i(E) = \mp \frac{1}{\pi} \int_{-\infty}^{\infty} dE \, \mathrm{Im} \, \{G_{ii}^{r,a}(E)\} = 1. \qquad (5.23)$$

Property 5. As one can easily see from Eq. (5.13), the following simple relation between G^r and G^a holds:

$$\mathbf{G}^r(E) = [\mathbf{G}^a(E)]^{\dagger}. \qquad (5.24)$$

This means in practice that we only need to compute one of these two type of functions.

Property 6. As a last issue, let us consider the case in which the Hamiltonian $\mathbf{H}$ can be written as

$$\mathbf{H} = \mathbf{H}_0 + \mathbf{V}, \qquad (5.25)$$

where $\mathbf{H}_0$ is the Hamiltonian of a problem for which the Green's functions are known, $\mathbf{g}^{r,a}$, and $\mathbf{V}$ is an arbitrary single-particle perturbation. We want to express the Green's functions of the full problem in terms of the unperturbed Green's functions. This can be easily done starting from the definition of Eq. (5.13)

$$\mathbf{G}^{r,a}(E) = [(E \pm i\eta)\mathbf{1} - \mathbf{H}_0 - \mathbf{V}]^{-1}. \qquad (5.26)$$

Taking into account that for the unperturbed problem we have

$$\mathbf{g}^{r,a}(E) = [(E \pm i\eta)\mathbf{1} - \mathbf{H}_0]^{-1}, \qquad (5.27)$$

it is easy to obtain the following relation (see Exercise 5.4)

$$\mathbf{G}^{r,a}(E) = \mathbf{g}^{r,a}(E) + \mathbf{g}^{r,a}(E)\mathbf{V}\mathbf{G}^{r,a}(E), \qquad (5.28)$$

The previous equation is known as *Dyson's equation* and it can also be derived in the interacting case, as we shall show in the next chapter. However, in the general case the operator $\mathbf{V}$ is replaced by a energy-dependent operator, $\mathbf{\Sigma}(\mathbf{E})$, known as *self-energy*. Dyson's equation is extremely useful to compute the Green's functions in different situations, as we shall illustrate in the next section. We shall also show that it is possible to have a energy-dependent self-energy in single-particle problems when one deals with a subspace of the full Hilbert space of the problem.

5.3   Application to tight-binding Hamiltonians

In this section we shall apply what we have learned so far to the computation of the Green's functions of several simple electronic systems described in terms of tight-binding Hamiltonians.[1] Such Hamiltonians, as we shall see in the next chapters, play a fundamental role in the field of molecular electronics. A generic tight-binding Hamiltonian adopts the following form in the language of second quantization (see Appendix A)

$$\mathbf{H} = \sum_{i\sigma} \epsilon_i \mathbf{c}_{i\sigma}^{\dagger} \mathbf{c}_{i\sigma} + \sum_{i\neq j;\sigma} t_{ij} \mathbf{c}_{i\sigma}^{\dagger} \mathbf{c}_{j\sigma}. \tag{5.29}$$

Here, the indexes i and j run over of the sites (atoms) of the system and σ represents the electron spin ($\sigma = \uparrow, \downarrow$). The different operators have the following meaning. For instance, $\mathbf{c}_{i\sigma}^{\dagger}$ is the operator that creates an electron in the site i with spin σ, while $\mathbf{c}_{i\sigma}$ annihilates such an electron. For the sake of simplicity, we shall assume in this discussion that there is a single relevant orbital per site. The parameters ϵ_i are the on-site energies, while the *hoppings* t_{ij} describe the coupling between the different sites (see Appendix A for a precise definition of all these parameters).

Our goal is the calculation of the different Green's functions $G_{ij}^{r,a}(E)$ in this local basis representation. In principle, we have three methods at our disposal: (i) the definition of Eq. (5.13), (ii) the spectral representation of Eq. (5.14) and (iii) Dyson's equation, see Eq. (5.28). We shall illustrate the use of these different approaches with the analysis of three basic examples that will be frequently used in subsequent chapters.

5.3.1   *Example 1: A hydrogen molecule*

We describe a hydrogen molecule with the following two-sites tight-binding Hamiltonian (see Fig. 5.1)

$$\mathbf{H} = \epsilon_0 \sum_{\sigma} (\mathbf{n}_{1\sigma} + \mathbf{n}_{2\sigma}) + t \sum_{\sigma} (\mathbf{c}_{1\sigma}^{\dagger} \mathbf{c}_{2\sigma} + \mathbf{c}_{2\sigma}^{\dagger} \mathbf{c}_{1\sigma}). \tag{5.30}$$

Here, $\mathbf{n}_{i\sigma} = \mathbf{c}_{i\sigma}^{\dagger} \mathbf{c}_{i\sigma}$, ϵ_0 is the $1s$-level of the hydrogen atoms and t is the hopping connecting these two levels and it is assumed to be real. Our goal is to compute the retarded/advanced diagonal Green's function of site 1, i.e.

---

[1]The tight-binding approach is briefly described in Appendix A and it is explained in detail in Chapter 9. Here, we shall use the term tight-binding to refer to models or Hamiltonians where the electronic structure is described in terms a local (atomic-like) basis. We shall not discuss here how the matrix elements of such a Hamiltonian are actually computed, and we shall just use them as parameters.

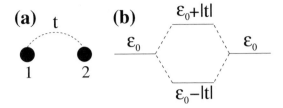

Fig. 5.1   (a) Model for the hydrogen molecule. We consider a single orbital per site (atom) with energy ϵ, and the coupling is described by a hopping t. (b) Level scheme of the hydrogen molecule in which the two orbitals hybridize to form the bonding and antibonding states with energies $\epsilon_0 \pm |t|$.

$G_{11}^{r,a}(E)$ (since the problem has spin degeneracy, we omit the spin indexes in the Green's functions). For symmetry reasons, this Green's function is equal to $G_{22}^{r,a}(E)$. In order to compute this function, we shall employ the three methods mentioned above:

Method 1: Direct definition. According to the definition of Eq. (5.13), the matrix Green's function can be simply calculated by inverting the Hamiltonian of Eq. (5.30). In the basis of the atomic states localized in the hydrogen atoms, $\{|1\rangle, |2\rangle\}$, this Hamiltonian adopts the following matrix form

$$\mathbf{H} = \begin{pmatrix} \epsilon_0 & t \\ t & \epsilon_0 \end{pmatrix}, \tag{5.31}$$

and therefore the matrix Green's function is given by

$$\mathbf{G}^{r,a}(E) = \begin{pmatrix} E^{r,a} - \epsilon_0 & -t \\ -t & E^{r,a} - \epsilon_0 \end{pmatrix}^{-1}, \tag{5.32}$$

where $E^{r,a} \equiv E \pm i\eta$, η being the infinitesimal imaginary part of the energy appearing in the definition of Eq. (5.13). Thus, the element $(1,1)$ that we are looking for reads

$$G_{11}^{r,a}(E) = \frac{E^{r,a} - \epsilon_0}{(E^{r,a} - \epsilon_0)^2 - t^2} = \frac{1/2}{E^{r,a} - (\epsilon_0 + t)} + \frac{1/2}{E^{r,a} - (\epsilon_0 - t)}. \tag{5.33}$$

One can show that this expression fulfills the different properties of a Green's function discussed in the previous section. Thus for instance, notice that Eq. (5.33) has precisely the form of the spectral representation of Eq. (5.14). The poles in this case are nothing else but the energies $\epsilon_\pm = \epsilon_0 \pm t$ of the bonding and antibonding orbitals of the hydrogen molecule,[2]

---

[2]The hooping t is indeed a negative quantity and thus $\epsilon_+ = \epsilon_0 + t$ corresponds to the lowest energy level (bonding state).

see Fig. 5.1. Notice also that the sum of the weights (coefficients appearing in the numerators) is equal to 1.

On the other hand, the density of states projected onto the site 1 is given in this case by

$$\rho_1(E) = \mp \frac{1}{\pi} \mathrm{Im} \left\{ G_{11}^{r,a}(E) \right\} = \frac{1}{2} \delta(E - \epsilon_+) + \frac{1}{2} \delta(E - \epsilon_-), \qquad (5.34)$$

i.e. it is a sum of delta functions evaluated at the molecular energies. This is a consequence of the fact that we are dealing with a finite system. In a similar way, one could demonstrate that the rest of the properties listed at the end of the previous section are satisfied. In particular, properties 4 and 5 are rather obvious from Eq. (5.33).

Method 2: Spectral representation. Let us now use the spectral representation of Eq. (5.14). To evaluate this expression we need both the eigenfunctions and the eigenvalues of the hydrogen molecule. For this purpose we just need to diagonalize the Hamiltonian of Eq. (5.31). The eigenfunctions are simply the bonding ($|\psi_+\rangle$) and antibonding ($|\psi_-\rangle$) states given by: $|\psi_\pm\rangle = (|1\rangle \pm |2\rangle)/\sqrt{2}$ with the corresponding eigenvalues $\epsilon_\pm$. Thus, the function $G_{11}^{r,a}(E)$ is then given by

$$G_{11}^{r,a}(E) = \langle 1|\mathbf{G}|1\rangle = \sum_{n=+,-} \frac{\langle 1|\psi_n\rangle\langle\psi_n|1\rangle}{E^{r,a} - \epsilon_n} = \sum_{n=+,-} \frac{|\langle 1|\psi_n\rangle|^2}{E^{r,a} - \epsilon_n}. \qquad (5.35)$$

Using the fact that $\langle 1|\psi_\pm\rangle = 1/\sqrt{2}$, we arrive immediately at the expression of Eq. (5.33). Obviously, this method is not very practical in general since it requires the knowledge of the eigenfunctions of the system, which are typically unknown.

Method 3: Dyson's equation. Now, our starting point is Eq. (5.28). The first thing to do is to divide the Hamiltonian of Eq. (5.30) into the unperturbed part $\mathbf{H_0}$ and the perturbation $\mathbf{V}$. The natural choice is that the perturbation be the coupling term between the two atoms (second term in Eq. (5.30)). Thus, these two parts of the Hamiltonian adopt the following matrix form

$$\mathbf{H_0} = \begin{pmatrix} \epsilon_0 & 0 \\ 0 & \epsilon_0 \end{pmatrix}; \quad \mathbf{V} = \begin{pmatrix} 0 & t \\ t & 0 \end{pmatrix}. \qquad (5.36)$$

To solve Dyson's equation we also need the Green's functions of the unperturbed system, $\mathbf{g}^{r,a}$. These functions are simply given by

$$\mathbf{g}^{r,a} = [E^{r,a}\mathbf{1} - \mathbf{H_0}]^{-1} = \begin{pmatrix} E^{r,a} - \epsilon_0 & 0 \\ 0 & E^{r,a} - \epsilon_0 \end{pmatrix}^{-1} = \frac{1}{E^{r,a} - \epsilon_0}\mathbf{1}. \qquad (5.37)$$

Now, we can determine the function $G_{11}^{r,a}(E)$ by taking the element $(1,1)$ in Eq. (5.28), i.e.

$$G_{11}^{r,a}(E) = g_{11}^{r,a}(E) + g_{11}^{r,a}(E)V_{12}G_{21}^{r,a}(E). \qquad (5.38)$$

Remember that $\mathbf{g}^{r,a}$ is diagonal, while $\mathbf{V}$ is purely off-diagonal. In order to get a closed equation for $G_{11}^{r,a}$, we still need an equation for $G_{21}^{r,a}$. Taking now the element $(2,1)$ in Eq. (5.28), we get

$$G_{21}^{r,a}(E) = g_{22}^{r,a}(E)V_{21}G_{11}^{r,a}(E). \qquad (5.39)$$

Substituting this expression now in Eq. (5.38), we arrive at

$$G_{11}^{r,a}(E) = g_{11}^{r,a}(E) + g_{11}^{r,a}(E)V_{12}g_{22}^{r,a}(E)V_{21}G_{11}^{r,a}(E). \qquad (5.40)$$

This equation can now be trivially inverted and using the explicit expression of the unperturbed Green's functions one arrives once more at the result of Eq. (5.33).

We can use the discussion above to illustrate the concept of self-energy, which was briefly mentioned at the end of the last section. In the previous equation, we can identify the following energy-dependent function

$$\Sigma_{11}^{r,a}(E) \equiv V_{12}g_{22}^{r,a}(E)V_{21} = t^2 g_{22}^{r,a}(E). \qquad (5.41)$$

This function describes how the properties of the atom 1 are modified via the interaction with the second atom. This can be better seen by rewriting Eq. (5.40) as

$$G_{11}^{r,a}(E) = \frac{1}{E^{r,a} - \epsilon_0 - \Sigma_{11}^{r,a}(E)}, \qquad (5.42)$$

where we have used the expressions of the unperturbed Green's functions. In this equation we see that the self-energy renormalizes dynamically (depending on the energy) both the position (ϵ_0) and the lifetime of the energy level in the atom 1 (this latter point will become clearer in the next examples). Notice that the self-energy depends both on the coupling to the second atom and on the electronic structure of this second atom. We shall see in the next examples that, no matter the problem, the concept of self-energy appears naturally and it describes the renormalization of the properties of a finite system due to its interaction with an external system. In particular, we shall show in the next chapter that the concept of self-energy remains valid even in the presence of interactions.

5.3.2   *Example 2: Semi-infinite linear chain*

As a first example of an infinite solid, we consider now a semi-infinite linear chain with only nearest-neighbor couplings. This system, which is schematically illustrated in Fig. 5.2(a), will be sometimes used in the next chapters as a model for a metallic electrode. The corresponding tight-binding Hamiltonian of this system reads

$$\mathbf{H} = \epsilon_0 \sum_{i\sigma} \mathbf{n}_{i\sigma} + t \sum_{i\sigma} \left(\mathbf{c}_{i\sigma}^{\dagger} \mathbf{c}_{i+1\sigma} + \mathbf{c}_{i+1\sigma}^{\dagger} \mathbf{c}_{i\sigma} \right), \qquad (5.43)$$

where $i = 1, 2, 3, ...$ represents the different sites starting from the surface. We shall carry out here the calculation of the surface Green's function, $G_{11}^{r,a}(E)$. As in the previous example, there are, in principle, three methods avaliable. However, the first two are rather impractical. The first one would require the inversion of an infinite matrix, while the second would need the calculation of the eigenfunctions and eigenvalues of this infinite (non-periodic) system. For these reasons, we shall resort to Dyson's equation. The first step in this method is to choose the unperturbed problem and the corresponding perturbation. One possible choice would be to select the uncoupled atoms as unperturbed system and the coupling between them as the perturbation. Such a legitimate choice would lead us to an infinite algebraic system, which is really difficult to solve (try it, just for fun!). There is an alternative "trick" that does the job in a few steps. The idea goes as follows. Let us consider that the unperturbed system is composed of two uncoupled systems, namely the atom 1 and the rest of the chain. Then, the perturbation is simply the coupling between these two subsystems, i.e.

$$\mathbf{V} = t \sum_{\sigma} \left(\mathbf{c}_{1\sigma}^{\dagger} \mathbf{c}_{2\sigma} + \mathbf{c}_{2\sigma}^{\dagger} \mathbf{c}_{1\sigma} \right). \qquad (5.44)$$

This means in practice that the only two non-zero elements of the perturbation are $V_{12} = V_{21} = t$.

Now, we can use Dyson's equation [Eq (5.28)] to obtain the equation for $G_{11}^{r,a}(E)$. Taking the element $(1, 1)$ we have

$$G_{11}(E) = g_{11}(E) + g_{11}(E)V_{12}G_{21}(E)$$
$$G_{21}(E) = g_{22}(E)V_{21}G_{11}(E),$$

where the second relation is necessary to obtain a closed equation for $G_{11}(E)$. Here, we have omitted again the spin index σ since there is spin degeneracy in this problem and we have also dropped the superindexes r, a because the equations are valid for both retarded and advanced functions. The unperturbed function g_{11} of the site $i = 1$ is simply given by

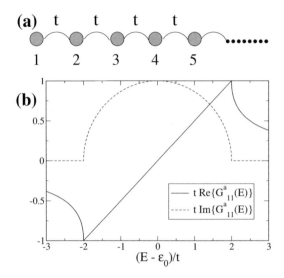

Fig. 5.2 (a) Semi-infinite linear chain with a single orbital per site and only nearest-neighbor couplings. (b) Real and imaginary parts of the advanced surface Green's function, G_{11}^a, of the semi-infinite chain as a function of the energy, see Eq. (5.46).

$g_{11}(E) = 1/(E - \epsilon_0)$. On the other hand, the unperturbed function g_{22} is nothing else but the surface Green's function of a semi-infinite chain,[3] which is precisely what we are looking for, i.e. $g_{22} = G_{11}$. This allows us to obtain the following closed equation for $G_{11}(E)$

$$(E - \epsilon_0)G_{11}(E) = 1 + t^2 G_{11}^2(E). \qquad (5.45)$$

This is a quadratic equation that possesses two possible solutions. In order to choose the "physical" one, it is necessary to take into account the boundary condition $E \to E^{r,a} = E \pm i\eta$ to distinguish between the retarded and advanced solutions. As a practical advice, remember that the imaginary part of these functions has a well-defined sign. The final solution adopts the following expression

$$G_{11}^{r,a}(E) = \frac{1}{t}\left(\frac{E^{r,a} - \epsilon_0}{2t} - \sqrt{\left(\frac{E^{r,a} - \epsilon_0}{2t}\right)^2 - 1}\right). \qquad (5.46)$$

The real and imaginary parts of the advanced function are depicted in Fig. 5.2(b). Notice that the imaginary part, and therefore the density of states, is only non-zero in the region $|E - \epsilon_0| < 2|t|$, which defines the

---

[3]The removal of an atom from the chain does not modify the fact that the remaining chain is again a semi-infinite chain.

energy band of the linear chain. In this region, the Green's function adopts the following form

$$G_{11}^{r,a}(E) = \frac{1}{t}\left(\frac{E - \epsilon_0}{2t} \mp i\sqrt{1 - \left(\frac{E - \epsilon_0}{2t}\right)^2}\right). \tag{5.47}$$

This expression can be written in a form that is very useful to do algebraic manipulations (see Exercise 5.5) by defining $\cos(\phi) \equiv (E - \epsilon_0)/2t$:

$$G_{11}^{r,a}(E) = \frac{1}{t}\exp(\mp i\phi) \tag{5.48}$$

The density of states in the surface atom of the chain can be then expressed as

$$\rho_1(E) = \frac{1}{\pi}\mathrm{Im}\left\{G_{11}^a(E)\right\} = \frac{1}{\pi t}\sqrt{1 - \left(\frac{E - \epsilon_0}{2t}\right)^2}, \quad |E - \epsilon_0| \leq 2|t|. \tag{5.49}$$

and it can be seen in Fig. 5.2(b). Contrary to the example of the hydrogen molecule, in this case there is an infinite number of states that are grouped in an energy band of width $4t$. Notice that we have not specified the actual occupation of this band. If we had an electron per site, the band would be half-filled (with the Fermi energy equal to ϵ_0) and there would be electron-hole symmetry.

It is worth mentioning that in Eq. (5.45) one can identify the self-energy $\Sigma_{11}^{r,a}(E) = t^2 G_{11}^{r,a}(E)$, which plays exactly the same role as in the case of the hydrogen molecule and it has the same functional form.

Let us say to conclude this discussion that one can check that the expression of Eq. (5.46) satisfies the different properties discussed in the previous section. The reader is encouraged to show, in particular, that

$$\lim_{E \to \infty} \mathrm{Re}\left\{G_{11}^{r,a}(E)\right\} = \frac{1}{E}, \tag{5.50}$$

and that the following sum rule is fulfilled

$$\int_{-\infty}^{\infty} dE\, \rho_1(E) = 1. \tag{5.51}$$

5.3.3   *Example 3: A single level coupled to electrodes*

We consider now the case of single energy level coupled to two infinite electrodes. This is a very important example that will teach us a couple of important lessons for molecular electronics. The system that we are

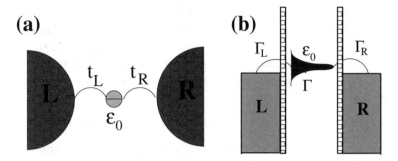

Fig. 5.3   (a) A single level of energy ϵ_0 is coupled to two infinite electrodes via the hoppings t_L and t_R. (b) The corresponding energy scheme where one can see the continuum of states in the electrodes filled up to the Fermi energy and the resonant level, which has acquired a half width at half maximum equal to $\Gamma = \Gamma_L + \Gamma_R$ due to the coupling to the reservoirs.

interested in is schematically represented in Fig. 5.3(a), and it is described by the following Hamiltonian

$$\mathbf{H} = \mathbf{H}_L + \mathbf{H}_R + \sum_{\sigma} \epsilon_0 \mathbf{c}_{0\sigma}^{\dagger} \mathbf{c}_{0\sigma} + \tag{5.52}$$

$$\sum_{\sigma} t_L \left(\mathbf{c}_{0\sigma}^{\dagger} \mathbf{c}_{L\sigma} + \mathbf{c}_{L\sigma}^{\dagger} \mathbf{c}_{0\sigma} \right) + \sum_{\sigma} t_R \left(\mathbf{c}_{0\sigma}^{\dagger} \mathbf{c}_{R\sigma} + \mathbf{c}_{R\sigma}^{\dagger} \mathbf{c}_{0\sigma} \right).$$

Here, the Hamiltonians $\mathbf{H}_L$ and $\mathbf{H}_R$ describe the left and right electrodes that are coupled to a single energy level. It will not be necessary for the present discussion to specify anything about the shape or concrete electronic structure of these two leads. The subindex 0 refers to the localized level, the energy of which is denoted by ϵ_0. This level is coupled to the electrodes via the hoppings t_L and t_R, which are assumed to be real. The subindexes L and R refer here to the outermost sites of the left and right electrodes (we have in mind again that there is a single relevant orbital per site in these leads).

The question that we want to address is: How is this level modified by the coupling to the electrodes? This question is very relevant for many different contexts. We have in mind the problem of a molecule (or atom) coupled to metallic leads, but it is also important for problems like the chemisorption of molecules on surfaces (in this case there would be only one electrode). In order to answer this question, we will compute the local density of states projected onto the level. This requires the calculation of the Green's function $G_{00}(E)$ (no matter whether it is retarded or advanced). For this purpose, we resort to Dyson's equation. Our choice for the unperturbed Hamiltonian $\mathbf{H}_0$ is the sum of the Hamiltonians of the three uncoupled subsystems, i.e. the right hand side of the first line of Eq. (5.52). Thus, the perturbation $\mathbf{V}$ is the term that describes the coupling between the localized level and the electrodes (second line in Eq. (5.52)). Notice that we are assuming that there is no direct coupling between the leads.

With this choice in mind, we take the element $(0,0)$ in Eq. (5.28) to obtain

$$G_{00}(E) = g_{00}(E) + g_{00}(E)V_{0L}G_{L0}(E) + g_{00}(E)V_{0R}G_{R0}(E), \qquad (5.53)$$

where $V_{0L} = t_L$ and $V_{0R} = t_R$ and $g_{00}(E) = 1/(E - \epsilon_0)$ is the unperturbed Green's function of the single-level system. As usual, to close this equation, we have to determine the functions G_{L0} and G_{R0}. This can be done by taking the corresponding elements in Dyson's equation, i.e.

$$G_{L0}(E) = g_{LL}(E)V_{L0}G_{00}(E)$$
$$G_{R0}(E) = g_{RR}(E)V_{R0}G_{00}(E),$$

where $V_{L/R0} = t_{L/R}$ and g_{LL} and g_{RR} are the Green's functions of the two outermost sites of the left and right electrodes, respectively. Substituting these expressions in Eq. (5.53), we obtain the following closed equation

$$G_{00}(E) = g_{00}(E) + g_{00}(E)V_{0L}g_{LL}(E)V_{L0}G_{00}(E) \qquad (5.54)$$
$$+ g_{00}(E)V_{0R}g_{RR}(E)V_{R0}G_{00}(E).$$

In this expression one can identify, as in the previous examples, the self-energy $\Sigma_{00}(E) = t_L^2 g_{LL}(E) + t_R^2 g_{RR}(E)$, which in this case is the sum of two contributions associated to the two leads. In terms of the self-energy we can express the function $G_{00}(E)$ as

$$G_{00}(E) = \frac{1}{E - \epsilon_0 - \Sigma_{00}(E)}, \qquad (5.55)$$

where we have used the expression of $g_{00}(E)$. Here, we see once more that the self-energy describes how the resonant level is modified by the interaction with the leads. In particular, its real part is responsible for the renormalization of the level position, which becomes $\tilde{\epsilon}_0 = \epsilon_0 + \mathrm{Re}\{\Sigma_{00}(E)\}$, while its imaginary part describes the finite energy "width" acquired by the level via the interaction with the leads. This latter point becomes more clear by using the following approximation. Let us assume that the Green's functions of the leads are imaginary for energies in the vicinity of ϵ_0 and that they do not depend significantly on energy in this region.[4] Thus, we can approximate these functions by $g_{LL,RR}^{r,a} \approx \mp i/W_{L,R}$, where

---

[4]This approximation is usually known as *wide-band approximation*.

$W_{L,R}$ are energy scales related to the density of states of the leads at the energy ϵ_0.[5] For instance, if we modeled the electrodes by the semi-infinite chains like in the previous example, $W_{L,R}$ would then be the bulk hopping element of these chains. Within this approximation, the self-energy becomes $\Sigma_{00}^{r,a} = \mp i\left(\Gamma_L + \Gamma_R\right)$, where we have defined the scattering rates $\Gamma_{L,R} \equiv t_{L,R}^2/W_{L,R}$. Obviously, with this approximation the level position remains unchanged (see Exercise 5.9). Finally, the function $G_{00}(E)$ adopts in this case the form

$$G_{00}^{r,a}(E) = \frac{1}{E^{r,a} - \epsilon_0 \pm i\left(\Gamma_L + \Gamma_R\right)}, \qquad (5.56)$$

Thus, the local density of states that we wanted to calculate is given by

$$\rho_0(E) = \mp\frac{1}{\pi}\mathrm{Im}\left\{G_{00}^{r,a}(E)\right\} = \frac{1}{\pi}\frac{\Gamma_L + \Gamma_R}{(E - \epsilon_0)^2 + (\Gamma_L + \Gamma_R)^2}, \qquad (5.57)$$

which is a Lorentzian function, where $\Gamma = \Gamma_L + \Gamma_R$ is the half-width at half-maximum (HWHM). This result shows clearly that the resonant level, which originally had zero width (it was an eigenstate of the isolated central system), acquires a finite width Γ via the coupling to the leads. This fact is illustrated in Figs. 5.3(b). It is worth stressing that the width depends both on the strength of the coupling to the electrodes (via $t_{L,R}^2$) and on the local electronic structure of the leads (via $W_{L,R}$ or, more generally, via $g_{LL,RR}$). The time scale $\hbar/\Gamma$ that can be interpreted as the finite lifetime of the resonant level due to the interaction with the leads, or in other words, as the time that an electron spends in the resonant level.

Thus, the take-home message of this example is that *when an isolated molecule (or an atom) is coupled to a continuum of states, its levels are, in general, shifted and they acquire a width that depends on the strength of the coupling and on the local electronic structure of the leads.*

Let us finally say that we hope that the reader has realized that all the calculations of this section involved simple algebraic manipulations. Indeed, we shall show in the next chapters that, as long as we deal with systems with only elastic interactions (described by mean-field Hamiltonians), the evaluation of the Green's functions, both in equilibrium and out of equilibrium, reduces to straightforward exercises of linear algebra. So maybe, this Green's function stuff is not so scary after all, don't you think?

For more detailed discussion of Green's functions in the framework of tight-binding models, we recommend the book of Ref. [181], as well as the exercises 5-9 at the end of this chapter.

---

[5]This energy scales are simply given by $W_{L,R} = 1/[\pi\rho_{L,R}(E = \epsilon_0)]$, where $\rho_{L,R}$ are the local densities of states of the two outermost sites of the leads.

5.4   Green's functions in time domain

The energy-dependent retarded and advanced Green's functions introduced in the previous section for single-particle problems can be considered as Fourier transforms of time-dependent Green's functions, the definition of which is much more general and they are still valid in the case of interacting systems. The utility of these new definitions will become apparent in the next chapter when we deal with the perturbation theory. Moreover, it will be clear that we need to introduce a new kind of function known as the causal Green's function.

Using the second quantization formalism and an arbitrary representation (or basis), the retarded Green's function that depends on two time arguments can be defined as follows

$$G_{ij}^r(t, t') = -i\theta(t - t')\langle\Psi_{\mathrm{H}}| \left\{\mathbf{c}_{i\sigma}(t), \mathbf{c}_{j\sigma}^\dagger(t')\right\} |\Psi_{\mathrm{H}}\rangle, \qquad (5.58)$$

where $|\Psi_{\mathrm{H}}\rangle = |\Psi_{\mathrm{S}}(0)\rangle$ is the wave function of the ground state of the system (that can include interactions) and the operators are in Heisenberg picture. We shall only include explicitly the spin index σ in G_{ij}^r in those problems where the spin symmetry is broken. In this definition, the step function, θ, ensures that $t > t'$ and the symbol $\{\,,\,\}$ stands for the anticommutator.

The Green's functions are often defined using the basis $\{|\mathbf{r}\rangle\}$ formed by the eigenfunctions of the position operator. The corresponding creation and annihilation operators in this representation are known as *field operators* and they are denoted by $\mathbf{\Psi}_\sigma^\dagger(\mathbf{r})$ and $\mathbf{\Psi}_\sigma(\mathbf{r})$, These operators are simply related to $\mathbf{c}_{i\sigma}^\dagger$ and $\mathbf{c}_{i\sigma}$ by the basis transformation

$$\mathbf{\Psi}_\sigma(\mathbf{r}) = \sum_i \phi_i(\mathbf{r})\mathbf{c}_{i\sigma} \text{ and } \mathbf{\Psi}_\sigma^\dagger(\mathbf{r}) = \sum_i \phi_i^*(\mathbf{r})\mathbf{c}_{i\sigma}^\dagger, \qquad (5.59)$$

where $\phi_i(\mathbf{r})$ are the basis wave functions of the discrete representation. These field operators satisfy the standard type of anticommutation relations, i.e.

$$\{\mathbf{\Psi}_\sigma(\mathbf{r}), \mathbf{\Psi}_{\sigma'}^\dagger(\mathbf{r}')\} = \delta(\mathbf{r} - \mathbf{r}')\delta_{\sigma,\sigma'}; \text{ etc.} \qquad (5.60)$$

In terms of these field operators, the retarded Green's function is defined as

$$G^r(\mathbf{r}t, \mathbf{r}'t') = -i\theta(t - t')\langle\Psi_{\mathrm{H}}| \left\{\mathbf{\Psi}_\sigma(\mathbf{r}, t), \mathbf{\Psi}_\sigma^\dagger(\mathbf{r}', t')\right\} |\Psi_{\mathrm{H}}\rangle, \qquad (5.61)$$

which is a complex function that depends on two spatial arguments and two time arguments.

The advanced Green's function has a similar definition, the only difference being that the propagation takes place backwards in time

$$G_{ij}^a(t, t') = i\theta(t' - t)\langle\Psi_H| \left\{ \mathbf{c}_{i\sigma}(t), \mathbf{c}_{j\sigma}^\dagger(t') \right\} |\Psi_H\rangle. \qquad (5.62)$$

Finally, it is convenient to define an additional Green's function, namely the one known as *causal Green's function*, which is defined as follows

$$G_{ij}^c(t, t') = -i\langle\Psi_H|\mathbf{T} \left[\mathbf{c}_{i\sigma}(t)\mathbf{c}_{j\sigma}^\dagger(t') \right] |\Psi_H\rangle, \qquad (5.63)$$

where $\mathbf{T}$ is the time-ordering operator. It acts on a product of time-dependent operators by ordering them chronologically from right to left. Thus for instance, the previous function has the following explicit form

$$G_{ij}^c(t, t') = \begin{cases} -i\langle\Psi_H|\mathbf{c}_{i\sigma}(t)\mathbf{c}_{j\sigma}^\dagger(t')|\Psi_H\rangle & t > t' \\ i\langle\Psi_H|\mathbf{c}_{j\sigma}^\dagger(t')\mathbf{c}_{i\sigma}(t)|\Psi_H\rangle & t' > t. \end{cases} \qquad (5.64)$$

Notice the sign change for $t' > t$ due to the anticommutation of fermion operators.

So far, our discussion in this section has been a bit mathematical and there are questions that arise naturally. The first one is: What is the physical meaning of the Green's functions? To answer this question notice that these functions contain factors like $\langle\Psi_H|\mathbf{c}_{i\sigma}(t)\mathbf{c}_{j\sigma}^\dagger(t')|\Psi_H\rangle$. Here, $\mathbf{c}_{j\sigma}^\dagger(t')|\Psi_H\rangle$ describes the creation (or injection) in the ground state of an electron at time t' in the state j. Then, the previous expectation value yields the probability amplitude of finding such an electron at a later time t in the state i. In other words, the Green's functions simply describe the probability amplitude of the occurrence of certain processes. The type of processes described depends on the arguments of these functions. Thus for instance, they can describe the propagation of electrons in time domain or in energy space, propagation in real space, in momentum space or simply in an atomic lattice.[6]

Another natural question is: What is the relation between this definition of the Green's functions and the one put forward in the previous section? At a first glance, it seems that there is no relation at all. However, we shall show below that if the system is noninteracting, the Fourier transform with respect to the time arguments of these new Green's functions fulfill Eqs. (5.13) and (5.14), i.e. these two type of functions are equivalent.

Simple example: degenerate electron gas. To illustrate the previous definitions, we consider now the example of a free electron gas at zero

---

[6]In this sense, it is not surprising that the elastic transmission of any real system can be naturally expressed in terms of these functions.

temperature, which is discussed in Exercise 5.1. As we know, the ground state of this noninteracting system is a Fermi sea, where the single-particle states are occupied up to the Fermi energy, E_F (or chemical potential μ). These states, $|\mathbf{k}\sigma\rangle$, are plane waves characterized by an energy $\epsilon_\mathbf{k} = k^2/2m$, where $\mathbf{k}$ is the electron momentum. In this case, it is easy to compute both the exact time evolution of the Heisenberg operators (see Exercise 5.1) and the expectation values over this ground state (Fermi sea). Thus for instance,

$$\langle \Psi_H | c^\dagger_{\mathbf{k}\sigma} c_{\mathbf{k}'\sigma} | \Psi_H \rangle = \delta_{\mathbf{k},\mathbf{k}'} \theta(k_F - k), \tag{5.65}$$

where k_F is the Fermi momentum.

Bearing these ideas in mind, it is easy to show that the retarded and advanced Green's functions defined in Eqs. (5.58) and (5.62) can be written in the $\mathbf{k}$-basis (momentum space) as

$$G^r(\mathbf{k}, t - t') = -i\theta(t - t')e^{-i\epsilon_k(t-t')} \tag{5.66}$$
$$G^a(\mathbf{k}, t - t') = +i\theta(t' - t)e^{-i\epsilon_k(t-t')},$$

while the causal function can be written as

$$G^c(\mathbf{k}, t - t') = \begin{cases} -i\theta(k - k_F)e^{-i\epsilon_k(t-t')} & t > t' \\ i\theta(k_F - k)e^{-i\epsilon_k(t-t')} & t < t'. \end{cases} \tag{5.67}$$

Notice first that these functions depend on the difference of the time arguments, which is a general property for equilibrium systems. Notice also that they are diagonal in k-space. Having in mind the physical meaning of the Green's functions, it is easy to understand why they have such a simple time dependence. Since we are injecting electrons in a state $|\mathbf{k}\sigma\rangle$, which is an eigenstate of the system, the probability of finding it at a later time in such state must be equal to one. This is precisely what the previous expressions illustrate.

It is instructive to make contact with the results of the previous section. For this purpose we must now Fourier transform the previous functions with respect to the time difference, i.e.

$$G^{r,a,c}(\mathbf{k}, E) = \int_{-\infty}^{\infty} dt \, G^{r,a,c}(\mathbf{k}, t)e^{-iE(t-t')}. \tag{5.68}$$

In the course of doing the Fourier transformations, one gets the impression that the time integrals diverge. This can be cured by introducing a small imaginary part in the energy $(E \to E \pm i\eta)$.[7] So finally, the retarded and advanced Green's functions in energy space are given by

$$G^{r,a}(\mathbf{k}, E) = \frac{1}{E - \epsilon_k \pm i\eta}. \tag{5.69}$$

This is exactly the result that one would have obtained directly from Eq. (5.13) in this plane wave basis.

On the other hand, the causal function adopts the form

$$G^c(\mathbf{k}, E) = \frac{\theta(k - k_{\mathrm{F}})}{E - \epsilon_k + i\eta} + \frac{\theta(k_{\mathrm{F}} - k)}{E - \epsilon_k - i\eta} = \frac{1}{E - \epsilon_k + i\,\mathrm{sgn}(k - k_{\mathrm{F}})\eta}. \tag{5.70}$$

Therefore, for the free electron gas, the causal Green's function is equal to the retarded one for $E > \mu$ and equal to the advanced one for $E < \mu$. This relation is true in general, as we shall show below.

5.4.1   *The Lehmann representation*

The goal is now to get an insight into the energy dependence of the Green's functions introduced above for a general interacting system. For this purpose, we shall derive here the spectral representation of a Green's function, which for the noninteracting case reduces to Eq. (5.14). We shall focus our analysis on the causal function defined in Eq. (5.63). In equilibrium, this function depends only on the difference of the time arguments. Choosing $t' = 0$ we have

$$G^c_{ij}(t) = -i\langle \Psi^N_0 | \mathbf{T}\left[\mathbf{c}_{i\sigma}(t)\mathbf{c}^\dagger_{j\sigma}(0) \right] |\Psi^N_0\rangle, \tag{5.71}$$

where we have added the superindex N in the ground state wave function, $|\Psi^N_0\rangle = |\Psi_{\mathrm{H}}\rangle$, to indicate the total number of electrons in the system. Writing explicitly the time-evolution of Heisenberg operators (see Eq. (5.5)) one has

$$G^c_{ij}(t) = -i\theta(t)\langle \Psi^N_0 | e^{i\mathbf{H}t} \mathbf{c}_{i\sigma} e^{-i\mathbf{H}t} \mathbf{c}^\dagger_{j\sigma} |\Psi^N_0\rangle \tag{5.72}$$
$$+ i\theta(-t)\langle \Psi^N_0 | \mathbf{c}^\dagger_{j\sigma} e^{i\mathbf{H}t} \mathbf{c}_{i\sigma} e^{-i\mathbf{H}t} |\Psi^N_0\rangle.$$

We now use the fact that $\mathbf{H}|\Psi^N_0\rangle = E^N_0|\Psi^N_0\rangle$, where E^N_0 is the ground state energy of the system with N electrons, to arrive at

$$G^c_{ij}(t) = -i\theta(t)\langle \Psi^N_0 | \mathbf{c}_{i\sigma} e^{-i\mathbf{H}t} \mathbf{c}^\dagger_{j\sigma} |\Psi^N_0\rangle e^{iE^N_0 t} \tag{5.73}$$
$$+ i\theta(-t)\langle \Psi^N_0 | \mathbf{c}^\dagger_{j\sigma} e^{i\mathbf{H}t} \mathbf{c}_{i\sigma} |\Psi^N_0\rangle e^{-iE^N_0 t}.$$

---

[7]A more rigorous way of solving this problem involves the introduction of the integral representation of the step function:

$$\theta(t - t') = -\int_{-\infty}^{\infty} \frac{dE}{2\pi i} \frac{e^{-iE(t-t')}}{E + i\eta}.$$

We now insert $\sum_m |\Psi_m^{N+1}\rangle\langle\Psi_m^{N+1}|$ in the part for $t > 0$ and $\sum_m |\Psi_m^{N-1}\rangle\langle\Psi_m^{N-1}|$ in the part for $t < 0$, where $|\Psi_m^{N+1}\rangle$ and $|\Psi_m^{N-1}\rangle$ are the eigenfunctions of the system with one more and one less electrons, respectively. The resulting expression reads

$$G_{ij}^c(t) = -i\theta(t)\sum_m \langle\Psi_0^N|\mathbf{c}_{i\sigma}|\Psi_m^{N+1}\rangle\langle\Psi_m^{N+1}|\mathbf{c}_{j\sigma}^\dagger|\Psi_0^N\rangle e^{-i(E_m^{N+1}-E_0^N)t}$$

$$+i\theta(-t)\sum_m \langle\Psi_0^N|\mathbf{c}_{j\sigma}^\dagger|\Psi_m^{N-1}\rangle\langle\Psi_m^{N-1}|\mathbf{c}_{i\sigma}|\Psi_0^N\rangle e^{-i(E_0^N-E_m^{N-1})t}.$$

We now Fourier transform with respect to the time argument to obtain the expression of the Green's function in energy space

$$G_{ij}^c(E) = \sum_m \frac{\langle\Psi_0^N|\mathbf{c}_{i\sigma}|\Psi_m^{N+1}\rangle\langle\Psi_m^{N+1}|\mathbf{c}_{j\sigma}^\dagger|\Psi_0^N\rangle}{E - (E_m^{N+1}-E_0^N) + i\eta} \tag{5.74}$$

$$+ \sum_m \frac{\langle\Psi_0^N|\mathbf{c}_{j\sigma}^\dagger|\Psi_m^{N-1}\rangle\langle\Psi_m^{N-1}|\mathbf{c}_{i\sigma}|\Psi_0^N\rangle}{E + (E_m^{N-1}-E_0^N) - i\eta},$$

which in the diagonal case adopts the form

$$G_{ii}^c(E) = \sum_m \frac{|\langle\Psi_m^{N+1}|\mathbf{c}_{i\sigma}^\dagger|\Psi_0^N\rangle|^2}{E - (E_m^{N+1}-E_0^N) + i\eta} + \sum_m \frac{|\langle\Psi_m^{N-1}|\mathbf{c}_{i\sigma}|\Psi_0^N\rangle|^2}{E + (E_m^{N-1}-E_0^N) - i\eta}. \tag{5.75}$$

This expression, referred to as *Lehmann or spectral representation*, shows clearly the pole structure of the Green's functions of a general electron system. The poles appear at the energy of the quasi-particles of the system, that is, at the energies that are necessary to add or remove an electron in the ground state of the system.[8] Before analyzing in more detail the properties of $G^c(E)$, let us see how the spectral representation of the retarded/advanced function looks like. One can repeat the process above to arrive at

$$G_{ij}^{r,a}(E) = \sum_m \frac{\langle\Psi_0^N|\mathbf{c}_{i\sigma}|\Psi_m^{N+1}\rangle\langle\Psi_m^{N+1}|\mathbf{c}_{j\sigma}^\dagger|\Psi_0^N\rangle}{E - (E_m^{N+1}-E_0^N) \pm i\eta} + \tag{5.76}$$

$$\sum_m \frac{\langle\Psi_0^N|\mathbf{c}_{j\sigma}^\dagger|\Psi_m^{N-1}\rangle\langle\Psi_m^{N-1}|\mathbf{c}_{i\sigma}|\Psi_0^N\rangle}{E + (E_m^{N-1}-E_0^N) \pm i\eta}.$$

The previous expressions of the Green's functions in energy space can be written in a slightly different way in the thermodynamical limit ($N \to \infty$). Let us focus on the expressions of the denominators. Considering first the

---

[8]Due to the factors $\pm i\eta$, the poles appear slightly shifted with respect to the real axis in the complex plane.

part of electrons, we can add and subtract the energy of the ground state with $N + 1$ electrons:

$$E - (E_m^{N+1} - E_0^N) = E - (E_0^{N+1} - E_0^N) - (E_m^{N+1} - E_0^{N+1}). \qquad (5.77)$$

The energy difference $E_0^{N+1} - E_0^N$ in the limit $N \to \infty$ is the chemical potential μ of the system, while $E_m^{N+1} - E_0^{N+1}$ is the energy of the excited state of the system with $N + 1$ electrons. Repeating the same operations for the hole part, one can finally write the Green's functions in the thermodynamic limit as (we only consider diagonal elements)

$$G_{ii}^c(E) = \sum_m \frac{|\langle \Psi_m^{N+1} | c_{i\sigma}^\dagger | \Psi_0^N \rangle|^2}{E - \mu - \epsilon_m^{N+1} + i\eta} + \sum_m \frac{|\langle \Psi_m^{N-1} | c_{i\sigma} | \Psi_0^N \rangle|^2}{E - \mu + \epsilon_m^{N-1} - i\eta} \qquad (5.78)$$

$$G_{ii}^{r,a}(E) = \sum_m \frac{|\langle \Psi_m^{N+1} | c_{i\sigma}^\dagger | \Psi_0^N \rangle|^2}{E - \mu - \epsilon_m^{N+1} \pm i\eta} + \sum_m \frac{|\langle \Psi_m^{N-1} | c_{i\sigma} | \Psi_0^N \rangle|^2}{E - \mu + \epsilon_m^{N-1} \pm i\eta}, \qquad (5.79)$$

where $\epsilon_m^{N+1} = E_m^{N+1} - E_0^{N+1}$ and $\epsilon_m^{N-1} = E_m^{N-1} - E_0^{N-1}$ are the excitation energies of the system with $N + 1$ and $N - 1$ electrons, respectively.

From the previous expressions one can show that the spectral representation reduces to Eq. (5.14) in the noninteracting case (this exercise is left to the reader). This is one way to establish the connection between the definitions introduced in this section and those of section 5.2.

From the general spectral representation, it is possible to derive the following important properties of the exact Green's functions of an arbitrary electronic system, which are practically identical to those of section 5.2:

Property 1. It is possible to define a spectral density related to the imaginary part of the Green's functions as (we only write the diagonal elements)

$$\rho_i(E) = \sum_m |\langle \Psi_m^{N+1} | c_{i\sigma}^\dagger | \Psi_0^N \rangle|^2 \delta(E - \mu - \epsilon_m^{N+1}) + \qquad (5.80)$$

$$\sum_m |\langle \Psi_m^{N-1} | c_{i\sigma} | \Psi_0^N \rangle|^2 \delta(E - \mu + \epsilon_m^{N-1}).$$

In a case in which i stands for a site index in a tight-binding problem, the previous expression represents the quasiparticle density of states of the system projected onto that site. The relation of the previous function to the imaginary part of the Green's functions is obvious. Comparing Eq. (5.80) with Eqs. (5.78) and (5.79), one obtains

$$\rho_i(E) = \mp \frac{1}{\pi} \text{Im} \{ G_{ii}^{r,a}(E) \} \qquad (5.81)$$

$$\rho_i(E) = -\text{sgn}(E - \mu) \frac{1}{\pi} \text{Im} \{ G_{ii}^c(E) \}. \qquad (5.82)$$

Property 2. The diagonal Green's functions satisfy in any basis that $\mathrm{Im}\{G_{ii}^r(E)\} \le 0$ and $\mathrm{Im}\{G_{ii}^a(E)\} \ge 0$.

Property 3. Due to the pole structure of the Green's functions in energy space, their real and imaginary parts are related through a Hilbert transformation:

$$\mathrm{Re}\left\{G_{ii}^{r,a}(E)\right\} = \mp \mathcal{P} \int_{-\infty}^{\infty} \frac{dE'}{\pi} \frac{\mathrm{Im}\left\{G_{ii}^{r,a}(E')\right\}}{E - E'} \tag{5.83}$$

$$\mathrm{Re}\left\{G_{ii}^c(E)\right\} = -\mathcal{P} \int_{-\infty}^{\infty} \frac{dE'}{\pi} \frac{\mathrm{Im}\left\{G_{ii}^c(E')\right\} \mathrm{sgn}(E' - \mu)}{E - E'}. \tag{5.84}$$

As in the single-particle case, it is possible to write the Green's functions in terms of the spectral density as

$$G_{ii}^{r,a}(E) = \int_{-\infty}^{\infty} dE' \frac{\rho_i(E')}{E - E' \pm i\eta} \tag{5.85}$$

$$G_{ii}^c(E) = \int_{-\infty}^{\infty} dE' \frac{\rho_i(E')}{E - E' + \mathrm{sgn}(E' - \mu)i\eta}. \tag{5.86}$$

Property 4. The previous expressions imply that

$$\lim_{E \to \infty} G_{ii}^{r,a}(E) = \lim_{E \to \infty} G_{ii}^c(E) = \frac{1}{E}, \tag{5.87}$$

where we have used the fact that the spectral density is normalized to 1.

Property 5. From the spectral representations, one can easily deduce the following relations

$$G_{ij}^a(E) = \left[G_{ji}^r(E)\right]^* \quad \text{and} \quad G_{ij}^c(E) = \begin{cases} G_{ij}^r(E), & \text{if } E > \mu \\ G_{ij}^a(E), & \text{if } E < \mu \end{cases}.$$

5.4.2   *Relation to observables*

So far, we have seen that the Green's functions provide important information such as the density of states of states (or the excitation spectrum). But the main reason for studying the Green's functions is that the expectation value of any one-electron operator in the ground state of the system can be expressed in terms of the functions that we have just introduced. Thus for instance, the electronic density $n(\mathbf{r})$ in the ground state is given by

$$n(\mathbf{r}) = \langle \mathbf{n}(\mathbf{r}) \rangle = \sum_\sigma \langle \Psi_\sigma^\dagger(\mathbf{r}) \Psi_\sigma(\mathbf{r}) \rangle, \tag{5.88}$$

which is directly related to the causal Green's function

$$G_\sigma^c(\mathbf{r}t, \mathbf{r}'t') = -i\langle \Psi_\mathrm{H} | \mathbf{T} \left[\Psi_\sigma(\mathbf{r}t) \Psi_\sigma^\dagger(\mathbf{r}'t') \right] | \Psi_\mathrm{H} \rangle, \tag{5.89}$$

by means of

$$n(\mathbf{r}) = -i \sum_\sigma G_\sigma^c(\mathbf{r}t, \mathbf{r}t^+),$$ (5.90)

where t^+ is an abbreviation that means that t' tends t from above.

Analogously, if we use a discrete basis $\{|i\rangle\}$, the occupation of the state i will be given by

$$\langle \mathbf{n}_{i\sigma} \rangle = -iG_{ii\sigma}^c(t, t^+).$$ (5.91)

For instance, for the free electron gas, the time-dependent Green's function is given by Eq. (5.67) and thus, the occupation of a state with wave vector $\mathbf{k}$ in the ground state (Fermi sphere) is

$$\langle \mathbf{n}_\mathbf{k} \rangle = \theta(k_F - k).$$ (5.92)

Let us now demonstrate the general statement made above. One-electron operators can be expressed generically in second quantized form as

$$\mathbf{V} = \sum_{ij\sigma} V_{ij} \mathbf{c}_{i\sigma}^\dagger \mathbf{c}_{j\sigma},$$ (5.93)

where $V_{ij} = \langle i|V(\mathbf{r})|j\rangle$.

Now, we want to compute the expectation value of this operator in the ground state, i.e.

$$\langle \mathbf{V} \rangle = \sum_{i,j,\sigma} V_{ij} \langle \Psi_H | \mathbf{c}_{i\sigma}^\dagger \mathbf{c}_{j\sigma} | \Psi_H \rangle.$$ (5.94)

The expectation values appearing in the previous expression can be related to the Green's functions. For instance, if we recall the definition of the causal Green's functions in the time representation, we have

$$G_{ij}^c(t) = -i\langle \Psi_H | \mathbf{T}[\mathbf{c}_{i\sigma}(t)\mathbf{c}_{j\sigma}^\dagger(0)] | \Psi_H \rangle.$$ (5.95)

If we evaluate this function at $t = 0^-$

$$G_{ij}^c(0^-) = -i\langle \Psi_H | \mathbf{c}_{j\sigma}^\dagger \mathbf{c}_{i\sigma} | \Psi_H \rangle,$$ (5.96)

and therefore

$$\langle \Psi_H | \mathbf{c}_{j\sigma}^\dagger \mathbf{c}_{i\sigma} | \Psi_H \rangle = -iG_{ij}^c(0^-).$$ (5.97)

On the other hand,

$$G_{ij}^c(0^-) = \int_{-\infty}^{\infty} \frac{dE}{2\pi} G_{ij}^c(E) e^{iE0^+}.$$ (5.98)

Making use of the spectral representation for $G^c_{ij}(E)$, we obtain

$$\langle \Psi_H | \mathbf{c}^\dagger_{j\sigma} \mathbf{c}_{i\sigma} | \Psi_H \rangle = \frac{1}{2\pi i} \oint dE \, G^c_{ij}(E) = \frac{1}{\pi} \int_{-\infty}^{\mu} dE \, \text{Im} \left\{ G^c_{ij}(E) \right\}. \quad (5.99)$$

Similar expressions can also be found in terms of the retarded and advanced functions.

Let us consider as an example the case in which the index i stands for a site in a tight-binding model. The average occupation per spin of this site is

$$\langle \mathbf{n}_{i\sigma} \rangle = \langle \Psi_H | \mathbf{c}^\dagger_{i\sigma} \mathbf{c}_{i\sigma} | \Psi_H \rangle = \frac{1}{\pi} \int_{-\infty}^{\mu} dE \, \text{Im} \left\{ G^c_{ii}(E) \right\}, \quad (5.100)$$

as it should be, since $\text{Im}\{G^c_{ii}(E)\}/\pi$ is nothing else than the local density of states projected onto the state i.

To conclude this subsection, let us say that in general the expectation in the ground state of two-electron operators, i.e. those containing two creation and two annihilation operators (see Appendix A), cannot be expressed in terms of the one-particle Green's functions that we have introduced in this chapter. However, a notable exception is the total energy of the system (for a discussion of this issue, see e.g. Ref. [173]).

5.4.3   *Equation of motion method*

So far we have discussed some of the properties of the "new" Green's functions and we have seen that they contain very important information. Now, let us discuss how they can be computed. In particular, we shall describe in this section a method referred to as *equation of motion*. Let us illustrate it in an example that is already familiar to us, namely in the case of an electron system described by a simple tight-binding Hamiltonian of the form

$$\mathbf{H} = \sum_{ij\sigma} t_{ij} \mathbf{c}^\dagger_{i\sigma} \mathbf{c}_{j\sigma}. \quad (5.101)$$

Here, the diagonal matrix elements $t_{i}i$ correspond to the on-site energies, ϵ_i, in the notation used in previous sections.

Our goal is the calculation of, for instance, the retarded Green's function

$$G^r_{ij,\sigma}(t) = -i\theta(t) \langle \Psi_H | \mathbf{c}_{i\sigma}(t) \mathbf{c}^\dagger_{j\sigma}(0) + \mathbf{c}^\dagger_{j\sigma}(0) \mathbf{c}_{i\sigma}(t) | \Psi_H \rangle. \quad (5.102)$$

For this purpose, let us calculate its time derivative

$$\frac{\partial}{\partial t} G^r_{ij,\sigma}(t) = -i\delta(t) \langle \Psi_H | \mathbf{c}_{i\sigma}(t) \mathbf{c}^\dagger_{j\sigma}(0) + \mathbf{c}^\dagger_{j\sigma}(0) \mathbf{c}_{i\sigma}(t) | \Psi_H \rangle \quad (5.103)$$

$$-i\theta(t) \langle \Psi_H | \frac{\partial}{\partial t} \mathbf{c}_{i\sigma}(t) \mathbf{c}^\dagger_{j\sigma}(0) + \mathbf{c}^\dagger_{j\sigma}(0) \frac{\partial}{\partial t} \mathbf{c}_{i\sigma}(t) | \Psi_H \rangle,$$

where we have used the fact that the derivative of the step function is a δ-function.

Now, in order to compute the time derivative of the annihilation operator appearing in the previous equation, we make use of the equation of motion for operators in the Heisenberg picture, see Eq. (5.6). Thus,

$$i\frac{\partial}{\partial t}\mathbf{c}_{i\sigma} = [\mathbf{c}_{i\sigma}, \mathbf{H}] = i\sum_{k} t_{ik}\mathbf{c}_{k\sigma}, \qquad (5.104)$$

where we have used Eq. (5.101) to obtain the last result. Substituting this expression in Eq. (5.103), we arrive at

$$i\frac{\partial}{\partial t}G^{r}_{ij,\sigma}(t) = \delta(t)\delta_{ij} + \sum_{k} t_{ik}G^{r}_{kj,\sigma}(t). \qquad (5.105)$$

It is now convenient to Fourier transform to energy space to convert this differential equation into an algebraic one. Thus, introducing

$$G^{r}_{ij,\sigma}(t) = \frac{1}{2\pi}\int_{-\infty}^{\infty} dE\, e^{-iEt}G^{r}_{ij,\sigma}(E) \; ; \; \delta(t) = \frac{1}{2\pi}\int_{-\infty}^{\infty} dE\, e^{-iEt} \qquad (5.106)$$

in Eq. (5.105), we obtain the following algebraic equation of the Green's function in energy space

$$EG^{r}_{ij,\sigma}(E) = \delta_{ij} + \sum_{k} t_{ik}G^{r}_{kj,\sigma}(E). \qquad (5.107)$$

This is nothing else but the element (i, j) of the matrix equation

$$\mathbf{G^{r}}(E) = [E\mathbf{1} - \mathbf{H}]^{-1}, \qquad (5.108)$$

which is precisely the expression that we used as a definition in section 5.2 [see Eq. (5.13)]. Thus, we have shown again the equivalence of the two type of definitions for the case of noninteracting electron systems.

It is important to emphasize that the equation-of-motion method illustrated above is by no means restricted to noninteracting system. However, if the Hamiltonian contains two-electron terms (with four creation/annihilation operators), in general there is no straightforward way to get a closed system of equations, as in the previous example. The problem is that the equation of motion for the one-particle Green's function couples this function to higher-order ones containing an increasing number of operators and the resulting algebraic system has, strictly speaking, an infinite dimension. In practice, one has to find an appropriate way of truncating the system, which is not an easy task in general.

In order to illustrate what we meant in the previous paragraph, let us consider the Anderson model that describes the interaction of a single level (including the electron-electron interaction in this level) with a continuum of states. This model can describe, for instance, a magnetic impurity in a metal or a quantum dot (or a molecule) coupled to metallic reservoirs. The Hamiltonian of this model adopts the form (see Appendix A)

$$\mathbf{H} = \sum_{\mathbf{k},\sigma} \epsilon_\mathbf{k} \mathbf{n}_{\mathbf{k}\sigma} + \sum_{\mathbf{k},\sigma} \left(V_{\mathbf{k}0} \mathbf{c}_{\mathbf{k}\sigma}^\dagger \mathbf{c}_{0\sigma} + V_{0\mathbf{k}} \mathbf{c}_{0\sigma}^\dagger \mathbf{c}_{\mathbf{k}\sigma} \right) + \sum_\sigma \epsilon_0 \mathbf{n}_{0\sigma} + U \mathbf{n}_{0\uparrow} \mathbf{n}_{0\downarrow},$$

(5.109)

where the subindex 0 refers to the correlated level and $\mathbf{k}$ to the metallic states in the reservoirs. Our goal is to compute the (retarded or advanced) Green's function $G_{00,\sigma}(E)$ in the impurity. For this purpose, we proceed as above and determine the time derivative of this function. This calculation requires the evaluation of the time derivative of the operator $\mathbf{c}_{0\sigma}(t)$, which in turn requires the determination of the commutator of this operator with the Hamiltonian. The novel term, as compared with the tight-binding example above, is $U \mathbf{n}_{0\uparrow} \mathbf{n}_{0\downarrow}$ and the corresponding commutator with it is

$$[\mathbf{c}_{0\sigma}, U \mathbf{n}_{0\uparrow} \mathbf{n}_{0\downarrow}] = U \mathbf{c}_{0\sigma} \mathbf{n}_{0\bar\sigma},$$

(5.110)

where we have used the notation $\bar\sigma = -\sigma$. Inserting this term in the equation of motion, it is straightforward to show that one arrives at (after Fourier transforming to energy space)

$$(E - \epsilon_0) G_{00\sigma}(E) = 1 + \sum_\mathbf{k} V_{0\mathbf{k}} G_{\mathbf{k}0}(E)$$

(5.111)

$$-iU\theta(t) \langle \Psi_H | \{ \mathbf{c}_{0\sigma}(t) \mathbf{n}_{0\bar\sigma}(t), \mathbf{c}_{0\sigma} \} | \Psi_H \rangle,$$

where { } stands for the anticommutator. Here, the novelty with respect to Eq. (5.107) is the appearance of the term in the second line that contains four operators. To close the equation, we need now an equation for this new expectation value. The reader can convince himself, that such an equation would generate terms containing expectation values of six operators. Then, the equation for these functions would involve terms with eight operators and so on and so forth. So, the only way to solve these equations in practice is to truncate the system with sensible arguments, but in most cases it is not clear how to do it. In the next chapter we shall discuss a more systematic approach to obtain the Green's functions in interacting problems.

There is one limit in which it is possible to obtain the exact Green's function, namely in the limit where the coupling to the reservoirs tends to zero ($V_{0\mathbf{k}} \to 0$ with U finite). In this case the equation of motion can be truncated and one obtains (see Problem 5.11)

$$G_{00\sigma}(E) = \frac{1 - \langle \mathbf{n}_{0\bar\sigma} \rangle}{E - \epsilon_0} + \frac{\langle \mathbf{n}_{0\bar\sigma} \rangle}{E - \epsilon_0 - U},$$

(5.112)

where $\langle \mathbf{n}_{0\sigma} \rangle$ is the occupation of the level ϵ_0 for spin σ, which in turn has to be calculated with the full Green's function of Eq. (5.112). Thus, in this limit the Green's functions exhibit poles at energies equal to ϵ_0 and $\epsilon_0 + U$. This tells us in particular that U is the energy that one has to supply to accommodate a second electron in the level. The expression of Eq. (5.112) can be used as an starting point to analyze the so-called Coulomb blockade in quantum dots or molecular transistors (see Exercise 8.9).

Let us conclude this section by recommending Chapter 9 of Ref. [185] for a more detailed discussion about the equation-of-motion method.

5.5 Exercises

5.1 Time evolution of the operators in Heisenberg picture:
(a) Let us consider a free electron gas described by the Hamiltonian

$$\mathbf{H} = \sum_{k,\sigma} \epsilon_k \mathbf{c}_{k\sigma}^{\dagger} \mathbf{c}_{k\sigma}.$$

Show that the time evolution of the operators $\mathbf{c}_{k\sigma}^{\dagger}$ and $\mathbf{c}_{k\sigma}$ in Heisenberg picture is given by

$$\mathbf{c}_{k\sigma}^{\dagger}(t) = \mathbf{c}_{k\sigma}^{\dagger}(0) e^{i\epsilon_k t} \quad \text{and} \quad \mathbf{c}_{k\sigma}(t) = \mathbf{c}_{k\sigma}(0) e^{-i\epsilon_k t}.$$

(b) Let us consider a diatomic molecule described by the following two-sites tight-binding Hamiltonian

$$\mathbf{H} = \epsilon_0 \sum_{\sigma} (\mathbf{n}_{1\sigma} + \mathbf{n}_{2\sigma}) + t \sum_{\sigma} (\mathbf{c}_{1\sigma}^{\dagger} \mathbf{c}_{2\sigma} + \mathbf{c}_{2\sigma}^{\dagger} \mathbf{c}_{1\sigma}).$$

Obtain the temporal evolution of the operators $\mathbf{c}_{1\sigma}$ and $\mathbf{c}_{2\sigma}$ in Heisenberg picture.

5.2 Green's function of a free electron in 1D: Let us consider the Schrödinger equation of a free electron in a 1D potential

$$\left[-\frac{1}{2m} \frac{\partial^2}{\partial x^2} + V_0 \right] \Psi(x) = E \Psi(x),$$

where V_0 is a spatially constant potential. Show that the electron Green's function is given by the expressions detailed in section 5.2.

5.3 Equivalence of expressions (5.13) and (5.14): Show the equivalence of Eq. (5.13) and Eq. (5.14). Hints: (i) Multiply both sides of Eq. (5.13) by $[(E \pm i\eta) - \mathbf{H}]$. (ii) Introduce then the closure relation $\sum_n |\psi_n\rangle\langle\psi_n| = \mathbf{1}$, where ψ_n are the eigenfunctions of $\mathbf{H}$. (iii) Use $\mathbf{H}|\psi_n\rangle = \epsilon_n |\psi_n\rangle$ and (iv) multiply by the inverse of the operator on the left hand side of the Green's function to obtain Eq. (5.14).

5.4 Dyson's equation: Starting from Eq. (5.26), show that the Green's functions fulfill the Dyson's equation (5.28).

5.5 Semi-infinite tight-binding chain: Let us consider the Hamiltonian of Eq. (5.43) for a semi-infinite chain. Calculate the off-diagonal retarded Green's functions G_{n1}^r of the chain (where 1 is the first site and n an arbitrary one) and demonstrate that it is given by the following expression for $|E - \epsilon_0| < 2|t|$:

$$G_{n1}^r(E) = \frac{e^{-in\phi}}{t} \quad \text{where} \quad \cos\phi = (E - \epsilon_0)/2t.$$

5.6 Infinite tight-binding chain: Let us consider an infinite chain of identical atoms with only nearest-neighbor hoppings, t.

(a) Making use of the eigenvalues of this problem, $\epsilon_k = \epsilon_0 + 2t\cos(ka)$, where a is the lattice constant, and the corresponding eigenfunctions, demonstrate that the advanced Green's functions $G_{ij}^a(E)$ are given by

$$G_{ij}^a(E) = \frac{i}{|t|} e^{-i\phi|i-j|} \cos\phi \quad \text{for} \quad |E - \epsilon_0| < 2|t|.$$

(b) An infinite chain can be viewed as two coupled semi-infinite chains. In this sense, consider the coupling between the semi-infinite chains as a perturbation and use Dyson's equation to obtain the diagonal advanced Green's functions in a site of the chain and demonstrate that it coincides with the result derived in (a) for $i = j$.

5.7 Tight-binding chain with a defect: Let us consider an infinite chain as in the previous problem in which a diagonal perturbation is introduced in one of the sites, let us say in site i, such that its on-site energy becomes $\epsilon_0 + \Delta$. Calculate the local density of states in the site i and, in particular, investigate the possibility of having a localized state outside the band. Study also the spatial extension of such a state by calculating the occupation of this state in different sites away from the one in which the defect is located.

5.8 Finite tight-binding chain: Let us consider a finite chain with N sites and only nearest-neighbor interactions. Calculate the advanced Green's function $G_{n1}^a(E)$, where 1 refers to the atom in one of the extremes of the chain and n to an arbitrary site. Demonstrate in particular that for $|E - \epsilon_0| < 2|t|$

$$G_{n1}^a(E) = \frac{1}{t} \frac{\sin[(N - n + 1)\phi]}{\sin[(N + 1)\phi]}.$$

5.9 Resonant level coupled to metallic electrodes: In the example 3 of section 5.3 we considered a single site with energy ϵ_0 connected to two electron reservoirs. We computed the local density of states in the wide-band approximation, see Eq. (5.57). Assume now that the electrodes are modeled by the semi-infinite linear chain of the example 2 of section 5.3 with on-site energy equal to zero and a hopping integral t. Study the local density of states in the central site as a function of the values of ϵ_0 and the coupling elements t_L and t_R. Discuss in particular how the level position is renormalized.

5.10 Time-dependent Green's functions: Make use of the expressions of the time dependence of the creation and annihilation operators of the two-sites problem of Exercise 5.1.(b) to compute the time-dependent retarded Green's functions. Show that the energy-dependent Green's functions that can be obtained from the previous solution coincide with the result of Eq. (5.13).

5.11 Equation of motion: Atomic limit of the Anderson's model: Let us consider the Anderson's Hamiltonian given in Eq. (5.109). Use the equation-of-motion method to show that in the atomic limit ($V_{0\mathbf{k}} \to 0$) the Green's function of the level can indeed be written as in Eq. (5.112).

Chapter 6

Green's functions and Feynman diagrams

In the previous chapter we have seen that the calculation of the zero-temperature Green's functions of a non-interacting system in equilibrium reduces to solve an algebraic linear system, summarized in Dyson's equation. This is practically all we need to tackle the problem of the determination of the elastic transmission of realistic systems. However, if we want to go beyond and treat systems where the electron correlations or inelastic interactions play a major role, we need many-body techniques. For this reason, we present in this chapter a systematic perturbative approach for the calculation of zero-temperature equilibrium Green's functions.[1] This formalism is valid for any type of system and interaction and constitutes the most general method for the computation of Green's functions. Moreover, the nonequilibrium formalism introduced in the next chapter follows closely the perturbative approach that we are about to describe.

The perturbative (or diagrammatic) approach is nicely explained in different many-body textbooks (see e.g. Refs. [173–175, 182–185]) and for this reason, our description here will be rather brief.[2] This approach is conceptually rather simple, but it contains several technical points that usually make it rather obscure. In the spirit of this monograph, we shall avoid very formal discussions and we shall provide instead simple plausibility arguments or we shall simply refer the reader to the adequate literature.

Before the trees do not let us see the forest, let us give a brief overview of what we are about to see. First, we shall learn how to write down a perturbative series for the Green's functions, i.e. how to express systematically the corrections to the Green's function due to a perturbation such

---

[1] In some sense, this approach is simply a generalization of the perturbation theory for the wave functions that one studies in elementary courses of quantum mechanics.

[2] This chapter is mainly based on Chapter 3 of Ref. [173].

as an external potential, electron-electron interaction, etc. Then, we shall discuss how these contributions can be "visualized" with the help of the so-called Feynman diagrams. These diagrams will in turn help us to organize and simplify the perturbative series. Finally, we shall show that this series can be formally resumed and cast in the Dyson's equation, which we have already introduced for case of non-interacting systems. Dyson's equation is expressed in terms of the concept of self-energy. This concept was also introduced in the previous chapter and in this one its precise meaning will be clarified.

So, it is time get started. The general problem that we want to tackle in this chapter is the analysis of an electron system in equilibrium that is described by a Hamiltonian of the following form

$$\mathbf{H} = \mathbf{H}_0 + \mathbf{V}, \qquad (6.1)$$

where $\mathbf{H}_0$ is a single-particle Hamiltonian and $\mathbf{V}$ is a perturbation that may contain an external potential and any type of interaction. Our goal is the compute the Green's functions of the system in terms of the unperturbed Green's functions, i.e. those associated with the Hamiltonian $\mathbf{H}_0$, which are supposed to be known. For this purpose, we shall develop a systematic perturbation theory, but before doing that we shall now introduce a convenient representation of quantum mechanics, known as the *interaction picture*, that will be very useful in what follows.

6.1   The interaction picture

Let us consider a system described by the Hamiltonian of Eq. (6.1). We define the interaction picture starting from the Schrödinger one by means of the following unitary transformation[3]

$$\Psi_{\mathrm{I}}(t) = e^{i\mathbf{H}_0 t}\Psi_{\mathrm{S}}(t) \ \text{ and } \ \mathbf{O}_{\mathrm{I}}(t) = e^{i\mathbf{H}_0 t}\mathbf{O}_{\mathrm{S}}(t)e^{-i\mathbf{H}_0 t}. \qquad (6.2)$$

Notice that, contrary to the case of the Schrödinger and Heisenberg pictures, in the interaction picture both wave functions and operators depend explicitly on time.

Let us analyze first the time evolution of the operators. It is obvious from Eq. (6.2) that the operators in this representation are the Heisenberg operators of the unperturbed system. Taking the derivative with respect to time in the definition of an operator in the interaction picture, one obtains

$$i\frac{\partial}{\partial t}\mathbf{O}_{\mathrm{I}} = [\mathbf{O}_{\mathrm{I}}, \mathbf{H}_0]. \qquad (6.3)$$

---

[3]In this chapter we shall also set $\hbar = 1$.

Therefore, the dynamics of the operators in this representation is governed by $\mathbf{H}_0$ and it is thus known.

Turning to the wave functions, we can make use of the evolution of the wave function in Schrödinger picture to obtain

$$\Psi_{\mathrm{I}}(t) = e^{i\mathbf{H}_0 t}\Psi_{\mathrm{S}}(t) = e^{i\mathbf{H}_0 t}e^{-i\mathbf{H}t}\Psi_{\mathrm{S}}(0). \qquad (6.4)$$

Let us remind that

$$e^{i\mathbf{H}_0 t}e^{-i\mathbf{H}t} \neq e^{-i\mathbf{V}t},$$

since, in general, $[\mathbf{H}_0, \mathbf{H}] \neq 0$.

In order to find the equation that describes the time evolution of the wave function in this picture, we now take the derivative with respect to time in Eq. (6.2)

$$i\frac{\partial}{\partial t}\Psi_{\mathrm{I}}(t) = -\mathbf{H}_0 e^{i\mathbf{H}_0 t}\Psi_{\mathrm{S}}(t) + ie^{i\mathbf{H}_0 t}\frac{\partial}{\partial t}\Psi_{\mathrm{S}}(t), \qquad (6.5)$$

and making use of the Schrödinger equation on the right hand side of the previous expression, one obtains

$$i\frac{\partial}{\partial t}\Psi_{\mathrm{I}}(t) = e^{i\mathbf{H}_0 t}(\mathbf{H} - \mathbf{H}_0)\Psi_{\mathrm{S}}(t) = e^{i\mathbf{H}_0 t}\mathbf{V}e^{-i\mathbf{H}_0 t}e^{i\mathbf{H}_0 t}\Psi_{\mathrm{S}}(t), \qquad (6.6)$$

which can be simply written as

$$i\frac{\partial}{\partial t}\Psi_{\mathrm{I}}(t) = \mathbf{V}_{\mathrm{I}}(t)\Psi_{\mathrm{I}}(t). \qquad (6.7)$$

This equation plays the role of the standard Schrödinger equation in this new picture. Notice that the dynamics of the wave functions is governed by the perturbation. This is very important because it makes possible, by means of an adiabatic hypothesis in which the perturbation is adiabatically switched on, to relate the perturbed and unperturbed ground states of the system by means of the evolution of the wave function in this picture. Due to this fact, the operator that describes the time evolution of the wave functions is of special interest and it will be discussed in detail in the next section.

To end this section, let us discuss now the relation between the Heisenberg picture and the interaction picture. Using the definitions of Eq. (6.2), one can easily show that

$$\Psi_{\mathrm{I}}(t) = e^{i\mathbf{H}_0 t}e^{-i\mathbf{H}t}\Psi_{\mathrm{H}} \qquad (6.8)$$

$$\mathbf{O}_{\mathrm{I}}(t) = e^{i\mathbf{H}_0 t}e^{-i\mathbf{H}t}\mathbf{O}_{\mathrm{H}}(t)e^{i\mathbf{H}t}e^{-i\mathbf{H}_0 t}.$$

The inverse transformation is obviously given by

$$\Psi_{\mathrm{H}}(t) = e^{i\mathbf{H}t}e^{-i\mathbf{H}_0 t}\Psi_{\mathrm{I}}(t) \qquad (6.9)$$

$$\mathbf{O}_{\mathrm{H}}(t) = e^{i\mathbf{H}t}e^{-i\mathbf{H}_0 t}\mathbf{O}_{\mathrm{I}}(t)e^{i\mathbf{H}_0 t}e^{-i\mathbf{H}t}.$$

6.2   The time-evolution operator

We define the time-evolution operator in the interaction picture as

$$\Psi_I(t) = \mathbf{S}(t, t_0)\Psi_I(t_0). \tag{6.10}$$

It is easy to find a formal expression for the operator $\mathbf{S}$ in terms of the system Hamiltonian. From the definition of the interaction picture one has

$$\Psi_I(t) = e^{i\mathbf{H}_0 t}\Psi_S(t). \tag{6.11}$$

Making use of the expression of the time evolution of the wave function in the Schrödinger picture we can write

$$\Psi_I(t) = e^{i\mathbf{H}_0 t}e^{-i\mathbf{H}(t-t_0)}\Psi_S(t_0). \tag{6.12}$$

Transforming the wave function $\Psi_S(t_0)$ to the interaction picture, one has finally

$$\Psi_I(t) = e^{i\mathbf{H}_0 t}e^{-i\mathbf{H}(t-t_0)}e^{-i\mathbf{H}_0 t_0}\Psi_I(t_0). \tag{6.13}$$

Comparing this expression with the definition of Eq (6.10), we can identify

$$\mathbf{S}(t, t_0) = e^{i\mathbf{H}_0 t}e^{-i\mathbf{H}(t-t_0)}e^{-i\mathbf{H}_0 t_0}. \tag{6.14}$$

From the definition of the time-evolution operator or from its formal expression, one can easily show the following properties:

- The operator $\mathbf{S}$ is unitary, i.e. $\mathbf{S}^{-1} = \mathbf{S}^\dagger$.
- $\mathbf{S}(t, t) = 1$.
- $\mathbf{S}(t, t')\mathbf{S}(t', t'') = \mathbf{S}(t, t'')$.
- $\mathbf{S}(t, t') = \mathbf{S}^\dagger(t', t)$.

The time-evolution operator is also related to the unitary transformation that relates Heisenberg and interaction pictures. From Eq. (6.14) one has

$$\mathbf{S}(0, t) = e^{i\mathbf{H}t}e^{-i\mathbf{H}_0 t}. \tag{6.15}$$

Comparing now with Eq. (6.9), we can write

$$\Psi_H = \mathbf{S}(0, t)\Psi_I(t) \tag{6.16}$$

$$\mathbf{O}_H(t) = \mathbf{S}(0, t)\mathbf{O}_I(t)\mathbf{S}(t, 0).$$

The operator $\mathbf{S}$ satisfies its own equation of motion, which is very similar to the equation for the wave functions in this representation. Taking the derivative with respect to time in Eq. (6.14) one has

$$i\frac{\partial}{\partial t}\mathbf{S}(t, t_0) = \mathbf{V}_I(t)\mathbf{S}(t, t_0). \tag{6.17}$$

Finally, the time-evolution operator can be expressed as a perturbative series in the interaction $\mathbf{V}_I(t)$. This can be shown either by solving iteratively the previous equation or by using the equation for the wave function $\Psi_I(t)$. We choose the second option and write Eq. (6.7) as an integral equation

$$\Psi_I(t) = \Psi_I(t_0) - i \int_{t_0}^{t} dt' \, \mathbf{V}_I(t')\Psi_I(t'). \tag{6.18}$$

This equation can now be solved iteratively. To zero order we have

$$\Psi_I(t) = \Psi_I(t_0). \tag{6.19}$$

Substituting this zero-order result in Eq. (6.18) we obtain the first-order result

$$\Psi_I(t) = \left[1 - i \int_{t_0}^{t} dt_1 \, \mathbf{V}_I(t_1)\right] \Psi_I(t_0). \tag{6.20}$$

Iterating we can arrive at

$$\Psi_I(t) = \left[1 + \sum_n (-i)^n \int_{t_0}^{t} dt_1 \, \mathbf{V}_I(t_1) \times \right. \tag{6.21}$$
$$\left. \int_{t_0}^{t_1} dt_2 \, \mathbf{V}_I(t_2) \cdots \int_{t_0}^{t_{n-1}} dt_n \, \mathbf{V}_I(t_n)\right] \Psi_I(t_0).$$

The expression inside the brackets is just the time-evolution operator $\mathbf{S}(t, t_0)$ expanded as a power series in the operator $\mathbf{V}_I(t)$. This expression is not very inconvenient because the upper and lower limits of the time integrals are different. It is possible to rewrite the previous expression in more adequate manner by noticing that the integration variables fulfill $t > t_1 > t_2 > \cdots > t_n > t_0$. This makes possible to rewrite the time-evolution operator in the interaction picture as

$$\mathbf{S}(t, t_0) = \sum_{n=0}^{\infty} \frac{(-i)^n}{n!} \int_{t_0}^{t} dt_1 \int_{t_0}^{t} dt_2 \cdots \int_{t_0}^{t} dt_n \, \mathbf{T}\left[\mathbf{V}_I(t_1)\mathbf{V}_I(t_2) \cdots \mathbf{V}_I(t_n)\right],$$
$$\tag{6.22}$$

where the $n = 0$ term is the unit operator and $\mathbf{T}$ is the time-ordering operator that we introduced in the last chapter. The demonstration of this last step is left to the reader as an exercise.

6.3   Perturbative expansion of causal Green's functions

Our goal now is the calculation of a generic causal Green's function, which in a discrete basis is given by

$$G_{ij}(t,t') = \frac{\langle \Psi_{\mathrm{H}} | \mathbf{T} \left[\mathbf{c}_{i\sigma}(t) \mathbf{c}_{j\sigma}^{\dagger}(t') \right] | \Psi_{\mathrm{H}} \rangle}{\langle \Psi_{\mathrm{H}} | \Psi_{\mathrm{H}} \rangle}. \tag{6.23}$$

Here, the expectation value is evaluated in the ground state of the system described by the Hamiltonian of Eq. (6.1) and the operators are written in Heisenberg picture. Notice that we omit the superindex c to abbreviate the notation and we introduce the denominator for normalization reasons that will become clear later on.

As explained in the previous section, it is convenient to use the interaction picture. We first transform the operators:

$$G_{ij}(t,t') = \frac{\langle \Psi_{\mathrm{H}} | \mathbf{T} \left[\mathbf{S}(0,t) \mathbf{c}_{i\sigma}^{(0)}(t) \mathbf{S}(t,t') \mathbf{c}_{j\sigma}^{(0)\dagger}(t') \mathbf{S}(t',0) \right] | \Psi_{\mathrm{H}} \rangle}{\langle \Psi_{\mathrm{H}} | \Psi_{\mathrm{H}} \rangle}. \tag{6.24}$$

Here, we have used the superindex (0) to emphasize that the operators in the interaction picture correspond to Heisenberg operators of the unperturbed system. We now transform the wave function by using

$$|\Psi_{\mathrm{H}}\rangle = \mathbf{S}(0,t)|\Psi_{\mathrm{I}}(t)\rangle, \tag{6.25}$$

where t is an arbitrary time. Now, we want to relate the state $|\Psi_{\mathrm{I}}(t)\rangle$ with the unperturbed ground state (for $\mathbf{V} = 0$), $|\phi_0\rangle$. This can be done using the so-called adiabatic hypothesis. In this hypothesis, one assumes that if the perturbation is switched on at an initial time, let us say $t = -\infty$, and grows slowly to its actual value at $t = 0$, the physics is not modified. This adiabatic switch on is achieved by replacing the perturbation $\mathbf{V}$ by $\mathbf{V}e^{-\epsilon|t|}$, where ϵ is an infinitesimally small positive parameter. In this way, at $t = \pm\infty$ the perturbation vanishes and the system tends to the unperturbed ground state

$$|\Psi_{\mathrm{H}}\rangle = \mathbf{S}(0,-\infty)|\phi_0\rangle. \tag{6.26}$$

This procedure is not completely well-defined and one can show that during the evolution of the ground state from $t = -\infty$ to $t = 0$ with the operator $\mathbf{S}$, the wave function acquires a phase that diverges as ϵ tends to zero. These phase factors are finally canceled by the terms in the denominator of the expectation value. The rigorous statement of this fact is known as the Gell-Mann and Low theorem and for more information we refer the reader to the book of Fetter and Walecka [173].

We now make use of Eq. (6.26) to write the causal Green's function as follows

$$G_{ij}(t,t') = \frac{\langle \phi_0 | S(\infty, 0) \mathbf{T} \left[\mathbf{S}(0,t) \mathbf{c}_{i\sigma}^{(0)}(t) \mathbf{S}(t,t') \mathbf{c}_{j\sigma}^{(0)\dagger}(t') \mathbf{S}(t',0) \right] S(0,-\infty) | \phi_0 \rangle}{\langle \phi_0 | \mathbf{S}(\infty, -\infty) | \phi_0 \rangle}.$$
(6.27)

Here, we have used the time symmetry of the problem that implies in particular that the ground state wave function is recovered at $t = +\infty$ (apart from a phase factor). On the other hand, it is obvious that in the previous expression we can introduce the time-evolution operators appearing next to the wave functions inside the time-ordered products. Thus, the expectation value now reads

$$G_{ij}(t,t') = \frac{\langle \phi_0 | \mathbf{T} \left[\mathbf{c}_{i\sigma}^{(0)}(t) \mathbf{c}_{j\sigma}^{(0)\dagger}(t') \mathbf{S}(\infty, -\infty) \right] | \phi_0 \rangle}{\langle \phi_0 | \mathbf{S}(\infty, -\infty) | \phi_0 \rangle},$$
(6.28)

where we have grouped all the pieces of the operator $\mathbf{S}$ since the operator $\mathbf{T}$ ensures the proper ordering. Now, we use the expansion of Eq. (6.22) for the operator $\mathbf{S}$ to write the expectation value as a perturbative expansion

$$G_{ij}(t,t') = \frac{1}{\langle \phi_0 | \mathbf{S}(\infty, -\infty) | \phi_0 \rangle} \left[\sum_{n=0}^{\infty} \frac{(-i)^n}{n!} \int_{-\infty}^{\infty} dt_1 \dots dt_n \times \right.$$
(6.29)
$$\left. \langle \phi_0 | \mathbf{T} \left[\mathbf{c}_{i\sigma}^{(0)}(t) \mathbf{c}_{j\sigma}^{(0)\dagger}(t') \mathbf{V}^{(0)}(t_1) \cdots \mathbf{V}^{(0)}(t_n) \right] | \phi_0 \rangle \right],$$

where the zero-order term ($n = 0$) corresponds to the unperturbed Green's function, which we shall denote as $G_{ij}^{(0)}(t,t')$. The previous expression is the central result of this section.

The perturbative expansion adopts the same form, irrespectively of the basis used. Thus for instance, if one uses a spatial representation, the previous expression becomes

$$G(\mathbf{r}t, \mathbf{r}'t') = \frac{1}{\langle \phi_0 | \mathbf{S}(\infty, -\infty) | \phi_0 \rangle} \left[\sum_{n=0}^{\infty} \frac{(-i)^n}{n!} \int_{-\infty}^{\infty} dt_1 \cdots dt_n \times \right.$$
(6.30)
$$\left. \langle \phi_0 | \mathbf{T} \left[\mathbf{\Psi}_{\sigma}^{(0)}(\mathbf{r}t) \mathbf{\Psi}_{\sigma}^{(0)\dagger}(\mathbf{r}'t') \mathbf{V}^{(0)}(t_1) \cdots \mathbf{V}^{(0)}(t_n) \right] | \phi_0 \rangle \right].$$

6.4 Wick's theorem

With the perturbative formalism that we have developed so far, the problem of calculating a Green's function or any expectation value of an operator in the ground state reduces to the calculation of expectation values in the unperturbed ground state of the following type

$$\langle\phi_0|\mathbf{T}\left[\mathbf{c}_{i\sigma}^{(0)}(t)\mathbf{c}_{j\sigma}^{(0)\dagger}(t')\mathbf{V}^{(0)}(t_1)\cdots\mathbf{V}^{(0)}(t_n)\right]|\phi_0\rangle. \qquad (6.31)$$

This is something that we can, in principle, calculate in an exact manner because we know the evolution of the operators in the unperturbed problem. However, in practice, the direct calculation of expectation values like the one in Eq. (6.31) is extremely cumbersome. Fortunately, *Wick's theorem* simplifies enormously this task.

Wick's theorem is the mathematical expression of the fact that the electrons in the unperturbed problem are uncorrelated. Before stating the theorem, let us illustrate it with a simple example. Let us consider the following two-sites tight-binding Hamiltonian

$$\mathbf{H} = \sum_\sigma \epsilon_0\left(\mathbf{n}_{1\sigma} + \mathbf{n}_{2\sigma}\right) + t\sum_\sigma\left(\mathbf{c}_{1\sigma}^\dagger\mathbf{c}_{2\sigma} + \mathbf{c}_{2\sigma}^\dagger\mathbf{c}_{1\sigma}\right). \qquad (6.32)$$

Let us also assume that we have two electrons in total. If $|\phi_0\rangle$ is the wave function of the noninteracting problem, it seems natural that

$$\langle\phi_0|\mathbf{n}_{1\uparrow}\mathbf{n}_{1\downarrow}|\phi_0\rangle = \langle\phi_0|\mathbf{n}_{1\uparrow}|\phi_0\rangle\langle\phi_0|\mathbf{n}_{1\downarrow}|\phi_0\rangle, \qquad (6.33)$$

since in the absence of interactions the probability of finding two electrons simultaneously in $|1\downarrow\rangle$ and in $|1\uparrow\rangle$ must be equal to the product of the probabilities (see Exercise 6.1).

Wick's theorem generalizes this result to the expectation value in a non-interacting ground state of a product of an arbitrary number of operators. Without many-body interactions, an average like the one in Eq. (6.31) look like

$$\langle\phi_0|\mathbf{T}\left[\mathbf{c}_{i\sigma}^{(0)\dagger}(t)\mathbf{c}_{j\sigma}^{(0)\dagger}(t')\cdots\mathbf{c}_{k\sigma}^{(0)\dagger}(t_1)\cdots\mathbf{c}_{l\sigma}^{(0)}(t_n)\right]|\phi_0\rangle. \qquad (6.34)$$

Wick's theorem establishes that such an expectation value is equal to the sum of all possible factorizations of averages of two operators. Since in our case the operators are fermionic and therefore anticommute, one has to follow the usual criterion, i.e. the factorization that respects the original order does not contain any minus sign, whereas the factorization that differs by an odd number of permutations from the original configuration introduces a minus sign. Thus for instance, the following expectation value of the product of four operators can be decomposed as follows

$$\langle\phi_0|\mathbf{T}\left[\mathbf{c}_{i\sigma}^{(0)}(t)\mathbf{c}_{j\sigma}^{(0)\dagger}(t')\mathbf{c}_{k\sigma}^{(0)}(t_1)\mathbf{c}_{l\sigma}^{(0)\dagger}(t_2)\right]|\phi_0\rangle = \qquad (6.35)$$

$$\langle\phi_0|\mathbf{T}\left[\mathbf{c}_{i\sigma}^{(0)}(t)\mathbf{c}_{j\sigma}^{(0)\dagger}(t')\right]|\phi_0\rangle\langle\phi_0|\mathbf{T}\left[\mathbf{c}_{k\sigma}^{(0)}(t_1)\mathbf{c}_{l\sigma}^{(0)\dagger}(t_2)\right]|\phi_0\rangle$$

$$-\langle\phi_0|\mathbf{T}\left[\mathbf{c}_{i\sigma}^{(0)}(t)\mathbf{c}_{l\sigma}^{(0)\dagger}(t_2)\right]|\phi_0\rangle\langle\phi_0|\mathbf{T}\left[\mathbf{c}_{k\sigma}^{(0)}(t_1)\mathbf{c}_{j\sigma}^{(0)\dagger}(t')\right]|\phi_0\rangle.$$

Notice that in the previous factorization one could have had additional terms containing expectation values like for instance

$$\langle\phi_0|\mathbf{T}\left[\mathbf{c}^{(0)}_{i\sigma}(t)\mathbf{c}^{(0)}_{k\sigma}(t_1)\right]|\phi_0\rangle, \ \langle\phi_0|\mathbf{T}\left[\mathbf{c}^{(0)\dagger}_{j\sigma}(t')\mathbf{c}^{(0)\dagger}_{l\sigma}(t_2)\right]|\phi_0\rangle$$

$$\text{or} \ \langle\phi_0|\mathbf{T}\left[\mathbf{c}^{(0)}_{i\sigma}(t)\mathbf{c}^{(0)\dagger}_{j,\bar\sigma}(t')\right]|\phi_0\rangle.$$

However, they usually vanish for different reasons. In the first two cases, the combinations of operators do not conserve the number of electrons. The third expectation value vanishes, unless the ground state is magnetic. Thus, usually the only terms that survive are those with the following form: $\langle\phi_0|\mathbf{T}\left[\mathbf{c}^{(0)}_{i\sigma}(t)\mathbf{c}^{(0)\dagger}_{j\sigma}(t')\right]|\phi_0\rangle$, i.e. those with a combination of a creation and an annihilation operator. As a convention, we shall always place the creation operator on the right hand side in these factors.

To end this section, notice that the basic factor appearing in the decomposition that results from Wick's theorem is closely related to a single-particle Green's function of the unperturbed system

$$\langle\phi_0|\mathbf{T}\left[\mathbf{c}^{(0)}_{i\sigma}(t)\mathbf{c}^{(0)\dagger}_{j\sigma}(t')\right]|\phi_0\rangle = iG^{(0)}_{ij\sigma}(t,t'). \tag{6.36}$$

Thus for instance, the expectation value of the previous example can be written as

$$\langle\phi_0|\mathbf{T}\left[\mathbf{c}^{(0)}_{i\sigma}(t)\mathbf{c}^{(0)\dagger}_{j\sigma}(t')\mathbf{c}^{(0)}_{k\sigma}(t_1)\mathbf{c}^{(0)\dagger}_{l\sigma}(t_2)\right]|\phi_0\rangle = \tag{6.37}$$

$$-G^{(0)}_{ij\sigma}(t,t')G^{(0)}_{kl\sigma}(t_1,t_2) + G^{(0)}_{il\sigma}(t,t_2)G^{(0)}_{kj\sigma}(t_1,t').$$

6.5   Feynman diagrams

Feynman diagrams are a graphical representation of the different contributions of the perturbative expansion of a Green's function, which result from the application of Wick's theorem. Let us recall that Green's functions can be interpreted as the propagation amplitude of an electron from one state to another. In this sense, the Feynman diagrams turn out to have a simple interpretation in terms of processes that contribute to the total amplitude of propagation of an electron. Moreover, apart from the physical insight that these diagrams provide, they also help in classifying and identifying the contributions resulting from the application of Wick's theorem.

Before describing the Feynman diagrams, we need a "dictionary" that assigns a convenient graphical representation to the different functions that appear in the perturbation theory. Thus for instance, the unperturbed causal Green's functions, which appear in the perturbative expansion through the application of Wick's theorem, will be represented by a solid line. This is shown in Fig. 6.1(a) for the function $G^{(0)}(\mathbf{r}t, \mathbf{r}'t')$ in real space. For this case, the arrow points from the second set of arguments (or event) to the first one (indicating the propagation of an electron from $\mathbf{r}'t'$ to $\mathbf{r}t$). If the problem depends explicitly on the spin, we would have to label the different events with the corresponding spin. If we use a discrete basis, the corresponding line will look like in Fig. 6.1(b).

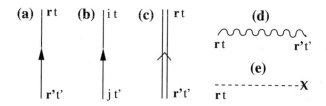

Fig. 6.1   Basic elements of Feynman diagrams. (a) Propagator line between the events $\mathbf{r}'t'$ to $\mathbf{r}t$. (b) Propagator line between the states $j\sigma'$ and $i\sigma$. (c) Full propagator line. (d) Interaction line between the events $\mathbf{r}'t'$ to $\mathbf{r}t$. (e) Interaction line for an external potential.

The full (or dressed) Green's function that corresponds to the total amplitude for the electron propagation will be represented as a double line, as shown in Fig. 6.1(c). On the other hand, the electron-electron interaction between two events will be represented by a wavy line, as in Fig. 6.1(d). Notice that, in general, the interaction is instantaneous and therefore $U(\mathbf{r}t, \mathbf{r}'t') \propto \delta(t - t')$. In the case in which the perturbation is an external potential, $V(\mathbf{r})$, this will then be represented by a dashed line, see Fig. 6.1(e).

The structure of perturbative series and the corresponding Feynman diagrams depends on the type of perturbation under study. In what follows, we shall illustrate the diagrammatic approach with the analysis of two examples where the perturbation is (i) the electron-electron interaction and (ii) an external static potential.

6.5.1   *Feynman diagrams for the electron-electron interaction*

Let us analyze the case of an electron system in which the electron-electron interaction is considered to be the perturbation. In this case the Hamiltonian has the following generic form in first quantization

$$\mathbf{H} = \mathbf{H}_0 + \mathbf{V} = \sum_{n=1}^{N} h(\mathbf{r}_i) + \frac{1}{2}\sum_{i\neq j}^{N} U(\mathbf{r}_i, \mathbf{r}_j), \qquad (6.38)$$

where $h(\mathbf{r})$ is single-electron Hamiltonian and $U(\mathbf{r}, \mathbf{r}')$ is the electron-electron (Coulomb) potential. Using the second quantization language and the basis of the eigenfunctions of the position operator $\{|\mathbf{r}\rangle\}$, the previous Hamiltonian can be expressed in terms of the field operators as follows

$$\mathbf{H} = \sum_{\sigma} \int d\mathbf{r}\ \boldsymbol{\Psi}_\sigma^\dagger(\mathbf{r})h(\mathbf{r})\boldsymbol{\Psi}_\sigma(\mathbf{r}) \qquad (6.39)$$

$$+\frac{1}{2}\sum_{\sigma\sigma'}\int d\mathbf{r}\int d\mathbf{r}'\ \boldsymbol{\Psi}_\sigma^\dagger(\mathbf{r})\boldsymbol{\Psi}_{\sigma'}^\dagger(\mathbf{r}')U(\mathbf{r},\mathbf{r}')\boldsymbol{\Psi}_{\sigma'}(\mathbf{r}')\boldsymbol{\Psi}_\sigma(\mathbf{r}).$$

Thus, the perturbation $\mathbf{V}$ appearing in the perturbative expansion of the causal Green's function of Eqs. (6.30) is given by

$$\mathbf{V}^{(0)}(t) = \frac{1}{2}\sum_{\sigma\sigma'}\int d\mathbf{r}\int d\mathbf{r}'\ \boldsymbol{\Psi}_\sigma^{(0)\dagger}(\mathbf{r}t)\boldsymbol{\Psi}_{\sigma'}^{(0)\dagger}(\mathbf{r}'t)U(\mathbf{r},\mathbf{r}')\boldsymbol{\Psi}_{\sigma'}^{(0)}(\mathbf{r}'t)\boldsymbol{\Psi}_\sigma^{(0)}(\mathbf{r}t).$$

$$(6.40)$$

Using this expression in Eq. (6.30) and applying Wick's theorem, we arrive at the following expression for the first-order correction for the causal Green's function[4]

$$\delta G^{(1)}(\mathbf{x}, \mathbf{x}') = \frac{1}{2}\int d\mathbf{x}_1 \int d\mathbf{x}_1'\ U(\mathbf{x}_1, \mathbf{x}_1')\ \{ \qquad (6.41)$$

$$n^{(0)}(\mathbf{r}_1')G^{(0)}(\mathbf{x},\mathbf{x}_1)G^{(0)}(\mathbf{x}_1,\mathbf{x}') + iG^{(0)}(\mathbf{x},\mathbf{x}_1)G^{(0)}(\mathbf{x}_1,\mathbf{x}_1')G^{(0)}(\mathbf{x}_1',\mathbf{x}')$$

$$+iG^{(0)}(\mathbf{x},\mathbf{x}_1')G^{(0)}(\mathbf{x}_1',\mathbf{x}_1)G^{(0)}(\mathbf{x}_1,\mathbf{x}') + n^{(0)}(\mathbf{r}_1)G^{(0)}(\mathbf{x},\mathbf{x}_1')G^{(0)}(\mathbf{x}_1',\mathbf{x}')$$

$$-iG^{(0)}(\mathbf{x},\mathbf{x}')G^{(0)}(\mathbf{x}_1',\mathbf{x}_1)G^{(0)}(\mathbf{x}_1,\mathbf{x}_1') - n^{(0)}(\mathbf{r}_1)n^{(0)}(\mathbf{r}_1')G^{(0)}(\mathbf{x},\mathbf{x}')\ \},$$

where we have used the shorthand $\mathbf{x} \equiv \mathbf{r}t$ to simplify the notation. In Eq. (6.41) it was necessary to write the causal Green's function with equal time arguments, i.e. $G^{(0)}(t, t)$, which has an ambiguous mathematical expression. We have used the following criterion that provides the correct result: $G^{(0)}(t, t^+)$, i.e. in Eq. (6.41) we have used

$$G^{(0)}(\mathbf{x}, \mathbf{x}) = G^{(0)}(\mathbf{r}t, \mathbf{r}t^+) \qquad (6.42)$$

$$= i\langle\phi_0|\boldsymbol{\Psi}_\sigma^{(0)\dagger}(\mathbf{r}t)\boldsymbol{\Psi}_\sigma^{(0)}(\mathbf{r}t)|\phi_0\rangle = in^{(0)}(\mathbf{r}). \qquad (6.43)$$

Now, we can use the graphical conventions introduced in Fig. 6.1 to represent the six different contributions to the first-order correction of the

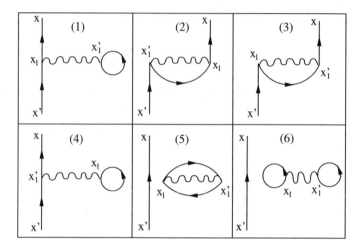

Fig. 6.2   First-order Feynman diagrams for the electron-electron interaction.

causal Green's function. This can be seen in Fig. 6.2, where we have numbered the terms from 1 to 6 following the order of Eq. (6.41).

Let us summarize some of the main features of these diagrams, which are also found in higher-order contributions:

- The only thing that matters in the diagrams is their topology, i.e. the way in which the different events are connected.
- The Green's functions with equal time arguments are represented by a closed loop and their value is equal to $in^{(0)}(\mathbf{r})$. If we used a local representation $\{|i\rangle\}$, then we would have

$$G_{ii}^{(0)}(t, t^+) = i\langle \mathbf{n}_{i\sigma}^{(0)}\rangle. \tag{6.44}$$

- Notice that all the intermediate events are linked by an interaction line and they have an incoming and an outgoing propagator, which correspond to the scattering process that the electron undergoes due to the electron-electron interaction. These intermediate events are known as vertexes (see Fig. 6.3).
- In Fig. 6.2 there are diagrams that have parts that are not connected to the the rest of the diagram and, in particular, to the initial and final events. Since there is an integration over the intermediate arguments appearing in these disconnected parts, they

---

[4]We assume here that there is spin symmetry in the unperturbed problem. Thus, all the Green's functions are diagonal in spin space and we will not write explicitly their spin index to abbreviate the notation.

Fig. 6.3   Vertex: point where two propagator lines and an interaction line meet.

simply give a constant that multiplies the contribution of the rest of the diagram. More importantly, one can show that these type of diagrams do not contribute to the final expansion because they are exactly canceled by the denominator of the full Green's functions. For a demonstration of this fact we refer the reader to the Exercise 6.3.

• As we can see in Fig. 6.2, several diagrams are topologically equivalent (e.g. diagrams 1 and 3 or 2 and 4) and the only difference is the order in which the arguments appear. However, since there are integrations over such intermediate variables, see Eq. (6.41), all these equivalent diagrams give exactly the same contribution. This happens indeed at any order of the perturbative expansion. Thus, at order n, any topologically connected diagram appears $2^n n!$ times. The factor $1/2$ in the expression of $V^{(0)}$ together with the factor $1/n!$ in the perturbative expansion (see Eq. (6.30)) cancel exactly this multiplicity. Therefore, we need to consider the topologically connected diagrams only once.

Summarizing, the series of diagrams that contribute to the expansion of the causal Green's function are formed by the topologically distinct connected diagrams. Moreover, the denominator in Eq. (6.30) drops. Therefore, we can finally write the diagrammatic series of Eq. (6.30) as

$$G(\mathbf{r}t, \mathbf{r}'t') = G^{(0)}(\mathbf{r}t, \mathbf{r}'t') + \sum_{n=1}^{\infty} (-i)^{n+1} \int_{-\infty}^{\infty} dt_1 \cdots \int_{-\infty}^{\infty} dt_n \times \quad (6.45)$$

$$\langle \phi_0 | \mathbf{T} \left[\mathbf{\Psi}_\sigma^{(0)}(\mathbf{r}t) \mathbf{V}^{(0)}(t_1) \cdots \mathbf{V}^{(0)}(t_n) \mathbf{\Psi}^{(0)\dagger}(\mathbf{r}'t') \right] | \phi_0 \rangle_{\text{connected}},$$

where only the contribution of the topologically distinct connected diagrams is considered. Of course, there would be a similar expression for the Green's functions in a discrete representation (or basis).

It is a very useful exercise to find the 10 topologically distinct connected Feynman diagrams that contribute to the second-order correction of the causal Green's function (see Exercise 6.4). In Fig. 6.4 we show some of these diagrams.

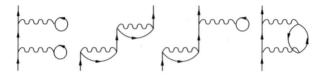

Fig. 6.4   Some of the 10 second-order topologically distinct connected Feynman diagrams for the electron-electron interaction.

The Feynman diagrams provide a very intuitive way of evaluating the different contributions to the perturbative expansion of a causal Green's function. In this sense, one proceeds sometimes by identifying directly the relevant diagrams rather than calculating the systematic perturbative series. Indeed, one can derive simple rules to quantify the contribution of the different diagrams. For the sake of completeness, we state here these rules for obtaining diagrammatically the contribution at a given order n to the causal Green's function in the case of the electron-electron interaction:

(1) Draw all the topologically distinct connected diagrams containing n interaction lines and $2n + 1$ propagator lines between the initial and the final events.

(2) Every event must be labeled with its corresponding space-time coordinate $\mathbf{r}t$ (or it, if one works with a discrete basis $|i\rangle$). All the events, apart from the initial and final ones, contain a vertex as the one of Fig. 6.3.

(3) Every propagator line connecting the events $\mathbf{x}_2 = \mathbf{r}_2 t_2$ and $\mathbf{x}_1 = \mathbf{r}_1 t_1$ contributes with a factor $G^0(\mathbf{x}_1, \mathbf{x}_2)$.

(4) Every interaction line connecting the events $\mathbf{x}_2 = \mathbf{r}_2 t_2$ and $\mathbf{x}_1 = \mathbf{r}_1 t_1$ introduces a factor $U(\mathbf{x}_1, \mathbf{x}_2) = U(\mathbf{r}_1, \mathbf{r}_2)\delta(t_1 - t_2)$. In the case of a discrete basis, this factor would be U_{ijkl} (corresponding matrix element of the Coulomb potential).

(5) One has to include integrals over all intermediate variables.

(6) Every diagram of order n contains a pre-factor i^n.

(7) Finally, there is a sign $(-1)^F$, where F is the number of closed loops in the diagram. The closed loop can be formed either by a single propagator or by a combination of several of them. Moreover, a Green's function with equal time variables must be interpreted as $G^{(0)}(\mathbf{x}t, \mathbf{x}'t^+)$.

As an illustration of these rules, let us compute the contribution corresponding to the last diagram in Fig. 6.4. This second-order contribution is equal to

$$-i^2 \int d\mathbf{x}_1 \int d\mathbf{x}_1' \int d\mathbf{x}_2 \int d\mathbf{x}_2'\, G^{(0)}(\mathbf{x},\mathbf{x}_1)U(\mathbf{x}_1,\mathbf{x}_1')G^{(0)}(\mathbf{x}_1',\mathbf{x}_2')G^{(0)}(\mathbf{x}_2',\mathbf{x}_1')$$
$$G^{(0)}(\mathbf{x}_1,\mathbf{x}_2)U(\mathbf{x}_2,\mathbf{x}_2')G^{(0)}(\mathbf{x}_2,\mathbf{x}').$$

6.5.2 Feynman diagrams for an external potential

Now, we assume that the electrons are subjected to an external time-independent perturbation of the form

$$\mathbf{V} = \sum_{i=1}^{N} V(\mathbf{r}_i), \qquad (6.46)$$

which in second quantization can be written as (in the interaction picture)

$$\mathbf{V}^{(0)} = \sum_{\sigma} \int d\mathbf{r}\, \mathbf{\Psi}_\sigma^{(0)\dagger}(\mathbf{r}t)V(\mathbf{r})\mathbf{\Psi}_\sigma^{(0)}(\mathbf{r}t). \qquad (6.47)$$

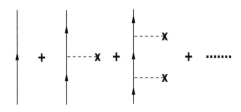

Fig. 6.5 Diagrammatic series for the propagator in the case of an external potential.

For the sake of simplicity, we have assumed that the potential does not depend on the electron spin. In this case, the diagrammatic series is very simple. Applying Wick's theorem to Eq. (6.30), one obtains the diagrammatic series shown in Fig. 6.5. This means that in the propagation of the electron from the initial instance to the final one, one simply has a series of sequential scattering events with the external potential. The rules for computing the contribution to the nth-order correction of the causal Green's functions are very simple in this case:

(1) Draw the sequential diagrams like in Fig. 6.5 with $n+1$ propagators and n interaction lines.
(2) Associate the corresponding Green's function to every propagator line.
(3) Assign the corresponding external potential to every interaction line.
(4) Integrate over the intermediate variables.

(5) The prefactor is 1.

Due to the simplicity of the diagrammatic series in this case, it is often possible to sum up all the contributions up infinite order (notice that the diagrammatic expansion leads to a geometrical series). As an illustration of the previous rules, the second-order diagram in Fig. 6.5 gives a contribution equal to

$$\int dx_1 \int dx_2 \; G^{(0)}(x, x_1) V(r_1) G^{(0)}(x_1, x_2) V(r_2) G^{(0)}(x_2, x'). \qquad (6.48)$$

6.6   Feynman diagrams in energy space

In spite of all the simplifications that we have introduced in the last section, it is still very difficult to compute the different terms of the perturbative series. This is due to the presence of the integrals over the intermediate arguments. Thus for instance, a diagram of order 1 for the electron-electron interaction contains up to six integrals.

The problem can be simplified by noticing first that in an equilibrium situation the Green's functions depend exclusively on the difference of the time arguments. Thus, we can Fourier transform with respect to time and work in the energy space. The introduction of the Fourier transformation modifies the Feynman diagrams and we now study how this occurs in detail.

On the other hand, if the system is spatially homogeneous, the problem can be simplified even further since then the Green's functions depend only on the difference of the space coordinates. We shall first discuss this case and later on, we shall generalize the results to an arbitrary non-homogeneous system.

As we have just said, if the system is spatially homogeneous and in equilibrium, the Green's functions satisfy

$$G(rt, r't') = G(r - r', t - t'), \qquad (6.49)$$

or, using the four-dimensional notation $(x \equiv rt)$, $G(x, x') = G(x - x')$. If we assume that the interaction potential also satisfies $U(x, x') = U(x - x')$, we can then Fourier transform

$$G(rt) = \int \frac{dk}{(2\pi)^3} \int \frac{dE}{2\pi} e^{i(k \cdot r - Et)} G(k, t). \qquad (6.50)$$

In what follows, we shall use the following simplified notation: $p \equiv (k, E)$ and $p \cdot x = kr - Et$. With this notation, the different Fourier transforms read

$$G(\mathbf{x}) = \int \frac{d\mathbf{p}}{(2\pi)^4} e^{i\mathbf{p}\mathbf{x}} G(\mathbf{p}); \ U(\mathbf{x}) = \int \frac{d\mathbf{p}}{(2\pi)^4} e^{i\mathbf{p}\mathbf{x}} U(\mathbf{p}), \qquad (6.51)$$

where $d\mathbf{p} \equiv d^3 k dE$ is the volume element in $(\mathbf{k}, E)$-space.

In order to illustrate how the diagrams are modified in energy space, we choose a first-order diagram for the electron-electron interaction, namely diagram 2 in Fig. 6.2. The contribution of this diagram, which we shall denote as $D(\mathbf{x} - \mathbf{x}')$, is given by

$$D(\mathbf{x} - \mathbf{x}') = i \int d\mathbf{x}_1 \int d\mathbf{x}'_1 \ G^{(0)}(\mathbf{x} - \mathbf{x}_1) U(\mathbf{x}_1 - \mathbf{x}'_1) \qquad (6.52)$$
$$G^{(0)}(\mathbf{x}_1 - \mathbf{x}'_1) G^{(0)}(\mathbf{x}'_1 - \mathbf{x}').$$

Substituting in the right hand side of this expression the Fourier transform of $G^{(0)}$ and U, one has

$$D(\mathbf{x} - \mathbf{x}') = i \int d\mathbf{x}_1 \int d\mathbf{x}'_1 \int \frac{d\mathbf{p}}{(2\pi)^4} \int \frac{d\mathbf{q}}{(2\pi)^4} \int \frac{d\mathbf{p}'}{(2\pi)^4} \int \frac{d\mathbf{q}'}{(2\pi)^4} \qquad (6.53)$$
$$G^{(0)}(\mathbf{p}) U(\mathbf{q}) G^{(0)}(\mathbf{q}') G^{(0)}(\mathbf{p}') e^{i\mathbf{p}(\mathbf{x}-\mathbf{x}_1)} e^{i\mathbf{q}(\mathbf{x}_1-\mathbf{x}'_1)} e^{i\mathbf{q}'(\mathbf{x}_1-\mathbf{x}'_1)} e^{i\mathbf{p}'(\mathbf{x}'_1-\mathbf{x}')}.$$

This expression can be greatly simplified in the following way. First, we regroup the exponential terms as follows

$$e^{i\mathbf{p}\mathbf{x}} e^{i\mathbf{x}_1(-\mathbf{p}+\mathbf{q}+\mathbf{q}')} e^{i\mathbf{x}'_1(\mathbf{p}'-\mathbf{q}-\mathbf{q}')} e^{-i\mathbf{p}'\mathbf{x}'}. \qquad (6.54)$$

Now, we integrate over the variables $\mathbf{x}_1$ and $\mathbf{x}'_1$:

$$\int d\mathbf{x}_1 \ e^{i\mathbf{x}_1(-\mathbf{p}+\mathbf{q}+\mathbf{q}')} = (2\pi)^4 \delta(\mathbf{p} - \mathbf{q} - \mathbf{q}') \Rightarrow \mathbf{q}' = \mathbf{p} - \mathbf{q} \qquad (6.55)$$

$$\int d\mathbf{x}'_1 \ e^{i\mathbf{x}'_1(\mathbf{p}'-\mathbf{q}-\mathbf{q}')} = (2\pi)^4 \delta(\mathbf{p}' - \mathbf{q} - \mathbf{q}') \Rightarrow \mathbf{p}' = \mathbf{q} + \mathbf{q}' = \mathbf{p}.$$

The previous equations simply express the conservation of the four-dimensional moment (momentum and energy) in every vertex, as we illustrate in Fig. 6.6, where the momentum lost by the electron in the scattering process is carried by the interaction line. If we now substitute Eq. (6.55) in Eq. (6.52), we obtain

$$D(\mathbf{x} - \mathbf{x}') = i \int \frac{d\mathbf{p}}{(2\pi)^4} e^{i\mathbf{p}(\mathbf{x}-\mathbf{x}')} \int \frac{d\mathbf{q}}{(2\pi)^4} U(\mathbf{q}) G^{(0)}(\mathbf{p}) G^{(0)}(\mathbf{p} - \mathbf{q}) G^{(0)}(\mathbf{p}).$$
$$(6.56)$$

This implies that the Fourier transform of the diagram can be written as

$$D(\mathbf{p}) = i \int \frac{d\mathbf{q}}{(2\pi)^4} U(\mathbf{q}) G^{(0)}(\mathbf{p}) G^{(0)}(\mathbf{p} - \mathbf{q}) G^{(0)}(\mathbf{p}). \qquad (6.57)$$

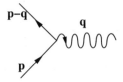

Fig. 6.6   Energy and momentum conservation in a vertex.

The previous derivation would be similar for any diagram. The key idea is that the energy and the momentum are conserved in every vertex. Thus, one can view the diagrams as flow diagrams in which the propagator lines and the interaction lines carry momentum and energy. The momentum **k** and the energy E carried by the initial propagator are also carried by the final one, due to the conservation of momentum and energy in every vertex of the diagram. This is illustrated in Fig. 6.7 with two first-order diagrams and a second-order one. Notice that, since the interaction lines carry both momentum and energy, one has to assign to them a direction, which is indicated by an arrow in the diagram.

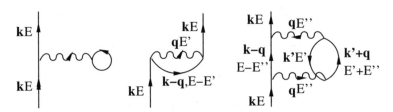

Fig. 6.7   Feynman diagrams in momentum and energy space.

As in the case of real space, it is possible to establish the diagrammatic rules for computing the perturbative expansion of the causal Green's function in energy space. Those rules for the nth-order correction now read:

(1) Draw all the topologically distinct connected diagrams with n interaction lines and $2n+1$ propagator lines. These diagrams are the same as in the ones in $(\mathbf{r}, t)$-space.
(2) Assign the flow direction (arrows) of the momentum and energy to every interaction and propagator line.
(3) The momentum and the energy must be conserved in every vertex.
(4) Every propagator with momentum **k** and energy E contributes with a factor that is equal to the unperturbed causal Green's function, which

for a homogeneous electron gas has the form

$$G^{(0)}(\mathbf{k}, E) = \frac{1}{E - \epsilon_{\mathbf{k}} - i\eta \mathrm{sgn}(k - k_F)}. \tag{6.58}$$

(5) Every interaction line with momentum $\mathbf{k}$ introduces an interaction potential in momentum space. For the homogenous system and for the Coulomb potential, it has the form

$$U(\mathbf{k}) = \frac{4\pi e^2}{k^2}. \tag{6.59}$$

(6) We have to integrate over all intermediate momenta and energies (for a non-homogeneous systems only over the energies).

(7) As a consequence of the previous rule, there is a factor for a diagram of order n equal to $1/(2\pi)^{4n}$ (equal to $1/(2\pi)^n$, if one only needs to integrate over the energies). Moreover, there is a factor i^n, as in the case of real space.

(8) As in the case of real space, there is a sign $(-1)^F$, where F is the number of closed loops.

(9) Finally, let us remind that for the diagrams in real space, there was an ambiguity that occurs when the time arguments of the causal Green's function are equal. This problem was solved with the criterion $G^{(0)}(t, t) = G^{(0)}(t, t^+)$. The consequence of this choice when we Fourier transform is the introduction of a convergence factor $\exp(iE\eta)$, which must appear associated to every propagator that forms a closed loop and to those that are connected by an interaction line (if the interaction is instantaneous).

As an example, let us write the contribution of the second-order diagram in Fig. 6.7. The result is

$$\int \frac{d\mathbf{q}}{(2\pi)^3} \int \frac{d\mathbf{k}'}{(2\pi)^3} \int \frac{dE''}{2\pi} \int \frac{dE'}{2\pi} U^2(\mathbf{q}) G^{(0)}(\mathbf{k}, E) G^{(0)}(\mathbf{k} - \mathbf{q}, E - E'') \times$$
$$G^{(0)}(\mathbf{k}', E') G^{(0)}(\mathbf{k}' + \mathbf{q}, E' + E'') G^{(0)}(\mathbf{k}, E).$$

To conclude this section, it is convenient to generalize the results obtained so far to the case of non-homogeneous systems. Indeed, this generalization is quite simple. Since the momentum is not a good quantum number, it makes no sense to Fourier transform with respect to the spatial coordinates. However, since the system is in equilibrium, one can still introduce the Fourier transform with respect to the time arguments. This is done exactly in the way explained above for the homogeneous system.

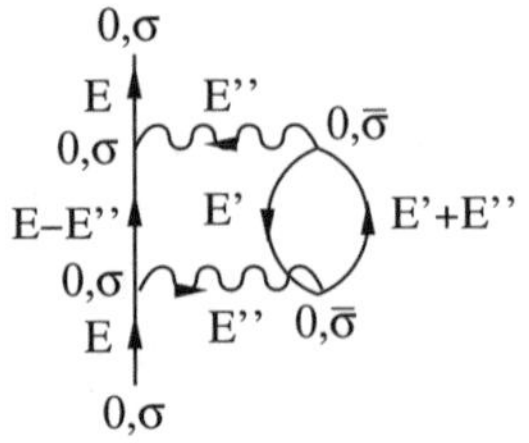

Fig. 6.8   Second-order Feynman diagrams in energy space for the Anderson model.

As an example, let us calculate the contribution of second-order diagram of Fig. 6.8 for the Anderson model that we discussed in section 5.4.3:

$$U^2 \int \frac{dE''}{2\pi} \int \frac{dE'}{2\pi} \, G_{00\sigma}^{(0)}(E) G_{00\sigma}^{(0)}(E - E'') G_{00\bar\sigma}^{(0)}(E') G_{00\bar\sigma}^{(0)}(E' + E'') G_{00\sigma}^{(0)}(E).$$

Here, the index 0 refers to the impurity level.

6.7   Electronic self-energy and Dyson's equation

In the previous sections we have analyzed the structure of the diagrammatic series of an electronic Green's function. In this section we shall show that it is possible to sum formally the diagrams up to infinite order, leading to the Dyson's equation. But before describing this further simplification of the perturbative expansion, let us introduce the concept of self-energy.

In Fig. 6.9 we show again the diagrammatic expansion for the Green's function in the cases in which the perturbation is an external potential and the electron-electron interaction. Notice that in both cases the diagrams have the same type of structure in the following sense. They are formed by an initial and a final Green's function (the same in all diagrams) and by a central part where one can find all the scattering processes. Obviously, this latter part is the interesting one. This structure of the diagrammatic series allows us to define the (improper) electronic self-energy as the sum of the central part of the diagrams to all orders (Σ_I in Fig. 6.10). Thus, the diagrammatic series for the self-energy insertion has the form shown in Fig. 6.11 for the cases of an external potential and the electron-electron interaction.

Notice that in the previous discussion we have neither specified the representation nor the space (time/energy). In this sense, the result discussed in the previous paragraphs is quite general. The diagrammatic expansion

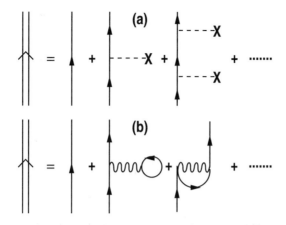

Fig. 6.9 Diagrammatic expansion for the propagator for (a) an external potential and (b) the electron-electron interaction.

of Fig. 6.9 can be summarized in the following equation in real space ($\mathbf{r}$-representation)

$$G(\mathbf{x}, \mathbf{x}') = G^{(0)}(\mathbf{x}, \mathbf{x}') + \int d\mathbf{x}_1 \int d\mathbf{x}_2 \, G^{(0)}(\mathbf{x}, \mathbf{x}_1)\Sigma_I(\mathbf{x}_1, \mathbf{x}_2)G^{(0)}(\mathbf{x}_2, \mathbf{x}').$$
(6.60)

The equation in momentum-energy space (for a homogeneous case) reads as follows

$$G(\mathbf{k}, E) = G^{(0)}(\mathbf{k}, E) + G^{(0)}(\mathbf{k}, E)\Sigma_I(\mathbf{k}, E)G^{(0)}(\mathbf{k}, E).$$
(6.61)

In the case of a localized basis (like in a tight-binding model), the previous equation adopts the form:

$$G_{ij}(E) = G_{ij}^{(0)}(E) + \sum_{kl} G_{ik}^{(0)}(E)\Sigma_{I,kl}(E)G_{lj}^{(0)}(E).$$
(6.62)

To avoid explicit reference to any particular representation or space, we shall write the previous equation in matrix form:

$$\mathbf{G} = \mathbf{G}^{(0)} + \mathbf{G}^{(0)}\boldsymbol{\Sigma}_I\mathbf{G}^{(0)},$$
(6.63)

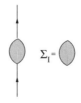

Fig. 6.10   Self-energy insertion.

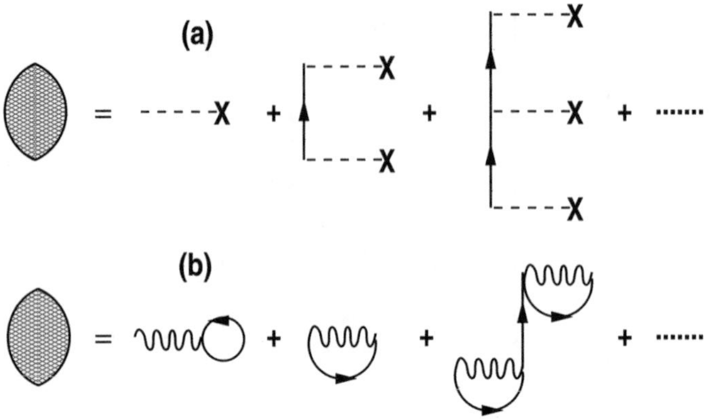

Fig. 6.11   Diagrammatic expansion for the self-energy insertion. (a) External potential. (b) Electron-electron interaction.

where the internal integrals and sums are implicitly assumed. It is possible to write this equation in a more convenient way by inspection of the perturbative series of $\mathbf{G}$ or $\boldsymbol{\Sigma}_I$. Let us illustrate this fact first with the example of an external potential. As we explained in previous sections, the diagrammatic expansion has in this case the form of a geometrical series where the diagram of order n is simply the repetition of n identical pieces. If we define in this case the *proper self-energy*, Σ, as the part of the diagram that includes only a single scattering process, which in this case is simply the external potential, we have the following identity

$$\boldsymbol{\Sigma}_I \mathbf{G}^{(0)} = \boldsymbol{\Sigma} \mathbf{G}. \tag{6.64}$$

This is evident when it is expressed diagrammatically as in Fig. 6.12.

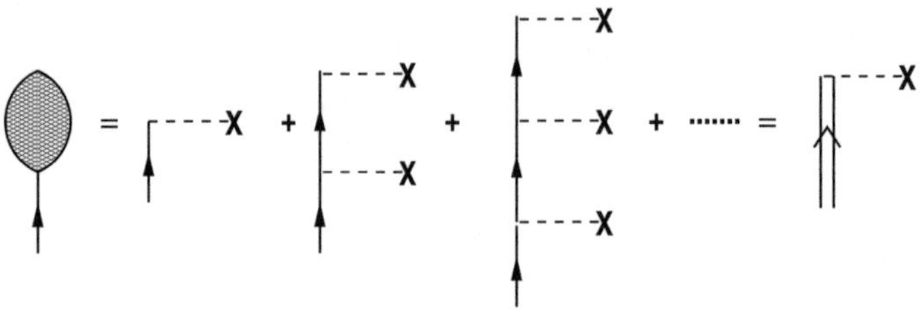

Fig. 6.12   Relation between the self-energy insertion, $\boldsymbol{\Sigma}_I$ and the proper self-energy, $\boldsymbol{\Sigma}$.

The proper self-energy, or from now on just self-energy, does not contain repetitions of the same process, but only one scattering event. Then,

Eq. (6.63) can be written in terms of the self-energy as

$$\mathbf{G} = \mathbf{G}^{(0)} + \mathbf{G}^{(0)}\mathbf{\Sigma G}, \qquad (6.65)$$

which constitutes the so-called *Dyson's equation* and was first obtained by F. Dyson in 1949 in the context of the quantum electrodynamics.

Let us now discuss the derivation of this result in the case of the electron-electron interaction. Notice first that in this case the diagrams that contribute to the self-energy insertion to all orders can be classified in two different ways. On the one hand, we have diagrams that cannot be separated in two parts by cutting a propagator line, i.e. they do not contain repetitions of the same elementary process. These diagrams are called *irreducible* [see Fig. 6.13(a)]. On the other hand, we have diagrams that can be divided into parts of lower order by cutting a propagator line, these are called *reducible* diagrams [see Fig. 6.13(b)].

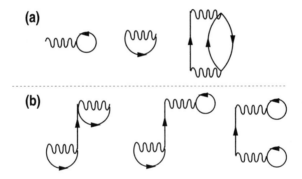

Fig. 6.13   (a) Examples of irreducible self-energy diagrams for the electron-electron interaction. (b) Reducible diagrams.

We define the proper self-energy (or simply self-energy) in this case as the sum of all the irreducible self-energy diagrams. With this definition, the Dyson's equation is also verified in this case. The proof is more complicated than in the case of an external potential and it will not be detailed here.

The Dyson's equation can be represented graphically as shown in Fig. 6.14. Notice that the due to the symmetry of the diagrammatic series, we could have chosen to close the Dyson's equation in an alternative way:

$$\mathbf{G} = \mathbf{G}^{(0)} + \mathbf{G}\mathbf{\Sigma G}^{(0)}. \qquad (6.66)$$

On the other hand, notice that the Dyson's equation obtained in the previous chapter for single-electron problems, see Eq. (5.28), is just a particular example of Eq. (6.65), which is valid for any electronic system.

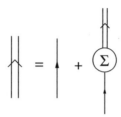

Fig. 6.14   Pictorial representation of the Dyson's equation.

For systems in equilibrium it is convenient to write the Dyson's equation in energy space

$$\mathbf{G}(E) = \mathbf{G}^{(0)}(E) + \mathbf{G}^{(0)}(E)\mathbf{\Sigma}(E)\mathbf{G}(E), \qquad (6.67)$$

which will be our starting point for the description of the equilibrium properties of any system.

Taking into account the definition of the single-particle Green's function in energy space introduced in the previous chapter, we can rewrite the previous Dyson's equation as

$$\left[\mathbf{G}^{(0)}(E)\right]^{-1}\mathbf{G}(E) = \mathbf{1} + \mathbf{\Sigma}(E)\mathbf{G}(E) \qquad (6.68)$$

$$[E\mathbf{1} - \mathbf{H}_0]\,\mathbf{G}(E) = \mathbf{1} + \mathbf{\Sigma}(E)\mathbf{G}(E),$$

which allows us to write the Green's function matrix of the full system as

$$\mathbf{G}(E) = [E\mathbf{1} - \mathbf{H}_0 - \mathbf{\Sigma}(E)]^{-1}. \qquad (6.69)$$

From this expression, one can interpret the self-energy as the matrix whose elements renormalize dynamically the matrix elements of the unperturbed system. Thus for instance, for the homogeneous electron gas with electron-electron interaction, the problem is diagonal in the plane wave basis that diagonalizes $\mathbf{H}_0$ and the previous Dyson's equation becomes

$$G(\mathbf{k}, E) = \frac{1}{E - \epsilon_{\mathbf{k}} - \Sigma(\mathbf{k}, E)}. \qquad (6.70)$$

In summary, the perturbative analysis reduces to the evaluation of the proper self-energy (or just self-energy) of the electronic system. For the two cases considered in the last sections, namely external potential and electron-electron interaction, this implies to calculate the diagrammatic series depicted in Fig. 6.15.

Finally, let us conclude this section with some comments and the main analytical properties of the electronic self-energy:

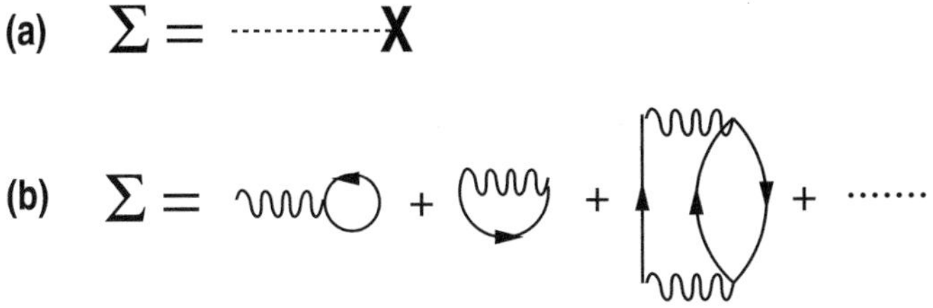

(a) $\Sigma = \cdots\cdots\cdots\mathbf{X}$

(b) $\Sigma = $ $+ $ $+ $ $+ \cdots\cdots$

Fig. 6.15 Diagrammatic expansion for the proper self-energy. (a) External potential and (b) electron-electron interaction.

- The Dyson's equation relates directly the self-energy with the full Green's function. Therefore, the analytical properties of $\Sigma(E)$ can be derived from those of $\mathbf{G}(E)$.
- One can interpret Eq. (6.69) as a definition of $\Sigma(E)$ in terms of $\mathbf{G}(E)$. Thus, it is also possible to define a retarded and advanced self-energy.
- From Lehmann's representation of the Green's functions, one can deduce the following properties that we state here without any proof:

$$\text{Im}\left\{\Sigma_{ii}^r(E)\right\} \leq 0 \; ; \; \text{Im}\left\{\Sigma_{ii}^a(E)\right\} \geq 0 \qquad (6.71)$$

$$\text{Im}\left\{\Sigma_{ii}^c(E)\right\} \geq 0 \; \text{ if } \; E < \mu \; ; \; \text{Im}\left\{\Sigma_{ii}^c(E)\right\} \leq 0, \; \text{ if } \; E > \mu.$$

- $\text{Im}\Sigma_{ii}(E)$ and $\text{Re}\Sigma_{ii}(E)$ are related through a Hilbert transformation:

$$\text{Re}\left\{\Sigma_{ii}^{r,a}(E)\right\} = \mp \mathcal{P}\int \frac{dE'}{\pi} \frac{\text{Im}\left\{\Sigma_{ii}^{r,a}(E')\right\}}{E - E'} \qquad (6.72)$$

$$\text{Re}\left\{\Sigma_{ii}^c(E)\right\} = -\mathcal{P}\int \frac{dE'}{\pi} \frac{\text{Im}\left\{\Sigma_{ii}^c(E')\right\}\,\text{sgn}(E' - \mu)}{E - E'}.$$

6.8 Self-consistent diagrammatic theory: The Hartree-Fock approximation

Apart from the Dyson's equation, there exist other ways to include certain diagrams in the expansion of the self-energy up to infinite order. By inspection of the set of diagrams that contribute to the self-energy, it is possible to distinguish two types of diagrams. On the one hand, there are diagrams, like the one shown in Fig. 6.16, in which in one of the propagators there is a self-energy insertion. On the other hand, there exist diagrams that do not contain insertions and they are called *skeleton diagrams*. An example of a second-order skeleton diagram is shown in Fig. 6.15(b).

Analyzing the diagrammatic series of the self-energy, one realizes that if we consider any skeleton diagram, there appear diagrams at higher orders

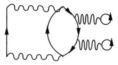

Fig. 6.16 Example of diagram with a self-energy insertion in one of the propagators.

with the same structure (or skeleton), but with all possible self-energy insertions in their propagators. This fact makes possible to sum up to infinite order all the diagrams that share the same skeleton, which leads to effective diagrams like the one depicted in Fig. 6.17. Here, we have taken into account the fact that by adding all the diagrams with the same structure, the propagator in the skeleton diagram can be replaced by the full (dressed) propagator.

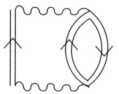

Fig. 6.17 Second-order skeleton diagram.

The previous result implies that it is possible to write the self-energy as an expansion that contains exclusively skeleton diagrams, where the propagators are the full ones (they are sometimes referred to as dressed or renormalized propagators). This is illustrated in Fig. 6.18.

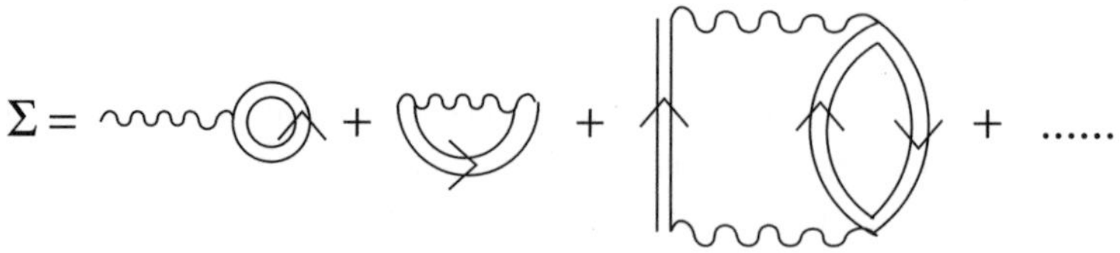

$$\Sigma = \text{\raisebox{-0.5em}{\includegraphics}} + \text{\raisebox{-0.5em}{\includegraphics}} + \text{\raisebox{-0.5em}{\includegraphics}} + \quad ......$$

Fig. 6.18 Expansion of the self-energy in terms of skeleton diagrams.

It is worth stressing that the propagators that appear in these skeleton diagrams are the perturbed ones, which are unknown and they have to be determined by solving the Dyson's equation. This means that the expansion of Fig. 6.18, together with the corresponding Dyson's equation provide two equations that have to be solved in a self-consistent manner. The most common practice is to include just a few diagrams in the expansion of Fig. 6.18. An interesting example that illustrates this procedure is the Hartree-Fock approximation, which from a diagrammatic point of view, is given by the approximation for the self-energy schematized in Fig. 6.19.

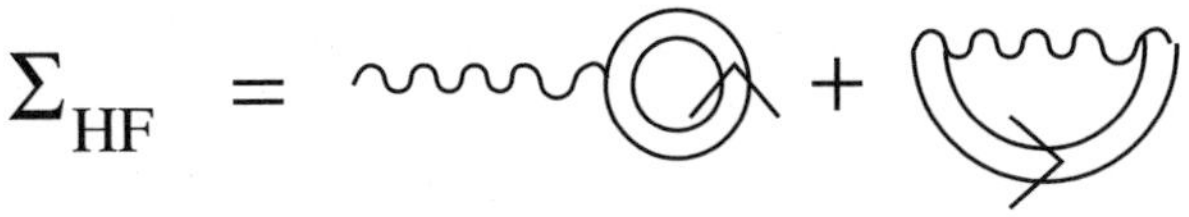

Fig. 6.19   Hartree-Fock approximation for the self-energy.

Let us show now this approximation is indeed equivalent to the well-known Hartree-Fock approximation in the more standard wavefunction-based language (see section 10.1.3). The diagram that contains the bubble (Hartree diagram) has the following expression in the representation $|\mathbf{r}\rangle$

$$\Sigma_\sigma^H(\mathbf{r}) = \sum_{\sigma'} \int d\mathbf{r}'\, U(\mathbf{r} - \mathbf{r}')G_{\sigma'}(\mathbf{r}'t', \mathbf{r}'t'^+) \tag{6.73}$$

$$= \sum_{\sigma'} \int d\mathbf{r}'\, U(\mathbf{r} - \mathbf{r}')n_{\sigma'}(\mathbf{r}') = \sum_{\sigma'} \int d\mathbf{r}'\, \frac{e^2 n_{\sigma'}(\mathbf{r}')}{|\mathbf{r} - \mathbf{r}'|},$$

which is nothing else but the Hartree potential, where $n_\sigma(\mathbf{r})$ is the perturbed electron density with spin σ that has to be determined self-consistently.

Analogously, the second diagram in Fig. 6.19 is given by (in the representation $|\mathbf{r}\rangle$)

$$\Sigma_\sigma^X(\mathbf{r}, \mathbf{r}') = iU(\mathbf{r} - \mathbf{r}')G_\sigma(\mathbf{r}t, \mathbf{r}'t^+). \tag{6.74}$$

One can show that this expression leads to the known nonlocal (Fock) exchange potential. For this purpose, one just needs to expand the field operators in the previous expression in terms of an arbitrary single-electron basis and take into account that the ground state is noninteracting. This leads to

$$\Sigma_\sigma^X(\mathbf{r}, \mathbf{r}') = -\sum_i \frac{e^2 \phi_{i\sigma}(\mathbf{r}')\phi_{i\sigma}(\mathbf{r})}{|\mathbf{r} - \mathbf{r}'|}. \tag{6.75}$$

As an additional illustration of the Hartree-Fock approximation, we discuss now the calculation of the energy bands in this approximation of a homogeneous electron gas (see Exercise 6.5). In this case, it is not necessary to do the self-consistency because it is automatically guaranteed due homogeneity of the system with a constant density $n = N/V$. Instead of using the expressions derived above, we compute now the self-energy in this approximation in the $(\mathbf{k}, E)$-space. Evaluating the Hartree-Fock diagrams in this space, one arrives at

$$\Sigma_\sigma^{\mathrm{H}} = U(q=0) \sum_{\sigma'} \int \frac{d\mathbf{k}'}{(2\pi)^3} \int \frac{dE'}{2\pi} G_{\sigma'}(\mathbf{k}', E') e^{iE'\eta}. \qquad (6.76)$$

Since the Fourier transform of the Coulomb potential, $U(q) = 4\pi e^2/q^2$, diverges at $q = 0$, we replace the potential $U(q)$ by $\lim_{\mu \to 0} 4\pi e^2/(q^2 + \mu^2)$, which allows us to control the divergence. This new expression is simply the Fourier transform of a Yukawa-like potential $\exp(-\mu r)/r$. Thus, if one computes the integral in the expression of $\Sigma_\sigma^{\mathrm{H}}$, one obtains

$$\Sigma_\sigma^{\mathrm{H}} = \frac{4\pi e^2}{\mu^2} n. \qquad (6.77)$$

Although this result diverges when $\mu \to 0$, it is exactly canceled in the jellium model by the potential created by the uniform background of positive charge. Thus, the only remaining contribution is the exchange one that can be expressed as

$$\Sigma_\sigma^{\mathrm{X}}(\mathbf{k}) = i \int \frac{d\mathbf{q}}{(2\pi)^3} \int \frac{d\nu}{2\pi} U(q) G_\sigma(\mathbf{k}-\mathbf{q}, E-\nu) = -\int \frac{d\mathbf{k}'}{(2\pi)^3} \frac{4\pi e^2}{|\mathbf{k} - \mathbf{k}'|} \langle n_{\mathbf{k}'\sigma} \rangle. \qquad (6.78)$$

Now using the Dyson's equation in this representation, $G(\mathbf{k}, E) = [E - \epsilon_\mathbf{k} - \Sigma(\mathbf{k}, E)]^{-1}$, we see that the energy bands in the Hartree-Fock approximation are given by $\epsilon_{\mathbf{k},\mathrm{HF}} = \epsilon_\mathbf{k} + \Sigma^{\mathrm{X}}(\mathbf{k})$. The explicit expression of the dispersion relation is computed in Exercise 6.5.

6.9   The Anderson model and the Kondo effect

The goal of this section is two-fold. On the one hand, we shall use the Anderson model, already discussed in section 5.4.3 and Appendix A, to illustrate the perturbative approach described in this chapter. On the other hand, we shall use this model to get a flavor of the Kondo effect. This is a many-body phenomenon which can appear in molecular junctions and it will be described in much more detail in Chapter 15.

The Anderson model describes the interaction of a localized level with electron-electron interaction with the continuum of states of a metallic system. It was introduced by Anderson to describe a magnetic impurity in a metal host, but it can also be used to describe a metal-molecule-metal junction, which is the problem that we are interested in. In this model, the

Hamiltonian is given by Eq. (5.109), where in particular, the U-term describes the electron-electron interaction in this level. In the absence of this interaction, this model reduces to the resonant tunneling model of section 5.3.3.

Our goal now is to study the influence of the electron-electron interaction in the equilibrium properties of a molecular junction, with special attention to the local density of states. For this purpose, we shall make use of the perturbative approach described in this chapter. In this approach we shall consider the entire system without electron-electron interaction as the unperturbed system and this interaction, i.e. the last term in Eq. (5.109), will be considered as the perturbation. The unperturbed Green's functions projected onto the localized level were already obtained in section 5.3.3, see Eq. (5.56). In particular, the causal function adopts the following form in the wide-band approximation[5]

$$G_{00}^{(0)}(E) = \frac{1}{E - \epsilon_0 - i\mathrm{sgn}(E - \mu)\Gamma}, \qquad (6.79)$$

where μ is the chemical potential of the system and $\Gamma = \Gamma_L + \Gamma_R$ is the total broadening of the level acquired via the interaction with the metal electrodes. In what follows, we shall only consider symmetric situations ($\Gamma_L = \Gamma_R$). As we saw in section 5.3.3, in this approximation the density of states in the localized level is a Lorentzian with Γ as its half width at half maximum.

In the rest of this section, and in order to study the effect of the electron-electron interaction, we shall first discuss the so-called Friedel sum rule, which is an exact result that relates the local density of states at the Fermi energy to the occupation of the level, and then we shall do a perturbative analysis up to second order in the interaction U.

6.9.1 *Friedel sum rule*

We discuss now an important exact result, known as Friedel' sum rule, which is a consequence of the Fermi liquid properties of the system described by the Anderson model.[6] This sum rule can be derived as follows. The effect of the electron-electron interaction in the localized level can be included via the exact self-energy of the problem, $\Sigma_{00,\sigma}(E)$.[7] The (retarded) full Green

---

[5]Notice that this function is independent of the spin.

[6]Although we have not discussed the Fermi liquid theory in this book, we find important to introduce this discussion about Friedel sum rule because it provides a simple way to understand the appearance of the Kondo effect.

[7]Notice that we have now included the spin index σ in the self-energy.

function projected onto the level can written in terms of the self-energy as

$$G^r_{00,\sigma}(E) = \frac{1}{E - \epsilon_0 + i\Gamma - \Sigma^r_{00,\sigma}(E)}. \tag{6.80}$$

Taking now into account that the density of states in the level is given by $\rho_{0\sigma}(E) = -(1/\pi)\mathrm{Im}G^r_{00,\sigma}(E)$, the corresponding occupation can be expressed as

$$\langle n_{0\sigma} \rangle = \int_{-\infty}^{\mu} dE \, \rho_{0\sigma}(E) = -\frac{1}{\pi} \int_{-\infty}^{\mu} dE \, \frac{1}{E - \epsilon_0 + i\Gamma - \Sigma^r_{00,\sigma}(E)}. \tag{6.81}$$

We can now use the relation

$$\frac{1}{E - \epsilon_0 + i\Gamma - \Sigma^r_{00,\sigma}(E)} = \frac{\partial}{\partial E} \ln \left[E - \epsilon_0 + i\Gamma - \Sigma^r_{00,\sigma}(E) \right] +$$

$$\frac{\partial \Sigma^r_{00,\sigma}(E)/\partial E}{E - \epsilon_0 + i\Gamma - \Sigma^r_{00,\sigma}(E)} \tag{6.82}$$

together with the Ward identity (see Exercise 6.6)

$$\int_{-\infty}^{\mu} dE \, G^r_{00,\sigma}(E) \frac{\partial \Sigma^r_{00,\sigma}(E)}{\partial E} = 0, \tag{6.83}$$

to write the occupation as

$$\langle n_{0\sigma} \rangle = -\frac{1}{\pi} \mathrm{Im} \int_{-\infty}^{\mu} dE \, \frac{\partial}{\partial E} \ln \left[E - \epsilon_0 + i\Gamma - \Sigma^r_{00,\sigma}(E) \right]. \tag{6.84}$$

Integrating this expression we arrive at

$$\langle n_{0\sigma} \rangle = \frac{1}{2} - \frac{1}{\pi} \tan^{-1} \left[\frac{\epsilon_0 - \mu - \mathrm{Re}\Sigma^r_{00,\sigma}(\mu)}{\Gamma} \right]. \tag{6.85}$$

Here, we have used the fact that in a Fermi liquid $\mathrm{Im}\Sigma^r_{00,\sigma}(\mu) = 0$, which physically means that the quasiparticles have an infinite lifetime at the Fermi energy.

Thus, we can write the local density of states as

$$\rho_{0\sigma}(E) = \frac{1}{\pi} \frac{\Gamma + \mathrm{Im}\Sigma^r_{00,\sigma}(E)}{\left[E - \epsilon_0 - \mathrm{Re}\Sigma^r_{00,\sigma}(\mu) \right]^2 + \left[\Gamma + \mathrm{Im}\Sigma^r_{00,\sigma}(E) \right]^2}. \tag{6.86}$$

Using Eq. (6.85), we can relate the exact density of states at the Fermi energy with the occupation of the level as follows

$$\rho_{0\sigma}(\mu) = \frac{1}{\pi\Gamma} \sin^2 \left[\pi \langle n_{0\sigma} \rangle \right], \tag{6.87}$$

which is known as *Friedel sum rule*. In a case with electron-hole symmetry and $\langle n_{0\sigma} \rangle = 1/2$, the previous expression reduces to

$$\rho_{0\sigma}(\mu) = \frac{1}{\pi\Gamma}. \tag{6.88}$$

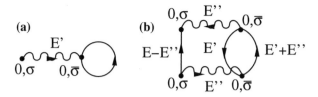

Fig. 6.20 First (a) and second (b) order self-energy diagrams in the Anderson model.

Notice that this equation implies that in the symmetric case, the density of states at the Fermi energy coincides with the corresponding one in the unperturbed problem, i.e. $\rho_{0\sigma}(\mu) = \rho_{0\sigma}^{(0)}(\mu)$.

Friedel sum rule implies the appearance of a narrow peak in the density of states in the limit $U/\Gamma \to 0$. Let us discuss how this comes about. In section 5.4.3 we saw that the level Green's function in the limit $U/\Gamma \to 0$ (atomic limit) is given by Eq. (5.112). This equation suggests that when $U \gg \Gamma$, the density of states consists mainly of two subbands (of width $\sim \Gamma$) around ϵ_0 and $\epsilon_0 + U$, which have most of the total spectral weight. However, Eq. (6.88) tells us that there is a finite density at the Fermi energy. Therefore, the exact density of states must exhibit a narrow peak at the Fermi energy, known as *Kondo peak* or *Kondo resonance*, the width of which tends to zero in the limit $U/\Gamma \to 0$. Indeed, it can be shown that this weight decays exponentially in this limit.

6.9.2 *Perturbative analysis*

We now want to calculate the properties of the system via a perturbative expansion of the Green's functions. For this purpose, we need an approximation for the self-energy, which can be obtained from the lowest-order diagrams. Expanding up to second order in U, one finds only two self-energy diagrams that give a finite contribution, namely those depicted in Fig. 6.20. The first-order diagram, see Fig. 6.20(a), is the Hartree diagram and it yields the following contribution

$$\Sigma_{00,\sigma}^{(1)}(E) = U \int_{-\infty}^{\infty} \frac{dE'}{2\pi} G_{00,\sigma}^{(0)}(E') e^{iE'\eta} = U \langle n_{0\bar{\sigma}} \rangle. \qquad (6.89)$$

The standard Hartree approximation requires to determine the occupation $\langle n_{0\bar{\sigma}} \rangle$ in a self-consistent manner, i.e. by dressing the Green's function line in the Hartree diagram.

The level Green's function can then be written within this approximation as

$$G_{00,\sigma}(E) = \frac{1}{E - \epsilon_0 + i\Gamma \text{sgn}(E) - U\langle n_{0\bar{\sigma}}\rangle}, \qquad (6.90)$$

where we have set $\mu = 0$. Notice that the role of the interaction is to shift the position of the resonant level, which moves to $\epsilon_0 + U\langle n_{0\bar{\sigma}}\rangle$. In the special case in which $\epsilon_0 = -U/2$, known as the symmetric case, the self-consistent solution, assuming that there is no magnetic solution, is $\langle n_{0\sigma}\rangle = \langle n_{0\bar{\sigma}}\rangle = 1/2$. The problem exhibits in this case electron-hole symmetry around $\mu = 0$ and the density of states is still described by a Lorentzian of width Γ.

Let us now analyze the contribution of the second-order diagram, see Fig. 6.20(b). Such contribution is given by

$$\Sigma_{00,\sigma}^{(2)}(E) = U^2 \int_{-\infty}^{\infty} \frac{dE''}{2\pi} \int_{-\infty}^{\infty} \frac{dE'}{2\pi} \, G_{00\sigma}^{(0)}(E - E'')G_{00\sigma}^{(0)}(E')G_{00\bar{\sigma}}^{(0)}(E' + E'').$$

$$(6.91)$$

This expression is not easy to evaluate, but the main features of this self-energy can be reproduced in a simple analytical calculation in which one assumes a constant density of states for the unperturbed problem (see Exercise 6.7).

If in the diagram of Fig. 6.20(b) the Green's function line is dressed with the Hartree diagram and one considers the symmetric case ($\epsilon_0 = -U/2$), the second-order approximation preserves the electron-hole symmetry around $\mu = 0$ and one has $\langle n_{0\sigma}\rangle = \langle n_{0\sigma}^{(0)}\rangle$. Moreover, in this case one can show that $\text{Re}\Sigma_{00,\sigma}^{(2)}(\mu) = \text{Im}\Sigma_{00,\sigma}^{(2)}(\mu) = 0$. This implies that $\rho_{0\sigma} = \rho_{0\sigma}^{(0)}$ and therefore the Friedel sum rule is satisfied. This is one of the reasons why this second-order approximation gives an excellent description in the symmetric case, even if U is not too small in comparison with Γ.

In order to illustrate the effect of the electron-electron interaction in the density of states, we have computed it numerically in the symmetric case using the second-order self-energy of Eq. (6.91). The results for different values of the ratio U/Γ are shown in Fig. 6.21.[8] As one can see, as the U/Γ increases, the density of states exhibits two subbands around ϵ_0 and $\epsilon_0 + U$ and a narrow peak at the Fermi energy (the Kondo peak). Notice that the height of this peak remains constant and it is equal to $1/(\pi\Gamma)$, as in the case without electron-electron interaction. The appearance of this peak at

---

[8]In this figure we explore cases in which U is considerably larger than Γ, which in principle should be out of the scope of this second-order approximation. However, as stated above, this approximation works nicely in the symmetric case and it reproduces the main features of the exact solution [651].

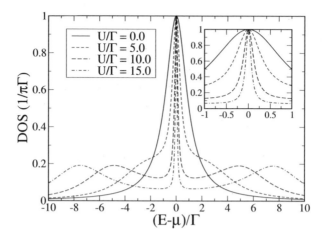

Fig. 6.21 Density of states projected onto the localized level as a function of the energy in the Anderson model for $\epsilon_0 = -U/2$ and different values of the ratio U/Γ. The calculation has been done including the self-energy diagrams up to second order. The inset shows a blow-up of the energy region close to the Fermi energy.

the Fermi energy has very important consequences for the low-temperature transport properties of molecular junctions. This will be discussed in detail in section 15.6.2.

6.10   Final remarks

In this chapter we have presented a systematic perturbative approach to compute zero-temperature Green's functions of an electronic system. The next natural step in most textbooks is to discuss the generalization of this approach to finite temperatures. However, we shall skip this extension and jump in the next chapter to the nonequilibrium formalism in which the temperature will enter in a natural manner. Anyway, the reader is now in position to study the finite-temperature formalism, which can be found in different textbooks, see e.g. Refs. [173, 174, 182, 185].

It is worth stressing that in this chapter we have focused on the description of electronic systems, but a similar perturbative approach can be extended to other types of systems. For instance, in nanoscale junctions phonons or local vibrations play an important role both in the electronic and thermal transport properties. In this sense, it is interesting to learn how the diagrammatic formalism described in this chapter can be applied to phonons and other bosonic degrees of freedom. This subject will not be address in this monograph and for those readers interested in this topic we recommend Refs. [173, 174, 182, 185].

Finally, we would like to emphasize that at this stage the reader is ready to study many important topics in solid state physics which are out of the scope of this book. For instance, the formalism detailed in this chapter is the starting point to understand the Fermi liquid theory, which is very important to get a deeper insight into the physics of metals. The reader is now also prepared to study the physics of the homogeneous electron gas, which is a model system where one can learn many important lessons related to the relevance of electronic correlations. Again, Refs. [173, 174, 182, 185] are very recommendable for studying these topics.

6.11 Exercises

6.1 Wick's theorem I: Let us consider the two-sites tight-binding Hamiltonian of Exercise 5.1(b). Compute the ground state wave function, $|\phi_0\rangle$, for the case in which there are 2 electrons in the system. Then, show that the following relations hold:

$$\langle\phi_0|\mathbf{n}_{1\uparrow}\mathbf{n}_{1\downarrow}|\phi_0\rangle = \langle\phi_0|\mathbf{n}_{1\uparrow}|\phi_0\rangle\langle\phi_0|\mathbf{n}_{1\downarrow}|\phi_0\rangle$$

$$\langle\phi_0|\mathbf{n}_{2\uparrow}\mathbf{n}_{2\downarrow}|\phi_0\rangle = \langle\phi_0|\mathbf{n}_{2\uparrow}|\phi_0\rangle\langle\phi_0|\mathbf{n}_{2\downarrow}|\phi_0\rangle.$$

6.2 Wick's theorem II: Starting from the results of Exercise 5.1(b) about the time evolution of the creation and annihilation operators of the two-sites system, show without applying Wick's theorem that

$$\langle\phi_0|\mathbf{T}\left[\mathbf{c}_{1\sigma}(t)\mathbf{c}^\dagger_{2\bar{\sigma}}(t)\mathbf{c}^\dagger_{2\sigma}(t')\mathbf{c}_{1\bar{\sigma}}(t')\right]|\phi_0\rangle = -G^{(0)}_{12\sigma}(t-t')G^{(0)}_{12\bar{\sigma}}(t'-t),$$

which is the result that one obtains using Wick's theorem.

6.3 Cancellation of the disconnected diagrams: Compute the denominator of the Green's function, $\langle\phi_0|\mathbf{S}|\phi_0\rangle$ up to first order for the electron-electron interaction and show that it exactly cancels the contribution of the disconnected diagrams that appear in the numerator of the Green's function (see Fig. 6.2). Hint: Show that $\langle\phi_0|\mathbf{S}|\phi_0\rangle$ has the following diagrammatic expansion up to first order:

Fig. 6.22   Diagrammatic expansion of the denominator of the Green's function up to first order in the electron-electron interaction.

6.4 Feynman diagrams for the electron-electron interaction: Let us consider a system of interacting electrons with the electron-electron interaction as a perturbation. Use Wick's theorem to compute the different contributions of the 10 second-order topologically distinct diagrams. Check that the rules presented in section 6.5.1 reproduce these results.

6.5 Hartree-fock approximation for the homogeneous electron gas: Derive the expression for the exchange potential of an interacting electron gas and demonstrate that the energy dispersion relation in this case is equal to

$$\epsilon_{\mathbf{k},HF} = \frac{\hbar k^2}{2m} - \frac{2e^2 k_F}{\pi} \left[\frac{1}{2} - \frac{1-k_0}{4k_0} \ln \left| \frac{1+k_0}{1-k_0} \right| \right],$$

where $k_0 \equiv k/k_F$. Show also that the derivative of the dispersion relation exhibits a logarithmic divergence at $k = k_F$.

6.6 Ward identity: Demonstrate the Ward identity of Eq. (6.83).

6.7 Density of states and Kondo resonance in the Anderson model: Compute the second-order contribution to the retarded self-energy in the Anderson model, see Eq. (6.91), in the symmetric $\epsilon_0 = -U/2$ by assuming that the unperturbed density of states adopts the form

$$\rho_{0\sigma}^{(0)}(E) = \begin{cases} 1/W, & -W/2 < E < W/2 \\ 0, & |E| > W/2 \end{cases}$$

where W is a constant. Use this result to plot the density of states in the level as a function of energy for different values of the ratio U/Γ. Hint: Use first the spectral representation to write the unperturbed Green's function appearing in Eq. (6.91) in terms of the density of states $\rho_{0\sigma}^{(0)}$.

Chapter 7

Nonequilibrium Green's functions formalism

So far we have shown how the Green's function techniques can help us to understand the physics of systems in equilibrium. Since our goal is the analysis of the transport properties of different nanocontacts, we have to generalize those techniques to deal with situations in which the systems are driven out of equilibrium. This is precisely the goal of this chapter in which we shall discuss the so-called *nonequilibrium Green's function formalism* (NEGF). This formalism was developed independently by Kadanoff and Baym [186] and Keldysh [187] in the early 1960's. Here we shall follow Keldysh formulation of this approach and we shall refer to it as the *Keldysh formalism*. This formalism is a natural extension of the diagrammatic theory that we have presented in the previous chapter. The importance of the Keldysh formalism lies in the fact that it allows us to go beyond the usual linear response in a systematic manner. Since its appearance, it has been used in a great variety of topics (see Refs. [188, 189] and references therein). In particular, it has been applied to the study of electronic transport in many types of nanoscale devices and it constitutes a basic tool that will be used throughout the rest of the book.

Apart from the original paper [187], there exist a number of excellent reviews devoted to the Keldysh formalism in the literature [188–191]. We try to explain it here in a didactic manner, concentrating ourselves on its application to the problems of molecular electronics that we have in mind, rather than entering into very technical discussions about its foundation. Bearing this in mind, we have organized this chapter as follows. We first present the general ideas of the Keldysh formalism. Then, we shall briefly discuss how to perform the diagrammatic expansion within this formalism. We shall finish the formal discussion by reviewing both the main properties of the functions appearing in this nonequilibrium formalism and the main practical equations. Finally, the last part of this chapter is devoted to the application of the Keldysh formalism to some simple transport problems.

7.1 The Keldysh formalism

In an out-of-equilibrium situation the perturbative approach detailed in the previous chapter is not applicable. However, its generalization to nonequilibrium situations is straightforward. Let us consider an electron system that is described by the following Hamiltonian

$$\mathbf{H} = \mathbf{H}_0 + \mathbf{V}(t), \tag{7.1}$$

where $\mathbf{H}_0$ is a noninteracting Hamiltonian and $\mathbf{V}(t)$ is a time-dependent perturbation that can contain external potentials and interaction terms.

As in the equilibrium case, we are interested in the calculation of expectation values of operators like the following one

$$\langle \mathbf{A} \rangle = \frac{\langle \Psi_H | \mathbf{A}_H(t) | \Psi_H \rangle}{\langle \Psi_H | \Psi_H \rangle}, \tag{7.2}$$

where, for the sake of clarity, we consider the expectation value of a single operator rather than the usual product of two of them.

We now change to the interaction picture, where this expectation value becomes

$$\langle \mathbf{A} \rangle = \frac{\langle \Psi_I | \mathbf{A}_I(t) | \Psi_I \rangle}{\langle \Psi_I | \Psi_I \rangle}. \tag{7.3}$$

Although the perturbation in this case may depend on time, one can still assume that the interaction is adiabatically switched on and off at $t = -\infty$ and $t = \infty$, respectively. As usual, this can be done by the replacement $\mathbf{V}(t) \rightarrow \exp(-\epsilon|t|)\mathbf{V}(t)$, where ϵ is an infinitesimally small positive parameter. In the equilibrium case, the time symmetry is preserved and at time $t = \infty$ we recover the same noninteracting state $|\phi_0\rangle$ that we had at $t = -\infty$ (apart from a phase factor). However, out of equilibrium this symmetry is in general broken and the starting point for the perturbative expansion must be the following one

$$\langle \mathbf{A} \rangle = \frac{\langle \phi_0 | \mathbf{S}(-\infty, t) \mathbf{A}_I(t) \mathbf{S}(t, -\infty) | \phi_0 \rangle}{\langle \phi_0 | \mathbf{S}(-\infty, t) \mathbf{S}(t, -\infty) | \phi_0 \rangle}. \tag{7.4}$$

At a first glance, one might think that now the perturbative expansion becomes very cumbersome because we cannot group all the pieces of the time-evolution operator into a single one. Keldysh showed that one can

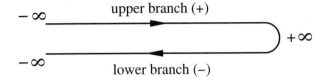

Fig. 7.1 The Keldysh contour.

still order the time arguments along a modified time contour. This contour is referred to as *the Keldysh contour* and it is depicted in Fig. 7.1.

On this contour, the time runs from $-\infty$ to $+\infty$ in the upper branch, whereas it does it backwards in the lower one, i.e. from $+\infty$ to $-\infty$. In order to indicate in which branch the time arguments lie, we introduce a subindex that will be equal to $+$ for the upper branch and $-$ for the lower one. With this notation, we can write now the expectation value of Eq. (7.4) as

$$\langle \mathbf{A} \rangle = \frac{\langle \phi_0 | \mathbf{S}_-(-\infty, \infty) \mathbf{S}_+(\infty, t) \mathbf{A}_\mathrm{I}(t) \mathbf{S}_+(t, -\infty) | \phi_0 \rangle}{\langle \phi_0 | \mathbf{S}_-(-\infty, \infty) \mathbf{S}_+(\infty, t) \mathbf{S}_+(t, -\infty) | \phi_0 \rangle}, \qquad (7.5)$$

if t lies in the upper branch or

$$\langle \mathbf{A} \rangle = \frac{\langle \phi_0 | \mathbf{S}_-(-\infty, t) \mathbf{A}_\mathrm{I}(t) \mathbf{S}_-(t, \infty) \mathbf{S}_+(\infty, -\infty) | \phi_0 \rangle}{\langle \phi_0 | \mathbf{S}_-(-\infty, t) \mathbf{S}_-(t, \infty) \mathbf{S}_+(\infty, -\infty) | \phi_0 \rangle}, \qquad (7.6)$$

if t lies in the lower one. Defining the operator $\mathbf{T}_c$ that orders the time arguments along the Keldysh contour, we can rewrite the expectation value as

$$\langle \mathbf{A} \rangle = \frac{\langle \phi_0 | \mathbf{T}_c \left[\mathbf{A}_\mathrm{I}(t) \mathbf{S}_-(-\infty, \infty) \mathbf{S}_+(\infty, -\infty) \right] | \phi_0 \rangle}{\langle \phi_0 | \mathbf{S}_-(-\infty, \infty) \mathbf{S}_+(\infty, -\infty) | \phi_0 \rangle}. \qquad (7.7)$$

This expression can be in turn rewritten in a more familiar way by defining the operator that describes the time-evolution along the Keldysh contour

$$\mathbf{S}_c(\infty, -\infty) \equiv \mathbf{S}_-(-\infty, \infty) \mathbf{S}_+(\infty, -\infty). \qquad (7.8)$$

With this definition we can finally write the expectation value $\langle \mathbf{A} \rangle$ as follows

$$\langle \mathbf{A} \rangle = \frac{\langle \phi_0 | \mathbf{T}_c \left[\mathbf{A}_\mathrm{I}(t) \mathbf{S}_c(\infty, -\infty) \right] | \phi_0 \rangle}{\langle \phi_0 | \mathbf{S}_c(\infty, -\infty) | \phi_0 \rangle}. \qquad (7.9)$$

Analogously, one can express the expectation value of any operator product.

The expectation value of Eq. (7.9) has formally the same structure as in an equilibrium situation. The main difference is the fact that one has to keep track of the branch in which the time arguments lie (t_+ and t_-).

This implies that when defining the propagators in this formalism, there are four different possibilities depending on the two time arguments. These definitions are analogous to those of the causal function in the equilibrium formalism

$$G_{ij}(t_\alpha, t'_\beta) = -i \frac{\langle \Psi_H | \mathbf{T}_c \left[\mathbf{c}_{i\sigma}(t_\alpha) \mathbf{c}^\dagger_{j\sigma}(t'_\beta) \right] | \Psi_H \rangle}{\langle \Psi_H | \Psi_H \rangle} \qquad (7.10)$$

$$G(\mathbf{r}t_\alpha, \mathbf{r}'t'_\beta) = -i \frac{\langle \Psi_H | \mathbf{T}_c \left[\mathbf{\Psi}_\sigma(\mathbf{r}t_\alpha) \mathbf{\Psi}^\dagger_\sigma(\mathbf{r}'t'_\beta) \right] | \Psi_H \rangle}{\langle \Psi_H | \Psi_H \rangle}, \qquad (7.11)$$

depending on whether we use the representation $|i\rangle$ or $|\mathbf{r}\rangle$. The subindexes α and β take the values $+$ and $-$ and indicate in which branch the time arguments lie. Let us now discuss in detail the expression for the four possible functions:

(1) $t = t_+$ and $t' = t'_+$:

In this case both time arguments lie in the upper branch and the corresponding Green's function reads (for a discrete representation)

$$G^{++}_{ij}(t, t') = -i \langle \mathbf{T} \left[\mathbf{c}_{i\sigma}(t) \mathbf{c}^\dagger_{j\sigma}(t') \right] \rangle, \qquad (7.12)$$

where, from now on, the subindexes $\alpha, \beta = +, -$ will appear as superindexes of the Green's functions. Moreover, in order to simplify the notation, we shall drop the wave functions in the expectation values and we shall not include the denominator $\langle \Psi_H | \Psi_H \rangle$, which indeed turns out to be equal to 1 (see discussion below). Notice that this function is nothing else but the causal Green's function.

(2) $t = t_+$ and $t' = t'_-$:

In this case, since any time in the lower branch of the Keldysh contour is "larger" than any time in the upper branch, one has

$$G^{+-}_{ij}(t, t') = i \langle \mathbf{c}^\dagger_{j\sigma}(t') \mathbf{c}_{i\sigma}(t) \rangle. \qquad (7.13)$$

This function plays a fundamental role in the nonequilibrium Green's functions theory and, as we shall see later, it contains information about the distribution function of the electrons.

(3) $t = t_-$ and $t' = t'_+$:

In this case we have

$$G^{-+}_{ij}(t, t') = -i \langle \mathbf{c}_{i\sigma}(t) \mathbf{c}^\dagger_{j\sigma}(t') \rangle. \qquad (7.14)$$

This function contains essentially the same information as $G^{+-}_{ij}(t, t')$.

(4) $t = t_-$ and $t' = t'_-$

In this last possibility, both time arguments lie in the lower branch, where the arguments are ordered in an antichronological way. Therefore, this new function reads

$$G_{ij}^{--}(t,t') = -i\langle \bar{\mathbf{T}}\left[\mathbf{c}_{i\sigma}(t)\mathbf{c}_{j\sigma}^{\dagger}(t')\right]\rangle, \qquad (7.15)$$

where the operator $\bar{\mathbf{T}}$ orders the time arguments in the opposite way as compared with the usual time-ordering operator $\mathbf{T}$, i.e. in a antichronological order.

The four Green's functions defined above can be grouped in a matrix as follows

$$\check{\mathbf{G}} = \begin{pmatrix} \mathbf{G}^{++} & \mathbf{G}^{+-} \\ \mathbf{G}^{-+} & \mathbf{G}^{--} \end{pmatrix}, \qquad (7.16)$$

where the check symbol ($\check{\ }$) indicates that we are dealing with a 2×2 matrix in *Keldysh space*. The perturbative expansion couples the different components of this matrix, which effectively leads to an enlargement of the propagator space in a factor of 2. This enlargement is indeed quite natural since in an out-of-equilibrium situation we have to determine not only the states, the information of which is contained in the causal function, but also the distribution function that describes how such states are occupied. This latter information is provided by the off-diagonal functions in Eq. (7.16).

Formally speaking, the perturbative expansion is very similar to the equilibrium one, and one has only to keep track of the matrix structure. A additional complication is that in time-dependent problems, the products are replaced by convolutions over intermediate arguments, which makes the calculations considerably more complicated. Fortunately, transport problems often admit a stationary solution and then, the application of the nonequilibrium formalism is not more complicated than the equilibrium one.

As stated above, apart from the matrix structure introduced by the Keldysh formalism, the rest of the perturbative approach is very similar to the equilibrium one. To derive the perturbative expansion of the matrix propagator of Eq. (7.16), one can use the expression of Eq. (7.9) and expand the operator $\mathbf{S}_c$. Let us recall that $\mathbf{S}_c(\infty, -\infty) \equiv \mathbf{S}_-(-\infty, \infty)\mathbf{S}_+(\infty, -\infty)$ and the perturbative expansions of both time-evolution operators are given by

$$\mathbf{S}_+(\infty, -\infty) = \sum_{n=0}^{\infty} \frac{(-i)^n}{n!} \int_{-\infty}^{\infty} dt_1 \cdots \int_{-\infty}^{\infty} dt_n \, \mathbf{T} \left[\mathbf{V}_\mathrm{I}(t_1) \cdots \mathbf{V}_\mathrm{I}(t_n) \right] \quad (7.17)$$

$$\mathbf{S}_-(-\infty, \infty) = \sum_{n=0}^{\infty} \frac{(-i)^n}{n!} \int_{\infty}^{-\infty} dt_1 \cdots \int_{\infty}^{-\infty} dt_n \, \bar{\mathbf{T}} \left[\mathbf{V}_\mathrm{I}(t_1) \cdots \mathbf{V}_\mathrm{I}(t_n) \right].$$

After expanding the operators $\mathbf{S}_+$ and $\mathbf{S}_-$, one applies the Wick's theorem in the standard way. Therefore, the resulting diagrammatic structure is analogous to the one in equilibrium, the main difference being the enlargement of the space that is encoded in the indexes α and β. We shall discuss the peculiarities of the nonequilibrium diagrammatic expansion in the next section.

Finally, since the structure of the diagrammatic expansion is identical to the equilibrium one, such an expansion can be also summarized in a Dyson's equation, which in the nonequilibrium case has the following matrix form

$$\check{\mathbf{G}}(t, t') = \check{\mathbf{g}}(t, t') + \int dt_1 \int dt_2 \, \check{\mathbf{g}}(t, t_1) \check{\mathbf{\Sigma}}(t_1, t_2) \check{\mathbf{G}}(t_2, t'). \quad (7.18)$$

Here, we have denoted the unperturbed propagators by $\check{\mathbf{g}}$ instead of $\check{\mathbf{G}}^{(0)}$ to simplify the notation. Here, the self-energy has a 2×2 matrix structure in Keldysh space analogous to Eq. (7.16). In general, the functions appearing in Eq. (7.18) depend on two time arguments and the Dyson's equation is an integral equation. However, in many stationary situations, both the propagators and the self-energies depend on the time difference and, after Fourier transforming, Eq. (7.18) recovers its standard equilibrium form of an algebraic equation with the frequency as the argument, i.e.

$$\check{\mathbf{G}}(E) = \check{\mathbf{g}}(E) + \check{\mathbf{g}}(E) \check{\mathbf{\Sigma}}(E) \check{\mathbf{G}}(E). \quad (7.19)$$

7.2   Diagrammatic expansion in the Keldysh formalism

Let us discuss now some of the peculiarities of the diagrammatic expansion in the Keldysh formalism. One of them is the fact that in this formalism the denominator of the Green's functions does not play any role (indeed $\langle \phi_0 | \mathbf{S}_c | \phi_0 \rangle = 1$, see Exercise 7.1). One can show that in the expansion of $\mathbf{S}_c$ the terms of order higher than zero cancel each other order by order. One might think that this fact creates a problem related to the cancellation of the disconnected diagrams. However, this is not the case because, as it is easy to show by applying Wick's theorem, these diagrams also cancel each other. Therefore, as in equilibrium, one needs to consider the topologically distinct diagrams only once.

Let us discuss the diagrammatic structure in two situations of interest:

- **Case 1: Time-dependent external potential.**

  Let us consider a system with N noninteracting electrons subjected to an external potential that can be time-dependent. The Hamiltonian in first quantization reads in this case

  $$\mathbf{H} = \mathbf{H}_0 + \mathbf{V}(t), \qquad (7.20)$$

  where

  $$\mathbf{V}(t) = \sum_{i=1}^{N} V(\mathbf{r}_i, t). \qquad (7.21)$$

  The diagrams in this case are trivial because, as in the case of a static potential, they consist of the repetition of identical scattering events. The matrix self-energy is therfore given by (see Exercise 7.2)

  $$\check{\Sigma}(\mathbf{r}, t) = \begin{pmatrix} V(\mathbf{r}, t) & 0 \\ 0 & -V(\mathbf{r}, t) \end{pmatrix}. \qquad (7.22)$$

  It is interesting to note that for this single-electron perturbation the components Σ^{+-} and Σ^{-+} vanish. The existence of off-diagonals components of the self-energies in the Keldysh space is only possible in the case of inelastic mechanisms such as electron-electron interaction or electron-phonon interaction (see next case).

- **Case 2: Electron-electron interaction.**

  Let us consider an electronic system where the electron-electron interaction is assumed to be the perturbation. The system might be out of equilibrium due to, for instance, the presence of a current. For the sake of concreteness, let us assume that the unperturbed system can be described by a tight-binding Hamiltonian and the interaction is Hubbard-like (see Appendix A)

  $$\mathbf{H} = \mathbf{H}_0 + \sum_i U \mathbf{n}_{i\uparrow} \mathbf{n}_{i\downarrow}. \qquad (7.23)$$

  The diagrams are topologically identical to the equilibrium ones and the only difference is the fact that one has to indicate where the time arguments reside on the Keldysh contour. In this respect, every equilibrium diagram gives rise to several diagrams for the different components of the self-energy in Keldysh space. We illustrate this fact in Fig. 7.2,

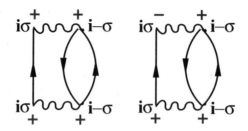

Fig. 7.2   Examples of second-order self-energy diagrams in the Keldysh space for the electron-electron interaction. The indexes $+$ and $-$ indicate in which branch the time arguments lie.

where we show the self-energy diagrams of second order in U for the components Σ^{++} and Σ^{+-}. The expression of the self-energy Σ_{ii}^{+-}, for instance, would be (ignoring the spin dependence)

$$\Sigma_{ii}^{+-}(t,t') = U^2 \left[g_{ii}^{+-}(t,t') \right]^2 g_{ii}^{-+}(t',t). \qquad (7.24)$$

7.3   Basic relations and equations in the Keldysh formalism

In the previous section we have seen that the Dyson's equation has acquired an additional 2×2 matrix structure, which gives the impression that one has to solve four times more equations than in the equilibrium case. Indeed, one can show that the different functions in the 2×2 matrix of Eq. (7.16) are not independent and the number of equations that one has to solve in practice can be reduced to only two. In this sense, the goal of this section is to derive those equations and to discuss the general properties of the Keldysh-Green's functions.

7.3.1   *Relations between the Green's functions*

Let us explore the different relations between the functions appearing in the Keldysh formalism. We start by showing that the four Green's functions G^{++}, G^{+-}, G^{-+} and G^{--} are not independent, but satisfy

$$\mathbf{G}^{++} + \mathbf{G}^{--} = \mathbf{G}^{+-} + \mathbf{G}^{-+}. \qquad (7.25)$$

This is a direct consequence of the definition of these functions. Thus for instance,

$$\begin{aligned} G_{ij}^{++}(t,t') &= -i\theta(t-t')\langle \mathbf{c}_{i\sigma}(t)\mathbf{c}_{j\sigma}^{\dagger}(t') \rangle + i\theta(t'-t)\langle \mathbf{c}_{j\sigma}^{\dagger}(t')\mathbf{c}_{i\sigma}(t) \rangle \\ &= \theta(t-t')G_{ij}^{-+}(t,t') + \theta(t'-t)G_{ij}^{+-}(t,t'). \end{aligned} \qquad (7.26)$$

Analogously,

$$G_{ij}^{--}(t,t') = \theta(t-t')G_{ij}^{+-}(t,t') + \theta(t'-t)G_{ij}^{-+}(t,t'). \qquad (7.27)$$

Adding these two equations, we obtain the relation stated above.

On the other hand, from this relation and using the Dyson's equation in Keldysh space, see Eq. (7.18), one can show the following relation between the different elements of the self-energy matrix in Keldysh space (see Exercise 7.3)

$$\mathbf{\Sigma}^{++} + \mathbf{\Sigma}^{--} = -\left(\mathbf{\Sigma}^{+-} + \mathbf{\Sigma}^{-+}\right). \qquad (7.28)$$

Other important relations are those between the Keldysh-Green's functions and the advanced and retarded functions G^a and G^r. Such relations can be found as follows. Using the expression of Eq. (7.26), one obtains

$$G_{ij}^{++}(t,t') - G_{ij}^{+-}(t,t') = -\theta(t-t')\left[G_{ij}^{+-}(t,t') - G_{ij}^{-+}(t,t')\right], \qquad (7.29)$$

and using the definitions of G^{+-} and G^{+-}, we arrive at

$$G_{ij}^{++}(t,t') - G_{ij}^{+-}(t,t') = -i\theta(t-t')\langle \mathbf{c}_{i\sigma}(t)\mathbf{c}_{j\sigma}^{\dagger}(t') + \mathbf{c}_{j\sigma}^{\dagger}(t')\mathbf{c}_{i\sigma}(t)\rangle$$
$$= G_{ij}^{r}(t,t') \qquad (7.30)$$

Proceeding in an analogous way, one can show the following relations

$$\mathbf{G}^{r} = \mathbf{G}^{++} - \mathbf{G}^{+-} = \mathbf{G}^{-+} - \mathbf{G}^{--} \qquad (7.31)$$

$$\mathbf{G}^{a} = \mathbf{G}^{++} - \mathbf{G}^{-+} = \mathbf{G}^{+-} - \mathbf{G}^{--}. \qquad (7.32)$$

These relations are crucial for the discussion of next section.

7.3.2 *The triangular representation*

As we have seen above, there are redundancies in the Green's functions and in that sense it is natural to try to get rid of them to simplify the equations as much as possible. In what follows, we shall try to eliminate G^{++} and G^{--} in favor of G^r and G^a. For this purpose, we will apply a unitary transformation to perform the following change

$$\begin{pmatrix} \mathbf{G}^{++} & \mathbf{G}^{+-} \\ \mathbf{G}^{-+} & \mathbf{G}^{--} \end{pmatrix} \longrightarrow \begin{pmatrix} \mathbf{0} & \mathbf{G}^{a} \\ \mathbf{G}^{r} & \mathbf{G}^{K} \end{pmatrix}, \qquad (7.33)$$

where $\mathbf{G}^{K} = \mathbf{G}^{++} + \mathbf{G}^{--} = \mathbf{G}^{+-} + \mathbf{G}^{-+}$ is known as the Keldysh function.

It is easy to show that the unitary transformation has the form

$$\check{\mathbf{R}} = \frac{1}{\sqrt{2}} \begin{pmatrix} 1 & -1 \\ 1 & 1 \end{pmatrix} = \frac{1}{\sqrt{2}}(\check{\mathbf{1}} - i\check{\sigma}_y), \qquad (7.34)$$

where $\check{\sigma}_y$ is the corresponding Pauli matrix. The representation above is known as the *triangular representation* and it is important from the practical point of view. Let us now denote the standard Keldysh matrix by $\check{\mathbf{G}}$ and the corresponding matrix in the triangular representation as $\tilde{\mathbf{G}}$. They are related by $\tilde{\mathbf{G}} = \check{\mathbf{R}}\check{\mathbf{G}}\check{\mathbf{R}}^{-1}$. Applying the transformation $\check{\mathbf{R}}$ to the Dyson's equation[1]

$$\check{\mathbf{G}} = \check{\mathbf{g}} + \check{\mathbf{g}}\check{\mathbf{\Sigma}}\check{\mathbf{G}}, \tag{7.35}$$

we obtain the corresponding Dyson's equation in the triangular representation

$$\tilde{\mathbf{G}} = \tilde{\mathbf{g}} + \tilde{\mathbf{g}}\tilde{\mathbf{\Sigma}}\tilde{\mathbf{G}}, \tag{7.36}$$

where the self-energy in this representation has the form

$$\tilde{\mathbf{\Sigma}} = \begin{pmatrix} \mathbf{\Sigma}^K & \mathbf{\Sigma}^r \\ \mathbf{\Sigma}^a & 0 \end{pmatrix}. \tag{7.37}$$

Here, the new self-energy components are expressed in terms of those of the original representation as follows

$$\mathbf{\Sigma}^K = \mathbf{\Sigma}^{++} + \mathbf{\Sigma}^{--} = -\left(\mathbf{\Sigma}^{+-} + \mathbf{\Sigma}^{-+}\right) \tag{7.38}$$

$$\mathbf{\Sigma}^r = \mathbf{\Sigma}^{++} + \mathbf{\Sigma}^{+-} = -\left(\mathbf{\Sigma}^{--} + \mathbf{\Sigma}^{-+}\right) \tag{7.39}$$

$$\mathbf{\Sigma}^a = \mathbf{\Sigma}^{++} + \mathbf{\Sigma}^{-+} = -\left(\mathbf{\Sigma}^{--} + \mathbf{\Sigma}^{+-}\right). \tag{7.40}$$

From Eqs. (7.36) and (7.37) one can show that the advanced and retarded Green's functions satisfy independent Dyson's equations, i.e.

$$\mathbf{G}^{r,a} = \mathbf{g}^{r,a} + \mathbf{g}^{r,a}\mathbf{\Sigma}^{r,a}\mathbf{G}^{r,a}. \tag{7.41}$$

Notice that this equation is formally identical to the equilibrium one. In the case in which the perturbation is an external potential, as we showed in the previous section, the corresponding self-energies reduce to $\Sigma^a(\mathbf{r}, t) = \Sigma^r(\mathbf{r}, t) = V(\mathbf{r}, t)$, i.e. like in equilibrium.

On the other hand, the Keldysh function $\mathbf{G}^K$ fulfills the following equation

$$\mathbf{G}^K = \mathbf{g}^K + \mathbf{g}^K\mathbf{\Sigma}^a\mathbf{G}^a + \mathbf{g}^r\mathbf{\Sigma}^r\mathbf{G}^K + \mathbf{g}^r\mathbf{\Sigma}^K\mathbf{G}^a. \tag{7.42}$$

Notice now that $\mathbf{G}^K$ is coupled to $\mathbf{G}^{r,a}$ and this equation requires to solve first Dyson's equation for these latter functions. Let us recall that the retarded and advanced functions are related, which in practice means that

---

[1] In this equation, as in the next ones, the integrations over the intermediate arguments are implicitly assumed.

there are only two functions to be determined, as we stated at the beginning of this section.

The previous equation can be written in a more symmetric way as follows. We first group on the left hand side all the terms containing $\mathbf{G}^K$ and then we multiply from the left by $(1 - \mathbf{g}^r \mathbf{\Sigma}^r)^{-1}$ on both sides of the equation to arrive at

$$\mathbf{G}^K = (1 - \mathbf{g}^r \mathbf{\Sigma}^r)^{-1} \mathbf{g}^K \left(1 + \mathbf{\Sigma}^a \mathbf{G}^a\right) + (1 - \mathbf{g}^r \mathbf{\Sigma}^r)^{-1} \mathbf{g}^r \mathbf{\Sigma}^K \mathbf{G}^a. \quad (7.43)$$

Then, using the Dyson's equation for the retarded function, we finally obtain

$$\mathbf{G}^K = (1 + \mathbf{G}^r \mathbf{\Sigma}^r) \mathbf{g}^K \left(1 + \mathbf{\Sigma}^a \mathbf{G}^a\right) + \mathbf{G}^r \mathbf{\Sigma}^K \mathbf{G}^a. \quad (7.44)$$

In this book, we shall mainly use the function $\mathbf{G}^{+-}$, rather than the Keldysh function $\mathbf{G}^K$. For this reason, we now proceed to derive the corresponding equation for $\mathbf{G}^{+-}$. We first take the element $+-$ in the Dyson's equation, i.e.

$$\mathbf{G}^{+-} = \mathbf{g}^{+-} + (\mathbf{g}\mathbf{\Sigma}\mathbf{G})^{+-}. \quad (7.45)$$

Then, we make use of the relations derived above between the different functions to arrive at (see Exercise 7.3)

$$\mathbf{G}^{+-} = \mathbf{g}^{+-} + \mathbf{g}^{+-} \mathbf{\Sigma}^a \mathbf{G}^a + \mathbf{g}^r \mathbf{\Sigma}^r \mathbf{G}^{+-} - \mathbf{g}^r \mathbf{\Sigma}^{+-} \mathbf{G}^a. \quad (7.46)$$

The function G^{-+} fulfills a similar equation that can be obtained from the previous one by exchanging $+$ by $-$ and vice versa. Eq. (7.46) for $\mathbf{G}^{+-}$ can be written in a more symmetric way, in analogy with what we did for the function $\mathbf{G}^K$. Thus, we obtain finally

$$\mathbf{G}^{+-} = (1 + \mathbf{G}^r \mathbf{\Sigma}^r) \mathbf{g}^{+-} \left(1 + \mathbf{\Sigma}^a \mathbf{G}^a\right) - \mathbf{G}^r \mathbf{\Sigma}^{+-} \mathbf{G}^a. \quad (7.47)$$

The function $\mathbf{G}^{-+}$ satisfies a similar equation given by

$$\mathbf{G}^{-+} = (1 + \mathbf{G}^r \mathbf{\Sigma}^r) \mathbf{g}^{-+} \left(1 + \mathbf{\Sigma}^a \mathbf{G}^a\right) - \mathbf{G}^r \mathbf{\Sigma}^{-+} \mathbf{G}^a. \quad (7.48)$$

7.3.3 *Unperturbed Keldysh-Green's functions*

In the Keldysh formalism the time dependence is introduced through the perturbation and the unperturbed Hamiltonian $\mathbf{H}_0$ must correspond to a noninteracting electron system in equilibrium. Thus, all unperturbed Green's functions depend only on the time difference and they are easy to obtain in energy space. The form and properties of the unperturbed retarded, advanced and causal functions in energy space were studied in detail in Chapter 5, whereas the properties of the functions $\mathbf{g}^{--}(E)$ can be easily deduced from those of $\mathbf{g}^{++}(E)$. Thus, we concentrate now on the analysis of the functions $\mathbf{g}^{+-}(E)$ and $\mathbf{g}^{-+}(E)$. From its definition in the time domain (and in a discrete basis)

$$G_{ij}^{+-}(t) = i \langle \mathbf{c}_{j\sigma}^{\dagger}(0) \mathbf{c}_{i\sigma}(t) \rangle, \tag{7.49}$$

it is obvious that this function is related to the electron distribution in equilibrium. Although the temperature does not appear explicitly in the Keldysh formalism, one uses the previous fact to introduce it. Thus, the previous expression for $t = 0$ and $i = j$ reads

$$G_{ii}^{+-}(0) = i \langle \mathbf{n}_{i\sigma} \rangle = \int_{-\infty}^{\infty} \frac{dE}{2\pi} G_{ii}^{+-}(E). \tag{7.50}$$

This implies that $G_{ii}^{+-}(E) = 2\pi i \rho_i(E) f(E)$, where $f(E)$ is the Fermi function and $\rho_i(E)$ is the local density of states in the site i. In the same way, one can show that $G_{ii}^{-+}(E) = -2\pi i \rho_i(E)[1 - f(E)]$. Taking into account this result, it is clear that $G^{+-} \propto f(E)$ and $G^{-+} \propto 1 - f(E)$. This fact together with the general relation

$$\mathbf{G}^a(t) - \mathbf{G}^r(t) = \mathbf{G}^{+-}(t) - \mathbf{G}^{-+}(t), \tag{7.51}$$

leads to the following relations

$$\mathbf{G}^{+-}(E) = [\mathbf{G}^a(E) - \mathbf{G}^r(E)] f(E) \tag{7.52}$$

$$\mathbf{G}^{-+}(E) = -[\mathbf{G}^a(E) - \mathbf{G}^r(E)] [1 - f(E)]. \tag{7.53}$$

It is worth stressing that we have written the previous expressions using capital letters to indicate that these expressions are always valid in equilibrium, even in an interacting case. In the Keldysh formalism the unperturbed system is moreover non-interacting, which implies that in a basis $|i\rangle$ one has

$$g_{ij}^{+-}(E) = [g_{ij}^a(E) - g_{ij}^r(E)] f(E) \tag{7.54}$$

$$g_{ij}^{-+}(E) = -[g_{ij}^a(E) - g_{ij}^r(E)] [1 - f(E)].$$

As a consequence, these functions are proportional to the spectral densities and to the thermal distribution function. The way in which we have introduced the temperature in the Keldysh formalism is certainly not very satisfactory. However, one can show that a rigorous derivation leads exactly to the result that we have just described (see for instance Ref. [192]).

7.3.4 Some comments on the notation

The notation used here for the different Keldysh-Green's functions is not shared by all the authors. In this sense, it is important to devote a few lines to make contact with other texts where the Keldysh formalism is described.

Frequently, the functions G^{+-} and G^{-+} are denoted by $G^<$ and $G^>$, respectively. Sometimes, the Keldysh function G^K is denoted by G^F or simply by F. On the other hand, the triangular representation is often written in a slightly different way. One first defines a new matrix function as $\bar{\mathbf{G}} = \sigma_z \check{\mathbf{G}}$, where σ_z is the Pauli matrix, and then the unitary transformation of Eq. (7.34) is applied. This leads to a 2×2 matrix with the form

$$\begin{pmatrix} \mathbf{G}^r & \mathbf{G}^K \\ \mathbf{0} & \mathbf{G}^a \end{pmatrix}, \tag{7.55}$$

which is often used in the field of superconductivity [193].

7.4 Application of Keldysh formalism to simple transport problems

In this section we shall illustrate the utility of the Keldysh formalism by applying it to the description of the electronic transport in some simple situations of special interest. Our goal is two fold. First, we want to illustrate how this formalism is used in practice and second, we want to show how the elastic transmission can be computed from an atomistic point of view.

Most of the systems that we have in mind (atomic contacts, molecular junctions, etc.) are conveniently described by a tight-binding Hamiltonian of the following form

$$\mathbf{H} = \sum_{i\sigma} \epsilon_i \mathbf{n}_{i\sigma} + \sum_{ij\sigma} t_{ij} \left(\mathbf{c}_{i\sigma}^\dagger \mathbf{c}_{j\sigma} + \mathbf{c}_{j\sigma}^\dagger \mathbf{c}_{i\sigma} \right), \tag{7.56}$$

where we have assumed, without loss of generality, that the hopping elements t_{ij} are real. Our first task is to derive an expression for the electrical current operator in this local basis. For this purpose, we first consider the simple case of a tight-binding chain with only nearest-neighbor hoppings, denoted by t. Such a chain is schematically represented in Fig. 7.3. Let us compute now the current between the sites k and $k + 1$. Without doing any calculation, one can guess that the operator must adopt somehow the following form[2]

$$\mathbf{I} \propto t \sum_{\sigma} \left[\mathbf{c}_{k\sigma}^{\dagger}(t)\mathbf{c}_{k+1\sigma}(t) - \mathbf{c}_{k+1\sigma}^{\dagger}(t)\mathbf{c}_{k\sigma}(t) \right], \qquad (7.57)$$

where the first term in the sum represents the current flowing in one direction and second one corresponds to the current flowing in the opposite one. Let us see if a rigorous calculation confirms our intuition.

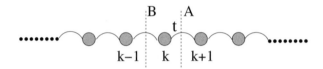

Fig. 7.3   Schematic representation of a linear chain with only nearest-neighbor hoppings.

The current operator must be obtained from the continuity equation that describes the charge conservation. Such equation can be written in a discrete representation as

$$\mathbf{I}_A - \mathbf{I}_B + \frac{\partial \rho_k}{\partial t} = 0, \qquad (7.58)$$

where A represents a point between the sites k and $k+1$ and B a point between $k-1$ and k, see Fig. 7.3. Here, ρ_k is the operator that describes the charge in the site k

$$\rho_k = e \sum_{\sigma} \mathbf{c}_{k\sigma}^{\dagger}\mathbf{c}_{k\sigma} \qquad (7.59)$$

and satisfies the equation of motion of Heisenberg operators

$$\frac{\partial \rho_k}{\partial t} = -\frac{i}{\hbar} \left[\rho_k, \mathbf{H} \right]. \qquad (7.60)$$

Notice that we have reintroduced $\hbar$, and we shall write it explicitly from now on. Using the expression of Eq. (7.56) for the homogeneous chain that we are considering, it is straightforward to compute the commutator that appears in the previous equation of motion and thus, one arrives at

$$\frac{\partial \rho_k}{\partial t} = \frac{-iet}{\hbar} \sum_{\sigma} \left\{ \mathbf{c}_{k\sigma}^{\dagger}\mathbf{c}_{k+1\sigma} - \mathbf{c}_{k+1\sigma}^{\dagger}\mathbf{c}_{k\sigma} + \mathbf{c}_{k\sigma}^{\dagger}\mathbf{c}_{k-1\sigma} - \mathbf{c}_{k-1\sigma}^{\dagger}\mathbf{c}_{k\sigma} \right\}.$$

Rewriting this expression in the form of the continuity equation, see Eq. (7.58), we can identify the current operator, which at point A takes the form

$$\mathbf{I}_A(t) = \frac{iet}{\hbar} \sum_{\sigma} \left\{ \mathbf{c}_{k\sigma}^{\dagger}(t)\mathbf{c}_{k+1\sigma}(t) - \mathbf{c}_{k+1\sigma}^{\dagger}(t)\mathbf{c}_{k\sigma}(t) \right\}. \qquad (7.61)$$

---

[2]We believe that no confusion can arise between the hopping t and the time appearing as an argument in the creation and annihilation operators.

Notice that this has exactly the intuitive form that we had anticipated above.

This expression can be easily generalized to any 3D system described by a tight-binding Hamiltonian as in Eq. (7.56). The electrical current through an arbitrary surface that separates two regions A and B is given by

$$\mathbf{I}(t) = \frac{ie}{\hbar} \sum_{i \in A; j \in B} \sum_{\sigma} t_{ij} \left\{ \mathbf{c}_{i\sigma}^{\dagger}(t)\mathbf{c}_{j\sigma}(t) - \mathbf{c}_{j\sigma}^{\dagger}(t)\mathbf{c}_{i\sigma}(t) \right\}. \tag{7.62}$$

Let us now compute the expectation value of the current operator, for instance, for the case of the chain. According to Eq. (7.61), one can write (dropping the subindex A)

$$\langle \mathbf{I}(t) \rangle = \frac{iet}{\hbar} \sum_{\sigma} \left\{ \langle \mathbf{c}_{k\sigma}^{\dagger}(t)\mathbf{c}_{k+1\sigma}(t) \rangle - \langle \mathbf{c}_{k+1\sigma}^{\dagger}(t)\mathbf{c}_{k\sigma}(t) \rangle \right\}. \tag{7.63}$$

The expectation values appearing in the previous equation can be expressed in terms of the Keldysh functions G^{+-} as follows

$$\langle \mathbf{I}(t) \rangle = \frac{e}{\hbar} t \sum_{\sigma} \left\{ G_{k+1,k}^{+-}(t,t) - G_{k,k+1}^{+-}(t,t) \right\}, \tag{7.64}$$

and there is a similar expression for the most general case of Eq. (7.62). In many situations, for instance when there is a constant voltage applied in a junction, the problem admits a stationary solution and the Green's functions depend exclusively on the difference of the time arguments. In those cases, Eq. (7.64) can be written in terms of the Green's functions in energy space as

$$\langle \mathbf{I} \rangle = \frac{e}{\hbar} t \sum_{\sigma} \int_{-\infty}^{\infty} \frac{dE}{2\pi} \left\{ G_{k+1,k}^{+-}(E) - G_{k,k+1}^{+-}(E) \right\}. \tag{7.65}$$

We are now in position to discuss the electronic transport in some simple examples of special interest.

7.4.1 *Electrical current through a metallic atomic contact*

As a first example, we consider an atomic constriction. As we learned in the first part of this book, such contacts can nowadays be fabricated with the scanning tunneling microscope or with the mechanically controllable break-junctions. For the sake of simplicity, we consider the case of a metal described by a tight-binding Hamiltonian with a single relevant atomic orbital per site. We assume that the two electrodes forming the atomic junction are only coupled through their outermost atoms, denoted

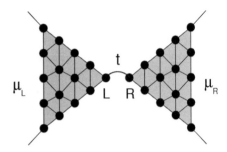

Fig. 7.4 Schematic representation of a single-channel atomic contact. The electrodes are coupled via the hopping element t that describes the coupling between the two outermost atoms of both leads, denoted by L and R. There is a bias voltage applied across the system giving rise to a difference in the chemical potential of the electrodes: $eV = \mu_L - \mu_R$.

as L and R, via a single hopping element t. This situation is schematically represented in Fig. 7.4. Here, the specific shape of the electrodes is irrelevant for our discussion. As it will become clear later, this is a model for a contact with a single conduction channel and if everything is consistent, we should arrive at the Landauer formula. However, contrary to the scattering approach, we will now be able to obtain a microscopic expression for the transmission coefficient in terms of the coupling element t and the local electronic structure of the electrodes.

This model system is described by the following tight-binding Hamiltonian

$$\mathbf{H} = \mathbf{H}_L + \mathbf{H}_R + \sum_{\sigma} t \left(\mathbf{c}_{L\sigma}^{\dagger} \mathbf{c}_{R\sigma} + \mathbf{c}_{R\sigma}^{\dagger} \mathbf{c}_{L\sigma} \right), \qquad (7.66)$$

where $\mathbf{H}_L$ and $\mathbf{H}_R$ are the Hamiltonians describing the left and right electrodes, respectively. We assume that there is a bias voltage V applied across the contact and that the potential drops abruptly in the interface region. The task in this example is to compute the current-voltage characteristics. According to Eqs. (7.63-7.65), the current evaluated at the interface between the electrodes is given by[3]

$$I = \langle \mathbf{I} \rangle = \frac{2et}{h} \int_{-\infty}^{\infty} dE \left[G_{RL}^{+-}(E) - G_{LR}^{+-}(E) \right], \qquad (7.67)$$

---

[3]We assume that the voltage is time-independent and therefore the problem admits a stationary solution. This allows us to write the current in terms of the Fourier transform of the Green's functions with respect to the difference of the time arguments.

where the factor 2 is due to the spin degeneracy in this problem. At this stage the problem is to determine the Green's functions appearing in Eq. (7.67). For this purpose, we employ the perturbative method that we have just described in the previous sections. Therefore, the first thing that we need to do is to choose the perturbation. Let us remind that in the Keldysh formalism the unperturbed system has to be in equilibrium. One possibility would be to introduce the voltage as a perturbation, but this is not very convenient because such a perturbation is extended over the whole system and the calculation would be rather cumbersome. The most convenient choice is to treat the coupling term in Eq. (7.66) as the perturbation and include the voltage in the unperturbed Hamiltonians by shifting the corresponding chemical potential (e.g. $\mu_L = eV$ and $\mu_R = 0$).[4]

With this choice, the retarded and advanced self-energies associated to this single-particle perturbation adopt the form

$$\Sigma_{LR}^{r,a} = \Sigma_{RL}^{r,a} = t, \tag{7.68}$$

while the Keldysh self-energies vanish: $\Sigma^{+-} = \Sigma^{-+} = 0$ (there are no inelastic interactions). Now, the functions G_{LR}^{+-} and G_{RL}^{+-} appearing in the expression of the current can be determined in terms of the Green's functions of the uncoupled electrodes (unperturbed functions) using Eq. (7.47). But before doing so, we can simplify the algebra by writing the current in terms of the diagonal Green's functions of both electrodes. For this purpose, we compute G_{LR}^{+-} making use of Eq. (7.46) by writing it as (remember that $\Sigma^{+-} = 0$ in this problem)

$$\mathbf{G}^{+-} = \mathbf{g}^{+-} + \mathbf{g}^{+-}\mathbf{\Sigma}^a\mathbf{G}^a + \mathbf{g}^r\mathbf{\Sigma}^r\mathbf{G}^{+-}, \tag{7.69}$$

while we compute G_{RL}^{+-} using this equation, but written in the following alternative form:

$$\mathbf{G}^{+-} = \mathbf{g}^{+-} + \mathbf{G}^{+-}\mathbf{\Sigma}^a\mathbf{g}^a + \mathbf{G}^r\mathbf{\Sigma}^r\mathbf{g}^{+-}. \tag{7.70}$$

It is important to emphasize that these equations are algebraic equations in energy space and we shall often omit, as we have just done, the energy argument of the Green's functions, E, to abbreviate the notation.

Using the last two equations, we can write G_{LR}^{+-} and G_{RL}^{+-} as

$$G_{LR}^{+-} = g_{LL}^{+-}\Sigma_{LR}^a G_{RR}^a + g_{LL}^r\Sigma_{LR}^r G_{RR}^{+-}, \tag{7.71}$$

$$G_{RL}^{+-} = G_{RR}^{+-}\Sigma_{RL}^a g_{LL}^a + G_{RR}^r\Sigma_{RL}^r g_{LL}^{+-}. \tag{7.72}$$

---

[4]This does not mean that the unperturbed system is out of equilibrium since in the absence of coupling, there is no current and the electron distributions in both leads are the equilibrium one.

Substituting now G_{LR}^{+-} and G_{RL}^{+-} in Eq. (7.67) and using the general relation $\mathbf{G}^a - \mathbf{G}^r = \mathbf{G}^{+-} - \mathbf{G}^{-+}$, one arrives at

$$I = \frac{2e}{h} t^2 \int_{-\infty}^{\infty} dE \left[g_{LL}^{+-}(E) G_{RR}^{-+}(E) - g_{LL}^{-+}(E) G_{RR}^{+-}(E) \right]. \qquad (7.73)$$

We now compute the functions G_{RR}^{+-} and G_{RR}^{-+} by using Eqs. (7.47) and (7.48)

$$G_{RR}^{+-/-+} = (1 + G_{RL}^r \Sigma_{LR}^r) g_{RR}^{+-/-+} (1 + \Sigma_{RL}^a G_{LR}^a) + \qquad (7.74)$$

$$G_{RR}^r \Sigma_{RL}^r g_{LL}^{+-/-+} \Sigma_{LR}^a G_{RR}^a. \qquad (7.75)$$

Introducing these expressions in Eq. (7.73) we obtain

$$I = \frac{2e}{h} t^2 \int_{-\infty}^{\infty} dE \, |1 + t G_{RL}^r(E)|^2 \left[g_{LL}^{+-}(E) g_{RR}^{-+}(E) - g_{LL}^{-+}(E) g_{RR}^{+-}(E) \right]. \qquad (7.76)$$

Here, we have used the explicit expression of the self-energies, see Eq. (7.68), and the fact that $\mathbf{G}^a(E) = [\mathbf{G}^r(E)]^\dagger$ (thus e.g., $G_{LR}^a(E) = [G_{RL}^r(E)]^*$).

To complete the calculation we still have to determine the retarded function $G_{RL}^r(E)$. This can be done, very much like in equilibrium, using its Dyson's equation, see Eq. (7.41). Taking the element (R, L) we arrive at

$$G_{RL}^r = g_{RR}^r \Sigma_{RL}^r G_{LL}^r. \qquad (7.77)$$

To close this equation, we need now an equation for G_{LL}^r, which is obtained by taking the element (L, L) in the Dyson's equation, i.e.

$$G_{LL}^r = g_{LL}^r + g_{LL}^r \Sigma_{LR} G_{RL}^r. \qquad (7.78)$$

Substituting back into the equation for G_{RL}^r, we obtain finally

$$G_{RL}^r = \frac{t g_{RR}^r g_{LL}^r}{1 - t^2 g_{RR}^r g_{LL}^r} \quad \text{and} \quad 1 + t G_{RL}^r = \frac{1}{1 - t^2 g_{RR}^r g_{LL}^r}. \qquad (7.79)$$

Before coming back to the expression of current, let us remind that the unperturbed Keldysh functions $\mathbf{g}^{+-/-+}$ can be expressed in terms of the retarded and advanced ones using Eq. (7.54). Thus, the functions appearing in Eq. (7.76) can be written as

$$g_{LL}^{+-}(E) = [g_{LL}^a(E - eV) - g_{LL}^r(E - eV)] f(E - eV) \qquad (7.80)$$
$$= 2\pi i \rho_L(E - eV) f(E - eV)$$
$$g_{LL}^{-+}(E) = - [g_{LL}^a(E - eV) - g_{LL}^r(E - eV)] [1 - f(E - eV)]$$
$$= -2\pi i \rho_L(E - eV) [1 - f(E - eV)]$$
$$g_{RR}^{+-}(E) = [g_{RR}^a(E) - g_{RR}^r(E)] f(E) = 2\pi i \rho_R(E) f(E)$$
$$g_{RR}^{-+}(E) = - [g_{RR}^a(E) - g_{RR}^r(E)] [1 - f(E)] = -2\pi i \rho_R(E) [1 - f(E)],$$

where $f(E)$ is the Fermi function and $\rho_{L/R}$ is the local density of states of the leads projected onto the sites L and R. Notice that we have already taken into account the relative shift of the chemical potentials due to the bias voltage V.

Using Eqs. (7.79) and (7.80), we can finally write the current as follows[5]

$$I = \frac{2e}{h} \int_{-\infty}^{\infty} dE \, \frac{4\pi t^2 \rho_L(E - eV)\rho_R(E)}{|1 - t^2 g_{LL}(E - eV)g_{RR}(E)|^2} \, [f(E - eV) - f(E)].$$
(7.81)

Notice that Eq. (7.81) has exactly the form of the Landauer formula, i.e.

$$I = \frac{2e}{h} \int_{-\infty}^{\infty} dE \, T(E, V) \, [f(E - eV) - f(E)], \qquad (7.82)$$

where we can identify $T(E, V)$ as an energy and voltage-dependent transmission probability given by

$$T(E, V) = \frac{4\pi t^2 \rho_L(E - eV)\rho_R(E)}{|1 - t^2 g_{LL}(E - eV)g_{RR}(E)|^2}. \qquad (7.83)$$

As it can be seen, the transmission depends primarily on the coupling element t and the local electronic structure of the leads.

For sufficiently low voltages, there is a linear regime where the current is proportional to the voltage. In this limit, the conductance is given by $G = (2e^2/h)T(E_F, V = 0)$, where $T(E_F, V = 0)$ is the zero-bias transmission at the Fermi energy given by

$$T(E_F, V = 0) = \frac{4\pi t^2 \rho_L(E_F)\rho_R(E_F)}{|1 - t^2 g_{LL}(E_F)g_{RR}(E_F)|^2}. \qquad (7.84)$$

One can often consider that the Green's functions are constant around the Fermi energy and one can also neglect their real part (this is the wide-band approximation introduced in Chapter 5). This means that the lead Green's functions can be approximated by

$$g_{LL} \approx \frac{i}{W}, \qquad (7.85)$$

where $W = 1/\pi\rho_{L/R}(E_F)$ (we are assuming a symmetric contact ($g_{LL} = g_{RR}$) for simplicity). Within this approximation, one obtains the following expression for the transmission

$$T = \frac{4t^2/W^2}{(1 + t^2/W^2)^2}. \qquad (7.86)$$

---

[5]This expression for the current was first derived in Ref. [194] for a more realistic model.

This expression illustrates the transition from the tunnel regime, when the electrodes are separated by a large distance, to the contact regime at small distances. In the former limit, the transmission given in Eq. (7.86) can be approximated by $4t^2/W^2$. This means that the dependence of the transmission on the distance between the electrodes, and therefore that of the linear conductance, is determined by t^2. At large distances, a hopping element is roughly proportional to the overlap of the atomic orbitals and decays exponentially with the distance between the corresponding atoms. This is how the exponential length dependence, which we already discussed in section 4.4, comes about from an atomistic point of view. From the scattering approach, see section 4.4, we concluded that the length dependence of a metallic tunnel junction is determined by the metal work function. However, with this simple model, we get the impression that such a dependence is governed by a local property, namely the coupling between the outermost orbitals of the electrodes. These two pictures, which at first glance look contradictory, can indeed be reconciled. This is, however, a subtle issue that is out of the scope of this book and we refer the reader to Ref. [195] for a discussion of this question.

When the electrodes approach each other the hopping t becomes of the same order as the energy scale W and the transmission can reach unity and in turn the conductance approaches the quantum of conductance $G_0 = 2e^2/h$. The transition from tunnel to contact was first discussed within this type of atomistic models in Ref. [194] in connection with the first experiment that explored such a transition [55]. For an overview on recent experiments exploring the tunnel-to-contact transition both in single atoms and molecules, see Refs. [196, 197].

Let us now study in more detail the tunnel limit $(t \to 0)$. In this case, the non-linear current of Eq. (7.81) can be approximated by

$$I = \frac{8\pi e}{h} t^2 \int_{-\infty}^{\infty} dE \, \rho_L(E - eV)\rho_R(E) \left[f(E - eV) - f(E) \right], \qquad (7.87)$$

which tell us that the current in this limit is determined by the convolution of the local density of states of both electrodes. This well-known expression is a fundamental result for the theory of STM and provides a simple interpretation of the STM images. Assuming that the left electrode represents a STM tip with a constant density of states around the Fermi energy, the differential conductance at low temperatures is simply given by

$$G(V) = \frac{dI}{dV} = \frac{2e^2}{h} 4\pi t^2 \rho_L(E_{\mathrm{F}})\rho_R(E_{\mathrm{F}} + eV), \qquad (7.88)$$

i.e. the conductance is a measure of the local density of states of the sample (or right electrode in our case).

7.4.2 *Shot noise in an atomic contact*

Another interesting transport property that can easily be calculated with the Keldysh formalism is the shot noise (or nonequilibrium current fluctuations), which was introduced in Chapter 4. Let us consider the model for an atomic contact discussed in the previous subsection. Our goal now is the calculation of the current fluctuations in the zero-temperature limit and finite bias (shot noise).

The noise is characterized by the fluctuation spectral density that is defined as

$$P(\omega) = \hbar \int_{-\infty}^{\infty} dt \; e^{i\omega t} \langle \delta \mathbf{I}(t) \delta \mathbf{I}(0) + \delta \mathbf{I}(0) \delta \mathbf{I}(t) \rangle, \qquad (7.89)$$

where $\delta \mathbf{I}(t) = \mathbf{I}(t) - \langle \mathbf{I}(t) \rangle$.

We are specially interested in the zero-frequency noise, $P(0)$,

$$P(0) = \hbar \int_{-\infty}^{\infty} dt \; \langle \delta \mathbf{I}(t) \delta \mathbf{I}(0) + \delta \mathbf{I}(0) \delta \mathbf{I}(t) \rangle. \qquad (7.90)$$

If we now substitute the expressions for $\mathbf{I}(t)$ and $\langle \mathbf{I}(t) \rangle$ for an atomic contact and we write the result in terms of the Green's functions, we obtain

$$P(0) = \frac{2e^2}{h} \int_{-\infty}^{\infty} dE \; [G_{LR}^{+-}(E)G_{RL}^{-+}(E) + G_{RL}^{+-}(E)G_{LR}^{-+}(E) -$$
$$G_{LL}^{+-}(E)G_{RR}^{-+}(E) - G_{RR}^{+-}(E)G_{LL}^{-+}(E)]. \quad (7.91)$$

Here, in order to obtain this expression, we have made use of Wick's theorem to decouple the averages of four operators (let us remind that this is valid since our electron system is noninteracting).

At this stage the calculation of the shot noise has been reduced to the computation of the different Keldysh-Green's functions that appear in Eq. (7.91). These functions can be calculated following exactly the same procedure detailed in the previous subsection. If we now assume zero temperature and use the wide-band approximation of Eq. (7.85) for the unperturbed Green's functions, we can obtain the following expression (see Exercise 7.5)

$$P(0) = \frac{4e^2}{h} T(1-T)V, \qquad (7.92)$$

which is the result derived in section 4.7 using the scattering approach.

7.4.3   *Current through a resonant level*

Let us now discuss the calculation of the current for the resonant level model discussed in section 5.3.3. Let us remind that in this model a single quantum level is coupled to two metallic electrodes and the corresponding Hamiltonian is given by

$$\mathbf{H} = \mathbf{H}_L + \mathbf{H}_R + \sum_\sigma \epsilon_0 \mathbf{n}_{0\sigma} + \qquad (7.93)$$

$$\sum_\sigma t_L \left(\mathbf{c}^\dagger_{L\sigma} \mathbf{c}_{0\sigma} + \mathbf{c}^\dagger_{0\sigma} \mathbf{c}_{L\sigma} \right) + \sum_\sigma t_R \left(\mathbf{c}^\dagger_{R\sigma} \mathbf{c}_{0\sigma} + \mathbf{c}^\dagger_{0\sigma} \mathbf{c}_{R\sigma} \right),$$

where ϵ_0 is the position of the resonant level, which in principle can also depend on the bias voltage, and $t_{L,R}$ are the matrix elements describing the coupling to the reservoirs. Here, L and R denote the outermost sites of the left and right electrodes, respectively. On the other hand, we now assume that there is a constant bias voltage across the system and our task is to compute the current-voltage characteristics.

We start by evaluating the current at the interface between the left electrode and the level, which in terms of the Green's functions G^{+-} can be written as follows

$$I = \frac{2et_L}{h} \int_{-\infty}^{\infty} dE \left[G^{+-}_{L0}(E) - G^{+-}_{0L}(E) \right]. \qquad (7.94)$$

In order to determine the Green's functions in the previous expression, we use again the Keldysh formalism and we treat the coupling terms between the level and the electrodes, i.e. the second line in Eq. (7.93), as a perturbation. With this choice the only non-vanishing elements of the self-energy are: $\Sigma^{r,a}_{L0} = \Sigma^{r,a}_{0L} = t_L$ and $\Sigma^{r,a}_{R0} = \Sigma^{r,a}_{0R} = t_R$.

Following now the same steps as in section 7.4.1, we can write the current in terms of diagonal elements of the Green's functions as

$$I = \frac{2et_L}{h} \int_{-\infty}^{\infty} dE \left[g^{+-}_{LL}(E) G^{-+}_{00}(E) - g^{-+}_{LL}(E) G^{+-}_{00}(E) \right]. \qquad (7.95)$$

Now, to determine the full Green's functions, we use the Dyson's equation, Eq. (7.47), to write

$$G^{+-/-+}_{00} = (1 + \mathbf{G}^r \mathbf{\Sigma}^r)_{00}\, g^{+-/-+}_{00}\, (1 + \mathbf{\Sigma}^a \mathbf{G}^a)_{00} + \qquad (7.96)$$

$$G^r_{00} \Sigma^r_{0L} g^{+-/-+}_{LL} \Sigma^a_{L0} G^a_{00} + G^r_{00} \Sigma^r_{0R} g^{+-/-+}_{RR} \Sigma^a_{R0} G^a_{00}.$$

If we now substitute this expression into the current formula, the term containing $g^{+-/-+}_{LL}$ is canceled. Moreover, the term proportional to $g^{+-/-+}_{00}$ does not contribute either. The reason is that $g^{+-/-+}_{00}(E) \propto \delta(E - \epsilon_0)$ and the prefactor of this term vanishes at $E = \epsilon_0$.[6] Thus, the current can now be expressed as

$$I = \frac{2e}{h} 4\pi^2 t_L^2 t_R^2 \int_{-\infty}^{\infty} dE \, \rho_L(E)\rho_R(E)|G_{00}^r(E)|^2 \left[f_L(E) - f_R(E) \right], \quad (7.97)$$

where it is implicitly assumed that the density of states (and distribution function) of the left electrode is shifted by eV. Notice that we have already used the expression of the lead Green's functions in terms of the local density of states and Fermi functions.

At this point, the only remaining task is the calculation of $G_{00}^r(E)$, but this is something that we have already done in section 5.3.3 and we just recall here the result

$$G_{00}^r(E) = \frac{1}{E - \epsilon_0 - t_L^2 g_L^r(E) - t_R^2 g_R^r(E)}. \quad (7.98)$$

Therefore, the current adopts again the form of the Landauer formula

$$I = \frac{2e}{h} \int_{-\infty}^{\infty} dE \, T(E, V) \left[f(E - eV) - f(E) \right], \quad (7.99)$$

where this time the transmission $T(E, V)$ is given by

$$T(E, V) = \frac{4\pi^2 t_L^2 t_R^2 \rho_L(E - eV)\rho_R(E)}{|E - \epsilon_0 - t_L^2 g_L^r(E - eV) - t_R^2 g_R^r(E)|^2}. \quad (7.100)$$

To simplify this expression, we use now as in section 5.3.3 the wide-band approximation and neglect the energy dependence introduced by the leads. This way, $g_{L/R}^r \approx -i\pi\rho_{L/R}(E_F)$ and we define the scattering rates $\Gamma_{L/R} = \pi t_{L/R}^2 \rho_{L/R}(E_F)$. In this approximation the transmission can be written as

$$T(E, V) = \frac{4\Gamma_L \Gamma_R}{(E - \epsilon_0)^2 + (\Gamma_L + \Gamma_R)^2}. \quad (7.101)$$

In this case, the voltage dependence of the transmission may only stem from the eventual voltage dependence of the level position. This expression is the well-known Breit-Wigner formula that was derived in Chapter 4 within the scattering approach (see Exercises 4.5 and 4.8) and it will be used extensively in later chapters.

Again, in the linear regime the low-temperature conductance is simply given by $G = (2e^2/h)T(E_F, 0)$. This expression shows that the maximum conductance is reached when $E_F = \epsilon_0$, which is the resonant condition. In the symmetric case ($\Gamma_L = \Gamma_R$), this maximum is equal to $G_0 = 2e^2/h$, irrespectively of the value of the scattering rates. These facts are illustrated in Fig. 7.5. The non-linear current-voltage characteristics of this model will be discussed in detail in Chapter 15.

---

[6]Physically speaking, it is quite reasonable that this term does not contribute to the current. It makes no sense that the current depends on the occupation of the level before being coupled to the electrodes.

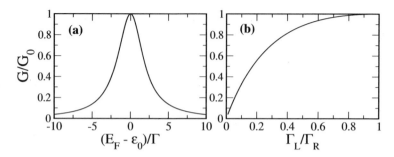

Fig. 7.5   Zero-temperature linear conductance in the resonant tunneling model. (a) Linear conductance (normalized by $G_0 = 2e^2/h$) as a function of the level position, ϵ_0 for a symmetric contact $\Gamma_L = \Gamma_R = \Gamma$. (b) Linear conductance at resonance ($\epsilon_0 = E_{\mathrm{F}}$) as a function of the ratio between the scattering rates.

7.5   Exercises

7.1 Diagrammatic expansion in the Keldysh formalism: Show explicitly that $\langle \phi_0 | \mathbf{S}_c | \phi_0 \rangle = 1$ by using the expansion of the operator $\mathbf{S}_c$. For this purpose, expand $\mathbf{S}_c$ up to second order and show that the contributions of order higher than zero cancel.

7.2 Time-dependent external potential: Let us consider a system with N noninteracting electrons subjected to a time-dependent external potential:

$$\mathbf{V}(t) = \sum_{i=1}^{N} V(\mathbf{r}_i, t). \tag{7.102}$$

Apply Wick's theorem to demonstrate that the self-energy is given by Eq. (7.22).

7.3 Properties of the Keldysh-Green's functions:
   (a) Demonstrate the property of Eq. (7.28).   Hint: Use the property of Eq. (7.25) and the Dyson's equation in Keldysh space.
   (b) Demonstrate Eq. (7.46).

7.4 Shot noise in a single-channel point contact:
   Derive the expression of the zero-frequency shot noise of a single-channel point contact following the discussion of the example of section 7.4.2 and demonstrate that it is given by

$$P(0) = \frac{4e^2}{h} T(1 - T) eV,$$

where T is the energy-independent transmission coefficient of the contact given by Eq. (7.86).

7.5 Electrical current through a linear chain: Consider the electronic transport in a finite one-dimensional system formed by a tight-binding chain with N

sites such that the site 1 is connected to the left electrode through a hopping t_L and the site N is connected to the right electrode with a hopping t_R. Show that the current formula in this case is given by

$$I = \frac{2e}{h} 4\pi^2 t_L^2 t_R^2 \int_{-\infty}^{\infty} dE \, \rho_L(E - eV)\rho_R(E)|G_{1N}^r(E)|^2 \left[f(E - eV) - f(E) \right].$$

For the sake of simplicity, consider that in the chain there are only hoppings between nearest-neighbor atoms, t, and that the on-site energy is given by ϵ_0. Study the linear conductance of this system as a function of the number of sites N in the chain and show that it may exhibit parity oscillations, depending on whether N is even or odd.

7.6 Thermopower of a single-channel point contact: Using the model of section 7.4.1, derive the expression for the thermopower for a single-channel contact and show that it coincides with the result obtained with the scattering approach in section 4.8.

Chapter 8

Formulas of the electrical current: Exploiting the Keldysh formalism

In the previous chapter we showed how the Keldysh formalism can be combined with simple Hamiltonians to compute the current in model systems. In this chapter we shall exploit this technique and derive some general expressions for the electrical current that can be combined with realistic methods for the determination of the electronic structure. To be precise, we shall address three basic issues:

(1) Derivation of Landauer formula in the framework of the non-equilibrium Green's function techniques. Here, the goal is the determination of the microscopic expression for the elastic transmission valid for any atomic and molecular junction.
(2) Generalization of Landauer formula to include inelastic and correlation effects.
(3) Description of the current in atomic-scale junctions subjected to time-dependent potentials.

This chapter is rather technical and it can be skipped by those who are not so interested in the algebra behind the current formulas. Anyway, we recommend to read the next section about the derivation of the Landauer formula, since the expression obtained there for the elastic transmission will be frequently used in subsequent chapters.

8.1 Elastic current: Microscopic derivation of the Landauer formula

In section 7.4 we discussed two simple examples of atomic-scale contacts. In both cases we ended up with a Landauer-like formula for the elastic current,

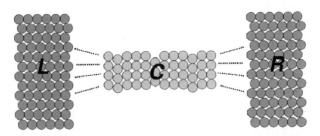

Fig. 8.1   Schematic representation of an atomic-scale contact of arbitrary geometry. We divide this system into three parts: a central region, C, and the two leads, L and R.

the only difference being the expression for the transmission coefficient. In this section we shall demonstrate that this was not a coincidence and we shall derive a general expression for the elastic current valid for any type of atomic and molecular junction.

Let us consider a contact with arbitrary geometry like the one depicted in Fig. 8.1. Such a contact can be either an atomic contact or a molecular junction. Since we shall ignore inelastic interaction in this discussion, one can describe the system in terms of the following generic tight-binding Hamiltonian

$$\mathbf{H} = \sum_{ij,\alpha\beta,\sigma} h_{i\alpha,j\beta}\ \mathbf{c}^{\dagger}_{i\alpha,\sigma}\mathbf{c}_{j\beta,\sigma}, \qquad (8.1)$$

where i, j run over the atomic sites and α, β denote the different atomic orbitals. The number of orbitals in each site can be arbitrary. For the sake of simplicity, we assume that the local basis is orthogonal. Later in this section, we shall generalize the results to the case of nonorthogonal basis sets. Notice also that we are assuming that matrix elements are independent of the spin, i.e. for the moment we do not consider magnetic situations.

We now distinguish three different parts in this contact: the reservoirs L and R, and a central region that can have arbitrary size and shape. In principle, the reservoirs L and R could also have an arbitrary shape and we assume that an electron in these subsystems has a well-defined temperature and chemical potential. In other words, these regions play the role of electron reservoirs, in the spirit of the scattering approach of Chapter 4. The separation of the contact in these three subsystems is somewhat arbitrary, especially in the linear response regime, and one can play with that, as we shall discuss below. We also assume that there is no direct coupling between the reservoirs. With this assumption the Hamiltonian above can be written in the following matrix form

$$\mathbf{H} = \begin{pmatrix} \mathbf{H}_{LL} & \mathbf{t}_{LC} & 0 \\ \mathbf{t}_{CL} & \mathbf{H}_{CC} & \mathbf{t}_{CR} \\ 0 & \mathbf{t}_{RC} & \mathbf{H}_{RR} \end{pmatrix}, \tag{8.2}$$

where the diagonal terms $\mathbf{H}_{XX}$ with $X = L, C, R$ are the Hamiltonian of the three subsystems and the $\mathbf{t}$'s describe the coupling between them.

Our aim is to determine the current through the contact induced by a constant bias voltage, $eV = \mu_L - \mu_R$. For this purpose, we first evaluate the current at the interface between the left lead L and central region C, which in the tight-binding representation adopts the form (see section 7.4)

$$I = \frac{ie}{\hbar} \sum_{i \in L; j \in C; \alpha, \beta, \sigma} \left(h_{i\alpha, j\beta} \langle c_{i\alpha, \sigma}^{\dagger} c_{j\beta, \sigma} \rangle - h_{j\beta, i\alpha} \langle c_{j\beta, \sigma}^{\dagger} c_{i\alpha, \sigma} \rangle \right), \tag{8.3}$$

where i runs over the atoms of the left electrode which are connected with the atoms in the central region C, and j runs over the atoms of the central region coupled to the left electrode (in principle, all of them). The indexes α and β indicate the different atomic orbitals in every site.

Following the ideas of the last section of the previous chapter, we make use of nonequilibrium Green's function techniques to calculate the current. First of all, we express the expectation values appearing in the current expression in terms of the Keldysh-Green's function G^{+-}. This function gives information about the distribution function of the system and in a local basis it adopts the following form

$$G_{i\alpha, j\beta}^{+-, \sigma\sigma'}(t, t') = i \langle c_{j\beta, \sigma'}^{\dagger}(t') c_{i\alpha, \sigma}(t) \rangle. \tag{8.4}$$

Using this expression one can write the current as

$$I = \frac{e}{\hbar} \sum_{i \in L; j \in C; \alpha, \beta, \sigma} \left[t_{i\alpha, j\beta} G_{j\beta, i\alpha}^{+-, \sigma\sigma}(t, t) - t_{j\beta, i\alpha} G_{i\alpha, j\beta}^{+-, \sigma\sigma}(t, t) \right]. \tag{8.5}$$

The current can be expressed in a more compact way in terms of the hopping matrices $\mathbf{t}_{LC}$ and $\mathbf{t}_{CL}$ [see Eq. (8.2)] whose elements are given by

$$(\mathbf{t}_{LC})_{i\alpha, j\beta} = h_{i\alpha, j\beta} \text{ with } i \in L; \ j \in C \tag{8.6}$$
$$(\mathbf{t}_{CL}) = (\mathbf{t}_{LC})^{\dagger}.$$

Analogously, one can define similar matrices for the Green's functions G^{+-}. With this new notation, one can express the current as

$$I = \frac{2e}{\hbar} \text{Tr} \left[\mathbf{G}_{CL}^{+-}(t, t) \mathbf{t}_{LC} - \mathbf{t}_{CL} \mathbf{G}_{LC}^{+-}(t, t) \right], \tag{8.7}$$

where Tr denotes the trace over atoms and orbitals in the central region C. The prefactor 2 comes from the sum over spins, since for the moment we do not consider any magnetic situation. For the same reason, we drop the superindex σ in the Green's functions.

This transport problem admits a stationary solution and therefore, the different Green's functions only depend on the difference of time arguments. Thus, we can Fourier transform with respect to the difference of the time arguments and write the current as

$$I = \frac{2e}{h} \int_{-\infty}^{\infty} dE \, \mathrm{Tr} \left[\mathbf{G}_{CL}^{+-}(E) \mathbf{t}_{LC} - \mathbf{t}_{CL} \mathbf{G}_{LC}^{+-}(E) \right]. \tag{8.8}$$

Notice that the current is expressed in terms of the trace of a matrix whose dimension is the number of orbitals in the central region, which we denote as N_C. At this stage, the problem has been reduced to the determination of the functions G^{+-} in terms of matrix elements of the Hamiltonian of Eq. (8.1). We shall calculate these functions considering the coupling terms between the electrodes and the central region as a perturbation. Then, starting from the Green's functions for the three isolated systems, we shall determine the corresponding functions for the whole system. With this choice, the self-energies of the problem are the hopping matrices defined in Eq. (8.6) and the equivalent ones for the interface between the central region and the right electrode R.

We now follow the ideas of section 7.4.3 and make use of Dyson's equation in Keldysh space, see Eq. (7.46), to write the functions G^{+-} as follows[1]

$$\mathbf{G}_{LC}^{+-} = \mathbf{g}_{LL}^{+-} \mathbf{t}_{LC} \mathbf{G}_{CC}^{a} + \mathbf{g}_{LL}^{r} \mathbf{t}_{LC} \mathbf{G}_{CC}^{+-} \tag{8.9}$$
$$\mathbf{G}_{CL}^{+-} = \mathbf{G}_{CC}^{+-} \mathbf{t}_{CL} \mathbf{g}_{LL}^{a} + \mathbf{G}_{CC}^{r} \mathbf{t}_{CL} \mathbf{g}_{LL}^{+-},$$

where $\mathbf{g}_{XX}^{r;a}$ are the (retarded, advanced) Green's functions of the uncoupled reservoirs $(X = L, R)$. Introducing this equation in the current expression and making use of the relation $\mathbf{G}^{+-} - \mathbf{G}^{-+} = \mathbf{G}^{a} - \mathbf{G}^{r}$, we obtain

$$I = \frac{2e}{h} \int_{-\infty}^{\infty} dE \, \mathrm{Tr} \left[\mathbf{G}_{CC}^{-+} \mathbf{t}_{CL} \mathbf{g}_{LL}^{+-} \mathbf{t}_{LC} - \mathbf{G}_{CC}^{+-} \mathbf{t}_{CL} \mathbf{g}_{LL}^{-+} \mathbf{t}_{LC} \right]. \tag{8.10}$$

Then, we determine $\mathbf{G}^{+-/-+}$ by means of the relation

$$\mathbf{G}^{+-/-+} = (1 + \mathbf{G}^r \mathbf{t}) \, \mathbf{g}^{+-/-+} \, (1 + \mathbf{t} \mathbf{G}^a). \tag{8.11}$$

Taking the element (C,C) in this expression we obtain

$$\mathbf{G}_{CC}^{+-/-+} = \mathbf{G}_{CC}^{r} \hat{t}_{CL} \mathbf{g}_{LL}^{+-/-+} \mathbf{t}_{LC} \mathbf{G}_{CC}^{a} + \mathbf{G}_{CR}^{r} \mathbf{t}_{CR} \mathbf{g}_{RR}^{+-/-+} \mathbf{t}_{RC} \mathbf{G}_{CC}^{a}. \tag{8.12}$$

---

[1] In order to abbreviate the notation, we do not write the energy argument E explicitly. Moreover, since there are no inelastic processes involved in this model, the self-energies Σ^{+-} associated with them vanish.

Notice that there is an additional contribution containing $\mathbf{g}_{CC}^{+-/-+}$ that was left out in the previous expression. The reason for this is that, in analogy with our discussion of the resonant tunneling model in section 7.4.3, one can show that such a term does not contribute to the final expression of the current.

Substitution of the previous equation in the expression of the current yields

$$I = \frac{2e}{h} \int_{-\infty}^{\infty} dE \, \mathrm{Tr} \left[\mathbf{G}_{CC}^{r} \mathbf{t}_{CR} \mathbf{g}_{RR}^{-+} \mathbf{t}_{RC} \mathbf{G}_{CC}^{a} \mathbf{t}_{CL} \mathbf{g}_{LL}^{+-} \mathbf{t}_{LC} - \right.$$
$$\left. \mathbf{G}_{CC}^{r} \mathbf{t}_{CR} \mathbf{g}_{RR}^{+-} \mathbf{t}_{RC} \mathbf{G}_{CC}^{a} \mathbf{t}_{CL} \mathbf{g}_{LL}^{-+} \mathbf{t}_{LC} \right]. \quad (8.13)$$

Let us recall that the unperturbed functions $\mathbf{g}^{+-}$ and $\mathbf{g}^{-+}$ satisfy the following relations[2]

$$\begin{aligned} \mathbf{g}^{+-} &= (\mathbf{g}^{a} - \mathbf{g}^{r}) \, f = 2i \, \mathrm{Im} \, (\mathbf{g}^{a}) \, f \\ \mathbf{g}^{-+} &= (\mathbf{g}^{a} - \mathbf{g}^{r}) \, (f - 1) = 2i \, \mathrm{Im} \, (\mathbf{g}^{a}) \, (f - 1), \end{aligned} \quad (8.14)$$

where f is the Fermi function. Thus, the current can be expressed as

$$I = \frac{8e}{h} \int_{-\infty}^{\infty} dE \, \mathrm{Tr} \left[\mathbf{G}_{CC}^{r} \mathbf{t}_{CR} \mathrm{Im} \left\{ \mathbf{g}_{RR}^{a} \right\} \mathbf{t}_{RC} \mathbf{G}_{CC}^{a} \mathbf{t}_{CL} \mathrm{Im} \left\{ \mathbf{g}_{LL}^{a} \right\} \mathbf{t}_{LC} \right]$$
$$\times (f_{L} - f_{R}). \quad (8.15)$$

Here, $f_{L/R}$ is the Fermi function of the corresponding electrode, which takes into account the shift of the chemical potential induced by the voltage.

One can further simplify the expression of the current by defining

$$\boldsymbol{\Sigma}_{L}^{r,a} = \mathbf{t}_{CL} \mathbf{g}_{LL}^{r,a} \mathbf{t}_{LC} \quad \text{and} \quad \boldsymbol{\Sigma}_{R}^{r,a} = \mathbf{t}_{CR} \mathbf{g}_{RR}^{r,a} \mathbf{t}_{RC}, \quad (8.16)$$

These matrices are nothing else but the self-energies of this problem for the subspace of the central region. These self-energies describe the influence of the reservoir in the central region and they depend both on the coupling between the reservoirs and the central region and on the local electronic structure of the leads. Notice that these matrices have a dimension equal to the number of orbitals in the central region. Using these definitions, the current can now be rewritten in the following familiar form

$$I = \frac{2e}{h} \int_{-\infty}^{\infty} dE \, T(E,V) \, (f_{L} - f_{R}), \quad (8.17)$$

where $T(E,V)$ is the energy- and voltage-dependent total transmission probability of the contact given by

$$T(E,V) \equiv 4\mathrm{Tr} \left[\boldsymbol{\Gamma}_{L} \mathbf{G}_{CC}^{r} \boldsymbol{\Gamma}_{R} \mathbf{G}_{CC}^{a} \right]. \quad (8.18)$$

[2]Notice that in Eq. (8.14) we have assumed that that Hamiltonian is real, i.e. there is time reversal symmetry. One can easily show that this implies that $\mathbf{g}^{r}(E) = [\mathbf{g}^{a}(E)]^{*}$.

where we have defined the scattering rate matrices as $\Gamma_{L,R} \equiv \text{Im}\{\Sigma_{L,R}^a\}$.[3] The voltage dependence of the transmission comes through the scattering rates (i.e. via the leads), but also through the possible voltage dependence of the Hamiltonian matrix elements of the central region.

We can further symmetrize this expression by using the cyclic property of the trace and write $T(E,V) = \text{Tr}\left[\mathbf{t}(E,V)\mathbf{t}^\dagger(E,V)\right] = \text{Tr}\left[\mathbf{t}^\dagger(E,V)\mathbf{t}(E,V)\right]$, where

$$\mathbf{t}(E,V) = 2\Gamma_L^{1/2}\mathbf{G}_{CC}^r\Gamma_R^{1/2} \qquad (8.19)$$

is the transmission matrix of the system. The existence of $\Gamma^{1/2}$ as a real matrix is warranted by Γ being positive definite (see Exercise 8.1).

Finally, the current adopts the form

$$I = \frac{2e}{h}\int_{-\infty}^{\infty} dE\ \text{Tr}\left[\mathbf{t}^\dagger(E,V)\mathbf{t}(E,V)\right](f_L - f_R), \qquad (8.20)$$

valid for arbitrary bias voltage. In the linear regime this expression reduces to the standard Landauer formula for the zero-temperature conductance

$$G = \frac{2e^2}{h}\text{Tr}\left[\mathbf{t}^\dagger(E_{\text{F}},0)\mathbf{t}(E_{\text{F}},0)\right] = \frac{2e^2}{h}\sum_{i=1}^{N} T_i, \qquad (8.21)$$

where T_i are the eigenvalues of $\hat{\mathbf{t}}^\dagger\mathbf{t}$ (or $\mathbf{t}\hat{\mathbf{t}}^\dagger$) at the Fermi level. As one can see, in principle the number of channel would be N_C, which is the dimension of the matrix $\mathbf{t}^\dagger\mathbf{t}$. However, as we stated at the beginning of this section, the separation in three subsystems in somewhat arbitrary and one can evaluate the current at any point. Thus, it is evident that the actual number of channels is controlled by the narrowest part of the junction. This fact will be very important in our discussion of the conduction channels in metallic single-atom contacts, see section 11.5. Notice also that in this formulation, the conduction channels , defined as the eigenfunctions of $\mathbf{t}^\dagger\hat{\mathbf{t}}$, are linear combinations of the atomic orbitals in the central system.

As a result of the discussion above, we have not only re-derived the Landauer formula, but more importantly, we have also obtained an explicit formula for the transmission as a function of the microscopic parameters of the system. As one can see in Eq. (8.18) or in Eq. (8.19), the determination of the transmission requires the calculations of both the retarded/advanced Green's functions of the central system and the scattering rate matrices. These functions can be determined from their Dyson's equation

$$\mathbf{G}_{CC}^a = (\mathbf{G}_{CC}^r)^\dagger = \left[(E - i0^+)\mathbf{1} - \mathbf{H}_{CC} - \Sigma_L^a - \Sigma_R^a\right]^{-1}, \qquad (8.22)$$

---

[3] We have assumed without loss of generality that the hopping matrix elements are real.

where $\mathbf{H}_{CC}$ is the Hamiltonian of the central region and the self-energies $\boldsymbol{\Sigma}_X$ ($X = L, R$) are given by Eq. (8.16).

On the other hand, the calculation of the scattering rate matrices, which are the imaginary part of the self-energies of Eq. (8.16), requires the knowledge of the Green's functions of the uncoupled reservoirs, $\mathbf{g}_{XX}$ (with $X = L, R$). The leads are semi-infinite systems and thus they cannot possess in practice a very complicated geometry. A typical option is to describe these leads as ideal surfaces of the corresponding material and the unperturbed Green's functions are then computed using special recursive techniques like the so-called *decimation* [198].

Let us end this section with a brief technical discussion. The quantity $\mathbf{t}(E, V)$ appearing in Eq. (8.19) has been called transmission matrix without a real justification. We should demonstrate that this matrix fulfills the properties of a transmission matrix. In particular, we should at least prove that the eigenvalues of $\mathbf{tt}^\dagger$ are bounded between 0 and 1. Indeed, this property can be shown using a few algebraic manipulations (see Exercise 8.2).

Another way of showing that $\mathbf{t}(E, V)$ in Eq. (8.19) is indeed the transmission matrix of the contact is via the so-called Fisher-Lee relation [199], which expresses the elements of the scattering matrix in terms of Green's functions. For the readers interested in this route, we recommend the original work of Ref. [199] and the discussion on this matter in Chapter 3 of Ref. [50].

8.1.1 *An example: back to the resonant tunneling model*

As an application of the general formula derived above and in order to illustrate its use, let us now re-derive the current formula for the resonant tunneling model considered in section 7.4.3.

Our starting point is the expression for the transmission of Eq. (8.18). We need first to compute the retarded/advanced Green's functions of the central region. In this case this region consists of a single site with an on-site energy ϵ_0. Therefore, the Green's functions of the central region are scalars with the following form

$$G_{CC}^{r,a} = \left[E \pm i0^+ - \epsilon_0 - \Sigma_L^{r,a} - \Sigma_R^{r,a} \right]^{-1}, \tag{8.23}$$

where the self-energies are the scalars $\Sigma_{L/R}^{r,a} = t_{L/R}^2 g_{LL/RR}^{r,a}$. Assuming as in section 7.4.3 that the local Green's functions $g_{LL/RR}^{r,a}$ are purely imaginary and independent of the energy around E_F, the advanced self-energies reduce to $\Sigma_{L/R}^a = i\Gamma_{L/R}$, where $\Gamma_{L/R} = t_{L/R}^2 \text{Im}\{g_{LL/RR}^a(E_F)\}$ are the scattering rates at the Fermi energy. Substituting now Eq. (8.23) and the expressions of the self-energy in Eq. (8.18), we arrive again at the well-known Breit-Wigner formula

$$T(E) = \frac{4\Gamma_L\Gamma_R}{(E - \epsilon_0)^2 + (\Gamma_L + \Gamma_R)^2}. \tag{8.24}$$

Analogously, we can easily re-derive all the different formulas obtained in the last chapter like for instance, the current expression of the Exercise 7.5 for a linear tight-binding chain (see Exercise 8.3).

8.1.2   *Nonorthogonal basis sets*

In the context of molecular electronics the use of nonorthogonal local basis is quite common. In this sense, we have to discuss how to generalize the current formula derived above for this type of bases. We shall address this issue using a simple argument put forward by Emberly and Kirczenow [200] and we refer the reader to different entry points in the literature for more rigorous discussions.

In an orthogonal basis set, the overlap between the different basis states is: $\langle i|j \rangle = S_{ij} = \delta_{ij}$, while the corresponding secular equation that provides the eigenstates of the systems reads: $\mathbf{H}_O - E\mathbf{1} = 0$. Here, the subindex O indicates that we are working with an orthogonal basis set. Finally, the Green's functions are simply obtained by inverting the Hamiltonian in the usual way, i.e. $\mathbf{G}_O = [E\mathbf{1} - \mathbf{H}_O]^{-1}$.

For a nonorthogonal basis, the overlap matrix differs from the unity and the secular equation adopts the form: $\mathbf{H}_N - E\mathbf{S} = 0$. Here, N denotes nonorthogonal basis set. The left hand side of the secular equation can be rewritten as follows

$$\mathbf{H}_N - E\mathbf{S} = \mathbf{H}_N - E(\mathbf{S} - \mathbf{1}) - E\mathbf{1} \equiv \mathbf{H}_N' - E\mathbf{1}. \tag{8.25}$$

Notice that the secular equation has now the same form as in the orthogonal case, but with an effective energy-dependent Hamiltonian: $\mathbf{H}_N' \equiv \mathbf{H}_N - E(\mathbf{S} - \mathbf{1})$. In this Hamiltonian, the on-site energies remain unchanged, as compared with the original one, whereas the hopping matrix elements become energy dependent: $h_{ij}' = h_{ij} - ES_{ij}$. This argument suggests that the only effect that the nonorthogonal basis introduces is the renormalization of the hopping elements and therefore, the current formula is identical to the one derived above after replacing the orthogonal parameters by the nonorthogonal ones. Additionally, the retarded/advanced Green's functions appearing in the expression of the transmission have to be calculated by means of the following Dyson's equation

$$\mathbf{G}_{CC}^{r,a} = \left[(E \pm i0^+)\mathbf{S}_{CC} - \mathbf{H}_{CC} - \mathbf{\Sigma}_L^{r,a} - \mathbf{\Sigma}_R^{r,a} \right]^{-1}, \qquad (8.26)$$

where $\mathbf{H}_{CC}$ is now the nonorthogonal Hamiltonian of the central region and $\mathbf{S}_{CC}$ is the sector of the overlap matrix corresponding to the central region. On the other hand, in the expression of the self-energies we have to replace the hopping matrices $\mathbf{t}_{XC}$ by $\mathbf{t}_{XC} - E\mathbf{S}_{XC}$, where $X = L, R$.

There is another way of deriving the result above [201]. The idea is to transform every quantity from an orthogonal representation to a nonorthogonal one via the so-called Löwdin's transformation. This transformation is defined by $\mathbf{S}^{-1/2}$, where $\mathbf{S}$ is the overlap matrix and it transforms an operator $\mathbf{M}_O$ in the orthogonal basis to the corresponding one in the nonorthogonal basis, $\mathbf{M}_N$, as follows

$$\mathbf{M}_N = \mathbf{S}^{1/2}\mathbf{M}_O\mathbf{S}^{1/2}. \qquad (8.27)$$

Inserting $1 = \mathbf{S}^{-1/2}\mathbf{S}^{1/2}$ in the current formula in the orthogonal representation, we arrive after some straightforward algebra at the same conclusions as those stated above. For more detailed discussion of the derivation of this result, we recommend Refs. [202, 203].

8.1.3 *Spin-dependent elastic transport*

So far we have only considered situations where there was spin symmetry. We proceed now to generalize the Landauer formula derived above to situations where the spin symmetry is broken. Those situations include very prominent examples in molecular electronics such as the transport through ferromagnetic atomic-sized contacts (see Chapter 12) and molecular junctions with ferromagnetic leads.

For the sake of concreteness, let us first consider the case of a metallic atomic-sized contact made of a ferromagnetic material (like Fe, Co or Ni). It is customary to analyze the transport properties of these junctions within the two-current model put forward by N.F. Mott [204, 205]. Mott realized that at sufficiently low temperature, where the magnon scattering in a ferromagnet becomes vanishingly small, electrons of majority and minority spin, with magnetic moment parallel and antiparallel to the magnetization, respectively, do not mix in the scattering processes. This means in practice that the total current can then be expressed as the sum of two independent contribution coming from the two different spin projections, which implies that in ferromagnets the current is spin polarized. Therefore, the Landauer formula of Eq. (8.20) adopts now the form

$$I = \frac{e}{h} \int_{-\infty}^{\infty} dE \, T(E, V) \, (f_L - f_R), \qquad (8.28)$$

where $T(E, V)$ is the total transmission sum of the transmissions of the two spin bands

$$T = \sum_{\sigma=\uparrow,\downarrow} T_\sigma = \sum_\sigma \mathrm{Tr} \left\{ \mathbf{t}_\sigma^\dagger \mathbf{t}_\sigma \right\} = \sum_{n,\sigma} T_{n\sigma}, \qquad (8.29)$$

where $\mathbf{t}_\sigma$ is the transmission matrix of the spin sector σ and $T_{n\sigma}$ are the corresponding transmission coefficients. The transmission $\mathbf{t}_\sigma$ is given by Eq. (8.19), where all the quantity are referred to the spin band σ.

The previous current formula describes any (elastic) situation where there is no mixing of the two spin bands. This is what occurs in most of the atomic-scale junctions that we have in mind, where the system size is clearly smaller than the spin-diffusion length. However, this is no longer true if, for instance, there is a small domain wall of atomic size in the junction or a strong spin-orbit interaction is present. Let us show how the formula for the elastic current is modified in those situations.

A system in which the majority and minority spin bands are mixed can be generically described by the following tight-binding Hamiltonian

$$\mathbf{H} = \sum_{ij\alpha\beta\sigma\sigma'} h_{i\alpha,j\beta}^{\sigma\sigma'} \mathbf{c}_{i\alpha\sigma}^\dagger \mathbf{c}_{j\beta\sigma'}, \qquad (8.30)$$

where i, j run over the atomic sites, α, β denote the different atomic orbitals, and $\sigma = \uparrow, \downarrow$ the spin. Within this model, the current can be computed following the same steps as in the case with spin symmetry and we only sketch here the main idea and the final result. Briefly, the atomic-scale contact is divided into three parts, a central region C containing the constriction and the left/right (L/R) leads. The retarded Green's functions of the central part read[4]

$$\mathbf{G}_{CC}^r = [E\mathbf{S}_{CC} - \mathbf{H}_{CC} - \mathbf{\Sigma}_L^r - \mathbf{\Sigma}_R^r]^{-1}, \qquad (8.31)$$

where $\mathbf{\Sigma}_X^r = (\mathbf{t}_{CX} - E\mathbf{S}_{CX})\mathbf{g}_{XX}^r(\mathbf{t}_{CX} - E\mathbf{S}_{CX})^\dagger$ are the lead self-energies $(X = L, R)$. Here, $\mathbf{t}_{CX}$ and $\mathbf{S}_{CX}$ are the hoppings and overlaps between the C region and the lead X, and $\mathbf{g}_{XX}^r$ is a lead Green's function. Notice that the dimension of all the matrices in the previous equation is equal to the total number of orbitals in the central region multiplied by two. This factor

---

[4]Notice that we take into account the possibility of using non-orthogonal basis sets.

two comes from the structure in spin space. As before, the transmission matrix is given by $\mathbf{t} = 2\mathbf{\Gamma}_L^{1/2}\mathbf{G}_{CC}^r\mathbf{\Gamma}_R^{1/2}$, but this time the scattering rate matrices are given by where $\mathbf{\Gamma}_X = i[\mathbf{\Sigma}_X^r - (\mathbf{\Sigma}_X^r)^\dagger]/2$. The reason for this is that, in general, the Hamiltonian is not real and $\mathbf{g}_{XX}^r = (\mathbf{g}_{XX}^a)^\dagger$. Finally, the current then adopts the standard Landauer form of Eq. (8.28), but now the trace includes not only a sum over the orbitals in the central part, but also over spins. Finally, the low-temperature linear conductance can be written as $G = (e^2/h)\sum_n T_n$, where T_n are the transmission coefficients, i.e. the eigenvalues of $\mathbf{t}^\dagger\mathbf{t}$ at Fermi energy.

8.2 Current through an interacting atomic-scale junction

As we explained in previous chapters, one of the main advantages of the Green's functions techniques with respect to the scattering approach is the possibility to describe the influence of correlation and inelastic effects in the transport characteristics. The goal of this section is to show how the Landauer formula derived in the previous section is modified when such effects are present in an atomic-scale junction. The derivation of the current formula for an interacting system in the framework of Hamiltonian written in a local basis was first done by Caroli and coworkers [209]. Later, Meir and Wingreen re-derived this formula to express the current in a more appealing way [210]. Although this latter formula is widely used in the context of mesoscopic physics, its simplest form is not generally valid for atomic-scale systems (see discussion below). We follow now the formulation of Caroli and coworkers [209] and then discuss the Meir-Wingreen formula in section 8.2.2.

Let us consider again the generic junction of Fig. 8.1. For the sake of simplicity, we assume that the interactions (such as electron-electron or electron-phonon interactions) are restricted to the central region. The calculation of the current is identical to that of the elastic case up to Eq. (8.10). At this point we have to determine the functions $\mathbf{G}^{+-/-+}$ of the central region, which can be done using the general Keldysh relations [see Eqs. (7.47) and (7.48)]

$$\mathbf{G}^{+-/-+} = (1 + \mathbf{G}^r\mathbf{t})\,\mathbf{g}^{+-/-+}\,(1 + \mathbf{t}\mathbf{G}^a) - \mathbf{G}^r\mathbf{\Sigma}^{+-/-+}\mathbf{G}^a, \qquad (8.32)$$

where $\mathbf{\Sigma}^{+-/-+}$ are the Keldysh components of the self-energy describing the inelastic effects. Notice that the last term was absent in the elastic case. We now take the block-element (C,C) in the previous equation and obtain

$$\mathbf{G}_{CC}^{+-/-+} = \mathbf{g}_{CC}^{+-/-+} + \mathbf{G}_{CC}^r \hat{t}_{CL} \mathbf{g}_{LL}^{+-/-+} \mathbf{t}_{LC} \mathbf{G}_{CC}^a + \tag{8.33}$$
$$\mathbf{G}_{CR}^r \mathbf{t}_{CR} \mathbf{g}_{RR}^{+-/-+} \mathbf{t}_{RC} \mathbf{G}_{CC}^a - \mathbf{G}_{CC}^r \boldsymbol{\Sigma}_{CC}^{+-/-+} \mathbf{G}_{CC}^a.$$

Here we have used the fact that the interactions are restricted to the central region, which in practice means that the inelastic self-energies $\boldsymbol{\Sigma}^{+-/-+}$ have only a (CC) component. Introducing now these Green's functions in Eq. (8.10), one can readily show that the current can be written as the sum of two contributions: $I = I_{el} + I_{inel}$, where

$$I_{el} = \frac{8e}{h} \int_{-\infty}^{\infty} dE\, \mathrm{Tr}\left[\mathbf{G}_{CC}^r \boldsymbol{\Gamma}_R \mathbf{G}_{CC}^a \boldsymbol{\Gamma}_L\right] (f_L - f_R) \tag{8.34}$$

$$I_{inel} = \frac{4ie}{h} \int_{-\infty}^{\infty} dE\, \mathrm{Tr}\left\{\mathbf{G}_{CC}^a \boldsymbol{\Gamma}_L \mathbf{G}_{CC}^r \left[(f_L - 1)\boldsymbol{\Sigma}_{CC}^{+-} - f_L \boldsymbol{\Sigma}_{CC}^{-+}\right]\right\} \tag{8.35}$$

Again, the trace in these expressions has to be understood as a sum over all the orbitals in the central region. The first term, I_{el}, represents the elastic current and it has the same form as the Landauer formula derived in the previous section. The second term, I_{inel}, which we call inelastic current, is the new contribution due to the inelastic interactions. Notice that this term has a rather asymmetric form, which is a consequence of our choice of computing the current in the left interface. If wanted, one can symmetrize this expression by combining it with the inelastic current evaluated in right interface[5] and using current conservation to define the inelastic current as $I_{inel} = (I_{inel}^L + I_{inel}^R)/2$.

From Eq. (8.35) it is not obvious that the inelastic current vanishes at zero bias. However, this can be shown by using the general relations for a system in equilibrium

$$\boldsymbol{\Sigma}^{+-}(E) = (\boldsymbol{\Sigma}^r - \boldsymbol{\Sigma}^a)\, f(E); \quad \boldsymbol{\Sigma}^{-+}(E) = (\boldsymbol{\Sigma}^r - \boldsymbol{\Sigma}^a)\,(f(E) - 1), \tag{8.36}$$

where $f(E)$ is the Fermi function.

It is important to emphasize that the retarded and advanced Green's functions of the central region are computed through a Dyson's equation that now also includes the new inelastic self-energies

$$\mathbf{G}_{CC}^a = (\mathbf{G}_{CC}^r)^\dagger = [(E - i0^+)\mathbf{1} - \mathbf{H}_{CC} - \boldsymbol{\Sigma}_L^a - \boldsymbol{\Sigma}_R^a - \boldsymbol{\Sigma}_{CC}^a]^{-1}, \tag{8.37}$$

---

[5] Such expression reads

$$I_{inel}^R = -\frac{4ie}{h} \int_{-\infty}^{\infty} dE\, \mathrm{Tr}\left\{\mathbf{G}_{CC}^a \boldsymbol{\Gamma}_R \mathbf{G}_{CC}^r \left[(f_R - 1)\boldsymbol{\Sigma}_{CC}^{+-} - f_R \boldsymbol{\Sigma}_{CC}^{-+}\right]\right\}.$$

where $\mathbf{\Sigma}_{CC}^a$ is the advanced component of the self-energy describing the inelastic interactions in the central region and $\mathbf{\Sigma}_{L,R}^a$ are given by Eq. (8.16).

The precise form of the inelastic self-energies and in turn of the contribution of the inelastic term to the total current depends on the specific nature of the inelastic interactions. In order to illustrate the use of this new current formula, we present in the next subsection an important example concerning the role of the electron-phonon interaction in molecular junctions.

8.2.1 *Electron-phonon interaction in the resonant tunneling model*

In most transport experiments in molecular junctions, there is no certainty that the current is indeed flowing through a molecule. Thus, it is to find unambiguous signatures of the presence of the molecule, for instance, in the current-voltage (I-V) characteristics. As we shall discuss extensively in Chapter 16, presently the most convincing signatures are those related to the excitation of vibration modes of the molecules used to form the junctions. For this reason, it has become very important to understand how the local interaction between the conduction electrons and molecular vibrations is manifested in the I-V curves. We shall address this issue here with a toy model that will also serve us to illustrate the use of the inelastic current formula derived above.

Let us consider the resonant tunneling model that was already discussed in section 7.4.3. Let us recall that in this model an electronic level with energy ϵ_0 is coupled to two metallic reservoirs via hopping elements t_L and t_R, where L and R denote the left and right leads, respectively. Now, we assume that this resonant level is also coupled to a single local vibrational mode of energy $\hbar\omega$. This model is schematically represented in Fig. 8.2. Our goal is to compute the current-voltage characteristics when a constant bias voltage, V, is applied. In particular, we shall pay special attention to the correction of the current due to the electron-vibration interaction.

The Hamiltonian of the system that we have just described has the following form

$$\mathbf{H} = \mathbf{H}_e + \mathbf{H}_{vib} + \mathbf{H}_{e-vib}, \tag{8.38}$$

where $\mathbf{H}_e$ describes the electronic part of this problem as it is given by Eq. (7.93). The vibrational mode is described as a simple harmonic oscil-

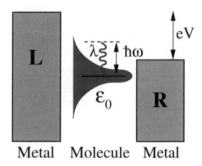

Metal　　Molecule　　Metal

Fig. 8.2　Schematic representation of the resonant tunneling model where the electronic level is coupled to a single vibrational mode of frequency ω with an electron-phonon coupling constant λ.

lator of energy $\hbar\omega$ by

$$\mathbf{H}_{vib} = \hbar\omega \left(\mathbf{b}^\dagger \mathbf{b} + \frac{1}{2} \right), \qquad (8.39)$$

where the creation and annihilation operators $\mathbf{b}^\dagger$ and $\mathbf{b}$ satisfy the bosonic commutation relations, e.g. $[\mathbf{b}, \mathbf{b}^\dagger] = 1$. Finally, the interaction between the vibration mode and the conduction electrons is described by the following Hamiltonian [174]

$$\mathbf{H}_{e-vib} = \lambda \mathbf{c}_0^\dagger \mathbf{c}_0 (\mathbf{b}^\dagger + \mathbf{b}), \qquad (8.40)$$

where λ is the electron-vibration coupling constant and $\mathbf{c}_0^\dagger$ and $\mathbf{c}_0$ are the fermionic operators related to the electronic level.[6]

In this simple model, the central region consists of a single site and therefore the Green's functions, scattering rates and self-energies appearing in the current formulas of Eqs. (8.34) and (8.35) are just scalars. Such formulas reduce to the following expressions

$$I_{el} = \frac{8e}{h} \Gamma_L \Gamma_R \int_{-\infty}^{\infty} dE \, |G^r|^2 (f_L - f_R), \qquad (8.41)$$

$$I_{inel} = \frac{4ie}{h} \Gamma_L \int_{-\infty}^{\infty} dE \, |G^r|^2 \left[(f_L - 1)\Sigma_{e-vib}^{+-} - f_L \Sigma_{e-vib}^{-+} \right]. \qquad (8.42)$$

Here, the Green's function $G^r(E)$ refers to the central site or resonant level. Moreover, as usual, we have assumed that the scattering rates, $\Gamma_{L,R}$, that describe the strength of the coupling between the resonant level and the leads are independent of the energy. Now, we have to determine the

---

[6]The spin does not play any role in this problem and we have dropped it in the previous expression.

self-energy associated to the electron-vibration interaction, $\check{\Sigma}_{e-vib}$. The simplest approximation for this self-energy can be obtained by applying perturbation theory and keeping only the lowest order correction. Physically, this means that one only takes single-phonon processes into account. As it is shown in Exercise 8.5, the first non-vanishing correction to the self-energy is proportional to λ^2 and its different components are given by[7]

$$\Sigma^r_{e-vib}(E) = i\lambda^2 \int_{-\infty}^{\infty} \frac{dE'}{2\pi} \left\{ D^r(E')\tilde{G}^{-+}(E-E') + D^{+-}(E')\tilde{G}^r(E-E') \right\},$$

$$\Sigma^{+-}_{e-vib}(E) = -i\lambda^2 \int \frac{dE'}{2\pi} D^{+-}(E')\tilde{G}^{+-}(E-E'),$$

$$\Sigma^{-+}_{e-vib}(E) = -i\lambda^2 \int \frac{dE'}{2\pi} D^{-+}(E')\tilde{G}^{-+}(E-E'). \tag{8.43}$$

Here, the functions with tilde are the electronic Green's functions of the resonant site where the coupling to the leads is taken into account and the electron-vibration is not included, i.e. these are, loosely speaking, the unperturbed functions of this problem, which are given by

$$\tilde{G}^r(E) = [(E+i\eta) - \epsilon_0 + i(\Gamma_L + \Gamma_R)]^{-1},$$
$$\tilde{G}^{+-}(E) = 2i|\tilde{G}^r(E)|^2 [\Gamma_L f_L + \Gamma_R f_R],$$
$$\tilde{G}^{-+}(E) = -2i|\tilde{G}^r(E)|^2 [\Gamma_L(1-f_L) + \Gamma_R(1-f_R)], \tag{8.44}$$

where $\eta = 0^+$.

On the other hand, the D's are the phonon Green's functions of this problem and their general definitions can be found in Exercise 8.4. Assuming that the vibration mode is in thermal equilibrium at the temperature of the electrodes, these functions are given by (see Exercise 8.4)

$$D^r(E) = \frac{1}{E - \hbar\omega + i\eta} - \frac{1}{E + \hbar\omega + i\eta},$$
$$D^{+-}(E) = -2\pi i \left\{ (n_B+1)\delta(E+\hbar\omega) + n_B\delta(E-\hbar\omega) \right\},$$
$$D^{-+}(E) = -2\pi i \left\{ (n_B+1)\delta(E-\hbar\omega) + n_B\delta(E+\hbar\omega) \right\}, \tag{8.45}$$

where $n_B = 1/[\exp(\beta\hbar\omega) - 1]$, with $\beta = 1/k_B T$ is the Bose function that describes the thermal occupation of the vibration mode.

---

[7]There is an additional contribution to $\Sigma^r_{e-vib}(E)$ which is equal to

$$-i\lambda^2 D^r(0) \int_{-\infty}^{\infty} \frac{dE'}{2\pi} \tilde{G}^{+-}(E').$$

This gives a constant contribution that simply renormalizes the position of the resonant level and we ignore it in what follows.

Now, we expand the current to lowest order in the coupling constant λ. To do so, in the inelastic term of Eq. (8.42) we just need to introduce the expressions of Σ_{e-vib} and replace the full G^r by $\tilde{G}^r$. In the elastic term of Eq. (8.41) we have to insert the lowest order correction of the Green's functions, i.e. $G^r \approx \tilde{G}^r + \tilde{G}^r \Sigma^r_{e-vib} \tilde{G}^r$, and collect all the terms up to second order in λ. Doing this, the current can be expressed as the sum of three terms: $I = I^0_{el} + \delta I_{el} + I_{inel}$, where the different contributions are given by

$$I^0_{el} = \frac{8e}{h}\Gamma_L\Gamma_R \int_{-\infty}^{\infty} dE \, |\tilde{G}^r(E)|^2 \left[f_L(E) - f_R(E)\right], \qquad (8.46)$$

$$\delta I_{el} = \frac{16e}{h}\Gamma_L\Gamma_R \int_{-\infty}^{\infty} dE \, |\tilde{G}^r(E)|^2 \times$$
$$\text{Re}\left\{\tilde{G}^r(E)\Sigma^r_{e-vib}(E)\right\}\left[f_L(E) - f_R(E)\right], \qquad (8.47)$$

$$I_{inel} = \frac{8e\lambda^2}{h}\Gamma_L\Gamma_R \int_{-\infty}^{\infty} dE \, \left\{(n_B+1)|\tilde{G}^r(E)\tilde{G}^r(E-\hbar\omega)|^2 \times\right.$$
$$\left[f_L(E)(1-f_R(E-\hbar\omega)) - f_R(E)(1-f_L(E-\hbar\omega))\right]$$
$$+n_B|\tilde{G}^r(E)\tilde{G}^r(E+\hbar\omega)|^2 \times$$
$$\left.\left[f_L(E)(1-f_R(E+\hbar\omega)) - f_R(E)(1-f_L(E+\hbar\omega))\right]\right\}. \quad (8.48)$$

The first contribution, I^0_{el}, is nothing else but the elastic current in the absence of electron-vibration interaction that we have studied in section 7.4.3, see Fig. 8.3(a). The third term, I_{inel}, is the inelastic contribution coming from the emission and absorption of a vibrational mode. Notice that the term in I_{inel} proportional to n_B corresponds to the contribution of processes assisted by the absorption of a mode, see Fig. 8.3(b), whereas the term proportional to $(n_B + 1)$ is the contribution of tunneling processes mediated by the stimulated and spontaneous emission of a mode, see Fig. 8.3(c). At temperatures much lower than $\hbar\omega/k_B$, the second one dominates. Moreover, it is easy to see that at low temperatures the emission term has a threshold voltage equal to the vibration energy $(\hbar\omega/e)$ below which it vanishes. Above this voltage this term gives always a positive contribution, which means that it gives rise to a step up in the conductance. The second term, δI_{el}, has a less obvious interpretation. It is an elastic term that involves the emission and absorption of a virtual vibrational mode, see Fig. 8.3(d). This term will be referred to as *elastic correction*.

It is easy to evaluate numerically the different contributions to the current for an arbitrary range of parameters. However, in order to gain some insight, we concentrate here on a limiting case that can be worked out analytically. Let us assume that the energy dependence in the retarded

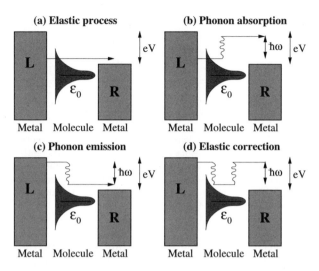

Fig. 8.3 Schematic representation of the elastic (a) and inelastic (b-d) tunneling processes that can occur in the model in which an electronic level is coupled to a single vibration mode. Here, we have assumed that the electron-phonon interaction is weak and the processes (b-d) are responsible for the inelastic correction to the elastic current up to order λ^2.

electronic Green's functions can be neglected, i.e. $\tilde{G}^r(E) = \tilde{G}^r(E_{\mathrm{F}})$. This means in practice that we assume that both the local density of states and the transmission are energy-independent. This is a good approximation in two cases: (i) when the coupling to the leads is so strong that $\Gamma_L + \Gamma_R >> \hbar\omega, eV, |E_{\mathrm{F}} - \epsilon_0|$ and (ii) when the resonant level is far away from the Fermi energy, i.e. $|E_{\mathrm{F}} - \epsilon_0| \gg \Gamma_{L,R}, eV, \hbar\omega$. With this approximation the different terms can be computed analytically. At temperatures well below the vibrational energy, the correction to the elastic current is a competition between the emission term in I_{inel} and the elastic correction δI_{el}. Assuming a symmetric junction, $\Gamma_L = \Gamma_R = \Gamma$, the three contributions to the zero-temperature differential conductance are given by (see Exercise 8.6)

$$\frac{G_{el}^0}{G_0} = T,$$

$$\frac{\delta G_{el}(V)}{G_0} = \frac{\lambda^2}{\Gamma^2} \left\{ \begin{array}{ll} T^2(1-T)/2; & |eV| \leq \hbar\omega \\ T^2(1-2T)/2; & |eV| > \hbar\omega \end{array} \right.$$

$$\frac{G_{inel}(V)}{G_0} = \frac{\lambda^2}{\Gamma^2} \left\{ \begin{array}{ll} 0; & |eV| \leq \hbar\omega \\ T^2/4; & |eV| > \hbar\omega \end{array} \right. , \tag{8.49}$$

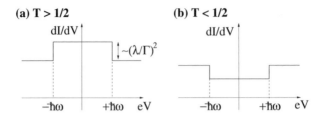

Fig. 8.4   Signature of a vibration mode in the zero-temperature differential conductance of a resonant level. (a) For transmissions greater than $1/2$, the differential conductance exhibits a step down at $eV = \pm\hbar\omega$ due to electron-vibration interaction. The height of the step is mainly determined by the ratio λ^2/Γ^2 and it has been exaggerated for clarity. (b) The signature of the vibration mode in the differential conductance for transmissions less than $1/2$ is a step up at $eV = \pm\hbar\omega$.

where $T = 4\Gamma^2|\tilde{G}^r(E_F)|^2$ is the elastic transmission in the absence of electron-vibration interaction. Notice that δG_{el} has a discontinuity (step down) at $eV = \pm\hbar\omega$ proportional to $-T^3/2$, while the emission term contributes to this jump as $\sim +T^2/4$. This means that the sign of the conductance jump depends on the junction transmission and it is given by: $(\lambda^2/\Gamma^2)T^2(1-2T)/4$. Notice that the magnitude is determined by the ratio of the two relevant coupling constants, λ and Γ, which has been assumed to be small. On the other hand, the conclusion of this analysis is that the electron-vibration interaction in this simple model is reflected in the appearance of a jump in the low-temperature conductance at $eV = \pm\hbar\omega$. This jump is seen as a step up in conductance for $T < 1/2$ and as step down for $T > 1/2$. This conclusion is summarized schematically in Fig. 8.4. The signature of the vibration modes can be seen more clearly in the second derivative of the current, d^2I/dV^2, where it appears as a peak or as a dip depending on the junction transmission.[8] The results of this model will be discussed in much more detail in section 17.1.1.

8.2.2   *The Meir-Wingreen formula*

As we mentioned in the introduction of this section, Meir and Wingreen proposed in 1992 [210] an alternative form for the formula of the current through an interacting region. This formula has been widely used in mesoscopic physics and, in particular, for studying the transport through all kind of quantum dots and molecular transistors. For this reason and for the sake

---

[8]The signature is antisymmetric with respect to the voltage polarity in the sense that if it appears as peak for positive bias, it appears as dip for negative one.

of completeness, we include here a short discussion of the derivation of this formula. Further technical details can be found in Exercise 8.7.

Once more, we consider the system of Fig. 8.1, where the central part represents an interacting region. The current evaluated at the left interface is given by Eq. (8.8). Now, to determine the Keldysh-Green's functions appearing in that expression we make use of Dyson's equation in Keldysh space as follows

$$\mathbf{G}_{LC}^{+-} = \mathbf{g}_{LL}^{++}\mathbf{t}_{LC}\mathbf{G}_{CC}^{+-} - \mathbf{g}_{LL}^{+-}\mathbf{t}_{LC}\mathbf{G}_{CC}^{--} \qquad (8.50)$$
$$\mathbf{G}_{CL}^{+-} = \mathbf{G}_{CC}^{++}\mathbf{t}_{CL}\mathbf{g}_{LL}^{+-} - \mathbf{G}_{CC}^{+-}\mathbf{t}_{CL}\mathbf{g}_{LL}^{--}.$$

Using the general relations $\mathbf{G}^{+-}+\mathbf{G}^{-+} = \mathbf{G}^{++}+\mathbf{G}^{--}$ and $\mathbf{G}^{+-}-\mathbf{G}^{-+} = \mathbf{G}^a - \mathbf{G}^r$, it is straightforward to show that the current evaluated at the left interface, I_L, is given by

$$I_L = \frac{4ie}{h} \int_{-\infty}^{\infty} dE \, \mathrm{Tr} \left\{ \mathbf{\Gamma}_L \left[\mathbf{G}_{CC}^{+-} + (\mathbf{G}_{CC}^r - \mathbf{G}_{CC}^a)f_L \right] \right\}, \qquad (8.51)$$

where the scattering rate $\mathbf{\Gamma}_L$ is defined in the usual way.

Analogously, one can obtain the expression of the current, evaluated this time at the right interface, I_R. Writing then the current in a more symmetric manner as $I = (I_L + I_R)/2$, one arrives at

$$I = \frac{2ie}{h} \int_{-\infty}^{\infty} dE \, \mathrm{Tr} \left\{ (\mathbf{\Gamma}_L - \mathbf{\Gamma}_R)\mathbf{G}_{CC}^{+-} + (f_L\mathbf{\Gamma}_L - \mathbf{\Gamma}_R f_R)(\mathbf{G}_{CC}^r - \mathbf{G}_{CC}^a) \right\}.$$
$$(8.52)$$

This is the Meir-Wingreen formula in its most general form. It is completely equivalent to the expression derived above and in the non-interacting case it reduces to the Landauer formula (see Exercise 8.7). The "popularity" of this formula is due to the fact that it takes an appealing form in the case in which the couplings to the leads differ only by a constant factor, $\mathbf{\Gamma}_L(E) = \lambda\mathbf{\Gamma}_R(E)$. In this case, the current reads

$$I = \frac{8e}{h} \int_{-\infty}^{\infty} dE \, \mathrm{Tr} \left\{ \mathbf{\Gamma}\mathbf{A} \right\} (f_L - f_R), \qquad (8.53)$$

where $\mathbf{\Gamma} \equiv \mathbf{\Gamma}_L\mathbf{\Gamma}_R/(\mathbf{\Gamma}_L + \mathbf{\Gamma}_R)$ and $\mathbf{A} \equiv i(\mathbf{G}_{CC}^r - \mathbf{G}_{CC}^a)/2$ is the spectral function of the central region. The division in the expression of $\mathbf{\Gamma}$ has to be understood as multiplication by the inverse of the matrix appearing in the denominator. The nice thing about this formula is that the current is expressed in terms of the spectral function, $\mathbf{A}$. Unfortunately, the condition of proportionality of the scattering rates is quite restrictive and most cases it is not really fulfilled. For applications of this latter formula, see Exercises 8.9 and 8.10.

8.3   Time-dependent transport in nanoscale junctions

Up to now we have only considered stationary situations where the current was time-independent. In this section we shall illustrate the use of the Keldysh formalism for computing the transport properties of a system that is subjected to an externally applied time-dependent drive.

As a model problem, which will be very important for Chapter 20, we consider here the calculation of the current in an atomic or molecular contact under the presence of an oscillating bias voltage: $V(t) = V + V_{ac}\sin(\omega t)$, where V is the dc part of the bias and V_{ac} and ω are the amplitude and the frequency of this periodic potential, respectively. This ac field can be simply due to an applied chemical potential difference, but one can also imagine that it is induced in the junction by the application of an external radiation, which is a situation is of special interest for us. The question of how the current through an atomic-scale junction can be modified by irradiation is a very important subject in molecular electronics [211]. In other contexts, like for instance in the case of superconducting tunnel junctions, this problem has a long history [212]. From the theory side, the "photon-assisted" transport has been traditionally addressed following the seminal work of Tien and Gordon (TG) [213], where this phenomenon was described by a harmonic voltage at the radiation frequency ω applied to one of the leads of a junction. Such a simple approach have been quite successful in gaining a qualitative understanding of radiation-induced currents in many situations like superconducting systems [212], semiconductor heterostructures [214], STM [215], and and other mesoscopic systems [211]. Our discussion in this section provides the basis to address similar problems in the context of atomic and molecular junctions.

Different theoretical approaches have been applied to the problem that we are about to tackle such as the scattering approach [165] or Floquet theory [211, 214]. We shall follow here the nonequilibrium Green's function formalism used in the previous sections of this chapter (see also Refs. [216–220]). This approach allows us to describe the photo-transport in realistic atomic and molecular contacts, in the sense that it can be combined by advanced electronic structure methods.

Let us consider the generic geometry of Fig. 8.1, which again represents an atomic or molecular contact of arbitrary shape. For simplicity, we assume that the correlation and inelastic effects do not play a mayor role in this case. In other words, we assume that the transport in the absence of the ac drive is coherent. We describe the system with the following time-dependent Hamiltonian: $\mathbf{H}(t) = \mathbf{H}_0 + \mathbf{H}_1(t)$. Here, $\mathbf{H}_0$ is the Hamiltonian of Eq. (8.1) that contains the full microscopic information about the system in the absence of dc and ac voltages.[9] The time-dependent part $\mathbf{H}_1(t)$ describes the driving potential and it can be written generically as

$$\mathbf{H}_1(t) = \sum_{ij} W_{ii}(t) c_i^\dagger c_i, \qquad (8.54)$$

where $W_{ii}(t) = U_i^{dc} + U_i^{ac} \sin(\omega t)$ describe the shifts in the on-site energies induced by the dc and ac parts of voltage. Here, the U_i's are the amplitudes of the local potential at site i and equal for all the orbitals in the same atom. We assume that the potential is spatially constant in the L and R leads and equal to $U_X(t) = U_X^{dc} + U_X^{ac} \sin(\omega t)$, and $X = L, R$. The applied dc voltage is $V = (U_L^{dc} - U_R^{dc})/e$ and the corresponding ac part is $V_{ac} = (U_L^{ac} - U_R^{ac})/e$. We shall calculate the current for an arbitrary potential profile in the central region (encoded in the functions $U_i(t)$), the actual shape of which should in principle be obtained self-consistently [165].

In order to derive the current formula in this situation, we shall follow the same steps taken in section 8.1 and we shall emphasize here only the main differences with respect to that calculation. Our starting point is the expression of the time-dependent current evaluated at the left interface, which can be written in terms of the Green's functions as follows

$$I(t) = \frac{2e}{\hbar} \mathrm{Tr} \left[\mathbf{G}_{CL}^{+-}(t,t) \mathbf{t}_{LC} - \mathbf{t}_{CL} \mathbf{G}_{LC}^{+-}(t,t) \right]. \qquad (8.55)$$

To determine the Green's functions we follow the same perturbative approach as in section 8.1. The essential difference now is that the Green's functions depend explicitly on two time arguments (rather than on their difference), which introduces an extra complication, as we are about to show. Using the Dyson's equation [see Eq. (7.46)] we can express the functions appearing in the current as[10]

$$\mathbf{G}_{LC}^{+-}(t,t') = \left\{ \mathbf{g}_{LL}^{+-} \circ \mathbf{t}_{LC} \circ \mathbf{G}_{CC}^a + \mathbf{g}_{LL}^r \circ \mathbf{t}_{LC} \circ \mathbf{G}_{CC}^{+-} \right\}(t,t') \quad (8.56)$$

$$\mathbf{G}_{CL}^{+-}(t,t') = \left\{ \mathbf{G}_{CC}^{+-} \circ \mathbf{t}_{CL} \circ \mathbf{g}_{LL}^a + \mathbf{G}_{CC}^r \circ \mathbf{t}_{CL} \circ \mathbf{g}_{LL}^{+-} \right\}(t,t'),$$

where the product $\circ$ is defined by $(\mathbf{A} \circ \mathbf{B})(t,t') = \int dt_1\, \mathbf{A}(t,t_1)\mathbf{B}(t_1,t')$, i.e. it is a convolution over the intermediate time arguments. This means that any Dyson's equation is no longer an algebraic equation as before, but rather an integral equation. Anyway, if we handle carefully this non-commutative

---

[9] We shall assume throughout this discussion that this Hamiltonian is written is a local orthogonal basis.

[10] Here, the time-dependent hopping matrices are defined, for instance, as: $\mathbf{t}_{LC}(t,t') = \mathbf{t}_{LC}\delta(t-t')$.

product, the derivation still follows the same steps as in section 8.1. Thus, we can easily arrive at the following expression for the current

$$I(t) = \frac{2e}{\hbar} \text{Tr} \left[\mathbf{G}_{CC}^r \circ \mathbf{\Sigma}_R^{-+} \circ \mathbf{G}_{CC}^a \circ \mathbf{\Sigma}_L^{+-} - \right.$$
$$\left. \mathbf{G}_{CC}^r \circ \mathbf{\Sigma}_R^{+-} \circ \mathbf{G}_{CC}^a \circ \mathbf{\Sigma}_L^{-+} \right] (t, t), \qquad (8.57)$$

which is the analog of Eq. (8.13). Here, we have define the "lead self-energies"

$$\mathbf{\Sigma}_X^{+-/-+}(t, t') = \left[\mathbf{t}_{CX} \circ \mathbf{g}_{XX}^{+-/-+} \circ \mathbf{t}_{XC} \right] (t, t'), \qquad (8.58)$$

where $X = L, R$.

The lead Green's functions have now a more complicated time dependence. Due to the ac voltage they oscillate on time as follows[11]

$$\mathbf{g}_X^c(t, t') = e^{-i\phi_X(t)} \mathbf{g}_X^c(t - t') e^{i\phi_X(t')}, \qquad (8.59)$$

where $c = r, a, +-, -+$. Here, $\partial\phi_X(t)/\partial t = \mu_X(t)/\hbar$, where $\mu_X(t)$ is the chemical potential of the corresponding electrode. Therefore, $\phi_X(t) = (U_X^{dc}/\hbar)t + \alpha_X \cos(\omega t)$, with $\alpha_X = U_X^{ac}/(\hbar\omega)$.

As usual, it is more convenient to work in energy space and for this reason we now Fourier transform with respect to the two time arguments

$$\mathbf{g}_X^c(t, t') = \frac{1}{2\pi} \int dE \int dE' \, e^{-iEt/\hbar} e^{iE't'/\hbar} \mathbf{g}_X^c(E, E'). \qquad (8.60)$$

From Eq. (8.59) it is easy to show that the lead Green's functions admit a Fourier expansion of the form

$$\mathbf{g}_X^c(t, t') = \sum_m e^{im\omega t'} \int \frac{dE}{2\pi} e^{-iE(t-t')/\hbar} \mathbf{g}_X^c(E, E + m\hbar\omega). \qquad (8.61)$$

In other words, the functions $\mathbf{g}_X^c(E, E')$ satisfy the following relation

$$\mathbf{g}_X^c(E, E') = \sum_n [\hat{\mathbf{g}}_X^c]_{0,n}(E)\delta(E - E' + n\hbar\omega), \qquad (8.62)$$

where $[\hat{\mathbf{g}}_X^c]_{0,n}(E) \equiv \mathbf{g}_X^c(E, E+n\hbar\omega)$. Other Fourier components are related by $[\mathbf{g}_X^c]_{n,m}(E) = [\mathbf{g}_X^c]_{0,m-n}(E + n\hbar\omega)$. These Fourier components can be seen as the matrix elements of the Green's functions in energy space. We

---

[11]This time dependence can be shown by solving the Dyson's equation for the lead Green's function, which e.g. for the retarded component reads:

$$\left(i\hbar\frac{\partial}{\partial t} - \mathbf{H}_{XX} - \mathbf{W}_{XX}(t) \right) \mathbf{g}_{XX}^r(t, t') = \hbar\delta(t - t'),$$

where $\mathbf{W}_{XX}(t)$ is nothing else but a spatially constant term equal to the chemical potential of the electrode.

denote the matrices in this space with a "hat" symbol. The previous relation is the mathematical expression of the fact that all physical quantities in this problem oscillate in time with the driving frequency and all its harmonics.

With the help of the relation

$$e^{i\alpha \cos(\omega t)} = \sum_m i^m J_m(\alpha) e^{im\omega t}, \qquad (8.63)$$

where J_m is the Bessel function of first kind of order m, one can show that the Fourier components of the lead Green's functions are given by

$$[\hat{\mathbf{g}}_X^c]_{n,m}(E) = i^{m-n} \sum_l J_{n-l}(\alpha_X) J_{m-l}(\alpha_X) g_X^{c,\text{eq}}(E - U_X^{dc} + l\hbar\omega), \quad (8.64)$$

where $\mathbf{g}_X^{c,eq}$ are the equilibrium Green's functions of the lead X, i.e. the usual lead Green's function for us. With these expressions, it is straightforward to show that the self-energies, like the ones in Eq. (8.58), and the corresponding scattering rates are related to the corresponding equilibrium quantities as follows

$$[\hat{\mathbf{\Sigma}}_X^c]_{m,n} = \sum_l [\hat{\mathbf{\Sigma}}_X^{c(l)}]_{m,n}, \quad [\hat{\mathbf{\Gamma}}_X]_{m,n} = \sum_l [\hat{\mathbf{\Gamma}}_X^{(l)}]_{m,n}, \qquad (8.65)$$

where we define the components

$$[\hat{\mathbf{\Gamma}}_X^{(l)}]_{n,m}(E) = i^{m-n} J_{n-l}(\alpha_X) J_{m-l}(\alpha_X) \mathbf{\Gamma}_X^{\text{eq}}(E - U_X^{dc} + l\hbar\omega), \qquad (8.66)$$

with a similar equation for $\hat{\mathbf{\Sigma}}_X^{c(l)}(E)$.

The full Green's functions in the central region have a similar structure in energy space and their different Fourier components are given by the following matrix Dyson's equation

$$[\hat{\mathbf{G}}_{CC}^{r,a}]^{-1} = \hat{\mathbf{E}} - \mathbf{H}_{CC}\hat{\mathbf{I}} - \hat{\mathbf{W}}_{CC} - \hat{\mathbf{\Sigma}}_L^{r,a} - \hat{\mathbf{\Sigma}}_R^{r,a}. \qquad (8.67)$$

Here $[\hat{\mathbf{E}}]_{n,m} = (E + n\hbar\omega)\delta_{m,n}\mathbf{1}$, $[\hat{\mathbf{W}}_{CC}]_{n,m} = \mathbf{W}_{CC}^{dc}\delta_{n,m} + \mathbf{W}_{CC}^{ac}(\delta_{n-1,m} + \delta_{n+1,m})/2$. This means that the Fourier components of the Green's functions can be obtained by inverting the usual matrix, but this time in an extended space. This is a $(\infty \times \infty)$ matrix that has to be truncated and its actual dimension is determined by the amplitude of the ac voltage.

Now, we can bring all these results into the current expression, see Eq. (8.57). The first thing to notice is that, as we have already pointed out, all the quantities in this problem, Green's functions, self-energies, etc., admit a Fourier expansion of the form of Eq. (8.61). It is easy to show that the convolution (or $\circ$-product) of two quantities with this property is a function that also fulfills this property. Therefore, it is obvious that the current in Eq. (8.57), in which the two time arguments are equal, has the following time dependence

$$I(t) = \sum_m I_m e^{im\omega t}, \tag{8.68}$$

i.e. as anticipated, it oscillates with the external frequency and all its harmonics. We are only interested here in the dc component, I_0, which from now on we will simply denote as I.

Using the generic Fourier expansion of Eq. (8.61) for all the quantities appearing in the current expression, see Eq. (8.57), it is easy to show that the dc current can be written in terms of the different Fourier components in energy space defined above as

$$I = \frac{8e}{h} \int_{-\infty}^{\infty} \sum_{k,l,m,n,n'} \mathrm{Tr}\left\{ [\hat{\mathbf{G}}^r]_{0,k} [\hat{\mathbf{\Gamma}}_R^{(n)}]_{k,l} [\hat{\mathbf{G}}^a]_{l,m} [\hat{\mathbf{\Gamma}}_L^{(n')}]_{m,0} \right\} (f_L^{(n')} - f_R^{(n)}), \tag{8.69}$$

where $f_X^{(n)}(E) = f(E - U_X^{dc} + n\hbar\omega)$. At this stage it is already obvious that in the absence of an ac field, this formula reduces to the Landauer formula derived in section 8.1. We can write the current in numerous ways by changing summation indices and the integration variable. Thus for instance, it is not difficult to show that the dc current can be expressed as follows[12]

$$I(V;\alpha,\omega) = \frac{2e}{h} \sum_{k=-\infty}^{\infty} \int_{-\infty}^{\infty} dE\, [T_{RL}^{(k)}(E,V;\alpha,\omega) f_L(E) - \tag{8.70}$$

$$T_{LR}^{(k)}(E,V;\alpha,\omega) f_R(E)],$$

where $f_X(E) = f(E - U_X^{dc})$, the parameter $\alpha = \alpha_L - \alpha_R = eV_{ac}/\hbar\omega$ is the strength of the ac drive and the coefficients appearing inside the energy integral are given by

$$T_{RL}^{(k)}(E) = 4\mathrm{Tr}_\omega[\hat{\mathbf{G}}^r(E)\hat{\mathbf{\Gamma}}_R^{(k)}(E)\hat{\mathbf{G}}^a(E)\hat{\mathbf{\Gamma}}_L^{(0)}(E)], \tag{8.71}$$

$$T_{LR}^{(k)}(E) = 4\mathrm{Tr}_\omega[\hat{\mathbf{G}}^a(E)\hat{\mathbf{\Gamma}}_L^{(k)}(E)\hat{\mathbf{G}}^r(E)\hat{\mathbf{\Gamma}}_R^{(0)}(E)], \tag{8.72}$$

where trace Tr_ω includes a summation over the "harmonic" indexes, i.e. over the Fourier components in energy space, and over the usual site and orbital indexes of the central region. Here $T_{RL}^{(k)}(E)$ can be interpreted as a transmission coefficient that describes processes taking an electron from left (L) to right (R), under the absorption of a total of k energy quanta

---

[12]For the sake of clarity, we make explicit the dependence of the current on the dc voltage, V, the frequency, ω, and the strength of the ac drive, $\alpha = \alpha_L - \alpha_R = eV_{ac}/\hbar\omega$.

$\hbar\omega$. The coefficient $T_{LR}^{(k)}(E)$ has a similar interpretation. By the way, these interpretations are the reason why one usually talks about photo-assisted processes in this problem, although there is indeed no quantized electromagnetic field interacting with the conduction electrons in our model.

Let us summarize the discussion above. The current of Eq. (8.70) describes the dc current in the presence of an oscillating potential and it adopts a form similar to the standard Landauer formula. The main difference is that all the quantities have now a matrix structure in an extended Hilbert space, which includes both the orbital and the energy space. The appearance of off-diagonal elements in energy space is a natural consequence of the occurrence of the inelastic processes that take place in this problem. In those inelastic tunneling processes, a certain number of energy quanta (multiples of $\hbar\omega$) can be either absorbed or emitted. The retarded/advanced Green's functions appearing in the current formula are determined by solving the matrix equation (8.67), while the scattering rates are given by Eq. (8.66). All these matrices have, in principle, an infinite dimension in energy space, but they can be truncated in practice and their actual dimension is governed by the amplitude of the ac drive, α.

The formalism above has been recently used to discuss both the photon-assisted transport in atomic [221] and molecular wires [222]. This formalism is a bit cumbersome and numerically demanding due to the large size of the matrices involved. However, the current formula above can be greatly simplified in the case in which we can ignore the energy dependence in the leads, which is frequently a very good approximation. In this situation the self-energies $\hat{\Sigma}_X$ become diagonal (see Exercise 8.8)

$$[\hat{\Sigma}_X]_{n,m}(E) = \Sigma_X(E)\delta_{n,m}. \tag{8.73}$$

If in addition we assume that the ac potential profile is such that it is constant in the central region (i.e. the drops occur at the interfaces), the current formula reduces to [165, 211, 214]

$$I(V;\alpha,\omega) = \frac{2e}{h} \sum_{l=-\infty}^{\infty} \left[J_l\left(\frac{\alpha}{2}\right) \right]^2 \int dE\, T(E+l\hbar\omega)[f_L(E) - f_R(E)], \tag{8.74}$$

where $T(E)$ is the transmission in the absence of ac drive.[13] Moreover, we have assumed here that the ac potential drops symmetrically at both interfaces, i.e. $\alpha_{L,R} = \pm\alpha/2$. The result of Eq. (8.74) is quite remarkable and it tell us that the current under a periodic time-dependent field depends primarily on the energy dependence of the elastic transmission. This becomes

---

[13]This transmission may include the dc part of the voltage.

even more apparent in the case of the conductance. At low temperatures and in the linear response regime (vanishing dc bias), the conductance, which will be referred to as *photoconductance*, takes the particularly simple form

$$G(V = 0; \alpha, \omega) = G_0 \sum_{l=-\infty}^{\infty} \left[J_l \left(\frac{\alpha}{2} \right) \right]^2 T(E_{\mathrm{F}} + l\hbar\omega), \qquad (8.75)$$

where $T(E)$ is the zero-bias equilibrium transmission. Let us remind that here l can be interpreted as the number of absorbed or emitted photons, $J_l(x)$ is a Bessel function of the first kind (of order l), and $\alpha = eV_{ac}/\hbar\omega$ is the dimensionless parameter describing the strength of the ac drive. Note that if the transmission does not depend on energy in the range explored by the inelastic processes, the conductance reduces to the conductance in the absence of drive, i.e. $G_0 T(E_{\mathrm{F}})$.

In the limit $\alpha \ll 1$ and frequencies small in comparison with the energy scale in which transmission changes significantly, we can expand $T(E)$ and the Bessel functions in Eq. (8.75) to leading order in these small quantities, yielding $G(\omega) = G_0 T(E_{\mathrm{F}}) + G_0 (\alpha\hbar\omega)^2 T''(E_{\mathrm{F}})/16$, where T'' denotes the second derivative respect to energy. Defining then the induced conductance correction $\Delta G(\omega) = G(\omega) - G(\omega = 0)$, where $G(\omega = 0) = G = G_0 T(E_{\mathrm{F}})$, the relative correction becomes

$$\frac{\Delta G(\alpha, \omega)}{G} = \frac{(\alpha\hbar\omega)^2}{16} \frac{T''(E_{\mathrm{F}})}{T(E_{\mathrm{F}})}. \qquad (8.76)$$

We thus see that this quantity gives experimental access to the second derivative of the transmission function at $E = E_{\mathrm{F}}$. Note that in this approximation, which can be seen as an adiabatic or "classical" limit [212], the conductance correction depends only on the driving field through the ac amplitude $V_{ac} = \alpha\hbar\omega/e$.

Finally, let us mention that Eq. (8.74) may equally well be written in the form [213, 212]

$$I(V; \alpha, \omega) = \sum_{l=-\infty}^{\infty} \left[J_l \left(\frac{\alpha}{2} \right) \right]^2 I_0(V + 2l\hbar\omega/e), \qquad (8.77)$$

where $I_0(V)$ is the I-V characteristic in the absence of light.

The main assumption leading to these simplified formulas is the fact that the profile is flat across the central part of the constriction. However, it has been shown in Refs. [221, 222] that the detailed shape of the profile does not change significantly the main results, unless the ac amplitude is very large.

8.3.1 *Photon-assisted resonant tunneling*

In order to illustrate the previous time-dependent formalism, let us now apply it to the resonant tunneling model (see section 7.4.3). This problem has been analyzed by Jauho *et al.* [218]. As we have discussed many times by now, this simple model gives useful insight into the conduction through a single-molecule junction. Now, the question is: How is the resonant transport modified in the presence of radiation? Following the discussion above, we assume here that an electromagnetic field simply induces an ac voltage of frequency ω across the junction. If, as usual, we neglect the energy dependence of the scattering rates, we can analyze this problem in terms of the simplified formulas presented at the end of the previous section.

The first issue that we want to address is the modification of the non-linear conductance. For this purpose, we use the expression of Eq. (8.74) to determine the current-voltage characteristics and in turn the differential conductance $dI(V; \alpha, \omega)/dV$, where V is the dc voltage. In this formula we make use of the expression of Eq. (8.24) for the elastic transmission through the resonant level in the absence of radiation. We assume that both the bias voltage and the ac drive drop symmetrically at the interfaces. This means in practice that the chemical potentials of both electrodes are shifted by $\pm eV/2$, while the resonant level is not shifted by the bias. An example of the zero-temperature I-V characteristics for different values of α is shown in Fig. 8.5(a). In this example, the level position (measured with respect to the equilibrium chemical potential of the electrodes) is $\epsilon_0 = 5\hbar\omega$ and $\Gamma_L = \Gamma_R = 0.1\hbar\omega$. The corresponding differential conductance as a function of the bias voltage is shown in Fig. 8.5(b). Notice that in the absence of the external ac field, the conductance is simply given by a Breit-Wigner resonance centered around $2\epsilon_0$ (see curve for $\alpha = 0$). The factor two is due to the symmetric voltage profile adopted here. When the radiation is applied, one can see the appearance of additional steps in the current and satellite peaks in the conductance with a regular spacing equal to $2\hbar\omega$. In the case of the conductance, the peaks on the left hand side of the central elastic resonance are due to the photon absorption, i.e. due to tunneling processes in which an incoming electron with energy $E = \epsilon_0 - eV/2 - k\hbar\omega$ absorbs k photons and crosses the level exactly at resonance. Similarly, the peaks on the right hand side are due to emission processes in which an electron loses energy emitting a certain number of photons. The number of satellite peaks (or side bands) depends on the strength of the ac drive, α, which is basically a measure of the local field intensity at the junction.

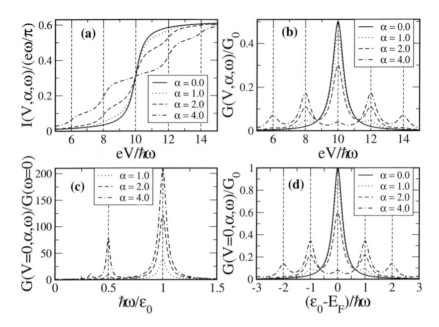

Fig. 8.5 Photon-assisted transport in the resonant tunneling model. In this example we consider a symmetric junction with $\Gamma_L = \Gamma_R = 0.1\hbar\omega$ and all the results are obtained at zero temperature. (a) Current as a function of the dc bias voltage V for $\epsilon_0 - E_F = 5\hbar\omega$ and different values of $\alpha = eV_{ac}/\hbar\omega$. (b) The differential conductance corresponding to the I-V curves of panel (a). (c) Photoconductance normalized by the conductance in the absence of radiation as a function of the radiation frequency for different values of α. (d) Photoconductance versus level position measured with respect to the Fermi energy.

An important quantity for us is the photoconductance $G(V = 0; \alpha, \omega)$, i.e. the conductance when the dc voltage is infinitesimally small. In Fig. 8.5(c) we show an example of this quantity as a function of the radiation frequency. The fact that we want to illustrate here is that when the frequency matches the distance between the Fermi energy and the level position, one observes a huge enhancement of the conductance that can reach up to several orders of magnitude. The additional peaks that one can see in Fig. 8.5(c) are due to multi-photon processes. Finally, in some situations the position of the resonant level can be tuned by means of, for instance, a gate voltage. Therefore, it is interesting to know what is the expected dependence of the photoconductance on the level position. This can be seen in Fig. 8.5(d), where one can observe that in this case the inelastic tunneling events give rise to satellite peaks that are separated by an energy equal to $\hbar\omega$.

8.4 Exercises

8.1 Scattering rate matrices: Show that the scattering rate matrices defined in section 8.1 as $\mathbf{\Gamma}_X = \mathrm{Im}\{\mathbf{\Sigma}_X\}$ ($X = L, R$), where $\mathbf{\Sigma}_X$ are the self-energies of Eq. (8.16), are positive definite and therefore their square roots are well-defined.

8.2 Transmission matrix: The goal of this exercise is to show that the matrix defined in Eq. (8.19) has indeed the basic properties of a transmission matrix. For this purpose, it must be shown that the eigenvalues of $\mathbf{tt}^\dagger$ are bounded between 0 and 1. Demonstrate this property following the next steps:

(i) Using the result of the previous exercise, show that $\mathbf{tt}^\dagger$ is positive definite and therefore all its eigenvalues are real and positive.

(ii) Use the definition of the scattering rate matrices and Dyson's equation for the retarded and advanced Green's functions to prove the following relation

$$\mathbf{G}^r_{CC}\left[\mathbf{\Gamma}_L + \mathbf{\Gamma}_R\right]\mathbf{G}^a_{CC} = \frac{i}{2}\left[\mathbf{G}^r_{CC} - \mathbf{G}^a_{CC}\right].$$

(iii) Use the previous relation to demonstrate the following relation

$$\mathbf{1} = \mathbf{rr}^\dagger + \mathbf{tt}^\dagger,$$

where $\mathbf{r}$ is the reflection matrix given by

$$\mathbf{r} = \mathbf{1} - 2i\mathbf{\Gamma}_L^{1/2}\mathbf{G}^r_{CC}\mathbf{\Gamma}_L^{1/2}.$$

(iv) Using this last relation, show that the all eigenvalues of $\mathbf{tt}^\dagger$ are less than (or equal to) one.

8.3 Formula for the current through an atomic chain: Consider the model for an atomic chain described in Exercise 7.5. Use the general expression of Eq. (8.20) to re-derive the formula for the electrical current obtained in that exercise.

8.4 Phonon Green's functions: The phonon Green's functions are defined in analogy with the electronic ones as

$$D^r(t,t') = -i\theta(t-t')\langle\left[\mathbf{A}(t), \mathbf{A}^\dagger(t')\right]\rangle, \quad D^a(t,t') = -i\theta(t'-t)\langle\left[\mathbf{A}(t), \mathbf{A}^\dagger(t')\right]\rangle,$$

$$D^{+-}(t,t') = i\langle\mathbf{A}^\dagger(t')\mathbf{A}(t)\rangle, \quad D^{-+}(t,t') = -i\langle\mathbf{A}(t)\mathbf{A}^\dagger(t')\rangle,$$

where $\mathbf{A} = \mathbf{b} + \mathbf{b}^\dagger$ and the creation and annihilation operators $\mathbf{b}^\dagger$ and $\mathbf{b}$ satisfy the bosonic commutation relations (see Appendix A). Show that for the case of a free phonon (or vibration) mode, described by the Hamiltonian of Eq. (8.39), these functions are by given Eq. (8.45). Hint: Compute first the time evolution of the bosonic operators by solving the equation of motion of an operator in the Heisenberg picture.

8.5 Lowest order expansion of the electron-vibration self-energy: The goal of this exercise is to demonstrate that Eq. (8.43) gives the correct expression for the lowest order correction to the electronic self-energy in the problem of section 8.2.1. For this purpose, follow the next steps:

(i) Use the Hamiltonian of Eq. (8.40) as the perturbation in this problem. With this choice, show that the second order correction in λ of an electronic Green's function is equal to

$$G_c^{(2)}(t_\alpha, t_\beta) = \lambda^2 \frac{(-i)^3}{2!} \int_c dt_1 \int_c dt_2 \times$$

$$\langle \mathbf{T}_c \left[\mathbf{c}(t_\alpha)\mathbf{c}^\dagger(t_1)\mathbf{c}(t_1)[\mathbf{b}^\dagger(t_1) + \mathbf{b}(t_1)]\mathbf{c}^\dagger(t_2)\mathbf{c}(t_2)[\mathbf{b}^\dagger(t_2) + \mathbf{b}(t_2)]\mathbf{c}^\dagger(t_\beta) \right] \rangle.$$

Here, the subindex c indicates that the Green's functions can be any of the four components in Keldysh space depending upon where the time arguments, t_α and t_β ($\alpha, \beta = +, -$), lie on the Keldysh contour. The integrations above have to be understood as follows

$$\int_c dt_i = \int_{-\infty}^{\infty} dt_{i,+} - \int_{-\infty}^{\infty} dt_{i,-}.$$

(ii) Apply Wick's theorem to the previous expression and keep only the contributions of topologically distinct connected diagrams. Show that the only two relevant self-energy diagrams are the ones shown in Fig. 8.6.

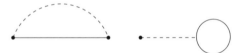

Fig. 8.6 Lowest-order electronic self-energy diagrams associated to the electron-phonon interaction in the resonant tunneling model. The solid lines represent electronic Green's functions, while the dashed ones correspond to phonon Green's functions.

(iii) Evaluate the contribution of the diagrams of Fig. 8.6 to the different components of the self-energy in energy space. Show that the contributions coming from the diagram on the left hand side lead to the results of Eq. (8.43). Discuss also the relevance of the contributions coming from the other diagram.

8.6 Signature of a vibrational mode in the differential conductance: Consider the model used in section 8.2.1 to understand the signature of a vibration mode in the current through a single resonant level. Assume that the density of states in that level and the corresponding transmission are energy-independent and show that the zero-temperature differential conductance is given by Eq. (8.49) in the case of a symmetric junction.

Hint: The only complicated term in the expression for the current is the elastic correction, which contains the self-energy Σ_{e-vib}^r. Separating the contributions of the real and imaginary part of the retarded phonon Green's function D^r, this self-energy can be written as

$$
\Sigma^r_{e-vib}(E)/\lambda^2 = i \int_{-\infty}^{\infty} \frac{dE'}{2\pi} \left\{ \frac{2\hbar\omega}{(E')^2 - (\hbar\omega)^2} \tilde{G}^{-+}(E - E') \right\} +
$$
$$
\frac{1}{2}\left[\tilde{G}^{-+}(E - \hbar\omega) - \tilde{G}^{-+}(E + \hbar\omega) \right] +
$$
$$
(n_B + 1)\tilde{G}^r(E + \hbar\omega) + n_B \tilde{G}^r(E - \hbar\omega).
$$

The first term, which has to be understood as principle value, does not contribute to the conductance in the case of a symmetric junction, while the others are responsible for the contribution of the elastic correction to Eq. (8.49).

8.7 The Meir-Wingreen formula:

(i) Follow the steps indicated in section 8.2.2 to show that the current through an interacting region is given by Eq. (8.52).

(ii) Show that the current given by Eq. (8.52) vanishes in equilibrium.

(iii) Demonstrate that in the noninteracting case the Meir-Wingreen formula of Eq. (8.52) reduces to the Landauer formula derived in section 8.1.

(iv) Assume that the scattering rates fulfill $\Gamma_L(E) = \lambda \Gamma_R(E)$ and prove that the Meir-Wingreen formula adopts the form given in Eq. (8.53).

8.8 Photo-current formula in the wide-band approximation:

(i) Show that the general formula of Eq. (8.70) for the current in a nanocontact under an ac field reduces to Eq. (8.74) when (i) the energy dependence of the density of states in the leads can be neglected (wide-band approximation) and (ii) the ac potential is assumed to be flat in the central region of the system.

(ii) Starting from Eq. (8.75), show that in the limit of $\alpha \ll 1$ and small frequencies, the conductance correction induced by the ac drive is a measure of the second derivative of the transmission around the Fermi energy, i.e. demonstrate Eq. (8.76).

Hint: Use the following properties of Bessel's functions

$$
\sum_{l=-\infty}^{\infty} J_{n+l}(x)J_{m+l}(x) = \delta_{nm}, \qquad \sum_{l=-\infty}^{\infty} [J_l(x)]^2 = 1,
$$

$$
J_l(x \ll 1, l > 0) \approx \frac{(\pm x/2)^l}{l!} - \frac{(\pm x/2)^{l+2}}{(l+1)!}.
$$

8.9 Linear conductance in the Coulomb blockade regime:
As we shall explain in Chapter 15, the Coulomb blockade is a transport phenomenon that takes place in weakly coupled quantum dots and molecular junctions. The signatures of Coulomb blockade in the linear conductance (i.e. at vanishingly small bias voltage) are: (i) the appearance of peaks as a function of the gate voltage (or chemical potential) known as Coulomb oscillations and (ii) a characteristic temperature dependence that is described by the derivative of the Fermi function with respect to energy. The goal of this exercise is to explain these two signa-

tures by combining the Meir-Wingreen formula of Eq. (8.53) and the single-level Anderson model of Eq. (5.109). For this purpose, carry out the following tasks:

(i) Adapt the Meir-Wingreen formula to the case of a single-level Anderson model and derive an expression for the linear conductance in terms of the spectral function in the resonant level.

(ii) Use the approximation of Eq. (5.112) to compute the spectral function in the weak coupling limit.

(iii) Combine the results of (i) and (ii) to obtain the gate voltage and temperature dependence of the linear conductance and show that this model reproduces the two signatures described above.

Hint: This problem was addressed by Meir *et al.* in Ref. [632].

8.10 Kondo effect in molecular transistors: Unitary limit. The Kondo effect in molecular junctions is manifested in the appearance of a pronounced resonance in the density of states at the Fermi energy. This many-body effect is usually described with the help of the Anderson model (see section 6.9). Apply the Meir-Wingreen formula to this model and show that in the Kondo regime the low-temperature linear conductance in a symmetric junction ($\Gamma_L = \Gamma_R$) is equal to the conductance quantum (G_0). This is referred to as the unitary limit. Hint: Use the Friedel sum rule discussed in section 6.9.1.

Chapter 9

Electronic structure I: Tight-binding approach

In the previous chapters we have shown how to compute the transport properties of an atomic-scale junction once the corresponding Hamiltonian is known. Therefore, in order to make our theoretical background self-contained, at least to a certain extent, we need to discuss how those Hamiltonians are determined in practice. In other words, we have to describe adequate methods for the description of the electronic structure of atomic and molecular junctions. Such methods are based on the standard approaches for the calculation of the electronic structure of atoms, molecules and solids that are used in atomic physics, theoretical chemistry and solid state physics. There is a great variety of electronic structure methods and, obviously, we cannot review all of them here. We shall focus our attention on the two methods that have had the largest impact so far in the field of molecular electronics. First, in this chapter we shall discuss the tight-binding approach, which is a very intuitive empirical or semi-empirical method that has been crucial to elucidate the physics of, in particular, metallic atomic-sized contacts. Then, the next chapter is devoted to the density functional theory (DFT), which is the most widely used approach among the so-called *ab initio* methods.

The tight-binding approach is reviewed in several textbooks and we recommend in particular Refs. [223–226] to the physics-oriented readership and Ref. [227] for a chemistry view on this subject.

9.1 Basics of the tight-binding approach

The main idea of the tight-binding approach was already introduced in Appendix A and indeed it has been extensively used in the previous chapters devoted to the Green's function techniques. Anyway, let us now define more precisely what we mean by tight-binding approach or by a tight-binding model. The problem that we are interested in is the determination the electronic structure of a system composed of a collection of atoms that are located in different positions denoted by $\mathbf{R}_i$. The corresponding Hamiltonian, $\mathbf{H}$, of this system can be written in a local basis, i.e. in a basis formed by single-particle wave functions that are localized around the different atomic positions. This is the spirit of the method known as linear combination of atomic orbitals (LCAO), which is so popular in theoretical chemistry. The first approximation in the tight-binding approach is to assume that the Hamiltonian adopts the form of Eq. (A.67), which in first quantization language reads (using Dirac's notation)[1]

$$\mathbf{H} = \sum_{ij,\alpha\beta} H_{i\alpha,\beta j} |i\alpha\rangle\langle j\beta|, \tag{9.1}$$

where $|i\alpha\rangle$ denotes the state that corresponds to the localized orbital α that is centered around $\mathbf{R}_i$, i.e. $\langle\mathbf{r}|i\alpha\rangle = \phi_{i\alpha}(\mathbf{r}) = \phi_\alpha(\mathbf{r} - \mathbf{R}_i)$. This generic form for the Hamiltonian implies that either the many-body interactions such as the electron-electron interaction are neglected or they are taken into account in a mean field manner by an appropriate choice of the matrix elements. In the former case, the matrix elements are rigorously defined as

$$H_{i\alpha,j\beta} = \int d\mathbf{r}\, \phi_\alpha^*(\mathbf{r} - \mathbf{R}_i) \left[-\frac{\hbar^2}{2m}\nabla^2 + V(\mathbf{r}) \right] \phi_\beta(\mathbf{r} - \mathbf{R}_j), \tag{9.2}$$

where $V(\mathbf{r})$ is the potential that describes the Coulomb interaction between the electrons and ions. Finally, in the tight-binding approach, as it is used in this book, the matrix elements are not determined from first principles, i.e. from a direct evaluation of the integral in Eq. (9.2), but they are used merely as parameters that may be derived approximately or may be fitted to experiment or to other theories. Thus, by tight-binding model we mean here a model in which the system is described in terms of a single-particle Hamiltonian written in a local basis, the elements of which are determined in a empirical or semi-empirical way. The different tight-binding models differ in the way in which these parameters are obtained.

There are two situations where the wave function associated to a tight-binding model can be determined in a straightforward manner. The first one corresponds to the case of a small finite system such as a molecule and

---

[1]This Hamiltonian in our usual second quantization language reads

$$\mathbf{H} = \sum_{ij,\alpha\beta} H_{i\alpha,\beta j} c_{i\alpha}^\dagger c_{j\beta}.$$

the second one corresponds to the case of an infinite periodic system. In the first case, the Hamiltonian can be diagonalized by writing first the wave function as a combination of the localized orbitals:

$$\Phi(\mathbf{r}) = \sum_{j\beta} c_{i\alpha,j\beta}\phi_{j\beta}(\mathbf{r}). \tag{9.3}$$

This leads immediately to the following set of equations for the coefficients (see Exercise 9.1)

$$\sum_{j\beta} [H_{i\alpha,j\beta} - ES_{i\alpha,j\beta}] c_{i\alpha,j\beta} = 0, \tag{9.4}$$

where E is the energy and

$$S_{i\alpha,j\beta} = \int d\mathbf{r} \, \phi_\alpha^*(\mathbf{r} - \mathbf{R}_i)\phi_\beta(\mathbf{r} - \mathbf{R}_j), \tag{9.5}$$

is the overlap between the states $|i\alpha\rangle$ and $|j\beta\rangle$. Here, we have taken into account the possibility that the localized orbitals centered in different atoms can be non-orthogonal. These equations have non-trivial solutions if

$$\det(\mathbf{H} - E\mathbf{S}) = 0, \tag{9.6}$$

where the symbol "det" denotes the determinant of the matrix appearing inside the brackets. The roots of this secular equation yield the eigenenergies or energy levels of the finite problem and the eigenfunctions are the corresponding waves functions (or molecular orbitals) of this system. The dimension of the matrices in Eq. (9.6) is simply the total number of localized orbitals in the problem and therefore, the solution of the generalized eigenvalue problem of Eq. (9.6) requires typically to resort to numerics.

In the case of an infinite periodic system, typical of solid state physics, one can diagonalize the Hamiltonian making use of Bloch's theorem (see for instance Ref. [223]). The idea goes as follows. Consider a periodically replicated unit cell, where the lattice vectors are denoted as $\mathbf{R}_m$, with a set of atoms i located at positions $\mathbf{b}_i$ in each unit cell. Associated with each atom is a set of atomic-like orbitals $\phi_{i\alpha}$, where α denotes both the orbital and angular quantum number of the atomic state. The Hamiltonian can be easily diagonalized in reciprocal space as follows. We first construct the following wavefunctions (Bloch sums)

$$\Phi_{\mathbf{k}i\alpha}(\mathbf{r}) = \frac{1}{\sqrt{N}} \sum_n \exp(i\mathbf{k} \cdot \mathbf{R}_n)\phi_{i\alpha}(\mathbf{r} - \mathbf{R}_n - \mathbf{b}_i), \tag{9.7}$$

where $\mathbf{k}$ is the Bloch wave vector, which is restricted to the Brillouin zone, and N is the number of unit cells in the sum. The solution to Schrödinger equation for wave vector $\mathbf{k}$ then requires the diagonalization of the Hamiltonian matrix using the basis functions of Eq. (9.7). Since the Hamiltonian has the periodicity of the lattice, this basis will block-diagonalize the Hamiltonian, with each block having a single value of $\mathbf{k}$. Within one of these blocks, the matrix elements can be written in the form

$$H_{i\alpha,j\beta}(\mathbf{k}) = \sum_n \exp(i\mathbf{k} \cdot \mathbf{R}_n) \int \phi_{i\alpha}^*(\mathbf{r} - \mathbf{R}_n - \mathbf{b}_i) H \phi_{j\beta}(\mathbf{r} - \mathbf{b}_j) d^3\mathbf{r}, \quad (9.8)$$

where we have used the translation symmetry of the lattice to remove one of the sums over the lattice vector $\mathbf{R}$ (see Exercise 9.2). In the same way, one can also define the overlap matrix in reciprocal space where the different elements adopt the form

$$S_{i\alpha,j\beta}(\mathbf{k}) = \sum_n \exp(i\mathbf{k} \cdot \mathbf{R}_n) \int \phi_{i\alpha}^*(\mathbf{r} - \mathbf{R}_n - \mathbf{b}_i) \phi_{j\beta}(\mathbf{r} - \mathbf{b}_j) d^3\mathbf{r}. \quad (9.9)$$

The corresponding secular equation reads this time

$$\det\left(\mathbf{H}(\mathbf{k}) - E\mathbf{S}(\mathbf{k})\right) = 0. \quad (9.10)$$

The solution of this generalized eigenvalue problem yields the different energy bands, $\epsilon_\mu(\mathbf{k})$ of the solid and the corresponding eigenvectors $\mathbf{Q}_\mu(\mathbf{k})$. Notice that the number of bands, i.e. the number of solutions of Eq. (9.10), equals the number of atoms in the unit cell times the number of orbitals per atom. Thus, in some simple cases the solution can be found analytically and, in general, this problem can be easily solved numerically.

An important quantity for many purposes is the density of states (DOS) per unit energy E. The local DOS projected onto a given atom, orbital and spin (summarized by the index ν) is defined in terms of the energy bands $\epsilon_\mu(\mathbf{k})$ as follows

$$\rho_\nu(E) = \frac{1}{N_k} \sum_{\mathbf{k},\mu} |Q_{\nu,\mu}(\mathbf{k})|^2 \delta(\epsilon_\mu(\mathbf{k}) - E) \quad (9.11)$$

$$= \frac{\Omega_{cell}}{(2\pi)^d} \sum_\mu \int_{BZ} d\mathbf{k} \, |Q_{\nu,\mu}(\mathbf{k})|^2 \delta(\epsilon_\mu(\mathbf{k}) - E),$$

where BZ denotes the Brillouin zone, Ω_{cell} is the volume of the unit cell and d is the dimensionality of the system.

In the case of infinite non-periodic systems, like the atomic-scale junctions that we are interested in, the determination of the wavefunction is literally impossible. However, the use of the Green's function techniques described in Chapter 5 allows to extract most of the relevant information about the electronic structure from a tight-binding Hamiltonian.

9.2 The extended Hückel method

The history of quantum chemistry is plagued with examples of approximations in the framework of the LCAO method, which fall into our definition of tight-binding approach. One of the oldest and most familiar of such approaches in quantum chemistry is the "extended Hückel approximation" [228]. Let us explain briefly the idea behind this approach. It was developed by Roald Hoffmann in 1963 [228] to describe the electronic structure of a variety of organic molecules. It is based on the Hückel method [229–232] but, while the original Hückel method only considers π-orbitals, the extended method also includes the σ-orbitals. The idea goes as follows. We seek matrix elements of the Hamiltonian between atomic orbitals on adjacent atoms, $\langle i|\mathbf{H}|j\rangle$. If $|j\rangle$ were an eigenstate of the Hamiltonian, we could replace $\mathbf{H}|j\rangle$ by $\epsilon_j|j\rangle$, where ϵ_j, the on-site energy of the atom j, is the eigenvalue. Then, if the overlap $\langle i|j\rangle$ is written S_{ij}, the matrix element becomes $\epsilon_j S_{ij}$. This, however, treats the two orbitals differently, so we might use the average instead of ϵ_j. Finding that this does not give good values, we introduce a scale factor K, to be adjusted to fit the properties of heavy molecules (a value of $K = 1.75$ is usually taken); this leads to the extended Hückel formula[2]

$$\langle i|\mathbf{H}|j\rangle = K S_{ij}(\epsilon_i + \epsilon_j)/2. \qquad (9.12)$$

In the extended Hückel method, only valence electrons are considered; the core electron energies and wave functions are supposed to be more or less constant between atoms of the same type. The method uses a series of parameterized energies calculated from atomic ionization potentials or theoretical methods to fill the diagonal of the Hamiltonian matrix. After filling the non-diagonal elements (with the formula above) and diagonalizing the resulting Hamiltonian matrix, the energies (eigenvalues) and wavefunctions (eigenvectors) of the valence orbitals are found.

The extended Hückel approximation and a wide range of methods that may be considered as descendents of it have enjoyed considerable success in theoretical chemistry. This method can be used for determining the molecular orbitals, but it is not very successful in determining the structural geometry of an organic molecule. It can however determine the relative energy of different geometrical configurations. It is common in quantum chemistry to use the extended Hückel molecular orbitals as a first guess in

---

[2]This formula is indeed due to M. Wolfsberg and L.J. Helmholtz [233].

the determination of the molecular orbitals by ab initio quantum chemistry methods.

9.3   Matrix elements in solid state approaches

In the context of solid state physics most of the semi-empirical tight-binding applications are largely based on the seminal work of Slater and Koster (SK) [234], in which they proposed a modified LCAO method to interpolate the results of first-principles electronic structure calculations. At that time (1954), it was computationally impossible to directly evaluate the large number of integrals occurring in the LCAO method. However, since this approach shows all the correct symmetry properties of the energy bands as well as providing solutions of the single-particle Schrödinger equation at arbitrary points in the Brillouin zone, they suggested that these integrals could be considered as adjustable constants to be determined from the results of other, more efficient, calculations. In order to understand the basis of the simplified LCAO/tight-binding method proposed by Slater and Koster, we need first to discuss in certain detail the nature of the matrix elements that appear in the tight-binding approach. Thus, the explanation of the SK-method will be postponed until the next section.

The tight-binding approach benefits from the consideration of the symmetries of the basis orbitals and the crystal or molecule. On each site of the physical system, the atomic-like functions can be written as radial functions multiplied by spherical harmonics,

$$\phi_{nlm}(\mathbf{r}) = \phi_{nl}(r)Y_{lm}(\hat{\mathbf{r}}), \qquad (9.13)$$

where $r = |\mathbf{r}|$, $\hat{r} = \mathbf{r}/r$ and n indicates different functions with the same angular momentum. We shall work frequently with real basis functions that can be defined using the real angular functions $S_{lm}^{+} = (Y_{lm} + Y_{lm}^{*})/\sqrt{2}$ and $S_{lm}^{-} = (Y_{lm} - Y_{lm}^{*})/(i\sqrt{2})$. The examples of real s ($l = 0$), p ($l = 1$) and d ($l = 2$) orbitals are given in Fig. 9.1. The analytical expressions of the angular dependence of these real orbitals can be found in many textbooks, see e.g. Chapter 1 of Ref. [224] or Chapter 3 of Ref. [227].

The key problem in a tight-binding model is the determination of the matrix elements (or integrals) that appear both in Eq. (9.8) and Eq. (9.9). Those matrix elements can be divided into one-, two-, and three-center terms. The simplest is the overlap matrix in Eq. (9.9), which involves only one center if the two orbitals are on the same site and two centers otherwise.

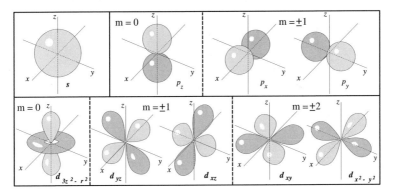

Fig. 9.1 Boundary surfaces for real s-, p-, and d-orbitals. The index m indicates the quantum number corresponding to the z-component of the orbital angular momentum.

The Hamiltonian matrix elements appearing in Eq. (9.8) consist of kinetic and potential terms

$$\mathbf{H} = -\frac{\hbar^2}{2m}\nabla^2 + \sum_{nk} V_k(\mathbf{r} - \mathbf{R}_n - \mathbf{b}_k), \qquad (9.14)$$

where the first term is the usual kinetic energy and the second is the potential decomposed into a sum of spherical terms centered on each site k in the unit cell. The kinetic part of the Hamiltonian matrix element always involves one or two centers. However, the potential terms may depend upon the positions of other atoms; they can be divided into the following.

- One-center, where both orbitals and the potential are centered on the same site. These terms have the same symmetry as an atom in free space.
- Two-center, where the orbitals are centered on different sites and the potential is on one of the two. These terms have the same symmetry as other two-center terms.
- Three-center, where the orbitals and the potential are all centered on different sites. These terms can also be classified into various symmetries based upon the fact that three sites define a triangle.
- A special class of two-center terms with both orbitals on the same site and the potential centered on a different site. These terms add to the one-center terms above, but depend upon the crystal symmetry.

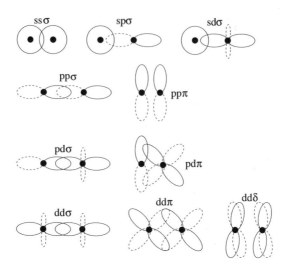

Fig. 9.2   The 10 irreducible SK-parameters for the s, p and d orbitals, which are classified by the angular momentum about the axis with the notation σ $(m = 0)$, π $(m = 1)$ and δ $(m = 2)$. The orbitals shown are the real combinations of the angular momentum eigenstates. Positive and negative lobes are denoted by solid and dashed lines, respectively.

9.3.1   *Two-center matrix elements*

Two-center matrix elements play a special role in most practical tight-binding approaches and are considered here in more detail. The analysis applies to all overlap terms and to any Hamiltonian matrix element that involves only orbitals and potential on two sites. For these integral the problem is the same as for a diatomic molecule in free space with cylindrical symmetry. The orbitals can be classified in terms of the azimuthal angular momentum about the line between the centers, i.e. the value of m with the axis chosen along the line, and the only non-zero matrix elements are between orbitals with the same m. If $K_{lm,l'm'}$ denotes an overlap or two-center Hamiltonian matrix element for states lm and $l'm'$, then in the standard form with orbitals quantized about the axis between the pair of atoms, the matrix elements are diagonal in mm' and can be written as $K_{lm,l'm'} = K_{ll'm}\delta_{m,m'}$. The quantities $K_{ll'm}$ are independent matrix elements that are irreducible, i.e. they cannot be further reduced by symmetry. By convention the states are labeled with l or l' denoted by s, p, d, ..., and $m = 0, \pm 1, \pm 2, ...$, denoted by σ, π, δ, ..., leading to the notation $K_{ss\sigma}$, $K_{sp\sigma}$, $K_{pp\pi}$, etc.

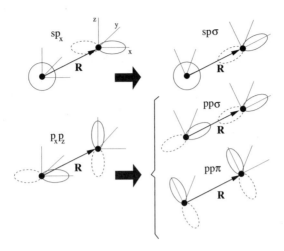

Fig. 9.3 Schematic representation of two examples of two-center matrix elements of s and p orbitals for atoms separated by a displacement vector $\mathbf{R}$. Matrix elements are related to σ and π integrals by the transformation to a combination of orbitals that are aligned along $\mathbf{R}$ and perpendicular to $\mathbf{R}$. The top figure illustrates the transformation to write a real matrix element K_{s,p_x} in terms of $K_{sp\sigma}$: the s orbital is unchanged and the p_x orbital is written as a sum of the σ orbital, which is shown, and the π orbitals, which are not shown because there is no $sp\pi$ matrix element. The lower figure illustrates the transformation needed to write K_{p_x,p_z} in terms of $K_{pp\sigma}$ and $K_{pp\pi}$. The coefficients of the transformation for all s and p matrix elements are given in Table 9.1.

In Fig. 9.2 we show the orbitals for the non-zero σ, π, and δ matrix elements for s, p, and d orbitals. The orbitals shown are actually the real basis functions $S_{lm}^{\pm}$ defined as combinations of the $\pm m$ angular momentum eigenstates. These are oriented along the axes defined by the line between the neighbors and two perpendicular axes. All states except the s state have positive and negative lobes. Note that states with odd l are odd under inversion. Their sign must be fixed by convention (typically one chooses the positive lobe along the positive axis). The direction of the displacement vector is defined to lie between the site denoted by the first index and that denoted by the second index. For example, in Fig. 9.2, the $K_{sp\sigma}$ matrix element in the top center has the negative lobe of the p function oriented toward the s function. Interchange of the indices leads to $K_{ps\sigma} = -K_{sp\sigma}$ and, more generally, to $K_{ll'm} = (-1)^{l+l'} K_{l'lm}$.

An actual set of basis functions is constructed with the quantization axis fixed in space, so that the functions must be transformed to utilize the standard irreducible form of the matrix elements. Examples of two-center matrix elements of s and $p_i = p_x, p_y, p_z$ orbitals for atoms separated by

Table 9.1   Table of two-center matrix elements for either the overlap or the Hamiltonian, with real orbitals s and p_x, p_y, p_z. The vector $\mathbf{R}$ between sites, as shown in Fig. 9.3, is defined to have direction components $\hat{\mathbf{R}} \equiv x, y, z$. The matrix elements are then expressed in terms of these coordinates and the four irreducible matrix elements: $K_{ss\sigma}$, $K_{sp\sigma}$, $K_{pp\sigma}$ and $K_{pp\pi}$. Other matrix elements can be found by permuting elements.

Element	Expression
$K_{s,s}$	$K_{ss\sigma}$
K_{s,p_x}	$x K_{sp\sigma}$
K_{p_x,p_x}	$x^2 K_{pp\sigma} + (1 - x^2) K_{pp\pi}$
K_{p_x,p_y}	$xy(K_{pp\sigma} - K_{pp\pi})$
K_{p_x,p_z}	$xz(K_{pp\sigma} - K_{pp\pi})$

the displacement vector $\mathbf{R}$ are shown in Fig. 9.3. Each of the orbitals on the left-hand side can be expressed as a linear combination of orbitals that have the standard form oriented along the rotated axes, as shown on the right. An s orbital is invariant and a p orbital is transformed to a linear combination of p orbitals. The only non-zero matrix elements are the σ and π matrix elements, as shown. The top row of the figure illustrates the transformation of the p_x orbital needed to write the matrix element K_{s,p_x} in terms of $K_{sp\sigma}$ and the bottom row illustrates the relation of K_{p_x,p_z} to $K_{pp\sigma}$ and $K_{pp\pi}$. Specific relations for all s and p matrix elements are given in Table 9.1. Expressions for d orbitals are given in Refs. [234, 224, 225].

9.4   Slater-Koster two-center approximation

Now we are in position to describe the Slater and Koster approach [234]. These authors proposed that the Hamiltonian matrix elements can be approximated with the two-center form and fitted to theoretical calculations (or empirical data) as a simplified way of describing and extending calculations of electronic bands. Within this approach, all matrix elements have the same symmetry as for two atoms in free space (see Fig 9.3 and Table 9.1). This is a great simplification that leads to an extremely useful approach to understanding electrons in materials.

Slater and Koster gave extensive tables for matrix elements, including the s and p matrix elements given in Table 9.1. In addition, they presented expressions for the d states and analytical formulas for bands in several crystal structures. Examples of the latter are presented in the next section to illustrate the useful information that can be derived. However, the primary use of the SK approach in electronic structure has become the description of complicated systems, including the bands, total energies, and forces for relaxation of structures and molecular dynamics. These applications have very different requirements that often lead to different choices of SK parameters.

For the bands, the parameters are usually designed to fit selected eigenvalues for a particular crystal structure and lattice constant. For example, the extensive tables derived by Papaconstantopoulos [235] are very useful for interpolation of results of more expensive methods. It has been pointed out by Stiles [236] that for a fixed ionic configuration, effects of multi-center integrals can be included in two-center terms that can be generated by an automatic procedure. This makes it possible to describe any band structure accurately with a sufficient number of matrix elements in SK form. However, the two-center matrix elements are not transferable to different structures.

On the other hand, any calculation of total energies, forces, etc., requires that the parameters be known as a function of the position of the atoms. Thus, the choices are usually compromises that attempt to fit a large range of data. Such models are fit to structural data and, in general, are only qualitatively correct for the bands. Since the total energy depends only upon the occupied states, the conduction bands may be poorly described in these models. Of particular note, Harrison [224, 225] has introduced a table that provides parameters for any element or compound. The forms are chosen for simplicity, generality, and ability to describe many properties in a way that it is instructive and useful. The basis is assumed to be orthonormal, i.e. $S_{mm'} = \delta_{mm'}$. The diagonal Hamiltonian matrix elements are given in a table for each atom. Any Hamiltonian matrix element for orbitals on neighboring atoms separated by a distance R is given by a factor times $1/R^2$ for s and p orbitals and $1/R^{l+l'}$ for $l > l'$.

Many other SK parameterizations have been proposed, each tailored to particular elements and compounds. Some additional examples can be found in Chapter 14 of Ref. [226].

9.5 Some illustrative examples

Let us illustrate the tight-binding approach with the analysis of some simple situations.

9.5.1   *Example 1: A benzene molecule*

Often in molecular electronics one wants to establish a relation between the transport properties of a molecular junction and the corresponding the electronic structure of an isolated molecule. Let us illustrate how this electronic structure can be described by means of simple tight-binding models. For this purpose, we consider here the case of a benzene molecule, which was introduced in section 3.2. Benzene is an emblematic example of a molecule in which the relevant electronic structure is determined by a conjugated π-system. This means that the electrons in the highest occupied orbitals reside in π orbitals, which in this case are formed by the $2p_z$ orbitals of the six carbon atoms (here z is the direction perpendicular to the plane of the molecule). The word conjugated refers to the fact that this π-system extends over several neighboring atoms, which is the way in which the binding energy is increased in these molecules. In molecular orbital theory in chemistry, benzene is often described within the simplified *Hückel approximation* [229, 232, 227]. This approximation is based on the following basic assumptions: (i) only π-orbitals are considered (the σ-orbitals are much more strongly bound and they can be ignored), i.e. only one orbital per carbon atom is taken into account, (ii) the overlap integrals between different orbitals are set to zero: $S_{ij} = \delta_{ij}$, (iii) all the diagonal matrix elements of the Hamiltonian are ascribed the same value: $H_{ii} = \epsilon_0$, and (iv) the off-diagonal elements are set equal to zero except for those between neighboring atoms, all of which are set equal to $-t$, where t is positive. This model for benzene is summarized schematically in Fig. 9.4(a).

Following our discussion on finite systems in section 9.1, see in particular Eq. (9.6), the energy levels in this model are the roots of the following secular equation

$$\begin{vmatrix} \epsilon_0 - E & -t & 0 & 0 & 0 & -t \\ -t & \epsilon_0 - E & -t & 0 & 0 & 0 \\ 0 & -t & \epsilon_0 - E & -t & 0 & 0 \\ 0 & 0 & -t & \epsilon_0 - E & -t & 0 \\ 0 & 0 & 0 & -t & \epsilon_0 - E & -t \\ -t & 0 & 0 & 0 & -t & \epsilon_0 - E \end{vmatrix} = 0. \qquad (9.15)$$

In this determinant we have followed the order sketched in Fig. 9.4(a) and the π-orbitals in the different C atoms are denoted by $|i\rangle$, with $i = 1, ..., 6$. This equation can be solved analytically (see Exercise 9.3) and the different eigenenergies are given by

$$E_1 = \epsilon_0 - 2t; E_2 = E_3 = \epsilon_0 - t; E_4 = E_5 = \epsilon_0 + t; E_4 = \epsilon_0 + 2t, \qquad (9.16)$$

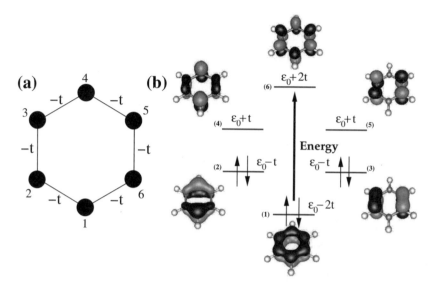

Fig. 9.4 (a) Schematic representation of the Hückel model for the benzene molecule, as described in the text. (b) Energy level diagram of benzene as obtained from this approximation. The levels are labeled from 1 to 6 following Eq. (9.16). We also show charge-density plots of the molecular orbitals obtained from a density-functional-theory calculation to show that indeed the Hückel approximation reproduces the character of the orbitals, see Eq. (9.17). The two colors indicate different signs of the wavefunctions. The ground state is obtained by doubly occupying the three lowest energy levels.

and the corresponding molecular orbitals (eigenfunctions) read

$$\phi_1 = \frac{1}{\sqrt{6}} \left(|1\rangle + |2\rangle + |3\rangle + |4\rangle + |5\rangle + |6\rangle \right)$$

$$\phi_2 = \frac{1}{\sqrt{12}} \left(2|1\rangle + |2\rangle - |3\rangle - 2|4\rangle - |5\rangle + |6\rangle \right)$$

$$\phi_3 = \frac{1}{2} \left(|2\rangle + |3\rangle - |5\rangle - |6\rangle \right)$$

$$\phi_4 = \frac{1}{\sqrt{12}} \left(2|1\rangle - |2\rangle - |3\rangle + 2|4\rangle - |5\rangle - |6\rangle \right)$$

$$\phi_5 = \frac{1}{2} \left(|2\rangle - |3\rangle + |5\rangle - |6\rangle \right)$$

$$\phi_6 = \frac{1}{\sqrt{6}} \left(|1\rangle - |2\rangle + |3\rangle - |4\rangle + |5\rangle - |6\rangle \right) \qquad (9.17)$$

These orbitals indeed describe correctly the symmetry and extension of the molecular orbitals that one obtains with more sophisticated methods, as one can see in Fig. 9.4(b), where we show the orbitals as obtained from a density-functional-theory calculation. The ground state of benzene is that in which the six π electrons are occupying the three lowest energy levels, as shown in Fig. 9.4.(b) and it has a total energy of $(6\epsilon_0 - 8t)$. Notice that the gap between the highest occupied molecular orbital (HOMO) and the lowest unoccupied molecular orbital (LUMO) is equal to $2t$. This result can be used to obtain the value of the hopping parameter t by comparing with spectroscopical data or ab initio calculations. A fit to a density-functional-theory calculation gives a value of around 2.6 eV.

It is worth stressing that the molecule gains stability or binding energy by delocalizing the electrons over the entire molecule (*conjugation*). This fact can be quantified in the following manner. If the molecule were described as having three unconjugated π-bonds, its total π-electron energy would have been $6(\epsilon_0 - t)$. By means of the conjugation, the molecule has gained an energy equal to $-2t$ (this gain is sometimes called delocalization energy).

On the other hand, notice that the form of the orbitals is determined solely by the symmetry of the molecule. Notice also that the six electrons just complete the molecular orbitals with net bonding effect, leaving unfilled the orbitals with net antibonding character. Another feature of the energy of levels of benzene is that the array of levels is symmetrical: to every bonding level there corresponds an antibonding level. This symmetry is a characteristic feature of alternant hydrocarbons and can be traced to the topology of the molecules.

This has been a simple example of the insightful molecular orbital theory, which is widely used in theoretical chemistry. For more examples, see for instance Chapter 8 of Ref. [227].

9.5.2 *Example 2: Energy bands in line, square and cubic Bravais lattices*

In molecular electronics it is important to know the bulk electronic structure of typical metals that are used in molecular junctions. For this reason, we consider now a bulk solid with a single atom per unit cell. The simplest possible example of bands is that of a lattice in which we have a single relevant orbital per site with s-symmetry. As a further simplification, we consider the case of orthogonal basis states and non-zero Hamiltonian matrix elements $\langle i|\mathbf{H}|j\rangle = t$ only if i and j are nearest neighbors. The on-site energy can be chosen to be zero, $\langle i|\mathbf{H}|i\rangle = 0$. There are three cases (line, square and cubic lattices) that can be treated together. For the cubic lat-

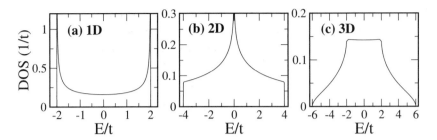

Fig. 9.5 Density of states per spin (DOS) vs. energy for an s-band in (a) a one-dimensional line, (b) a two-dimensional square, and (c) a three dimensional simple cubic lattice with nearest neighbor interactions t.

tice with spacing a the general expressions (9.8) and (9.10) reduce to (see Exercise 9.4)

$$\epsilon(\mathbf{k}) = H(\mathbf{k}) = 2t \left[\cos(k_x a) + \cos(k_y a) + \cos(k_z a)\right]. \qquad (9.18)$$

The bands for the square lattice in the xy-plane are given by this expression, omitting the k_z term; for a line (or chain) in the x-direction, only the k_x term applies. From this expression one can easily deduce several interesting consequences. First, the bands are symmetric about $\epsilon(\mathbf{k}) = 0$ in the sense that every state at $+\epsilon$ has a corresponding state at $-\epsilon$. This can be seen in Fig. 9.5, where we show the density of states (DOS) for one, two and three dimensions. The shapes can be found analytically in this case (see Exercise 9.6). Notice that the bandwidth is determined by the hopping element t. In the case of a metal, this parameter has a value of around 1 eV.

In the square lattice, the energy $\epsilon(\mathbf{k}) = 0$ at a face zone $\mathbf{k} = (\pi/a, 0)$. This is a saddle point since the slope vanishes and the bands curve upward and downward in different directions. This leads to a density of states with a logarithmic divergence at $\epsilon = 0$. Furthermore, for a half-filled band (one electron per cell), the Fermi surface is at $\epsilon(\mathbf{k}) = 0$. This leads to the result that the Fermi surface is a square rotated by $\pi/4$ with half the volume of the Brillouin zone, and the density of states diverges at $\epsilon = E_F$ as shown in Fig. 9.5(b). If there are second-neighbor interactions, the symmetry of the bands in $\pm\epsilon$ is broken and the Fermi surface is no longer square.

Let us assume now that the states are no longer orthogonal, but the overlap between nearest neighbors is equal to s. Then the solution for the bands, Eq. (9.18), is generalized to (Exercise 9.7)

$$\epsilon(\mathbf{k}) = \frac{H(\mathbf{k})}{S(\mathbf{k})} = \frac{2t \left[\cos(k_x a) + \cos(k_y a) + \cos(k_z a)\right]}{1 + 2s \left[\cos(k_x a) + \cos(k_y a) + \cos(k_z a)\right]}. \qquad (9.19)$$

In this case, the symmetry about $\epsilon = 0$ is broken, so that the conclusions on bands and the Fermi surface no longer apply. In fact s has an effect like longer range Hamiltonian matrix elements, indeed showing strictly infinite range but rapid exponential decay.

9.5.3 *Example 3: Energy bands of graphene*

As an example of a lattice with more than one atom per unit cell, we consider now the case of graphene. Graphene is a two-dimensional system formed by a single sheet of carbon atoms. Although the graphene band structure was already discussed theoretically more than 50 years ago [237], only recently it has been shown to exist in reality [238] and its physical properties are attracting a great attention [239–241]. Graphene has the planar honeycomb structure shown in Fig. 9.6(a). The corresponding Brillouin zone is a hexagon. Full calculations show that the band of graphitic systems at the Fermi energy are π bands, composed of electronic states that are odd in reflection in the plane. For graphene the π bands are well represented as linear combinations of p_z orbitals of the C atoms, where z is perpendicular to the plane. Since graphene has two atoms per cell, the p_z states form two bands. If there is a nearest neighbor Hamiltonian matrix element t, the bands are given by (Exercise 9.8)

$$|\mathbf{H}(\mathbf{k}) - \epsilon(\mathbf{k})| = \begin{vmatrix} -\epsilon(\mathbf{k}) & H_{12}(\mathbf{k}) \\ H_{12}^*(\mathbf{k}) & -\epsilon(\mathbf{k}) \end{vmatrix} = 0, \tag{9.20}$$

where (with the lattice oriented as in Fig. 9.6(a))

$$H_{12}(\mathbf{k}) = t \left[e^{ik_y a/\sqrt{3}} + 2e^{-ik_y a/2\sqrt{3}} \cos(k_x a/2) \right], \tag{9.21}$$

and a is the lattice constant. This is readily solved to yield the bands

$$\epsilon(\mathbf{k}) = \pm t \left[1 + 4\cos(\sqrt{3}k_y a/2)\cos(k_x a/2) + 4\cos^2(k_x a/2) \right]^{1/2}. \tag{9.22}$$

The most remarkable feature of the graphene bands is that they touch at the corners of the hexagonal Brillouin zone, e.g. the points denoted $\mathbf{K}_\pm = (k_x = \pm 4\pi/3a, k_y = 0)$. Note also that the bands are symmetric in $\pm\epsilon$. Since there is one π electron per atom, the band is half-filled and the bands touch with finite slope at the Fermi energy, i.e. a Fermi surface consisting of points. Indeed, one can show (see Exercise 9.8) that the dispersion relation around the points $\mathbf{K}_\pm$ (Dirac points) is linear, i.e. $\epsilon(\mathbf{q}) = \hbar v |\mathbf{q}|$, where $\mathbf{q} = \mathbf{k} - \mathbf{K}_\pm$ with a velocity given by $v = (\sqrt{3}a/2\hbar)t$. This linear dispersion relation resembles that of Dirac's massless fermions and it is the origin of

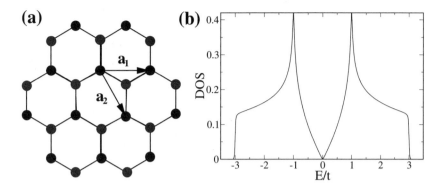

Fig. 9.6 (a) Honeycomb lattice for a graphene sheet: the lattice is triangular and there are two atoms per unit cell. Two primitive vectors are $\vec{a}_1 = a(1, 0)$ and $\vec{a}_2 = a/2(1, -\sqrt{3})$, where a is the lattice constant. (b) Local density of states (per spin) projected onto an atom of the unit cell as a function of the energy normalized by the hopping parameter t. Notice that the DOS vanishes at $E = 0$ and there are van Hove singularities at $E = \pm t$.

the extraordinary properties of this material [239–241]. Finally, if we have a look at the density of states, see Fig. 9.6(b), we can see that (undoped) graphene is a zero-bandgap semiconductor.

9.6   The NRL tight-binding method

There is a basic difficulty in generating tight-binding models that can describe very different structures. In models that have only two-center matrix elements, the values of the matrix elements must take into account effects of three-center terms. These effects change drastically between structures. There are two primary approaches toward making tight-binding models that are transferable between different structures. One is to define environment-dependent matrix elements, the values of which depend upon the presence of other neighbors. The other approach involves non-orthogonal tight-binding, which is more transferable than orthogonal forms.

The goal of this section is to describe in certain detail a sophisticated tight-binding parameterization in the spirit of the SK approach that meets the two requirements discussed in the previous paragraph. Moreover, this parameterization, which has been developed by Cohen, Mehl, and Papaconstantopoulos [242, 243], also allows us the computation of total energies and related quantities. This method has been widely used, in particular, for the analysis of the transport properties of metallic atomic-sized contacts, as we will show in later chapters. This parameterization is referred to as NRL[3] and it is nicely described in several review articles [244, 245].

Up to this point we have mainly discussed tight-binding parameterizations of the band structure alone.[4] Total-energy information is typically not given by these calculations, although a band energy can be readily determined from the sum of the eigenvalues over the occupied states. However, in single-particle band theory this sum is only a partial contribution to the total energy. In the Kohn-Sham single particle density functional theory (DFT) Ansatz, which will be explained in the next chapter, the total energy is given by

$$E = \int \frac{d^3\mathbf{k}}{(2\pi)^3} \sum_n \epsilon_n(\mathbf{k}) + F[n(\mathbf{r})], \qquad (9.23)$$

where the integral is over the first Brillouin zone, the sum is over occupied states, and $F[n(\mathbf{r})]$ is a functional of the density which includes the repulsion of the ionic cores, correlation effects, and part of the Coulomb interaction. Note that the value of the integral depends upon the choice of zero for the Kohn-Sham potential $v_{KS}(\mathbf{r})$ which generates the eigenvalue spectrum:

$$-\nabla^2\psi_n(\mathbf{r}) + v_{KS}(\mathbf{r})\psi_n(\mathbf{r}) = \epsilon_n\psi_n(\mathbf{r}). \qquad (9.24)$$

This choice is arbitrary. In the method developed at NRL [242, 243], the potential v_{KS} in the previous equation is shifted by an amount

$$V_0 = F[n(\mathbf{r})]/N_e, \qquad (9.25)$$

where N_e is the number of electrons in the unit cell. Then, the total energy of the system is

$$E = \int \frac{d^3\mathbf{k}}{(2\pi)^3} \sum_n \epsilon_n(\mathbf{k}) + F[n(\mathbf{r})] = \int \frac{d^3\mathbf{k}}{(2\pi)^3} \sum_n \epsilon_n(\mathbf{k}) + N_e V_0$$

$$= \int \frac{d^3\mathbf{k}}{(2\pi)^3} \sum_n [\epsilon_n(\mathbf{k}) + V_0], \qquad (9.26)$$

If we now define a shifted eigenvalue: $\epsilon'(\mathbf{k}) = \epsilon_n(\mathbf{k}) + V_0$, then to get the total energy we just sum the shifted eigenvalues of the occupied states:

$$E = \int \frac{d^3\mathbf{k}}{(2\pi)^3} \sum_n \epsilon'(\mathbf{k}). \qquad (9.27)$$

---

[3]NRL stands for Naval Research Laboratory, which is located in Washington D.C. For more practical information about this parameterization, visit the web page: http://cst-www.nrl.navy.mil/bind/.

[4]The only exception was the method put forward by Harrison that was briefly mentioned in section 9.4.

Note that V_0 depends upon the structure of the crystal, as well as the original method for determining the energy zero. Notice also that the $\epsilon'(\mathbf{k})$ are in some sense "universal". That is, if any two band structure methods are sufficiently well converged, they will give the same total energy, and the eigenvalues derived from the two methods will differ by only a constant. Then the definition of V_0 for each method will be such that the shifted eigenvalues $\epsilon'(\mathbf{k})$ are identical.

In the NRL method, the authors construct a first-principles[5] database of eigenvalues $\epsilon(\mathbf{k})$ and total energies E for several crystal structures at several volumes. Then, they find V_0 for each system, and shift the eigenvalues. Next, they attempt to find a set of parameters which will generate non-orthogonal, two-center SK Hamiltonians which will reproduce the energies and eigenvalues in the database.

Let us now describe how the TB parameters for elemental systems are constructed. One assumes that the on-site terms are diagonal and sensitive to the environment. For single-element systems one assigns atom i in the crystal an embedded-atom-like "density"

$$\rho_i = \sum_j \exp(-\lambda^2 R_{ij}) F(R_{ij}), \tag{9.28}$$

where the sum is over all the atoms j within a range R_c of atom i; λ is the first fitting parameter, squared to ensure that the contributions are greater from the nearest neighbors; and $F(R)$ is a cut-off function,

$$F(R) = \theta(R_c - R) / \left\{ 1 + \exp\left[(R - R_c)/l + 5\right] \right\}, \tag{9.29}$$

where $\theta(z)$ is the step function. Typically one takes R_c between 10.5 and 16.5 Bohr and l between 0.25 and 0.5 Bohr.

One then defines the angular-momentum-dependent on-site terms by

$$h_{il} = a_l + b_l \rho_i^{2/3} + c_l \rho_i^{4/3} + d_l \rho_i^2, \tag{9.30}$$

where $l = s, p$, or d. These $(a, b, c, d)_l$ form the next 12 fitting parameters.

In the spirit of the two-center approximation, one assumes that the hopping integrals depend only upon the angular momentum of the orbitals and the distance between the atoms. As we showed in section 9.3.1, all the two-center (spd) hopping integral can then be constructed from ten independent parameters, the SK parameters, $H_{ll'm}$, where

$$(ll'm) = ss\sigma, sp\sigma, pp\sigma, pp\pi, sd\sigma, pd\sigma, pd\pi, dd\sigma, dd\pi, \text{ and } dd\delta. \tag{9.31}$$

---

[5]The first-principle methods used by the authors are typically the augmented plane wave method (APW) or the linearized augmented plane wave method (LAPW).

Then, it is assumed the following polynomial × exponential form for these parameters:

$$H_{ll'm}(R) = (e_{ll'm} + f_{ll'm}R + g_{ll'm}R^2)\exp(-h_{ll'm}^2 R)F(R), \qquad (9.32)$$

where R is the separation between these atoms and $F(R)$ is the cut-off function defined above. The parameters $(e_{ll'm}, f_{ll'm}, g_{ll'm}, h_{ll'm})$ constitute the next 40 fitting parameters.

Since this is a non-orthogonal calculation, one must also define a set of SK overlap functions. These represent the overlap between two orbitals separated by a distance R. They have the same angular momentum behavior as the hopping parameters:

$$S_{ll'm}(R) = (p_{ll'm} + q_{ll'm}R + r_{ll'm}R^2)\exp(-s_{ll'm}^2 R)F(R), \qquad (9.33)$$

The parameters $(p_{ll'm}, q_{ll'm}, r_{ll'm}, s_{ll'm})$ make up the final 40 fitting parameters for a monoatomic system, giving in total 93 fitting parameters that are chosen to reproduce the contents of the first-principles database, as noted above.[6]

So in summary, this parameterization uses an analytical set of two-center integrals, nonorthogonal parameters and on-site parameters that depend on the local environment. The method reproduces not only the band structure, but also the total energy of the system. It has been demonstrated that this method reproduces very well structural energy differences, elastic constants, phonon frequencies, vacancy formation energies, and surface energies for both transition metal and noble metals.

As an application of this parameterization, we have computed the bulk density of states of six different metals that play an important role in molecular electronics.[7] The results can be seen in Fig. 9.7. Notice that in the cases of Ag and Au (noble metals), the Fermi energy lies in the region where the DOS is dominated by the s band. In the case of Al and Pb, the s and p bands dominate the DOS around the Fermi energy. The main difference between these two metals is that Pb has 4 valence electrons and therefore, the Fermi energy lies well inside the p band. Finally, Nb and Pt are examples of transition metals, where the d band dominates the DOS at the Fermi energy and for this reason, the d orbitals play a fundamental in the transport properties of these metals.

---

[6]These parameters for many different elementary solids can be found in the following web page: http://cst-www.nrl.navy.mil/bind/.

[7]In particular, we shall analyze in Chapter 11 the conductance of single-atom contacts of these six materials.

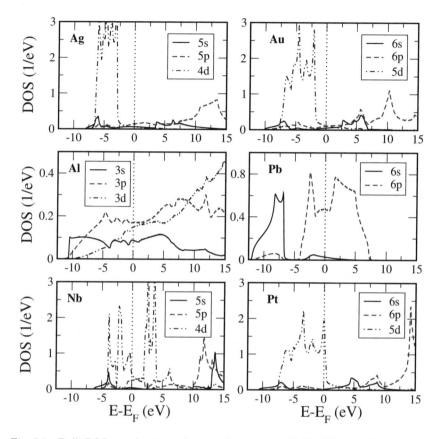

Fig. 9.7 Bulk DOS as a function of energy for Ag, Au, Al, Pb, Nb, and Pt computed using the NRL-tight-binding parameterization. The DOS is projected onto the s, p, and d orbitals that give rise to the bands around the Fermi energy (E_F).

9.7 The tight-binding approach in molecular electronics

In this final section we shall explain how the tight-binding approach is used in practice to describe the transport properties of atomic-scale junctions and we shall also review very briefly the impact of this approach in the field of molecular electronics to date.

9.7.1   *Some comments on the practical implementation of the tight-binding approach*

In the previous chapters we have learned how the Green's function techniques can be combined with the knowledge of the Hamiltonian matrix elements in a local basis to describe both equilibrium and transport properties of atomic-scale junctions. Thus, the application of the tight-binding approach to the description of the physics of atomic or molecular junctions is now rather straightforward. Anyway, let us mention some of technical issues and difficulties that one encounters in the practical application of this method.

One of the most common theoretical problems in molecular electronics is the calculation of the elastic current of an atomic-scale junction. Such calculation proceeds in a series of steps that we now proceed to describe.

Step 1: Geometry of the contact. As a first step one has to define the geometry of the junction. This geometry includes a central part of arbitrary shape and two ideal leads or electrodes that, for practical reasons, must have a regular structure (see Fig. 8.1). Ideally, one should determine the junction geometry by doing, for instance, molecular dynamics simulations or geometry optimization. In principle, this is possible with sophisticated tight-binding approaches like the NRL-method described in section 9.6 which allows us to compute the forces between the atoms or the total energy of the system. In practice, the determination of the geometry with tight-binding models has been restricted to the case of atomic contacts and for molecular junctions one needs to resort to more sophisticated methods like density function theory (see next chapter).

Step 2: Hamiltonian matrix elements. Once the geometry is defined, one proceeds to the determination of the matrix elements of the Hamiltonian, as explained in the previous sections. For instance, in the approaches based on the two-center approximation (see section 9.4), one can construct those matrix elements between two neighboring atoms by projecting the irreducible SK-parameters according to their relative position.

Step 3: Calculation of the Green's functions. The retarded and advanced Green's functions of the central part of the junction contain all the relevant information about both the equilibrium and transport properties of the system. These functions are computed via their Dyson's equation, see Eq. (8.22). This requires to previously calculate the self-energies and in turn the Green's functions of the leads. This is indeed the most complicated step in the whole calculation. The leads are semi-infinite systems and the lack of periodicity complicates the calculation of their Green's functions. There are different solutions for this problem. For instance, one can describe the electrodes with simple structures like Bethe lattices [246]. A more satisfactory solution to avoid artificial interface resistances is to describe the leads as ideal surfaces and compute the Green's functions with recursive methods like the decimation technique of Ref. [198].[8]

Step 4: Computation of the current. The final step is the calculation of the current from the knowledge of the Green's functions, which is done using Eqs. (8.18-8.20).

In general, the "recipe" described above has to be carried out numerically, but the computer codes can be developed by a single person in a few weeks. Moreover, the tight-binding approach is extremely efficient, computationally speaking, and the calculations of the transport properties of realistic systems can be done in standard PC's. Of course, the level of accuracy of these calculations depends on the quality tight-binding parameters, which in turn depends on the system under study.

9.7.2   *Tight-binding simulations of atomic-scale transport junctions*

The tight-binding approach has been used to describe a great variety of problems related to the electronic transport in atomic-scale junctions. Our goal in this subsection is to mention very briefly some these applications and we refer the reader to Todorov's review [248] for a more detailed discussion and for a more complete list of references.

One of the first applications of the tight-binding approach was the analysis of the operation of the STM and the interpretation of images taken with this instrument [194, 249–252]. This approach has been very important to elucidate the role of the tip-substrate distance and it allows to identify the characteristic signature of many different adsorbates.

In the context of molecular junctions, at the end of the 1980's Sautet and Joachim pioneered the use of the tight-binding approach (within the extended Hückel approximation) to compute the current and conductance of single-molecule junctions [253, 254]. Later, Ratner and coworkers used simple tight-binding models to address a number of issues such as the dependence of the conductance on the length of the molecular wires [255–259].[9]

---

[8] An extension of the decimation technique to the case of non-orthogonal basis sets can be found in the appendix C of Ref. [247].

[9] In Chapter 13 we shall make use of simple tight-binding models to discuss basic issues

Also in the middle of the 1990's, Datta and coworkers employed the tight-binding approach to describe the current-voltage characteristics of different organic molecules and to establish a detailed comparison with the experiments [260, 261].

In the context of metallic atomic-sized contacts,[10] the tight-binding formalism was first used, in combination with molecular dynamics simulations, to elucidate the origin of the conductance jumps observed during the formation of these nanowires [262]. Tight-binding models were then used to establish the relation between the conduction channels of single-atom contacts and the detailed chemistry of the metal atoms [263, 264]. The tight-binding approach has also been extended to calculate changes to interatomic forces under electrical current flow in atomic-scale conductors [265] and this formalism has been used to model electromigration and current-induced fracture of atomic wires [266, 267].

9.8   Exercises

9.1 Secular equation for a finite system: Use the Schrödinger equation to show that the coefficients of the expansion of Eq. (9.3) satisfy the set of equations given by Eq. (9.6).

9.2 Bloch's theorem: Using the translational invariance in a Bravais lattice, show that matrix elements of the Hamiltonian with basis functions $\Phi_{\mathbf{k}i\alpha}$ and $\Phi_{\mathbf{k}'j\beta}$ are non-zero only for $\mathbf{k} = \mathbf{k}'$, and derive the expression of Eq. (9.8).

9.3 Energy spectrum of benzene: Solve analytically the secular equation (9.15) and show that within the Hückel approximation the energy levels and the corresponding molecular orbitals are given by Eq. (9.16) and Eq. (9.17), respectively.

9.4 Molecular orbital structure of butadiene: The 1,3-butadiene molecule (C_4H_6) shown in Fig. 9.8 is a simple conjugated diene, i.e. it is a hydrocarbon which contains two double bonds. Determine its energy levels and molecular orbitals using the Hückel approximation.

9.5 Energy bands of s-bands in line, square and cubic Bravais lattices: Show that for an s-band in a line, square lattice, and simple cubic lattice with only nearest neighbor Hamiltonian matrix elements, the energy bands are given by Eq. (9.18).

9.6 Density of states of s-bands in line, square and cubic Bravais lat-

---

concerning the coherent transport through molecular junctions.

[10]The physics of these metallic nanowires is described in Part 3, where in particular we shall make extensive use of the tight-binding approach.

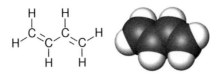

Fig. 9.8   Composition and structure of the 1,3-butadiene molecule.

tices: Reproduce the results of Fig. 9.5 for the density of states of s-band in a line, square lattice, and simple cubic lattice with only nearest neighbor Hamiltonian matrix elements.

9.7 Energy bands of s-bands in line, square and cubic Bravais lattices in a non-orthogonal model: Show that the expression for bands with non-orthogonal basis orbitals, Eq. (9.19), is correct. Why are the bands in this case no longer symmetric about $\epsilon = 0$?

9.8 Electronic structure of graphene: Consider the model for graphene detailed in section 9.5.3. Carry out the following tasks: (i) Determine the Brillouin zone of the honeycomb lattice, (ii) show that the energy bands are given by Eq. (9.22), (iii) demonstrate that the dispersion relation around the Dirac's point is linear, and (iv) compute the local density of states and show that it is given by the result of Fig. 9.6(b).

9.9 The NRL tight-binding method: An interesting project for graduate students and advanced undergraduate students is to write a computer code (in whatever language) to calculate the energy bands and bulk density of states [see Fig. 9.7] of elementary solids within the NRL tight-binding method described in section 9.6.

Chapter 10

Electronic structure II: Density functional theory

This second chapter about electronic structure calculations provides a basic introduction to the density functional theory (DFT). This theory is presently the most successful (and also the most promising) approach for computing the electronic structure of matter. Its applicability ranges from atoms, molecules and solids to nuclei and quantum and classical fluids. Thus for instance, in chemistry DFT is widely used to predict a great variety of molecular properties: molecular structures, vibrational frequencies, atomization energies, ionization energies, electric and magnetic properties, reaction paths, etc. Originally, DFT was designed to provide the electron density and total energy of the ground state of (non-magnetic) electronic systems. However, meanwhile the theory has been generalized to deal with many different situations: spin polarized systems, multicomponent systems such as nuclei and electron hole droplets, free energy at finite temperatures, superconductors with electronic pairing mechanisms, relativistic electrons, time-dependent phenomena and excited states, bosons, molecular dynamics, etc.

More importantly for the scope of this book, DFT is at the moment the theoretical approach with the largest impact in molecular electronics. In this sense, we believe that DFT should be part of the general culture of the researchers working in this field. For this reason, we have included here a concise introduction to the standard formulation of DFT, which can be read almost independently of the rest of the book. Our goals are: (i) To explain what kind of information this theory can provide, (ii) to describe how it is used in molecular electronics, and (iii) to discuss its advantages and limitations. For those readers who want to get deeper into the subtleties and performance of this theory, the following entry points into the literature are recommended. First of all, one has of course the original papers [268, 269] and Kohn's Nobel lecture [270]. Among the DFT reviews, we recommend the one of Ref. [271]. Finally, let us say that this chapter is based on the monographs of Refs. [272, 226] and specially on that of Ref. [273]. Those readers familiar with DFT who only want to know how it is applied in molecular electronic are advised to jump directly to the last section of this chapter.

10.1   Elementary quantum mechanics

In order to pave the way for the understanding of the basic formulation of density function theory, we start reminding some basic issues in quantum mechanics.[1]

10.1.1   *The Schrödinger equation*

The ultimate goal of most theoretical approaches in solid state physics and quantum chemistry is the solution of the time-independent, non-relativistic Schrödinger equation[2]

$$\mathbf{H}\Psi_i(\vec{x}_1, ..., \vec{x}_N, \vec{R}_1, ..., \vec{R}_M) = E_i\Psi_i(\vec{x}_1, ..., \vec{x}_N, \vec{R}_1, ..., \vec{R}_M), \qquad (10.1)$$

where $\mathbf{H}$ is the Hamiltonian for a system consisting of M nuclei and N electrons. In the absence of external fields, $\mathbf{H}$ has the following form:

$$\mathbf{H} = -\frac{1}{2}\sum_{i=1}^{N}\nabla_i^2 - \frac{1}{2}\sum_{A=1}^{M}\frac{1}{M_A}\nabla_A^2 - \sum_{i=1}^{N}\sum_{A=1}^{M}\frac{Z_A}{r_{iA}} + \sum_{i=1}^{N}\sum_{j>i}^{N}\frac{1}{r_{ij}} + \sum_{A=1}^{M}\sum_{B>A}^{M}\frac{Z_A Z_B}{R_{AB}}.$$
$$(10.2)$$

Here, A and B run over the M nuclei while i and j denote the N electrons in the system. The first two terms describe the kinetic energy of the electrons and nuclei. The other three terms represent the attractive electrostatic interaction between the nuclei and the electrons and repulsive potential due to the electron-electron and nucleus-nucleus interactions. Let us stress that, following the common practice in most textbooks, we shall use atomic units throughout this chapter.[3]

---

[1]Throughout this chapter we shall be using the more standard first quantization formulation of quantum mechanics.

[2]In this chapter the operators will be written in boldface, while the vector character of a variable will be indicated by an arrow on top of it.

[3]In this system of units the masses are measured in units of the electron mass, the charges in units of the electron charge, $\hbar$ is the unit of action, the energy is measured in Hartrees (27.211 eV) and the length unit is the Bohr (0.52910 Å).

The Schrödinger equation [Eq. (10.1)] can be simplified using the Born-Oppenheimer approximation. Due to their masses the nuclei move much slower than the electrons, which implies that we can consider the electrons as moving in the field of fixed nuclei, i.e. the nuclear kinetic energy is zero and their potential energy is merely a constant. Thus, the electronic Hamiltonian reduces to

$$\mathbf{H}_{elec} = -\frac{1}{2}\sum_{i=1}^{N}\nabla_i^2 - \sum_{i=1}^{N}\sum_{A=1}^{M}\frac{Z_A}{r_{iA}} + \sum_{i=1}^{N}\sum_{j>i}^{N}\frac{1}{r_{ij}} = \mathbf{T} + \mathbf{V}_{Ne} + \mathbf{V}_{ee}. \quad (10.3)$$

The solution of the Schrödinger equation with $\mathbf{H}_{elec}$ is the electronic wave function Ψ_{elec} and the electronic energy E_{elec}. The total energy E_{tot} is then the sum of E_{elec} and the constant nuclear repulsion term E_{nuc}, i.e.

$$\mathbf{H}_{elec}\Psi_{elec} = E_{elec}\Psi_{elec}, \quad (10.4)$$

and

$$E_{tot} = E_{elec} + E_{nuc} \text{ where } E_{nuc} = \sum_{A=1}^{M}\sum_{B>A}^{M}\frac{Z_A Z_B}{R_{AB}}. \quad (10.5)$$

In principle, our main problem now is to solve Eq. (10.4), which is simply impossible to accomplish in general.[4]

10.1.2 *The variational principle for the ground state*

A general strategy for the search of the lowest-energy solution of Schrödinger equation [Eq (10.4)] is provided by the variational principle. This idea goes as follows. When a system is in the state Ψ, the expectation value of the energy is given by

$$E[\Psi] = \frac{\langle\Psi|\mathbf{H}|\Psi\rangle}{\langle\Psi|\Psi\rangle} \text{ where } \langle\Psi|\mathbf{H}|\Psi\rangle = \int \Psi^*\mathbf{H}\Psi \, d\vec{x}. \quad (10.6)$$

The variational principle states that *the energy computed from a guessed* Ψ *is an upper bound to the true ground-state energy* E_0. Full minimization of the functional $E[\Psi]$ with respect to all allowed N-electrons wave functions will give the true ground state Ψ_0 and energy $E[\Psi_0] = E_0$; that is

$$E_0 = \min_{\Psi\to N} E[\Psi] = \min_{\Psi\to N} \langle\Psi|\mathbf{T} + \mathbf{V}_{Ne} + \mathbf{V}_{ee}|\Psi\rangle, \quad (10.7)$$

where $\Psi \to N$ indicates that Ψ is an allowed N-electron wave function.

---

[4]From now one we shall only consider the electronic problem and subscript "elec" will be dropped.

For a system of N electrons and given nuclear potential V_{ext}, the variational principle defines a procedure to determine the ground-state wave function Ψ_0, the ground-state energy $E_0[N, V_{ext}]$, and other properties of interest. In other words, the ground state energy is a functional of the number of electrons N and the nuclear potential V_{ext}

$$E_0 = E[N, V_{ext}]. \tag{10.8}$$

10.1.3   *The Hartree-Fock approximation*

Although the variational principle offers a strategy for finding the ground state wave function, it is simply impossible to solve Eq. (10.4) by searching through all acceptable many-body wave functions. We need to define a suitable subset, which offers a reasonable approximation to the exact wave function without being unmanageable in practice. The Hartree-Fock approximation provides the simplest, yet physically sound, solution to this problem. Let us briefly explain the basic idea behind this approach.

Suppose that the ground state wave function, Ψ_0, is approximated as an antisymmetrized product of N orthonormal spin orbitals $\psi_i(\vec{x})$, each a product of a spatial orbital $\phi_k(\vec{r})$ and a spin function $\sigma(s)$, the Slater determinant

$$\Psi_0 \approx \Psi_{\mathrm{HF}} = \frac{1}{\sqrt{N!}} \begin{vmatrix} \psi_1(\vec{x}_1) & \psi_2(\vec{x}_1) & \dots & \psi_N(\vec{x}_1) \\ \psi_1(\vec{x}_2) & \psi_2(\vec{x}_2) & \dots & \psi_N(\vec{x}_2) \\ \vdots & \vdots & & \vdots \\ \psi_1(\vec{x}_N) & \psi_2(\vec{x}_N) & \dots & \psi_N(\vec{x}_N) \end{vmatrix}. \tag{10.9}$$

The Hartree-Fock approximation is the method whereby the orthogonal orbitals ψ_i are found that minimize the energy for this determinantal form of Ψ_0

$$E_{\mathrm{HF}} = \min_{(\Psi_{\mathrm{HF}} \to N)} E\left[\Psi_{\mathrm{HF}}\right]. \tag{10.10}$$

The expectation value of the Hamiltonian operator with Ψ_{HF} is given by (Exercise 10.1)

$$E_{\mathrm{HF}} = \langle \Psi_{\mathrm{HF}} | \mathbf{H} | \Psi_{\mathrm{HF}} \rangle = \sum_{i=1}^{N} H_i + \frac{1}{2} \sum_{i,j=1}^{N} (J_{ij} - K_{ij}), \tag{10.11}$$

where

$$H_i \equiv \int \psi_i^*(\vec{x}) \left[-\frac{1}{2}\nabla^2 + V_{ext}(\vec{x})\right] \psi_i(\vec{x}) \, d\vec{x} \tag{10.12}$$

defines the contribution due to the kinetic energy and the electron-nucleus attraction and

$$J_{ij} = \int \int \psi_i(\vec{x}_1)\psi_i^*(\vec{x}_1) \, \frac{1}{r_{12}} \, \psi_j^*(\vec{x}_2)\psi_j(\vec{x}_2) d\vec{x}_1 d\vec{x}_2, \qquad (10.13)$$

$$K_{ij} = \int \int \psi_i^*(\vec{x}_1)\psi_j(\vec{x}_1) \, \frac{1}{r_{12}} \, \psi_i(\vec{x}_2)\psi_j^*(\vec{x}_2) d\vec{x}_1 d\vec{x}_2. \qquad (10.14)$$

The integrals are all real, and $J_{ij} \geq K_{ij} \geq 0$. The J_{ij} are called *Coulomb integrals*, the K_{ij} are called *exchange integrals*. We have the property $J_{ii} = K_{ii}$.

The variational freedom in the expression of the energy [Eq. (10.11)] is in the choice of the orbitals. The minimization of the energy functional with the normalization conditions $\int \psi_i^*(\vec{x})\psi_j(\vec{x})d\vec{x} = \delta_{ij}$ leads to *the Hartree-Fock differential equations* (see Exercise 10.2)

$$\mathbf{f} \, \psi_i = \epsilon_i \, \psi_i \, , i = 1, 2, ..., N. \qquad (10.15)$$

These N equations have the appearance of eigenvalue equations, where ϵ_i are the eigenvalues of the operator $\mathbf{f}$. The Fock operator $\mathbf{f}$ is an effective one-electron operator defined as

$$\mathbf{f} = -\frac{1}{2}\nabla_i^2 - \sum_A^M \frac{Z_A}{r_{iA}} + \mathbf{V}_{\mathrm{HF}}(i). \qquad (10.16)$$

The first two terms are the kinetic energy and the potential energy due to the electron-nucleus attraction. $\mathbf{V}_{\mathrm{HF}}(i)$ is the *Hartree-Fock potential*, the average repulsive potential experienced by the i-th electron due to the remaining N-1 electrons, and it is given by

$$\mathbf{V}_{\mathrm{HF}}(\vec{x}_1) = \sum_j^N (\mathbf{J}_j(\vec{x}_1) - \mathbf{K}_j(\vec{x}_1)), \qquad (10.17)$$

$$\mathbf{J}_j(\vec{x}_1) = \int |\psi_j(\vec{x}_2)|^2 \frac{1}{r_{12}} \, d\vec{x}_2. \qquad (10.18)$$

The Coulomb operator $\mathbf{J}$ represents the potential that an electron at position $\vec{x}_1$ experiences due to the average charge distribution of another electron in spin orbital ψ_j.

The second term in Eq. (10.17) is the exchange contribution to the HF potential. It has no classical analog and it describes the modification of the energy that can be ascribed to the effects of spin correlation. It is defined through its effect when operating on a spin orbital

$$\mathbf{K}_j(\vec{x}_1) \, \psi_i(\vec{x}_1) = \int \psi_j^*(\vec{x}_2)\frac{1}{r_{12}}\psi_i(\vec{x}_2) \, d\vec{x}_2 \, \psi_j(\vec{x}_1). \qquad (10.19)$$

Two important remarks to conclude this section. First, the HF potential is non-local and it depends on the spin orbitals. Thus, the HF equations must be solved self-consistently. Second, the Koopman's theorem [274] provides a physical interpretation of the orbital energies: it states that the orbital energy ϵ_i is an approximation of minus the ionization energy (IE) associated with the removal of an electron from the orbital ψ_i, i.e. $\epsilon_i \approx E_N - E_{N-1}^i = -IE(i)$.

10.2   Early density functional theories

In this section we shall introduce the electron density, which is the fundamental quantity in DFT, and we shall briefly review some early attempts to develop a density functional theory.

The electron density is defined as the integral over the spin coordinates of all electrons and over all but one of the spatial variables ($\vec{x} \equiv \vec{r}, s$)

$$\rho(\vec{r}) = N \int \ldots \int |\Psi(\vec{x}_1, \vec{x}_2, \ldots, \vec{x}_N)|^2 ds_1 d\vec{x}_2 \ldots d\vec{x}_N. \tag{10.20}$$

The electron density $\rho(\vec{r})$ determines the probability of finding any of the N electrons within volume element $d\vec{r}$. Clearly, $\rho(\vec{r})$ is a non-negative function of only the three spatial variables which vanishes at infinity and integrates to the total number of electrons, i.e.

$$\rho(\vec{r} \to \infty) = 0; \quad \int \rho(\vec{r}) d\vec{r} = N. \tag{10.21}$$

Moreover, unlike the wave function, the electron density is an observable and it can be measured experimentally, e.g. by X-ray diffraction.

At this stage, one may wonder whether the central role of the complicated N-electron wave function, which depends on 3N spatial variables, could be played by a simpler function such as the electron density. As early as in the 1920's, several authors conjectured that indeed the total energy could be a functional of the electronic density alone. Probably, the most famous example of such an early density functional theory is the so-called *Thomas-Fermi model*, which was put forward in 1927. Based on the uniform electron gas, Thomas and Fermi proposed independently the following functional for the kinetic energy

$$T_{TF}[\rho(\vec{r})] = \frac{3}{10}(3\pi^2)^{2/3} \int \rho^{5/3}(\vec{r}) d\vec{r}. \tag{10.22}$$

This functional was combined with the classical expression for the electron-nuclei potential and the electron-electron potential to write down the following functional for the energy of an atom

$$E_{TF}\left[\rho(\vec{r})\right] = \frac{3}{10}(3\pi^2)^{2/3} \int \rho^{5/3}(\vec{r})d\vec{r}$$

$$-Z \int \frac{\rho(\vec{r})}{r}d\vec{r} + \frac{1}{2} \int \int \frac{\rho(\vec{r}_1)\rho(\vec{r}_2)}{r_{12}}d\vec{r}_1 d\vec{r}_2. \quad (10.23)$$

Notice that the energy is given completely in terms of the electron density.

In order to determine the correct density to be included in Eq. (10.23), they employed a variational principle. They assumed that the ground state of the system is connected to the $\rho(\vec{r})$ for which the energy is minimized under the constraint of $\int \rho(\vec{r})d\vec{r} = N$. The obvious question at this stage is: does this variational principle make sense? The Hohenberg-Kohn theorems discussed in the next section will prove that this approach can be rigorously justified.

10.3 The Hohenberg-Kohn theorems

Density functional theory as we know it today is founded in the so-called Hohenberg-Kohn theorems that were put forward in 1964 [268]. In this section we present these theorems and discuss some of their basic implications. The proofs of these theorems will not be detailed here and they can be found in any of the references given at the beginning of this chapter.

The first Hohenberg-Kohn theorem states that the electron density uniquely determines the Hamiltonian operator and thus all the properties of the system. To be precise, this theorem states that *the external potential $V_{ext}(\vec{r})$ is (to within a constant) a unique functional of $\rho(\vec{r})$; since, in turn $V_{ext}(\vec{r})$ fixes* **H** *we see that the full many particle ground state is a unique functional of $\rho(\vec{r})$.*

Thus, $\rho(\vec{r})$ determines N and $V_{ext}(\vec{r})$ and hence all the properties of the ground state, for example the kinetic energy $T[\rho]$, the potential energy $V[\rho]$, and the total energy $E[\rho]$. Now, we can write the total energy as

$$E[\rho] = E_{Ne}[\rho] + T[\rho] + E_{ee}[\rho] = \int \rho(\vec{r})V_{Ne}(\vec{r})d\vec{r} + F_{\mathrm{HK}}[\rho]; \quad (10.24)$$

$$F_{\mathrm{HK}}[\rho] = T[\rho] + E_{ee}[\rho]. \quad (10.25)$$

Here, we have separated the contributions that depend on the actual system, i.e. the potential energy due to the electron-nuclei attraction,

$E_{Ne}[\rho] = \int \rho(\vec{r})V_{Ne}(\vec{r})d\vec{r}$, from those which are universal, $F_{\mathrm{HK}}[\rho]$. This functional $F_{\mathrm{HK}}[\rho]$ is the holy grail of density functional theory. If it was known, we would be able to solve the Schrödinger equation exactly and for any system. This functional contains the functional for the kinetic energy $T[\rho]$ and that for the electron-electron interaction, $E_{ee}[\rho]$. The explicit form of both these functionals is unknown. However, from the latter we can extract at least the classical part $J[\rho]$,

$$E_{ee}[\rho] = \frac{1}{2} \int \int \frac{\rho(\vec{r}_1)\rho(\vec{r}_2)}{r_{12}} d\vec{r}_1 d\vec{r}_2 + E_{ncl} = J[\rho] + E_{ncl}[\rho]. \qquad (10.26)$$

E_{ncl} is the *non-classical* contribution to the electron-electron interaction: self-interaction correction, exchange and Coulomb correlation. The explicit form of the functionals $T[\rho]$ and $E_{ncl}[\rho]$ is the major challenge of DFT.

Let us now address the following question: how can we be sure that a certain density is the ground-state density that we are looking for? **The second Hohenberg-Kohn theorem** answers this question. This theorem states that $F_{\mathrm{HK}}[\rho]$, *the functional that delivers the ground state energy of the system, delivers the lowest energy if and only if the input density is the true ground state density*. This is nothing but the variational principle

$$E_0 \leq E[\tilde{\rho}] = T[\tilde{\rho}] + E_{Ne}[\tilde{\rho}] + E_{ee}[\tilde{\rho}]. \qquad (10.27)$$

In other words, this means that for any trial density $\tilde{\rho}(\vec{r})$, which satisfies the necessary boundary conditions such as $\tilde{\rho}(\vec{r}) \geq 0$, $\int \tilde{\rho}(\vec{r})d\vec{r} = N$, and which is associated with some external potential $\tilde{V}_{ext}$, the energy obtained from the functional of Eq. (10.24) represents an upper bound to the true ground state energy E_0. E_0 results if and only if the exact ground state density is inserted in Eq. (10.27).

Let us summarize what we have learned so far and some basic consequences of the previous theorems:

- All the properties of a system defined by an external potential V_{ext} are determined by the ground state density. In particular, the ground state energy associated with a density ρ is available through the functional

$$\int \rho(\vec{r})V_{ext}d\vec{r} + F_{\mathrm{HK}}[\rho]. \qquad (10.28)$$

- This functional attains its minimum value with respect to all allowed densities if and only if the input density is the true ground state density, i.e. for $\tilde{\rho}(\vec{r}) \equiv \rho(\vec{r})$.
- The applicability of the variational principle is limited to the ground state. Hence, we cannot easily transfer this strategy to the problem of excited states.

- The explicit form of the functional $F_{HK}[\rho]$ is unknown and this remains as the major challenge of DFT.

10.4   The Kohn-Sham approach

We have seen that the ground state energy of a system can be written as

$$E_0 = \min_{\rho \to N} \left(F_{HK}[\rho] + \int \rho(\vec{r}) V_{Ne} d\vec{r} \right), \qquad (10.29)$$

where the universal functional $F_{HK}[\rho]$ contains the contributions of the kinetic energy, the classical Coulomb interaction and the non-classical portion

$$F_{HK}[\rho] = T[\rho] + J[\rho] + E_{ncl}[\rho]. \qquad (10.30)$$

Of these, only $J[\rho]$ is known. The main problem is to find the expressions for $T[\rho]$ and $E_{ncl}[\rho]$. The Thomas-Fermi model of section 10.2 provides an example of density functional theory. However, its performance is really bad due to the poor approximation of the kinetic energy. To solve this problem Kohn and Sham proposed in 1965 [269] the following approach. They suggested to calculate the exact kinetic energy of a non-interacting reference system with the same density as the real, interacting one

$$T_S = -\frac{1}{2} \sum_i^N \langle \psi_i | \nabla^2 | \psi_i \rangle, \quad \rho_S(\vec{r}) = \sum_i^N \sum_s |\psi_i(\vec{r}, s)|^2 = \rho(\vec{r}), \quad (10.31)$$

where the ψ_i are the orbitals of the non-interacting system. Of course, T_S is not equal to the true kinetic energy of the system. Kohn and Sham accounted for that by introducing the following separation of the functional $F_{HK}[\rho]$

$$F_{HK}[\rho] = T_S[\rho] + J[\rho] + E_{XC}[\rho], \qquad (10.32)$$

where E_{XC}, the so-called *exchange-correlation energy* is defined through Eq. (10.32) as

$$E_{XC}[\rho] \equiv (T[\rho] - T_S[\rho]) + (E_{ee}[\rho] - J[\rho]). \qquad (10.33)$$

The exchange and correlation energy E_{XC} is the functional that contains everything that is unknown.

Now the question is: How can we uniquely determine the orbitals in our non-interacting reference system? In other words, how can we define a potential V_S such that it provides us with a Slater determinant which is characterized by the same density as our real system? To solve this problem, we write down the expression for the energy of the interacting system in terms of the separation described in Eq. (10.32)

$$E[\rho] = T_S[\rho] + J[\rho] + E_{XC}[\rho] + E_{Ne}[\rho], \tag{10.34}$$

where

$$
\begin{aligned}
E[\rho] &= T_S[\rho] + \frac{1}{2} \int \int \frac{\rho(\vec{r}_1)\rho(\vec{r}_2)}{r_{12}} d\vec{r}_1 d\vec{r}_2 + E_{XC}[\rho] + \int V_{Ne}\rho(\vec{r})d\vec{r} \\
&= -\frac{1}{2} \sum_i^N \langle \psi_i | \nabla^2 | \psi_i \rangle + \frac{1}{2} \sum_i^N \sum_j^N \int \int |\psi_i(\vec{r}_1)|^2 \frac{1}{r_{12}} |\psi_j(\vec{r}_2)|^2 d\vec{r}_1 d\vec{r}_2 \\
&\quad + E_{XC}[\rho] - \sum_i^N \int \sum_A^M \frac{Z_A}{r_{1A}} |\psi_i(\vec{r}_1)|^2 d\vec{r}_1.
\end{aligned}
\tag{10.35}
$$

The only term for which no explicit form can be given is E_{XC}. We now apply the variational principle and ask: What condition must the orbitals $\{\psi_i\}$ fulfill in order to minimize this energy expression under the usual constraint $\langle \psi_i | \psi_j \rangle = \delta_{ij}$? The resulting equations are the **Kohn-Sham equations**:

$$\left(-\frac{1}{2}\nabla^2 + V_{eff}(\vec{r}_1) \right) \psi_i = \epsilon_i \psi_i, \tag{10.36}$$

where the effective potential $V_{eff}(\vec{r}_1)$ is given by

$$V_{eff}(\vec{r}_1) = \int \frac{\rho(\vec{r}_2)}{r_{12}} d\vec{r}_2 + V_{XC}(\vec{r}_1) - \sum_A^M \frac{Z_A}{r_{1A}}. \tag{10.37}$$

Thus, once we know the various contributions in Eq. (10.37), we can insert the potential V_{eff} into the one-particle equations, which in turn determine the orbitals and hence the ground state density and the ground state energy employing Eq. (10.35). Notice that V_{eff} depends on the density, and therefore the Kohn-Sham equations have to be solved iteratively.

One term in the above equations needs some additional comments. The exchange-correlation potential, V_{XC} is defined as the functional derivative of E_{XC} with respect to ρ, i.e. $V_{XC} = \delta E_{XC}/\delta\rho$. It is very important to realize that if the exact forms of E_{XC} and V_{XC} were known, the Kohn-Sham strategy would lead to the exact energy.

A question of special relevance for the use of DFT in molecular electronics is: Do the Kohn-Sham orbitals and eigenvalues mean anything? It is often said that the Kohn-Sham orbitals and eigenvalues have no physical meaning. In particular, the eigenvalues are not the energies to add or subtract electrons from the interacting many-body system. There is only one exception [275]: The highest eigenvalue in a finite system, which is minus the ionization energy.[5] Anyway, several authors have lately pointed to the interpretative power of the Kohn-Sham orbitals in traditional qualitative molecular orbital schemes (see section 5.3.3 in Ref. [273]) and in solid state physics it is customary to use these orbitals as an approximation for the true spectrum of an electronic system, see Ref. [271]. After all, these orbitals are not only associated with a one-electron potential which includes all non-classical effects, they are also consistent with the exact ground state density.

Let us close this section by saying that the Kohn-Sham (KS) approach provides a practical strategy to find both the electron density and the total energy of the ground state of any electronic system. However, there are still two important issues that we have to address. First, we need to find reasonable approximations for the exchange-correlation functional and second, we have to discuss how to solve in practice the Kohn-Sham equations. These two issues are the subject of the next sections.

10.5 The exchange-correlation functionals

The genius of the Kohn-Sham approach described in the previous section is two-fold. First, the Ansatz leads to tractable single-particle equations that hold the hope of solving interacting many-body problems. Second, by explicitly separating the independent-particle kinetic energy and the long-range Hartree terms, the remaining exchange-correlation functional $E_{XC}[\rho]$ can be reasonable approximated as a local or nearly local functional of the density. Even though the exact functional $E_{XC}[\rho]$ must be very complex, great progress has been made with remarkably simple approximations. This sections is devoted to a brief description of some of those approximations. For more details about this topic, see for instance Chapter 8 of Ref. [226] and Chapter 6 of Ref. [273].

10.5.1 *LDA approximation*

The *local density approximation* (LDA) is the basis of all approximate exchange-correlation functionals. At the center of this model is the idea of a uniform electron gas. This is a system in which electrons move on

---

[5]The asymptotic long-range density of a bound system is governed by the occupied state with the highest eigenvalue; since the density is assumed to be exact, so must the eigenvalue be exact. No other eigenvalue is guaranteed to be correct.

a positive background charge distribution such that the total ensemble is neutral.

The central idea of LDA is the assumption that we can write E_{XC} in the following form

$$E_{XC}^{\text{LDA}}[\rho] = \int \rho(\vec{r}) \epsilon_{XC}(\rho(\vec{r})) \, d\vec{r}. \tag{10.38}$$

Here, $\epsilon_{XC}(\rho(\vec{r}))$ is the exchange-correlation energy per particle of a uniform electron gas of density $\rho(\vec{r})$. This energy per particle is weighted with the probability $\rho(\vec{r})$ that there is an electron at this position. The quantity $\epsilon_{XC}(\rho(\vec{r}))$ can be further split into exchange and correlation contributions,

$$\epsilon_{XC}(\rho(\vec{r})) = \epsilon_X(\rho(\vec{r})) + \epsilon_C(\rho(\vec{r})). \tag{10.39}$$

The exchange part, ϵ_X, which represents the exchange energy of an electron in a uniform electron gas of a particular density, was originally derived by Bloch and Dirac in the late 1920's and it is given by

$$\epsilon_X = -\frac{3}{4} \left(\frac{3\rho(\vec{r})}{\pi} \right)^{1/3}. \tag{10.40}$$

No such explicit expression is known for the correlation part, ϵ_C. However, highly accurate numerical quantum Monte-Carlo simulations of the homogeneous electron gas are available from the work of Ceperly and Alder [276]. On the basis of these results various authors have presented analytical expressions of ϵ_C based on sophisticated interpolation schemes.

Up to this point the local density approximation was introduced as a functional depending solely on $\rho(\vec{r})$. If we extend the LDA to an unrestricted case, i.e. to a case without spin symmetry, we arrive at the *local spin-density approximation*, or LSDA, where the two spin densities, $\rho_\uparrow(\vec{r})$ and $\rho_\downarrow(\vec{r})$, with $\rho(\vec{r}) = \rho_\uparrow(\vec{r}) + \rho_\downarrow(\vec{r})$, are employed as the central input. In this approximation, instead of Eq. (10.38) one now writes

$$E_{XC}^{\text{LSDA}}[\rho_\uparrow, \rho_\downarrow] = \int \rho(\vec{r}) \epsilon_{XC}(\rho_\uparrow(\vec{r}), \rho_\downarrow(\vec{r})) \, d\vec{r}. \tag{10.41}$$

As for the simple spin-symmetric situation, there are related expressions for the exchange and correlation energies per particle of the uniform electron gas characterized by $\rho_\uparrow(\vec{r}) \neq \rho_\downarrow(\vec{r})$, the so-called spin polarized case. In the following we do not differentiate between the local and the local spin-density approximation and use the abbreviation LDA for both.

We conclude this subsection with some brief comments about the performance of LDA, especially in the context of molecular physics:

- The accuracy of the LDA for the exchange energy is typically within 10%, while the normally much smaller correlation energy is generally overestimated by up to a factor 2. The two errors typically cancel partially.
- Experience has shown that the LDA gives ionization energies of atoms, dissociation energies of molecules and cohesive energies with a fair accuracy of typically 10-20%. However, the LDA gives bond lengths of molecules and solids typically with an astonishing accuracy of $\sim 2\%$.
- The moderate accuracy that LDA delivers is insufficient for most applications in chemistry. For this reason, for many years, where LDA was the only approximation for the exchange-correlation functional, DFT was mostly used by solid-state physicists and it hardly had any impact in quantum chemistry.

10.5.2 *The generalized gradient approximation*

The first step beyond the local approximation is a functional of the magnitude of the gradient of the density $\nabla\rho(\vec{r})$ as well as the value of $\rho(\vec{r})$ at each point. Such a *gradient expansion approximation* (GEA) was already suggested in the original paper of Kohn and Sham. The low-order expansion of the exchange and correlation energies is known. However, the GEA does not lead to consistent improvement over the LDA. It violates exact sum rules and other relevant conditions and, indeed, often leads to worse results. The basic problem is that gradients in real systems can be so large that the expansion breaks down.

The term *generalized gradient approximation* (GGA) denotes a variety of ways proposed for functionals that modify the behavior at large gradients in such way as to preserve the desired properties. These functionals are the workhorses of current DFT and they can be generically written as

$$E_{XC}^{\mathrm{GGA}}[\rho_\uparrow, \rho_\downarrow] = \int f(\rho_\uparrow, \rho_\downarrow, \nabla\rho_\uparrow, \nabla\rho_\downarrow)\, d\vec{r}. \qquad (10.42)$$

In practice, E_{XC}^{GGA} is usually split into its exchange and correlation contributions, $E_{XC}^{\mathrm{GGA}} = E_X^{\mathrm{GGA}} + E_C^{\mathrm{GGA}}$, and approximations for the two terms are sought separately. With respect to the exchange part E_X^{GGA}, it can be written as

$$E_X^{\mathrm{GGA}} = E_X^{\mathrm{LDA}} - \sum_{\sigma=\uparrow,\downarrow} \int F(s_\sigma)\rho_\sigma^{4/3}(\vec{r})\, d\vec{r}. \qquad (10.43)$$

The argument of the function F is the reduced density gradient for spin σ

$$s_\sigma = \frac{|\nabla \rho_\sigma(\vec{r})|}{\rho_\sigma^{4/3}}. \tag{10.44}$$

Numerous forms for the function F above have been given. We just mention here three of the most widely used ones that were proposed by Becke in 1986 (B86) [277], Perdew also in 1986 (P) [278], and Perdew, Burke and Ernzerhof in 1996 (PBE) [279]. In all these cases, F is a complicated rational function of the reduced density gradient that we shall not write here explicitly.

The corresponding gradient-corrected correlation functionals have even more complicated analytical forms and cannot be understood by simple physically motivated reasoning. Among the most widely used choices is the correlation counterpart of the 1986 Perdew exchange functional [278], usually referred to as P or P86. This functional employs an empirical parameter, which was fitted to the correlation energy of the neon atom. A few years later Perdew and Wang [280] refined their correlation functional, leading to the parameter free PW91. Another, nowadays even more popular correlation functional is due to Lee, Yang, and Parr (LYP) [281]. This functional was derived from an expression for the correlation energy of the helium atom. The LYP functional contains one empirical parameter and it differs from other GGA functionals in that it contains some local components.

In principle, each exchange functional could be combined with any of the correlation functionals, but only a few combinations are currently in use. The exchange part is usually chosen to be Becke's functional which is either combined with Perdew's 1986 correlation functional or the LYP one. These combinations are termed BP86 and BLYP, respectively. Sometimes also the PW91 correlation functional is employed, corresponding to BPW91. It is worth stressing that these combinations lead to results that are of very similar quality.

As a general statement about the performance of GGA-based functionals, let us say that they have reduced the LDA errors of, in particular, atomization energies of standard set of small molecules by a factor 3-5. This improved accuracy has made DFT one of the most widely used tools in quantum chemistry.

10.5.3 *Hybrid functionals*

Usually the exchange contributions are significantly larger than the corresponding correlation effects. Therefore, an accurate expression for the exchange functional is a prerequisite for obtaining meaningful results from density functional theory. In this sense, it is important to remind that the exchange energy of a Slater determinant can be computed exactly (see discussion of the Hartree-Fock (HF) approximation in section 10.1.3). This fact has motivated the construction of functionals called *hybrid* because they are a combination of orbital-dependent Hartree-Fock and an explicit density functional. These are the most accurate functionals available as far as the energetics is concerned and are the method of choice in the quantum chemistry community.

The hybrid functionals differ in the way in which the exchange HF energy is mixed with the exchange-correlation energy of a density functional. Becke [282] has argued that the total exchange-correlation energy can be approximated by

$$E_{XC} = \frac{1}{2} \left(E_X^{\mathrm{HF}} + E_{XC}^{\mathrm{DFA}} \right), \tag{10.45}$$

where DFA denotes an LDA or GGA functional. Later Becke presented parameterized forms that are accurate for many molecules, such as "B3P91" [282, 283], a three-parameter functional that mixes Hartree-Fock exchange, the exchange functional of Becke (B88), and correlation from Perdew and Wang (PW91).

Currently the most popular hybrid functional is the so-called B3LYP [284] that uses the LYP correlation functional. In this case the definition of the exchange-correlation energy is

$$E_{XC} = E_{XC}^{\mathrm{LDA}} + a_0 \left(E_X^{\mathrm{HF}} - E_X^{\mathrm{DFA}} \right) + a_x E_X^{\mathrm{Becke}} + a_c E_C, \tag{10.46}$$

where the three coefficients a_i are empirically adjusted to fit atomic and molecular data.

10.6 The basic machinery of DFT

In this section we shall address how the Kohn-Sham single-particle equations are solved in practice. There are three main types of methods that are applied to this problem with their own advantages and disadvantages. The first one are the *plane wave and grid methods* that provide general approaches for the solution of differential equations, including the Schrödinger and Poisson equations. A second family is formed by the *atomic sphere methods* that are the most general methods for precise solution of the Kohn-Sham equations. The basic idea is to divide the electronic structure problem, providing efficient representation of atomic-like features that are rapidly varying near each nucleus and smoothly varying functions between atoms. Finally, the third type of methods is based on *localized atomic-(like) orbitals (LCAO)* that provide a basis that captures the essence of the atomic-like features of solids and molecules. They provide a satisfying, localized description of electronic structure widely used in chemistry. Since this latter method is, in principle, better adapted to the type of systems studied in molecular electronics, we shall devote the rest of this section to describe how it is actually used to solve the Kohn-Sham equations. The other two types of methods are extensively discussed in Ref. [226].

10.6.1   *The LCAO Ansatz in the Kohn-Sham equations*

Recall the central ingredient of the Kohn-Sham (KS) approach to density functional theory, i.e. the one-electron KS equations,

$$\left(-\frac{1}{2}\nabla^2 + \left[\sum_j^N \int \frac{|\psi_j(\vec{r}_2)|^2}{r_{12}} d\vec{r}_2 + V_{XC}(\vec{r}_1) - \sum_A^M \frac{Z_A}{r_{1A}} \right] \right) \psi_i = \epsilon_i \, \psi_i.$$

$$(10.47)$$

The term in square brackets defines the Kohn-Sham one-electron operator and Eq. (10.47) can be written more compactly as

$$\mathbf{f}^{KS} \, \psi_i = \epsilon_i \, \psi_i. \qquad (10.48)$$

Most of the applications in chemistry of the Kohn-Sham density functional theory make use of the LCAO expansion of the Kohn-Sham orbitals. Indeed, the way to proceed is almost identical to the case of the tight-binding approach that we discussed in the previous chapter. Let us assume that we are dealing with a finite system and introduce a set of L predefined basis functions $\{\eta_\mu\}$ and linearly expand the K-S orbitals as

$$\psi_i = \sum_{\mu=1}^{L} c_{\mu i} \eta_\mu. \qquad (10.49)$$

We now insert Eq. (10.49) into Eq. (10.48) and obtain

$$\mathbf{f}^{KS}(\vec{r}_1) \sum_{\nu=1}^{L} c_{\nu i} \eta_\nu(\vec{r}_1) = \epsilon_i \sum_{\nu=1}^{L} c_{\nu i} \eta_\nu(\vec{r}_1). \qquad (10.50)$$

If we now multiply this equation from the left with an arbitrary basis function η_μ and integrate over space we get L equations[6]

$$\sum_{\nu=1}^{L} c_{\nu i} \int \eta_\mu(\vec{r}_1) \mathbf{f}^{\mathrm{KS}}(\vec{r}_1) \eta_\nu(\vec{r}_1) d\vec{r}_1 = \epsilon_i \sum_{\nu=1}^{L} c_{\nu i} \int \eta_\mu(\vec{r}_1) \eta_\nu(\vec{r}_1) d\vec{r}_1, \quad (10.51)$$

where i runs from 0 to L.

The integrals on the left hand side of this equation define the Kohn-Sham matrix, $\mathbf{F}^{\mathrm{KS}}$, with the corresponding elements defined as

$$F_{\mu\nu}^{\mathrm{KS}} = \int \eta_\mu(\vec{r}_1) \mathbf{f}^{\mathrm{KS}}(\vec{r}_1) \eta_\nu(\vec{r}_1) d\vec{r}_1, \quad (10.52)$$

and on the right hand side we can identify the overlap matrix, $\mathbf{S}$, the elements of which are given by

$$S_{\mu\nu} = \int \eta_\mu(\vec{r}_1) \eta_\nu(\vec{r}_1) d\vec{r}_1. \quad (10.53)$$

Both matrices are $L \times L$ dimensional. The previous equation can be rewritten compactly as a matrix equation

$$\mathbf{F}^{\mathrm{KS}} \mathbf{C} = \mathbf{S} \mathbf{C} \epsilon. \quad (10.54)$$

Hence, through the LCAO expansion we have translated the non-linear optimization problem into a linear one, which can be expressed in the language of standard algebra.

By expanding $\mathbf{f}^{\mathrm{KS}}$ into its components, the individual elements of the KS matrix become

$$F_{\mu\nu}^{\mathrm{KS}} = \int \eta_\mu(\vec{r}_1) \left(-\frac{1}{2}\nabla^2 - \sum_{A}^{M} \frac{Z_A}{r_{1A}} + \int \frac{\rho(\vec{r}_2)}{r_{12}} d\vec{r}_2 + V_{XC}(\vec{r}_1) \right) \eta_\nu(\vec{r}_1) d\vec{r}_1. \quad (10.55)$$

The first two terms describe the kinetic energy and the electron-nuclear interaction, and they are usually combined since they are one-electron integrals[7]

$$h_{\mu\nu} = \int \eta_\mu(\vec{r}_1) \left(-\frac{1}{2}\nabla^2 - \sum_{A}^{M} \frac{Z_A}{r_{1A}} \right) \eta_\nu(\vec{r}_1) d\vec{r}_1. \quad (10.56)$$

For the third term we need the charge density ρ which takes the following form in the LCAO scheme

$$\rho(\vec{r}) = \sum_{i}^{L} |\psi_i(\vec{r})|^2 = \sum_{i}^{N} \sum_{\mu}^{L} \sum_{\nu}^{L} c_{\mu i} c_{\nu i} \eta_\mu(\vec{r}) \eta_\nu(\vec{r}). \quad (10.57)$$

---

[6]We assume without loss of generality that the basis functions are real.

[7]These are the hooping matrix elements introduced in the frame of the tight-binding approach.

The expansion coefficients are usually collected in the so-called *density matrix* **P** with elements

$$P_{\mu\nu} = \sum_{i}^{N} c_{\mu i} c_{\nu i}. \tag{10.58}$$

Thus, the Coulomb contribution in Eq. (10.55) can be expressed as

$$J_{\mu\nu} = \sum_{\lambda}^{L} \sum_{\sigma}^{L} P_{\lambda\sigma} \int \int \eta_{\mu}(\vec{r}_1)\eta_{\nu}(\vec{r}_1) \frac{1}{r_{12}} \eta_{\lambda}(\vec{r}_2)\eta_{\sigma}(\vec{r}_2) d\vec{r}_1 d\vec{r}_2. \tag{10.59}$$

Up to this point, exactly the same formulas also apply in the Hartree-Fock case. The difference is only in the exchange-correlation part. In the Kohn-Sham scheme this is represented by the integral

$$V_{\mu\nu}^{XC} = \int \eta_{\mu}(\vec{r}_1) V_{XC}(\vec{r}_1)\eta_{\nu}(\vec{r}_1) d\vec{r}_1, \tag{10.60}$$

whereas the Hartree-Fock exchange integral is given by

$$K_{\mu\nu} = \sum_{\lambda}^{L} \sum_{\sigma}^{L} P_{\lambda\sigma} \int \int \eta_{\mu}(\vec{x}_1)\eta_{\lambda}(\vec{x}_1) \frac{1}{r_{12}} \eta_{\nu}(\vec{x}_2)\eta_{\sigma}(\vec{x}_2) d\vec{x}_1 d\vec{x}_2. \tag{10.61}$$

The $L^2/2$ one-electron integrals contained in $h_{\mu\nu}$ can be easily computed. The computational *bottle-neck* is the calculation of the $\sim L^4$ two-electron integrals in the Coulomb and exchange-correlation terms. For a discussion about efficient ways of computing these latter integrals, see Ref. [273].

10.6.2   *Basis sets*

In order to complete our discussion of the LCAO approach in DFT, we shall now describe the main types of localized basis functions that are used. A first type of orbitals are the **Gaussian-type-orbitals (GTOs)**, which have been inherited from wave-functions-based methods like Hartree-Fock. The GTO basis functions have the following general form

$$\eta^{GTO} = N x^l y^m z^n \exp\left[-\alpha r^2\right]. \tag{10.62}$$

Here, N is a normalization factor which ensures that $\langle \eta_{\mu} | \eta_{\mu} \rangle = 1$, but note that the η_{μ} are not orthogonal. The orbital exponent α determines how compact or how diffusive the resulting function is. $L = l + m + n$ is used to classify the GTO as s-functions ($L = 0$), p-functions ($L = 1$), etc. The advantage of this type of basis functions lies in the existence of very efficient algorithms for calculating analytically the huge number of multi-center integrals appearing in the Coulomb and exchange-correlation terms.

On the other hand, from a physical point of view, **Slater-type-orbitals (STO)** seem to be the natural choice for basis functions. They are exponential functions that mimic the exact eigenfunctions of the hydrogen atom. A typical STO is expressed as

$$\eta^{\text{STO}} = N r^{n-1} \exp\left[-\beta r\right] Y_{lm}(\Theta, \phi). \tag{10.63}$$

Here, n corresponds to the principal quantum number, the orbital exponent is termed β and Y_{lm} are the usual spherical harmonics. Unfortunately, many-center integrals are very difficult to compute with STO basis, and they do not play a major role in the DFT community.

The so-called **contracted Gaussian functions (CGF)** try to combine the advantages of the two previous type of orbitals. In this case, several primitive Gaussian functions are combined in a fixed linear combination:

$$\eta^{\text{CGF}}_{\tau} = \sum_{a}^{A} d_{a\tau} \eta^{\text{GTO}}_{a}. \tag{10.64}$$

The original motivation for contracting was that the contraction coefficients $d_{a\tau}$ can be chosen in a way that the CGF resembles as much as possible a single STO function. In density functional theory, CGF basis sets enjoy a strong popularity.

A fourth type of basis functions are the **numerical basis functions**. In this case, the orbitals are represented numerically on atomic centered grids. These functions can be generated, for instance, by numerically solving the atomic KS equations with a given approximation for the exchange-correlation functional. Obviously, in this approach the different integrals are computed numerically.

Irrespective of the type of functions used, the basis sets can be classified in the following simple way that already gives a hint about their quality. The simplest (and smallest) basis functions are those that use a single basis function for each atomic orbital up to and including the valence orbitals. These basis sets are called, for obvious reasons, *minimal basis sets*. A typical representative is the STO-3G basis set, in which three primitive GTO functions are combined into one CGF. For carbon, this basis set consists of five functions, one describing the $1s$ atomic orbital, another one for the $2s$ orbital and three more for the $2p$ shell. One should expect no more than only qualitative results from minimal sets and nowadays they are hardly used anymore.

In the next level of sophistication are the *double-zeta* basis sets. Here, the set of functions is doubled, i.e. there are two functions for each orbital. If only the valence orbitals are doubled, and each core atomic orbitals is still described by a single function, the resulting basis set is called *split-valence* basis set. Typical examples are the 3-21G or 6-31G Gaussian basis sets. In most applications, such basis sets are augmented by *polarization functions*, i.e. functions of higher angular momentum than those occupied in the atom. Polarized double-zeta or split valence basis sets are the mainstay of routine quantum chemical applications since usually they offer a balance compromise between accuracy and efficiency. Finally, it is obvious how these schemes can be extended by increasing the number of functions in the various categories. This results in triple- or quadruple-zeta basis sets which are augmented by several sets of polarization functions.

If the molecules or solids of interest contain elements heavier than, say krypton, one usually employs *effective core potentials*, also called *pseudopotentials*, to model the core electrons. For a detailed discussion of the theory of pseudopotentials see Ref. [226].

10.7 DFT performance

Our discussion about density functional theory would not be complete without answering, at least partially, the most obvious question at this stage, namely: how much should one trust DFT? In other words, what is the accuracy of DFT at present, i.e. with the existent approximations for the exchange-correlation functional? A detailed answer to this question is out of the scope of this book and we just pretend to give here a flavor about DFT's performance in the case of the systems of interest in molecular electronics.

Let us remind again that the standard DFT, as presented here, gives only results for the ground state energy and density of a system and related properties. In the context of chemistry (or molecular physics), this means in practice that one can expect from DFT information about the structure of molecules, vibrational frequencies, atomization energies, dipole moments, reactions paths and other similar properties. In what follows, we shall illustrate DFT's performance with a very brief discussion of its predictions for some basic molecular properties. We follow here Ref. [273], where an excellent discussion of "goodness" of DFT in the context of quantum chemistry can be found.

Table 10.1 Calculated and experimental bond lengths for different bonding situations [Å]. The LDA calculations were done with the 6-31G(d) basis set and the GGA ones with the 6-311++G(d,p) basis set.

Bond		LDA	BLYP	BP86	BPW91	Experiment
H-H	R_{H-H}	0.765	0.748	0.752	0.749	0.741
H_3C-CH_3	R_{C-C}	1.510	1.542	1.535	1.533	1.526
	R_{C-H}	1.101	1.100	1.102	1.100	1.088
H_2C=CH_2	R_{C-C}	1.331	1.339	1.337	1.336	1.339
	R_{C-H}	1.098	1.092	1.094	1.092	1.085
HC≡CH	R_{C-C}	1.203	1.209	1.210	1.209	1.203
	R_{C-H}	1.073	1.068	1.072	1.070	1.061

Molecular structures: DFT calculations provide the electronic part of the energy of a molecule. If this information is combined with the classical nuclear energy in Eq. (10.5), the total energy of the molecule can be minimized with respect to the position of the nuclei to find the most stable structure. This is one of the main applications of DFT, which gives the bond lengths of a large set of molecules with a precision of 1-2%. Gradient-corrected and hybrid functionals have improved the LDA results, which for this property are already surprisingly good. The degree of accuracy is illustrated in Table 10.1, where we show the calculated bond distances for several basic covalently bound molecular structures with different functionals and their comparison with experimental results (data taken from Ref. [273]).

Vibrational frequencies: The frequencies of molecular vibrational modes can be calculated by evaluation of second derivatives of the total energy with respect to Cartesian coordinates. DFT predicts the vibrational frequencies of a broad range of molecules within 5-10% accuracy.

Scott and Radom [285] have investigated the performance of a variety of gradient-corrected and hybrid functionals for predicting vibrational frequencies of a large set of 122 test molecules. By fitting computed data to a basis of 1066 experimental vibrations, they obtained scaling factors relating the computed frequencies to experimental values. Some of the results are reproduced in Table 10.7.

Atomization energies: The most common way of testing the performance of new functionals is the comparison with the experimental atomization energies (the energies needed to break up a molecule into its constituent atoms) of well-studied sets of small molecules. These comparisons

Table 10.2   Vibrational frequencies of a set of 122 molecules: functional, frequency scaling factor (f), root mean square (RMS) error after scaling in cm^{-1} and percentage of frequencies that fall outside the experimental values by more than 10%.

Functional	f	RMS	10%
BLYP	0.9945	45	10
BP86	0.9914	41	6
B3LYP	0.9614	34	6
B3P86	0.9558	38	4
B3PW91	0.9573	34	4

have established the following hierarchy of functionals:

$$\text{LDA} < \text{GGA} < \text{hybrid functionals} \ .$$

The hybrid functionals are progressively approaching the desired accuracy in the atomization energies, and in many cases they deliver results comparable with highly sophisticated post-HF methods.

Ionization and affinity energies: The energies needed to remove (IE) or to add an electron (EA) can be determined with hybrid functionals with an average error of around 0.2 eV for a large variety of molecules.

The discussion above shows the impressive accuracy that DFT is achieving in many situations. However, it is worth stressing that DFT (with the present approximations) is still failing in situations where the density is not a slowly varying function. An important example of the failure of DFT is the description of systems where the binding is dominated by van der Waals interactions, which is something essential for supramolecular chemistry. Another example is the description of electronic tails evanescing into the vacuum near the surfaces of bounded electronic systems, which is a key problem in the context of the STM.

10.8   DFT in molecular electronics

In this final section we shall discuss how DFT is used in practice in the field of molecular electronics. For this purpose, we shall first discuss how DFT can be combined with the nonequilibrium Green's function (NEGF) techniques presented in previous chapters to describe the electronic transport in atomic-scale junctions. Then, we shall end this section with some com-

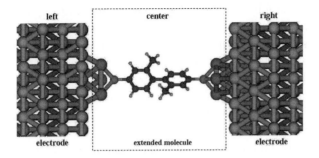

Fig. 10.1 Schematic representation of a molecular junction. We distinguish regions: the left (L) and right (R) semi-infinite electrodes and the central region or "extended molecule" that contains the molecule and part of the leads.

ments about the advantages and limitations of such a combination. For a more detailed discussion of the use of DFT to compute the transport properties of nanostructures, we recommend the excellent review of Pecchia and Di Carlo [286].

10.8.1 *Combining DFT with NEGF techniques*

In section 10.6 we have shown how DFT is applied to the description of the electronic structure of finite systems like molecules or atomic clusters, in particular within the LCAO approach. For periodic systems like infinite solids, one proceeds in a similar manner, but in this case the Kohn-Sham equations are solved in reciprocal space. In both types of systems the dimension of the problem, i.e. the number of Kohn-Sham equations, is finite. In the first case this dimension is mainly determined by the number of atoms in the system, whereas in the latter it is governed by the size of the unit cell. In molecular electronics we are interested in the description of the electronic structure and transport properties of atomic-scale junctions, like the one depicted in Fig. 10.1. These junctions are neither finite nor periodic, which makes more complicated the application of DFT. Moreover, we are also interested in situations in which these systems are driven out of equilibrium, for instance by the application of a bias voltage. Such situations are out of the scope of the standard ground state DFT. The goal of this subsection is to show how DFT can be combined with the Green's function techniques of Chapters 5-8 to describe the equilibrium and transport properties of nanoscale junctions.

When applying DFT to systems like the one in Fig. 10.1, one is confronted with the following two questions: (i) how to compute the charge density? and (ii) how to make finite the dimension of the problem? Both questions can be answered with the help of Green's function methods as follows. First, we divide the junction into three parts: the left (L) and right (R) electrodes and a central part or "extended molecule" that contains the narrowest part of the junction (the molecule in Fig. 10.1) and part of the electrodes.[8] Second, within the LCAO approach, the charge density is computed in terms of the density matrix, see Eqs. (10.57) and (10.58), which in turn can be computed in terms of Green's functions in the following way. Let us assume that the system is in equilibrium. The retarded and advanced Green's functions $G_{\mu\nu}^{r,a}$ referred to the local basis functions μ and ν can be written via their spectral representation [see Eq. (5.14)] as follows

$$G_{\mu\nu}^{r,a}(E) = \sum_i \frac{c_{\mu i} c_{i\nu}^*}{E \pm i\eta - E_i}. \tag{10.65}$$

Here, the c's are the coefficients of the expansion of the system eigenfunctions (or molecular orbitals) in terms of local orbitals and E_i are the corresponding eigenenergies. Notice that $\sum_i c_{\mu i} c_{i\nu}^*$ is nothing but the element $P_{\mu\nu}$ of the density matrix, see Eq. (10.58).[9] Therefore, the density matrix in the central part of the junction can be calculated from the retarded or advanced Green's function matrix of this part of the system as[10]

$$\mathbf{P} = \mp \frac{1}{\pi} \int_{-\infty}^{\infty} dE \, \mathrm{Im} \left\{ \mathbf{G}^{r,a}(E) \right\} f(E), \tag{10.66}$$

where $f(E)$ is the Fermi function that ensures that only the occupied states contribute to the electron density. Now, these Green's functions can be computed via their Dyson's equation [see Eq. (8.26)]

$$\mathbf{G}^{r,a} = \left[(E \pm i\eta)\mathbf{S} - \mathbf{H} - \mathbf{\Sigma}_L^{r,a} - \mathbf{\Sigma}_R^{r,a} \right]^{-1}, \tag{10.67}$$

where $\mathbf{S}$ is the overlap matrix, $\mathbf{H}$ is the one-electron Kohn-Sham Hamiltonian of the central part and $\mathbf{\Sigma}_{L/R}^{r,a}$ are the left and right self-energies (see section 8.1). The calculation of these self-energies requires the computation of the Hamiltonian and Green's functions of the electrodes. This issue will be discussed in detail below.

This discussion shows that DFT can be applied to describe nanoscale junctions by using Eq. (10.66) for determining the density matrix, rather

---

[8]The reason for dividing the system in this way will become clear below.

[9]In Eq. (10.58), we assumed that the c's were real, but in principle, these coefficients can be complex numbers and then, $\sum_i c_{\mu i} c_{i\nu}^*$ is the most general definition of $P_{\mu\nu}$.

[10]In what follows, we shall not write explicitly the subindexes CC to refer to the central part of the junction, as we did, for instance, in section 8.1.

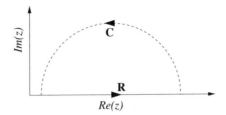

Fig. 10.2　The integral of a retarded Green's function $G^r(z)$, considered as a function of a complex variable z, is the same along the contour C and along the real axis R. However, $G^r(z)$ is much smoother away from the real axis and for this reason, it is advantageous to integrate the Green's functions in Eq. (10.66) along a contour like C. The lower limit of this contour has to be below the lowest lying states of the system, while the upper limit should be the chemical potential of the system.

than solving the Kohn-Sham equations. The evaluation of the density matrix requires the calculation of the Green's functions via Eq. (10.67) from the knowledge of the Kohn-Sham Hamiltonian of the central part. Since this Hamiltonian depends on the charge density (or density matrix), Eqs. (10.66) and (10.67) are coupled and they have to be solved in a self-consistent manner. Finally, when these equations are solved, one can compute the different equilibrium properties of a junction such as charge density, total energy, local density of states, etc.

A technical comment is pertinent at this point. Usually Green's functions vary rapidly as a function of energy, which complicates the integration appearing in Eq. (10.66). One can get around this problem by making use of the fact that the Green's functions are analytical functions and they can be extended into the complex plane. This means in practice that the integral in Eq. (10.66) can be done by integrating along a contour in the complex plane, see Fig. 10.2, where these functions are very smooth. Thus, one needs a much smaller number of points to carry out the numerical integration.

The previous discussion also suggests a straightforward way of generalizing this approach to nonequilibrium situations. In this case, the density matrix can be expressed in terms of the Keldysh-Green's functions as

$$\mathbf{P} = \frac{1}{2\pi i} \int_{-\infty}^{\infty} dE \, \mathbf{G}^{+-}(E), \qquad (10.68)$$

where $\mathbf{G}^{+-}$ can be computed in terms of the retarded and advanced functions as [see Eq. (8.12)]

$$\mathbf{G}^{+-}(E) = 2i\mathbf{G}^r \left[\mathbf{\Gamma}_L f_L + \mathbf{\Gamma}_R f_R \right] \mathbf{G}^a. \qquad (10.69)$$

Here, the scattering rates $\Gamma_{L/R}$ are the imaginary part of the self-energies $\Sigma^a_{L/R}$ (see section 8.1) and $f_{L/R}$ are the Fermi functions of the left and right electrodes that include the energy shift caused by the applied bias voltage. In this latter equation, the retarded and advanced functions can be calculated from a Dyson's equation like Eq. (10.67) taking into account the presence of the bias voltage. Again, Eqs. (10.69) and (10.69) are coupled and they have to be solved self-consistently. Once this is done, the different transport properties can be computed as described in Chapter 8. It is worth mentioning that again the integration in Eq. (10.68) can be done more efficiently in the complex plane, although this time the integration close to the Fermi energy requires to modify the contour shown in Fig. 10.2 (see Ref. [287] for details).

The key step to make our generic problem finite was the division of the system into three parts, see Fig. 10.1. In this division one assumes that the electrodes are not perturbed by the central part and therefore, their Hamiltonians and charge densities can be obtained from a separate (bulk-like) calculation, which only needs to be done once. This assumption is based on the idea that deep inside a solid the Kohn-Sham potential approaches the bulk potential. This approximation is often referred to as the screening approximation and it provides natural boundary conditions for the potential of the open system. In any calculation, it should be checked that the potential of the central part actually matches that of the bulk calculation. Such a check defines in practice the size of the central part and this size depends on the nature of the electrodes.

The practical implementations of the DFT-NEGF combination differ mainly in the way in which the electrodes Green's functions are determined and how the potential and Hamiltonian of the central part are forced to match the corresponding ones in the leads. Roughly speaking, one can grouped all the existent approaches into the following two families:

1.- Methods based on quantum chemistry software. In order to take advantage of the powerful and well-tested existent quantum chemistry codes, several groups have implemented the DFT-NEGF approach as follows. The diagonalization in these codes of the Kohn-Sham Hamiltonian of the finite central system is replaced by Eqs. (10.68) and (10.69), which are solved self-consistently. The self-energies required to compute the retarded and advanced Green's functions appearing in Eq (10.69) are obtained from a separate calculation, from which one extracts the bulk Hamiltonian as well as the coupling matrix elements between the central part and the leads. This separate calculation can be done at different levels of sophistication.

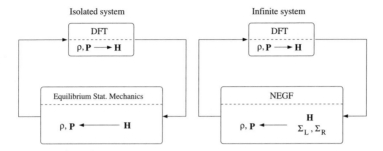

Fig. 10.3 Schematic description of the self-consistent loop in DFT for the determination of the electronic structure of a finite system (left panel) and an infinite non-periodic system (right panel) [293]. For an isolated system, the density matrix is constructed by occupying the states of the Kohn-Sham Hamiltonian H with N electrons. For an infinite system, the density matrix is computed from the nonequilibrium Green's functions, see Eq. (10.68), which requires the determination of the self-energies from a separate calculation.

Thus for instance, some authors describe the leads in terms of simple parameterized tight-binding methods [288–292] and others extract the bulk Hamiltonian from DFT calculations of finite clusters [293, 294, 247]. The bulk parameters are then used to construct surface Green's functions using recursive methods like those described in Refs. [198, 295, 296]. Following Damle *et al.* [293], we summarize in Fig. 10.3 this approach and emphasize the main differences with the standard method used for finite equilibrium systems.

In this approach it is implicitly assumed that the central Hamiltonian is a functional of the charge density only in the central system. This is indeed the case for the contributions coming from the kinetic energy and the electron-nuclei interaction, see Eq. (10.56). It is also true for the exchange-correlation potential, see Eq. (10.59), but it is not really the case for the classical Coulomb contribution, see Eq. (10.60). In this latter term, there are non-local contributions coming from the leads, which are not easy to describe correctly in the approach discussed in the previous paragraph. The lack of these non-local contributions causes sometimes severe problems in the convergence procedure in this method. However, it seems that, as shown in Ref. [247], if the central system is sufficiently large, those additional contributions do not play a major role in the physical quantities of interest, and the self-consistent loop is not crucial in equilibrium systems.

2.- Methods based on solid state software. In the implementations of the DFT-NEGF method based on the computer codes specially designed for the description of (infinite) solid states systems, the Hartree potential

V_H, or classical Coulomb term of Eq. (10.59), is obtained in the central region as a solution of the Poisson equation (in Hartree atomic units)

$$\nabla^2 V_H = -4\pi \rho(\vec{r}). \tag{10.70}$$

This equation needs to be solved with appropriate boundary conditions given by the contact potentials, which are obtained from a separate bulk calculation. This Poisson equation can be solved with different strategies. For instance, in the TRANSIESTA code [287], which is an extension of the SIESTA code for equilibrium systems, this equation is solved via a fast Fourier transformation algorithm by constructing a periodic supercell. In other implementations, this equation is solved in real space with 3D multigrid algorithms [297, 298].

In this approach, the Hamiltonian of the leads, necessary for the construction of the self-energies, is determined with recursive methods similar to those employed in the method described above. This second approach

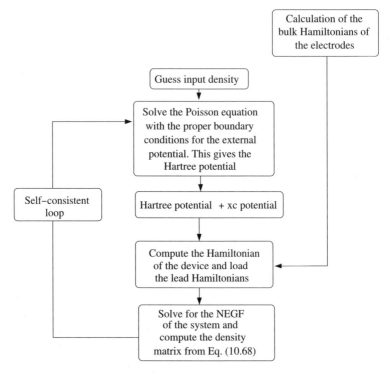

Fig. 10.4   Flowchart of the self-consistent loop for the solution of the nonequilibrium transport problem based on the solution of the Poisson equation [286].

Table 10.3 List of implementations of the combination of DFT and Green's function techniques for the description of equilibrium and nonequilibrium properties of nanoscale junctions. We provide the name of the code, if any, some characteristics, the reference equilibrium DFT code in which it is based on, and a reference where details about it can be found. Methods 1 and 2 refer to the methods described in the text, BC means boundary conditions and TB corresponds to tight-binding.

Name	Key features	Basis code	Ref.
McDCAL	Method 2, real space, non-linear transport	SIESTA	[297]
TRANSIESTA	Method 2, periodic BC, non-linear transport	SIESTA	[287]
—	complex band structure method, linear transport	FIREBALL	[299]
—	Periodic BC + Wannier functions, linear transport	Dacapo	[300]
SMEAGOL	Method 2, periodic BC, non-linear and spin-dependent transport	SIESTA	[301]
ALACANT	Method 1, TB for the leads	GAUSSIAN	[290]
—	Method 1, DFT for the leads, non-linear transport	GAUSSIAN	[294]
—	Method 1, TB for the leads, non-linear transport	GAUSSIAN	[292]
Cluster-based method	Method 1, DFT for the leads, linear transport	TURBOMOLE	[247]

is summarized in the flowchart of Fig. 10.4.

We conclude this discussion by listing a few implementations of the combination of DFT and Green's function techniques, see Table 10.3. This list is by no means complete and it is included here just to give some entry points into the literature where one can find the technical details that we have skipped in our discussion of the use of DFT for transport problems.

10.8.2 *Pluses and minuses of DFT-NEGF-based methods*

Let us end this section with some brief comments about the advantages and drawbacks of the DFT-NEGF approach for the description of the properties of nanoscale junctions. As we have seen in previous sections, DFT was designed to describe the ground state energy and related equilibrium properties of a system. In this sense, from DFT we can expect to obtain an excellent description of, for instance, contact geometries, breaking forces, vibration modes and electron-phonon coupling constants. This constitutes a very valuable information for the understanding of issues like the structure and formation mechanisms of molecular junctions as well as for the description of the vibration-assisted inelastic transport (see Chapter 17) or the phonon contribution to heat conduction in these systems.

On the other hand, the use of DFT for transport problems has clear limitations. As it is used in these problems, DFT just provides a mean field approach which is unable to describe strong electronic correlations like the ones that give rise to phenomena such as Coulomb blockade or the Kondo effect (see Chapter 15). Furthermore, the use of Kohn-Sham orbitals as the molecular orbitals of a system is just an approximation that in some cases leads to large errors in the position of the relevant energy levels responsible for the transport. In the case of molecular junctions, the transport often proceeds through the tails of the molecular orbitals closest to the Fermi energy. Thus, small errors in the position of those levels can lead to big errors in the transport properties. It is also worth stressing that the use of DFT in nonequilibrium situations, as it was described above, is just a reasonable Ansatz, but it is not really justified at the same level as the corresponding ground state theory. For all these reasons, the DFT-NEGF combination should be seen as a first step towards a quantitative theory of transport in nanoscale junctions.

In summary, DFT provides crucial information for the description of the transport properties of atomic and molecular junctions, but it is important to be aware of its limitations. In Part 4, we shall discuss in detail the performance of this theory when applied to different aspects of the electronic and thermal transport of molecular junctions.

10.9   Exercises

10.1 Energy in the Hartree-Fock approximation: Show that the expectation value of the electronic Hamiltonian of Eq. (10.3) in the Hartree-Fock approximation [see Eq. (10.9)] is given by Eq. (10.11).

10.2 Hartree-Fock equations: Use the variational principle to derive the Hartree-Fock equations [see Eqs. (10.15) and (10.16)].

10.3 Kohn-Sham equations: Use the variational principle to derive the Kohn-Sham equations [see Eq. (10.36)].

PART 3
Metallic atomic-size contacts

Chapter 11

The conductance of a single atom

In order to understand the electrical and thermal conduction through molecular junctions, which is the main goal of this monograph, it is necessary to first understand the corresponding properties of the metallic atomic contacts that are used as electrodes in these nanoscale circuits. The conduction through atomic-scale wires constitutes a field of its own that started at the beginning of the 1990's and it has reached maturity in the last years. Metallic wires of atomic dimensions have become a marvelous playground where many basic concepts of quantum transport have been tested [15]. The physics of these nanocontacts and the progress made in this field up to 2003 have been reviewed in a magnificent article by Agraït, Levy Yeyati and van Ruitenbeek [15].[1] For this reason, we shall not make any attempt to provide a historical revision of this field or to give a complete list of references. Instead, we shall present here a short elementary introduction to some basic aspects that will be useful in subsequent chapters where the physics of molecular transport junctions is described.

With this idea in mind, we initiate here a series of two chapters devoted to the electrical conduction through metallic atomic-size contacts. In this first chapter, we shall focus our attention on the conduction through non-magnetic contacts, with special emphasis in the simplest structures, namely single-atom junctions and monoatomic chains. Our main goal here is to establish the relation between the transport characteristics of these nanowires and the quantum properties of those atoms used as building blocks. The next chapter will be devoted to the spin-dependent transport through magnetic atomic-size contacts. This is a topic in which a lot of progress has been made in recent years and these advances are not covered in the review

---

[1]A brief introduction to the transport properties of metallic atomic contacts can be found in [302].

of Ref. [15].

11.1   Landauer approach to conductance: brief reminder

Before discussing the experimental results for the conductance of atomic contacts, it is convenient to say a few words about how this transport property is described theoretically. The metallic point contacts and nanowires that we are considering here have characteristic dimensions that are much smaller than the typical elastic and inelastic scattering lengths of metals. In particular, the electron mean free path for elastic scattering on defects and impurities near the contact is usually much larger than the contact size.[2] The main source of (elastic) scattering in these nanocontacts are the walls forming the boundary of the system. Thus, the transport through atomic contacts is phase-coherent and it can be described within the framework of the scattering or Landauer approach, which was extensively described in Chapter 4.

Within this approach, the low-temperature linear conductance G is given by Landauer formula

$$G = G_0 T(E_F) = G_0 \operatorname{Tr} \left\{ \mathbf{t}^\dagger \mathbf{t} \right\} (E_F), \qquad (11.1)$$

where $T(E) = \operatorname{Tr}\{\mathbf{t}^\dagger\mathbf{t}\}(E)$ is the energy-dependent total transmission of the structure, E_F is the Fermi energy and $G_0 = 2e^2/h = 77.5$ nS $= (12.9$ k$\Omega)^{-1}$ is the conductance quantum.[3] Here, $\mathbf{t}$ is the transmission matrix of the contact whose elements t_{mn} give the probability amplitude for an electron wave in mode n on the left of the contact to be transmitted into mode m on the right of the contact. Since the trace is an invariant, one can choose to write Eq. (11.1) in the basis that diagonalizes the matrix $\mathbf{t}^\dagger\mathbf{t}$ and then the conductance expression adopts the following simplified form

$$G = G_0 \sum_{n=1}^{N_c} T_n(E_F), \qquad (11.2)$$

where T_n (with $0 \leq T_n \leq 1$) are the transmission coefficients defined as the eigenvalues of $\mathbf{t}^\dagger\mathbf{t}$. A simple estimate of the number channels N_c in 3D metallic contact is given by $N_c \approx (\pi R/\lambda_F)^2$, where R is the radius of the contact radius and λ_F the Fermi wavelength of the conduction electrons. For one-atom thick contacts this number is between 1 and 3 for most metals.

[2]Moreover, the spin-diffusion and the phase-breaking lengths are also typically much larger than the contact dimensions.

[3]The factor 2 is due to the spin degeneracy in non-magnetic situations.

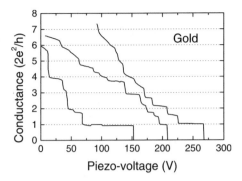

Fig. 11.1 Three typical recordings of the conductance G measured in atomic-size contacts for gold at helium temperatures, using the MCBJ technique. The electrodes are pulled apart by increasing the piezo-voltage. The corresponding displacement is about 0.1 nm per 25 V. After each recording the electrodes are pushed firmly together, and each trace has new structure. Reprinted with permission from Ref. [74].

As we shall see later in this chapter, the actual number channels that give a significant contribution to the conductance depends on the geometry of the narrowest part of the contacts and on the number of valence orbitals of the atoms of the corresponding metal.

11.2 Conductance of atomic-scale contacts

The first question that we want to address is: what is the conductance of a metallic atomic contact? As we discussed in Chapter 2, a metallic contact of atomic size can be fabricated with various techniques, but the most widely used ones are the scanning tunneling microscope (STM) and the mechanically controllable break-junction (MCBJ). In Fig. 11.1 one can see some typical examples of the conductance measured during breaking of a gold contact at low temperatures, using a MCBJ device.[4] Notice that the conductance decreases by sudden jumps, separated by "plateaus", which have a negative slope, the higher conductance the steeper. Some of the plateaus are remarkably close to multiples of the conductance quantum, G_0; in particular the last plateau before loosing contact is nearly flat and very close to $1\,G_0$.[5] This behavior resembles the conductance quantization

---

[4]In these atomic contacts the current-voltage characteristics are typically linear at low voltages (below, let us say, 100 mV) and for this reason we shall mainly talk about the linear conductance as the central transport property.

[5]As it will become clear later in this chapter, the last conductance plateaus most likely correspond to contacts with one atom in cross section and, in particular, long plateaus,

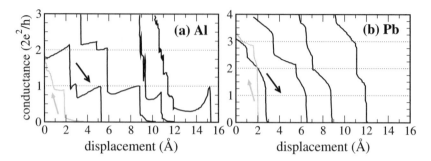

Fig. 11.2   Evolution of conductance vs tip-sample relative displacement for several representative nanocontacts of Al and Pb in STM experiments at low temperatures (4.2 K for Al and 1.5 K for Pb). The black and grey curves correspond to elongation (opening of the contact) and contraction (closing of the contact), respectively. Adapted with permission from [264]. Copyright 1998 by the American Physical Society.

that occurs in point contacts defined in 2D electron gases (2DEG), see section 4.6.1 and in particular Fig. 4.11. Indeed, different authors interpreted the step-like evolution of the conductance as an evidence of conductance quantization in atomic contacts. However, closer inspection of Fig. 11.1 shows that many plateaus cannot be identified with integer multiples of the quantum unit, and the structure of the steps is different for each new recording. Also, the height of the steps is of the order of the quantum unit, but they can vary by more than a factor of 2, where both smaller and larger steps are found.

The conductance traces not only change from realization to realization, but they are also clearly distinct for different metals. In Fig. 11.2 we show several examples of conductance curves for aluminum and lead wires obtained in the last stages of the breaking of contacts formed with a STM at low temperatures. In the case of aluminum, one finds that many plateaus have an anomalous slope: the conductance increases when pulling the contact, in contrast to the results for gold. For aluminum, the last plateau before breaking is still close to the quantum conductance, but one frequently observes the conductance diving below this value, and then recovering to nearly $1\,G_0$, before contact is lost. Lead, on the other hand, has a last conductance value, which is clearly above $1\,G_0$ and the slope is positive, i.e. the conductance is reduced upon stretching.

It is worth mentioning that, as one can see in the examples of Fig. 11.2,

---

like the one in the left curve in Fig. 11.1, are a signature of the formation of a monoatomic chain.

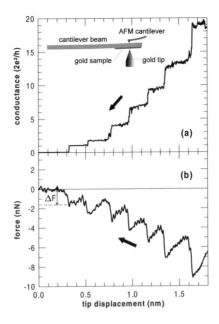

Fig. 11.3 Simultaneous measurement of force and conductance on atom scale point contacts for Au. The sample is mounted on a cantilever beam and the force between tip and sample is measured by the deflection of the beam using an AFM. The measurements are done in air at room temperature. Reprinted with permission from [58]. Copyright 1996 by the American Physical Society.

the conductance traces recorded when opening the contacts differ from those recorded during the closing of the contacts. The reason for this lies in the different atomic arrangements which can be achieved when stretching as opposed to the ones when pushing the electrodes together. Furthermore, as we shall explain later, the shape of the conductance traces also depends on the technique used for fabricating the nanowires.

The previous examples raise several basic questions. The main one is related to the origin of the conductance steps. In the case of point contacts defined in 2DEGs, these steps are due to continuous change in the number of conduction channels as the width varies. In that case the abruptness of the jumps depends in particular on the shape of the confinement potential. However, in the case of atomic contacts the cross section cannot be changed continuously. Early molecular dynamic simulations [303, 304, 262] already suggested that these jumps could be due to sudden atomic rearrangements. The idea goes as follows. Upon stretching of the contact, the stress accumulates elastic energy in the atomic bonds over the length of a plateau. This energy is suddenly released in a transition to a new atomic configuration, which will typically have a smaller contact size.

The direct proof of the relation between atomic rearrangements and conductance steps was provided in an experiment by Rubio, Agraït and Vieira [58], where the conductance for atomic-size gold contacts was measured simultaneously with the force on the contacts, see Fig. 11.3. Notice that the stress accumulation on the plateaus and the coincidence of the stress relief events with the jumps in the conductance can be clearly distinguished.

11.3   Conductance histograms

As shown in the previous section, the conductance of an atomic contact changes from realization to realization, but there are features that are certainly reproducible, like the last plateau in gold contacts. In order investigate objectively the intrinsic conductance of atomic junctions, several authors introduced a method [305, 306], which consists in recording histograms of conductance values encountered in a large number of runs.[6] The most studied metal has been gold, which has been investigated with various techniques and under very different conditions, see e.g. Refs. [60, 70, 307–315].[7] In Fig. 11.4 we show conductance histograms of gold contacts that were measured at different temperatures and different bias voltages using the MCBJ technique [315].[8] Notice that the histograms are largely dominated by the presence of a peak located very close to $1\,G_0$, while some additional peaks close 2 and $3\,G_0$ are also visible. Similar histograms to that of gold are found for the other two noble metals and an example for silver can be seen in the upper left panel of Fig. 11.6.

It is worth stressing that, although it does not seem to be very obvious in the case of gold, the conductance histograms are in general rather sensitive to experimental conditions such as temperature, voltage, breaking speed, environmental conditions, etc., see e.g. Ref. [315]. To illustrate this point, let us briefly discuss here the influence of the experimental technique. As mentioned in the previous section, the shape of the conductance traces

---

[6]This method was adopted later by Xu and Tao to study the conductance of single-molecule junctions [549].

[7]This metal plays a very important role in molecular electronics since it is by far the most common material used for the electrodes in molecular junctions.

[8]The Ph.D. thesis of A.I. Yanson [315] contains the most systematic study of the conductance histograms of various metals published to date (see in particular Chapter 4 of this work). Another systematic analysis can be found in Ref. [313].

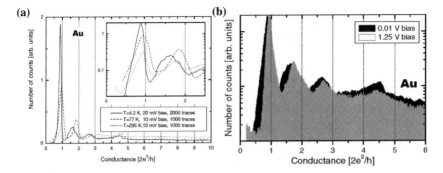

Fig. 11.4 (a) Conductance histograms of gold at 4.2, 77 and 295 K using the notched-wire MCBJ technique. The inset shows the first peaks on the expanded scale. (b) Conductance histograms of gold built from 2000 traces recorded at 1.25 V bias and 12 K (gray). The low temperature histogram (4.2 K) from the left panel is shown for comparison (black). Note that the vertical axis is in logarithmic scale. Reprinted with permission from [315].

depends on the sample fabrication method and this variation is reflected in the histograms. This fact can be easily understood with the help of a mechanical model of the atomic contact and its leads [316]. The nanowire can be modeled as a series of the atomic contact between the left and the right lead. The leads are modeled as one effective spring with a spring constant K_s. Obviously, when the spring constant of the leads is smaller than the effective spring constant of the atomic contact, only a small fraction of the total applied stretching force is concentrated at the atomic contact. When pulling the electrodes apart, the leads are elongated and the atomic contact remains almost unchanged. The conductance trace would thus display horizontal plateaus. The plateaus however do in fact not correspond to values which are favored electronically but by the minimum force. When however, the spring representing the atomic contact is softer than the lead springs (in Fig. 11.5 this is modeled as atomic contact without spring, thus an infinite spring constant of the lead), the majority of the strain is concentrated at the contact. It has to respond to the changing separation of the leads resulting in plateaus with rich substructure and rather broad conductance peaks in the histogram. When comparing the various techniques with respect to this property, we can deduce the following rule of thumb: Techniques with short free-standing electrodes such as the lithographic MCBJ technique are supposed to give rise to plateaus with fine structure and wide histogram peaks, while STM techniques and notched-wire MCBJs are expected to show straight plateaus and narrower peaks for the same metal.

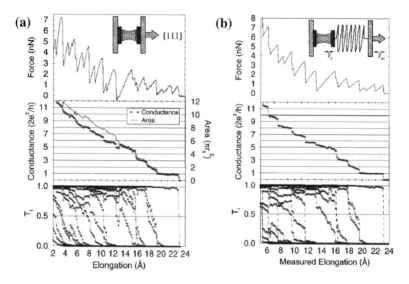

Fig. 11.5   (a) Molecular dynamics simulation of the elongation of a Au nanocontact. Inset: Initial atomic configuration. Top panel: Tensile force on the contact during elongation: Elastic straining of the metallic bonds interrupted by mechanical instabilities and processes. Middle panel: Conductance and minimum cross section area (r_s 51.6 Å is the Wigner-Seitz radius for Au). Lower panel: Transmissions coefficients of conductance channels. For smaller contacts, stages appear where the transmission is carried by a few almost fully transmitting channels, giving rise to a conductance plateau slightly downshifted from an integer value ($G = \{1, 3, 6\}G_0$). (b) The same as in the left panel but for a contact in series with a spring, in order to account for finite stiffness of the experimental setup. The spring constant is here taken to be $K_s = 25$ N/m corresponding to a typical value for contacts fabricated with an STM. Reprinted with permission from [316]. Copyright 1997 by the American Physical Society.

For alkali metals (Na, K, etc.) one finds histograms at low temperatures with peaks near 1, 3, 5 and 6 times G_0 [306]. An example for sodium is shown in the upper right panel of Fig. 11.6. The fact that peaks near 2 and $4\,G_0$ are absent points at an interpretation in terms of a smooth, near-perfect cylindrical symmetry of the sodium contacts. The alkali metals can be described to a good approximation as free electron systems. Within this framework, it can be shown that in smooth cylindrical contacts with continuously adjustable contact diameter [143, 317], the conductance increases from zero to $1\,G_0$ as soon as the diameter is large enough, so that the first conductance mode is occupied. When increasing the diameter further, the conductance increases by two units because the second and third modes are degenerate. In a similar way, one can explain the absence of a peak at $4\,G_0$

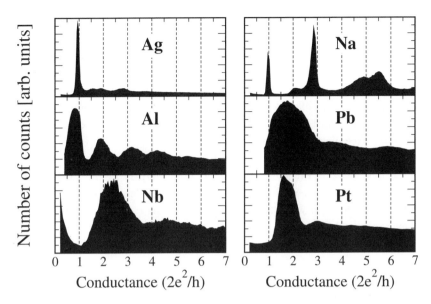

Fig. 11.6 Conductance histograms of several metals obtained using the MCBJ technique. All the histograms were recorded at 4.2 K, except for Nb, which was obtained at 16 K. The conductance was measured at 20 mV for Ag and Nb, 10 mV for Na and Al and 100 mV for Pb and Pt. Adapted with permission from [315].

and the presence of peaks at 5 and $6\,G_0$.

The analysis of atomic contacts of monovalent metals (alkali and noble metals) suggests that there is certain tendency to observe quantized values of the conductance, at least in the very last stages of the formation of these atomic contacts. However, this tendency is by no means universal. Indeed, most multivalent metals only show a rather broad first peak, which reflects the conductance of a single-atom contact (see discussion below). This peak can generally not be identified with an integer value of the conductance. This is illustrated in Fig. 11.6 for Pb, Nb and Pt, which exhibit peaks at roughly 1.7, 2.3 and $1.6\,G_0$, respectively. On the other hand, there are a few examples of multivalent metals, which show pronounced peaks in the histograms, like for instance Al [318] (see Fig. 11.6), Zn [315, 319, 320] and Mg [321]. As we shall discuss in the next section, the histogram for Al throws doubt upon a straightforward interpretation of the histogram peaks in terms of conductance quantization.

To conclude this section and following Ref. [315], we can summarize the findings concerning the conductance histograms of atomic contacts in the following way:

- With the exception of alkali metals, the highest peak is always lying at the lowest conductance value.
- The position of this peak for all the elements falls in the range between 0.7 and $2.3\,G_0$. There is no structure related to metallic conductance in the histograms below the position of the first peak.
- For free electron-like alkali metals the first peak is extremely sharp and is located almost exactly at $1\,G_0$. This statement also extends to the almost free electron-like noble metals.
- For divalent metals (zinc, magnesium) and trivalent ones (aluminum) the first peak is rather sharp and located slightly below $1\,G_0$. Other multivalent metals, and in particular transition metals, exhibit a broad first peak located well above $1\,G_0$ and in some case like niobium it lies even above $2\,G_0$.

11.4   Determining the conduction channels

As the Landauer formula indicates [see Eqs. (11.1) and (11.2)], the conductance measurements gives us only access to the total transmission $T = \sum_n T_n$ at the Fermi energy. Obviously, the experimental determination of the individual transmission coefficients, T_n, could provide a valuable insight into the origin of the differences between atomic contacts of different metals. From a mathematical point of view, it is clear that the extraction of the set $\{T_n\}$ requires the analysis of transport properties that depend on the transmission coefficients on a non-linear manner. As we saw in section 4.7, the shot noise is an example of such a quantity. Indeed, the experimental study of shot noise has provided very important information about the conduction channels of both atomic contacts and single-molecule junctions. This is discussed in detail in Chapter 19.

In this section we shall focus our attention on the first method that was used to extract the individual transmission coefficients of an atomic contact and which continues to be the most precise one. This method was put forward by Scheer *et al.* [77] and it is based on the analysis of the subgap structure in superconducting contacts. Let us explain this idea in certain detail. Many simple metals, like Al, Pb, Nb, etc., are superconducting below a critical temperature of the order of a few K. In the superconducting state these metals exhibit a gap in density of states, Δ, which is typically between 0.1 and 1 meV. This gap strongly influences the transport properties of superconducting contacts (including atomic junctions) leading to

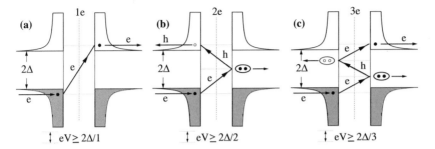

Fig. 11.7 Schematic representation of the multiple Andreev reflection (MAR) that take place in a contact between two superconductors with gap Δ. We have sketched the density of states of both electrodes, which exhibits a singularity at the gap edges. In order to simplify these graphical representations, we have not shifted the DOS of the leads with bias voltage, but equivalently we have taken into account the fact the quasiparticles gain an energy eV every time they cross the junction. (a) This panel describes the process in which a single electron tunnels through the system overcoming the gap due to a voltage $eV \geq 2\Delta$. (b) Andreev reflection process in which an electron is reflected as a hole transferring a Cooper pair to the other electrode. This process has a threshold voltage equal to Δ/e and its probability is proportional to T^2. (c) MAR of order 3 in which a quasiparticle is reflected twice before it finds an available state in the right electrodes. In this process three electron charges are transferred across the junction, the threshold voltage is $2\Delta/3e$ and its probability is proportional to T^3. Higher-order processes with contributions proportional to T^n can also occur when the bias voltage is larger than $2\Delta/ne$ (with n integer).

highly non-linear current-voltage (I-V) characteristics. In order to understand why this is so, let us consider a junction with a single conduction mode of transmission T. In the limit $T \ll 1$, we have essentially a tunnel junction and the I-V characteristic for a superconducting tunnel junction is known to directly reflect the gap [702]. As illustrated in Fig. 11.7(a), no current flows until the applied bias exceeds $2\Delta/e$, after which the current jumps to approximately the normal-state resistance line. For $eV > 2\Delta$ single quasiparticles can be transferred from the occupied states at $E_F - \Delta$ on the low voltage side of the junction to empty states at $E_F + \Delta$ at the other side. For $eV < 2\Delta$ this process is forbidden, since there are no states available in the gap.

If the transmission of the junction is not too low, one can still have current for voltages smaller than $2\Delta/e$ due to higher-order tunnel processes. Figure 11.7(b) illustrates a process, known as *Andreev reflection*, in which an electron is reflected as hole leading to the transfer of a Cooper pair to the other side of the junction.[9] The Andreev process is allowed for $eV > \Delta$

---

[9]This process can also be viewed as the simultaneous tunneling of two quasiparticles.

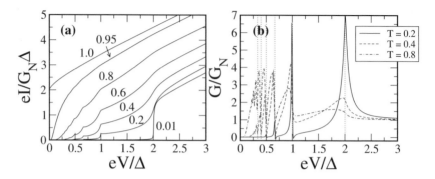

Fig. 11.8   (a) Zero-temperature I-V characteristics of a single channel superconducting quantum point contact for different values of the normal transmission coefficient (indicated in the graph). Notice that the current has been normalized with the normal state conductance $G_N = G_0 T$ to see all the curves in the same scale. (b) The corresponding differential conductance $G = dI/dV$ for three different values of the transmission. As a guide for the eyes, the vertical dotted lines indicate the position $eV = 2\Delta/n$ with $n = 1, \ldots, 6$. From Ref. [326].

and its onset causes a step in the current at $V = \Delta/e$. The height of the current step is smaller than the step at $2\Delta/e$ by a factor T, since the probability for two particles to tunnel is T^2. Depending on the junction transparency, similar processes of order n involving the transfer of n particles can occur. These processes give rise to current onsets at $eV = 2\Delta/n$ with a step height proportional to T^n. An example for $n = 3$ is illustrated in Fig. 11.7(c). These processes are referred to as *multiple Andreev reflections* (MARs) [323].[10] The microscopic theory of MARs for a single-channel point contact was developed in the late 1980's and in the 1990's by several groups independently [324–328]. In Fig. 11.8 we show the zero-temperature I-V curves and the corresponding differential conductance for different values of the normal transmission coefficient. Notice the appearance of a pronounced structure in the I-Vs close to voltages $V = 2\Delta/ne$ (with n integer) as a result of the onset of the different MAR processes. This structure, which is known as *subharmonic gap structure*, is more clearly seen in the differential conductance a series of maxima, see Fig. 11.8(b).

The subharmonic gap structure had been measured in the context of atomic contacts by several authors [69, 329, 330], but Scheer *et al.* [77]

---

[10]These multiple processes were first described by Schrieffer and Wilkins [322] in the limit of low transparent junctions. These authors coined the name multiple particle tunneling (MPT) for these tunnel events. It is now understood that the concepts of MAR and MPT are indeed equivalent.

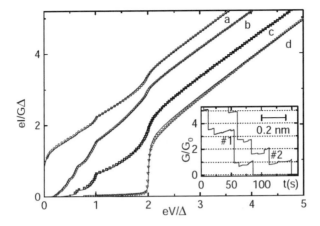

Fig. 11.9 Current–voltage characteristics for four atom-sized contacts of aluminum using a lithographically fabricated MCBJs at 30 mK (*symbols*). The *right inset* shows
the typical variation of the conductance, or total transmission $T = G/G_0$, as a function of the displacement of the electrodes, while pulling. The data in the main panel
have been recorded by stopping the elongation at the last stages of the contact (**a–c**) or
just after the jump to the tunneling regime (**d**) and then measuring the current while
slowly sweeping the bias voltage. The current and voltage are plotted in reduced units,
$eI/G\Delta$ and eV/Δ, where G is the normal state conductance for each contact and Δ is
the measured superconducting gap, $\Delta/e = (182.5 \pm 2.0)\mu V$. The *solid lines* have been
obtained by adding several theoretical curves for a single channel contact and optimiz-
ing the set of transmission values. The curves are obtained with: (**a**) three channels,
T_1=0.997, T_2=0.46, T_3=0.29 with a total transmission $\sum T_n$ =1.747, (**b**) two channels,
T_1=0.74, T_2=0.11, with a total transmission $\sum T_n$ =0.85, (**c**) three channels, T_1=0.46,
T_2=0.35, T_3=0.07 with a total transmission $\sum T_n$ =0.88. (**d**) In the tunneling range a
single channel is sufficient, here $\sum T_n = T_1$=0.025. Reprinted with permission from [77].
Copyright 1997 by the American Physical Society.

were the first to realize that the highly non-linear dependence of the superconducting I-Vs on the transmission coefficient offers the possibility to
extract the transmission coefficients of a few-atom thick contacts. The
principle is illustrated in Fig. 11.9. Using lithographic MCBJs, Al atomic
contacts were formed at very low temperatures (30 mK). During the breaking of the Al wires, I-V at low bias ($\lesssim$ 1 mV) were recorded along the
conductance plateaus. Examples of these I-Vs can be seen in the main
panel of Fig. 11.9 for different realizations of the contacts. Notice in particular that curves (b) and (c) correspond to similar values of the normal
state conductance (i.e. for voltages much larger than the Al gap). This
indicates that while these two junctions are almost indistinguishable in
the normal state, they exhibit clearly distinct superconducting I-Vs, which

means that their set of transmission coefficients are very different. The
I-V curves were fitted very accurately with the single-channel I-V curves of
Fig. 11.8(a) using as adjustable parameters both the number of conduction
channels and the transmission coefficients. The authors of Ref. [77] showed
that for the smallest contacts, the set of transmission probabilities can be
unambiguously determined.

The most important finding in these experiments was that in the last
"plateau" in the conductance, just before the breaking of the contact, typically three channels with different T's are required for a good description.
This is surprising since the conductance for such contacts is typically below
$1\,G_0$ (see Fig. 11.2(a) and the Al histogram in Fig. 11.6), and it would in
principle require only a single conductance channel. Contacts at the verge
of breaking are expected to consist of a single atom, and this atom would
then admit three conductance channels, but each of the three would only be
partially open, adding up to a conductance close to $1\,G_0$. This very much
contradicts a simple picture of quantized conductance in atomic-size contacts, and poses the question as to what determines the number of channels
through a single atom.

11.5   The chemical nature of the conduction channels of one-atom contacts

In order to answer the question posed at the end of the previous section,
Cuevas, Levy Yeyati and Martin-Rodero [263] put forward a minimal model
to compute the conductance of atomic contacts within the framework of
Landauer approach. This model is based on a combination of a simple tight-
binding (TB) model and nonequilibrium Green's functions techniques, in
the spirit of what we have discussed in Chapters 7-9, and it contains the following three basic ingredients. First, a proper description of the electronic
structure of atomic contacts, and in turn of their transport properties, requires the inclusion in the TB model of at least the atomic orbitals that
give the major contribution to the bulk density of states at the Fermi energy. As one can see in Fig. 9.7, this means in practice to include the s
orbitals for alkali and noble metals, the s and p orbitals for metals like Al
and Pb and the s and d orbitals in the case of transition metals. Second,
since often we do not have direct information about the geometry of the
atomic contacts, it is important to study the influence of the precise atomic
arrangements. Finally, metals often exhibit local charge neutrality due to

their small screening length. In this respect, it is important to impose such neutrality in the TB model via a self-consistent determination of the on-site energies.

With this minimal model, the authors of Ref. [263] focused on the analysis of the conductance of one-atom thick contacts like the one shown in Fig. 11.10(a). In such geometries the current proceeds mainly through the central atom. Thus, it is convenient to compute the transmission matrix at the central atom, where its dimension is just the number of orbitals included in the basis set.[11] This simple idea already tell us that *the number of channels is determined by the number of valence orbitals of the central atom.* This means in practice that the number of conduction channels for monovalent metals (alkali and noble ones) is limited to one, this number is at maximum four in the case of *sp*-like metals like Al or Pb and it can be up to 6 for transition metals due to the contribution of the *s* and *d* bands. It is worth stressing that this rule of thumb should be taken as an upper limit since some of the channels may be closed for symmetry reasons. The case of Al nicely illustrates this fact. Al in its atomic form has an electronic configuration $[Ne]3s^23p^1$, and a total of four orbitals would be available for current transport: one *s* orbital and three *p* orbitals (p_x, p_y and p_z). The calculations of Ref. [263] showed that for a single-atom Al contact there are three channels that give a significant contribution adding up to a total conductance of the order of $1\,G_0$. There is a dominant channel that originates from a combination of the *s* and p_z orbitals of the central atom (where the *z* coordinate is taken in the current direction), and two smaller identical contributions coming from the p_x and p_y orbitals.[12] The degeneracy of these two channels is due to the symmetry of the geometry considered [see Fig. 11.10(a)], and it can be lifted by changing the local environment for the central atom. The fourth possible channel, an antisymmetric combination of *s* and p_z, is found to have a negligible transmission probability.[13] Thus, these calculations explained the experimental observation by Scheer *et al.* [77] that three channels contribute to the conductance for a single

---

[11]Strictly speaking, this is only true if the hopping elements in the TB Hamiltonian are restricted to first nearest-neighbors. In general, it is a good approximation as long as the direct coupling between atoms on the left and on the right of the central atom is weak. The technical details concerning this discussion can be found in section 8.1.

[12]The conduction channels, defined as the eigenfunctions of $t^\dagger t$, can be expressed in the approach of Ref. [263] as a linear combination of the atomic orbitals of the central atom.

[13]In simple terms, this antisymmetric combination in the central atom is almost orthogonal to the incoming states from the leads which results in a very weak effective coupling and the corresponding negligible contribution to the total conductance.

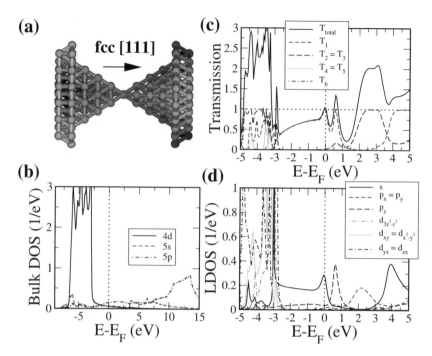

Fig. 11.10   (a) Ideal geometry of a single-atom Ag contact grown along the [111] direction (taken as the z-axis). The distances are set to bulk distances and the last two
layers on both sides correspond to those atoms in the infinite surfaces used to model
the leads that are coupled to the atoms in the constriction. (b) Bulk density of states
(DOS) projected onto the s, p and d orbitals as a function of energy (measured with
respect to the Fermi energy E_F). (c) Total transmission and transmission coefficients
of the contact of panel (a) as a function of energy. (d) Local density of states (LDOS)
at the central atom projected onto the different atomic orbitals as a function of energy.
Courtesy of Michael Häfner [331].

aluminum atom. Moreover, the results for the number of channels were
shown to be robust against changes in the atomic configuration, whereas
the total conductance was found to vary depending on the exact atomic
geometry. Finally, this analysis was extended to the case of transition metals (in particular Nb) showing that for these metals up to 5 channels can
be expected for a single-atom contact. Again, the sixth channel that could
potentially contribute in a transition metal is actually closed for symmetry
reasons.

Before turning to the analysis of the experiments that confirmed these
ideas, we now want to illustrate them in more detail. In what follows, we
shall make use of the NRL tight-binding method of section 9.6 and the

formulas derived in section 8.1. The NRL method provides a very accurate TB parameterization of the bulk properties of elementary solids that is also well suited for low-dimensional structures (see discussion below). Moreover, this parameterization takes into account long range hopping matrix elements and it includes up 9 orbitals in the basis set (the s, p and d closest to the Fermi energy). Thus, this parameterization is more accurate than that used in Ref. [263] and it serves us to test the conclusions drawn above. Let us start by analyzing the conductance of an ideal single-atom contact of Ag. The geometry of this ideal contact is shown in Fig. 11.10(a). It is constructed by starting from a central atom and including the nearest neighbors in the successive layers of the fcc lattice along the [111] direction. The leads are modeled as two infinite surfaces grown along the same direction. The bulk density of states (DOS) of this metal computed from this TB parameterization is shown in Fig. 11.10(b). Notice that the d bands are filled, while the p bands have little weight at the Fermi energy. Therefore, one expects the s band to dominate the transport properties of this monovalent metal. Moreover, from the arguments above, one also expects to have a single conduction channel in the case of one-atom contacts. This is indeed confirmed by the calculations, as one can see in Fig. 11.10(c). This figure shows both the total transmission and individual transmission coefficients as a function of energy for this geometry. Notice in particular that the transmission at the Fermi energy, which determines the conductance, is largely dominated by a single channel. One can get insight into the nature of the conductance channels of single-atom contacts by analyzing the corresponding local density of states (LDOS) at the central atom. This LDOS projected onto the different atomic orbitals for the geometry of panel (a) can be seen in Fig. 11.10(d). The first thing to notice is the presence of true energy bands that, although are narrower than those of the bulk solid, have widths of several electronvolts. This illustrates the fact that the central atom is strongly coupled to the electrodes and there is huge hybridization between its orbitals and those of the leads. Notice also that there is a clear correlation between the energy dependence of the transmission and that of the LDOS. In particular, one can see that the transmission at the Fermi energy arises from a resonance of the s band, as expected from the arguments above.

On the other hand, we can use this example to anticipate the results for single-atom contacts of other metals in the periodic table. The idea goes as follows. Most metals have similar energy bands and the main difference is the position of the Fermi level, which is determined by the number of valence electrons. Thus, the energy dependence of the transmission shown in Fig. 11.10(c) for an Ag contact can be used to understand what happens in the case of other metals by simply imagining that the Fermi level is located in a different position. To figure out what to expect for a transition metal, we can concentrate on energies 3 eV below the Fermi energy, which is the region dominated by the d bands, see Fig. 11.10(d). In this case one can see that up to 5 channels can give a significant contribution to the total transmission depending on the energy. This agrees with the predictions described above for transition metals. On the other hand, the region dominated by the p orbitals, a few eV above E_F in Fig. 11.10(c), can give us a hint about the expectations for metals like Al or Pb. In this region one can see that three channels dominate the transport and their relative contribution depends on the energy. Two of the channels are degenerate as a result of the symmetry of the contact. This degeneracy is also reflected in the LDOS, see Fig. 11.10(d), where the p_x and p_y bands are identical, while the p_z one has been shifted down due to the stronger hybridization of the p_z orbitals with the states in the leads. These results again agree with arguments described above.

To confirm these ideas, we present in Fig. 11.11 the results for the transmission coefficients for six different metals. This time we have chosen a geometry for one-atom thick contacts that contains a dimer in its central part. Different molecular dynamics simulations suggest that this type of geometry is the most frequently realized in the last stages of the breaking of the contacts [332, 333, 342, 343]. The results for Ag are similar to those of Fig. 11.10(c), the main difference being that the channels arising from the the p_x and p_y orbitals have been partially suppressed due to the reduction of the effective coupling of these orbitals to those in the leads. In the case of Au (another monovalent metal) the transmission is also dominated by a single channel, although the second and third one give a larger contribution than in Ag. Turning now to the sp-like metals Al and Pb, we see that the three channels give a significant contribution at the Fermi energy. Notice that two of them are degenerate in this highly symmetric contact due to the reasons explained above. As anticipated in the previous paragraph, the total conductance for Pb $(1.67\,G_0)$ is higher than for Al $(1.16\,G_0)$ because the former has one more valence electron and therefore the Fermi level lies in the middle of the p bands, rather than in their tails as in the case of Al. In the case of the transition metals Nb and Pt the d orbitals play a crucial role (see their bulk DOS in Fig. 9.7) leading to the opening of additional channels and conductances well above G_0. In

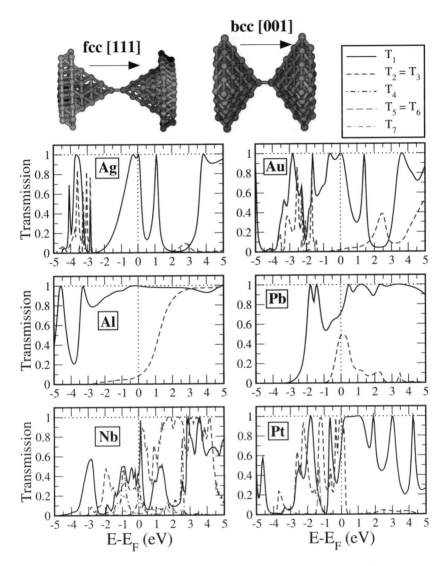

Fig. 11.11 Individual transmission coefficients as a function of energy for dimer contacts of Ag, Au, Al, Pb, Nb, and Pt computed with the NRL TB parameterization. The structures of the contacts are shown in the upper part of the graph. For all the cases the geometries are grown along the [111] direction of a fcc lattice (see left structure), except for Nb, which is grown along the [001] direction of a bcc lattice. Courtesy of Michael Häfner [331].

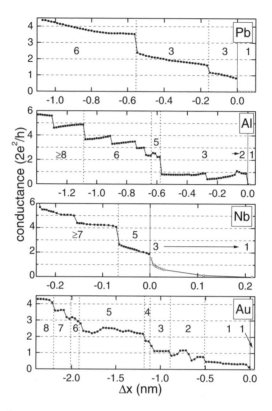

Fig. 11.12   Conductance curves measured as a function of contact elongation for Al, Pb, Nb and Au. The number of channels contributing to the conductance was determined at each point in the curves by recording the current-voltage relation and fitting the curves with the theory for superconducting subgap structure. The numbers along the curves in the figure indicate the number of channels obtained in this way. The number is constant over a plateau, and usually jumps to a smaller value at the steps in the conductance. Reprinted by permission from Macmillan Publishers Ltd: Nature [335], copyright 1998.

the case of Nb the total conductance ($1.56\,G_0$) is due to the contribution of approximately 5 partially open channels. In the case of Pt, however, three channels contribute to its conductance ($2.6\,G_0$). The difference between these two transition metals is due to the fact that Pt has five more valence electrons than Nb and the Fermi level lies on the edge of the d bands, while for Nb is almost in the center of those bands, see Fig. 9.7.

Turning now to the experiments, the method described in the previous section to extract the information on the conduction channels based on the analysis of the superconducting I-V curves has been extended to other metals. In particular, in a collaboration between three different laboratories atomic contacts of Pb, Al, Nb and Au were analyzed [335]. Fig. 11.12 shows conductance curves for Pb, Al, Nb and Au, where at each point in the figure I-V curves as in Fig. 11.9 were recorded and fitted in order to determine the number of channels involved. The number is constant over a plateau in the conductance, where the transmission probability for every mode changes gradually. At the steps in the conductance the number of channels involved is usually found to jump to a smaller number. In tunneling range, when the contact is broken and the distance is larger than 0.2 nm, the I-V characteristics can in all cases be described by a single channel, with a transmission probability which is given by the tunneling resistance. The number of channels found for the smallest contacts, just before the jump to tunneling is 1 for Au, 3 for Al and Pb, and 5 for Nb. The case of gold deserves a special comment. This metal is not a superconductor, and a special device was fabricated which allowed the use of proximity induced superconductivity [335, 334]. The device is a nanofabricated MCBJ having a thick superconducting Al layer forming a bridge with a gap of about 100 nm. This small gap was closed by a thin Au film in contact with the aluminum. Superconducting properties were thereby induced in the Au film, and by breaking this film and adjusting an atomic-size contact, the same subgap analysis could be performed.[14] Both the Al and Au junctions were measured at temperatures of 100 mK, far below the superconducting transition temperatures. Pb and Nb were measured at 1.5 K.

The number of channels and the total conductance at the last plateau before breaking found in the experiment agree very well with the theory detailed above and this can be seen as the confirmation of the fact that *the number of conduction channels in single-atom contacts is mainly determined by the number of valence orbitals*. Notice also that the relative conductances of Al and Pb, which are metals of the same group, also agrees with the theoretical predictions that say that Pb should be more conductive because of the larger contribution of the p_x-p_y channels (see also the conductance histograms of Fig. 11.6). Another important conclusion from the work of Ref. [335] was that *the smallest contacts produced by the different experimental techniques are indeed one-atom thick contacts since the number of channels in the last conductance plateau never exceeds the number of valence orbitals*. That was an important conclusion at that time since

---

[14]Notice that in Fig. 11.12 the conductance for Au in the last plateau is a factor 2–3 smaller than usually found, see Figs. 11.1 and 11.6. This was tentatively attributed to the strong scattering in the nanofabricated device.

there were no means to obtain direct information about the contact geometries. Later, it became possible to directly image atomic-size contacts by means of high resolution transmission electron microscopy (HR-TEM), see e.g. Refs. [63, 336, 62, 65]. This technique has allowed to confirm the existence of single-atom contacts and the formation of monoatomic chains (see below).

11.6   Some further issues

The ensemble of results presented in the previous section illustrates the very good level of understanding achieved in this field. Anyway, there are aspects of the transport properties of atomic contacts that deserve further discussion. From the theory side, in spite of excellent overall agreement with the experiments, one may wonder whether TB models based on parameterizations of bulk properties can provide a quantitative description of the conductance of atomic contacts. In recent years, many different groups have applied ab initio methods to the description of the transport properties of these metallic nanowires, see e.g. [337, 290, 287, 338, 339, 247] and references therein. Most of these methods are based on the density functional theory (DFT) and are described in Chapter 10. As an illustrative example, we present in Fig. 11.13 a comparison of the transmission of a single-atom Al contact computed with the NRL-TB method and with the cluster-based DFT method of Ref. [247]. As one can see, the agreement on the total conductance and transmission channels is quite satisfactory. Such an agreement is very important because for many problems involving large contacts or the analysis of a great number of configurations, the ab initio calculations are either extremely time-consuming or simply not possible. However, those problems can nowadays be tackled with, for instance, the relatively inexpensive NRL-TB method.

A more important issue is the role of the mechanical properties in the conductance of atomic contacts. Since these properties determine the geometry of these metallic wires, they obviously have a major impact in the transport properties. In this sense, a complete theoretical description should ideally determine also the possible contact geometries realized in the experiments. For this reason, different authors have combined conductance calculations with geometry optimizations or molecular dynamics simulations of different levels of sophistication, see for instance [332, 333, 340–344] and references therein. Such combinations have allowed to tackle the following

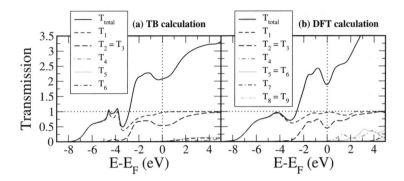

Fig. 11.13 Total transmission and individual transmission coefficients as a function of energy for an ideal single-atom Al contact in [111]-direction, as depicted in Fig. 11.10(a). (a) Calculation done with the NRL-TB method [331]. Courtesy of Michael Häfner. (b) DFT calculation from Ref. [247].

two important problems that we now proceed to describe.

The first one concerns the different slopes observed in the conductance traces of different metals (see section 11.2). This issue was addressed in Ref. [264], where it was argued that the slopes depend primarily on the evolution of the local density of states at the contact region upon stretching. This was further investigated with the help of first-principle simulations for the case of Al by Jelínek *et al.* [332]. Their main results are reproduced in Fig. 11.14 where one can see a typical evolution of an Al wire in their simulations of the stretching process and the corresponding total conductance and the contribution of the individual channels. As one can see, these results nicely reproduce the main findings of the experiments of Ref. [77]: (i) the anomalous positive slope of the conductance plateaus, (ii) the fact that three channels contribute to the conductance in the last plateau and (iii) the fact that this last plateau has a conductance below $1\,G_0$, but it raises to almost $1\,G_0$ on the verge to break.

Maybe, the main problem that remains without a fully satisfactory solution is that of the origin of the peaks in the conductance histograms of the different metals, see Fig. 11.6. From the discussion in the last section, it is clear that those peaks cannot longer be interpreted as signatures of conductance quantization, even if they appear close to integers of G_0 as in the case of Al (see Fig. 11.6). In some cases it has been understood that these peaks are the result of the interplay between mechanical and electrical properties. Thus for instance, in the case of alkali metals it has been understood that the peaks are associated to the existence of exceptionally stable configura-

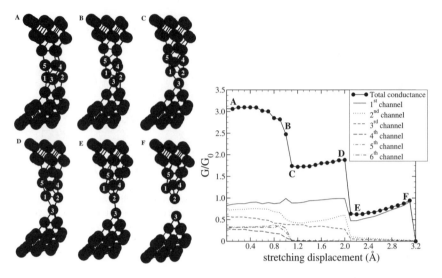

Fig. 11.14   Left panel: Snapshots of the structure of an Al nanowire during a stretching process. The atoms involved in the important bonding rearrangements related to discontinuous changes in total energy, force, and conductance are labeled 1-5. Right panel: Total conductance in units of G_0 and the channel contribution along the stretching path. Adapted with permission from [332]. Copyright 2003 by the American Physical Society.

tions due to both electronic and atomic shell effects [72, 345, 346].[15] In the case of gold, it has become clear that the pronounced peak close to $1\,G_0$ is related to the formation of monoatomic chains which sustain a single almost fully open channel (see section 11.8). Several suggestions for the origin of the peaks in the low-temperature histograms of some multivalent metals have been made [347].[16] An interesting idea was put forward by Hasmy *et al.* [351] who performed molecular dynamics simulations to study the histograms of the minimum cross section for Al contacts. At low temperatures they obtained peaks at multiples of the cross section of a single atom, which led them to an interpretation of the conductance histogram peaks based on preferential geometrical arrangements of nanocontact necks. Dreher *et al.* [342, 343] have corroborated the existence of well-defined peaks in the minimum cross section histograms. However, they have shown that those peaks

---

[15]This is a very interesting topic that will not be further discussed here because our interest is focused on the smallest contacts. For a detailed discussion of the shell effects we recommend Refs. [15, 315].

[16]Let us mention that more recently room-temperature conductance histograms of Al and noble metals have been interpreted as an evidence of electronic and atomic shell effect in these metals [348–350].

are not necessarily reflected in the corresponding conductance histograms. Their theoretical analysis of conductance histograms of Au, Ag, Pt and Ni contacts shows that the lowest peak is related to the formation of single-atom contacts and monoatomic chains in the cases of Au and Pt. The origin of the multi-peak structure of metals like Al or Zn remains however to be understood.

11.7 Conductance fluctuations

The method described in section 11.4 to extract the set of transmission coefficients of an atomic contact can only be easily applied in the case of superconducting metals. In this sense, it is important to have other methods that give access to the conduction channels. As already mentioned, the shot noise and the thermopower are two valuable transport properties in this respect and we shall discuss them in detail in Chapter 19. In this section we shall focus on the analysis of the so-called conductance fluctuations. This discussion is based on Ref. [302].

The elastic scattering of conduction electrons on defects and/or impurities near atomic contacts leads to interference effects that are clearly visible in the second derivative of the current with respect to bias voltage, i.e. in dG/dV. This is similar to the universal conductance fluctuations in diffusive mesoscopic conductors [50]. This phenomenon is well-known in point contacts with dimensions much larger than those of an atomic contact [352–355]. The conductance fluctuations on atomic contacts were studied systematically by Ludoph *et al.* [312]. In these experiments lock-in amplifiers were used to measure simultaneously the conductance and its derivative during the breaking of the nanowires. By repeating this operation many times one can construct a conductance histogram together with the average properties of dG/dV. Typical results for gold contacts can be seen in Fig. 11.15, where the upper panel shows the standard deviation of the derivative of the conductance with bias voltage $\sigma_{GV} = \langle (dG/dV)^2 \rangle$ as a function of the conductance. The conductance histogram for the same set of data is shown in the lower panel. As one can see in this figure, the data for σ_{GV} display pronounced minima for G near multiples of G_0.

The authors of Ref. [312] offered a very appealing explanation for this quantum suppression of the conductance fluctuations, which is sketched in the inset of Fig. 11.15. The idea goes as follows. The atomic contact is modeled by a ballistic central part, which is described by a set of transmission coefficients, sandwiched between diffusive banks, where electrons are scattered by defects characterized by an elastic scattering length l_e. An electron wave of a given mode impinging on the contact is transmitted with probability amplitude t and it is partially reflected back to the contact by the diffusive medium, into the same mode, with probability amplitude $a_n \ll 1$. This back-scattered wave is then reflected again at the contact with probability amplitude r_n, where $T_n = |t_n|^2 = 1 - |r_n|^2$. The latter wave interferes with the original transmitted wave. This interference

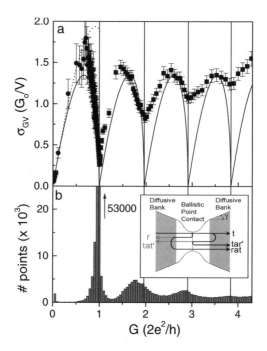

Fig. 11.15   (a) Standard deviation of the voltage dependence of the conductance versus conductance for 3500 curves for gold measured with the notched-wire MCBJ technique at 4.2 K. All data points in the set were sorted as a function of the conductance after which the rms value of dG/dV was calculated from a fixed number of successive points. The circles are the averages for 300 points, and the squares for 2500 points. The solid and dashed curves depict the calculated behavior for a single partially-open channel and a random distribution over two channels respectively. The vertical grey lines are the corrected integer conductance values (see text). (b) Conductance histogram obtained from the same data set. The peak in the conductance histogram at G_0 extends to 53000 on the vertical scale. The insets shows a schematic diagram of the configuration used in the analysis. The dark lines with arrows show the paths, which contribute to the conductance fluctuations in lowest order. Reprinted with permission from [312]. Copyright 1999 by the American Physical Society.

depends on the phase difference between the two waves, and this phase difference depends on the phase accumulated by the wave during the passage through the diffusive medium. The probability amplitude a_n is a sum over all possible trajectories, and the phase for such a trajectory of total length L is simply kL, k being the wave vector of the electron. The wave vector can be influenced by increasing the voltage over the contact, thus launching the electrons into the other electrode with a higher speed. The interference of the waves changes as we change the bias voltage, and therefore the total transmission probability, or the conductance, changes as a function of V. This describes the dominant contributions to the conductance fluctuations, and from this description it is clear that the fluctuations are expected to vanish either when $t_n = 0$, or when $r_n = 0$.

Based on this model, Ludoph *et al.* [312] obtained the following analytical expression for σ_{GV},

$$\sigma_{GV} = \frac{2.71 \, e \, G_0}{\hbar k_F v_F \sqrt{1 - \cos\gamma}} \left(\frac{\hbar/\tau_e}{eV_m}\right)^{3/4} \sqrt{\sum_n T_n^2 (1 - T_n)} , \qquad (11.3)$$

where k_F and v_F are the Fermi wave vector and Fermi velocity, respectively, $\tau_e = l_e/v_F$ is the scattering time. The shape of the contact is taken into account in the form of the opening angle γ (see the inset in Fig. 11.15), and V_m is the applied voltage modulation amplitude. The solid lines in Fig. 11.15(a) are obtained from Eq. (11.3), assuming a single partially-open channel at any point, i.e. assuming that channels open one-by-one as the conductance increases. In agreement with the results discussed in previous sections, the conductance for the smallest gold contacts is very well described by this simple approximation. The amplitude of the curves is adjusted to fit the data, from which a value for the mean free path is obtained, $l_e = 5 \pm 1$ nm. Similar experiments [312, 313] for copper and silver and for sodium also show the quantum suppression of conductance fluctuations observed here for gold. However, this suppression is not observed in the cases of aluminum or niobium [313], which clearly indicates that the transport in these multivalent metals is governed by partially open channels. Thus, the conductance fluctuation measurements confirm the overall picture described in previous sections in which the transport in these nanowires is determined by the valence orbitals of the corresponding material.

11.8   Atomic chains: Parity oscillations in the conductance

As we have seen in previous sections, all evidence shows that for an one-atom thick contact of monovalent metals the current is carried by a single mode, with a transmission probability close to one. Guided by this knowledge in experiments on gold Yanson *et al.* [356] discovered that during the contact breaking process the atoms in the contact form stable chains of single atoms, up to 7 atoms long. Independently, Ohnishi *et al.* [63] discovered the formation of chains of gold atoms at room temperature using an instrument that combines a STM with a transmission electron microscope, where an atomic strand could be directly seen in the images. Similar results were also obtained in Refs. [336, 357].

Some understanding of the underlying mechanism can be obtained from molecular dynamics simulations. Already before the experimental observations, several groups had observed the spontaneous formation of chains of atoms in computer simulations of contact breaking [358, 359]. The authors argue that the interatomic potentials used in the simulation may not be reliable for this unusual configuration. However, the stability of these atomic wires has now been confirmed by various more advanced calculations [360–364].

Only three metals are known to form purely metallic atomic chains, namely Au, Pt, and Ir [365]. They are neighbors in the sixth period of the periodic table of the elements and they share another property: they make similar reconstructions of the surface atoms on clean [100], [110], and [111] surfaces. A common origin for these two properties has been suggested in terms of a relativistic contribution to the linear bond strength [365].

There are many interesting aspects of the physics of metallic atomic chains that could be discussed in detail such as the formation mechanism, their stability or the fundamental limits for their length. However, we shall focus our attention here on the analysis of their transport properties and, in particular, of the so-called parity oscillation of the conductance because it nicely illustrates how the electrical conduction takes place in these remarkable 1D systems.

Let us start our discussion by briefly describing the original observations of the parity oscillations reported by Smit *et al.* [366]. These authors investigated the changes of conductance in the process of pulling atomic chains of Au, Pt and Ir using a STM and MCBJs. In Fig. 11.16 we show a typical conductance trace obtained during the breaking of an Au contact. As we have discussed in previous sections, the last conductance plateau

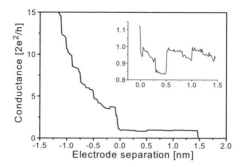

Fig. 11.16 Evolution of the conductance while pulling a contact between two gold electrodes (measured with the notched-wire MCBJ technique at 4.2 K). In the inset, an enlargement of the plateau of conductance at $\sim 1\,G_0$ is shown. Variations to lower conductance and back up by about 10–15% can be noticed when the atomic chain is stretched. Reprinted with permission from [366]. Copyright 2003 by the American Physical Society.

before rupture is in general due to a single-atom contact. The formation of an atomic wire results from further pulling of this one-atom contact, and its length can be estimated from the length of the last conductance plateau [356, 365, 367]. A histogram made of those lengths, see filled curves in Fig. 11.17, shows peaks separated by distances equal to the inter-atomic spacing in the chain. These peaks correspond to the lengths of stretching at which the atomic chain breaks, since at that point the strain to incorporate a new atom is higher than the one needed to break the chain [368]. This implies that a chain of atoms with a length between the position of the n-th and $(n + 1)$-th peak consists typically of $n + 1$ atoms.

As we learned in previous sections, the valence of the metal determines the number of electronic channels through the chain, and each channel contributes to the conductance with a maximum of G_0. For gold, a monovalent metal, both the one-atom contact and the chain have a conductance of about $1\,G_0$ with only small deviations from this value (see Fig. 11.16) suggesting that the single channel has a nearly perfect coupling to the banks. The small changes of conductance during the pulling of the wire shown in the inset in Fig. 11.16 are suggestive of an odd-even oscillation. The jumps result from changes in the connection between the chain and the banks when new atoms are being pulled into the atomic wire. In order to uncover possible patterns hidden in these changes the authors averaged many conductance traces starting from the moment that a single-atom contact is formed (defined here as a conductance dropping below $1.2\,G_0$) until the

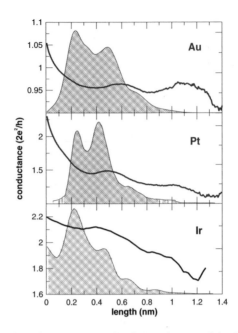

Fig. 11.17   Averaged conductance traces for chains of atoms of Au, Pt, and Ir (measured with the notched-wire MCBJ at 4.2 K). Each of the curves are made by the average of individual traces of conductance while pulling atomic contacts or chains. Histograms of the plateau lengths for the three metals obtained from the same set of data are shown by the filled curves. Reprinted with permission from [366]. Copyright 2003 by the American Physical Society.

wire is broken (conductance dropping below $0.5\,G_0$). In the upper panel of Fig. 11.17 it can be seen that the thus obtained average plateau shows an oscillatory dependence of the conductance with the length of the wire. The amplitude of the oscillation is small and differs slightly between experiments.[17]

The same procedure was repeated for Pt and Ir. These metals have s and d orbitals giving rise to several conduction channels. Each channel may have a different transmission that can be affected by the details of the contact and therefore the average plateau conductance is expected to show a more complicated behavior. A one-atom Pt contact has a conductance of about $2\,G_0$ while for a Pt atomic chain it is slightly smaller, $\sim 1.5\,G_0$ with variations during the pulling process that can be as large as $0.5\,G_0$. For Pt similar oscillations to those for Au were observed, which were compared

---

[17]This behavior is clearly at variance with that found in molecular junctions, where the conductance typically decays exponentially with the molecule length.

to the peak spacing in the length histogram in Fig. 11.17. The latter is obtained by taking as a starting point of the chain a conductance dropping below $2.4\,G_0$. Ir shows a similar behavior although somewhat less pronounced and it is more difficult to obtain good length histograms.

The simplicity of the atomic chains had stimulated numerical simulations of their transport properties well before their experimental observation [369]. In particular, various groups [369–374] had found oscillations in the conductance as a function of the number of atoms for calculations of sodium atomic chains, where this metal was selected because it has the simplest electronic structure. Sim *et al.* [370], using first-principles calculations and exploiting the Friedel sum rule, found that the conductance for an odd number of atoms is equal to G_0, independent of the geometry of the metallic banks, as long as they are symmetric for the left and right connections. On the other hand, the conductance is generally smaller than G_0 and sensitive to the lead structure for an even number of atoms. The odd-even behavior follows from a charge neutrality condition imposed for monovalent-atom wires. These predictions agree nicely with the results found for the Au chains.

As explained by the authors of Ref. [366], the odd-even behavior is essentially an interference effect and it can be easily understood in the frame of a simplified one-dimensional free-electron model, see Exercise 11.1 and Refs. [366, 375]. Instead, we shall provide here an argument from our usual "atomistic" point of view. We shall analyze the parity effect in gold chain with the help of the simple model described in Exercise 7.5, which is represented schematically in Fig. 11.18. In this model we describe the gold chain with a tight-binding Hamiltonian with a single orbital per atom and with hopping elements, t, only between nearest neighbors. We assume that the on-site energy, ϵ_0, is the same for all the atoms in the chain and we set it to zero. We describe the leads by two identical semi-infinite linear chains with, for simplicity, the same parameters as in the finite chain (bulk hopping t and on-site energy ϵ_0). Finally, the coupling between the chain and the leads is described by the hoppings t_L and t_R that can be different from the intra-chain hopping t.

The calculation of the transmission in this model, and therefore of the zero-bias conductance, is a simple exercise that we proceed to sketch.[18] From the general formula of Eq. 8.18, it is easy to show that the zero-bias

---

[18]It is not necessary to follow this calculation to understand the main conclusions that will be drawn from this toy model.

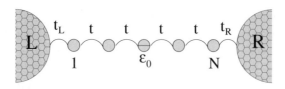

Fig. 11.18 Schematic representation of the simple tight-binding model used to analyze the parity effect in gold chains. In this model the chain has N atoms with a single orbital per site and with an on-site energy $\epsilon_0 = 0$. There is only coupling between nearest neighbors inside the chain, t. The coupling to the leads is given by the matrix elements $t_{L,R}$. The leads are modeled in practice by two identical semi-infinite chains with the same parameters as the finite central chain.

transmission in this model is given by (see also Exercise 7.5)

$$T(E) = 4\Gamma_L(E)\Gamma_R(E)|G^a_{1N}(E)|^2, \tag{11.4}$$

where the scattering rates are given by $\Gamma_{L,R} = \mathrm{Im}\{\Sigma^a_{L,R}\}$ and the self-energies can be by expressed as $\Sigma_{L,R} = t^2_{L,R}g^a_{L,R}$, where $g^a_{L,R}$ are the advanced Green's functions of the last atom of the two semi-infinite chains used to model the leads [see Eq. (5.46)]. Finally, we have to determine $G^a_{1N}(E)$, which is simply the element $(1, N)$ of the following matrix

$$\mathbf{G}^a(E) = [E^a - \mathbf{H}_{\text{chain}} - \Sigma^a_L - \Sigma^a_R]^{-1} \tag{11.5}$$

$$= \begin{pmatrix} E^a - \epsilon_0 - \Sigma_L & -t & 0 & \cdots \\ -t & E^a - \epsilon_0 & -t & 0 \\ \vdots & \vdots & \vdots & \vdots \\ 0 & t & E^a - \epsilon_0 & -t \\ \cdots & 0 & -t & E^a - \epsilon_0 - \Sigma_R \end{pmatrix}^{-1},$$

where $E^a = E - i0^+$. This tridiagonal matrix of dimension N can be inverted numerically or even analytically (see Exercise 11.2).

Let us illustrate the results of this simple model. In Fig. 11.19 we show the transmission as a function of energy for two chains with 4 and 5 atoms. Let us remind that the conductance is determined by the value of the transmission at the Fermi energy, which in this model is zero due to the inherent electron-hole symmetry (we have a single electron per atom). The different curves in both panels correspond to different values of the interface hopping $t_{L,R}$. Notice that if $t_{L,R} = t$ the system becomes an ideal infinite chain where there is no backscattering, which leads to a perfect transparency in the whole band ($|E| < 2t$). If the interface hopping is different from the intra-chain hopping (interface mismatch), the backscattering builds Fabry-Perot-like resonances in the transmission. In the case of the 4-atom chain

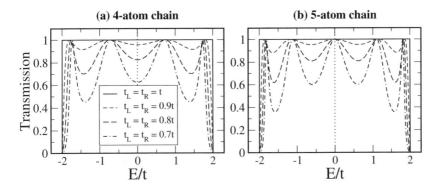

Fig. 11.19   Transmission as a function of energy (normalized by the intra-chain hopping t) for two chains with 4 (a) and 5 (b) atoms. The different curves correspond to different values of the interface hopping $t_{L,R}$, as indicated in the legend. The vertical dotted lines indicate the position of the Fermi energy, which is zero in this case.

(and in any chain with an even number of atoms), those resonances produce a minimum of the transmission at the Fermi energy, whereas they lead to a maximum for the chains with an odd number of atoms. This result explains qualitatively the parity effect discussed above for a monovalent metal.

The presence of transmission maxima at the Fermi energy for odd number of atoms in the chain and the minima for the chains with even number of atoms can be understood as follows (see Exercise 11.1). The maxima of the transmission appear at the position of the levels of the decoupled chain. In the case of odd N there is always a level in the chain spectrum exactly at the Fermi energy $(E = 0)$ for symmetry reasons, which together with the charge neutrality leads to a maximum of the linear conductance. On the contrary, when N is even there is no chain level at the Fermi energy and therefore these chains exhibit a lower conductance. To conclude this discussion, we show in Fig. 11.20 the transmission at the Fermi energy as a function of the number of chain atoms. As one can see, the amplitude of the even-odd oscillations depends on the quality of the interfaces.[19]

The simple explanation presented above can account qualitatively for the experimental behavior in the case of Au, characterized by a full $5d$ band and a nearly half-filled $6s$ band. However, for the case of Pt and Ir, in which the contribution of $5d$ orbitals to the conductance is important, there

---

[19]The conductance does not decay with length in this case because the Fermi energy lies inside the "band" formed by the states of the finite chain. For energies outside this energy window, the conductance decays exponentially with length. This is what happens in the case of molecular junctions (see discussion in section 13.4).

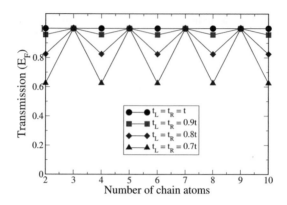

Fig. 11.20   Transmission at the Fermi energy as a function of the number of atoms in a linear chain. Notice the even-odd effect. The different curves correspond to different values of the interface hopping $t_{L,R}$, as indicated in the legend.

is no reason why this simple picture should still hold. The experiments of Smit *et al.* [366] triggered off new the theoretical analyses of the conductance of these monoatomic wires, see for instance Refs. [342, 343, 376–381]. In particular, de la Vega *et al.* [376] presented an appealing comparative study that we now proceed to describe. These authors studied the conductance of ideal chain geometries of Au, Pt and Ir, in which the atomic chain is connected to bulk electrodes represented by two semi-infinite fcc perfect crystals along the (111) direction. Using the Green's function techniques detailed in Chapters 7 and 8 and a parameterized self-consistent tight-binding model, they obtained the evolution of the conductance with the number of atoms in the chain depicted in Fig. 11.21(a). Notice that this evolution is rather sensitive to the elongation, especially in the case of Pt and Ir (for Au the conductance exhibits small amplitude even-odd oscillations, which remain practically unaffected upon stretching).

The main features and the differences between Au, Pt and Ir are more clearly understood by analyzing the local density of states and the energy dependence of the transmission, shown in Fig. 11.21(b) for a $N = 5$ chain of these metals at an intermediate elongation. The Au chains are characterized by a single conduction channel around the Fermi energy with predominant s character. The transmission of this channel lies close to one and exhibits small oscillations as a function of energy resembling the behavior of the single band TB model discussed above.

In the case of Pt the contribution from the almost filled $5d$ bands becomes important for the electronic properties at the Fermi energy. There

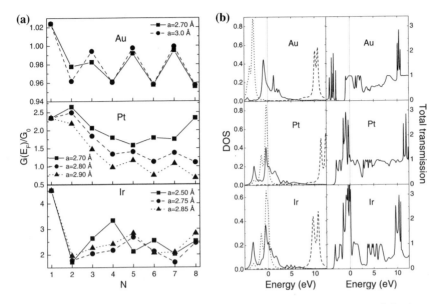

Fig. 11.21   (a) Evolution of the conductance with N for different values of the inter-atomic distance a. (b) Local density of states (LDOS) at the central atom and total transmission for Au, Pt and Ir chains $N = 5$ at an intermediate elongation. The LDOS is decomposed in s (full line), d (dotted line) and p (dashed line) orbitals with the same normalization in the three cases. Reprinted with permission from [376]. Copyright 2004 by the American Physical Society.

are three conduction channels with significant transmission at E_F: one due to the hybridization of s-p_z and d_{z^2} orbitals, and another two almost degenerate with p_x-d_{xz} and p_y-d_{yz} character, respectively (here z corresponds to the chain axis). The contribution of the $5d$ orbitals is even more important in the case of Ir where a fourth channel exhibits a significant transmission.

As discussed in Ref. [376], more insight into these results can be obtained by analyzing the band structure of the infinite chains. Fig. 11.22(a) shows the bands around the Fermi energy for Pt obtained from ab-initio calculations. Two main features are worth commenting: (i) Symmetry considerations allow to classify the bands according to the projection of the angular momentum along the chain axis, m. (ii) Close to E_F there is an almost flat filled two-fold degenerate band with d_{xy} and $d_{x^2-y^2}$ ($m = \pm 2$) character. The other partially filled and more dispersing bands have s-p_z-d_{z^2} ($m = 0$) and p_x-d_{xz} or p_y-d_{yz} ($m = \pm 1$) character (see labels in Fig. 11.22).

The close connection between this band structure and the conduction

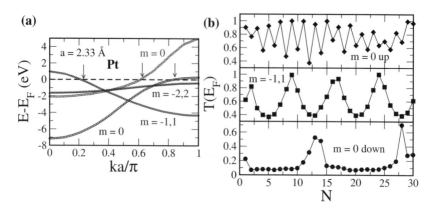

Fig. 11.22   (a) Band structure of the infinite Pt chain. The bands are classified by the quantum number m corresponding to the projection of the angular momentum on the chain axis. The arrows indicate the crossing of the Fermi level for the $m = 0$ and the $m = \pm 1$ bands. (b) Channel decomposition for Pt chains as a function of N. The legends indicate the symmetry of the corresponding bands in the infinite chain. Courtesy of A. Levy Yeyati.

channels of the chains is realized when analyzing the evolution of the conductance and its channel decomposition for even longer chains than in Fig. 11.21 ($N > 8$). This is illustrated in Fig. 11.22(b). As it can be observed, the decrease of the total conductance of Pt for $N < 7 - 8$ corresponds actually to a long period oscillation in the transmission of the two nearly degenerate channels associated with the $m = 1\pm$ bands. This period can be related to the small Fermi wave vector of these almost filled d bands, as indicated by the arrows in Fig. 11.22(a). In addition, the upper $m = 0$ band crossing the Fermi level is close to half-filling giving rise to the even-odd oscillatory behavior observed in the transmission of the channel with predominant s character. The lower $m = 0$ band tends to be completely filled and the corresponding channel is nearly closed for short chains. However, one can appreciate a very long period oscillation in its transmission, rising up to $\sim 0.5\,G_0$, for $N \sim 13$-14.

The general rule that emerges from the above analysis is that the transmission corresponding to each conduction channel oscillates as $\sim \cos^2(k_{F,i}Na)$, where $k_{F,i}$ is the Fermi wave vector of the associated band in the infinite chain. In the case of Pt the total conductance for short chains ($N < 7$-8) exhibits an overall decrease with superimposed even-odd oscillations in qualitative agreement with the experimental results. For even longer chains (not yet attainable in experiments) these calculations predict an increase of the conductance due to the contribution of conduction channels with d_{xz}, d_{yz} character.

11.9 Concluding remarks

As we have shown in this chapter, thanks to a close interaction between experiment and theory it has become possible to establish a coherent picture of the transport in metallic atomic contacts. Moreover, we have learned a few important lessons that are very useful for the field of single-molecule conduction. First of all, we now understand the close relation between the quantum properties of individual atoms used as building blocks and the macroscopic transport properties of the circuits in which they are embedded. This relation is nicely summarized in the connection between the number of channels of a single-atom contact and the valence orbitals of the corresponding atom. It has also become clear that a deep understanding of the electrical conduction in these nanocircuits can only be achieved by combining different experimental techniques and by studying a variety of transport properties.

Let us emphasize again that in this chapter we have addressed only a few basic issues concerning the very rich physics of (non-magnetic) atomic contacts. We have left out many important topics, a discussion of which can be found in the review of Ref. [15]. On the other hand, it is worth stressing that there are still many basic issues to be resolved. We have already mentioned some of them, like the problem of the origin of the peaks in the conductance histograms of multivalent metals, but there are many others. For instance, some materials like the semi-metals [382, 311, 384] exhibit very peculiar transport properties that are not yet fully understood. Concerning atomic chains, it has been recently discovered that the presence of impurity atoms can facilitate the formation of atomic chains, even in metals in which they are not formed in the pure state [385]. The fundamental limits for the lengths of these hybrid chains as well as their mechanical and transport properties still need to be investigated in further detail. With respect to the optical properties of atomic contacts, experiments measuring the laser-assisted transport are just beginning to be reported (see Chapter 20). On the other hand, experiments on light emission from atomic contacts are starting to reveal new information about the physics of these contacts [386]. So in short, atomic contacts will continue to be a marvelous source of new and fascinating physics.

11.10 Exercises

11.1 1D model for the parity effect in gold chains:

Let us model the gold monoatomic chains by the simple 1D potential shown in Fig. 11.23. The regions on the left and right of the potential step represent the electrodes, while the chain corresponds to the step region. The scattering at the chain-leads interfaces is taken into account via a mismatch in the wave vectors: $k_1 = k_3 \neq k_2$.

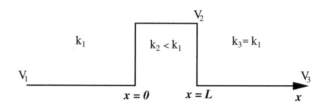

Fig. 11.23 One-dimensional model for the potential landscape describing an atomic chain of length L.

(i) Compute the transmission through this potential barrier for energies higher than the step height as a function of k_1 and k_2. Hint: the solution is given in Eq. (4.16).

(ii) Show that the transmission exhibits oscillations as a function of the chain length where the maxima are given by $T_{max} = 1$ and minima by $T_{min} = 4\gamma^2/(1 + \gamma^2)^2$, where $\gamma = k_2/k_1$.

(iii) To determine the value of k_2 relevant for the transport, one might be tempted to fix it to k_F of an infinite chain. Show that assuming that there is an electron per atom this Fermi wave vector is given by $k_F = \pi/(2a)$, where $a = L/N$ is the interatomic distance, N the number of atoms in the chain and L its length. Show also that with this choice for k_2 this 1D model predicts that the conductance maxima should appear for chains with even number of atoms, contrary to the model explained in section 11.8.

(iv) The problem found in (iii) can be solved by computing k_2 in the following more appropriate manner. Since the chain is finite, there is a limited set of possible values for k_2, namely $k_2^i = (\pi/a)i/N$ with $i = 1, ..., N$ (can you explain why?). Then, imposing charge neutrality one obtains $k_2 = (\pi/2a)N/(N+1)$ for the Fermi wave vector. Use this value for k_2 in the expression of the transmission to show that now the model reproduces the correct phase of the conductance oscillations.

11.2 Even-odd effect in gold atomic chains:

Let us consider the chain model discussed in section 11.8 to explain the even-odd effect in the conductance of gold atomic wires.

(i) Reproduce the results of Figs. 11.19 and 11.20.

(ii) Diagonalize the Hamiltonian of the uncoupled finite chain for different number atoms, N, to obtain its energy spectrum. Show that for N odd there is always an energy level of the chain at $E = 0$.

(iii) Compute the local density of states inside the chain for $N = 4, 5$ and study its relation with both level spectrum obtained in (iii) and the transmission function of Fig. 11.19.

(iv) Demonstrate that the transmission at the Fermi energy $(E_F = 0)$ is given by the following analytical expression

$$T(E_F) = \begin{cases} 4t^2\Gamma_L\Gamma_R/(t^2 + \Gamma_L\Gamma_R)^2 & \text{for even N} \\ 4\Gamma_L\Gamma_R/(\Gamma_L + \Gamma_R)^2 & \text{for odd N.} \end{cases} \tag{11.6}$$

Here, $\Gamma_{L,R}$ are the scattering rates at the Fermi energy given by $\Gamma_{L,R} = t_{L,R}$, where we have assumed that the semi-infinite chains describing the leads have the same hopping as the finite central chain.

Chapter 12

Spin-dependent transport in ferromagnetic atomic contacts

The use of the spin degree of freedom of the electron in conventional charge-based electronic devices has lead to the discovery of many fundamental effects and, in some cases, to new technological applications [387, 388]. The emblematic physical effects in this new field, already known as *spintronics*, like the giant magnetoresistance (GMR), tunneling magnetoresistance (TMR) or anisotropic magnetoresistance (AMR) stem from the spin-sensitivity of the scattering mechanisms that dominate the transport properties in electronic devices made of magnetic materials.[1] In recent years, a great effort has been devoted to understand how these fundamental effects are modified when the dimensions of a magnetic device are reduced of the way down to the atomic scale. Contrary to the case of non-magnetic atomic contacts, the physics of their ferromagnetic counterparts is not so well established and there are still basic open problems. The goal of this chapter is to provide a brief introduction to the transport properties of ferromagnetic atomic-size contacts and to draw the attention to problems that could be soon analyzed in the context of molecular junctions.

There are many different topics in this field that one could address. In order to illustrate the interesting physics of ferromagnetic atomic contacts, we have chosen to discuss three issues that are attracting a lot of attention. The first one concerns the conductance of these atomic contacts in the absence of domain or external magnetic fields and, in particular, the possibility of observing conductance quantization. The second problem is related to the magnetoresistance of these atomic-scale conductors, which has been shown to be enormous in comparison with the one found in larger devices made of the same materials. Finally, we shall address the issue of

---

[1] For a basic explanation of all these magnetoresistive effects, see Ref. [388] or Chapter 15 in Ref. [389].

the anisotropic magnetoresistance in ferromagnetic atomic contacts, which again is very different from the one found in bulk systems. After discussing these topics in the next three sections, we shall conclude this chapter with some final remarks and a brief discussion about the challenges and open problems.

12.1   Conductance of ferromagnetic atomic contacts

The first issue that we want to address is the conductance of ferromagnetic atomic-size contacts.[2] This question has been experimentally investigated by numerous groups [311, 313, 315, 390–406] which, in particular, have studied the conductance histograms of atomic contacts made of the $3d$ ferromagnetic metals (Ni, Co and Fe). To make a long story short, let us say that two type of contradictory results have been reported. On the one hand, several groups have observed peaks in the conductance histogram at half-integer multiples of G_0 [397–402]. This has been interpreted as a manifestation of half-integer conductance quantization [400], implying that only fully open channels contribute to the conductance. In this sense, a peak at $0.5\,G_0$ would then additionally mean the existence of a fully spin-polarized current. Furthermore, some authors have reported conductance histograms that are very sensitive to an external magnetic field [394].

On the other hand, another group of experiments, see e.g. Refs. [311, 313, 390, 403], show that the conductance histograms are either featureless at room temperature or they exhibit a single peak at conductances well above $1\,G_0$ at low temperatures. Let us mention in particular the work of Untiedt *et al.* [403] in which conductance histograms of Fe, Co, and Ni using notched-wire MCBJs under cryogenic vacuum conditions. These histograms are reproduced in Fig. 12.1. Notice the absence of fractional conductance quantization, even when a high external magnetic field was applied. Notice also that the histograms show broad peaks above $1\,G_0$, with only little weight below it. Furthermore these authors suggested that the differences between these two groups of experiments could be due to the fact that all room temperature experiments are performed under atmospheres that are considerably less pure than that provided by cryogenic vacuum and thus, one cannot disregard the possibility of atomic-scale contamination of the contact by foreign atoms or molecules. Indeed, these

---

[2]Here, we have in mind situations where there is a homogeneous magnetization in the junctions, i.e. no domains.

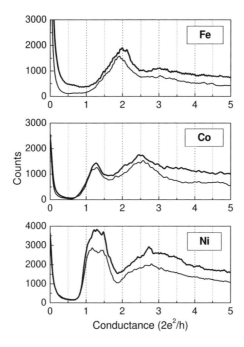

Fig. 12.1 Conductance histograms for Fe, Co and Ni atomic contacts obtained with the notched-wire MCBJ technique without magnetic field (thin curve) and when a magnetic field of 5 T parallel to the current direction was applied (thick curve). The conductance was measured using a dc bias voltage of 20 mV and a temperature of 4.2 K. Reprinted with permission from [403]. Copyright 2004 by the American Physical Society.

authors showed that the inclusion of hydrogen molecules significantly modifies the histograms of Fe, Co, and Ni.

The observation of any kind of quantization of ferromagnetic materials like Fe, Co or Ni is certainly surprising. These materials are transition metals in which, according to our discussions in the previous chapter, the d bands are expected to play a fundamental role in the transport properties contributing with partially open channels. The conductance of ferromagnetic atomic contacts has been analyzed theoretically by many different groups using a variety of methods, see e.g. Refs. [407–427]. The general picture that emerges from these works confirms the naive picture and clearly suggests that conductance quantization is not really expected in these ferromagnetic nanowires.

In what follows, we shall discuss in certain detail the results of Ref. [427], which illustrate the most commonly accepted picture of the transport in ferromagnetic contacts. In this work the calculations are based on the combination of the NRL tight-binding method of section 9.6 (see also Ref. [428]) and nonequilibrium Green's function techniques, which was already used in the last chapter.[3] Let us stress that in this discussion we shall neglect both the spin-orbit interaction and we shall assume that there are no domain walls present in the contacts.[4] With these assumptions, the transport properties of ferromagnetic contacts can be described in terms of two independent contributions coming from both spin bands. In particular, in the framework of Landauer's approach the linear conductance at low temperature can be expressed as follows

$$G = \frac{e^2}{h} \sum_{\sigma} T_\sigma(E_F),$$ (12.1)

where $T_\sigma(E)$ is the total transmission for spin $\sigma = \uparrow, \downarrow$ at energy E and E_{F} is the Fermi energy. We also define the spin-resolved conductances $G_\sigma = (e^2/h)T_\sigma(E_{\mathrm{F}})$, such that $G = G_\uparrow + G_\downarrow$. The transmissions are obtained as follows

$$T_\sigma(E) = \mathrm{Tr}[\mathbf{t}_\sigma^\dagger(E)\mathbf{t}_\sigma(E)] = \sum_{n} T_{n,\sigma}(E),$$ (12.2)

where $\mathbf{t}_\sigma(E)$ is the transmission matrix and $T_{n,\sigma}(E)$ are the individual transmission eigenvalues for each spin σ (see section 8.1.3 for more details on the calculation of these transmission matrices).

An important quantity in our discussion will be the spin polarization P of the current, which we define as

$$P = \frac{G_\uparrow - G_\downarrow}{G_\uparrow + G_\downarrow} \times 100\%.$$ (12.3)

Here, we shall assume that spin up denotes the majority spins, while spin down corresponds to the minority ones.

In order to understand the results described below, it is instructive to first discuss the bulk density of states (DOS). The spin- and orbital-resolved bulk DOS of these materials around E_{F}, as calculated from the NRL tight-binding method, is shown in Fig. 12.2. The common feature for the three ferromagnets is that the Fermi energy for the minority spins lies inside the d bands. This fact immediately suggests that the d orbitals may play an important role in the transport. For the majority spins the Fermi energy lies

---

[3]The technical details of the calculation of the current in ferromagnetic contacts have been discussed in section 8.1.3.

[4]Later in this chapter we shall discuss the role of these two factors.

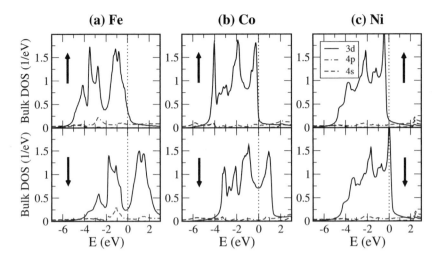

Fig. 12.2 Bulk density of states (DOS) of Fe, Co, and Ni, resolved with respect to the individual contributions of $3d$, $4s$, and $4p$ orbitals, as indicated in the legend. The upper panels show the DOS for the majority spins (spin up) and the lower ones the DOS for minority spins (spin down). The vertical dotted lines indicate the Fermi energy, set to zero. Reprinted with permission from [427]. Copyright 2008 by the American Physical Society.

close to the edge of the d band. The main difference between the materials is that for Fe there is still an important contribution of the d orbitals, while for Ni the Fermi level is in a region where the s and p bands become more important. The calculated values of the magnetic moment per atom (in units of the Bohr magneton) of 2.15 for Fe, 1.3 for Co, and 0.45 for Ni are reasonable agreement with the literature values [429].

Let us turn now to the analysis of the conductance of Fe, Co and Ni contacts. We consider ideal one-atom thick contact geometries with a central dimer as shown in the upper part of Fig. 12.3. In this figure we also present the total transmission for majority spins and minority spins as a function of energy as well as the individual transmission coefficients for those geometries. As one can see in Fig. 12.3(a), for the case of Fe one finds 3 channels for the majority spins, yielding $G_\uparrow = 1.24e^2/h$, while for the minority spins 3 channels contribute to $G_\downarrow = 0.70e^2/h$. The total conductance is $0.94\,G_0$ and the polarization $P = +28\%$. For the Co contact, see Fig. 12.3(b), one finds $G_\uparrow = 0.90e^2/h$ and $G_\downarrow = 2.23e^2/h$, summing up to a total conductance of $1.6\,G_0$. The transmission is formed by 3 channels for the majority spins (with one clearly dominant) and 6 channels for mi-

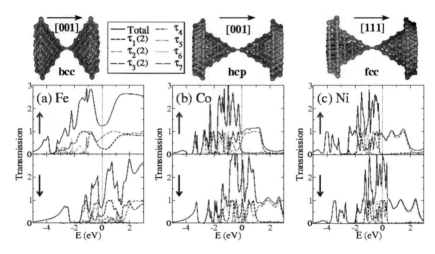

Fig. 12.3   Transmission for the three single-atom contacts of Fe, Co, and Ni containing a dimer in the central part of the contact. The geometries are shown in the upper graphs. The distances are set to bulk distances and the atoms of the last two layers correspond to the atoms of the leads (semi-infinite surfaces) that are coupled to the central atoms in the model. We present the total transmission (black solid line) for both majority spins and minority spins as well as the transmission of individual conduction channels that give the most important contribution at Fermi energy, which is indicated by a vertical dotted line. The channels corresponding to τ_1, τ_2, and τ_3 are two-fold degenerate. Reprinted with permission from [427]. Copyright 2008 by the American Physical Society.

nority spins and polarization is $P = -42\%$. Finally, for the Ni contact in Fig. 12.3(c), a single channel contributes to $G_\uparrow = 0.86e^2/h$ and 4 channels add up to $G_\downarrow = 2.66e^2/h$. This means that one has a total conductance of $1.8\,G_0$, while the current polarization adopts a value of $P = -51\%$.

From the analysis of Fig. 12.3 and many other one-atom thick geometries, the following basic conclusions were drawn in Ref. [427] concerning the conductance of a ferromagnetic single-atom contact. First, both spin bands contribute significantly to the transport. Second, the d orbitals give a very important contribution to conductance of the minority spins and they give rise to several channels (from 3 to 5 depending on the material). Third, for the majority spins there is a smaller number of channels ranging from 3 for Fe to 1 for Ni. This contribution is dominated by the d and s orbitals for Fe and only by the s orbitals for Co and Ni. The relative contribution and number of channels of the two spin species is a simple consequence of the position of the Fermi level and the magnitude of the spin splitting, see Fig. 12.2. In particular, notice that as we move from

Fe to Ni, the Fermi energy lies more and more outside of the d band for the majority spins, which implies that the number of channels is reduced for this spin species. In particular, for Ni a single majority spin channel dominates and in some sense, this material behaves as a monovalent and a transition metal combined in parallel. Finally, the conductance values for single-atom contacts lie typically above $1\,G_0$, in agreement with the experimental results of Fig. 12.1.

The analysis of Häfner *et al.* [427] also put forward two additional important conclusions. First, as a consequence of the contribution of the d bands, the value of the conductance and the current polarization are very sensitive to the contact geometry and to disorder. Second, in the tunneling regime one can have a much higher current polarization reaching in some cases values close to 100%.

These ideas and conclusions can be further illustrated with the results of Ref. [343] where conductance calculations were combined with classical molecular dynamics simulations to determine the contact geometries. In Fig. 12.4 we show the formation of a single-atom Ni contact containing a dimer in its central part just before rupture. Moreover, this figure shows the corresponding conductance and channel transmissions for both spin components, the strain force necessary to break the contact, the spin polarization of the current and the contact geometries. As one can see in this figure, in the last stages of the stretching the conductance is dominated by a single channel for the majority spins, while for the minority spin band there are still up to 4 open channels. In particular, in the very final stages (regions of 3 or 1 open channels for $G_\uparrow$) the spin-up conductance lies below $1.2e^2/h$, while for spin down it is close to $2e^2/h$, adding up to a conductance of around 1.2-1.6 G_0.

It is worth discussing the behavior of the spin polarization of the current, P. Notice that at the beginning of this contact evolution it takes a value around -40%, which is indeed close to the spin polarization of the bulk DOS at the Fermi energy (-40.5% in these model calculations). However, as the contact evolves, P fluctuates and even increases to positive values, which cannot be simply explained in terms of the bulk DOS. Notice also that P reaches the value of $+80\%$ in the tunneling regime, when the contact is broken. Such a huge value in this regime is due to the fact that the couplings between the d orbitals of the two Ni tips decrease much faster with distance than the corresponding s orbitals. As a result there is a great reduction of the spin-down conductance and in turn in a large positive value of P.

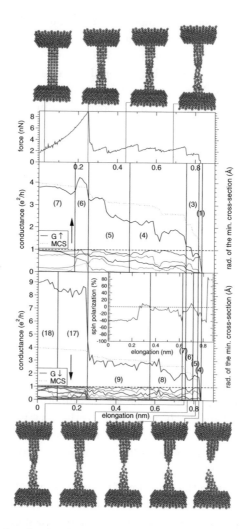

Fig. 12.4   Classical molecular dynamics simulation of the formation of a single-atom Ni contact at 4.2 K ([001]-direction). The upper panel shows the strain force as a function of the elongation of the contact. In the lower two panels the conductance G, the MCS (minimum cross-section) radius and the channel transmissions are displayed for the majority and minority spin components. Vertical lines separate regions with different numbers of open channels ranging from 7 to 1 and 18 to 4, respectively. The inset shows the evolution of the spin polarization of the current. Above and below these graphs snapshots of the stretching process are shown. Reprinted with permission from [343]. Copyright 2006 by the American Physical Society.

As a final comment on these results, let us to point out that the contribution of the minority spin component to the conductance is more sensitive to changes in the contact geometry, as one can see in Fig. 12.4. Again, this is a consequence of the fact that the minority spin contribution is dominated by the bands arising from the d orbitals, which are anisotropic and therefore more sensitive to disorder than the s states responsible for the conductance of the majority spins.

12.2 Magnetoresistance of ferromagnetic atomic contacts

Another aspect of the transport through ferromagnetic atomic contacts that has been extensively investigated in recent years is the magnetoresistance. In this case the resistance (or conductance) of a junction is measured for the two possible relative orientations (parallel and antiparallel) of the magnetization of the electrodes.[5] The way to quantify the resistance change is by means of the magnetoresistance defined as

$$\text{MR} = \frac{R(AP) - R(P)}{R(P)} \times 100\%, \qquad (12.4)$$

where $R(AP)$ is the resistance with an antiparallel orientation for the magnetizations in the electrodes and $R(P)$ is the resistance with parallel magnetizations. Normally, the antiparallel orientation exhibits a higher resistance and with this definition the magnetoresistance has not upper bound and, in particular, it can be larger than 100%.[6]

In the AP orientation there must be a domain wall somewhere in the contact and it can play an important role in the resistance of the sample. For large contacts (with diameter greater than tens of nm), the main contribution to the domain wall resistance is expected to come from the anisotropic MR, a difference in the resistivity of a magnetic material depending on whether the magnetic moment is oriented parallel or perpendicular to the current.[7] This contribution is relatively small, typically giving MR values of a few percent [430].

As the contact diameter is reduced, the width of the domain wall can be constrained by the geometry and decreases in proportion to the contact width [431]. Eventually a new mechanism of MR may become dominant

---

[5]This requires to design the geometry of the magnetic electrodes so that their moments can be controlled between reliably antiparallel and parallel configurations.

[6]Other definitions are also used in the literature as, for instance, MR = $[R(AP) - R(P)]/R(AP) \times 100\%$, which in the usual situations has an upper bound of 100%.

[7]The anisotropic MR will be the subject of the next section.

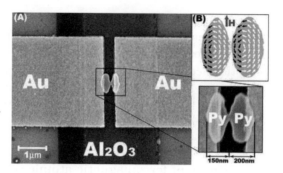

Fig. 12.5 (A) Scanning electron micrograph of a device where gold electrodes are used to contact two permalloy thin-film magnets (inset) on top of an oxidized aluminum gate. (B) Micromagnetic modeling showing antiparallel magnetic alignment across the tunneling gap in an applied magnetic field of H= 66 mT. Reprinted with permission from [438]. Copyright 2006 American Chemical Society.

if a domain wall is sufficiently narrow that the spin of a conduction electron cannot follow the direction of the local magnetization adiabatically [432]. In that case the domain wall can enhance the electron scattering in a similar way to what happens in the giant magnetoresistance effect in magnetic multilayers [433]. For contacts approaching the single-atom diameter regime, values of MR as large as 200% [434] to 100000% [435] have been reported, and ascribed to a "ballistic magnetoresistance" effect involving scattering of electrons from an atomically-abrupt domain wall. However, these large effects have not been reliably observed in well-controlled mechanical break-junctions [404], and it has been argued that the very large changes in resistance are due to the effects of magnetostriction or magnetostatic forces that cause the contact to break and reform as the magnetic field is varied [436].

Finally, if the contact diameter is reduced beyond the single-atom limit, it enters the tunneling regime. MR in that regime reflects the spin polarization of tunneling electrons and it is also expected to depend on the geometry of the contacts [437, 415].

As a representative example, we shall now describe the experiments of Bolotin *et al.* [438]. These authors fabricated two thin-film ferromagnets connected by a small magnetic constriction made of permalloy which can be controllably narrowed by electromigration from about 100×30 nm^2 to the atomic scale and finally to a tunnel junction, see Fig. 12.5. This allowed them to study the MR as the contact region between the two ferromagnets is progressively narrowed in a single sample. One additional

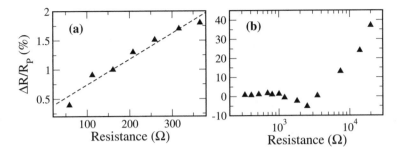

Fig. 12.6   (a) Magnetoresistance as a function of resistance in the range less than 400 Ω (device I). (b) Magnetoresistance as a function of resistance in the range 60 Ω - 15 kΩ (device II). Adapted with permission from [438]. Copyright 2006 American Chemical Society.

advantage of this device geometry is that the magnets are attached rigidly to a non-magnetic substrate with no suspended parts, so that the influence of magnetostriction and magnetostatic forces on the contact are expected to be negligible. Moreover, these experiments were conducted at low temperatures (4.2 K) to have the required thermal stability. Although it cannot be taken for granted that electromigration of alloys would maintain the stoichiometry down to the atomic scale, the magnetic properties seem not to have changed during the final phase of the electromigration process.

Let us now summarize the main findings of this work. When the resistance of a device is low (< 400 Ω), it increases smoothly as electromigration proceeds. The cross-section of the constriction varies from 100 $\times$ 30 nm^2 (60 Ω) to approximately 1 nm^2 (400 Ω). In this regime small ($< 3\%$) positive MR was found which increases as the constriction is narrowed, see Fig. 12.6(a), as expected from the semiclassical theory of Levy and Zhang [432]. In this theory, the resistance of the domain wall scales inversely with its width and the MR ranges typically from 0.7% to 3% for bulk ferromagnets.

In the resistance range from 400 Ω to 25 kΩ, corresponding to a crossover between transport through just a few atoms and tunneling, the value of MR exhibits pronounced dependence on the resistance of the device, see Fig. 12.6(b). The MR has a minimum for resistances above 1 kΩ, and typically changes sign here to give negative values. As the resistance is increased further into the kΩ range, the MR increases gradually to positive values of 10-20%. The observed MR values in the point contact regime are smaller than expected from scaling results of the semiclassical theory [432], which is not surprising since the current is transmitted through just a few channels.

Finally, in the tunnel regime, when the resistance of a device becomes greater than tens of kΩ, MR values in the range from -10% to a maximum of 85% where observed. These large fluctuations clearly indicate that the MR is sensitive to the details of the atomic structure near the tunnel gap. The tunneling current is flowing through just a few atoms on each of the electrodes, and the electronic structure at these atoms does not necessarily reflect the same degree of spin polarization as in the bulk of the ferromagnet.

Experiments like the ones just described raise several basic questions, most of them related to the role of a domain wall scattering in the magnetoresistance of atomic-size contacts and whether or not it can be responsible for the huge MR values reported in some experiments. These questions have been addressed theoretically by numerous authors, see e.g. Refs. [439–441, 409, 410, 413, 412, 414, 415]. In order to elucidate these issues a theory should incorporate three basic ingredients: (i) a proper description of the electronic structure of ferromagnetic atomic contacts, (ii) an adequate description of the domain wall or magnetization profiles that can appear in atomic-scale junctions and (iii) an analysis of realistic atomic geometries. One of the few works that meets these requirements is that of Jacob *et al.* [415] in which the authors studied the magnetoresistance of Ni atomic contacts using ab initio transport calculations. In Fig. 12.7 we reproduce results from this work for the transmission as a function of energy for a single-atom Ni contact for both P and AP orientations. These results were obtained using local spin density approximation (LSDA). In the AP case the self-consistent magnetization reverses abruptly between the tips atoms, i.e. this calculation confirms the possibility of having atomically-abrupt domain walls. However, despite this fact, the MR acquires a moderate value of 23%, which suggests that the domain wall scattering does not account for the large MR in Ni single-atom contacts.

The quantitative result above was found to be very sensitive to the functional used in the DFT calculations, but in no case very large MR values were found. According to the authors, the reason for the moderate MR values is two-fold. First, in the AP configuration the resistance is never too low because of the robust contribution of the *s* orbitals, which is of the order of G_0 for a single-atom contact. Second, in the P configuration the resistance never reaches the minimum value of the ballistic case because, as we saw in the previous section, the transport in these ferromagnetic contacts is not

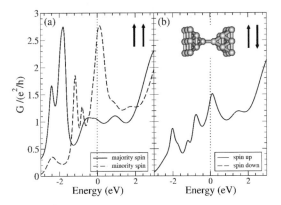

Fig. 12.7 Conductance per spin channel in the P configuration for the model nanocontact shown in the inset calculated with the local spin density approximation. (b) Same as in (a), but for the AP configuration. Reprinted with permission from [415]. Copyright 2005 by the American Physical Society.

really ballistic[8] and the d bands contribute with partially open channels. Another interesting finding of this work is the fact that the MR in atomic contacts can become negative, as in the experiments described above. This shows once more that the usual classical or semiclassical arguments do not apply to the transport in ferromagnetic atomic-size contacts.

12.3   Anisotropic magnetoresistance in atomic contacts

Lord Kelvin discovered in 1857 that the resistivity of bulk ferromagnetic metals depends on the relative angle between the electric current and the magnetization direction.[9] The importance of this phenomenon, known as *anisotropic magnetoresistance* (AMR), was recognized in the 1970's when AMR of a few percent at room temperature was found in a number of alloys based on Fe, Co, and Ni. This fact stimulated the development of AMR sensors for magnetic recording (for reviews on AMR see Refs. [389, 443]).

In the usual AMR effect, the resistivity of a ferromagnetic metal reaches a maximum when the current is parallel to the magnetization direction, $\rho_\parallel$, and a minimum when the current is perpendicular to the magnetization

---

[8]By ballistic transport we mean a situation where all the open conduction channels have a transmission equal to one.

[9]The experimental study of the anisotropic magnetoresistance requires the application a magnetic field high enough to saturate the magnetization of the system.

direction, $\rho_\perp$. The magnitude of AMR can be defined by

$$\text{AMR} = \frac{\rho_\parallel - \rho_\perp}{\rho_\perp}. \qquad (12.5)$$

As a function of the angle, θ, between the current and magnetization, the resistivity of a polycrystalline sample can often be described by

$$\rho(\theta) = \rho_\perp + (\rho_\parallel - \rho_\perp)\cos^2\theta. \qquad (12.6)$$

The origin of AMR stems from the anisotropy of scattering produced by the spin-orbit interaction [444]. The stronger scattering is expected for electrons traveling parallel to magnetization, resulting in larger resistivity $\rho_\parallel$ as compared to $\rho_\perp$ (see Refs. [389, 443, 444] for more details).

As usual, when the dimensions of a metallic wires are shrunk to the atomic scale, its transport properties (including its AMR) are significantly altered. Indeed, inspired by the work of Ref. [445] on Ni contacts, Bolotin *et al.* [446] investigated the AMR of permalloy electromigrated junctions and found that it can be considerably enhanced as compared with bulk samples and that it exhibits an angular dependence that clearly deviates from the $\cos^2\theta$ law of Eq. (12.6). These results are illustrated in Fig. 12.8. In panel (a) one can see, as a reference, the AMR signal of a large device exhibiting the $\cos^2\theta$-behavior. Panel (b) shows the AMR signal of a device as its cross section is reduced approaching atomic dimensions. Notice how the AMR signal progressively deviates from the bulk behavior. Finally, panel (c) shows results for the tunneling regime (when the contacts are already broken) exhibiting an amplitude of more than 10% to be compared with typical amplitudes of the bulk samples of less than 1%.

Additionally, these authors found a significant voltage dependence on the scale of millivolts, which led them to interpret the effect as a consequence of conductance fluctuations due to quantum interference [447]. Independently, Viret and coworkers [448] reported similar results in Ni contacts, but also the occurrence of conductance jumps upon rotation of the magnetization. Similar stepwise variations of the conductance have been found in Co nanocontacts [449], see Fig. 12.9.

These jumps have been interpreted as a manifestation of the so-called *ballistic* AMR (BAMR), a concept that we now proceed to explain. In 2005 Velev *et al.* [450] predicted that the conductance of a ferromagnetic ballistic conductor can change abruptly with the direction of magnetization. This prediction was based on ab initio calculations of the electronic band structure of infinite chains of Ni and Fe. One of those calculations for Ni is reproduced in Fig. 12.10, where one can see the band structure of an

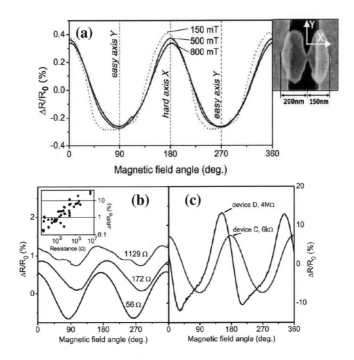

Fig. 12.8 (a) Zero-bias differential resistance vs angle of applied magnetic field at different field magnitudes at 4.2 K, illustrating bulk AMR for a permalloy constriction size of 30×100 nm^2 and resistance $R_0 = 70$ Ω. The inset shows a scanning electron micrograph of a typical device. (b) Evolution of AMR as the device resistance R_0 is increased from 56 to 1129 Ω. (c) AMR for a device with $R_0 = 6$ kΩ exhibiting 15% AMR, and a $R_0 = 4$ MΩ tunneling device, exhibiting 25% AMR. All measurements were made at a field magnitude of 800 mT at 4.2 K. Inset in panel (b): AMR magnitude as a function of R_0 for 12 devices studied into the tunneling regime. Adapted with permission from [446]. Copyright 2006 by the American Physical Society.

infinite chain in the absence of spin-orbit interaction [panel (a)] and in the presence of spin-orbit interaction for magnetizations both parallel to the chain axis [panel (b)] and perpendicular to it [panel (c)]. The key idea is that by rotating the magnetization one can change the number of bands crossing the Fermi energy, E_F. Since in a ballistic conductor the number of bands at E_F is equal to the number of conduction channels (all of them with perfect transparency), this change is reflected in an abrupt change of the corresponding linear conductance. In the particular example of Fig. 12.10, the conductance would change from $6\,e^2/h$ to $7\,e^2/h$ and back upon rotation of the magnetization.

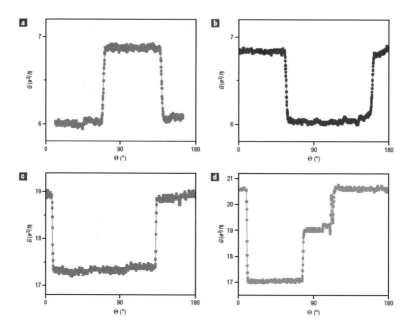

Fig. 12.9   Angular dependence of conductance of Co nanocontacts. (a-d) The angle Θ between the magnetic field and the sample plane changes from 0° to 180°. Results for four different samples exhibiting different sign and magnitude of AMR. Reprinted by permission from Macmillan Publishers Ltd: Nature Nanotechnology [449], copyright 2007.

The results of Fig. 12.9 definitively resemble the expected BAMR behavior described above, although in general the conductance jumps do not occur between quantized values, as can easily be seen in the representative traces However, as we have seen in previous sections, realistic ferromagnetic contacts made of transition metals are not ballistic and thus, the interpretation of the conductance jumps in terms of BAMR is at least questionable. Indeed, Shi and Ralph [451] have suggested that these jumps might originate from two-level fluctuations due to changes in atomic configurations [452].

From the above discussion, one can see that at present there is still a controversy about AMR in atomic contacts, concerning the origin of the enhanced amplitude, the anomalous angular dependence, the occurrence of conductance jumps and the voltage dependence. Different theoretical groups have tried recently to shed new light on this problem. Thus for instance, it has been proposed that the presence of resonant states local-

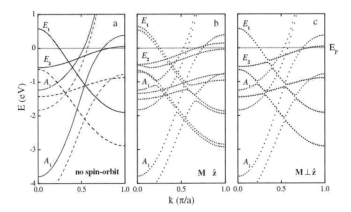

Fig. 12.10 Calculated electronic structure of monoatomic Ni chain with equilibrium interatomic distance in the absence of spin-orbit interaction (a) and in the presence of spin-orbit interaction for magnetization lying along the wire axis **M** $\parallel$ **ẑ** (b) and perpendicular to the wire axis **M** $\perp$ **ẑ** (c). The solid and dashed lines in (a) show the minority-spin and majority-spin bands, respectively. The labels stand for the irreducible representation of the group $C_{\infty v}$ and are displayed for minority-spin bands only. Reprinted with permission from [450]. Copyright 2005 by the American Physical Society.

ized in the electrodes near the junction break could give rise to a strong dependence of the conductance on the magnetization direction [453, 454]. This is an appealing explanation, but as we have seen in section 12.1, the transmission of ferromagnetic contacts is usually very smooth around the Fermi energy on the scale of a few meV. On the other hand, Autes *et al.* [455] have proposed an alternative explanation of the conductance jumps in terms of the existence of giant orbital moments in the contacts.

More recently, Häfner *et al.* [456] have put forward a simple explanation for the anomalous AMR in terms of the reduced symmetry of the atomic contacts as compared with bulk samples. The idea goes as follows. The AMR stems from the scattering between the s and d energy bands induced by the spin-orbit interaction and therefore, the AMR signal may reflect the symmetry of the lattice. In bulk samples the final signal is a result of the average over many impurities, but in the extreme case of an atomic-scale contact, such signal strongly depends on the local geometry. This is particularly clear in the case of a single-atom contact where all the current must flow through a single bond. Thus, it is not so strange to observe, depending on the contact realization, a large amplitude or an anomalous angular dependence as compared with bulk samples. This idea is illustrated in Fig. 12.11, where we reproduce the results of Ref. [456]. Here, one can

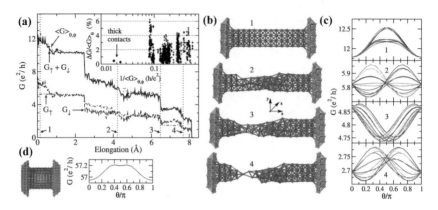

Fig. 12.11 Contact evolution of a Ni junction grown in the fcc [001] direction as obtained from classical molecular dynamics simulations. (a) Spin-projected, $G_{\uparrow,\downarrow}$, and total conductance in the absence of spin-orbit interaction and total conductance averaged over θ, ϕ in the presence of spin-orbit interaction. Vertical lines correspond to the contact geometries in (b). Inset: relative AMR amplitude $\Delta G/\langle G\rangle_\theta = (G_{\max,\theta}(\phi) - G_{\min,\theta}(\phi))/\langle G(\theta,\phi)\rangle_\theta$ vs. inverse averaged conductance. (c) Conductance vs. θ for the geometries in (b) and with ϕ in steps of $\pi/6$. (d) Same as (c) for the thick contact with 324 atoms shown in this panel. Reprinted with permission from [456]. Copyright 2009 by the American Physical Society.

see the evolution of the conductance and AMR signal[10] as a function of the polar angle θ and azimuthal angle ϕ during the formation of a Ni atomic contact. This formation was simulated by means of molecular dynamics (see Ref. [456] for technical details).

In Fig. 12.11(d) one can see that in the limit of thick contacts, these model calculations recover the bulk behavior with an AMR amplitude of 0.45%. However, in the case of small contacts, one can observe clear deviations from the $\cos^2 \theta$-behavior and an enhancement of the amplitude [see Fig. 12.11(a-c)]. Notice in particular that the signal in this case also depends strongly on the azimuthal angle ϕ, contrary to the bulk case. Finally, the statistical analysis of the data of these simulations reveals strong fluctuations in the AMR signal and an increase to 2% on average in the last steps before breaking, see inset of Fig 12.11(a).

On the other hand, in the analysis of these realistic geometries, Häfner *et al.* [456] did not find signs of BAMR or the presence of pronounced resonances in the local density of states of the electrodes. These authors argued finally that the voltage dependence observed in the experiments

---

[10]In this case the AMR signal is defined in terms of the conductance, see caption of Fig. 12.11.

of Refs. [446, 452] could be a result of the combination of the intrinsic anomalous AMR of atomic contacts with conductance fluctuations originating from impurities near the contacts [447].

12.4 Concluding remarks and open problems

As we have seen in the previous sections, the ferromagnetic atomic contacts exhibit a very rich phenomenology. Moreover, in spite of progress made in recent years, the level of understanding of the transport effects in these systems is not comparable to the corresponding one in non-magnetic contacts. With respect to the issues addressed in this chapter, there is an increasing evidence (both experimental and theoretical) that the electronic transport in these nanowires is not ballistic and phenomena like conductance quantization are not expected. Different theories show consistently that in ordinary ferromagnetic metals, like Fe, Co or Ni, the d bands give an important contribution to the transport, but the conduction channels originating from these states are in general only partially open.

While the results for the properties of ferromagnetic atomic contacts in the absence of field seem to be converging, there is not yet a similar consensus about the magnetoresistive effects. With respect to MR, it is becoming clear that the huge values reported in some of the first experiments are most likely due to magnetostriction. However, more controlled experiments are needed to establish the values of the MR for different materials as a function of the different system parameters (contact size, field, temperature, etc.). From the theory side, more work is required to elucidate the questions related to the existence and properties of domain walls in these systems, as well as their influence in the transport characteristics. In some cases, very sophisticated calculations have been performed for academic geometries like atomic chains, which often results in misleading conclusions that do not apply to the systems explored experimentally.

The situation is very similar in the case of AMR of these magnetic contacts. New experiments are needed to, in particular, identify the origin of the abrupt steps observed in some experiments in the conductance as a function of the angle between the magnetization and the current direction. More theoretical calculations for realistic geometries are highly desirable to find out the origin of the anomalous amplitude and angular dependence of the AMR in these systems.

Finally, let us say the phenomena described in this chapter constitute the starting point for the investigation the spin-dependent transport through single-molecule junctions. As we shall see in the next part of book, experiments of that kind have been already reported. Some of them are exploring the spin injection in molecules with the use of ferromagnetic electrodes, while others investigate how the molecular magnetism is reflected in the transport properties of molecular junctions with non-magnetic electrodes.

PART 4

Transport through molecular junctions

Chapter 13

Coherent transport through molecular junctions I: Basic concepts

As we have just seen in Part 3, the level of understanding achieved in the field of metallic atomic-size contacts is certainly remarkable. However, it is also clear that such metallic nanowires are not very "flexible" in many respects. Thus for instance, their conductance can hardly be changed with a gate voltage and often their current-voltage characteristics are simply linear, which hinders the implementation of interesting electronic functionalities. Thus, it seems natural to investigate the use of molecules as possible building blocks of nanoscale circuits. Molecules are still small enough to take advantage of their size, and the great variety of their physical properties make them ideal not only to mimic ordinary components of today's microelectronics, but also to provide new electronic functions.[1] For these reasons, the analysis of the transport properties of molecular junctions is attracting a lot of attention and this will be the subject of the rest of this book.

The study of the transport properties of molecular junctions constitutes a formidable challenge. As we discussed in Chapter 3, there are still many basic problems to be solved from the experimental side: reproducibility of the results, stability of the contacts, external control, mass production, etc. On the other hand, the theoretical description of the electrical conduction in molecular circuits is, in general, considerably more complicated than in the case of atomic wires for various reasons. First, a molecule has a more complicated electronic structure than an atom, simply because it is composed of several atoms of, in general, different species. The accurate description of the interaction between those atoms which leads, among other things, to substantial charge transfer between them requires sophis-

---

[1]The basic properties of the main molecules explored so far in molecular electronics, as well as their possible functionalities, are described in section 3.2.

ticated ab initio methods.[2] Second, in the case of a molecular junction, a molecule may have a weak chemical interaction with the metallic electrodes, which implies that the charge carriers can spend a long time in the molecule. This may in turn lead to the appearance of correlation effects well-known in mesoscopic physics such as the Coulomb blockade or the Kondo effect. Third, molecules possess internal degree of freedom, in particular vibrations modes, which can be excited by the transport electrons leading to a modification of the current-voltage (I-V) characteristics. Obviously, the probability to excite a vibration depends on various factors like the strength of the electron-vibration interaction, the quality of metal-molecule interfaces and, of course, the length of the molecule. Depending on this latter factor, the vibrations can produce weak signals in the I-V curves in the case of short molecules or they can completely dominate the transport characteristics like in long DNA strands. Fourth, a molecule can undergo conformation changes due to, for instance, the high electric fields applied in the contacts, mechanical stress, an external field (electromagnetic radiation) or the local environment (red-ox reactions).

Due to the very rich phenomenology of molecular transport junctions, it is not an easy task to organize the existent material in the literature. Since this is not merely a review, we shall not follow a chronological order. Instead, we find didactic to organize the huge amount of results concerning the physics and chemistry of molecular junctions according to the dominant transport mechanism. Thus, we shall first discuss the coherent transport through molecular wires, in which electrons flow elastically through the molecules without exchanging energy. The main goal in this case is to understand the relation between the electronic structure of individual molecules and the transport properties of the junctions in which they are embedded. This discussion will be divided into two parts. In the first one, which is covered in this chapter, we shall discuss several coherent transport phenomena which can be understood in the light of simple toy models or handwaving arguments. These phenomena cover issues like the shape of the current-voltage characteristics, their temperature dependence, their symmetry or the dependence of the conductance on the length of the molecules. Then, in the Chapter 14, we shall address similar issues, but

---

[2]Let us remind the reader at this stage that empirical methods like extended Hückel and its descendents have played a fundamental role in quantum chemistry, but one cannot expect these methods to give quantitative answers to the key questions in molecular electronics, such as the position of the molecular levels, hybridization with the extended states of the metallic leads, metal-molecule charge transfer, etc.

this time from a more quantitative point of view. In particular, we shall discuss the transport through short molecules which serve as test-beds for molecular electronics and we shall try to establish to what extend their electronic transport properties are quantitatively understood at present.

In Chapter 15 we shall discuss the transport through weakly coupled molecules, where correlation effects such as the Coulomb blockade and the Kondo effect play an essential role. Then, we present in Chapters 16 and 17 a thorough discussion of the role of vibration modes in the current through short molecules, while the incoherent or hooping transport regime in long molecules will be deferred until Chapter 18. Chapter 19 is devoted to the analysis of transport properties (different from the electrical conductance) that provide very valuable information about the transport in different types of junctions. In particular, we address in that chapter the thermal transport in molecular circuits. In Chapter 20 we shall discuss the optical properties of current-carrying molecular junctions. Finally, in Chapter 21 we shall briefly mention some of the topics in molecular electronics that are not addressed in this monograph.

We want to stress that, as the previous part of this book, the remaining chapters have bee written in such a way that most sections are accessible for both theorists and experimentalists. Our main goal has been to give a didactic introduction to the basic concepts in molecular electronics, but at the same time we have made an effort to review the most relevant contributions to the different topics in this field. Let us finally say that, as stated in the introductory part of this book, we shall mainly focus our attention on single-molecule junctions.

13.1   Identifying the transport mechanism in single-molecule junctions

As explained in the introduction, in this chapter we want to discuss the coherent transport through molecular junctions. Let us stress that by coherent transport (or tunneling) we mean the transport regime in which the information about the phase of the wavefunction of conduction electrons is preserved along the molecular bridges and the inelastic interactions take only place well inside the electrodes.[3] The first question that we want to

---

[3]In Chapter 4 we have presented an introduction to the scattering approach, which is the most popular and appealing theoretical formalism for the description of phase-coherent transport in nanoscale junctions. If you are not familiar with this approach, we recommend you to read that chapter at least up to section 4.4.

Table 13.1   Possible conduction mechanisms. Here, J is the current density, V is the bias voltage, φ_B is the barrier height, d is the barrier length and T the temperature.

Conduction mechanism	Characteristic behavior	Temperature dependence	Voltage dependence
Direct tunneling	$J \sim V \exp\left(-\frac{2d}{\hbar}\sqrt{2m\varphi_B}\right)$	none	$J \sim V$
Fowler-Nordheim tunneling	$J \sim V^2 \exp\left(-\frac{4d\sqrt{2m}\varphi_B^{3/2}}{3q\hbar V}\right)$	none	$\ln(\frac{J}{V^2}) \sim \frac{1}{V}$
Thermionic emission	$J \sim T^2 \exp\left(-\frac{\varphi_B - q\sqrt{qV/4\pi\epsilon d}}{k_B T}\right)$	$\ln\left(\frac{J}{T^2}\right) \sim \frac{1}{T}$	$\ln(J) \sim V^{1/2}$
Hopping conduction	$J \sim V \exp\left(-\frac{\varphi_B}{k_B T}\right)$	$\ln\left(\frac{J}{V}\right) \sim \frac{1}{T}$	$J \sim V$

address is: How do we know that the transport in a particular junction is coherent? Or more generally, how can we identify the transport mechanism from the experimental results? There is no unique answer to these questions, but certainly both the shape of the I-V characteristics and, specially, their temperature dependence are very useful in this respect. Following the instructive work of Reed's group (see Ref. [130]), we list in Table 13.1 some possible conduction mechanisms along with their characteristic temperature and voltage dependence of the current.[4] This list is by no means complete and some other mechanisms will be discussed in later chapters, but it constitutes a good starting point. The mechanisms listed in Table 13.1 have been extensively studied in the context of metal tunnel junctions and semiconductor devices.

The first two conduction mechanisms, direct tunneling and Fowler-Nordheim tunneling, are two manifestations of coherent tunneling through a potential barrier. The explicity voltage and temperature dependence are taken from the Simmons model that we discussed in detail in section 4.4. Direct tunneling refers to what happens at low bias, when the voltage is much smaller than the barrier height, whereas Fowler-Nordheim tunneling occurs when voltage is larger than the average barrier height and it is similar to field emission. Both mechanisms (or regimes, to be more precise)

[4]This list was adapted from Ref. [457].

have in common that the I-V's are rather insensitive to temperature and they only differ in the voltage dependence.

The third mechanism, thermionic emission, is a process that takes place when the electrons are excited over a potential barrier, as opposed to tunneling through it. This clearly has a very strong temperature dependence, and will become significant when the potential barrier is relatively small. Notice that, strictly speaking, thermionic emission is also a coherent mechanism since the electrons proceed elastically through the barrier without losing their phase memory.

Hopping conduction is a mechanism in which electrons are localized at certain points within the molecule, and can hop between those points. This will also be a thermally activated process.[5] This mechanism dominates the transport properties of long molecules, except in some remarkable cases such as carbon nanotubes.

Based on whether thermal activation is involved, the conduction mechanisms fall into two distinct categories: (i) thermionic or hopping conduction, which has temperature-dependent I-V characteristics, and (ii) direct tunneling or Fowler-Nordheimer tunneling, which does not exhibit temperature-dependent I-V curves. According to this slightly oversimplified discussion, one can conclude that if the I-V curves are temperature independent, the dominant conduction mechanism is (coherent) tunneling. Moreover, the transport regime can be discriminated by the analysis of the shape of the I-V characteristics. It is important to recall that most experimental techniques, especially those designed to work with single molecules, are not suitable for temperature-dependent measurements. Thus, it may not be easy to carry out the test proposed above to elucidate the transport mechanism.

The working principle stated in the previous paragraph has been used in many different investigations to establish the conduction mechanism. In Fig. 13.1 one can see an example taken from Ref. [130]. In this case, the authors studied the transport through thiolated alkanes of different length using the nanopore technique (see section 3.5.1). In this experiment the transport through a self-assembled monolayer (SAM) was investigated. Although our main interest is on single-molecule junctions, this experiment is specially illustrative and it will be used several times in this chapter. As one can see in Fig. 13.1(b,c), the current is rather insensitive to the temperature and thus it was concluded that the conduction mechanism

---

[5]This transport mechanism will be discussed in detail in Chapter 18.

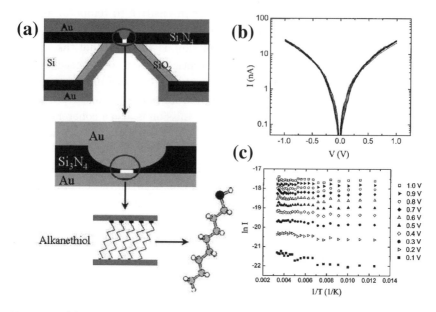

Fig. 13.1   (a) Schematics of a nanometer-scale device used in the experiments [130]. The structure of octanethiol is shown as an example. (b) Temperature-dependent I-V characteristics of dodecanethiol (C12). I-V data at temperature from 300 to 80 K with 20 K steps are plotted on a logarithmic scale. (c) Arrhenius plot generated from the I-V data in panel (b), at voltages from 0.1 to 1.0 V with 0.1 V steps. Reprinted with permission from [130]. Copyright 2003 by the American Physical Society.

through alkanethiols is tunneling, i.e. the electronic transport is coherent.

Once it has been established that coherent tunneling is the dominant transport mechanism, one can use, for instance, the Simmons model to understand the shape of the I-V characteristics. As we explained in section 4.4, see in particular Eq. (4.17), the current in this model is given in terms of different parameters like the electron mass, m, the barrier width, d, the barrier height, φ_B, and a dimensionless parameter called α. This parameter is of the order of 1 for a rectangular barrier and bare electron mass. It is sometimes used as a fitting parameter to account for the possibility of non-rectangular barriers or an effective mass, m^*, different from the bare electron mass. Eq. (4.17), with φ_B and α as adjustable parameters, was used in Ref. [130] to fit the I-V curves of different thiolated alkanes. An example of such fits is shown in Fig. 13.2, where the I-V curve of a dodecanethiol (C12) was fitted. The best nonlinear least-square fitted was performed with $\varphi_B = 1.42$ eV and $\alpha = 0.65$.

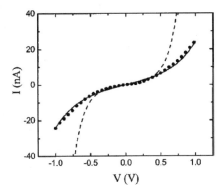

Fig. 13.2 Measured current-voltage characteristics (circular symbols) of a nanopore junction with dodecanethiols compared with calculations based on Simmons model (solid line) using $\varphi_B = 1.42$ eV and $\alpha = 0.65$. The calculated I-V from a simple rectangular model ($\alpha = 1$) with $\varphi_B = 0.65$ eV is also shown as dashed curve. Reprinted with permission from [130]. Copyright 2003 by the American Physical Society.

In spite of the quality of the fit shown in Fig. 13.2, there are a few things that are not very satisfactory. First, an attempt to fit the results with a rectangular barrier fails to describe the high-bias regime, see dashed line in Fig. 13.2. This conclusion has been drawn in several analyses of the transport through alkanethiol [458, 459]. This is the reason why α was used above as an adjustable parameter, although its physical meaning is not really clear. Second, the value obtained for the barrier height is certainly small as compared with the expectations. This height, φ_B, is in principle the distance between the Fermi energy of the electrodes and the nearest molecular energy level in the molecule. For the combination of Au contacts and alkanes, this distance is expected to be between 4 and 5 eV [299]. A possible way out for these problems has been pointed out by Akkerman *et al.* [460]. These authors have shown that the description of the transport through SAMs of alkenedithiols can be improved by including the effect of image charges in the Simmons model (see brief discussion of the role of image charges in section 4.4). They were able to describe the transport in their experiments up to 1 V by using a single effective mass and a barrier height. The barrier heights found were in the order of 4-5 eV and, irrespective of the length of the molecules, an effective mass of $0.28\ m$ was determined in agreement with theoretical predictions [299].

Simmons model has been used in many other examples in molecular electronics to interpret the observed I-V characteristics. For instance, another beautiful example can be seen in Ref. [461], where the authors used this model to explain the measured I-V curves in metal-molecule-metal junctions formed from π-conjugated thiols, which were consistent with a change in transport mechanism from direct tunneling to field emission.

Tunneling models, like Simmons one, borrowed from the field of metallic tunnel junctions and semiconductor devices will continue to play an important role in molecular electronics. However, their use is at least questionable. For instance, one may argue that one should use at least a double-barrier model to describe a metal-molecule-metal junction since we have two interfaces. Of course, such models are available, as we showed in Chapter 4. However, one could still argue that the bound states of a simple double-barrier structure do not necessarily resemble those of a molecule. One could go on trying to refine even further such barrier models, but it seems more natural to use models that already incorporate the molecular features right from the start. This is precisely the strategy that we are going to follow in the rest of this chapter, where we shall introduce simple molecular-based models to describe the transport in molecular junctions. In particular, we shall start in the next section by studying the main conclusions that can be drawn from the simple resonant tunneling model.

13.2 Some lessons from the resonant tunneling model

When the coherent transport through a metal-molecule-metal contact is discussed, one typically thinks of the molecular orbitals of the molecule within the junction. These orbitals are occupied up to the highest occupied molecular orbital (HOMO), which for a characteristic molecule could be roughly -7 eV. This has to be compared with the Fermi level of the metal, which for a noble material is around -5 eV.[6] Due to the interaction between the molecule and the metal electrodes, some charge flow, charge rearrangements, and geometric reorganization will occur. After this process, the simplest viewpoint is expressed by the level scheme depicted in Fig. 13.3(a). Here, the Fermi energy of the electrodes lies somewhere within the HOMO-LUMO (lowest unoccupied molecular orbital) gap of the molecule. Moreover, due to the hybridization of the molecular orbitals and the metallic states, the former ones acquire a finite broadening that depends on the strength of the metal-molecule coupling, i.e. the original molecular states have now a finite lifetime.

---

[6]These energies are measured with respect to the vacuum level, which is set to zero.

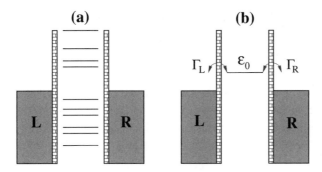

Fig. 13.3   (a) Level scheme of a molecular junction. The molecule has a series of sharp resonances corresponding to the different molecular orbitals, whereas the metal possess a continuum of states that is filled up to the Fermi energy of the metal. (b) The same as in panel (a) for a situation where the transport is dominated by a single level, ϵ_0.

In principle, different molecular orbitals can participate in the electron transport simultaneously. However, there are many situations where one level (HOMO or LUMO) lies closest to the Fermi level of the metal and therefore dominates the transport in a certain voltage range. In this case, the situation is better represented by the scheme of Fig. 13.3(b). This is precisely the situation that we will be considering throughout this section. Such situation can be described with the (single-level) resonant tunneling model considered, for instance, in section 7.4.1.[7] In this model, the level position is denoted by ϵ_0 and we measure it with respect to the Fermi energy of the electrodes, which we set to zero. At finite bias, this position depends on the voltage applied across the junction (and on the way the voltage drops at the interfaces) and to indicate it explicitly we shall write $\epsilon_0(V)$. The other key parameters of this model are the scattering rates $\Gamma_{L,R}$, which describe the strength of the coupling to the metal electrodes (L, R). These parameters have dimensions of energy and they determine the lifetime or broadening of the resonant level. Such broadening, to be precise the half-width at half-maximum, is simply given by $\Gamma = \Gamma_L + \Gamma_R$. The different parameters of the model will be considered as phenomenological parameters, but they could in principle be obtained from a fit to the experimental results or they can be calculated from ab initio methods.

As we have seen in the previous chapters, see Chapter 4 and section

---

[7]The word "resonant" in the name of this model is maybe a bit misleading since it may suggest that the transport takes place on resonance. This is actually not the case and, as we shall see in this section, this model describes in a unified manner different regimes within the coherent tunneling picture.

7.4.1, following the spirit of the Landauer approach, the I-V characteristics in this model can be computed from the following expression

$$I(V) = \frac{2e}{h} \int_{-\infty}^{\infty} dE \, T(E, V) \left[f(E - eV/2) - f(E + eV/2) \right], \qquad (13.1)$$

where the factor 2 is due to the spin symmetry of the problem, $f(E)$ is the Fermi function and $T(E, V)$ is the energy- and voltage-dependent transmission coefficient given by the Breit-Wigner formula

$$T(E, V) = \frac{4\Gamma_L \Gamma_R}{[E - \epsilon_0(V)]^2 + [\Gamma_L + \Gamma_R]^2}. \qquad (13.2)$$

Here, the scattering rates are assumed to be energy- and voltage-independent. This assumption can be easily relaxed, but it is usually a good approximation for noble metals like gold with a rather flat density of states around the Fermi energy. Notice also that we assume that the voltage is applied symmetrically between the left and right electrode. Obviously, this is irrelevant and the current only depends on the different of the chemical potentials. The previous simple expressions will be our starting point to discuss a few basic issues in the next subsections.

13.2.1   *Shape of the I-V curves*

The first obvious issue to be discussed is the shape of the I-V characteristics. Let us assume for the moment that the voltage drops symmetrically in both interfaces and therefore $\epsilon_0(V) = \epsilon_0$. This is the situation expected when the molecule is equally coupled to both electrodes ($\Gamma_L = \Gamma_R$). In the zero-temperature limit the integral of Eq. (13.1) can be done analytically and the current adopts the following form

$$I(V) = \frac{2e}{h} \frac{4\Gamma_L \Gamma_R}{\Gamma} \left[\arctan\left(\frac{eV/2 - \epsilon_0}{\Gamma} \right) + \arctan\left(\frac{eV/2 + \epsilon_0}{\Gamma} \right) \right], \quad (13.3)$$

where $\Gamma \equiv \Gamma_L + \Gamma_R$. From this expression one can see that the current at sufficiently large voltages saturates to a value given by $I_{\text{sat}} = (2e/h)4\pi\Gamma_L\Gamma_R/\Gamma$, which one can show to be independent of the temperature. This simple result illustrates how the scattering rates determine the order of magnitude of the current.

In order to have an idea about how the I-V curves looks like, we show in Fig. 13.4 the current vs. bias voltage and the corresponding differential conductance ($G = dI/dV$) for different values of the scattering rates (symmetric situation), a level position of $\epsilon_0 = 1$ eV and room temperature. Notice that the current is symmetric with respect to voltage inversion and it

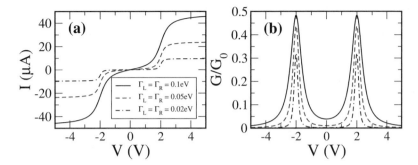

Fig. 13.4 (a) Current vs. bias voltage in the resonant tunneling model for a level position $\epsilon_0 = 1$ eV (measured with respect to the Fermi energy of the electrodes) and at room temperature ($k_B T = 0.025$ eV). The different curves correspond to different values of the scattering rates that are assumed to be equal for both interfaces. (b) The corresponding differential conductance $G = dI/dV$ normalized by $G_0 = 2e^2/h$.

has a characteristic shape where one can distinguish three different regions. We focus on the positive bias part. The first region is at low bias, when the voltage is much smaller than $|\epsilon_0|$, see Fig. 13.5(a). In this case the current is quite low, specially if Γ is rather small. The second region is defined by the resonant condition: $eV/2 = \epsilon_0(V)$, i.e. $eV = 2\epsilon_0$, see Fig. 13.5(b), where the level is aligned with the chemical potential of one of the electrodes. Here, when the voltage approaches this condition, the current is greatly enhanced. Finally, when the voltage is larger than $2|\epsilon_0| + \Gamma$, the current saturates to the value given by I_{sat} obtained above, see Fig. 13.5(c).

As one can see in Fig. 13.4(b), the corresponding differential conductance, $G = dI/dV$, exhibits two peaks at the resonant conditions

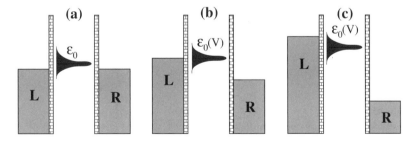

Fig. 13.5 Voltage dependence of the level alignment in the resonant tunneling model for symmetric coupling. (a) Zero bias region, (b) resonant situation where the level is aligned with the chemical potential of one of the electrodes and (c) large bias region where the current saturates. The level has a finite broadening given by $\Gamma = \Gamma_L + \Gamma_R$.

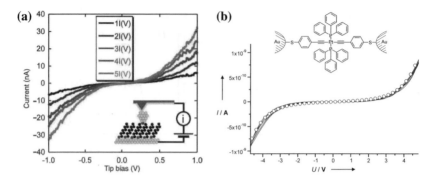

Fig. 13.6   (a) I-V curves of octanedithiol-based junctions measured with the appara-
tus shown in the inset. Here octanedithiol molecules are placed inside a monolayer of
octyl chains on top a gold surface. The molecules are contacted with a gold nanoparticle, which in turn is contacted by the gold tip of a conducting AFM. From [125].
Reprinted with permission from AAAS. (b) I-V curves (solid lines) measured in molecular junctions formed with the break-junction technique at room temperature where a
trans-platinum(II) complex is contacted with gold electrodes (see inset). The curves
were fitted with a model for a rectangular barrier of height 2.5 eV (circles). Reproduced
with permission from [462]. Copyright Wiley-VCH Verlag GmbH & Co. KGaA.

$eV = \pm 2\epsilon_0$. The width of these peaks is determined by the largest energy scale between Γ and $k_B T$. In the example of Fig. 13.4, the width is
mainly determined by Γ and the conductance at low temperatures would
reach a value close to G_0 at the resonant conditions. Then, in the plot of
the conductance vs. bias voltage one can read off at low temperatures the
parameter Γ, which determines the strength of the metal-molecule coupling.

13.2.2   *Molecular contacts as tunnel junctions*

In Fig. 13.2 one can see an example of the I-V characteristics of a molecular junction that resembles those that typically are reported in tunnel
junctions. These type of curves are encountered quite frequently in the literature and we show two more examples in Fig. 13.6. The first one, see panel
(a), was obtained by measuring the current through an alkane thiol ad-layer
with a gold cluster attached to a conductive AFM tip [125]. In the second
example, see panel (b), the current through a trans-platinum complex was
measured making use of the microfabricated MCBJ technique [462]. As
discussed in the previous section, this type of curves can be described with
standard tunneling models. Indeed, in the example of Fig. 13.6(b), the
authors were able to fit quite well the I-V curves using the model of a
rectangular potential barrier of height 2.5 eV.

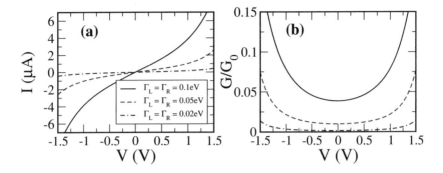

Fig. 13.7 The same as in Fig. 13.4 for low bias ($|eV| < \epsilon_0(V)$).

Such tunneling curves can also be described with the resonant tunneling model. If in Fig. 13.4 we focus on the low bias regime, i.e. before the resonant condition is reached, one obtains the current and conductance shown in Fig. 13.7. The similarity with the experimental curves is rather obvious and by adjusting the parameters ϵ_0 and Γ, one can in principle fit those I-V curves. Anyway, it is important to emphasize that what it is usually called a tunnel-like curve is nothing but a cubic function of the form: $I(V) = AV + BV^3$, where A and B are constants. Almost any tunneling model that produces symmetric I-V curves gives rise to such a voltage-dependence at low bias and therefore it is suitable for fitting the I-V characteristics in this regime.[8] For this reason, if the I-V curves have no much structure, one must be careful in interpreting the fits and one should make sure that the values of the parameters obtained from the fits are sensible.

13.2.3 *Temperature dependence of the current*

As we discussed in the previous section, the temperature dependence of the current is a key issue for identifying the transport mechanism. In particular, we concluded that temperature-independent I-V curves are a signature of coherent tunneling. In this subsection we shall show that this conclusion is basically supported by the resonant tunneling model, although we shall show that coherent tunneling can also give rise to temperature-dependent I-V curves.

Quite generally, if the transport is coherent, the Landauer formula, see

---

[8]We showed in section 4.4 that the I-V curves in Simmons model has this cubic dependence in an intermediate voltage range.

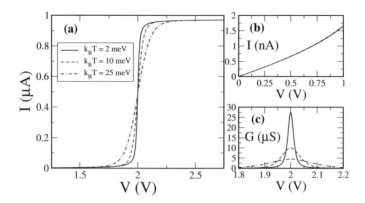

Fig. 13.8 (a) Current-voltage characteristics computed with the resonant tunneling model for different temperatures. The parameter values are: $\Gamma_L = \Gamma_R = 2$ meV and $\epsilon_0 = 1$ eV. (b) Blow-up of the low bias region. Notice that the current is independent of the temperature. (c) The corresponding differential conductance vs. voltage.

Eq. (13.1), tell us that the temperature dependence of the current or of the conductance is determined by the energy dependence of the transmission coefficient, which is usually not very pronounced. Thus, the temperature dependence in the coherent regime, if any, is typically a power law, which is clearly at variance with, for instance, the exponential behavior in the incoherent hopping regime that takes place in very long molecules. In the particular case of the resonant tunneling model, it is easy to see that if the transmission is fairly energy-independent in the energy window controlled by the voltage, then the current is insensitive to the temperature. This is precisely what occurs at low bias when the level lies well above (or below) the equilibrium Fermi energy of the system. Therefore, we can conclude that the current (and also the conductance) is temperature independent in an off-resonant situation.

The situation changes when the transport takes place on resonance. In this case, if the temperature is comparable or larger than Γ, the current depends on temperature. This is illustrated in Fig. 13.8 where we show the I-V curves and the corresponding differential conductance for temperatures larger than the width of the resonance. As one can see, the current and conductance depend on temperature for voltages around the resonant condition, while at low bias they are insensitive to its value, see Fig. 13.8(b).

To be more precise, let us now study the temperature dependence of the conductance in the linear regime. From Eqs. (13.1) and (13.2), one can show that the linear conductance is given by

$$G(T) = \left(\frac{2e^2}{h}\right) \frac{1}{4k_BT} \int_{-\infty}^{\infty} dE \left[\frac{4\Gamma_L\Gamma_R}{(E - \epsilon_0)^2 + \Gamma^2}\right] \frac{1}{\cosh^2(\beta E/2)}, \quad (13.4)$$

where $\beta = 1/k_BT$. There are two limiting cases in which we can get a simple analytical expression. First, if we are in an off-resonant situation, where $|\epsilon_0| \gg \Gamma, k_BT$, then the conductance is temperature independent and it is given by

$$G = \left(\frac{2e^2}{h}\right) \frac{4\Gamma_L\Gamma_R}{\epsilon_0^2}. \quad (13.5)$$

On the other hand, in a weak coupling situation ($\Gamma \ll k_BT$) the linear conductance can be expressed as

$$G(T) = \left(\frac{2e^2}{h}\right) \frac{\pi\Gamma_L\Gamma_R}{\Gamma} \frac{1}{k_BT \cosh^2(\beta\epsilon_0/2)}. \quad (13.6)$$

This means that in this limit the conductance increases as the temperature decreases. Such temperature dependence is illustrated in Fig. 13.8(c).

13.2.4   *Symmetry of the I-V curves*

The symmetry of the I-V characteristics with respect to voltage inversion has played a prominent role in the history of molecular electronics. As we discussed in section 1.2, Aviram and Ratner suggested in their seminal paper [8] that a single molecule with a donor-spacer-acceptor structure would behave as a diode when placed between two electrodes.

Rectifying behavior was already observed in 1990 and 1993 by two groups using a monolayer of hexadecylquinolinium tricyanoquin-odimethanide sandwiched between dissimilar metal electrodes (magnesium and platinum) [463, 464] and then confirmed later in 1997 and 2001 by Metzger and coworkers, who used identical metals (first aluminum, then gold) [14, 465, 466]. These papers use Langmuir-Blodgett monolayers (one molecule thick), with maybe 10^{14} to 10^{15} molecules measured in parallel. About nine similar rectifiers of vastly different structure have been found by Metzger's group between 1997 and 2006 [467]. Rectification has also been studied at the level of single-molecule contacts, see for instance Ref. [468].

Let us see now what the resonant tunneling model can teach us about the symmetry of the I-V characteristics. This model suggests that a possible rectification mechanism is related to the voltage profile across the junction. Let us consider an asymmetric situation, where the molecule is differently

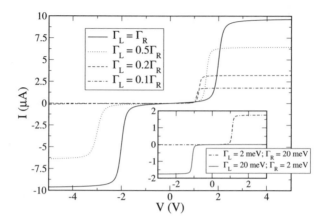

Fig. 13.9   Current-voltage characteristics in the resonant tunneling model for an asymmetric situation for $\epsilon_0 = 1$ eV, $\Gamma_R = 20$ meV and at room temperature ($k_BT = 25$ meV). The different curves correspond to different values of the left scattering rate. The inset shows very asymmetric situations where the scattering rates have bee interchanged. Notice the that the *I-V* curves exhibit a clear rectification behavior.

coupled to the left and right electrodes. If the scattering rates Γ_L and Γ_R are different, it is reasonable to assume that the voltage drops at the interfaces accordingly to the ratio of the scattering rates. This can be simply modeled by assuming that the voltage dependence of the level position is of the form: $\epsilon_0(V) = \epsilon_0 + (eV/2)(\Gamma_L - \Gamma_R)/\Gamma$. This expression simply reflects the fact that if one of the rates is much greater than the other, the level follows the shift of the chemical potential of the electrode that is better coupled.

With this simple model, we can now compute the I-V curves and an example is shown in Fig. 13.9. Here, the different curves correspond to different values of the ratio Γ_L/Γ_R. As we can see, when this ratio clearly differs from one, the I-V curves become very asymmetric and the desired rectification behavior becomes apparent. Notice that the polarity of the curves can be controlled by exchanging the values of the scattering rates in an asymmetric situation, as it is shown in the inset of Fig. 13.9.

It is easy to understand the shape of the I-V curves in Fig. 13.9. For instance, if we focus on the situation where $\Gamma_L \ll \Gamma_R$, the level is shifted with the bias as $\epsilon_0(V) = \epsilon_0 - eV/2$, i.e. it follows the chemical potential of the right electrode. Then, the resonant condition is reached for positive voltages when the Fermi energy of the left electrode is aligned with the level, i.e. when $eV/2 = \epsilon_0 - eV/2$, which implies $\epsilon_0 = eV$. For negative voltages, since the level follows the right electrode, the resonant condition is never reached and then the current for this polarity is much lower than for positive voltages. These arguments explain the curve in Fig. 13.9 for $\Gamma_L = 0.1\Gamma_R$. Using similar arguments, one can easily explain the other curves in this figure.

It is worth pointing out that the asymmetry in the coupling can be due to extrinsic factors, like a different coupling between left and right due to an asymmetric configuration of the molecular junction, or it can be due to something intrinsic, like the geometry of the molecule under investigation. Thus for instance, an asymmetric molecule has molecular orbitals with an asymmetric charge distribution. This induces a different coupling with the electrodes, which can lead in turn to an asymmetric voltage profile. In both cases, the final result is the observation of asymmetric I-V curves. For an illustrative experimental example, we refer to the reader to Ref. [469].

13.2.5   *The resonant tunneling model at work*

After the extensive discussion of the previous subsections about the transport characteristics that can be deduced from the resonant tunneling model, the reader may be wondering whether this model actually works. The purpose of this subsection is to show that indeed it does.

The resonant tunneling model has been used by several authors to describe the experimental results in different types of molecular junctions. Thus for instance, Grüter *et al.* [470] used this model to obtain information about the tunneling rates in the transport through thiolated C_{60} molecules in a liquid environment. As we shall discuss in detail in section 18.2, this model was used by Poot *et al.* [471] to describe successfully the temperature dependence of I-V characteristics of three-terminal devices containing individual tercyclohexylidene molecules.

More recently, Zotti *et al.* [472] have shown that the I-V curves of single tolane molecules attached to gold electrodes via different anchoring groups can be accurately fitted with the resonant tunneling model. In Fig. 13.10 we show typical examples of those I-Vs for three different molecules together with the corresponding fits to this model. The curves in this figure correspond to symmetric I-Vs, but also asymmetric curves were fitted using the ideas of the previous subsection. It is worth mentioning that these I-V curves could not be so accurately described with other models like the Simmons one. On the other hand, as one can see in the figure caption, the values of the scattering rate ($\Gamma = \Gamma_L = \Gamma_R$) vary depending on the

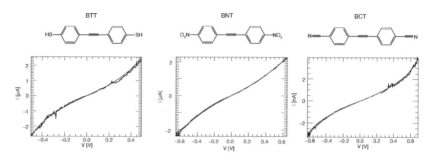

Fig. 13.10   I-V curves of tolane-based molecules junctions measured with the micro-fabricated MCBJ technique at room temperature and under liquid environment [472]. The molecules investigated are shown in the upper part: 4,4′-bisthiotolane (BTT), 4,4′-bisnitrotolane (BNT) and 4,4′-biscyanotolane (BCT). The black lines in the different panels correspond to the experimental results, while the lighter lines are the fits to the resonant tunneling model. The parameters used in the fits of these symmetric curves ($\Gamma = \Gamma_L = \Gamma_R$) are: $\Gamma = 42$ meV and $\epsilon_0 = 404$ meV for BTT, $\Gamma = 93$ meV and $\epsilon_0 = 271$ meV for BNT and $\Gamma = 1.8$ meV and $\epsilon_0 = 558$ meV for BCT. Courtesy of Artur Erbe.

anchoring group used to bind the molecules. Let us also stress that Zotti *et al.* showed by means of ab initio DFT-based calculations that the use of the resonant tunneling model was justified. To be precise, they showed that the transport in these molecules is indeed dominated by a single molecular orbital that gives rise to a Breit-Wigner resonance close to the Fermi energy. In particular, the transport was found to be dominated by the HOMO in the case of the thiolated molecule, while the LUMO was found to be responsible for the conduction in the other two cases with nitro and cyano (or nitril) groups. The implications of this work for the role of anchoring groups in the transport through molecular junctions will be discussed in section 14.2.

13.3   A two-level model

In the previous section we have assumed that the coherent transport was completely dominated by a single molecular level. Of course, this is not always the case. For instance, the Fermi level may lie more or less in the middle of the HOMO-LUMO gap and then both molecular orbitals would contribute to the transport. In other situations, we can have other levels very close to the HOMO or to the LUMO contributing significantly to the transport. For these reasons, we want to refine the resonant tunneling model to include a second level. Our goal is to learn how the conductance

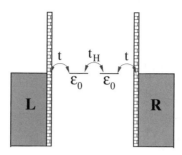

Fig. 13.11   Schematic representation of a two-level model where two sites with on-site energy ϵ_0 are coupled via a hopping t_H. Each site is coupled to its closest electrode by a hopping t (the same for both leads).

depends on the distance between the two levels and on the strength of the coupling to the electrodes.

The model that we are about to describe is inspired by an important example in molecular electronics, namely the transport through a hydrogen molecule [569]. As we shall see in section 14.1.3, Smit and coworkers [127] investigated the transport through hydrogen molecules with Pt contacts using the break junction technique. These authors concluded that a hydrogen molecule can form a stable bridge between Pt electrodes and that such a bridge has typically a conductance very close to the conductance quantum $G_0 = 2e^2/h$. Obviously, in this situation only two molecular levels can participate in the transport, namely the bonding and antibonding state of the hydrogen molecule.

With the hydrogen molecule in mind, we now proceed to analyze the transmission in the model represented schematically in Fig. 13.11. In this model we consider that the molecule is formed by two atoms with a single relevant orbital per site. The on-site energy is denoted by ϵ_0 and it is assumed to be the same in both sites. The two sites are connected by a hopping t_H, while the symmetric coupling to the electrodes is described by the hopping t. Notice that, for simplicity, we assume that the electrodes are only coupled to its closest atom. The hopping t_H is related to the splitting between the bonding (ϵ_+) and the antibonding state (ϵ_-) of the molecule, namely $\epsilon_\pm = \epsilon_0 \pm t_H$. Thus, the HOMO-LUMO gap is simply $2t_H$ in this case. Obviously, within this model the conductance is made up of a single channel because there is only one distinct path to cross the molecule.

The calculation of the zero-bias transmission is a simple exercise for those who have followed the theoretical background (see Exercise 13.1).

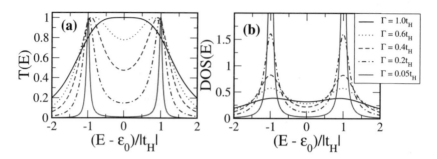

Fig. 13.12 (a) Transmission as a function of the energy for the two-level model. The different curves correspond to different values of the scattering rate Γ. (b) Total density of states (DOS) projected onto the molecule, i.e. the sum of the local DOS in both sites vs. energy.

We just state here the final result that reads[9]

$$T(E) = \frac{4\Gamma^2 t_H^2}{[(E - \tilde{\epsilon}_+)^2 + \Gamma^2]\,[(E - \tilde{\epsilon}_-)^2 + \Gamma^2]}. \tag{13.7}$$

Here, $\tilde{\epsilon}_\pm = \epsilon_0 \pm t_H + t^2 \mathrm{Re}\{g^a\}$ are the renormalized molecular levels, $g^a(\epsilon)$ being the advanced Green function which describes the local electronic structure of the leads. The scattering rate Γ, which determines the broadening of the molecular levels, is given by $\Gamma(\epsilon) = t^2 \mathrm{Im}\{g^a\} = \pi t^2 \rho(\epsilon)$, where $\rho(\epsilon)$ is the LDOS of the metallic contacts. For the sake of simplicity, we now assume that Γ is independent of the energy and that the levels are not renormalized ($\tilde{\epsilon}_\pm = \epsilon_\pm$). In Fig. 13.12(a) we show the transmission as a function of energy for different values of Γ in units of t_H. We also show in Fig. 13.12(b) the corresponding total density of states (DOS) projected onto the molecule.[10] Let us recall that the linear conductance is finally determined by the value of the transmission at the Fermi energy, which we have not yet specified.

As one can see in Fig. 13.12(a), the energy dependence of the transmission depends crucially on the ratio between the scattering rate and the hopping t_H. In a weak coupling situation, where $\Gamma \ll t_H$, the molecular levels are clearly resolved and there is a pronounced pseudo-gap between them. On the other hand, as Γ becomes of the order of t_H, and therefore of the order of the distance between the molecular levels, the gap is filled with states, see Fig. 13.12(b). In this case one can reach a transmission close to

---

[9]It is not important to understand the meaning of all the functions appearing in this formula to appreciate the main conclusion that we want to draw.

[10]This DOS is given by $\rho_+ + \rho_-$, where $\pi \rho_\pm = \Gamma / \{(\epsilon - \tilde{\epsilon}_\pm)^2 + \Gamma^2\}$.

one even in the energy region between the two molecular states. The first limit describes the typical situation in many organic molecules in which the Fermi energy lies somewhere in the HOMO-LUMO gap and the broadening of the levels (0.01-0.5 eV) is clearly smaller than the gap (3-8 eV). This is the reason why most organic molecules, even those with delocalized orbitals, are poorly conductive. The opposite limit describes the situation that occurs in strongly coupled systems such as the hydrogen molecule [127] and other short organic molecules coupled to transition metals [473, 474], where the linear conductance can be as high as $1\,G_0$. In these cases the strong hybridization between the molecules and the electrodes (made of Pt) provides a broadening to the molecular levels of several electronvolts, which is in some cases comparable to the gap of the molecules or it simply facilitates the resonant condition for the relevant orbital for transport (see sections 14.1.3 and 14.1.4). Thus, almost irrespective of the exact position of the Fermi energy, the transmission reaches a value close to unity. This is, in simple terms, the explanation for the high conductance observed in those examples.

Another simple two-level model is that in which the transmission is assumed to be the sum of two independent Lorentzian functions. We shall make use such a model in section 19.3 in our discussion of the thermopower of molecular junctions.

13.4   Length dependence of the conductance

One of the most studied issues in molecular electronics is the length dependence of the conductance of molecular junctions. Typically the experiments are restricted to low bias, but there are also studies of the influence of a finite bias on this length dependence. Series of molecules like alkanes, oligophenylenes, oligothiophenes, etc., have been extensively studied with different techniques (see Ref. [41] for exhaustive list of references). The most common finding is that conductance decays exponentially with the length of the molecule, L, as

$$G(L) = Ae^{-\beta L}, \tag{13.8}$$

where the attenuation factor β depends on the particular type of molecule, the presence of side groups, eventually on the bias voltage and not so much on the anchoring group. Here, A is just a prefactor that determines the order of magnitude of the conductance. Typical values of β range from 0.2-0.4 Å^{-1} for conjugated molecules to 0.8-1.2 Å^{-1} for aromatic compounds.

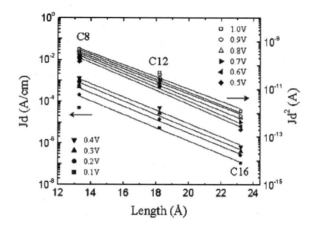

Fig. 13.13   Length dependence of the current through a self-assembled monolayer of alkane thiols measured for different bias voltages with the nanopore technique. The figure shows a log plot of the tunneling current densities multiplied by the molecular length, which is denoted by d in this graph, at low bias and by d^2 at high bias (symbols) vs. molecular lengths. The lines through the data points are linear fittings. Reprinted with permission from [130]. Copyright 2003 by the American Physical Society.

The exponential length dependence is expected in almost any tunneling model. Thus for instance, from the Simmons model (see section 4.4) one expects at low voltages a length dependence of the type $G \propto (1/L) \exp(-\beta_{\mathrm{LV}} L)$, where β_{LV} is a bias-independent decay coefficient given by

$$\beta_{\mathrm{LV}} = \frac{2\sqrt{2m}}{\hbar} \alpha \sqrt{\varphi_{\mathrm{B}}}, \qquad (13.9)$$

where let us recall that φ_{B} is the barrier height, m is the electron mass and α is a parameter that depends on the exact shape of the barrier. For higher voltages (HV) (i.e. $eV > \varphi_{\mathrm{B}}$), the attenuation factor depends on the bias as

$$\beta_{\mathrm{HV}} = \beta_{\mathrm{LV}} \left(1 - \frac{eV}{2\varphi_{\mathrm{B}}}\right)^{1/2}. \qquad (13.10)$$

We show a typical experimental example of this type length dependence in Fig. 13.13 taken from Ref. [130]. Let us recall that in this experiment the current through a self-assembled monolayer of alkanethiols was measured for different bias voltages with the nanopore technique. The data correspond to three different alkanethiols: $CH_3(CH_2)_{n-1}SH$ with $n = 8, 12, 16$, denoted as C8, C12 and C16. The current density has been normalized

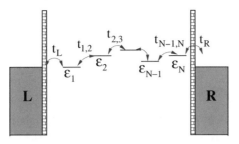

Fig. 13.14 Schematic representation of the bridge model to explain the exponential length dependence of the conductance. For further explanations, see text.

following the expectation of Simmons model and as it can be seen, the fit is satisfactory. On the other hand, in order to compare with other results reported in the literature, the authors also performed a fit to Eq. (13.8). They obtained a β value from 0.83 to 0.72 Å^{-1} in the bias range from 0.1 to 1.0 V, which is comparable to results reported previously with other techniques [458, 475, 476].

From an atomistic point of view, the exponential length dependence of the conductance can be understood using a simple tight-binding model, often used in the field of electron transfer [38]. Let us briefly explain the main idea. The model is schematically represented in Fig. 13.14. In this model a molecular bridge formed by N sites (or segments) with on-site energies ϵ_i (only one orbital per side) is coupled to two metallic leads via the hoppings $t_{L,R}$. In the bridge we only consider nearest-neighbor hoppings denoted by $t_{i,i+1}$. Notice that this model is simply the inhomogeneous version of the model that we have used to explain the even-odd effect in gold atomic chains in section 11.8.

Let us briefly remind how the transmission through the molecular bridge can be calculated.[11] Using the result of the Exercise 7.5 or the general formulas derived in section 8.1, the zero-bias transmission coefficient can be written as

$$T(E) = 4\Gamma_L(E)\Gamma_R(E)|G_{1N}^a(E)|^2, \qquad (13.11)$$

where the $\Gamma_{L,R}$ are the scattering rates determining the strength of the coupling to the metallic electrodes. Usually they do not have a very significant energy dependence and we assume here that they are constant. Moreover, G_{1N}^a is the (advanced) Green's function connecting the first and last site in

---

[11]Those readers not familiar with the Green's function techniques described in the second part of the book can skip this discussion and go directly to Eq. (13.14).

the molecular bridge. In this sense, $|G_{1N}^a(E)|^2$ can be seen as the probability for an electron to propagate along the molecular wire. This function can be calculated by taking the element $(1, N)$ of the following matrix (see section 11.8)

$$\mathbf{G}^a(E) = [E^a \mathbf{1} - \mathbf{H}_{\text{bridge}} - \mathbf{\Sigma}_L^a - \mathbf{\Sigma}_R^a]^{-1}, \qquad (13.12)$$

where $E^a = E - i0^+$ and $\mathbf{H}_{\text{bridge}}$ is the Hamiltonian of the molecular bridge. Here, the only non-vanishing elements of the matrix self-energies are $(\mathbf{\Sigma}_L^a)_{11} = t_L^2 g_L^a$ and $(\mathbf{\Sigma}_R^a)_{NN} = t_R^2 g_R^a$, where $g_{L,R}^a$ are the lead Green's functions (their exact expressions are irrelevant for our present discussion). The scattering rates are giving by $\Gamma_{L,R} = t_{L,R}^2 \text{Im}\{g_{L,R}^a\}$.

Rather than inverting exactly the previous $N \times N$ matrix, we compute the first non-vanishing contribution to G_{1N}^a. Obviously, this lowest-order contribution corresponds to the sequential tunneling along the bridge without any reflection. This is a good approximation to the exact expression in the weak coupling regime, where $\max\{t_{i,i+1}\} \ll \min\{|E - \epsilon_i|\}$. Mathematically, this contribution can be written as

$$G_{1N}^a(E) \approx \frac{1}{E^a - \epsilon_N} \prod_{i=1}^{N-1} \frac{t_{i,i+1}}{E^a - \epsilon_i}. \qquad (13.13)$$

For the sake of simplicity, we now assume that all bridge segments are identical, i.e. $t_{i,i+1} = t$ and $\epsilon_i = \epsilon$. Substituting the previous result into the expression of the transmission, one obtains for the homogeneous bridge

$$T(E) \approx \frac{4\Gamma_L \Gamma_R}{|t|^2} \left| \frac{t}{E - \epsilon} \right|^{2N}. \qquad (13.14)$$

This result implies a simple form for the attenuation parameter of Eq. (13.8)

$$\beta(E) = \frac{2}{a} \ln \left| \frac{E - \epsilon}{t} \right|, \qquad (13.15)$$

where a measures the segment size, so that the bridge length is Na. Notice that β is independent of the coupling to the leads and it is just determined by intrinsic properties of the molecular bridge. The exponential dependence on the bridge length is a manifestation of the tunneling character of this process. Again, remember that the relevant energy for the linear conductance is the Fermi energy, E_F. For typical values, e.g. $|(E_F - \epsilon)/t| = 10$ and $a = 5$ Å, Eq. (13.15) yields $\beta = 0.92$ Å^{-1}.

So in short, the general conclusion of our discussion is that the exponential length dependence of the conductance is a signature of coherent tunneling in an off-resonant situation. Things may be different if the transport occurs via a resonant molecular orbital. In this case, the conductance could be length independent or at least a non-monotonic function (see Exercise 13.4). Indeed, it is easy to show that if an electron is injected within the molecular bridge energy band, the conductance oscillates as a function of both the injection energy and of wire length (see Refs. [255–259] and Exercise 13.4). This is precisely the behavior found in the monoatomic chains in section 11.8. However, this situation seems to occur very rarely in the case of molecular junctions and it is reserved to "metallic" solid-like molecules like the carbon nanotubes.

13.5   Role of conjugation in π-electron systems

We want to address in this section the role of the conjugation in a delocalized π-electron system.[12] It is obvious that the "goodness" of the electrical conduction in a molecular junction depends crucially on the degree of delocalization of the molecular orbitals. After all, we have learned that the conductance is governed, among other things, by the strength of the metal-molecule coupling. In order to have a high current flowing through a molecular orbital, it has to be strongly coupled to both electrodes, which in turn implies that it has to be extended over the whole molecule. This is the reason why conjugated molecules are believed to be good candidates for molecular wires. A delocalized π-electron system in a conjugated molecule can be interrupted by the introduction of adequate side groups that rotate one part of the molecule with respect to the other. In this case the coupling of the two subsystems, which is mainly determined by a matrix element (or hopping) between two π-orbitals, decreases as the twist angle increases, and eventually it vanishes when the two orbitals are exactly orthogonal at an angle of $\pi/2$. This argument suggests a way of testing the role of the conjugation in the conductance of a molecular junction.

A beautiful experimental illustration of this simple idea was reported by Venkataraman and coworkers [477]. These authors investigated the transport through different biphenyl molecules using an STM-based break-junction technique. They studied in particular a series of biphenyl molecules with different ring substitutions that alter the twist angle of the molecules. They found that the conductance for this series decreases with increasing

---

[12]This type of electron systems was discussed in section 9.5.1 using benzene as an example.

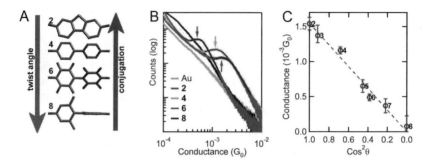

Fig. 13.15 (a) Structures of a subset of the biphenyl series studied, shown in order of increasing twist angle or decreasing conjugation. (b) Conductance histograms of the different molecules obtained with an STM at a bias voltage of 25 mV. (c) Position of the peaks for all the molecules studied plotted against $\cos^2 \theta$, where θ is the calculated twist angle for each molecule, see Ref. [477] for more details. Reprinted by permission from Macmillan Publishers Ltd: Nature [477], copyright 2006.

twist angle, consistent with a cosine-squared relation, which is expected in transport through π-conjugated biphenyl systems, see Fig. 13.15.

Let us briefly explain how the $\cos^2 \theta$-dependence comes about in these biphenyl compounds. For this purpose, let us use the model of the previous section. For simplicity, we assume that the bridge is composed of two identical segments linked by a hopping t. In an off-resonant situation, according to Eq. (13.14) the transmission is simply given by

$$T(E) \approx \frac{4\Gamma_L \Gamma_R}{|t|^2} \left| \frac{t}{E - \epsilon} \right|^4 = \frac{4\Gamma_L \Gamma_R}{|E - \epsilon|^4} |t|^2, \qquad (13.16)$$

i.e. the transmission is proportional to $|t|^2$. Here, t is the hopping between two π-orbitals that is simply proportional to $\cos \theta$, where θ is the angle between them (see Exercise 13.3). Thus, we arrive at the result that the transmission, and therefore the linear conductance, is expected to be proportional to $\cos^2 \theta$, as it was nicely observed in Ref. [477]. For a more rigorous discussion of this $\cos^2 \theta$-law, see Ref. [478].

13.6 Fano resonances

As we have discussed in sections 13.2 and 13.3, in most cases the coherent transport through molecular junctions is determined by Breit-Wigner resonances that originate from the different molecular orbitals. However, these are not the only transmission line shapes that can be expected in molec-

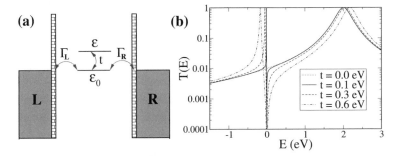

Fig. 13.16 (a) Schematic representation of a simple that illustrates the physics of Fano resonances. Here, the resonant tunneling model of section 13.2 is modified by introducing an additional level (ϵ) that is coupled to the resonant level, but not to the leads. (b) Zero-bias transmission as a function of the energy for the model of panel (a) for $\epsilon_0 = 2.0$ eV, $\epsilon = 0.0$ eV, $\Gamma_L = \Gamma_R = 0.1$ eV and different values of the coupling t.

ular junctions. In the last years, different authors have discussed the role of quantum interference [259, 479–482] and, in particular, Fano resonances [483–487] in the transport through molecular contacts. It has been shown that these phenomena can give rise of transmission line shapes that differ significantly from the standard Breit-Wigner resonances of section 13.2. As an example, in this section we shall briefly discuss the physics of Fano resonances in molecular wires.

In 1961 U. Fano showed that in the context of the excitation spectra of atoms and molecules, the interference of a discrete autoionized state with a continuum gives rise to characteristically asymmetric peaks [488], which are nowadays referred to as *Fano peaks or resonances*. The appearance of this type of resonances in transport experiments have been discussed in several contexts in mesoscopic physics ranging from one-dimensional waveguides to Kondo impurities. In the context of molecular junctions, a Fano resonance can appear in the transmission, for instance, due to the interplay between extended molecular orbitals and states that are localized in a side group of the molecule which is decoupled from the electrodes [484].

Following Ref. [484], we shall use the toy model schematically represented in Fig. 13.16(a) to explain the origin of Fano resonances. This model is based on the resonant tunneling model and the ingredient is the presence of an additional site (or energy level) that represents a side group that is not directly connected to the electrodes. The coupling to the resonant level is given by the hopping t and the level position of this "side group" is denoted by ϵ. The calculation of the zero-bias transmission in this model is a simple exercise (see Exercise 13.5) and the final result reads

$$T(E) = \frac{4\Gamma_L\Gamma_R}{[E - \epsilon_0 - t^2/(E - \epsilon)]^2 + \Gamma^2}, \tag{13.17}$$

where $\Gamma = \Gamma_L + \Gamma_R$. This equation reduces to the Breit-Wigner formula of Eq. (13.2) when the coupling element t vanishes. The main new feature in this model is the appearance of an antiresonance at $E = \epsilon$ where the transmission vanishes. This feature stems from a destructive quantum interference between the direct path crossing the resonant level and a path in which the electron "visits" the side group. Apart from this antiresonance, the transmission exhibits two maxima at $E = \epsilon_\pm$, where $\epsilon_\pm$ are given by

$$\epsilon_\pm = \frac{1}{2}\left\{(\epsilon + \epsilon_0) \pm \sqrt{(\epsilon - \epsilon_0)^2 + 4t^2}\right\}. \tag{13.18}$$

In the limit $t \ll |\epsilon - \epsilon_0|$, i.e. when the "side group" is weakly coupled to the central backbone, the transmission exhibits a Breit-Wigner resonance of width Γ in the vicinity of $E = \epsilon_0$. Moreover, a Fano peak occurs near the antiresonance ($E = \epsilon$) separated from it by a distance of approximately $t^2/|\epsilon - \epsilon_0|$. Thus, in this limit the hybridization with the weakly coupled side group leads to the appearance of a peculiar asymmetric structure formed by a peak followed by an antiresonance, which is the main fingerprint of this phenomenon. Examples of those asymmetric line shapes are shown in Fig. 4.73(b) in the limit of weak coupling (small t). Notice in particular the dramatic change in the transmission that can go from 1 all the way down to zero by changing slightly the energy. Obviously, in order to have an impact in the transport properties, the Fano resonances needs to be located close to the Fermi energy. If this is the case, they can give rise to a pronounced structure in the I-V curves [480] or they can significantly modify the thermoelectric properties of a molecular junction [487].

It is worth mentioning that an experimental situation that closely mimics our simple model was reported in Ref. [489]. In this work, an artificial quantum structure consisting of a single CO molecule adsorbed on a Au chain was assembled by manipulating single Au atoms on NiAl(110) at 12 K with a STM. It was shown that the CO disrupts the delocalization of electron density waves in the chain, as it suppresses the coupling between neighboring chain atoms. In a subsequent paper, Calzoni *et al.* [490] showed theoretically that the electronic properties of this system can be tuned by the selective adsorption of small molecules. In particular, they showed that a single CO group induces a quantum interference pattern that modulates the electronic wave functions and modifies the coherent transport properties of the system.

13.7   Negative differential resistance

As explained in Chapter 1, one of the goals of molecular electronics is to complement current Si technology. For this purpose, one must find molecular systems that, at least, mimic some of today's microelectronic components. In this respect, one of the most studied issues in the last years is the occurrence of negative differential resistance (NDR) in molecular junctions, which indeed has already been observed in several systems [18, 491–499]. NDR is the key feature in the I-V characteristics of the semiconductor device known as *resonant tunneling diode*, which was pioneered by Esaki and coworkers [500]. This device consists of two potential barriers in series, the barrier being formed by thin layers of a wide-gap material like AlGaAs sandwiched between layers of a material like GaAs having a smaller gap. Both barriers are thin enough for electrons to tunnel through. The NDR that occurs in this device forms the basis for practical applications as a switching device and in high frequency oscillators [501–503].

In the context of molecular junctions, several mechanisms for NDR have been suggested involving, for instance, charging and/or conformation changes [18, 504–508] or polaron formation [509]. Following the philosophy of this chapter, we are interested in the following question: Is it possible to induce NDR simply by means of coherent tunneling processes? With our analysis so far, based mainly on the resonant tunneling model, one might get the impression that this is not possible. However, it is well-known that the NDR in Esaki's resonant diode is explained in terms of coherent transport (for a didactic discussion of the essential physics of this device, see Chapter 6 of Ref. [50]). The NDR in that device is originated from the energy dependence of the electron injection rate (or scattering rate in our usual language), which is due to the band structure of the semiconducting leads. Thus, the take-home message is that coherent tunneling can lead to NDR, but one needs to have a pronounced energy dependence of the scattering rates.[13] This is not easy to achieve with metallic leads because they typically exhibit a rather flat density of states around the Fermi energy. An alternative is then to use of semiconductor electrodes (at least one of them). Indeed, many experiments have demonstrated the feasibility of attaching various organic molecules on Si substrates (see list of references in Refs. [510, 511]). The first theoretical analysis of the transport through metal-molecule-semiconductor junctions was carried out by Datta

---

[13]Let us recall that so far we have always assumed that the scattering rates were energy independent.

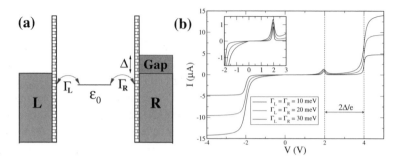

Fig. 13.17 (a) Schematic representation of a metal-molecule-semiconductor junction. In equilibrium, the right electrode has a gap in the energy window $E \in [E_F, E_F + \Delta]$, which simulates a heavily p-type doped semiconductor. The scattering rate Γ_R vanishes inside the gap of the semiconductor. (b) I-V characteristics of the metal-molecule-semiconductor junction of panel (a) for different values of the scattering rates and at room temperature (300 K). The gap is $\Delta = 1$ eV and the level position is $\epsilon_0 = -1$ eV (measured with respect to E_F). The value of Γ_R indicated in the legend refers to value outside the gap of the semiconductor. We have assumed that the voltage drops symmetrically at both interfaces. The vertical dotted lines indicate the voltage region where the resonant level lies inside the gap of the right electrode. The inset shows a blow up of the voltage region where the NDR occurs.

and coworkers [512]. These authors showed that indeed one can have NDR in these systems by means of coherent resonant tunneling. The presence of a semiconductor band-edge leads to NDR when the molecular levels are driven by the external potential into the semiconducting band-gap. We now proceed to illustrate this mechanism with a simple model.

Let us consider once more the resonant tunneling model of section 13.2. In order to describe a metal-molecule-semiconductor junction, we now assume that there is a gap in, let us say, the right electrode, see Fig. 13.17(a). The size of this gap is denoted by Δ. In a heterojunction like this one, it is important to describe correctly the band-bending in the semiconductor and the overall level alignment. We shall ignore these important details, in order to emphasize the basic conceptual issues. We assume that the equilibrium band-alignment is as shown in Fig. 13.17(a). Here, the Fermi energy lies near the semiconductor valence band-edge, i.e. we assume that the semiconductor is heavily p-type doped. The presence of a gap in the right electrode strongly modifies the scattering rate, which in particular vanishes inside the gap. We model this situation by a rate, Γ_R, that is constant outside the gap region and equal to zero at energies $E \in [E_F - eV/2, E_F - eV/2 + \Delta]$. Here, we have already taken into account the shift of the chemical potential of the right electrode induced by the bias voltage. The energy dependence of Γ_R is the only difference with respect to the standard resonant tunneling model.

In Fig. 13.17(b) we show examples of the I-V characteristics obtained with this simple model for a symmetric situation where the voltage drops equally at both interfaces. Here, we have assumed that in equilibrium the level lies 1 eV below the Fermi energy and the gap is $\Delta = 1$ eV. The most prominent feature is the appearance of NDR (a decrease in the current) at $V = +2$ V. This voltage corresponds to the bias at which the resonant level reaches the semiconductor valence band-gap and therefore the transmission drops abruptly. Another important feature is the strong asymmetry of the I-V's with respect to voltage inversion. In particular, notice that there is no NDR for negative bias. The reason is that for negative voltages the resonant level is shifted down with respect to the chemical potential of the right electrode and thus, it never "feels" (or reaches) the band-gap. The I-V curves of Fig. 13.17(b) reproduce qualitatively the line shapes obtained with a Hückel model and ab initio methods by Datta and coworkers [512, 513]. These authors also pointed out that in order to see NDR at negative voltages, one would need to use n-type semiconductors (see Exercise 13.6).

The first observation of NDR through individual organic molecules on silicon surfaces was reported by Hersam and coworkers [514]. This work reported room temperature charge transport measurements performed on individual organic molecules mounted on degenerately doped Si(100) surfaces using UHV STM. In particular, for 2,2,6,6-tetramethyl-1-piperidinyloxy (TEMPO) molecules, NDR was observed only for negative sample bias on n-type Si(100) and for positive sample bias on p-type Si(100). This unique behavior is consistent with the resonant tunneling mechanism described above. However, let us mention that the origin of the NDR in the n-type junction is not so clear and it has been attributed to possible vibronic interactions [513]. An example of the experimental results of Ref. [514] is shown in Fig. 13.18. Since this first observation, the conditions that give rise to electronic NDR on silicon within the coherent regime have been investigated at length both experimentally and theoretically [510, 511, 513, 515–520].

We conclude this section by saying that the mechanism described above is not the only possibility to obtain NDR in a situation where the transport is mainly coherent. The electrostatic potential profile across a molecular conductor is a key factor determining the shape of the I-V characteristics. It has been suggested by Liang *et al.* [521] that a complex potential profile might lead to NDR in a molecular junction.

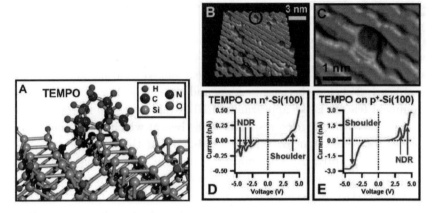

Fig. 13.18   Experimental observation of NDR in the transport through TEMPO molecules on Si(100)-2 × 1 surfaces probed with STM. (A) Molecular mechanics optimized structure of an individual TEMPO molecule on a truncated Si(100)-2 × 1 surface. (B) STM topography image of isolated TEMPO molecules on a degenerately n-type Si(100)-2 × 1 surface. (C) STM image of the isolated TEMPO molecule that is circled in part (B). (D) I-V curves of an isolated TEMPO molecule bound to *n*-type Si(100). At negative sample bias, three distinct NDR events are observed, while a shoulder is observed at positive sample bias. (E) Current-voltage plot of an isolated TEMPO molecule bound to *p*-type Si(100). At negative sample bias, a shoulder is observed, whereas two NDR events are detected at positive sample bias. Adapted with permission from [514]. Copyright 2004 American Chemical Society.

13.8   Final remarks

The goal of this chapter has been to describe and illustrate some basic concepts related to the coherent transport through molecular junctions. It is often believed in the context of molecular electronics that the theory is unable to reproduce the experimental observations. We hope to have shown that this judgment is unfair. We have been able to explain qualitatively a variety of effects by simply using toy models and handwaving arguments. A different story is our quantitative understanding that, as we shall see in the next chapter, is not yet that satisfactory.

We also want to stress that there are other basic issues related to the coherent transport that we have not covered in this chapter. Probably the most important one is the issue of the electrostatic potential profile. We have learned in this chapter that the position of the energy levels plays a crucial role determining the current through a molecular junction. At finite bias the energy levels are shifted with the voltage in a way that depends on the exact electrostatic profile across the junction. Therefore, the determination of such profile is crucial for the proper description of the current-voltage characteristics. The theoretical analysis of the electrostatic potential profile across atomic-scale junctions has been addressed by different authors with a variety of methods such as simple tight-binding models [246], model calculations based on a combination of the Schrödinger and Poisson equations [522], a simple Thomas-Fermi-type screening model [523] or ab initio approaches [293] mainly based on DFT (see section 10.8). For a detailed discussion about the electrostatic potential profile in molecular conductors we recommend Ref. [521] and references therein.

13.9 Exercises

13.1 Resonant tunneling model: Let us consider the resonant tunneling model of section 13.2 for a symmetric situation ($\Gamma_L = \Gamma_R = \Gamma/2$). Calculate the current at zero temperature up to third order in the bias voltage, i.e. determine the relation $I(V) = AV + BV^3$ at low bias and express the constants A and B as a function of the two parameters of the model, namely Γ and ϵ_0.

13.2 Two-level model: Let us consider the two-level model of section 13.3.
  (a) Use the general expressions derived in section 8.1, see Eq. (8.18) or (8.19), to show that the zero-bias transmission is given by Eq. (13.7). Discuss also under which conditions one recovers the expression of the transmission of the (single-level) resonant tunneling model.
  (b) Compute the I-V characteristics within the two-level model and discuss the results. Hint: Assume that there is no voltage drop inside the molecular bridge and that the scattering rates are independent of the energy.

13.3 The $\cos^2\theta$-law: Show that a matrix element (or hopping) between two π-orbitals is proportional to $\cos\theta$, where θ is the angle formed by the axes of the two orbitals. Hint: See discussion about two-center matrix elements in section 9.3.1.

13.4 Length and energy dependence of the transmission in molecular wires: In a series of papers Mujica and coworkers studied the conduction through molecular wires using an effective tight-binding Hamiltonian (equivalent to a Hückel model) [255–259]. They obtained the following interesting results for the linear conductance of a molecular junction:

(1) The conductance achieves large (but bounded) values in the vicinity of any of the wire energy eigenvalues.

(2) The conductance oscillates as a function of both injection energy and of wire length when the electron is injected within the wire's energy band.

(3) The conductance decreases exponentially with length when the electron is injected outside the band of the wire.

Use the model for a molecular bridge discussed in section 13.4, see Fig. 13.14, to demonstrate the previous conclusions. For this purpose, solve the model exactly inverting Eq. (13.12), rather than using perturbation theory as we did in our discussion in section 13.4. On the other hand, model the leads as semi-infinite linear chains and use Eq. (5.46) for the Green's functions of the outermost atoms of the chains that are coupled to the molecular bridge.

13.5 Fano resonances: Show that the transmission in the model introduced in section 13.6 is given by Eq. (13.17). Hints: (i) Use the general expression of Eq. (8.18) to show that the zero-bias transmission can be written as $T(E) = 4\Gamma_L\Gamma_R|G_{00}^a(E)|^2$, where $G_{00}^a(E)$ is the advanced Green's function in the resonant level. (ii) Show that $G_{00}^a(E) = 1/\{E - \epsilon_0 - t^2/(E - \epsilon) - i\Gamma\}$. This result leads directly to Eq. (13.17).

Finally, investigate the impact of Fano resonances on the current through a molecular junction by computing the I-V curves within this model. For simplicity, assume that the system is symmetrically coupled to the leads and that the voltage drops occur at the metal-molecule interfaces.

13.6 NDR in metal-molecule-semiconductor junctions: Using the model of section 13.7 show that one can encounter NDR at negative voltages using a heavily n-type doped semiconductor. For this purpose, (i) assume that in equilibrium the Fermi level lies near the edge of the conduction band of the semiconducting lead and (ii) assume that in equilibrium the resonant level lies above the semiconductor gap.

13.7 Transmission of a benzene junction: The goal of this exercise is to compute the transmission as a function of energy for a metal-benzene-metal junction. Use for this purpose the Hückel approximation for the benzene molecule described in section 9.5.1 with $\epsilon_0 = 0$ for the on-site energy of the π-orbital in each carbon atom and $t = -2.5$ eV for the hopping between neighboring atoms. Assume that the benzene molecule is coupled to the leads through a single carbon atom in each side [e.g. atoms 1 and 4 in Fig. 9.4(a)] and describe the strength of the coupling with a scalar and energy-independent scattering rate Γ (the same for both interfaces). Calculate the zero-bias transmission as a function of energy within this model for different values of Γ. Determine also the linear conductance assuming that the Fermi energy is $E_F = 0$ (i.e. it lies in the middle of the HOMO-LUMO gap of the benzene molecule) and estimate the value of Γ necessary to reach a conductance larger than $0.1\,G_0$.

Chapter 14

Coherent transport through molecular junctions II: Test-bed molecules

In the previous chapter we have learned that the coherent transport through molecular junctions is determined by the strength of the metal-molecule coupling as well as by the intrinsic properties of the molecules, including their length, conformation, the HOMO-LUMO gap and the alignment of this gap to the metal Fermi level. Moreover, we have shown that in many cases the experimental observations can be explained by means of very simple qualitative arguments. In this chapter we shall go on discussing the coherent transport in single-molecule junctions, but from a more quantitative point of view. Our goal is twofold. On the one hand, we want to calibrate our present level of understanding and for this purpose, we shall compare different experimental and theoretical results for various test systems. On the other hand, we shall illustrate some of the basic concepts discussed in the previous chapter in more quantitative terms.

Bearing these goals in mind, we shall discuss in the next section the results obtained so far for some test-bed molecules of special interest in molecular electronics. Then, we shall review recent advances in the understanding of the role of the metal-molecule interface and the efforts to chemically tune the conductance with the use of side-groups. Moreover, we shall briefly describe a set of controlled experiments performed with the STM, in which the junctions are fully characterized providing thus important test systems. We shall finish this chapter with a summary of the main conclusions and some comments about the future challenges and open problems.

Before getting started, let us say that the current status of the understanding of the electronic transport through molecular junctions has been reviewed several times in this decade. In particular, we recommend the following articles by Lindsay and collaborators [524–529].

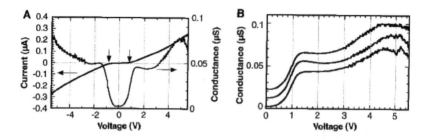

Fig. 14.1   I-V characteristics and differential conductance of Au-benzenedithiol-Au junctions measured with the MCBJ technique at room temperature. (A) Typical I-V curves, which illustrate a gap of 0.7 V; and the differential conductance $G(V) = dI/dV$, which shows a steplike structure. (B) Three independent $G(V)$ measurements, offset for clarity, illustrating the reproducibility of the conductance values. From [16]. Reprinted with permission from AAAS.

14.1   Coherent transport through some test-bed molecules

In order to establish to what extend we understand the electronic transport through single-molecule junctions, we shall review in this section several representative examples related to small molecules, in which the electrical conduction is believed to be dominated by coherent tunneling. These examples will also serve to illustrate in more detail some of the basic concepts discussed in the previous chapter.

14.1.1   *Benzenedithiol: how everything started*

As we discussed in our brief review of the history of molecular electronics in section 1.2, the experiment of Reed and coworkers in 1997 [16] is often considered as the beginning of the field of single-molecule conduction. This experiment was performed using a mechanically controllable break-junction (MCBJ) device working at room temperature, with the junction immersed in a solution of the organic compound of interest. The compound that they selected was 1,4-benzenedithiol (BDT), which has become a workhorse in this field. In the experiment the broken gold wire was allowed to interact with the molecules for a number of hours so that a self-assembled monolayer covered the surface. Next, the junction was closed and re-opened a number of times and I-V curves were recorded at the position just before contact was lost completely. The I-V curves showed some degree of reproducibility with a large energy gap feature of about 0.7 V, see Fig. 14.1, which was attributed to a metal-molecule-metal junction.

Transport through BDT molecules have been studied by several experimental groups with different setups [530–536]. The reported values for the linear conductance vary between $5 \times 10^{-5} G_0$ and $0.1 G_0$, i.e. they are scattered over more than 3 orders of magnitude. From the theory side, many authors have calculated the linear conductance of Au-BDT-Au junctions [537–546]. The typical values lie in the range of $(0.05\text{-}0.4) G_0$, which in general overestimate the observed linear conductance.

A certain level of disagreement between different experiments and different theories might be understandable because the transport depends on the microscopic details of the junction. Indeed, the conductance of BDT contacts has been found theoretically to be strongly dependent on the bonding site of the S atom [540, 543], while variations in the Au-S bond length only affects the transmission function weakly [541]. However, it is difficult to understand the differences found in the conductance histograms, where a statistical analysis is supposed to average out the microscopic details. Thus for instance, Xiao *et al.* [530] using a STM-based break-junction setup found that the most probable value for the room temperature linear conductance is $\sim 0.01 G_0$, while Lörtscher *et al.* [534] reported a value of $\sim 5 \times 10^{-5} G_0$ using microfabricated break-junctions. Also using this latter technique, Martin *et al.* [535, 536] found no distinct peaks in the histograms. Moreover, in Refs. [534, 535] I-V characteristics were reported that clearly differ from those of the original work of Reed and coworkers [16]. Both Lörtscher *et al.* [534] and Martin *et al.* [535] found non-linear I-Vs that were sensitive to temperature and, although some features like the gap at low bias were similar, there were significant differences in the magnitude of the measured current in these experiments. The origin of these discrepancies still remains unclear.

With respect to the theory, what is the origin of the differences between different theoretical results and why does the theory seem to overestimate the value of the linear conductance? There is no definite answer to these questions, but let us try to give some ideas. Most of the calculations mentioned above are based on the combination of nonequilibrium Green's function techniques (NEGF) and density functional theory (DFT), which was explained in detail in section 10.8. At this stage the technical details related to the implementation of this combination still matter and in some cases the deficiencies in the implementation of this approach lead to artificial results, which has nothing to do with the limitation of the NEGF-DFT approach (for a discussion of this issue, see Ref. [545]). On the other hand, the discrepancy between experiment and theory might be due to the intrinsic approximations made in the NEGF-DFT method. For instance, Delaney and Greer have suggested that the problem might lie in the insufficient description of the electronic correlation [547]. These authors claim that by reformulating the transport problem using boundary conditions suitable for correlated many-electron systems, one can obtain I-V curves for BDT that are close to experimental observations. Although, this method is not generally accepted (see Ref. [548] for severe objections to this approach), it is quite reasonable to believe that correlations beyond the scope of the NEGF-DFT approach play a fundamental role in molecular junctions. The development of those theoretical methods is presently one of the major challenges in the field.

Bearing in mind the limitations of the existent theories, let us try to give a simple picture of the expected transport mechanism in benzenedithiol. First of all, since this molecule is rather small and the Au-S bond is sufficiently strong, one does not expect the transport in Au-BDT-Au junctions to be dominated by vibronic degrees of freedom or correlation effects like Coulomb blockade. In other words, it is reasonable to assume that the transport in this case is coherent and therefore, it is probably determined by the electronic structure of the contact.[1] With respect to the molecule itself, it possesses an electronic structure that closely resembles that of benzene (see section 14.1.4). In Fig. 14.2(a) we show the frontier orbitals of this molecule, as obtained from a DFT calculation of the isolated molecule.[2] With respect to the vacuum level, the HOMO and the LUMO of the molecule lie at -4.95 eV and -1.42 eV, respectively. It is worth stressing that when the molecule is coupled to the electrodes, its levels are shifted and broadened depending on the strength of the interaction with the metal. Anyway, since the gold Fermi energy lies at approximately -5 eV and the Au-S bond is rather strong, one naively expects a rather high conductance dominated by the HOMO level. This picture is indeed confirmed by calculations based on the DFT-NEGF combination. Apart from the numerical discrepancies mentioned above, these calculations show that the transport proceeds through the tail of the HOMO of the molecule that lies at around 1 eV below the Fermi energy. An example of the zero-bias transmission as a function of energy of an Au-BDT-Au junction taken from Ref. [545] is

---

[1]The temperature dependence of the I-V curves in Refs. [534, 535] cannot be easily explained within a coherent transport picture, unless such changes are related to the thermal stability of the contacts.

[2]These DFT results were obtained with the code TURBOMOLE v5.7 [575] using a split valence polarization basis set and the BP86 exchange-correlation functional.

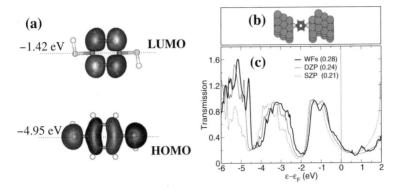

Fig. 14.2 (a) Frontier orbitals of a benzenedithiol (BDT) molecule as obtained from a DFT calculation (see footnote 2). (b) Supercell used to model the central region of a Au(111)-BDT-Au(111) junction with S at the fcc hollow site. (c) The calculated transmission functions with two different methods and different basis sets. The transmission at the Fermi level is indicated in the parentheses following the legends. Reprinted with permission from [545]. Copyright 2008, American Institute of Physics.

shown in Fig. 14.2(c). In this case the leads are ideal Au(111) surfaces and the S atoms were place at the minimum energy positions in the fcc hollow sites. The linear conductance obtained in this case is $\sim 0.28\, G_0$ in line with the naive expectation and clearly higher that in the experiments.

14.1.2 *Conductance of alkanedithiol molecular junctions: A reference system*

From our discussion about benzenedithiol in the previous section, one may infer that the level of agreement between experiments, theories, and experiment and theory is certainly disappointing. We shall see in this section that the situation is definitively improving and for this purpose we shall discuss the transport through alkanedithiols.

Alkanes[3] ($C_n H_{2n+2}$) are simple saturated chains of carbon atoms that constitute the most popular test-bed molecules for studies of the electrical conductance of molecular junctions in the last few years. Their chemical stability and large HOMO-LUMO gap (see below) make them ideal for investigating the contribution of the metal-molecule coupling to the conductance. In most cases thiol groups (SH) have been attached to the ends

---

[3]There are different types of alkanes such as branched alkanes or cyclic alkanes. Here, we restrict ourselves to the linear alkanes. A brief discussion of the properties of these molecules can be found in section 3.2.

of alkane molecules to investigate the transport with gold electrodes, making use of the well-known chemistry of the covalent Au-S bond.[4] Transport through thiolated alkanes has been studied extensively both at the level of single molecules [125, 549–556, 535] and self-assembled monoloyares (SAMs), indexself-assembled monoloyares (SAM) see Refs. [130, 41] and references therein. Furthermore, as we shall discuss later in this chapter, alkanes have also been used as a platform for testing the anchoring efficiency of different chemical groups [557, 126, 559].

It is presently acknowledged that a reliable measurement of the transport properties of single-molecules junctions requires a detailed statistical analysis. In this sense, the method introduced by Xu and Tao in 2003 [549] has been adopted by many authors, especially in the context of STM and break-junctions. In this statistical analysis, the conductance of a molecular contact is measured by repeatedly forming thousands of junctions. Often the corresponding conductance histograms reveal well-defined peaks at integer multiples of a fundamental conductance value, which is typically interpreted as the conductance of a single molecule. Xu and Tao presented in their seminal paper [549] the first statistical results on the conductance of alkanedithiols. In particular, they reported zero-bias resistances of 10.5 ± 0.5, 51 ± 5, and 630 ± 50 MΩ for hexanedithiol, octanedithiol, and decanedithiol. Moreover, the attenuation factor (β_N) for N-alkanedithiols was 1.0 ± 0.1 per carbon atom and was weakly dependent on the applied bias, which is in qualitative agreement with the values reported in SAMs by various authors (see Refs. [130, 460] and references therein).

Despite using the same statistical analysis and comparable experimental techniques, Xu and Tao [549] and Haiss *et al.* [550] obtained qualitatively different results for both the average conductance of an N-alkanedithiol and the length dependence. Further studies showed that the analysis of the conductance histogram would not yield a unique trace for the conductance, but rather several traces or peaks [551, 553]. The initial puzzling situation is by now resolved to a large extent. Presently, different groups agree that molecular junctions based on alkanedithiols are typically characterized by three conductance values. These can be labeled G^l (low), G^m (medium) and G^h (high). The authors explain that each G value corresponds to a single molecular junction of a different type, which is characterized by the atomic configuration at the molecule-electrode bond [551, 553, 555, 561]. Changes in the internal alkane conformation (from trans to gauche) can also

---

[4]When a thiolated molecule is adsorbed on gold surfaces the H of the thiol terminations desorbs and the sulfur atoms at each end bond strongly to the Au surfaces [560].

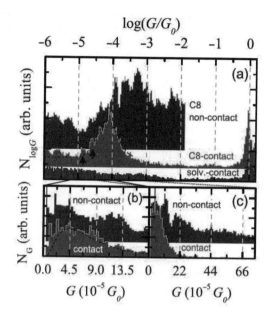

Fig. 14.3   Conductance histograms of Au-octanedithiol-Au junctions measured with a microfabricated MCBJ setup. (a) Log G-histograms built from sets of 100 G(z) traces (figure 1(c)). A peak structure is observed when C8 is in solution in contrast to the flat, pure solvent histogram. The broad peak in contact mode can be fitted to a Gaussian curve (dotted line). A finer, superimposed structure is observed (black arrows). In the non-contact mode, new peaks appear between - 4 and -3. (b) and (c) Linear G-histograms for C8 in two different G ranges. A peak centered at $2.2 \times 10^{-4} G_0$ appears only in the non-contact mode. Reprinted with permission from [556]. Copyright 2008 IOP Publishing Ltd.

result in different conductance values [553, 555, 561]. These assumptions are supported by several ab initio calculations that predict a significant conductance variation upon atomic rearrangement [562, 563, 555].

The present situation has been summarized by González *et al.* [556]. There is good agreement among the values assigned to G^l, G^m and G^h by different groups [551, 553, 555, 561]. Also, individual G values reported in initial experiments (where only a restricted conductance range was explored) [549, 550] are in good agreement with one of the three conductance values. The important exception was until recently G^h. While this value is reported by several groups working with STM break-junctions [549, 555, 561], no molecular signature was initially observed in that conductance range in MCBJ experiments [552]. In Ref. [556] González *et al.* studied the conductance of octanedithiol using a MCBJ setup and found for

Table 14.1   Values of the three main peaks (low, medium and high) of the conductance histograms of Au-alkanedithiol-Au junctions. The integer N indicates the number of C atoms in the molecule.*

alkanedithiol	G^l (G_0)	G^m (G_0)	G^h (G_0)
$N = 5$	2.45×10^{-5}	8.26×10^{-4}	-
$N = 6$	3.16×10^{-5}	2.58×10^{-4}	1.22×10^{-3}
$N = 8$	1.14×10^{-5}	5.68×10^{-5}	2.71×10^{-4}
$N = 9$	6.06×10^{-6}	2.58×10^{-5}	1.27×10^{-4}
$N = 10$	2.84×10^{-6}	5.81×10^{-6}	2.17×10^{-5}

* Taken from Ref. [555].

the three peaks. The first one was found at $G^l = 1.2 \times 10^{-5} G_0$,[5] which was attributed to the conductance of a single-molecule junction. The other two peaks appear at $G^m = 4.5 \times 10^{-5} G_0$, and $G^h = 2.3 \times 10^{-4} G_0$, see Fig. 14.3. They found that the G^m has the strongest statistical weight, whereas G^h is only observed in a non-contact mode, in which the electrodes do not get into contact before each new molecular junction formation. They proposed that these two values reflect the formation of several molecular junctions in parallel between the electrodes.

Then, what is the linear conductance of Au-alkanedithiol-Au? In Table 14.1 we have reproduced the experimental results of Wandlowski's group obtained with STM break-junctions for the three main peaks in the conductance histograms of alkanedithiol molecules of different length [555]. These values show an exponential decay of the linear conductance as a function of the number of C atoms (or length) for both the medium and the high peaks with exponents of 0.94 and 0.96 per carbon atom (β_N), respectively. However, the low peak does not exhibit such an exponential decay, see Ref. [555] for further details.

Let us discuss now how the transport takes place through alkane molecules. As we explained in the previous chapter, the analysis of the I-V characteristics in experiments involving alkane SAMs have shown clearly that the transport mechanism is coherent tunneling [41, 130]. This has also been confirmed in single-molecule experiments [557]. This is indeed what is naively expected from the electronic structure of these carbon chains. In Fig. 14.4 we have summarized some of the main features of such electronic structure, as obtained from DFT-based calculations (see footnote 2). As one can see, these molecules exhibit a very large HOMO-LUMO gap of

[5]This peak was followed by several ones at multiples of G^l.

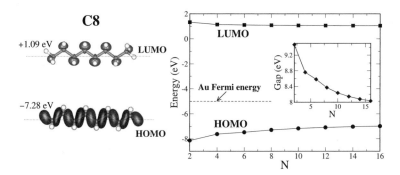

Fig. 14.4   Electronic structure of alkane molecules as computed from DFT (see text). (a) Frontier orbitals (HOMO and LUMO) of octane (C8). (b) HOMO and LUMO levels for alkanes of different length (N is the number of carbon atoms). The dashed line indicates the approximate position of the Fermi energy of gold. The inset shows the HOMO-LUMO gas vs. N.

more than 8 eV. The HOMO lies around 2-3 eV below the Fermi energy (or negative work function) of gold.[6] Thus, it is reasonable to assume that the transport in Au-alkanedithiols-Au junctions takes place through the tails of the HOMO of these molecules. This simple picture is basically confirmed by the existent DFT-based calculations of the linear conductance of these junctions [562, 563, 555]. However, there are still significant discrepancies between the different theoretical studies, as we now proceed to explain.

The DFT-based study of Ref. [555] indicates that the conductance of these junctions strongly depends on the binding geometry. These authors proposed values of 0.83 and 0.88 for the attenuation factor per C atom (β_N) for the medium and high conductance peaks, respectively, which is in fair agreement with the experimental results reported in that work. They also indicated that these exponents are sensitive to the functional used in the DFT calculations and differences up to 20% between functional can be expected. On the other hand, the estimates based on a complex band structure analysis, performed by Tomfohr and Sankey [299] and by Picaud *et al.* [564], suggested $\beta_N \approx 1.0$ and 0.9, respectively. However, their estimates for the tunneling barrier (distance between the HOMO of the molecule and the gold Fermi energy) of 3.5-5.0 eV exceed the values of Ref. [555] by a factor of 2. Another study by Müller [563] reported a comprehensive

---

[6]The inclusion of thiol groups at the end of the carbon chains introduces states close to the gold Fermi energy. These states are mainly localized in the sulfur atoms and therefore, they are not expected to play a role in the conduction, at least for long molecules. The situation may be different in the case of short alkanes.

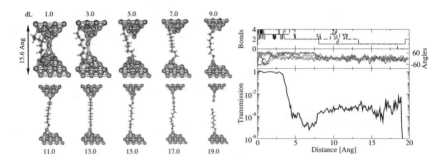

Fig. 14.5   (left) Snapshots of the formation of an octanedithiol molecular junction sim-ulated using DFT-based molecular dynamics. As the junction is being stretched, the molecule migrates into the junction and pulls out a short gold chain before finally break-ing. (right) Calculated electron transmission probability as a function of stretching distance. The number of Au-S bonds (defined by $r_{Au-S} < 3.3$ Å) and dihedral angles ($0° \sim$ straight molecule; $60° \sim$ gauche defect) for the $S-C_8-S$ chain are also shown. Reprinted with permission from [566]. Copyright 2009 American Chemical Society.

transport calculation using the TRANSIESTA package and also showed a strong dependence of the conductance on the contact geometry. How-ever, the obtained exponent of $\beta_N = 1.25$ is in clear disagreement with the other theoretical results. Moreover, the HOMO-LUMO gaps reported in that study were unrealistically large (17 eV). An attempt to go beyond the DFT approach for the conductance through alkanes was made by Fagas *et al.* [565] using a configuration interaction method. Unfortunately, the discrepancy between the obtained value, $\beta_N = 0.5$, and the experimental observation is even larger than the DFT-related uncertainty.

One of the main problems of ab initio theories in molecular electronics is the fact that the numerical calculations are so time-consuming that at the moment it is practically impossible to do a proper statistical analysis of the transport properties of a molecular junction. In this sense, most the-oretical studies are restricted to the analysis of a few idealized geometries and the comparison of these results with the experiment should be taken with caution. In some cases, it has been possible to perform some small molecular dynamic simulations to get an insight into the most probable con-tact geometries, see our discussion about hydrogen and benzene in the next subsections. In the case of alkanes, Paulsson *et al.* [566] have recently re-ported a study of the formation and conductance of alkanedithiol junctions using DFT-based molecular dynamics. This study provides a very valu-able insight into the formation mechanism of junctions based on thiolated molecules and gold electrodes. This work also shows that the conductance along the last "plateau" is very sensitive to the contact geometry, and one can observe upon stretching large variations in the conductance of an order of magnitude when gauche defects are present. We show an example of these simulations in Fig. 14.5 for octanedithiol (C8), where one can see both the evolution of the contact geometry upon stretching and the corresponding transmission probability. From these simulations, the authors constructed rudimentary conductance histograms from which they deduced a value of $\beta_N = 1.19$ and values of the conductance peaks of 2.2×10^{-3}, 3.1×10^{-4}, and 2.0×10^{-5} in units of G_0 for C6, C8, and C10, respectively. Notice that these values tend to overestimate the experimental results of Table 14.1 (see Ref. [566] for further details).

14.1.3 *The smallest molecular junction: Hydrogen bridges*

Since the goal of this section is to discuss the coherent transport through certain reference systems, it seems natural to include here the analysis of probably the simplest molecular junction that one can think of. Smit *et al.* [127] obtained molecular junctions of a hydrogen molecule between platinum leads using the MCBJ technique. In Fig. 14.6 we reproduce some of the results of this experiment. The inset shows a conductance curve for clean Pt (black) at 4.2 K, before admitting H_2 gas into the system. About 10000 similar curves were used to build the conductance histogram shown in the main panel (black, normalized by the area). After introducing hydrogen gas the conductance curves were observed to change qualitatively as illustrated by the gray curve in the inset. The dramatic change is most clearly brought out by the conductance histogram (gray, hatched). Clean Pt contacts show a typical conductance of $1.5 \pm 0.2\,G_0$ for a single-atom contact, as it can be inferred from the position and width of the first peak in the Pt conductance histogram. Below $1\,G_0$ very few data points are recorded, since Pt contacts tend to show an abrupt jump from the one-atom contact value into the tunneling regime towards tunnel conductance values well below $0.1\,G_0$. In contrast, after admitting hydrogen gas a lot of structure is found in the entire range below $1.5\,G_0$, including a pronounced peak in the histogram near $1\,G_0$.

Apart from the simplicity of the hydrogen molecule, what makes this system so special is the thorough characterization of these junctions that was carried out both in the original work and in subsequent papers. The information gathered in the different works can be summarized as follows:

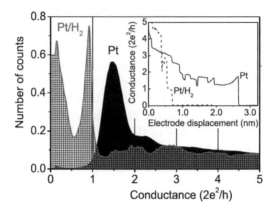

Fig. 14.6 Conductance curves and histograms for clean Pt, and Pt in a H_2 atmosphere. The inset shows a conductance curve for clean Pt (black) at 4.2 K recorded with a bias voltage of 10 mV, before admitting H_2 gas into the system. About 10000 similar curves are used to build the conductance histogram shown in the main panel (black). After introducing hydrogen gas the conductance curves change qualitatively as illustrated by the grey curve in the inset, recorded at 100 mV. This is most clearly brought out by the conductance histogram (grey; recorded with 140 mV bias). Reprinted by permission from Macmillan Publishers Ltd: Nature [127], copyright 2002.

- The presence of the molecules was confirmed with the signatures of vibration modes at energies between 40-70 meV in I-V characteristics [127]. Such signatures cannot be attributed to Pt, which has a Debye energy of around 20 meV.
- The shift in the vibrational energies upon isotope substitution of H_2 by D_2 and HD confirmed that the modes were indeed associated to a hydrogen molecule.
- Upon stretching of the contacts, the energy of the lowest modes increased, which indicates that these modes are transverse ones [567].
- The analysis of the conductance fluctuations [127] and especially shot noise measurements showed that the conductance in the range of $1\,G_0$ is largely dominated by a single channel [568].

The physics behind the signatures of vibration modes and the importance of transport properties like the shot noise will be discussed in detail in subsequent chapters, where we shall come back to this example.

All the observations detailed above offer a very stringent test to the theory, which should explain consistently all the experimental results. Before reviewing the existent work in the literature, it is interesting to discuss the

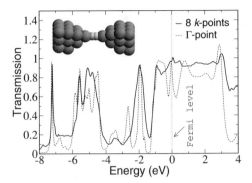

Fig. 14.7   Calculated transmission for the molecular hydrogen contact shown in the inset. For comparison both the k-point sampled transmission and the Γ-point transmission are shown. The wide plateau with $T \approx 1$ extending across the Fermi level indicates a single, robust conductance channel with nearly perfect transparency. Reprinted with permission from [571]. Copyright 2005 by the American Physical Society.

naive expectation for the conductance of a hydrogen molecule. Obviously, in the transport through this molecule, only the bonding and antibonding states (formed by the hybridization of the $1s$ orbitals of the H atoms) can contribute. A typical DFT calculation yields a HOMO-LUMO gap of about 10.5-11 eV, depending on the functional used, and the HOMO turns out to be located at ~ -10 eV with respect to the vacuum level. These results are for the equilibrium geometry, where the H-H distance is equal to 0.74 Å. This means that Fermi energy of Pt lies more or less in the middle of the gap. Thus, one might naively think that in view of the huge gap, H_2 should be poorly conductive, in clear contrast to the experiments. We shall see below that this naive picture fails because the molecule is significantly distorted in the Pt junction.

Different authors have studied theoretically the conductance of Pt-H_2-Pt junctions [127, 569–572]. The most satisfactory explanation so far has been proposed by Thygesen and Jacobsen [571], who presented conductance calculations based on density functional theory (DFT) showing that a hydrogen molecule bridging a pair of Pt contacts can have a conductance close to $1\,G_0$. In Fig. 14.7 we reproduce the main result of Ref. [571], where one can see the transmission as a function of energy for the junction shown in the inset. Notice in particular the presence of a plateau with a transmission close to 1 in an energy window of 4 eV around the Fermi level, suggesting the existence of a single conduction channel with nearly perfect transmission. The geometry of the inset was fully optimized and it is characterized by the bond lengths $d_{H-H} = 1.0$ Å and $d_{Pt-H} = 1.76$ Å. This means that in this stable configuration the molecule has been largely deformed and, in particular, the HOMO-LUMO gap is significantly reduced with respect to its value in vacuum. On the other hand, the authors showed that the vibration modes of the hydrogen molecule in this configuration are in fair agreement with the experimental results [567].

With respect to the physical mechanism, by performing a Wannier function analysis, they could establish that the transport is dominated by the antibonding state of the molecule. In particular, the transmission plateau in Fig. 14.7 is a result of a strong hybridization between the H_2 antibonding state and a combination of d- and s-like orbitals located on the neighboring Pt atoms. The antibonding orbital was found to be 0.1 eV above the Fermi energy (E_F), while the bonding orbitals lied 6.4 eV below E_F. Moreover, a coupling matrix element of 1.9 eV between the antibonding state and the leads was obtained. Therefore, the conclusion is that one has resonant transport through the antibonding orbital that has been largely broadened due to the strong hybridization with the Pt electrodes.

Other DFT calculations have been performed [569, 570, 572]. Using a slightly different approach García *et al.* [570] obtained a conductance well below $1 G_0$. They propose an alternative atomic arrangement to explain the high conductance for the Pt-H bridge, consisting of a Pt-Pt-bridge with two H atoms bonded to the sides in a perpendicular arrangement. However, this configuration gives rise to three conduction channels, which is excluded based on the analysis of shot noise and conductance fluctuations as discussed above. The origin of this discrepancy is still unclear (see Ref. [545] for some ideas).

Let us conclude this discussion by saying that conductance histograms recorded using Fe, Co and Ni electrodes in the presence of hydrogen also show a pronounced peak near $1 G_0$ [403], indicating that many transition metals may form similar single-molecule junctions. Also Pd seemed a good candidate, but Csonka *et al.* [573] find an additional peak at $0.5 G_0$ in the conductance histogram, and it was argued that hydrogen is incorporated into the bulk of the Pd metal electrodes.

14.1.4 *Highly conductive benzene junctions*

As we shall discuss in detail in the next section, the most common approach in fabrication of molecular junctions utilizes functional side groups attached to the main molecule structure as anchoring "arms" that chemically bind to metallic leads (e.g. thiol [16], amine [126] and carboxylic [557] groups). In many cases anchoring groups act as resistive spacers between the electrodes and the molecule. This leads to low conductance and sensitivity to different environmental effects such as neighbor adsorbed species [574]. In order to overcome these problems, Ruitenbeek and coworkers [473] have recently reported on a highly conductive molecular junction achieved by direct binding of a π-conjugated organic molecule (benzene) to metallic electrodes (Pt) without the use of anchoring groups. Again, the thorough analysis of the transport properties through these junctions makes Pt-benzene-Pt contacts a nice test system. In this sense, the goal of this section is to briefly describe the work of Ref. [473].

The measurements were performed using the MCBJ technique and they were conducted at 4 K. Following the formation of the Pt junction, the benzene was admitted using a leak valve via a heated capillary to the Pt junction while the latter is broken and formed repeatedly. During the benzene introduction, the typical Pt peak is observed to be suppressed, and a single peak appears near $1\,G_0$ accompanied with a low conductance tail (Fig. 14.8, filled curve). In some cases, the histogram exhibits a peak near $0.2\,G_0$ on top of the tail. These findings imply that after the introduction of benzene, the formation of pure Pt junctions is suppressed while new junctions with preferred conductance of $1\,G_0$ and sometimes $0.2\,G_0$ are formed while stretching the contact.

Following the spirit of the experiments on hydrogen, the presence of the molecule was identified by vibrational spectroscopy that revealed a well-defined mode at around 42 meV in the zero-bias-conductance region of 0.05-$0.4\,G_0$, which was rather insensitive to stretching of the contact. On the other hand, shot noise measurements showed that the number of channels is eventually reduced to one when the conductance is reduced to $0.2\,G_0$, while at higher conductance (also well below $1\,G_0$) multiple channels make up the transport across the junction.

What is the naive expectation for the conductance of a benzene molecule? As we discussed in section 9.5.1, the electronic structure of benzene is determined by a delocalized π-orbital system formed by 6 π-orbitals (the p-orbitals pointing out of the benzene plane), one per C atom. This

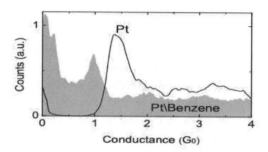

Fig. 14.8   Conductance histograms for a Pt junction (black), and for Pt after introducing benzene (filled) measured with the MCBJ technique. Each conductance histogram is constructed from more than 3000 conductance traces recorded with a bias of 0.1 V during repeated breaking of the contact. Reprinted with permission from [473]. Copyright 2008 by the American Physical Society.

simple picture is confirmed by DFT calculations (see footnote 2), which also predict a HOMO-LUMO gap of 5.14 eV with the two-fold degenerate HOMO lying at -6.26 eV (measured with respect to vacuum), see Fig. 9.4. Taking into account that the Fermi energy (or negative work function) of Pt is around -5.4 eV, one naively expects that if there is no substantial charge transfer, the transport must be dominated by the HOMO. With respect to the value of the conductance, it will depend crucially on the strength of the metal-molecule coupling (see Exercise 13.7).

The authors of Ref. [473] performed DFT structural simulations to determine the contact geometry and conductance calculations based on the method detailed in Ref. [576]. Their main conclusions are: (i) benzene can indeed form a stable bridge between Pt contacts with a conductance as high as $1\,G_0$ and (ii) stretching of the junction leads to tilting of the molecule which reduces both the conductance and the number of transmission channels across the junction as a consequence of sequential breaking of the Pt-C bonds. The main take-home message from the theoretical analysis is that the high conductance can be attributed to the strong hybridization of the benzene molecule with the Pt contacts. This can be seen in Fig. 14.9, where we show both the transmission and density of states (DOS) projected onto the frontier molecular orbitals as a function of energy for two geometries at different stages of the stretching process. Notice that when the conductance is close to $1\,G_0$ (see left panels), the transport is dominated by the HOMOs of the molecule, which are no longer degenerate because of the different coupling to the Pt electrodes. In this geometry, the conductance is due

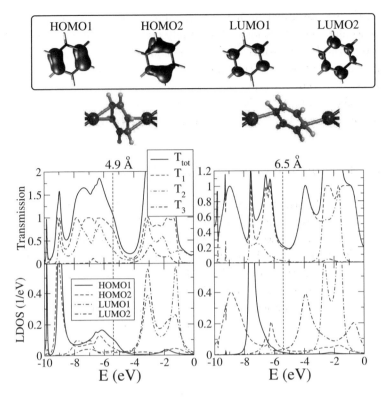

Fig. 14.9   Transmission and density of states (DOS) as a function of energy in a Pt-benzene-Pt junction as calculated in Ref. [473]. The left panels show the results for a geometry where the outermost Pt atoms were separated a distance of 4.9 Å, while the right ones show the corresponding results for a separation of 6.5 Å. The contact geometries are shown on top of these panels. The transmission plots show both the total transmission and its decomposition into individual transmission coefficients, T_i. The local DOS has been projected onto the four benzene frontier orbitals, which are shown in the upper part of the figure. The vertical dashed lines indicate the position of the Fermi energy (-5.4 eV). Courtesy of Sören Wohlthat.

to two channels and the frontier orbitals of the benzene acquired a large broadening due to the strong interaction with the metallic leads. When the elongation of the contact proceeds, the reduction of the metal-molecule coupling becomes apparent in the transmission curve with the appearance of a pseudo-gap around the Fermi energy (see right panels in Fig. 14.9). In this case, the transport is dominated by a single conduction channel. The reason for this is not really obvious from the information of the local DOS. Then, what determines the number of channels in this case? As a rule of thumb, an upper limit to the number of conduction channels when the molecular junction is formed, is simply given by the number of C atoms bonded to the Pt tip atoms. This can be understood as follows. Since each C atom has only one orbital taking part in the π-system of the benzene ring, each C atom can build at most one π-channel. This is nothing else than another example of the simple rule that we discussed in the context of the conductance of a single-atom contact (see section 11.5).

14.2   Metal-molecule contact: The role of anchoring groups

As we discussed in the previous chapter, one of the fundamental ingredients that determines the coherent transport through molecular junctions is the strength of the metal-molecule coupling. This strength can be tuned chemically, at least up to certain degree, by using appropriate anchoring groups to bind a molecule to metallic electrodes. How to choose the linker group? The choice depends primarily on the type of metal-molecule combination used to build the junctions and usually only a few anchoring groups are possible. On the other hand, the choice also depends on the functionality that one wants to implement in the system. If the goal is to achieve a high conductance, then the anchoring group is chosen to maximize the strength of the metal-molecule coupling.[7] Other important factors to bear in mind are the stability of the contact and the variability of the bonding between the terminal group and the metal, which can play a fundamental role in the reproducibility of the experimental results.

The majority of candidates for end-group/metal pairings for molecular electronics come from studies of self-assembled monolayers (SAMs) [577], such as thiolated molecules on gold surfaces. The combination of thiol as an end group and gold electrodes is by far the most studied metal-molecule binding motif in molecular electronics so far. In the previous section we have given several examples of this combination. Lately, it has been argued that the variability in the bonding between thiol groups and gold may be harmful for the reliability of electrical measurements on single molecules [126]. For this reason, different alternatives are currently being explored in many laboratories. An interesting possibility was put forward by Venkataraman and coworkers in Ref. [126], where the authors suggested the use of amine

---

[7]A strong coupling is not always the goal. In some cases, one may want to partially decouple the molecule from the leads, like in the case of the molecular transistors (see next chapter).

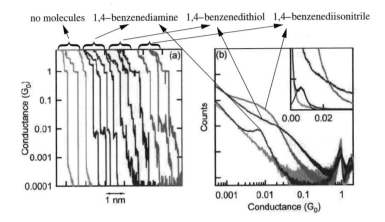

Fig. 14.10   (a) Sample conductance traces measured with STM Au break-junction without molecules and with 1,4-benzenediamine, 1,4-benzenedithiol, and 1,4-benzenediisonitrile shown on a semilog plot. (b) Conductance histograms constructed from over 3000 traces measured in the presence of the three molecules shown on a log-log plot. The control histogram of Au without molecules is also shown. Inset: same data on a linear plot showing a Gaussian fit to the peak (black curve). Adapted with permission from [126]. Copyright 2006 American Chemical Society.

(NH_2) groups to obtain well-defined values of the conductance of molecular junctions. In this work, the conductance of amine-terminated molecules was measured by breaking Au atomic contacts in a molecular solution at room temperature. It was found that the variability of the observed conductance for the diamine molecule-Au junctions is much less than the variability for diisonitrile- and dithiol-Au junctions. This narrow distribution enabled the authors to unambiguously determine the conductance of single molecules. The conductance histograms obtained in Ref. [126] for three differently substituted aromatics, 1,4-benzenedithiol, 1,4-benzenediisonitrile, and 1,4-benzenediamine, are shown in Fig. 14.10(b). Notice that in comparison to the data for the dithiol or the diisonitrile, the conductance histogram for 1,4-benzenediamine is particularly well-defined. From this histogram, a conductance value for this molecule of $0.0064 \pm 0.0004 \, G_0$ was deduced.

With the help of DFT-based calculations, it was suggested in Ref. [126] that the reproducible electrical characteristics result from the selective binding between the gold electrodes and amine link groups through a donor-acceptor bond to under-coordinated gold atoms. The amine end groups have been used by Venkataraman and coworkers to study the transport through alkanes [558], to analyze the role of the conjugation in the transport through biphenyl molecules [477] and to establish a detailed comparison with theory [578]. For more information, see Ref. [579], which provides a comprehensive review of single-molecule junction conductance measurements across families of molecules measured while breaking gold point contacts in a solution of molecules with amine end groups.

Tao and coworkers have systematically studied and compared the single-molecule conductance of alkanes terminated with dicarboxylic-acid (COOH), diamine, and dithiol anchoring groups [557]. The conductance values of these molecules were found to be independent of temperature, indicating coherent tunneling. For each anchoring group, the authors reported an exponential decay of the conductance with the molecular length, given by $G = A \exp(-\beta_N N)$, which also suggests the tunneling mechanism. The prefactor of the exponential function, A, a measure of contact resistance, turned out to be highly sensitive to the type of the anchoring group, which varies in the order Au-S > Au-NH$_2$ > Au-COOH. This dependence was attributed to the different coupling strengths provided by the different anchoring groups between the alkane and the electrodes. On the other hand, with respect to the spread of the peaks in the conductance histograms, there were no significant differences between thiols and amines. Something similar has also been reported by Martin *et al.* [535]. Using microfabricated gold break-junctions, these authors measured the conductance histogram for benzenediamine. In contrast to Ref. [477], they did not find a pronounced peak structure. According to these authors, the difference may be due to the absence of a solvent in their experiment and also to the fast rupture of the metal-molecule bond that must have reduced the probability of forming stable molecular junctions.

From the previous discussion it is obvious that the conductance values are not necessarily correlated with the selectivity of the binding that leads to narrow peaks in the conductance histograms. It would be desirable to find linker groups with the properties of amines, but providing a stronger coupling. With this goal in mind, Park *et al.* [559] have compared the low bias conductance of a series of alkanes terminated on their ends with dimethyl phosphines, methyl sulfides, and amines and found that junctions formed with dimethyl phosphine terminated alkanes have the highest conductance. Furthermore, they observed a clear conductance signature with these linker groups, indicating that the binding is well-defined and electronically selective.

As we discussed in a previous section devoted to the transport through benzene molecules, an interesting possibility is the use of other metals than

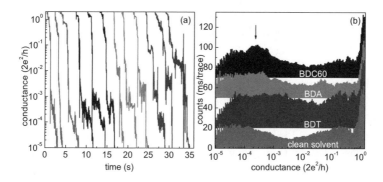

Fig. 14.11   (a) Conductance traces of 1,4-bis(fullero[c]pyrrolidin-1-yl)benzene (BDC60), 1,4-benzenediamine (BDA) and 1,4-benzenedithiol (BDT) measured using lithographic gold MCBJs. (b) Conductance histograms on a semilog scale, constructed from 400 consecutive traces. The arrow marks the typical junction conductance of BDC60. All curves are offset for clarity. Colors in panel (a) correspond to those in panel (b). Reprinted with permission from [536]. Copyright 2008 American Chemical Society.

gold. As we showed in that section, the use of a transition metal like Pt allows exploring the chemistry of unsaturated carbon bonds. In the case of benzene, this led to a very high conductance, of the order of $1\,G_0$, to be compared with the conductance of $0.0064 \pm 0.0004\,G_0$ reported for benzenediamine in Ref. [477]. This illustrates the fact that in many cases the anchoring groups are acting as spacers or potential barriers that diminish the conductance of the junctions. Of course, the use of other metals is often hindered by the oxidation of those metals, which can only be avoided working under UHV conditions.

The direct binding of carbon structures, like C_{60}, to gold electrodes has also been explored in the literature. C_{60} is known to hybridize strongly with gold surfaces [580], and in single-molecule junctions it can exhibit conductances on the order of one tenth of G_0 [128]. These results suggest that one could also use C_{60} as an anchoring group. Indeed, this possibility has been recently investigated by Martin *et al.* [536]. These authors have designed and synthesized a linear and rigid C_{60}-capped molecule, 1,4-bis(fullero[c]pyrrolidin-1-yl)benzene (BDC60), and compared the electrical characteristics to those of 1,4-benzenediamine (BDA) and 1,4-benzenedithiol (BDT) using lithographic MCBJs. The main conclusion of this work is the suitability of fullerene-anchoring for single-molecule transport measurements. In particular, compared to thiols the fullerene-anchoring leads to a considerably lower spread in low-bias conductance due to the higher junction stability that minimizes fluctuations due to atomic details at the anchoring site.

More recently, Zotti *et al.* [472] have studied both experimentally and theoretically the transport through tolane molecules attached to gold contacts via different anchoring groups. From the experimental side, they showed that the molecules with thiol and nitro groups can sustain a much higher current (see Fig. 13.10). From the theory side, and with the help of DFT-based calculations, they showed that the anchoring not only determine the strength of the metal-molecule coupling (i.e. width of the molecular resonances), but they also control the position of the molecular energy levels. In particular, they showed that in the case of thiol and amine groups, their electron-donating character is reflected in the fact that the HOMO of the molecules dominates the transport. On the contrary, nitro and cyano groups have an electron-withdrawing character, which means in practice that the LUMO is pushed closer to the gold Fermi energy and it dominates the electrical conduction. Moreover, these authors showed that there is no direct relation between the metal-molecule binding energies for different anchoring groups and the corresponding junction conductances. This is obvious in the case of molecules where the LUMO dominates the transport, since this orbital plays practically no role in the binding energy.

As a last comment, let us say that not only the type of anchoring group matters, but also its exact position. In a nice work, Mayor *et al.* [581] showed that the conductance of a thiol-terminated indexanchoring groups! thiol rod-like conjugated molecule depends crucially on the position of the thiol group. They showed that by placing the thiol group in the *meta* position of the last phenyl ring, the conjugation is partially interrupted and the current decreases significantly as compared with the case in which the thiol group is in the *para* position.

14.3   Tuning chemically the conductance: The role of side-groups

As it is clear from our discussions in the previous chapter, the coherent transport through a molecular junction depends crucially on the position of the relevant orbitals of the molecule with respect to the metal Fermi energy and also on their character (degree of delocalization). Thus, the internal electronic structure of a molecule plays a fundamental role and it can be chemically tuned to certain extend with the inclusion of appropriate side-groups or substituents . In principle, side-groups can have two main effects: (i) they can control the structure of a molecule which in turn determines the degree of conjugation (delocalization of the molecular orbitals) and (ii) they can tune the position of the frontier orbitals. Both effects can have an impact in the conductance of a junction. These effects are well-known in the field of electron transfer [582, 583], but so far they have been quite difficult to test systematically in molecular junctions.

The fact that the conformation must have a major impact on the conduction through a molecular junction has been predicted long ago [584, 585] and it is very easy to explain, as we saw in section 13.5. Such impact have been illustrated in different experiments [586–588], but probably the most illustrative example have been reported by Venkataraman and coworkers in Ref. [477]. As we explained in section 13.5, these authors carried out a detailed study of the conductance of a series of biphenyl molecules with different twist angles, θ, that were coupled to gold electrodes via amino linking groups. They showed that the conductance follows a $\cos^2 \theta$ dependence, as expected for transport through π-conjugated biphenyl systems (see Refs. [584, 260] and section 13.5).

As it was shown by Pauly *et al.* [478], ab initio calculations based on DFT show that the low-temperature conductance of biphenyl derivatives follows closely the $\cos^2 \theta$ law consistent with an effective π-orbital coupling model. A comparison between theory and the results of Ref. [477] has been reported by Finch *et al.* [589]. These authors studied the conductance of the series of 8 molecules shown in Fig. 14.12(a), with both thiol and amine anchoring groups. They showed that if the Fermi energy E_F lies within the HOMO-LUMO gap, then the experimental results are reproduced. More generally, however, if E_F is located within either the LUMO or HOMO states,[8] the presence of resonances destroys the linear dependence of the conductance on $\cos^2 \theta$ and gives rise to non-monotonic behavior associated with the level structure of the different molecules. These results are illustrated in Fig. 14.12(b).

It is worth mentioning that the conduction in the experiment of Ref. [477] is not completely suppressed when $\theta = \pi/2$. In this limit also σ-orbitals contribute to the effective coupling that allows a finite current

---

[8]In an experiment, E_F may differ from the computed value for a number of reasons, including the presence of a dielectric environment, such as air or water.

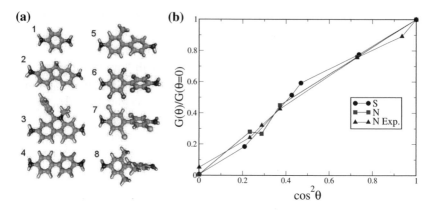

Fig. 14.12 Theoretical results on the conductance of biphenyl derivatives. (a) Molecules studied capped with NH_2. The dark vertex in the backbone of molecule 3 corresponds to N and the side groups of molecules 6 and 7 (other than H) correspond to F and Cl atoms, respectively. (b) Zero-bias conductance in a hollow configuration, for sulfur contact (circles), nitrogen contact (squares) and values from Ref. [477] (triangles). All cases have been normalized to the $\theta = 0$ value. Adapted with permission from [589]. Copyright 2008 IOP Publishing Ltd.

to flow through the system [478]. In this sense, Pauly *et al.* [546] have shown theoretically that the conductance of oligophenylenes of different length remains finite when the molecules are modified with methyl side-groups, although these substituents induce a rotation of the neighboring phenyl rings of about 90°. The typical reduction of the conductance, in comparison with the conjugated molecules, is about two order of magnitude. Recently, Lörtscher *et al.* [590] have shown experimentally that such non-conjugated molecules are still conductive.

The role of the conjugation in the conduction through molecular systems can also be illustrated without resorting to side-groups. Thus for instance, the comparison of the conductance through alkanes to that through proto-typical molecular wires with extended π-electron states, like oligophenyle-neethynylene (OPE) or oligophenylenevinylene (OPV), shows substantially higher conductance through the conjugated molecules and a rational dependence on the HOMO-LUMO gap [475, 591–593].

As mentioned above, the second main effect of side-groups is to shift the frontier orbitals of a molecule. In this sense, side-groups can be used to improve the usually bad alignment between the molecular levels and the Fermi energy of the metallic electrodes. In other words, and using terminology of semiconductor physics, one can use side-groups to "dope"

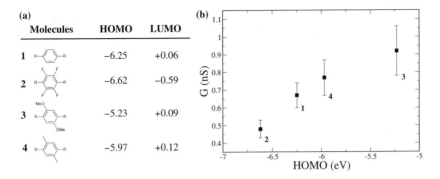

Fig. 14.13   Chemical control of double barrier tunneling in α, ω-dithiaalkane molecular wires. (a) Molecules used and their frontier orbital energies (in eV) as obtained from a DFT calculation. R = HS(CH$_2$)$_6$- in all cases. (b) Plot of conductances determined by $I(t)$ method (with standard deviations) against HOMO energy for molecules 1-4. From [594]. Reproduced by permission of The Royal Society of Chemistry.

molecular junctions. This effect has been studied experimentally by several groups [558, 594, 102], but it seems that it is still rather difficult to show this basic effect in a systematic manner. Let us briefly describe the work of Leary *et al.* [594], where the authors studied the low-bias conductance of 1,4-bis-(6-thiahexyl)-benzene derivatives using the STM-based $I(t)$ and $I(s)$ methods that will be discussed in the next section. In particular, they investigated the four benzene derivatives shown in Fig. 14.13(a). In order to achieve a contact with gold electrodes, these molecules contain radicals, which act a linking groups, consisting of thiolated alkyl chains [HS(CH$_2$)$_6$]. The central idea of this work was to study the correlation between the low-bias conductance and the position of the frontier orbitals. For this purpose, the authors determined theoretically the position of these orbitals by means of DFT calculations [see Fig. 14.13(a)]. In panel (b) of the same figure one can see the experimental results for the conductance plotted as a function of the theoretical position of the HOMO of the isolated molecules. This graph shows that the more electron-rich benzene rings (with a higher HOMO) give higher conductances, which is consistent with hole conduction (i.e. via the benzene HOMO). These results constitute a beautiful illustration of the doping effect, although the change in the conductance is still rather small (smaller than a factor 2). Anyway, let us stress that what really determines the conductance is the actual position of the frontier orbitals of the molecule in the junction, which in principle may differ from the corresponding ones in gas phase. In that sense, it would be highly desirable to obtain information in-situ about the level alignment with the electrode Fermi energy, along the lines of Ref. [461]. This is of course extremely challenging in the case of single-molecule junctions, although not impossible as we shall see in Chapter 20.

Another example of this doping effect was presented by Venkataraman *et al.* [558]. In this case, the authors studied the single molecule conductances of a series of very short conjugated molecules (substituted 1,4-diaminobenzenes) using an STM-based break-junction technique. They found that electron donating substituents resulted in higher molecular conductances, and there was an approximate correlation between the conductance and the Hammett σ_p parameter,[9] consistent with hole transport (i.e. transport dominated by the HOMO of the molecules). Another interesting example related to the influence of side-groups has been reported by Baheti *et al.* [102]. In this work the thermopower of molecular junctions based on several 1,4-benzenedithiol (BDT) derivatives was investigated. The BDT molecule was modified by the addition of electron-withdrawing or -donating groups such as fluorine, chlorine, and methyl on the benzene ring. It was found that the substituents on BDT generated small and predictable changes in conductance depending on their character. Moreover, the authors showed that by replacing the thiol end groups by cyanide end groups the transport changes radically and it turns out to be dominated by the LUMO of the molecule. These results will be discussed in more detail in Chapter 19 in the context of thermoelectricity in molecular junctions.

14.4   Controlled STM-based single-molecule experiments

One of the major problems in most of the experiments that we have discussed so far is the fact that it is not easy to prove that one is dealing with a single molecule. In principle, the STM constitutes an ideal tool to resolve this issue.[10] The STM can be utilized to perform controlled transport experiments through individual molecules that have been deposited onto metal surfaces by bringing the metallic tip into contact to the molecule. The obvious advantage of the STM is that the structure under investigation–a molecule along with its substrate–can be imaged with submolecular precision prior to and after taking conductance data. In this way, parameters

---

[9]Roughly speaking, the Hammett parameter (or constant) describes the change in reaction rates upon introduction of substituents. For a precise definition, see Ref. [595].
[10]The STM as a tool to fabricate molecular junctions has extensively described in section 3.4.4.

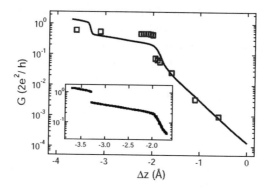

Fig. 14.14   Conductance of a C_{60} molecule deposited on a Cu(100) surface measured with a STM as a function of the tip displacement Δz. Data are an average of 500 measurements. Zero displacement corresponds to the tip position before freezing the feedback loop at $V = 300$ mV and $I = 3$ nA. The solid line correspond to the experimental data, while the square symbols correspond to calculations performed with the TRANSIESTA package. The inset shows a single conductance curve revealing a discontinuity at $\Delta z = 3.3$ Å. Reprinted with permission from [116]. Copyright 2007 by the American Physical Society.

such as molecular orientation or binding site can be monitored. Another advantage of STM is the possibility to characterize to some extent the status of the second electrode, the microscope tip, by recording conductance data on clean metal areas. Maybe the main disadvantage of the STM is its mechanical stability, which does not reach level of the break-junction methods. On the other hand, it is sometimes said that in this approach there is an inherent asymmetry in the contact and the strong substrate-molecule interaction that may distort some of the intrinsic properties of the molecule. In any case, the high degree of control in these experiments is extremely valuable at this stage in order to establish the basic transport mechanisms at the molecular scale. In this sense, it is somewhat surprising that STM data for molecular contacts are so scarce. In this section we shall briefly describe some illustrative examples of this type of controlled single-molecule experiments.

One of the first experiments of this kind was performed by Joachim *et al.* [115] who used a STM at room temperature to study the contact conductance of a C_{60} on Au(110). This experiment has been revisited more recently by Néel *et al.* [116], but at 8 K and under UHV conditions. In this experiment the molecules were deposited by sublimation onto a clean Cu(100) surface and were probed by a Cu-covered tip. The orientation of the molecules on top of the Cu(100) surface could be resolved, and only those molecules were selected that exposed a C-C bond between a hexagon and a pentagon at the top. When approaching the tip towards the molecule they observed a very reproducible jump into contact from about $G = 2.5 \times 10^{-2} G_0$ to $G = 0.25 G_0$ (see Fig. 14.14). When approaching the tip further towards the molecule a jump up to $G \approx G_0$ was observed. The detailed information provided in this experiment makes it ideal to compare with the theory. Indeed, in the same work a theoretical analysis of the linear conductance based on DFT yielded a satisfactory agreement, as one can see in Fig. 14.14. From the modeling the authors inferred that the controlled contact to a C_{60} molecule does not significantly deform its spherical shape and they also showed that the conductance around the tip-molecule contact formation is affected by a fluctuation between different microscopic configurations.

In the context of the STM, two important methods have been introduced by the Nichols' group for the measurement of single-molecule conductance [550, 596–599]. These methods are referred to as the $I(s)$ and $I(t)$ methods. In these methods the starting point for the measurements is the adsorption of a low coverage of the molecules under investigation on a Au surface. This condition typically results in flat-lying molecules and enables the formation of single-molecule wires with high probability. To attach a molecule to the STM tip, usually made also of Au, the tip is lowered onto the surface by fixing the tunneling current I_0 at relatively high values and then lifted, while keeping a constant position in the x-y plane. This procedure is illustrated in Fig. 14.15. The current decay shows distinctive current plateaus when molecular wires bridge the gap between the tip and substrate, whereas in the absence of wire formation the current simply decreases nearly exponentially with tip-sample separation, see Fig. 14.15.

The current plateaus obtained with this method have been related to electron tunneling through molecular wires bridging the STM tip and the substrate [550, 596]. Statistical analysis of the data using histogram plots has shown that the current-plateau values group themselves into discrete values, which are integer multiples of a lowest value. The lowest current peak in the histogram corresponds to a single molecule, whereas the next discrete conductance step has been assigned to conduction through two wires and so on.

An alternative method is the so-called $I(t)$ method [550]. This involves holding the Au STM tip at a given distance above the substrate while monitoring current jumps as molecular wires bridging the tip and substrate form

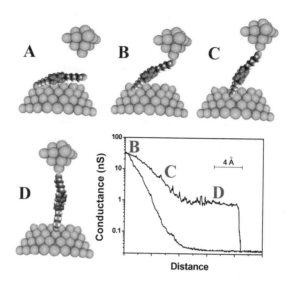

Fig. 14.15   Schematic illustration of the $I(s)$ STM method of forming molecular wires. (A) A low coverage of the studied molecule is formed on the Au(111) surface, and the set-point current is increased. (B) Attachment of the molecule at one end to the Au STM tip is achieved, and then (C-D) the tip is retracted from the surface while recording the current. The graph shows the conductance decay with distance for a clean Au substrate (lower curve) and for a molecule on Au (upper curve). In the latter curve the different stages of the contact formation are indicated (B, C and D). Notice the presence of a plateau before rupture. Courtesy of Edmund Leary.

and subsequently break. It has been shown that both the $I(s)$ and the $I(t)$ method result in the same single-molecule conductance for alkanedithiols [550]. A nice application of the $I(t)$ method can be found in Ref. [599], where the authors showed, in combination with ab initio transport calculations, that the tilt-angle dependence of the electrical conductance is a sensitive spectroscopic probe, providing information about the position of the Fermi energy.

The use of methods in the spirit of the $I(s)$ and the $I(t)$ ones are opening new ways of looking at molecular conductance. Thus for instance, Temirov *et al.* [119] have reported beautiful results on a complex system, PTCDA (4,9,10-perylenetetracarboxylic-dianhydrid), on a Ag(111) surface. They demonstrated that one can controllably contact the molecule to the STM tip at one of the four oxygen corner groups and peel the molecule gradually from the surface. The conductance clearly varies in the process of peeling, but when pulled to an upright position the conductance is approximately

$0.15\,G_0$. During the process an interesting Kondo-like resonance develops that can be tuned by the electrode position. Another spectacular example has been recently reported by Lafferentz *et al.* [600]. These authors have measured the conductance and mechanical characteristics of a single polyfluorene wire by pulling it up from a Au(111) surface with a STM tip, thus continuously changing its length up to more than 20 nm. They showed that the conductance not only decays exponentially but also exhibits characteristics oscillations as one molecular unit after another is detached from the surface during stretching.

14.5   Conclusions and open problems

Although in this chapter we have only talked about some concrete aspects of single-molecule conduction, we can already draw a few general conclusions and point out some of the main challenges for the near future. It is clear that in the last years a significant progress has been made in the experimental approaches to study single-molecule junctions as well as in the qualitative understanding of their transport properties. From the experimental side, the introduction of statistical methods to determine the conductance has partially eliminated the discrepancies between different experimental results which appear when only individual traces are compared. The use of new techniques to measure other transport properties such as shot noise or thermopower provides very valuable additional information that is not contained in the standard conductance measurements (see Chapter 19). The use of low temperatures and the improvement in the stability of the devices allow now making use of the inelastic tunneling spectroscopy (see Chapter 16), which gives an essential information about the presence of the molecules and the geometry of the junctions.

From the theory side, the development of ab initio methods makes now possible to study both the mechanical and the electrical properties in a much more reliable way. In particular, DFT-based calculations provide, for instance, a detailed information about the possible structure of the contacts, the relevant vibration modes and the conductance of the junctions. These theoretical methods are now able to describe the general experimental trends, see e.g. Ref. [579], although they still fail in general to describe quantitatively the transport results.

So in short, there are good reasons to be optimistic about the development of this field. However, it has to be acknowledged that there are still basic issues to be resolved. The challenges for the experiments concern, in the first place, the reproducibility of the results. We have seen that the statistical methods are not the panacea and the interpretation of the conductance histograms is not always straightforward. One of the main goals should be to find strategies to rigidly bind molecules to electrodes via selective anchoring groups or by means of new trapping techniques. It would also be desirable to improve the stability of the contacts to be able to extend the statistical analysis also to the I-V curves [534], which contain much more information. Most of these requirements are indeed met by the controlled STM experiments that we discussed in the previous section and, in this sense, there is no doubt that they will play an important role in the near future.

The theory has also to face several basic problems. One of the main things to do is to understand the origin of the discrepancies between the DFT-based methods which, in principle, are supposed to deliver the same results. Thus, systematic comparisons between different implementations of the DFT-NEGF approach are necessary, as proposed in Ref. [545]. On the other hand, as we explained in section 10.8, DFT as it is used in molecular electronics has clear limitations. Thus, the biggest challenge for the theory is the introduction of new methods to describe properly the role of the electronic correlations in the transport through these systems. DFT does not describe correctly the energy spectrum of a system and, in particular, it tends to give small values for the HOMO-LUMO gap, as compared with the experiment. This is the main reason behind its systematic overestimation of the low-bias conductance of molecular junctions. Moreover, DFT is not well-founded in an out-of-equilibrium situation and its use to describe the transport at finite bias is then doubtful.[11] Finally, so far most theoretical methods used to describe the I-V curves are not able to take into account the possible conformational changes that may appear when a molecule is subjected to a rather high electric field. This issue is certainly playing an important role in many experiments and it is presently out of the scope of most theories in molecular electronics.

---

[11]The reader has surely noticed that we have not presented or discussed any comparison of the I-V curves and we have focused our attention on the low-bias conductance. There are several reasons for that. First, there are very few statistical analyses of the I-V characteristics and there are no yet test systems in which different experiments agree on the shape of the I-Vs. Second, most of the existent theoretical methods fail to describe quantitatively the level spectrum of a molecule (or a molecular junction), and as consequence no quantitative agreement between theory and experiment has been obtained yet.

Chapter 15

Single-molecule transistors: Coulomb blockade and Kondo physics

15.1 Introduction

In the previous two chapters we have considered the coherent transport regime in which the electrons proceed elastically (without exchanging energy) through the junctions. What is the range of validity of the coherent picture? Intuitively, the coherent mechanism will be the dominant one as long as the time that an electron needs to cross the molecular bridge is smaller than the time that it takes to interact with other electrons or to excite vibronic degrees of freedom, i.e. the time that is needed for an electron to undergo an inelastic scattering event. A problem here is that the time that an electron spends in a junction, sometimes referred to as tunneling traversal time, is not easy to define unambiguously. Close to a resonant situation, i.e. when a molecular level is close to the Fermi energy of the leads, a measure of this time scale is $\hbar/\Gamma$, where Γ is the width of the molecular resonance due to the coupling to the electrodes. The scale $\hbar/\Gamma$ can be viewed as the lifetime of an electron for escaping into the leads. Away from the resonant condition, the traversal time, τ, is mainly determined by the injection gap, ΔE, which is the energy difference between the leads' Fermi energy and the relevant molecular orbital (HOMO or LUMO).[1] Büttiker and Landauer have shown that the traversal time obtained in the deep tunneling limit for a square barrier of energy height ΔE and width D is $\tau = D\sqrt{m/2\Delta E}$, where m is the electron mass [601]. If, instead, the bridge is described in terms of a one-dimensional lattice of N equivalent sites, this time is given by $\tau = \hbar N/\Delta E$ [602]. Then, for practical purposes we can use the unified expression $\tau = \hbar/(\sqrt{\Delta E^2 + \Gamma^2})$ as an estimate of the traversal

---

[1]In the resonant tunneling model, the injection gap is simply the energy ϵ_0 of the level, measured with respect to the Fermi energy.

time, which covers the different situations.[2] Thus, if $\hbar/\tau$ is larger than the energy scales associated with inelastic interactions like electron-electron, U, or electron-vibration, λ, then the transport is mainly coherent. At the contrary, if a molecule is weakly coupled to the electrodes ($\Gamma < \max\{U, \lambda\}$) and the system is brought close to resonance ($\Delta E \approx 0$), the transport will very likely be dominated by the Coulomb interaction in the molecule or by the excitation of internal degrees of freedom like vibrational modes.

There are by now many examples in nanophysics in which the electronic transport through a small object which is weakly coupled to metallic electrodes has been explored. Let us mention, for instance, the cases of semiconductor quantum dots, carbon nanotubes or metallic nanoparticles. In all these systems, the transport in the weak-coupling regime is governed by single-electron tunneling processes that lead to phenomena like the *Coulomb blockade effect*. Moreover, if the coupling is not so weak, other interesting many-body phenomena, like the *Kondo effect*, can show up at show temperatures. We shall see in this chapter that these phenomena also appear in single-molecule junctions. These effects have been understood in great detail in different devices with the help of a gate electrode. This third terminal is only capacitively coupled to the small object and it allows to tune its energy level spectrum and to explore different charge (or redox) states. The gate electrode allows us in turn to control the current that flows through the system with an external field, very much like in the case of field-effect transistors in microelectronics. Due to this analogy and also to the fact the transport is usually dominated by single-electron processes, these weakly coupled systems are known as single-electron transistors (SETs). In the last decade it has become possible to incorporate a gate electrode into single-molecule devices. We shall refer to these three-terminal molecular devices as *single-molecule transistors* (SMTs).

The goal of this chapter is to discuss the electronic transport through SMTs with special emphasis in the role of the Coulomb interaction in the molecules. The role of the vibrational modes in these systems will be discussed in the next chapter. With this idea in mind, we shall first review briefly the general conditions necessary to observe charging effects and we shall recall the basic signatures of these effects in the transport characteristics. Then, in section 15.3 we shall recall the main experimental techniques that have been used so far to fabricate SMTs. The experimental results in SMTs are often analyzed in the light of the "orthodox" theory of Coulomb

---

[2]See Ref. [38] and references therein for a detailed discussion about the tunneling traversal time.

blockade. For this reason, we have included a detailed description of this theory in section 15.4. The attempts to generalize the standard theory to the specific problem of SMTs are discussed in section 15.5. Section 15.6 is devoted to the intermediate transport regime (Γ not too small) and we shall pay special attention to the Kondo effect. In the last section we shall review some of the most representative experimental results obtained in the context of SMTs.

The physics, results and challenges related to STMs have been discussed in the reviews of Refs. [39, 603–607].

15.2   Charging effects in transport through nanoscale devices

In this section we examine the circumstances under which Coulomb charging effects are important in the transport through small devices and we briefly recall the main main signatures of these effects in the transport characteristics.

Following Ref. [608], we want to address first the following question: How small and how cold should a conductor be so that adding or subtracting a single electron has a measurable effect? To answer this question, let us consider the electronic properties of the generic conductor depicted in Fig. 15.1, which is coupled to three terminals. Particle exchange can occur with only two of the terminals. These source and drain terminals connect the small conductor to macroscopic current and voltage meters. The third terminal provides an electrostatic or capacitive coupling and can be used as a gate electrode. If we first assume that there is no coupling to the source and drain contacts, then the small conductor acts as an island for electrons. The number of electrons on this island is an integer N, i.e. the charge on the island is quantized and equal to Ne. If we now allow tunneling to the source and drain electrodes, then the number of electrons N adjusts itself until the energy of the whole circuit is minimized.

When tunneling occurs, the charge on the island suddenly changes by the quantized amount e. The associated change in the Coulomb energy is conveniently expressed in terms of the capacitance C of the island. An extra charge e changes the electrostatic potential by the charging energy $E_C = e^2/C$. This charging energy becomes important when it exceeds the thermal energy $k_B T$. A second requirement is that the barriers are sufficiently opaque such that the electrons are located either in the source,

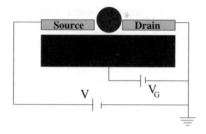

Fig. 15.1 Schematic representation of a generic three-terminal device. The sphere represents the dot (or island), which is weakly coupled to the source and drain electrodes by tunnel junctions. Finally, a third electrode (the gate) is capacitively coupled to the island.

in the drain, or on the island. This means that quantum fluctuations in the number N due to tunneling through the barriers are much less than one over the time scale of the measurement. (This time scale is roughly the electron charge divided by the current.) This requirement translates to a lower bound for the tunnel resistances R_t of the barriers. To see this, consider the typical time to charge or discharge the island $\Delta t = R_t C$. The Heisenberg uncertainty relation: $\Delta E \Delta t = (e^2/C)R_t C > h$ implies that R_t should be much larger than the resistance quantum $h/e^2 = 25.813$ kΩ in order for the energy uncertainty to be much smaller than the charging energy. To summarize, the two conditions for observing effects due to the discrete nature of charge are

$$R_t \gg h/e^2 \quad \text{and} \quad e^2/C \gg k_{\mathrm{B}}T. \tag{15.1}$$

The first criterion can be met by weakly coupling the small object (or dot) to the source and drain leads. The second criterion can be met by making the dot small or by lowering the temperature. Let us recall that the capacitance of an object scales with its radius R and for a sphere, $C = 4\pi\epsilon_0 R$. Thus for instance, the charging energy of a C_{60} molecule, which has a radius of ~ 4 Å, can be estimated to be $e^2/4\pi\epsilon_0 R \sim 3.6$ eV. This indicates that charging effects can in principle be readily observed in single-molecule junctions even at room temperature, as along as the molecules are weakly coupled to the electrodes.

The conditions summarized in Eq. (15.1) are met by many different nanoscale systems and for this reason charging effect have been observed, among other systems, in metallic islands [609, 610], semiconducting quantum dots [608, 611], nanoparticles [612], carbon nanotubes [613, 614], and semiconducting nanowire quantum dots [615, 616]. While the behavior of these type of quantum dots is fairly well understood, the properties of SMTs are much less established mainly because it is difficult to fabricate them in a reliable way (see next section).

When discussing charging effects, an important energy scale is the energy level spacing ΔE in the dot, i.e. the separation between the discrete energy states of the small conductor. To be able to resolve these levels, the spacing must be much larger than $k_B T$. The level spacing at the Fermi energy E_F for a box of size L depends on the dimensionality. Including spin degeneracy, we have

$$\Delta E = \frac{\hbar^2 \pi^2}{mL^2} \times \left\{ \begin{array}{ll} N/4 & \text{(1D)} \\ 1/\pi & \text{(2D)} \\ \left(1/3\pi^2 N\right)^{1/3} & \text{(3D)} \end{array} \right. , \qquad (15.2)$$

where m is the electron mass and N the number of electrons. The characteristic energy scale is thus $\hbar^2 \pi^2 / (mL^2)$. For a 1D box, the level spacing grows for increasing N, in 2D it is constant, while in 3D it decreases as N increases. The level spacing of a 100 nm 2D dot is ~ 0.03 meV, which is large enough to be observable at dilution refrigerator temperatures of ~ 100 mK. Thus, dots made in semiconductor heterostructures are true artificial atoms, with both observable quantized charge states and quantized energy levels. Using 3D metals to form a dot, one needs to make nanoparticles as small as ~ 5 nm in order to observe atom-like properties. In the case of molecular junctions, the spacing ΔE, which is basically the HOMO-LUMO gap, is typically of the order of several electronvolts. Therefore, level quantization should be easily observable in SMTs even at room temperature.

Now that we have identified the relevant scales for the occurrence of charging effects, let us now see how they are revealed in the transport characteristics. The tunneling of a single charge changes the electrostatic energy of the island by a discrete value, a voltage V_G applied to the gate (with capacitance C_G) can change the island's electrostatic energy in a continuous manner. In terms of charge, tunneling changes the island's charge by an integer while the gate voltage induces an effective continuous charge $q = C_G V_G$ that represents, in some sense, the charge that the dot would like to have. This charge is continuous even on the scale of the elementary charge e. If one sweeps V_G, the build up of the induced charge will be compensated in periodic intervals by tunneling of discrete charges onto the dot. This competition between continuously induced charge and discrete compensation leads to the so-called *Coulomb oscillations* in a measurement of the current (or conductance) as a function of gate voltage at a fixed source-drain voltage.

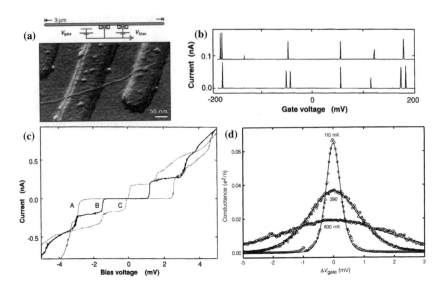

Fig. 15.2 Coulomb blockade in a single-wall carbon nanotube. (a) AFM image of a carbon nanotube on top of a Si/SiO$_2$ substrate with two 15-nm-thick Pt electrodes, and a corresponding circuit diagram. The total length of the tube is 3 μm, with a section of 140 nm between the contacts to which a bias (source-drain) voltage is applied. A gate voltage V_{gate} applied to the third electrode in the upper-left corner of the image is used to vary the electrostatic potential of the tube. (b) Current versus gate voltage at $V_{bias} = 30$ μV. Two traces are shown that were performed under the same conditions. (c) Current-voltage curves of the tube at a gate voltage of 88.2 mV (trace A), 104.1 mV (trace B) and 120.0 mV (trace C). (d) Conductance $G = I/V_{bias}$ versus ΔV_{gate} at low bias voltage $V_{bias} = 10$ μV and different temperatures. Solids lines are fits of $G \propto \cosh^{-2}(e\Delta V_{gate}/\alpha 2k_{\mathrm{B}}T)$, corresponding to the model of a single molecular level that is weakly coupled to two electrodes. The factor α is the gate coupling parameter (see text) and for this peak equals 16. Reprinted by permission from Macmillan Publishers Ltd: Nature [613], copyright 1997.

An example of these oscillations in a single-wall carbon nanotube weakly coupled to two metallic electrodes is shown in Fig. 15.2(b). As one can see, there appear a series of peaks or spikes in the current versus the gate voltage at very low bias (source-drain) voltage in a quasi-periodic fashion.[3] In the valley of the oscillations, the number of electrons in the nanotube is fixed and necessarily equal to an integer N. In the next valley to the right the number of electrons is increased to $N + 1$. At the crossover between two stable configurations N and $N + 1$, a charge degeneracy exists where the number can alternate between N and $N + 1$. This allowed fluctuation in

---

[3]In this case the bias voltage was quite low (linear regime) and the conductance exhibited the same peak structure as the current.

the number (i.e. according to the sequence $N \rightarrow N + 1 \rightarrow N \rightarrow \cdots$) leads to a current flow and results in the observed peaks.

An alternative measurement is performed by fixing the gate voltage, but varying the source-drain voltage V_{SD}. As shown in Fig. 15.2(c) for a carbon nanotube junction, one observes in this case a non-linear current-voltage characteristic exhibiting a series of steps. This characteristic structure is known as *Coulomb staircase*. A new current step occurs at a threshold voltage ($\sim e^2/C$) at which an extra electron is energetically allowed to enter the nanotube. It is seen in Fig. 15.2(c) that the threshold voltage can be modulated with the gate voltage until the low bias gap completely disappears, in accordance with the Coulomb oscillations. Finally, as one can see in Fig. 15.2(d), the conductance versus the gate voltage has a very characteristic temperature dependence close to a resonance, where the conductance maximum decreases as the temperature increases.

The origin of these peculiar transport characteristics will be analyzed in detail in detail in the next sections. Moreover, we shall show that one can obtain very valuable spectroscopic information about the charge state and energy levels of the dot by analyzing the precise shape of the Coulomb oscillations and the Coulomb staircase.

15.3   Single-molecule three-terminal devices

Most of the experiments described in the previous two chapters in our discussion of the coherent transport have been performed with two-terminal devices fabricated with the break-junction technique and the STM. These techniques have several advantages, but it is however very difficult to incorporate a gate electrode in their set-ups. This is an important drawback since, as discussed in the previous section, a gate electrode allows us to extract much more information about the junctions. Thus for instance, the gate makes possible to study the conduction through molecules in different transport regimes by bringing the energy levels into and out of resonance with the Fermi energy. This way, one can also probe excited states and different charge states can be accessed. Excited states can either be vibrational [22, 617, 678], electronic [618], or related to spin transitions [619, 620]. These excitations serve as a fingerprint of the molecule under study.

An important parameter in three-terminal devices is the gate coupling parameter, α. This parameter quantifies the shift of the orbital levels that can be induced with a gate electrode potential, V_G. In an experiment, the gate coupling should be as large as possible in order to access as many charge states as possible. The geometry plays an important role in the gate coupling and one should take care that the electrodes themselves do not screen the gate potential as this would decrease α. The electrode separation (and therefore the length of the molecule) and the breakthrough voltage of the gate oxide are other important parameters. Currently, two gate materials are frequently used: heavily doped silicon substrates with thermally grown SiO_2 on top and aluminum strips with a native Al_2O_3 oxide of only a few nanometers. For aluminum gates with an oxide thickness of 3 nm, the gate coupling is about 0.1 so that, with a typical breakthrough voltage of 4 V at low temperatures, the potential of the molecular levels can be shifted by ± 0.4 eV. On the other hand, in silicon devices with an SiO_2 thickness of 250 nm, the gate coupling is about 10^{-3}; with a typical breakthrough voltage of 100 V, the range over which the potential on the molecule can be varied equals $\pm$ 0.1 eV.

As we have seen in Chapter 3, three-terminal devices have been fabricated using different techniques. We follow here Ref. [606] and we now proceed to briefly describe the most successful approaches so far. They notably differ in the way the nanogap or the molecular junction is created. The most popular technique is electromigration, in which a large current density breaks a narrow and thin metal wire to form two physically separated electrodes [21]. Electromigration-induced nanogap formation has been imaged in situ by transmission (and scanning) electron microscopy [82, 621]. Several of these electrode pairs can be fabricated on top of a conducting substrate (coated by an insulating layer) which can then serve as a gate electrode. Although some control has been obtained over the electromigration process by using a feedback mechanism, the resulting nanogap geometry or size remains uncontrollable. The advantage of electromigrated devices on a Al/Al_2O_3 gate electrode is their large gate coupling ($\alpha_{max} \sim 0.27$). The planar geometry [see Fig. 15.3(a)] offers a large stability for systematic studies as a function of gate voltage, temperature and magnetic field. Molecules are deposited from solution either prior to gap formation or afterward.

A second technique involves the fabrication on top of a gate electrode of two gold electrodes using a shadow mask technique as illustrated in Fig. 15.3(b). If the tilt angle of evaporation is high there is no overlap between the source and drain shadows. Reducing the tilt angle decreases the source-drain gap. In situ measurements of the conductance allow for fine tuning of the gap distance when performed at low temperatures (~ 4.2

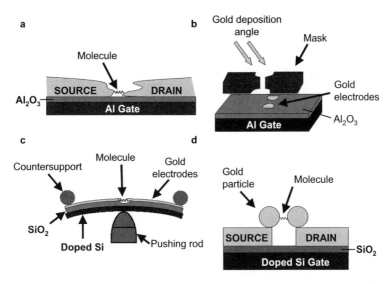

Fig. 15.3 Schematic diagrams of different three-terminal device techniques. (a) Electromigrated thin metal wire on top of a Al/Al_2O_3 gate electrode. (b) Angle evaporation technique to fabricate planar electrodes with nanometer separation on top of a Al/Al_2O_3 gate electrode. (c) Gated mechanical break junction. (d) The dimer contacting scheme (see text). Reprinted with permission from [606]. Copyright 2008 IOP Publishing Ltd.

K). Molecules are deposited by quench condensation without disruption of the vacuum [622, 623]. The advantages of this evaporation technique include all the ones from the electromigrated devices, plus the control over the gap distance and the ability for molecule deposition inside a clean environment. Typical gate coupling values are of the same order as the ones for electromigrated junctions.

Only recently it has been possible to integrate a gate electrode in MCBJs [86], see Fig. 15.3(c). So far it has been possible to place the gate electrode from the gap at a distance of 40 nm [86]. Although the gate coupling remains low as compared to other techniques with a planar geometry ($\alpha \sim$ 0.006 in Ref. [86]), MCBJs have the clear advantage of precise control over the gap distance; the reported breakthrough voltage [86] was 12 V. Molecule deposition is carried out from solution.

Another three-terminal approach was reported by Dadosh *et al.* [114]. Their method is based on synthesizing in solution a dimer structure consisting of two colloidal gold particles connected by a dithiolated molecule. The dimer is then electrostatically trapped between two gold electrodes defined on top of a gate electrode [see Fig. 15.3(d)]. According to the authors, this dimer-based contacting scheme provides several advantages such as the ability to fabricate single-molecule devices with high certainty in which the contacts to the molecule are well defined. The gold particles in this set-up, however, efficiently screen the gate potential. Moreover, at low temperatures spectroscopic features of the gold particles were sometimes observed to be superimposed on the characteristics of the molecule conduction.

15.4   Coulomb blockade theory: Constant interaction model

Most of the results obtained so far in single-molecule transistors (SMTs), i.e. in weakly coupled three-terminal molecular devices, have been analyzed with the help of the "orthodox" theory of Coulomb blockade [609], which has been very successful explaining the basic transport properties of semi-conductor quantum dots. For this reason, and before describing some of the main experiments reported to date, it is important to discuss in certain detail this theory, which is often referred to as the *constant interaction model*.

In the next subsections we present an introductory description of the standard theory of Coulomb blockade paying special attention to the relevant regime for molecular devices. In a later section we shall present an alternative formulation of this theory which is better adapted to SMTs, but it is much more involved. The next subsections are based on Refs. [624, 625] and on the didactic review of Ref. [605]. We also recommend the review on single-molecule junctions of Ref. [39].

15.4.1   *Formulation of the problem*

We consider a quantum dot or molecule,[4] which is weakly coupled via tunnel barriers to two metallic electrodes and it is also capacitively coupled to a gate electrode. The quantum dot has single-particle energy levels at E_p $(p = 1, 2, \cdots)$, labeled in ascending order and measured relative to the equilibrium chemical potential of the electrodes, which we set to zero $(E_F = 0)$. Each level contains either one or zero electrons. Spin degeneracy can be included by counting each level twice, and other degeneracies can be included similarly. Each electrode is considered to be in thermal equilibrium

---

[4]Although we have in mind molecular junctions, we shall use throughout this section the name quantum dot to refer generically to a small island weakly coupled to the source and drain.

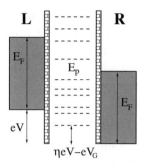

Fig. 15.4   Schematic drawing of the energy level diagram and electrostatic potential profile of a generic quantum dot. The dot possesses a single-particle spectrum with discrete levels, E_p. The Fermi levels in the left and right reservoirs are indicated. We measure the levels E_p with respect to E_F, which from now on we set to zero. The single particle spectrum may be shifted by the external potential. Here, η is the portion of the bias voltage that drops at the right interface and V_G corresponds to the gate voltage.

at temperature T and the continuum of states in the reservoirs is occupied according to the Fermi-Dirac distribution

$$f(E) = \left[1 + \exp\left(\frac{E}{k_B T} \right) \right]^{-1}. \tag{15.3}$$

In Fig. 15.4(a) we show schematically the energy level diagram of the quantum dot as well as the profile of the electrostatic potential.

Because in the weak coupling regime the number N of electrons localized in the dot can take integer values only, a charge imbalance, and hence a potential difference $V_{\text{dot}}(Q)$ can arise between the dot and reservoirs in equilibrium ($Q = -Ne$ is the charge on the dot). Following the orthodox model of the Coulomb blockade [609], one can express V_{dot} in terms of an effective N independent capacitance C between dot and the outside world,

$$V_{\text{dot}}(Q) = Q/C + V_{\text{ext}}, \tag{15.4}$$

where V_{ext} is a contribution from external charges (in particular those on a nearby gate electrode). The electrostatic energy $U(N) = \int_0^{-Ne} V_{\text{dot}}(Q) dQ$ then takes the form

$$U(N) = (Ne)^2 / 2C - NeV_{\text{ext}}. \tag{15.5}$$

Thus, the result for the total energy of a dot that contains N electrons, including the quantum energy due to the orbital energies is

$$E_{\text{dot}}(N) = U(N) + \sum_{p=1}^{N} E_p. \tag{15.6}$$

This expression for the total energy summarizes the constant interaction model, in which the capacitance C does not vary with N.

It is important to clarify the meaning of the external potential in Eq. (15.5). This can be done with the help of the equivalent circuit shown in Fig. 15.5. Elementary electrostatics gives the following relation between the different potentials and the charge Q on the island:

$$CV_{\text{dot}} - C_S V_S - C_D V_D - C_G V_G = Q, \qquad (15.7)$$

where $C = C_S + C_D + C_G$. Comparing this expression with Eq. (15.4) we arrive at the following result for the external potential

$$V_{\text{ext}} = (C_S V_S + C_D V_D + C_G V_G)/C. \qquad (15.8)$$

Thus, we see that the potential on the dot depends on the induced potential V_{ext} of the source, drain and gate. Notice that the change in the external potential due to a change in the gate voltage carries a factor $\alpha = C_G/C$, which is the gate coupling parameter that was mentioned in the previous section. On the other hand, assuming that the drain is grounded as in Fig. 15.5, the factor η introduced in Fig. 15.4 can now be simply expressed as the capacitance ratio $\eta = C_S/C$.

A key quantity determining whether the current can flow through the dot is its chemical potential, which is the minimum energy required to add an extra electron to the dot. From Eq. (15.6) it is easy to see that this chemical potential is given by

$$\mu_{\text{dot}}(N) = E_{\text{dot}}(N) - E_{\text{dot}}(N-1) = \left(N - \frac{1}{2}\right)\frac{e^2}{C} - eV_{\text{ext}} + E_N. \quad (15.9)$$

Before discussing the main predictions of this theory, it is important to be aware of the conditions for which the constant interaction model gives

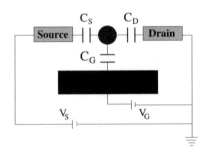

Fig. 15.5   Schematic representation of the capacitance model of a quantum dot. The dot is connected to source and drain electrodes with tunnel junctions and the gate electrode shifts the electrostatic potential of the dot. Here, we assume that drain electrode is grounded ($V_D = 0$).

a reliable description of the device. This is first of all weak coupling to the leads. A second condition is that the size of the device should be sufficiently large to make a description with single values for the capacitances possible. Finally, the single-particle spectrum E_p should not vary with the charge N residing on the dot. The constant interaction model works well for weakly coupled quantum dots for which it is very often used. However, the previous conditions are not fulfilled in general in molecular devices. Neither the charging energy nor the energy level spectrum are expected to be independent of the number of electrons in the molecule, specially for small ones. Thus, the constant interaction model should be used with caution in this case.

15.4.2 *Periodicity of the Coulomb blockade oscillations*

Let us now discuss the expected periodicity of the Coulomb oscillations, i.e. the distance between the peaks in the conductance as a function of the gate voltage in the limit of small (linear regime) source-drain voltage.

Electrons can flow from left to right when $\mu_{\rm dot}$ is between the potentials, μ_L and μ_R, of the leads (with $eV_{SD} = \mu_L - \mu_R$), i.e. $\mu_L > \mu_{\rm dot} > \mu_R$, see Fig. 15.6. For small bias voltages, $V_{SD} \approx 0$, the N-th Coulomb peak is a direct measure of the lowest possible energy state of an N-electron dot, i.e. the ground state electrochemical potential $\mu_{\rm dot}(N)$. From Eqs. (15.9) and (15.8) we obtain

$$\mu_{\rm dot}(N) = (N - 1/2)e^2/C - e\alpha V_G + E_N, \qquad (15.10)$$

where $\alpha = C_G/C$ is the gate coupling. The *addition energy* is given by

$$\Delta\mu(N) = \mu_{\rm dot}(N+1) - \mu_{\rm dot}(N) = \frac{e^2}{C} + E_{N+1} - E_N = \frac{e^2}{C} + \Delta E, \quad (15.11)$$

where $\Delta E = E_{N+1} - E_N$ is the level spacing mentioned in section 15.2.

In the absence of charging effects, the addition energy, $\Delta\mu(N)$, is determined by the irregular spacing ΔE of the single-electron levels in the quantum dot. The charging energy e^2/C *regulates* the spacing, once $e^2/C \gtrsim \Delta E$. If there is spin degeneracy of the levels, it is lifted by the charging energy. In the limit $(e^2/C)/\Delta E \to 0$, Eq. (15.11) is the usual condition for resonant tunneling. In the limit $(e^2/C)/\Delta E \to \infty$, Eq. (15.11) describes the periodicity of the classical Coulomb-blockade oscillations in metallic islands where the level spacing is negligible [609]. In molecular physics the related energies are defined as $A = E_{\rm dot}(N) - E_{\rm dot}(N+1)$ for the electron affinity

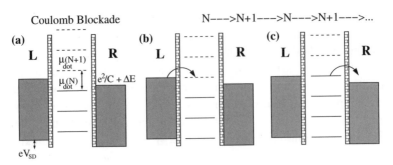

Fig. 15.6   Potential landscape through a quantum dot. The states in the contacts are filled up to the electrochemical potentials μ_L and μ_R, which are related by the external voltage $V_{SD} = (\mu_L - \mu_R)/e$. The discrete single-particle states in the dot are filled with N electrons up to $\mu_{dot}(N)$. The addition of one electron to the dot raises $\mu_{dot}(N)$ (i.e. the highest solid curve) to $\mu_{dot}(N + 1)$ (i.e. the lowest dashed curve). In (a) this addition is blocked at low temperatures. In (b) and (c) the addition is allowed since here $\mu_{dot}(N + 1)$ is aligned with the reservoir potentials by means of the gate voltage. (b) and (c) show two parts of the sequential tunneling process at the same gate voltage. (b) shows the situation with N and (c) with $N + 1$ electrons on the dot.

and $I = E_{dot}(N - 1) - E_{dot}(N)$ for the ionization energy. Their relation to the addition energy is $\Delta\mu(N) = I - A$.

From an experimental point of view, the Coulomb oscillations are measured as a function of the gate voltage. The peak spacing in terms of the gate voltage is given by

$$\Delta V_G = \Delta\mu(N)/e\alpha = (e^2/C + \Delta E)/e\alpha, \qquad (15.12)$$

while the condition $e\alpha V_G^N = (N - 1/2)e^2/C + E_N$ gives the gate voltage of the N-th Coulomb peak.

15.4.3   Qualitative discussion of the transport characteristics

Before discussing how to compute the transport properties in the Coulomb blockade regime, it is convenient to describe qualitatively the main expected features. Here, we shall follow Ref. [605] closely. As said in the previous subsection, in the weak coupling regime and at low temperature, the current is suppressed when all chemical potential levels lie outside of the bias window. As we can tune the location of these levels using the gate voltage, it is interesting to study the current and differential conductance of the device as a function of the bias and of the gate voltage. A two-dimensional plot of the current or conductance as a function of the two voltages is often referred to as *stability diagram*.

Let us now determine the line in the stability diagram (V_{SD}-V_G plane) that separates a region of suppressed current from a region with finite current. This line is given by the condition that the chemical potential of the source (or drain) is aligned with that of a level on the dot. We again assume the drain to be grounded as in Fig. 15.5. From the expression of Eq. (15.9) for the chemical potential and using the definition for V_{ext}, Eq. (15.8), we find the following condition for the dot chemical potential to be aligned with the source one ($\mu_L = eV$) keeping the dot's charge constant

$$V_{SD} = \beta(V_G - V_C),\tag{15.13}$$

where $\beta = C_G/(C_G + C_D)$ and $V_C = (N - 1/2)e/C_G + CE_N/(eC_G)$, which can be seen as the voltage corresponding to the chemical potential on the dot in the absence of an external potential. If the chemical potential is aligned with the drain ($\mu_R = 0$), we have

$$V_{SD} = \gamma(V_C - V_G),\tag{15.14}$$

with $\gamma = C_G/C_S$. The expressions given here are specific for a grounded drain electrode.[5]

Each dot resonance generates two straight lines in the V_{SD}-V_G plane, separating regions of suppressed current from those with finite current. For a sequence of resonances, one obtains generically the picture shown in Fig. 15.7(a). The diamond-shaped regions are traditionally called *Coulomb diamonds*, as they are often studied in the context of metallic dots, where the chemical potential difference of the levels is mainly made up of the Coulomb energy. The name is also used in molecular transport, although this is strictly speaking not justified since in this case the level spacing can be of the same order as the Coulomb interaction.

From the Coulomb diamond picture we can infer the values of some important quantities. Thus for instance, the addition energy, see Eq. (15.11), can be read off from the height of the diamond, see Fig. 15.7(a), or from the distance of the degeneracy points, although in this case one needs to know the gate coupling parameter, α. If the addition energy is dominated by the charging energy, we can find the total capacitance. Combining this with the slopes of the diamond sides, which give us the relative values of C_G, C_S and C_D, we can find all these capacitances explicitly.

---

[5]It is easily verified that, irrespective of the grounding, it holds that

$$\frac{C}{C_G} = \frac{1}{\alpha} = \frac{1}{\beta} + \frac{1}{\gamma}.$$

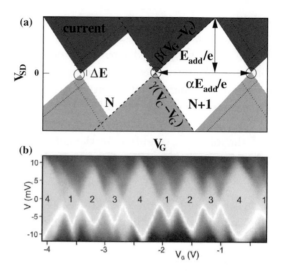

Fig. 15.7   (a) Generic two-dimensional plot of the current as a function of bias and gate voltage (*stability diagram*) for a quantum dot in the Coulomb blockade regime. For small bias current only flows in the three degeneracy points indicated with circles. Upper shadow region: positive currents. Lower shadowed region: negative currents. White: blockade, no current. The dotted lines indicate the presence of excitations (see text). (b) Measured stability diagram of a metallic single-walled carbon nanotube showing the expected fourfold shell filling. Blockade regime is white. Reprinted with permission from [626]. Copyright 2005 by the American Physical Society.

An interesting consequence of the previous analysis is that, if the capacitances do not depend on the particular state we are looking at, the height of successive Coulomb diamonds is constant. If, in addition to the Coulomb energy, the level splitting is significant, this homogeneity will be destroyed, as can be seen in Fig. 15.7(b), which shows the diamonds for a carbon nanotube [626]. The alternation of a large diamond with three smaller ones can be explained in terms of the electronic structure of the nanotubes [627]. In the case of transport through molecules there is no obvious underlying structure in the diamonds.

A stability diagram cannot only be used for finding addition energies, but it can also constitute a spectroscopic tool for revealing subtle excitations that arise on top of the ground state configurations of a dot with a particular number of electrons on it. This fact has been exploited in different contexts to study the level spectroscopy of a variety of systems such as metallic nanoparticles [612] or few-electron quantum dots [628]. The excitations appear as lines running parallel to the Coulomb diamond edges, see dotted lines in the upper panel of Fig. 15.7. At such a line, a new (electronically or vibrationally) excited state enters the bias window, creating an additional transport channel. The result is a step-wise increase of the current and a corresponding peak in the differential conductance. The energy of an excitation, ΔE in Fig. 15.7(a), can be determined by reading off the bias voltage of the intersection point between the excitation line and the Coulomb diamond edge through the same argument that we used for finding addition energies. The excitations correspond to the charge state of the Coulomb diamond they end up in [see Fig. 15.7(a)]. The width of the lines in the dI/dV_{SD} plot (or, equivalently, the voltage range over which the step-wise increase in the current occurs) is determined by the larger one of the energies $k_B T$ and Γ, which in the Coulomb blockade regime must be the first one. In practice, this means that sharp lines and thus accurate information on spectroscopic features are obtained at low temperatures and for weak coupling to the leads.

There are other important issues like the role of the asymmetry in the coupling that can be discussed at a qualitative level. For more details, recommend the review of Ref. [605].

15.4.4 *Amplitudes and line shapes: Rate equations*

We now want to put in more quantitative terms the statements of the previous subsection. For this purpose, we shall introduce here the so-called rate or master equations that allow us to compute, among other things, the amplitudes of the Coulomb blockade oscillations, the shape of the I-V curves (Coulomb staircase) and the stability diagrams. This section is based on Ref. [624].

In the weak coupling regime that we are interested in, the transport is determined by the tunnel rates from a given level p to the left and right reservoirs, see Fig. 15.4, which we shall denote by $\Gamma_L^{(p)}$ and $\Gamma_R^{(p)}$, respectively.[6] We assume here that $k_B T$, $\Delta E \gg (\Gamma_L + \Gamma_R)$ (for all levels participating in the conduction), so that the finite width $\Gamma = (\Gamma_L + \Gamma_R)$ of the transmission resonance through the quantum dot can be disregarded. This assumption allows us to characterize the state of the quantum dot by a set

---

[6]To be consistent with our notation in previous chapters, the rates $\Gamma_{L,R}^{(p)}$ have dimensions of energy. Thus, $\Gamma_{L,R}^{(p)}/\hbar$ gives the probability per unit of time of having a tunneling event that connects the level p of dot with the leads. The definition the rates in this chapter are a factor two larger than those used, for instance, in our discussion of the resonant tunneling model in Chapter 13.

of occupation numbers, one for each energy level. Notice that the restriction $k_B T$, $\Delta E \gg \Gamma$ results in the conductance being much smaller than e^2/h. We also assume conservation of energy in the tunneliing processes, thus neglecting contributions of higher order in Γ from tunneling via virtual intermediate states in the quantum dot. We finally assume that inelastic scattering takes place exclusively in the reservoirs — not in the quantum dot. The effect of inelastic scattering in the quantum dot is considered in Ref. [624] (see also Exercise 15.4).

Energy conservation upon tunneling from an initial state p in the quantum dot (containing N electrons) to a final state in the left reservoir at energy $E_p^{f,l}$ (in excess of the local electrostatic potential energy), requires that[7]

$$E_p^{f,l}(N) = E_p + U(N) - U(N-1) - (1-\eta)eV. \tag{15.15}$$

Here η is the fraction of the applied voltage V which drops over the right barrier,[8] see Fig. 15.4. The energy conservation condition for tunneling from an initial state $E_p^{i,l}$ in the left reservoir to a final state p in the quantum dot is

$$E_p^{i,l}(N) = E_p + U(N+1) - U(N) - (1-\eta)eV, \tag{15.16}$$

where N is the number of electrons in the dot *before* the tunneling event. Similarly, for tunneling between the quantum dot and the right reservoir one has the conditions

$$E_p^{f,r}(N) = E_p + U(N) - U(N-1) + \eta eV, \tag{15.17}$$

$$E_p^{i,r}(N) = E_p + U(N+1) - U(N) + \eta eV, \tag{15.18}$$

where $E_p^{i,r}$ and $E_p^{f,r}$ are the energies of the initial and final states in the right reservoir.

The stationary current through the left barrier equals that through the right barrier, and is given by

$$I = \frac{e}{\hbar} \sum_{p=1}^{\infty} \sum_{\{n_i\}} \Gamma_L^{(p)} P(\{n_i\}) \left[\delta_{n_p,0} f(E_p^{i,l}(N)) - \delta_{n_p,1} \bar{f}(E_p^{f,l}(N)) \right]. \tag{15.19}$$

where we have used the shorthand notation $\bar{f}(E) \equiv 1 - f(E)$. The second summation is over all realizations of occupation numbers $\{n_1, n_2, \ldots\} \equiv$

---

[7]Let us remind that the energies E_p are measured with respect to the Fermi energy of the leads.

[8]Notice that this definition differs from the one of Ref. [624]. This change has been introduced to preserve the convention our convention about the current direction and the sign of the bias voltage.

$\{n_i\}$ of the energy levels in the quantum dot, each with stationary probability $P(\{n_i\})$. here, the numbers n_i can take only the values 0 and 1. In equilibrium, this probability distribution is the Gibbs distribution in the grand canonical ensemble

$$P_{\text{eq}}(\{n_i\}) = \frac{1}{Z} \exp\left[-\frac{1}{k_B T}\left(\sum_{i=1}^{\infty} E_i n_i + U(N) - N E_F\right)\right], \qquad (15.20)$$

where $N \equiv \sum_i n_i$, and Z is the partition function given by

$$Z = \sum_{\{n_i\}} \exp\left[-\frac{1}{k_B T}\left(\sum_{i=1}^{\infty} E_i n_i + U(N) - N E_F\right)\right]. \qquad (15.21)$$

The non-equilibrium probability distribution P is a stationary solution of the kinetic equation

$$\hbar \frac{\partial}{\partial t} P(\{n_i\}) = 0$$

$$= -\sum_p P(\{n_i\}) \delta_{n_p,0} \left[\Gamma_L^{(p)} f(E_p^{i,l}(N)) + \Gamma_R^{(p)} f(E_p^{i,r}(N))\right]$$

$$-\sum_p P(\{n_i\}) \delta_{n_p,1} \left[\Gamma_L^{(p)} \bar{f}(E_p^{f,l}(N)) + \Gamma_R^{(p)} \bar{f}(E_p^{f,r}(N))\right]$$

$$+\sum_p P(n_1, \ldots n_{p-1}, 1, n_{p+1}, \ldots) \delta_{n_p,0}$$

$$\times \left[\Gamma_L^{(p)} \bar{f}(E_p^{f,l}(N+1)) + \Gamma_R^{(p)} \bar{f}(E_p^{f,r}(N+1))\right]$$

$$+\sum_p P(n_1, \ldots n_{p-1}, 0, n_{p+1}, \ldots) \delta_{n_p,1}$$

$$\times \left[\Gamma_L^{(p)} f(E_p^{i,l}(N-1)) + \Gamma_R^{(p)} f(E_p^{i,r}(N-1))\right], \qquad (15.22)$$

The kinetic equation, Eq. (15.22), for the stationary distribution function is equivalent to the set of detailed balance equations (one for each $p = 1, 2, \ldots$)

$$P(n_1, \ldots n_{p-1}, 1, n_{p+1}, \ldots)[\Gamma_L^{(p)} \bar{f}(E_p^{f,l}(\tilde{N}+1)) + \Gamma_R^{(p)} \bar{f}(E_p^{f,r}(\tilde{N}+1))]$$

$$= P(n_1, \ldots n_{p-1}, 0, n_{p+1}, \ldots)[\Gamma_L^{(p)} f(E_p^{i,l}(\tilde{N})) + \Gamma_R^{(p)} f(E_p^{i,r}(\tilde{N}))], \quad (15.23)$$

with the notation $\tilde{N} \equiv \sum_{i \neq p} n_i$. A similar set of equations formed the basis for the work of Averin, Korotkov, and Likharev on the Coulomb staircase in the non-linear I-V characteristics of a quantum dot [629–631].

Eq. (15.22), together with the normalization condition

$$\sum_{\{n_i\}} P(\{n_i\}) = 1, \qquad (15.24)$$

form a set of linear algebraic equations that can be easily solved numerically. For those readers not familiar with rate or master equations, we recommend Exercise 15.1, in which a single-level dot is considered.

15.4.4.1   *Linear response*

As shown by Beenakker in Ref. [624], in the linear response regime, the conductance can be calculated analytically, see also Exercise 15.2. The result can be expressed in terms of the equilibrium joint probability $P_{eq}(N, n_p = 1)$ that the quantum dot contains N electrons and that level p is occupied is

$$P_{eq}(N, n_p = 1) = \sum_{\{n_i\}} P_{eq}(\{n_i\}) \delta_{N, \sum_i n_i} \delta_{n_p, 1}. \qquad (15.25)$$

In terms of this probability distribution, the conductance is given by

$$G = \frac{e^2}{\hbar k_B T} \sum_{p=1}^{\infty} \sum_{N=1}^{\infty} \frac{\Gamma_L^{(p)} \Gamma_R^{(p)}}{\Gamma_L^{(p)} + \Gamma_R^{(p)}} P_{eq}(N, n_p = 1) \bar{f}(E_p^{f,l}(N)), \qquad (15.26)$$

where in this case $E_p^{f,l}(N) = E_p + U(N) - U(N-1)$ since the bias voltage is vanishingly small. This particular product of distribution functions expresses the fact that tunneling of an electron from an initial state p in the dot to a final state in the reservoir requires an occupied initial state and empty final state. The same formula was obtained independently by Meir, Wingreen, and Lee [632] by solving an Anderson model in the limit $k_B T \gg \Gamma$ (see Exercise 8.9).

   Eq. (15.26) is valid irrespective of the relative values of the temperature, charging energy and level splitting. The most relevant limit for molecular devices is $k_B T \ll e^2/C, \Delta E$. In this case Eq. (15.26) can be written in a simplified form. Now, the single term with $p = N = N_0$ gives the dominant contribution to the sum over p and N. If we consider that V_0 is the gate voltage at which the resonance associated with N_0 is reached, the dependence of linear conductance on the gate voltage, V_G, around V_0 is given by

$$G(V_G, T) = \left(\frac{e^2}{h}\right) \frac{\pi}{2k_B T} \frac{\Gamma_L^{(N_0)} \Gamma_R^{(N_0)}}{\Gamma_L^{(N_0)} + \Gamma_R^{(N_0)}} \cosh^{-2}\left(\frac{e\alpha(V_G - V_0)}{2k_B T}\right). \qquad (15.27)$$

This line shape is characterized by a maximum value of $G_{\max} = (e^2/2hk_B T)\Gamma_L^{(N_0)}\Gamma_R^{(N_0)}/(\Gamma_L^{(N_0)} + \Gamma_R^{(N_0)})$ attained when the gate voltage reaches the resonance V_0; this is the so-called *Coulomb peak*. The full-width at half maximum (FWHM) of this peak is $3.525 k_B T/(e\alpha)$, and the peak height decreases with temperature as $1/T$. Notice that Eq. (15.27) is nothing but the result that we obtained in section 13.2.3 in the context of the coherent tunneling through a single resonant level, see Eq. (13.6).[9] It is

---

[9]Let us remind that in Eq. (15.27) we do not consider spin degeneracy and the tunneling rates are a factor 2 larger than in Chapter 13.

natural to recover this result, since in the Coulomb blockade theory detailed in this section we have only considered elastic tunneling processes. Indeed, it is sometimes difficult to distinguish experimentally between Coulomb blockade effect and coherent transport through a weakly coupled system. From the theory side, there is an obvious difference. While in the coherent case the electrons tunnel through the single-particle levels of the dot, in the Coulomb blockade regime the resonances are also determined by the charging energy.

15.4.4.2 *Non-linear transport: A simple example*

In order to illustrate the main features of the non-linear transport characteristics of a quantum dot in the Coulomb blockade regime within the constant interaction model, we discuss now in detail an example of a two-level system, see Fig. 15.8(a). This system has two non-degenerate single-particle levels with energies $E_1 = 50$ meV and $E_2 = 80$ meV, which are measured with respect to the equilibrium chemical potential of the leads (set to zero). Level 2 plays the role of an excited state and we shall show how it is revealed in the different transport characteristics. We assume that the charging energy is $e^2/C = 100$ meV, which is larger than the excitation energy $\Delta E = E_2 - E_1 = 30$ meV. For simplicity, we assume that all tunneling rates are identical and equal to $\Gamma_{L,R}^{(p)} = 1$ meV ($p = 1, 2$). The temperature is $k_B T = 2.5$ meV (i.e. $T \approx 30$ K) and we take $\eta = 0.6$, where η is the parameter that describes the portion of the voltage that drops at the right barrier. Finally, we assume an arbitrary value for the gate coupling parameter, α, and in the different plots the gate voltage, V_G, will carry the factor α.

In this model there are four possible configurations for the dot: $\{n_i\} = (n_1, n_2)$, with $(0, 0)$, $(1, 0)$, $(0, 1)$ and $(1, 1)$. The first configuration has zero electrons in the dot, the second and the third ones correspond to one electron in the dot, and the fourth one to two electrons. To determine the different transport properties we have solved the stationary kinetic equation, see Eq. (15.22), to obtain the probabilities of the four configurations and we have then computed the current using Eq (15.19). The details of this simple calculation can be found in Exercise 15.3. Let us now proceed to describe the results of this model:

(i) *Coulomb oscillations:* As we explained in section 15.4.2, in the linear response regime ($V_{SD} \approx 0$), the current can flow when the chemical potential of the dot equals the equilibrium chemical potential of the

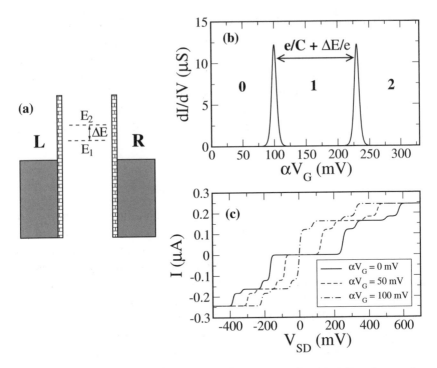

Fig. 15.8   (a) Two-level model to illustrate the transport characteristics of a quantum dot in the Coulomb blockade regime within the constant interaction model. The single-particle energies are $E_1 = 50$ meV and $E_2 = 80$ meV (measured with respect to the equilibrium chemical potential of the leads). The excitation energy is thus $\Delta E = E_2 - E_1 = 30$ meV. The charging energy is chosen to be $e^2/C = 100$ meV and all tunneling rates are assumed to be equal to $\Gamma = 1$ meV. The temperature is $k_B T = 2.5$ meV (i.e. $T \approx 30$ K) and $\eta = 0.6$. (b) Differential conductance vs. the gate voltage (including the gate coupling constant α) corresponding to the model of panel (a). The source-drain (or bias) voltage is 20 μV (linear regime). The numbers 0, 1 and 2 indicate the number of electrons in the different regions separated by the Coulomb peaks. (c) Corresponding non-linear current-voltage characteristics for several gate voltages.

leads (which we have set to zero): $\mu_{\text{dot}}(N, V_{SD} \approx 0) = (N - 1/2)e^2/C + E_N - e\alpha V_G = 0$. This implies that conductance peaks will appear at $e\alpha V_G^N = (N - 1/2)e^2/C + E_N$, which in this example correspond to $\alpha V_G^1 = 100$ mV and $\alpha V_G^2 = 230$ mV. This is illustrated in Fig. 15.8(b). Notice that the distance between the Coulomb peaks times the electron charge is equal to the addition energy $e^2/C + \Delta E = 130$ meV.

(ii) *Coulomb staircase:* In Fig. 15.8(c) we show the corresponding results for the current as a function of the source-drain or bias voltage (V_{SD}) for several values of the gate voltage. Notice that the I-V curves exhibit a series of steps, which correspond to the opening of new channels when the reservoir chemical potentials cross the different resonances in the dot. Thus for instance, at $V_G = 0$ and positive bias voltage, the current is blocked until $\mu_{\text{dot}}(N = 1)$ equals the chemical potential of the left reservoir, i.e. $\mu_{\text{dot}}(N = 1) = e^2/2C + E_1 + \eta eV_{SD} = eV_{SD}$. This occurs at $V_{SD} = 250$ mV. Then, the next step corresponds to the crossing of the excited state without changing the net charge in the dot. This requires an additional bias voltage equal to $\Delta E/e(1 - \eta) = 75$ mV, which explains the appearance of a step at $V_{SD} = 325$ mV. Following this line of reasoning, one can explain the position of all the steps in the I-V curves.

Two additional features in Fig. 15.8(c) are worth mentioning. First, notice that the gap in the low-bias voltage region can be completely closed by increasing the gate voltage, in accordance with the Coulomb oscillations. Second, notice that the I-V curves are not symmetric with respect to the inversion of the bias voltage. This is simply due to the fact that we have chosen an asymmetric electrostatic profile with $\eta \neq 0.5$.

(iii) *Stability diagrams:* As we discussed in section 15.4.3, the different energy scales and capacitances of the problem can be extracted from the so-called stability diagrams, where either the current or the differential conductance are plotted as a function of both the gate voltage and the bias voltage. In Fig. 15.9 we show the stability diagrams for our two-level example, which nicely illustrate the main conclusions of our qualitative discussion in section 15.4.3. In particular, notice that the addition energy can be extracted from the height of the middle diamond or from its width (distance between two consecutive degeneracy points). On the other hand, the energy of the excitation, $\Delta E = 30$ meV, can be read off from the bias voltage of the intersection point between the excitation line and the Coulomb diamond edge. The excitation line is particularly visible in the diagram of the differential conductance. Finally, notice that the diamonds are "inclined" due to the asymmetric potential profile ($\eta = 0.6$).

15.5   Towards a theory of Coulomb blockade in molecular transistors

There are two assumptions of the constant interaction model that are not met in general in SMTs. Neither the charging energy nor the level spectrum are expected to be independent of the number of electrons in a molecule. In this sense, the orthodox theory of Coulomb blockade may not be adequate

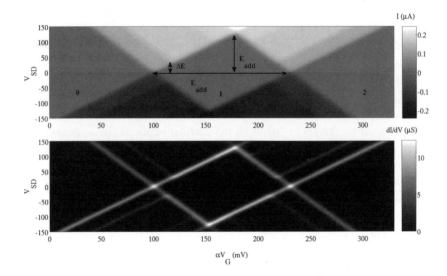

Fig. 15.9   Stability diagrams corresponding to the example of Fig. 15.8. (a) Current vs. gate voltage and source-drain (or bias) voltage. (b) Differential conductance vs. gate voltage and source-drain voltage.

for SMTs. The natural question is now how to generalize the standard theory to deal with molecular devices. It is worth mentioning that this question has also emerged in the other contexts like few-electron quantum dots [628] and ultrasmall metallic grains [633].

Part of the answer to this question is rather simple, at least conceptually speaking. Any theory of Coulomb blockade in SMTs should include an appropriate description of the molecular many-body spectrum as a function of the number of electrons in the molecule. In other words, we need as a starting point the ground state and excited states of the molecules not only for the neutral species, but also for the stable cations and anions. In principle, this requires the use of ab-initio (post Hartree-Fock) quantum chemistry methods like, for instance, the configuration interaction approach [176]. In practice, both model Hamiltonians and approximate methods like density functional theory[10] have been used for this purpose. Once the many-body spectrum for different number of electrons is known, one needs to solve a master equation to determine the occupation of the different states and finally their contribution to the current. By now many authors

---

[10]Density functional theory is not designed to give the level spectrum of a system and, although it may give reasonable results for neutral molecules, it has severe problems with charged species (anions and cations) [273].

have implemented such a procedure at different levels of sophistication, and we just mention here a few works from which it should be easy to trace back the entire literature [634–645].

Another important aspect that a theory of Coulomb blockade in SMTs should account for is the possible renormalization of the molecular levels due to the surrounding electrodes. As we shall discuss in the section 15.7, the addition energies found experimentally in SMTs are clearly much smaller than what is expected from the known ionization potential and the electron affinity of the molecules in gas phase. It has been suggested that this reduction of the addition energies is caused by image charges in the metallic electrodes [622], giving rise to a localization of the charges near the leads. This issue is a crucial one and surprisingly it has received little attention so far. Fortunately, some groups are starting to tackle this problem, see Ref. [646] and references therein, but this issue is by no means yet settled.

In the rest of this section, we shall sketch how more realistic single-tunneling theories for SMTs are formulated, and we shall present a simple example to illustrate those theories. The rest of this subsection is rather technical and it can be skipped in a first reading.

15.5.1   *Many-body master equations*

We describe in this subsection how to compute the current through a SMT in the weak-coupling regime within a model where the molecular part is described with a truly microscopic many-body approach. This discussion is based on Ref. [635], which we strongly recommend.

The starting point of our description of a molecular junction is the following model Hamiltonian: $\mathbf{H} = \mathbf{H}_M + \sum_r \mathbf{H}_r + \mathbf{H}_T$, which incorporates the molecule (M), reservoirs $r = L, R$ and the tunneling (T). The last two terms are given by

$$\mathbf{H}_r = \sum_{k\sigma} \epsilon_{k\sigma} \mathbf{a}_{k\sigma r}^\dagger \mathbf{a}_{k\sigma r} \tag{15.28}$$

$$\mathbf{H}_T = \sum_{ki\sigma r} t_i^r \mathbf{a}_{k\sigma r}^\dagger \mathbf{c}_{i\sigma} + \text{h.c.} \tag{15.29}$$

The molecular part H_M is chosen to contain the various strong interactions on the molecule which require an exact treatment. Generally, it is expressed in a basis of single-electron operators $\mathbf{c}_{i\sigma}$ for orbitals on the molecule labeled by i and spin projection σ. After diagonalization one can write $H_M = \sum_s E_s |s\rangle\langle s|$, where the discrete molecular many-body states have a summary label s which includes the total charge N, spin S, and other possible quantum numbers. Eq. (15.28) models the electrodes $r = L, R$ as non-interacting quasi-particle reservoirs which are fixed at electro-chemical potential $\mu_r = \mu \pm eV/2$ and temperature T. Here, $\mathbf{a}_{k\sigma r}$ are electron operators of electrode r labeled by k and spin σ. For simplicity, the density of states in the electrodes ρ_e is assumed to be flat around the Fermi-energy in order to focus on effects of the molecular part. The tunneling term, Eq. (15.29), describes charge transfer between electrode and molecule on a very small time scale. The level-dependent coupling strength is characterized by the intrinsic line width $\Gamma_i^r = 2\pi |t_i^r|^2 \rho_e$, where ρ_e is the density of states of the electrodes at the Fermi energy. $\Gamma = \max\{\Gamma_i^r\}$ denotes the overall coupling strength (in units of energy) between electrodes and the molecule and serves to define the scale of the current. The molecular states may additionally be coupled to the electromagnetic field (photons) and/or a mechanical environment which dissipate the energy accumulated on the molecule due to the tunneling. Whereas this does not change the charge on the molecule, it does have an effect on the non-equilibrium distribution of the molecular states and may thereby strongly influence the current, see Ref. [635] for more details. We do not explicitly discuss here the coupling to such bosonic reservoirs.[11]

Since we are particularly interested in the Coulomb blockade regime, it is important to treat the strong intramolecular interactions exactly, while the tunneling to the reservoirs is treated in a systematic perturbation theory in the electrode-molecule tunneling, Eq. (15.29). Assuming that $\Gamma \ll k_B T$, one only needs to compute the current to the lowest order in Γ. As in the constant interaction model, such a lowest order perturbation theory leads to master equations for the occupation probabilities p_s of molecular many-body states s. In principle, this approach can be improved systematically by going to higher orders in Γ. We summarize now the equations necessary to compute the current in lowest order perturbation theory in the coupling strengths Γ. Their systematic derivation using a diagrammatic technique has been discussed in [647, 648]. For time-independent applied bias, the time derivative of the probabilities p_s vanishes in the stationary state. The stationary nonequilibrium probabilities p_s^{st} are uniquely determined by the transition rates $W_{ss'}$ from state s' to s (forming a matrix $\hat{W}$) through the stationary master equation $\hat{W}\vec{p}^{\text{st}} = 0$ together with the normalization of

---

[11]This will be done in section 17.3.1 in the context of the study of the influence of the electron-phonon interaction in the transport properties of a molecular junction in the Coulomb blockade regime.

the distribution p_s for arbitrary times. We can then write

$$\vec{p}^{\,\mathrm{st}} = (\hat{\tilde{W}})^{-1}\vec{v}, \qquad (15.30)$$

where the matrix $\hat{\tilde{W}}$ is identical to $\hat{W}$ but with one (arbitrarily chosen) row s_0 replaced with $(\Gamma, \cdots, \Gamma)$ and $\vec{v}$ is a vector, $v_s = \Gamma\delta_{s,s_0}$. The transition rates $W_{ss'}$, with $s \neq s'$ (in the absence of bosonic coupling) are the sum $W_{ss'} = \sum_r W_{ss'}^r$ of the Golden rule rates for the tunneling of an electron to/from electrode $r = L, R$:

$$W_{ss'}^r = 2\pi\rho_e \sum_\sigma \begin{cases} f_r(E_s - E_{s'})\left|\sum_i t_i^r \langle s|c_{i\sigma}^\dagger|s'\rangle\right|^2 & N_{s'} < N_s \\ \bar{f}_r(E_{s'} - E_s)\left|\sum_i t_i^r \langle s|c_{i\sigma}|s'\rangle\right|^2 & N_{s'} > N_s \end{cases}, \qquad (15.31)$$

where $f(x) = 1/(\exp(x/k_\mathrm{B}T)+1)$ is the Fermi function, $f_r(x) = f(x - \mu_r)$ and $\bar{f}_r(x) = 1 - f(x - \mu_r)$. The decay rates are $W_{ss} = -\sum_{s' \neq s} W_{s's}$.

The stationary current $I = \langle \mathbf{I} \rangle$ is related to the current operator $\mathbf{I} = (\mathbf{I}_R - \mathbf{I}_L)/2$. We can use the symmetrized combination of currents $\mathbf{I}_r = -i(e/\hbar)\sum_{ik\sigma}(t_i^r \mathbf{a}_{k\sigma r}^\dagger c_{i\sigma} - \mathrm{h.c.})$ into electrode $r = L, R$ since in the stationary limit $\langle \mathbf{I}_R \rangle = -\langle \mathbf{I}_L \rangle$. The current can be explicitly calculated from the expression

$$I = \frac{e}{2\hbar} \vec{e}^{\,T} \hat{W}^I \vec{p}^{\,\mathrm{st}}. \qquad (15.32)$$

The vector $\vec{e}$ is given by $e_s = 1$ for all s. The rates entering the current are

$$W_{ss'}^I = \pm(W_{ss'}^R - W_{ss'}^L), \quad N_s \lessgtr N_{s'}. \qquad (15.33)$$

The inclusion of dissipative environments (photons, phonons) modifies only the rates $\hat{W}$ and thereby $\vec{p}^{\,\mathrm{st}}$, but the rates $\hat{W}^I$ are not affected. For an alternative, but equivalent, formulation of these master equations, see for instance Ref. [637].

15.5.2   *A simple example: The Anderson model*

In order to illustrate the formalism discussed above, let us now apply it to a simple example. We first need a Hamiltonian for the molecular part. Let us assume that its electronic structure can be described by a single-level Anderson model (see section 5.4.3 and Appendix A)

$$\mathbf{H}_M = \sum_\sigma \epsilon_0 \mathbf{n}_\sigma + U\mathbf{n}_\uparrow \mathbf{n}_\downarrow. \qquad (15.34)$$

Here, ϵ_0 is the energy of a (spin-degenerate) resonant level, $\mathbf{n}_\sigma = c_\sigma^\dagger c_\sigma$ is the number operator which describes the occupation of that level for spin σ

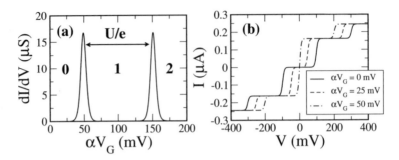

Fig. 15.10   (a) Differential conductance vs. the gate voltage (including the gate coupling constant α) for a molecular transistor described with the Anderson model. Here, $\Gamma_L = \Gamma_R = 1$ meV, $\epsilon_0 = 50$ meV and $U = 100$ meV. The temperature is $k_B T = 2.5$ meV (i.e. $T \approx 30$ K) and the bias voltage is 20 μV (linear regime). The numbers 0, 1 and 2 indicate the number of electrons in the different regions separated by the Coulomb peaks. (b) Corresponding non-linear current-voltage characteristics for several gate voltages. The voltage is assumed to drop symmetrically at the interfaces.

and U is the Coulomb repulsion energy on the molecule. This Hamiltonian has four eigenstates: $|1\rangle \equiv |00\rangle$ with $N = 0$, $|2\rangle \equiv |10\rangle$ and $|3\rangle \equiv |01\rangle$ with $N = 1$ and $|4\rangle \equiv |11\rangle$ with $N = 2$, where N is the number of electrons on the molecule. Here, we have used the notation $|n_\uparrow n_\downarrow\rangle$, where $n_\sigma = 0, 1$ is the occupation number of the single-particle level with spin σ. The corresponding eigenenergies are: $E_1 = 0$, $E_2 = E_3 = \epsilon_0$ and $E_4 = 2\epsilon_0 + U$. In this case, it is straightforward to compute the transition rates of Eq. (15.31) (see Exercise 15.5). They are determined by the scattering rates $\Gamma_r \equiv 2\pi \rho_e |t^r|^2$ $(r = L, R)$. Once the transition rates have been computed, one can obtain the stationary probabilities p_s $(s = 1, ..., 4)$ for the four states by solving numerically the 4×4 master equation, see Eq. (15.30). Finally, the current can be calculated from Eq. (15.32).

In Fig. 15.10 we present the results of this model for $\Gamma_L = \Gamma_R = 1$ meV, $\epsilon_0 = 50$ meV and $U = 100$ meV. Panel (a) of this figure shows the linear conductance versus the gate voltage. Here, this voltage has been introduced by shifting rigidly the level position, i.e. $\epsilon \rightarrow \epsilon - \alpha V_G$, where α is the gate coupling parameter. As one can see, the conductance exhibits Coulomb peaks at the degeneracy points. Notice that these peaks are separated by a distance U/e, which indicates that U plays here the role of the charging energy (the level splitting is zero in this case).

Fig. 15.10(b) shows the current as a function of the bias (source-drain) voltage for different values of the gate voltage. Here, we have assumed that the bias voltage drops symmetrically at both interfaces, i.e. $\eta = 0.5$. Due to the spin degeneracy, the I-V curves only exhibit two plateaus. Notice that the two steps (for a given voltage polarity) are separated by a distance $U/(e\eta)$.

In this example the transport characteristics are very similar to those obtained with the constant interaction model (orthodox theory). The differences become more pronounced when there are more charge states involved or additional quantum numbers play a fundamental role.

15.6 Intermediate coupling: Cotunneling and Kondo effect

So far in this chapter, we have focused on the limit where the tunnel coupling, Γ, is much smaller than any other energy scale in the problem. If this coupling strength is increased, higher-order tunneling processes begin to give a significant contribution to the transport properties [649]. In the opposite limit (strong coupling regime) where $\Gamma \gg e^2/C, \Delta E, k_B T$, the electronic states in the molecule and electrodes are strongly hybridized. In that case, as we have discussed in previous chapters, the elastic coherent tunneling dominates transport and signatures of the Coulomb blockade are washed out by quantum fluctuations of the molecular charge. Between the weak coupling and strong coupling regime one can identify a third regime which we shall refer to as the intermediate coupling regime. In this regime it is still possible to observe Coulomb diamonds, but higher-order processes lead to a non-negligible current inside the blockade regions. In this section we shall discuss three different types of higher-order tunneling processes: elastic and inelastic cotunneling and spin-flip cotunneling. This latter process is behind the appearance of the Kondo effect.

15.6.1 *Elastic and inelastic cotunneling*

The first process that we want to describe is the so-called *elastic cotunneling* process. This second-order process is illustrated in Fig. 15.11. In the situation depicted in this figure, energy conservation forbids the number of electrons to change as this would cost an energy ΔE, which is not available at the bias voltage considered in Fig. 15.11. Nevertheless, an electron can tunnel off the molecule, leaving it temporarily in a classically forbidden virtual state (middle diagram in Fig. 15.11). By virtue of Heisenberg's energy-time uncertainty principle this is allowed as long as another electron tunnels into the molecule in the same quantum process in order not to vi-

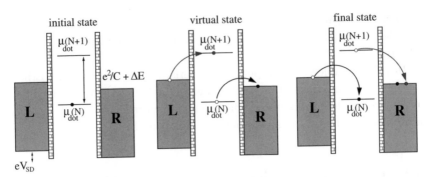

Fig. 15.11   Elastic cotunneling process. The Nth electron on the dot jumps to the drain (virtual state) to be immediately replaced (final state) by an electron from the source (black arrow sequence). A similar process involves the unoccupied state (light arrow sequence). In both examples, an electron is effectively transported from source to drain.

olate energy conservation. The final state then has the same energy as the initial one, but one electron has been transported through the molecule. This elastic cotunneling process is analogous to the superexchange mechanism in chemical electron transfer theory [582]. It occurs at arbitrarily low bias as the energy of the tunneling electron and the molecule are unchanged and leads to a nonzero background conductance in the blockade regions.

A cotunneling event that leaves the molecule in an excited state is called *inelastic*. An example of such a process is depicted in Fig. 15.12. As one can see, the onset of the cotunneling event occurs at $eV_{SD} = \Delta E$, the condition dictated by the energy conservation principle. In transport measurements, inelastic cotunneling appears in the stability diagram inside the Coulomb diamonds as two symmetric lines running parallel to the gate axis as represented by grey lines in Fig. 15.14(a). Their energy, ΔE, is the distance of the excitation to the zero-bias axis as illustrated in Fig. 15.14(a). Furthermore, the inelastic cotunneling line is expected to intersect, at the diamond boundary, the corresponding excitation line inside the single-electron tunneling region [650].

It is worth stressing that that higher-order coherent processes appear as sharp spectroscopic features as the conductance of the i-th order process is proportional to Γ^i, while for first-order incoherent single-electron tunneling, the current is proportional to Γ.

If the electron spin is taken into account, one can encounter another elastic cotunneling process connected to the Kondo effect. This will be analyzed in detail in the next subsection.

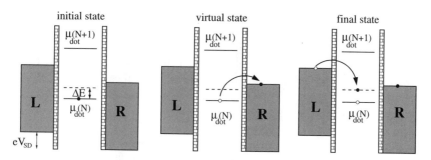

Fig. 15.12   Inelastic cotunneling process. For $eV_{SD} \geq \Delta E$, the Nth electron on the dot may jump from the ground state to the drain (virtual state) to be immediately replaced by an electron from the source (final state), which enters the excited state.

15.6.2   *Kondo effect*

The Kondo effect is a many-body phenomenon that occurs when a localized spin interacts with surrounding conduction electrons [651]. This effect is known to be the origin of the resistance increase at low temperatures in metals with magnetic impurities [652]. In recent years, there has been a renewed interest in the Kondo effect thanks to its observation in a variety of nanodevices [653]. Thus for instance, Kondo physics has been reported in the last decade in semiconductor quantum dots [654–656], in magnetic impurities on the surface of metals [657–659], and carbon nanotubes [660, 661]. More importantly for our discussion, this phenomenon has also been observed in single-molecule transistors [23, 24, 662–664] and for this reason, we shall briefly review in this section the basics of this effect. For a detailed discussion of the Kondo physics in different mesoscopic systems, see Refs. [653, 665–667].

In molecular transistors, and quantum dots in general, the Kondo effect can arise when the molecule has a net spin (magnetic moment). This, for example, occurs for an odd occupancy in the molecule (one electron is unpaired, $S = 1/2$). We shall mainly consider this simple case. The conduction mechanism for the Kondo effect involves spin-flip events such as the one illustrated in Fig. 15.13(a). The Heisenberg uncertainty principle allows the electron to tunnel out for only a short time of about $\hbar/|\epsilon_0|$, where ϵ_0 is the energy of the electron relative to the Fermi energy. During this time, another electron from the Fermi level at the opposite lead can tunnel onto the dot, keeping the total energy of the system conserved (elastic cotunneling). The exchange interaction causing the majority spin in the leads to be opposite to the original spin of the dot causes the probability

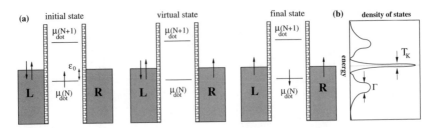

Fig. 15.13   (a) Spin-flip cotunneling. A spin-up electron jumps out of the dot (virtual state) to be immediately replaced by a spin-down electron (final state). (b) Kondo resonance in the density of states that appears as a consequence of the spin-flip tunneling processes.

for the new electron to have spin opposite to the first to be very high.

This spin exchange qualitatively changes the energy spectrum of the system. When many such processes are taken together, one finds that a new state, known as Kondo resonance, is generated exactly at the Fermi level, see Fig. 15.13(b).[12] In this situation, the localized spin is completely screened and the many-boy ground state turns out to be a singlet state $(S = 0)$. It is important to note that the Kondo state is always "on resonance" since it is fixed to the Fermi energy. Even though the system may start with an energy, ϵ_0, which is very far away from the Fermi energy, the Kondo effect alters the energy of the system so that it is always on resonance. For this reason, these many-body correlations can lead to a great enhancement of the conductance. The only requirement for this effect to occur is that the system is cooled to sufficiently low temperatures below the Kondo temperature T_K (see next paragraph).

The width of the Kondo resonance is proportional to the characteristic energy scale for Kondo physics, the so-called Kondo temperature T_K. For $\epsilon_0 \gg \Gamma$, T_K is given by [668]

$$k_B T_K = \frac{\sqrt{\Gamma U}}{2} \exp\left[\frac{\pi \epsilon_0 (\epsilon_0 + U)}{\Gamma U}\right]. \tag{15.35}$$

Here, Γ is the coupling strength and U can be seen as the charging energy, e^2/C. Typical Kondo temperatures are $T_K \sim 1$ K for semiconductor quantum dots [655], $T_K \sim 1-10$ K for carbon nanotubes [660, 661] and $T_K \sim 20-50$ K for molecular devices [23, 24, 662, 664]. This increase of T_K with decreasing dot size can be understood from the prefactor, which

---

[12]In section 6.9 we presented a discussion of the origin and description of the Kondo resonance in the framework of the Anderson model.

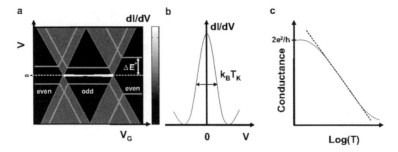

Fig. 15.14   Schematic representation of the main characteristics of the Kondo effect in electron transport through a molecular quantum dot. (a) In the stability diagram, the Kondo effect results in a zero-bias resonance (white line) for an odd number of electrons in the dot. Inelastic cotunneling excitations appear as lines running parallel to the gate axis at finite bias. (b) For $T \ll T_K$, the full width at half-maximum (FWHM) of the Kondo resonance is $\sim k_B T_K$. (c) Temperature dependence of the Kondo-peak height in the middle of the Coulomb diamond. Reprinted with permission from [606]. Copyright 2008 IOP Publishing Ltd.

contains the charging energy ($U = e^2/C$). Notice that the Kondo temperature depends on the position of the level and therefore it can be tuned in three-terminal devices by means of the gate voltage.

The theoretical description of the Kondo effect is very challenging. The reason is that below T_K, high order spin-flip processes contribute significantly to both the electronic structure and the transport properties. This implies that one needs to employ non-perturbative methods to describe properly this phenomenon. Different many-body methods have been used to account for the Kondo correlations in quantum dots and related structures. The description of such techniques is out of the scope of this book and in the rest of this subsection we shall concentrate ourselves on the discussion of its main transport characteristics and refer the interested reader to Refs. [651, 652, 669, 670, 665, 666] for more details about the theory.

The Kondo effect is manifested in the stability diagrams as a zero-bias resonance in the differential conductance, dI/dV_{SD}, versus V_{SD}, inside the Coulomb diamond connecting both degeneracy points as shown in Fig. 15.14(a). For an even number of electrons with all spins paired, $S = 0$ and there is no Kondo resonance. This even-odd asymmetry is very helpful in assigning the parity of the charge state which can then add extra information to the understanding of the spectroscopic features observed in the stability plots. In the low temperature limit ($T \ll T_K$), the full width at half-maximum (FWHM) of the Kondo resonance, as observed in a plot of the differential conductance versus V_{SD}, is of the order of $k_B T_K$, see Fig. 15.14(b). In the middle of the Coulomb diamond, the linear conductance exhibits a characteristic temperature dependence given by [670, 655]

$$G(T) = \frac{G_0}{\left[1 + (2^{1/s} - 1)(T/T_K)^2\right]^s},\qquad(15.36)$$

where $G_0 = 2e^2/h$ and $s = 0.22$ for $S = 1/2$. (Sometimes in this formula a temperature-independent offset is included.) This dependence is schematically drawn in Fig. 15.14(c). Notice that the conductance increases logarithmically with decreasing temperature and saturates at a value $2e^2/h$ at the lowest temperatures in the case of symmetric lead-dot coupling. The latter is commonly referred to as the Kondo effect in the unitary limit [671] (see Exercise 8.10).

In a magnetic field the Zeeman splitting of the Kondo resonance leads to the observation of two Kondo peaks symmetric in bias, separated by twice the Zeeman energy. On the other hand, although we have been only considering $S = 1/2$, it is important to note that other types of Kondo systems are possible owing to orbital degeneracies [672] or triplet states [673–675]; these can lead to a violation of the parity effect.

15.7 Single-molecule transistors: Experimental results

In this section we shall review some of most representative results obtained in single-molecule transistors (SMTs). We also recommend the following reviews on this subject [39, 603, 604, 606].

The question of whether charging effects play a major role in the conduction through molecular contacts arose immediately after the report of the first transport measurements in single-molecule junctions. Thus for instance, in the work of Reed and coworkers on benzenedithiol molecules [16], see section 14.1.1, one may argue that the conductance gap observed at low bias, see Fig. 14.1, is due to Coulomb blockade rather than to the HOMO-LUMO gap of this molecule.[13] Similar questions emerged in the early work of Kergueris *et al.* [676]. In this case, the authors reported room temperature measurements of the I-V characteristics of bisthiolterthiophene molecules using the technique of microfabricated MCBJ. Zero-bias conduc-

---

[13]The first maximum in the conductance in Fig. 14.1 at voltages above 1 V has a width that is larger than the temperature. This indicates that, strictly speaking, this system is not in the Coulomb blockade regime. Of course, this does not exclude that electronic correlations of some sort are at work.

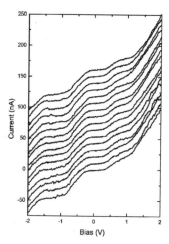

Fig. 15.15   I-V curves recorded at room temperature in a gold-bisthiolterthiophene-gold junction formed with the microfabricated MCBJ technique. Curves are shifted vertically for clarity. Reprinted with permission from [676]. Copyright 1999 by the American Physical Society.

tances were measured in the 10-100 nS range and different kinds of nonlinear I-V curves with steplike features were reproducibly obtained. An example of the results of these measurements can be seen in Fig. 15.15. Notice that the I-V curves resemble very much the Coulomb staircase observed in quantum dots. Indeed, the authors were able to fit the experimental results within the framework of the ortodox Coulomb blockade theory described in section 15.4, taking into account the discrete nature of the electronic spectrum of the molecule. Let us mention that charging energies of the order of 0.2 eV were used in the fits (see Ref. [676] for more details).

As we have discussed in previous sections, an unambiguous confirmation that characteristics like the ones shown in Fig. 15.15 are a consequence of the occurrence of charging effects requires the implementation of a gate electrode, which is very challenging in the case of MCBJs. To our knowledge, the first three-terminal single-molecule experiment was reported by Park *et al.* [22] in 2000. These authors prepared single-C_{60} junctions by depositing a dilute toluene solution of C_{60} onto a pair of connected gold electrodes fabricated using e-beam lithography. A gap of 1 nm between these electrodes was then created by electromigration [21]. The entire structure was defined on a SiO_2 insulating layer on top of a degenerately doped silicon wafer which served as a gate electrode that modulates the electrostatic po-

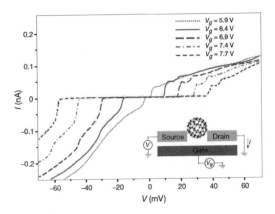

Fig. 15.16   I-V curves obtained at $T = 1.5$ K from a single-C_{60} transistor fabricated with the electromigration technique. The curves corresponds to five different gate voltages. The inset shows a schematic diagram of an idealized single-C_{60} transistor. Reprinted by permission from Macmillan Publishers Ltd: Nature [22], copyright 2000.

tential of C_{60}. A schematic diagram of an idealized single-C_{60} transistor is shown in the inset of Fig. 15.16.

Fig. 15.16 shows some typical I-V curves obtained in Ref. [22] at different gate voltages. Notice that the device exhibited a strongly suppressed conductance near zero bias voltage followed by step-like current jumps at higher voltages. The voltage width of the zero-conductance region (conductance gap) could be changed in a reversible manner by changing the gate voltage. These transport features clearly confirm that the conduction in this device is dominated by the Coulomb blockade effect and it can thus be stated that this experiment constitutes the first true example of a SMT reported in the literature.

A further confirmation of the underlying transport mechanism came from the analysis of the stability diagrams. In Fig. 15.17 we reproduce the results for the differential conductance as a function the bias and gate voltages. As one can see, two diamond-like regions can be identified corresponding to two charge states of the C_{60} molecules. In these plots, the peaks in the conductance, which correspond to the step-like features in Fig. 15.16, show up as lines. As seen clearly in Fig. 15.17, the size of he conductance gap and the peak positions evolve smoothly as the gate voltage is varied. As the gate voltage is varied further away in both positive and negative directions, the conductance gap continues to widen and exceeds 150 mV in some devices. This indicates that the charging energy of the C_{60}

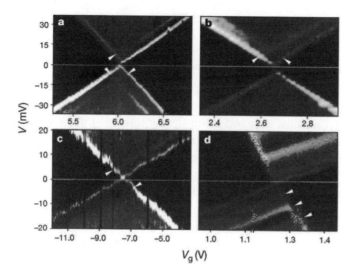

Fig. 15.17   Different conductance plots as a function of the bias voltage V and the gate voltage V_g obtained from four different devices. The dark triangular regions correspond to the conductance gap, and the bright lines represent peaks in the differential conductance. The arrows mark the point where the conductance lines intercept the conductance gap. Reprinted by permission from Macmillan Publishers Ltd: Nature [22], copyright 2000.

molecule in this geometry can exceed 150 meV. This value is much larger than in semiconductor quantum dots.

Notice that in the stability diagrams of Fig. 15.17 there are running lines that intersect the main diamonds or conductance gap regions. As we explained in previous sections, this indicates the presence of internal excitations of the C_{60} molecules. The energies of these excitations (of a few meV) are too small to correspond to electronic excitations. Moreover, some of these lines are observed for both charge states and multiple excitations with the same spacing are observed (see Fig. 15.16). These observations suggest that these lines may correspond to the excitation of vibration modes of the C_{60} molecules. The lowest-energy mode is known to have around 33 meV and this could explain some of the lines seen in the experiment. However, internal vibrational modes cannot account for the observed 5-meV features in Fig. 15.17. The authors of Ref. [22] suggested that this line could correspond to the excitation of the center-of-mass oscillation of C_{60} within the confinement potential that binds it to the gold surface.[14]

---

[14]The signatures of the excitation of vibration modes in the transport characteristics will be discussed in detail in the next two chapters.

After the first observation of the Coulomb blockade in molecular junctions, it was clear that to observe other single-electron tunneling phenomena was just a matter of time. Two years after the experiment on the C_{60} transistor, the observation of the Kondo effect was reported simultaneously and independently by two groups [23, 24]. In the work of Park *et al.* [23], two related molecules were examined containing a Co ion bonded to polypyridyl ligands, attached to insulating tethers of different length. The two molecules ($[Co(tpy-(CH_2)_5-SH)_2]^{2+}$ and $[Co(tpy-SH)_2]^{2+}$) differed by a five-carbon alkyl chain within the linker molecules. These molecules were selected because it is known from electrochemical studies that the charge state of the Co ion can be changed from 2+ to 3+ at low energy. The role of the linkers was to control the strength of the metal-molecule coupling. In this work the SMTs were fabricated using the electromigration technique with Au wires coated with the molecules. For the longer molecule, the transport results at temperatures of ~ 100 mK exhibited all the characteristic features of the Coulomb blockade effect. In particular, they observed two diamond-like regions which were associated to the two charge states of the Co ion.

For the shorter molecule, a significantly larger conductance owing to the shorter tether length was expected. The main results for this molecule are summarized in Fig. 15.18. The differential conductance for one such device is shown in Fig. 15.18(b). The most notable property is a peak at $V = 0$. The peak has a logarithmic temperature dependence between 3 and 20 K, see Fig. 15.18(c). The peak also splits in an applied magnetic field, as one can see in Fig. 15.18(d), with a splitting equal to $2g\mu_B H$, where $g \approx 2$ and μ_B is the Bohr magneton.

As we explained in section 15.6.2, all these observations are consistent with the occurrence of the ($S = 1/2$) Kondo effect. The observation of this effect is consistent with the fact that the Co^{2+} ion has $S = 1/2$. By setting the low-temperature full-width at half-maximum of the Kondo peak equal to $2k_B T_K/e$, where T_K is the Kondo temperature, the authors estimated that T_K in different devices varied between 10 and 25 K. These large Kondo temperatures indicate that the coupling between the localized state and the electrodes is strong, consistent with the high conductances found for the shorter linker molecule.

In the work of Liang *et al.* [24], the Kondo effect in SMTs was reported, where an individual divanadium molecule served as a spin impurity. These authors also used electromigrated break-junctions to form the molecular contacts. In Fig. 15.19 we show the results for the stability diagrams for

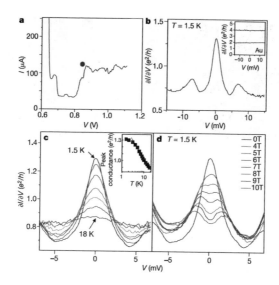

Fig. 15.18 Observation of the Kondo effect in a single-molecule transistor fabricated with the electromigration technique. (a) Breaking trace of a gold wire with adsorbed $[Co(tpy-SH)_2]^{2+}$ at 1.5 K. After the wire is broken the current level suddenly increases (dot) owing to the incorporation of a molecule in the gap. (b) Differential conductance of a $[Co(tpy-SH)_2]^{2+}$ device at 1.5 K showing a Kondo peak. The inset shows $\partial I/\partial V$ for bare gold point contacts for comparison. (c) The temperature dependence of the Kondo peak for the device shown in (b). The inset shows the $V = 0$ conductance as a function of temperature. The peak height decreases logarithmically with temperature and vanishes around 20 K. (d) Magnetic-field dependence of the Kondo peak. The peak splitting varies linearly with magnetic field. Reprinted by permission from Macmillan Publishers Ltd: Nature [23], copyright 2002.

two single-molecule devices, designated as D1 and D2. Two distinct characteristics are evident in the behavior of both devices. Each displays two conductance-gap regions, I and II, bounded by two broad $\partial I/\partial V$ peaks that slope linearly as a function of the gate voltage, V_g. These peaks cross at a gate voltage at which the conductance gaps vanish. Moving away from this point, the gaps in both regions continue to widen even beyond $V = 100\,\mathrm{mV}$. Most significantly, the devices also exhibit a sharp zero-bias $\partial I/\partial V$ peak in region I, whereas this peak is clearly absent in region II. This feature strongly suggests the occurrence of the Kondo effect. In order to confirm this impression, the authors carried out an analysis of the temperature and magnetic field dependence of the differential conductance. Most of the results were relatively well explained in terms of the $S = 1/2$ Kondo effect, but in particular the behavior of T_K suggested that maybe also the orbital

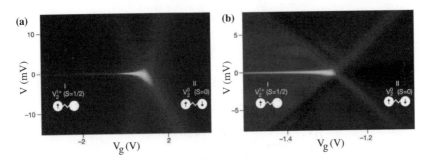

Fig. 15.19   Observation of the Kondo effect in a V_2 single-molecule transistor fabricated with the electromigration technique. The two panel show plots of differential conductance $\partial I/\partial V$ as a function of bias voltage (V) and gate voltage (V_g) obtained from two different single-V_2 transistors. Both measurements were performed at $T = 300$ mK. The values are represented by the color scale, which changes in (a), from dark (0) to bright ($1.55e^2/h$) and in (b), from dark (0) to bright ($1.3e^2/h$). The labels I and II mark two conductance-gap regions, and the diagrams indicate the charge and spin states of the V_2 molecule in each region. Reprinted by permission from Macmillan Publishers Ltd: Nature [24], copyright 2002.

degrees of freedom were playing an important role in the Kondo resonance (due to the V ion spin structure), see Ref. [24] for details.

Most three-terminal single-molecule experiments have been carried out with the electromigration technique. An interesting exception is the work of Kubatkin and coworkers of Ref. [622], where the angle evaporation technique that we described in section 15.3 was employed. These authors reported transport measurements through a single p-phenylenevinylene oligomer, which has five benzene rings connected through four double bonds (OPV5). The main experimental result of this work is reproduced in Fig. 15.20, where one can see up to eight different diamonds in the stability diagram. This suggests that the transport experiment had access to many different charge or redox states, which is very unusual in molecular transistors. Electrochemistry confirms however that this molecule can indeed have several stable redox states [677]. Even more surprising is the fact the addition energies extracted from the stability diagrams differ largely from those obtained from electrochemistry and computational methods. Specially dramatic is the deviation in the case of the neutral molecule. While the spectroscopic HOMO-LUMO gap for this molecule is of the order of 2.5 eV, the extracted one from the central Coulomb diamond was one order of magnitude smaller (~ 0.2 eV). The authors argued that this discrepancy is due to the fact that the intrinsic electronic levels of the molecules are

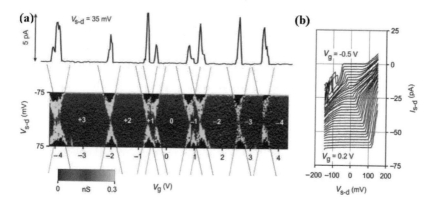

Fig. 15.20 Experimental results of the transport characteristic of an OPV5 single-molecule transistor. (a) Differential conductance as a function of bias voltage, V_{s-d}, and gate voltage, V_g. The full solid line at the top of the figure shows a representative trace of the current versus V_g. (b) Examples of the I-V curves at different gate potentials ($T = 4.2$ K). Curves are shifted vertically for clarity. Reprinted by permission from Macmillan Publishers Ltd: Nature [622], copyright 2003.

significantly altered in the metallic junction. In particular, they suggested that image charges generated in the source and drain electrodes by the charges on the molecule are probably the origin of this effect. This is a very interesting suggestion that may explain similar discrepancies in other experiments [678, 679].

Surprisingly, this important issue has not received much attention. A notable exception is the work of Kaasbjerg and Flensberg [646] in which a realistic description of the screening environment in a SMT was combined with quantum chemical calculations. These authors concluded that the addition energies in a junction are indeed strongly reduced as compared with naive expectations based on the ionization potentials and electron affinities of the molecules in gas phase. They explained that this is a consequence of both (a) a reduction of the electrostatic molecular charging energy and (b) polarization-induced level shifts of the HOMO and LUMO levels. These conclusions are at variance with most DFT-based calculations for two-terminal systems that suggest that the level spacing for small molecules inside the junctions is still rather large and comparable to the one of the isolated molecules. In this sense, it would be highly desirable to have further theoretical and experimental work to clarify this important issue.

SMTs have not only allowed to observe single-electron tunneling phenomena that were well-known in other nanodevices, but they have also made possible to access new transport regimes and to discover novel physical phenomena. In the context of Kondo physics, we would like to mention the work of Pasupathy *et al.* [663] where the Kondo effect in the presence of ferromagnetism has been reported for the first time. In this work the authors measured the transport through single-C_{60} transistors with ferromagnetic nickel electrodes. They showed that Kondo correlations persisted despite the presence of ferromagnetism, but the Kondo peak in the differential conductance was split by an amount that decreased (even to zero) as the spin polarizations in the two electrodes were turned from parallel to antiparallel alignment. Although, the reported splitting was too large to be explained by a local magnetic field, the voltage, temperature, and magnetic field dependence of the signal agreed with predictions for an exchange splitting of the Kondo resonance [680, 681].

SMTs have also allowed to study the interplay between Kondo physics and the electron-vibration interaction. The signatures of vibrational modes have been shown to persist in the Kondo regime [617, 664, 682] and we shall discuss this issue in certain detail in the next chapter. It is also worth mentioning that although most of the experiments on the Kondo effect in molecular junctions have been performed with the electromigration technique, the Kondo physics has also been studied with breakjunctions by Ralph's group [682]. Although in this case a gate electrode was not operative, the authors could tune the Kondo resonance in a single-C_{60} junction by adjusting the metal-molecule distance that is a capability of the breakjunction technique that is lacking in electromigration-based experiments. Changing the metal-molecule coupling the authors were able to tune the Kondo temperature and showed that the temperature dependence of the linear conductance agreed with the scaling function expected for the $S = 1/2$ Kondo problem [682].

SMTs have also been used to explore other basic aspects of the Kondo physics. Thus for instance, Roch *et al.* [684] have recently reported the observation of a quantum phase transition between a singlet and a triplet spin state at zero magnetic field in a single-C_{60} transistor. The analysis of the transport through three-terminal molecular devices has also allowed to study the fundamental scaling laws that govern the non-equilibrium the standard $S = 1/2$ Kondo effect [683].

Another aspect to which SMTs have contributed enormously is the understanding of the role of vibrational modes in the transport through single molecules. The signatures of the excitation of vibronic degrees of freedom

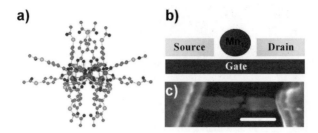

Fig. 15.21   (a) Side view of a Mn_{12} molecule with tailor made ligands containing acetyl-protected thiol end groups (R=C_6H_4). The diameter of the molecule is about 3 nm. (b) Schematic drawing of the Mn_{12} molecule (circle) trapped between electrodes. A gate can be used to change the electrostatic potential on the molecule enabling energy spectroscopy. (c) Scanning electron microscopy image of the electrodes. The gap is not resolvable. Scale-bar corresponds to 200 nm. Reprinted with permission from [619]. Copyright 2006 by the American Physical Society.

are specially visible in the transport characteristics in the limit of weak coupling between the molecule and the metallic electrodes [22, 685]. Moreover, in this regime the electron-vibration interaction can lead to a great variety of novel physical phenomena. This subject will be discussed in detail in the next chapter.

We now turn to a class of experiments where the transport through single-molecule magnets (SMMs) has been investigated (see Ref. [686] for a progress article on this subject). This type of molecules exhibits magnetic hysteresis due to their large spin and high anisotropy barrier, which hampers magnetization reversal [687, 688]. The first transport experiment on a SMM was performed by the group of van der Zant [619]. These authors studied the prototypical SMM, Mn_{12} acetate, which has a total spin $S = 10$ and an anisotropy barrier of about 6 meV. The molecules that were investigated were $[Mn_{12}O_{12}(O_2C\text{-}R\text{-}SAc)_{16}(H_2O)_4]$ (Mn_{12} from now on), where R=$\{C_6H_4, C_{15}H_{30}\}$, see Fig. 15.21(a). These molecules were designed with thiol groups in the outer ligand shell to ensure a strong affinity for gold surfaces. On the other hand, the ligands are believed to serve as tunnel barriers, so that the molecules are only weakly coupled electronically to the gold and their magnetic properties are preserved. The molecules were incorporated in a SMT geometry with gold electrodes using electromigration, see Fig. 15.21(b).

In Fig. 15.22(a) we reproduce some of the results of this work for the differential conductance as a function of gate (V_g) and bias voltage (V_b) for one of the devices ($T = 3$ K, R=C_6H_4). The lines separating the con-

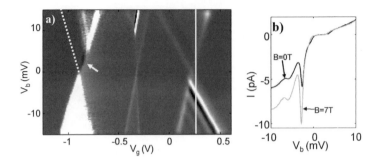

Fig. 15.22   (a) Differential conductance (gray-scale) as a function of gate voltage (V_g) and bias voltage (V_b) $(T = 3$ K, R=C$_6$H$_4$). A region of complete current suppression (left degeneracy point, arrow) and low-energy excitations with negative differential conductance (right degeneracy point) are observed. The dashed line near the left degeneracy point indicates the suppressed diamond edge. (Gray-scale from -0.8 nS [black] to 1.4 nS [white]). (b) $I - V_b$ at the gate voltage indicated in (a) with a line. NDC is clearly visible as a decrease in $|I|$ upon increasing $|V_b|$. Upon applying a magnetic field, current is increased for negative bias. Reprinted with permission from [619]. Copyright 2006 by the American Physical Society.

ducting regions from the diamond-shaped Coulomb blockade regions have different slopes for the three different charge transport regions. Within the orthodox Coulomb blockade theory this implies that the transport regions belong to different quantum dots, since the capacitance to the environment is assumed constant for each dot. However, for molecular quantum dots it is not possible to rule out that these three regions come from three different charge states of the same molecule.

The focus of the work of Ref. [619] was on transport features at low-energy ($\lesssim 5$ meV): a region of complete current suppression (CCS) and a strong negative differential conductance (NDC) excitation line in the stability diagrams. Both are visible in Fig. 15.22(a). At the left degeneracy point in this figure the current is fully suppressed at positive bias voltage above the left diamond edge (dashed line). Transport is restored beyond an excitation that lies at 5 meV. Remarkably, the right diamond edge does continue all the way down to zero bias, defining a narrow strip (~ 1 mV wide) where transport is possible. In the right conductive regime in Fig. 15.22(a), two excitations at an energy of 2 meV and 3 meV are the most pronounced features. The 2 meV excitation is visible as a bright line with positive differential conductance (PDC); the 3 meV excitation as a black line (NDC). The strength of the NDC is clearly visible in the I-V_b plot in Fig. 15.22(b).

The observations of CSS and NDC lines at low energy do not follow in a straightforward way from conventional Coulomb blockade theory. The authors of Ref. [619] explained qualitatively those features with the help of sequential tunneling model that takes into account the high-spin ground state and magnetic excitations of the molecule. They showed that sequential tunneling processes can result in spin blockade of the current, providing a possible explanation for the observed NDC and CCS. This effect is different from conventional spin-blockade [689], where there is no spin anisotropy.

The transport through similar Mn_{12}-based molecules, with short but weak binding ligands, were studied independently by Jo *et al.* [620]. In particular, these authors presented an extensive analysis of the magnetic-field dependence of the transport characteristics. They found two main signatures of magnetic molecular states and magnetic anisotropy in the data: an absence of energy degeneracy between spin states at zero magnetic field (B) and a nonlinear evolution of energy level positions with B. The magnitude of zero-field splitting between spin states was found to vary from device to device, and they interpreted this as evidence for magnetic anisotropy variations upon changes in molecular geometry and environment. On the other hand, they did not observe hysteresis in the electron-tunneling spectrum as a function of swept magnetic field, as one might expect to find in analogy to magnetization measurements on large ensembles of Mn_{12} molecules in bulk crystals. They pointed out that the absence of hysterisis might be due to the fact that sequential tunneling transitions can populate a sequence of excited magnetic levels that surmount the anisotropy barrier and enable rapid magnetic relaxation [690].

Another example of transport through individual magnetic molecules has been reported by Grose *et al.* [691]. In this case, the authors fabricated molecular transistors with individual molecules of the spin-3/2 endohedral fullerene $N@C_{60}$ and measured its spin excitations. $N@C_{60}$ is an attractive model system because of its simple spin structure and because of the possibility of doing control experiments with non-magnetic C_{60} molecules. $N@C_{60}$ molecules also have the advantage of being stable at the high temperatures present during the electromigration process by which the molecular junctions were formed in this work. In the experiments on SMMs that we have reviewed above, the molecular magnetism was usually destroyed during device fabrication. However, in the work of Ref. [691], it was observed that the $N@C_{60}$ devices exhibit clear magnetic character, meaning that they exhibit a spin-state transition as a function of applied magnetic field. The nature of this transition enabled the authors to identify the charge and spin states of the molecule inside the junctions. The spectra of N@C_{60} also exhibited low-energy excited states and signatures of non-equilibrium spin excitations predicted for this molecule [692]. The existence of a spin transition in N@C_{60} accessible at laboratory magnetic fields was associated with the scale of the exchange interaction between the nitrogen spin and electron(s) on the C_{60} cage.

As mentioned above, in the transport experiments on SMMs reported so far, no magnetic hysterisis has been observed, which is probably due to structural deformations of the molecules. In this sense, it would be interesting to test other (more robust) compounds. For instance, Mannini *et al.* [693] have shown recently that tailor-made Fe_4 complexes retain magnetic hysterisis on gold surfaces. These results demonstrate that isolated SMMs can be used for storing information and they open the way to address these molecules individually in their blocked magnetization state.

15.8   Exercises

15.1 Rate equations in a single-level model: For those who are not familiar with rate (or master) equations, it is convenient to start by analyzing the following situation. Let us consider a quantum dot with a single (non-degenerate) level of energy E_1, which is measured with respect to the equilibrium Fermi energy of the leads, which we set to zero. (The energy E_1 can depend on the gate voltage, the exact electrostatic profile and the charging energy). This dot has only two possible configurations with $n_1 = 0$ (empty dot) and $n_1 = 1$ (one electron in level E_1). We shall denote the corresponding probabilities as P_0 and P_1, respectively. As usual, we denote the left and right tunneling rates (in units of energy) as Γ_L and Γ_R, respectively, and we assume them to be energy-independent.

(a) Write down the kinetic equation for the probability distribution and show that in the stationary case the probabilities P_i are given by

$$P_0 = \frac{\Gamma_L \bar{f}_L + \Gamma_R \bar{f}_R}{\Gamma_L + \Gamma_R}, \quad P_1 = \frac{\Gamma_L f_L + \Gamma_R f_R}{\Gamma_L + \Gamma_R}.$$

Here, $f_{L,R} = f(E_1 \mp eV/2)$, where V is the bias voltage and $f(E)$ is the Fermi function.

(b) Use the previous solution to show that the current through the dot can be written as

$$I = \frac{e}{\hbar} \frac{\Gamma_L \Gamma_R}{\Gamma_L + \Gamma_R} [f_L - f_R].$$

Notice that this expression coincides with the expression for the current obtained in the single resonant tunneling model in the limit of weak coupling.

(c) Using the previous expression, show that the linear conductance is given by Eq. (15.26).

15.2 Linear conductance in the Coulomb blockade regime: Derive the formula of Eq. (15.26) for the linear conductance of a quantum dot in Coulomb blockade regime. For this purpose, follow the next steps:

(a) In the linear response regime the distribution function can be written as

$$P(\{n_i\}) \equiv P_{\mathrm{eq}}(\{n_i\}) \left(1 + \frac{eV}{k_{\mathrm{B}}T}\Psi(\{n_i\})\right).$$

Linearize the detailed balance equation (15.23) and solve it to show that Ψ can be written as

$$\Psi(\{n_i\}) = \mathrm{constant} + \sum_{i=1}^{\infty} n_i \left(\frac{\Gamma_R^{(i)}}{\Gamma_L^{(i)} + \Gamma_R^{(i)}} - \eta\right),$$

where the constant first term takes care of the normalization of P to first order in V and it does not need to be determined explicitly. Hint: Use the following relations:

$$1 - f(\epsilon) = f(\epsilon)e^{\epsilon/k_{\mathrm{B}}T}, \quad k_{\mathrm{B}}Tf'(\epsilon)(1 + e^{-\epsilon/k_{\mathrm{B}}T}) = -f(\epsilon),$$

$$P_{\mathrm{eq}}(n_1,\ldots n_{p-1},1,n_{p+1},\ldots) = P_{\mathrm{eq}}(n_1,\ldots n_{p-1},0,n_{p+1},\ldots)e^{-\epsilon/k_{\mathrm{B}}T},$$

where the prime symbol in the Fermi function stands for derivative with respect to its argument.

(b) Linearize the formula for the current in Eq. (15.19) and use the expression for $\Psi(\{n_i\})$ to obtain Eq. (15.26).

15.3 Coulomb oscillations, Coulomb staircase and stability diagrams: The goal of this exercise is to compute transport characteristics in the Coulomb blockade regime within the two-level model discussed in section 15.4.4.2.

(a) As a first step, compute the occupation probabilities of the four possible configurations of the dot. For this purpose, show that the stationary kinetic equation, Eq. (15.22), together with the normalization condition of Eq. (15.24) can be written in the following matrix form: $\hat{W}\vec{p} = \vec{v}$. Here, $\vec{p}$ is the column vector containing the probabilities of the four configurations of the dot, i.e. $\vec{p}^T = (P_1, P_2, P_3, P_4)$, where $1 \equiv (0,0)$, $2 \equiv (1,0)$, $3 \equiv (0,1)$ and $4 \equiv (1,1)$. The vector $\vec{v}$ is simply given by $\vec{v}^T = (1,0,0,0)$ and the different elements of the matrix $\hat{W}$

adopt the form

$$W_{1i} = 1 \ (i = 1, ..., 4), \ W_{23} = W_{32} = W_{41} = 0,$$

$$W_{21} = \Gamma_L^{(1)} f(E_1^{i,l}(N=1)) + \Gamma_R^{(1)} f(E_1^{i,r}(N=1)),$$

$$W_{22} = -\Gamma_L^{(1)} \bar{f}(E_1^{f,l}(N=1)) - \Gamma_R^{(1)} \bar{f}(E_1^{f,r}(N=1))$$
$$\qquad -\Gamma_L^{(2)} f(E_2^{i,l}(N=1)) - \Gamma_R^{(2)} f(E_2^{i,r}(N=1)),$$

$$W_{24} = \Gamma_L^{(2)} \bar{f}(E_2^{f,l}(N=2)) - \Gamma_R^{(2)} \bar{f}(E_2^{f,r}(N=2)),$$

$$W_{31} = \Gamma_L^{(2)} f(E_2^{i,l}(N=0)) + \Gamma_R^{(2)} f(E_2^{i,r}(N=0)),$$

$$W_{33} = -\Gamma_L^{(1)} f(E_1^{i,l}(N=1)) - \Gamma_R^{(1)} f(E_1^{i,r}(N=1))$$
$$\qquad -\Gamma_L^{(2)} \bar{f}(E_2^{f,l}(N=1)) - \Gamma_R^{(2)} f(E_2^{f,r}(N=1))$$

$$W_{34} = -\Gamma_L^{(1)} \bar{f}(E_1^{f,l}(N=1)) - \Gamma_R^{(1)} \bar{f}(E_1^{f,r}(N=1)),$$

$$W_{42} = \Gamma_L^{(2)} f(E_2^{i,l}(N=1)) + \Gamma_R^{(2)} f(E_2^{i,r}(N=1)),$$

$$W_{43} = \Gamma_L^{(1)} f(E_1^{i,l}(N=1)) + \Gamma_R^{(1)} f(E_1^{i,r}(N=1)),$$

$$W_{44} = -\Gamma_L^{(1)} \bar{f}(E_1^{f,l}(N=2)) - \Gamma_R^{(1)} \bar{f}(E_1^{f,r}(N=2))$$
$$\qquad -\Gamma_L^{(2)} \bar{f}(E_2^{f,l}(N=2)) - \Gamma_R^{(2)} \bar{f}(E_2^{f,r}(N=2)).$$

Here, the $\Gamma_{L,R}^{(p)}$ ($p = 1, 2$) are the tunneling rates, while the expressions for the energies appearing in the arguments of the Fermi functions can be found in section 15.4.4.

(b) Solve numerically the 4×4 system $\hat{W}\vec{p} = \vec{v}$ and use the expression of Eq. (15.19) to reproduce the results of Figs. 15.8 and 15.9.

(c) In a molecular transistor the level splitting ΔE may be larger than the charging energy, e^2/C. Study how the stability diagrams in this case differ from those shown in Figs. 15.9. Choose for instance $e^2/C = 30$ meV and $\Delta = 100$ meV, while keeping the other parameters equal to those in the example of section 15.4.4.2.

(d) An important experimental issue is that for a particular charge state lines are often only visible on one side of the Coulomb diamond. This is due to an asymmetry in the coupling. Illustrate this fact with the example of section 15.4.4.2 by choosing very different tunneling rates for the left and right barriers.

15.4 Effects of inelastic scattering in the Coulomb blockade regime: In the Coulomb blockade theory described in section 15.4.4 inelastic scattering was assumed to take place exclusively in the reservoirs. One of the effects of inelastic scattering in the dot is the thermalization of the electrons inside the dot. In the limiting case of full thermalization, the probability distribution function $P(\{n_i\})$ is given by the equilibrium expression of Eq. (15.20). Use this expression in the example of section 15.4.4.2 (and of the previous exercise) to study the effect of inelastic scattering in the different transport characteristics (Coulomb oscillations, Coulomb staircase and stability diagrams).

15.5 Coulomb blockade theory for single-molecule transistors: The goal of this problem is to compute the different transport characteristics of a SMT in the Coulomb blockade regime within the model of section 15.5.2. (a) Show first that the transition rates of Eq. (15.31) are given by

$$W_{11}^r = -\Gamma_r \left[f_r(E_2 - E_1) + f_r(E_3 - E_1) \right]$$
$$W_{12}^r = \Gamma_r \bar{f}_r(E_2 - E_1) = W_{13}^r$$
$$W_{21}^r = \Gamma_r f_r(E_2 - E_1)$$
$$W_{22}^r = -\Gamma_r \left[\bar{f}_r(E_2 - E_1) + f_r(E_4 - E_2) \right]$$
$$W_{24}^r = \Gamma_r \bar{f}_r(E_4 - E_2)$$
$$W_{31}^r = \Gamma_r f_r(E_3 - E_1)$$
$$W_{33}^r = -\Gamma_r \left[\bar{f}_r(E_3 - E_1) + f_r(E_4 - E_3) \right]$$
$$W_{34}^r = \Gamma_r \bar{f}_r(E_4 - E_3)$$
$$W_{42}^r = \Gamma_r f_r(E_4 - E_2) = W_{43}^r$$
$$W_{44}^r = -\Gamma_r \left[\bar{f}_r(E_4 - E_2) + \bar{f}_r(E_4 - E_3) \right]$$
$$W_{14}^r = 0 = W_{23}^r = W_{32}^r = W_{41}^r,$$

where $r = L, R$. The numbers 1 to 4 correspond to the four eigenstates of the molecular Hamiltonian, as defined in section 15.5.2.

(b) Using the numerical values chosen in section 15.5.2, reproduce the results of Fig. 15.10 and compute the corresponding stability diagram.

Chapter 16

Vibrationally-induced inelastic current I: Experiment

16.1 Introduction

In the previous chapter we discussed the transport phenomena that occur in molecular junctions when the conduction is dominated by the Coulomb interaction in the molecular bridge. We now want to focus on the corresponding effects that originate from another inelastic interaction, namely the electron-phonon interaction.[1] When an electron proceeds through a molecule, it can exchange energy by exciting its vibrational modes. Depending on the molecule, the energy of these modes ranges from a few meV to several hundreds of meV [694]. This is comparable to the excess energy of conduction electrons at the usual bias voltages applied in the junctions. Thus, these internal degrees of freedom may influence the transport properties of molecular junctions. Indeed, the interplay between electronic and nuclear dynamics does give rise to a great variety of transport phenomena, as we shall show in this chapter.

When is the electron-phonon interaction expected to play an important role in the electrical conduction through molecular junctions? As we explained in the introduction of the previous chapter, this will occur when the time needed to interact with a vibrational mode, $\hbar/\lambda$, becomes comparable to the traversal time, $\tau = \hbar/\sqrt{\Delta E^2 + \Gamma^2}$. Let us remind that here, λ is the electron-phonon coupling constant, ΔE is the injection energy and Γ is the width of the molecular resonance (or strength of the metal-molecule electronic coupling). In the limit of weak electron-phonon coupling, $\lambda \ll \sqrt{\Delta E^2 + \Gamma^2}$, the vibrational modes give rise to a small inelastic current that is superimposed in a background determined by the

---

[1] The term "phonon" in this chapter is used for vibrational modes associated with any nuclear motion.

elastic contribution. This inelastic current has typically well-defined signatures at energies that are basically the energies of the vibrational modes
of the neutral molecules inside the junctions. Thus, the analysis of the
inelastic current provides a local molecular spectroscopy and in turn, it
gives indirect information on the presence of the molecules, their structure,
orientation and coupling to the leads.

In the opposite limit of strong electron-phonon coupling, $\lambda \gg
\sqrt{\Delta E^2 + \Gamma^2}$, vibronic effects can dominate the transport characteristics of
molecular junctions. Thus for instance, in a resonant situation ($\Delta E \approx 0$)
and if the coupling Γ is not very large, as in the molecular transistors of the
previous chapter, the electron-phonon interaction can lead to pronounced
current steps, which contain valuable spectroscopic information about the
vibrational modes of the molecule in different charge states. On the other
hand, if the electron-phonon interaction is sufficiently strong, one can reach
a regime in which the vibrations make the electronic motion completely incoherent such that it can be described by successive classical rate processes,
usually referred to as *hopping*. The discussion of this hopping regime, where
the transport is mediated by thermally activated processes, will be deferred
until Chapter 18.

Apart from the energy scales mentioned in the previous paragraphs,
there are other important factors that determine the impact of vibrations
in the transport properties. Thus for instance, the temperature plays an
important role in determining the dominant transport mechanism. While
low temperatures favor the coherent transport, high temperatures reduce
drastically the inelastic scattering length by increasing the phonon popula-
tions and making the transport incoherent. On the other hand, the length
of the molecules is another important factor. Incoherent transport becomes
more important for longer molecules both because dephasing is more effective and (for off-resonant tunneling) because of the exponential fall off of
the coherent component.

We initiate here a series of two chapters in which we shall review our
present understanding of the role of molecular vibrations in the transport
properties of single-molecule junctions. In particular, we shall concentrate
on the analysis of their influence in the electrical current. The role of vibrational modes in other properties, including thermal transport, will be
discussed in the Chapter 19. We would like to remark that, following the
spirit of this monograph, we shall present a pedagogical introduction to this
subject, rather than a detailed review of the huge amount of work reported
in the last years. To be precise, after reading these two chapters the reader

should have a clear idea about: (i) what are the basic experimental signatures of vibrational modes in the current through single-molecule junctions, (ii) what are the physical mechanisms giving rise to those signatures and (iii) what are the main open problems related to this subject.

In this first chapter we shall describe some of the main experiments that have illustrated the role of vibrations in the electrical conduction through molecular junctions. We have grouped these experiments in three different categories. First of all, we shall discuss situations where the electron-phonon interaction is weak, in the sense explained above, and the electron tunneling is off-resonant. The analysis of the vibronic signatures in this regime is known as inelastic electron tunneling spectroscopy (IETS) for the historical reasons that will be explained in section 16.2. Then, we shall focus our attention in section 16.3 on the case of highly conductive junctions, where the electron-phonon interaction is also weak. In this regime, and again for historical reasons, the study of the vibrational modes is known as point-contact spectroscopy (PCS). Section 16.4 is devoted to a discussion of the relation between IETS and PCS. In section 16.5 we shall discuss the third group of experiments that correspond to the regime sometimes known as resonant inelastic electron tunneling. This regime corresponds to a situation where the transport is resonant and the electron-phonon interaction can be very strong. This regime is realized, in particular, in the molecular transistors described in the previous chapter. The discussion below will end with a brief summary of the main vibrational signatures that can be observed in the different transport regimes. If you are an impatient reader (as we are), please feel free to jump directly to section 16.6 and then come back to this point.

The recent progress in the understanding of vibrational effects in molecular transport junctions has been thoroughly described by Galperin, Ratner and Nitzan in the review of Ref. [695], which contains close to 500 references related to the main subject of these two chapters. For those who prefer a quick overview, we recommend them the shorter review of Ref. [696] of the same authors.

16.2 Inelastic electron tunneling spectroscopy (IETS)

The first studies of the influence of the electron-phonon interaction on the transport through molecules go back to the 1960's. In a pioneering work, Jaklevic and Lambe discovered in 1966 that vibrational spectra can be ob-

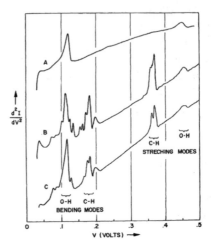

Fig. 16.1   Recorded traces of d^2I/dV^2 versus voltage for three Al-Al oxide-Pb junctions taken at 4.2 K. The zero of the vertical scale is shifted for each curve, and all three are normalized to the same arbitrary units. The largest peaks represent increases of 1% in the conductance. Also indicated are intervals associated with the energy of IR-active molecular vibrational modes. Curve A is obtained from a "clean" junction. Curves B and C are obtained from junctions exposed to propionic acid $[CH_3(CH_2)COOH]$ and acetic acid (CH_3COOH), respectively. The peaks positions are independent of the polarity. Reprinted with permission from [697]. Copyright 1966 by the American Physical Society.

tained from molecules adsorbed at the buried metal-oxide interface of a metal-oxide-metal tunneling junction [697]. In their experiment, the tunneling current I was measured as a function of the voltage V across the junction. Small, but sharp increases in the differential conductance, dI/dV, were observed when the energy of the tunneling electrons reached the energy of a vibrational mode for molecules in the junction. These increases represented changes in the differential conductance of about 1%. They were interpreted as the result of electrons losing their energies to the vibrational mode, giving rise to an inelastic tunneling channel, which is forbidden when tunneling electrons have energies below the quantized vibrational energy. In the experiment, a peak at each vibrational energy was observed in d^2I/dV^2, see Fig. 16.1. This method, known as *inelastic electron tunneling spectroscopy* (IETS), has been applied to a wide range of systems and has led to a better understanding of molecules in the adsorbed state [698–704].

It is convenient for our discussions below to briefly review some of the basic predictions of IETS theory concerning the following issues:

(1) **Tunneling mechanism:** The explanation for the appearance of the peaks in the tunneling spectra is the following [697, 705]. As shown schematically in Fig. 16.2, when the bias voltage applied to the junction is increased and crosses the threshold for excitation of a vibrational mode, electrons can tunnel either elastically or by emitting a vibrational mode. The opening of this latter inelastic channel is accompanied by an increase of the differential conductance (dI/dV) at $eV = \pm\hbar\omega$, where ω is the frequency of the excited mode. As mentioned above, this change is more clearly seen in the derivative of the conductance, d^2I/dV^2, where the signature related to the excitation of a vibrational mode is a peak (dip) for positive (negative) bias, see Fig. 16.2.

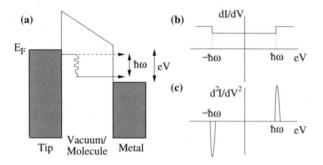

Fig. 16.2   (a) Schematic representation of the inelastic tunneling above the threshold for a vibrational excitation. An electron can tunnel losing part of its energy which is employed to excite a vibration mode of energy $\hbar\omega$. This process is only possible when $eV \geq \hbar\omega$. (b) The opening of the inelastic channel gives rise to an increase in the conductance at $eV = \pm\hbar\omega$. (c) The onset of the inelastic process is seen in the second derivative of the current, d^2I/dV^2, as a peak (dip) for positive (negative) bias.

(2) **Spectral linewidth:** The full width at half maximum (FWHM) of the d^2I/dV^2 vibrational peak is given by $W = [(1.7V_m)^2 + (5.4k_BT/e)^2 + W_I^2]^{1/2}$, where V_m is the modulation voltage in the lock-in technique, k_B is the Boltzmann constant, T is the temperature, and W_I is the intrinsic width (due to the finite phonon lifetime) [705, 706].

(3) **Selection rules:** Although there are no selection rules in IETS as there are in infrared (IR) and Raman spectroscopy, certain selection preferences have been established. According to the IETS theory [707, 708], molecular vibrations with net dipole moments perpendicular to the interface of the tunneling junction have larger peak intensities than vibrations with net dipole moments parallel to the interface (for dipoles close to the electrodes). For a more complete description of the propensity

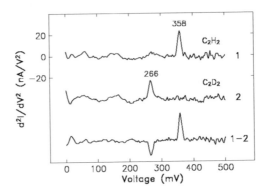

Fig. 16.3   STM inelastic tunneling spectra of acetylene molecules. The plot shows background difference d^2I/dV^2 spectra for C_2H_2 (1) and C_2D_2 (2), taken with the same STM tip. Notice the presence of peaks at 358 mV and 266 mV, respectively. The difference spectrum $(1-2)$ yields a more complete background subtraction. From [709]. Reprinted with permission from AAAS.

rules, see Ref. [704].

Soon after the invention of the STM, it was clear that this tool could serve to extend IETS all the way down to single molecules. However, this turned out to be very challenging since it requires the use of low temperatures ($\sim$ 4 K) and very high mechanical stability. The breakthrough came from Ho's group that reported in 1998 the first study of the vibrational spectra for a single molecule adsorbed on a solid surface [709]. To be precise, these authors measured the inelastic electron tunneling spectra for an isolated acetylene (C_2H_2) molecule adsorbed on the copper (100) surface using a STM under UHV conditions at a temperature of 8 K. They observed an increase in the tunneling conductance at 358 mV, which was attributed to the excitation of the C-H stretch mode. The increase in conductance is typically rather small (around 3-6% in these experiments depending on the tip) and for this reason the features related to the vibrational modes are better seen in the second derivative of the current, d^2I/dV^2, where they appear as peaks (for positive bias), very much like in the IETS in planar tunnel junctions. We show an example of the original data in Fig 16.3.

To confirm the interpretation of the origin of the peak in d^2I/dV^2, the authors used isotopic substitution, i.e. they replaced the hydrogen atoms by deuterium ones in the molecules. In the case of the deuterated acetylene (C_2D_2), they showed that the peak in d^2I/dV^2 is shifted to 266 mV, which corresponds to the expected change in energy of the C-H stretch mode.

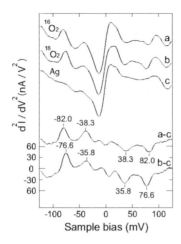

Fig. 16.4 Single-molecule vibrational spectra of oxygen molecules on an Ag(110) surface measured with a STM. Curve a corresponds to $^{16}O_2$, curve b to $^{18}O_2$ and curve c to a clean Ag(110) surface. The difference spectra (curve a-c, curve b-c) are also shown. Reprinted with permission from [711]. Copyright 2000 by the American Physical Society.

Indeed, these values are in close agreement with those obtained by electron energy loss spectroscopy (EELS). This experiment inspired an enormous amount of work in which the chemical sensitivity of the single-molecule IETS has been exploited. This has led in the last years to a better understanding and control of surface chemistry at the atomic level. The activities of the first years on STM-IETS have been reviewed by Ho in Ref. [710].

The first STM-IETS experiments raised several fundamental questions related, for instance, to the selection (or propensity) rules that apply in this case. With respect to the tunneling process that gives rise to the peaks seen in the spectra, it was believed that there is no fundamental difference with respect to the traditional IETS in oxide tunnel junctions. In other words, the process responsible for the vibrational signatures was believed to be the phonon emission process described in Fig. 16.2. However, it is worth stressing that the electron-phonon interaction in these systems does not always lead to an increase of the conductance at the phonon energies. Thus for instance, Ho and coworkers have reported in Ref. [711] STM-IETS studies that revealed two vibrational modes showing a *decrease* in the conductance at 682.0 and 638.3 mV for single oxygen molecules chemisorbed on the fourfold hollow sites of an Ag(110) surface at 13 K. These results can be seen in Fig. 16.4, where one can observe the presence of two well-defined dips at positive bias. It is worth remarking that in this case the change in

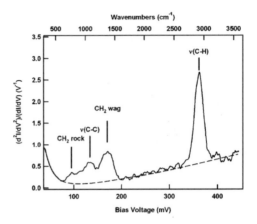

Fig. 16.5   IETS spectrum of a C11 junction formed with gold cross wires. The dashed line is a simple polynomial background and is presented as a guide to the eye. Mode assignments are from comparison to previous experimental results. Reprinted with permission from [712]. Copyright 2004 American Chemical Society.

the conductance at the vibrational energies continues to be rather small (a few percent). Let us mention that also more complicated line shapes have been observed in the context of STM-IETS studies [710].

The use of IETS to measure to the vibrational spectrum of metal-molecule-metal junctions relevant to molecular electronics was first reported in 2004 simultaneously by two different groups.[2] Kushmerick and coworkers presented in Ref. [712] in situ vibrational spectroscopy of metal-molecule-metal junctions containing prototypical molecular wires: C11 (an alkane chain with 11 carbon atoms), OPE, and OPV. The transport measurements were performed with a cryogenic crossed-wire tunnel junction, where one of the gold wires was coated with a monolayer of the molecule of interest. The experiments were conducted at 4 K and standard ac modulation techniques, along with two lock-in amplifiers, were utilized to measure directly both dI/dV and d^2I/dV^2. An example of the results for C11 is shown in Fig. 16.5. Here, the second derivative of the current is normalized by the conductance. Notice that, as in the traditional IETS, the signature of the molecular vibrations is a series of peaks, which were observed to have

---

[2]These experiments were not the first ones to investigate the role of vibronic coupling in molecular transport junctions, but they were the first ones that explored the regime discussed in this section, where the transport through the junctions takes place in a non-resonant manner and the current probes the vibrational modes of the ground state of the molecule.

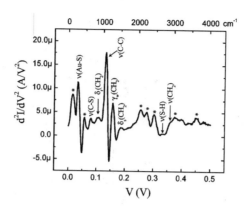

Fig. 16.6   IET spectrum of a C8 dithiol SAM measured with the nanopore technique. The spectrum was obtained from lock-in second-harmonic measurements with an ac modulation of 8.7 mV (rms value) at a frequency of 503 Hz ($T = 4.2$ K). Reprinted with permission from [713]. Copyright 2004 American Chemical Society.

the same height in the positive and negative bias polarity.

Based on previous infrared, Raman, and high-resolution electron energy loss spectroscopy studies of alkanethiolate monolayers, the authors were able to assign the observed peaks in the C11 junction to specific molecular vibrations. The C-H stretch at 362 mV is the most intense vibrational mode observed, but they also observed a number of lower energy vibrations in the region from 70 mV to 200 mV. An interesting observation in this work was the fact that most, although not all, of the modes that were identified corresponded to longitudinal molecular modes, which shows that this type of modes couples more strongly to the tunneling electrodes.

Reed's group reported simultaneously an IETS study of an alkanedithiol self-assembled monolayer (SAM) using the nanopore technique [713]. The second-harmonic signal d^2I/dV^2 was measured directly with a lock-in technique and an example of the results can be seen in Fig. 16.6. Notice that in this case the IET spectrum exhibits peaks with shapes that clearly differ from those of Fig. 16.5.[3] As in Ref. [712], the authors used known results from infrared, Raman, and high-resolution electron energy loss spectra of SAM-covered gold surfaces to identify some of the vibrational modes. Something remarkable in this work is the fact that the authors were able to verify

---

[3]Kushmerick and coworkers have argued that the discrepancies in the IET spectra between these two experiments could be due to the presence of metal nanoparticles in the nanopore devices of Wang *et al.* [713], see Ref. [714] for a detailed discussion of this issue.

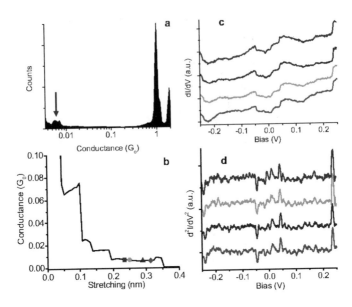

Fig. 16.7   Single-molecule IETS measurements using STM break-junctions. (a) Semilog conductance histogram with a peak at G_0, and an additional peak at $6 \times 10^{-3} G_0$, which is attributed to the conductance of propanedithiol. (b) A conductance curve with steps for a single molecule measurement. The four symbols represent four stretching distances where the bias was swept and the I-V and first derivative were recorded. (c) The corresponding four first derivative curves (offset for clarity), and (d) the corresponding IET spectra obtained numerically. The curves are antisymmetric, and certain features are very reproducible along the conductance plateau. Reprinted with permission from [722]. Copyright 2008 American Chemical Society.

that the observed spectra were indeed valid IETS data by examining the peak width as a function of temperature. This important test is usually very difficult to carry out with other techniques.

IETS has become quite popular in the field of molecular electronics over the last years and it has distinguished itself as a unique spectroscopic probe of molecular junctions. From comparison between experiments and computations, IETS can be useful for characterizing numerous aspects of molecular junctions such as the confirmation of the presence of the molecule, information on the nature of the interfaces, the orientation of the molecule and even electronic pathways can be identified. For further experimental examples of the use of the of IETS in the regime described in this section see Refs. [574, 715–721].

The experiments that we have just described correspond to situations where the transport is probed through an ensemble of molecules. In this sense, it is highly desirable to perform similar experiments, but with single-molecule junctions. However, such experiments in the off-resonant regime that we are discussing in this section are rather scarce. The main problem is to achieve the required stability, which can only be done by working at very low temperatures. Recently, this difficulty has been overcome by Hihath *et al.* [722] who reported IET spectra of a single 1,3-propanedithiol molecule using an STM break-junction at cryogenic temperatures. In particular, these authors were able to measure IET spectra at different stages of the formation of the molecular contacts, see Fig. 16.7. This allows them to correlate changes in the conductance with changes in the configuration of a single-molecule junction. Moreover, the authors were able to do a statistical analysis of the phonon spectra to identify the most relevant modes. Finally, the vibrational modes found for propanedithiol matched well with IR and Raman spectra and were described by a simple one-dimensional model. This type of experiment provides very important information about the formation of single-molecule junctions and in this sense, we are sure that many more experiments of this kind will be reported in the near future (for a more recent one see Ref. [723]).

16.3 Highly conductive junctions: Point-contact spectroscopy (PCS)

In this section we shall discuss the experimental signatures of the electron-phonon interaction in the case of molecular junctions with a high conductance (close to G_0). To be precise, the junctions discussed here are characterized by the presence of a broad electronic resonance around the Fermi energy with a width, Γ, considerably larger than the electron-phonon coupling constant, λ. The analysis of the electron-phonon interaction in this regime has its historical origin in the so-called *point contact spectroscopy* (PCS) [724–726]. Thus, we shall start this section by briefly explaining the basics of this technique.[4]

Many years before the rise of nanofabrication, ballistic metallic point contacts were widely studied [724–726]. The fabrication principle was introduced by Yanson in the 1970's [727] and later developed by his group and by Jansen *et al.* [728]. The technique has been worked out with various refinements for a range of applications, but essentially it consists of

---

[4]Our discussion follows closely Ref. [15].

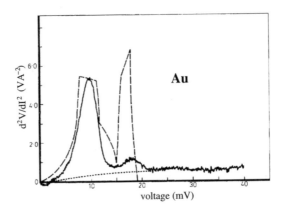

Fig. 16.8   An example of an electron-phonon spectrum measured for a gold point contact by taking the second derivative of the voltage with respect to the current. The long-dashed curve represents the phonon density of states obtained from inelastic neutron scattering. Reprinted with permission from [728]. Copyright 1980 IOP Publishing Ltd.

bringing a needle of a metal gently into contact with a metal surface. With this technique stable contacts are typically formed having resistances in the range from ~ 0.1 to ~ 10 Ω, which corresponds to contact diameters between $d \simeq 10$ and 100 nm. The elastic and inelastic mean free path can be much longer than this length d, when working with clean metals at low temperatures, and the ballistic nature of the transport in such contacts has been demonstrated in many experiments. The main application of the technique has been to study the electron-phonon interaction in metals. Here, one makes use of the fact that the (small but finite) probability for back-scattering through the contact is enhanced as soon as the electrons acquire sufficient energy from the electric potential difference over the contact that they are able to excite the main phonon modes of the material. The differential resistance, dV/dI, of the contact is seen to increase at the characteristic phonon energies of the material. Notice that this is at variance with the typical signature in IETS. A spectrum of the energy-dependent electron-phonon scattering can be directly obtained by measuring the second derivative of the voltage with respect to the current, d^2V/dI^2, as a function of the applied bias voltage. An example is given in Fig. 16.8. Peaks in the spectra are typically observed between 10 and 30 mV, and are generally in excellent agreement with spectral information about the phonons of the corresponding metal obtained from other experiments (e.g. neutron scattering), and with calculated spectra.

Traditionally, electron-phonon spectroscopy in large metallic contacts

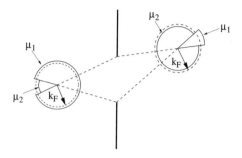

Fig. 16.9 Electron distribution function in the vicinity of the orifice. Here, k_F is the equilibrium Fermi wave vector, μ_1 and μ_2 are the chemical potentials for each side, which, far from the orifice and in the presence of an applied potential V, are equal to $E_F - eV/2$ and $E_F + eV/2$, respectively.

is described by considering the non-equilibrium electron distribution near the contact that results from the applied bias voltage, as illustrated in Fig. 16.9 [727–729]. Electrons that arrive at the left electrode, coming from the right, are represented in a Fermi surface picture by a cone with an angle corresponding to the solid angle at which the contact is viewed from that position in the metal. These electrons have eV more energy that the other Fermi surface electrons, and they can be scattered inelastically to all other angles outside the cone. Only those that scatter back into the contact will have a measurable effect on the current.

As the energy difference eV increases, this backscattering increases due to the larger phonon density of states, which will be observed as a decreasing conductance. Ignoring higher order processes, the decrease of the conductance comes to an end for energies higher than the top of the phonon spectrum, which is typically 20–30 meV. By taking the derivative of the conductance with respect to the voltage one obtains a signal that directly measures the strength of the electron-phonon coupling. An example for gold is illustrated in Fig. 16.8. Several authors have derived an expression for the spectrum [728, 730], which adopts the following form

$$\frac{d^2I}{dV^2} = \frac{4}{3\pi} \frac{e^3 m^2 v_F}{\hbar^4} a^3 \alpha^2 F_p(eV), \tag{16.1}$$

where a is the contact radius, v_F is the Fermi velocity, m the electron mass and the function $\alpha^2 F$ is given by

$$\alpha^2 F_p(E) = \frac{m^2 v_F}{4\pi h^3} \int d^2n \int d^2n' |g_{nn'}|^2 \eta(\theta(\mathbf{n}, \mathbf{n}')) \delta(E - \hbar\omega_{nn'}). \tag{16.2}$$

Here, the integrals run over the unit vectors of incoming and outgoing electron wave vectors $(\mathbf{n} = \mathbf{k}/k)$, $g_{nn'}$ is the matrix element for the electron-

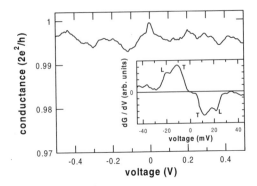

Fig. 16.10 Differential conductance as a function of the applied bias voltage for a one-atom Au contact at 4.2 K. The contact was tuned to have a conductance very close to $1\,G_0$, which suppresses the amplitude of the conductance fluctuations. This allows the observation of a phonon signal, which is seen as a maximum at zero bias. Inset: By taking the derivative of the conductance the transverse (T) and longitudinal (L) acoustic branches can be recognized symmetrically positioned around zero. Note the expanded scale of the voltage axis in the inset. Reprinted with permission from [731]. Copyright 2000 by the American Physical Society.

phonon interaction, and η is a function of the scattering angle that takes the geometry into account, such that only backscattering through the contact is effective, $\eta(\theta) = (1 - \theta/\tan\theta)/2$. From this expression, and by considering Fig. 16.9, one can see that the contribution of scattering events far away from the contacts is suppressed by the effect of the geometric angle at which the contact is seen from that point. The probability for an electron to return to the contact decreases as $(a/d)^2$, with a the contact radius and d the distance from the contact. This implies that the spectrum is dominantly sensitive to scattering events within a volume of radius a around the contact, thus the effective volume for inelastic scattering in the case of a clean opening (the contact) between two electrodes is proportional to a^3. Clearly, this effective volume must depend on the geometry of the contact. For a long cylindrical constriction, the electrons scattered within the constriction will have larger return probability, the effective volume, in this case, increases linearly with the length [724].

    The point-contact spectroscopy has been extended in recent years to atomic-sized contacts. As the contact becomes smaller, the signal comes from scattering on just a few atoms surrounding the contact. The spectrum no longer measures the bulk phonons, but rather local vibrational modes of the contact atoms. In attempting to measure the phonon signal for small

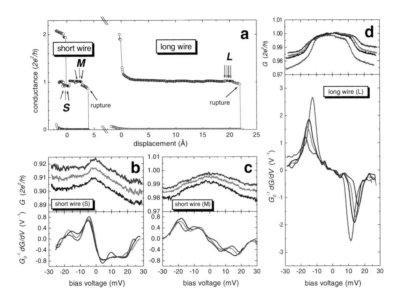

Fig. 16.11   Point contact spectroscopy of gold atomic chain. (a) Short and long atomic wire, ∼ 0.4 and ∼ 2.2 nm, respectively, as given by the length of the conductance plateau. Panels (b–d) show the differential conductance and its derivative at points S, M, and L, respectively. The various curves in (b–d) were acquired at intervals of 0.03, 0.03 and 0.05 nm, respectively. Note that the vertical scales for the last thee panels are chosen to be identical, which brings out the relative strength of the electron-phonon interaction for the longer chains. The wire in (d) has a length of about 7 atoms. Reprinted with permission from [732]. Copyright 2002 by the American Physical Society.

contact sizes one encounters the problem that the phonon signal intensity decreases, according to Eq. (16.2), while the amplitude of the conductance fluctuations[5] remains roughly constant, or slightly increases. The result is that the phonon signal is sometimes hidden in the conductance fluctuations for the smallest contacts. A solution to this problem is obtained for the special and interesting case of a contact made up of a single channel with nearly perfect transmission probability, where these fluctuations are suppressed. Under these conditions the features due to phonon scattering become clearly visible. This is illustrated in Fig. 16.10 where we show an example of the point contact spectrum of a gold one-atom contact [731]. Surprisingly, one observes a spectrum (see inset of the figure) that still closely resembles the bulk phonon spectrum, although the relative intensities of the features in the spectrum are different.

---

[5] These fluctuations were described in detail in section refsec-cond-fluct.

The point contact spectroscopy was pushed to study the phonon modes in Au atomic chains by Agraït and coworkers [732, 733]. As we described in section 11.8, atomic chains of certain metals can be formed with the STM and break junction techniques. These chains constitute in some sense the simplest molecules that one can think of. Thus, the PCS of gold chains of Refs. [732, 733] is of special interest for us. In these experiments, the differential conductance was measured using a lock-in detection with a 1 mV modulation voltage, from which dG/dV was calculated numerically. The energy resolution was limited by the temperature of 4.2 K to 2 meV. The results for the differential conductance and its derivative for a long atomic chain (~ 7 atoms) are shown in Fig. 16.11(d). Notice that at ± 15 mV bias the conductance exhibits a rather sharp drop by about 1%. In the second derivative d^2I/dV^2 this produces a pronounced single peak, point-symmetric about zero bias. The chains of Au atoms have the fortuitous property of having a single nearly perfectly transmitted conductance mode, which suppresses conductance fluctuations that would otherwise mask the phonon signal. Some asymmetry that can still be seen in the conductance curves is attributed to the residual elastic scattering and interference contributions.

The fact that only one conductance drop is clearly seen was interpreted by the authors as follows. By energy and momentum conservation the signal can only arise from electrons that are back-scattered, changing their momentum by $2k_F$. With $\hbar\omega_{2k_F}$ the energy for the corresponding phonon, the derivative of the conductance is expected to show a single peak at $eV = \pm\hbar\omega_{2k_F}$. The transverse phonon mode cannot be excited in this one-dimensional configuration and only the longitudinal mode is visible. We shall see in the next chapter that this argument is, strictly speaking, only valid for infinite chains, while it is approximate for the finite chains realized in the experiments.

Another interesting feature of the point-contact spectra of gold atomic chains is that the position of the peak in dG/dV shifts as a function of the strain in the wire. As one can see in Fig. 16.11(d), the frequency of the mode associated to the peak decreases as a function of the tension because of the decreasing bond strength between the atoms. However, the amplitude (peak height) increases, until an atomic rearrangement takes place, signaled by a small jump in the conductance (not shown here). At such points the amplitude and energy of the peak in dG/dV jump back to smaller and larger values, respectively. This is consistent with the phonon behavior of Au atomic chains found in ab initio calculations [360]. The growing

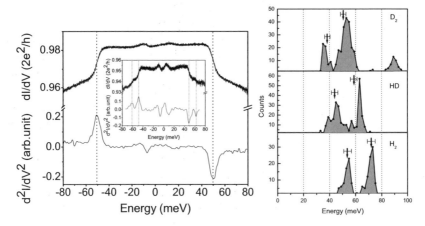

Fig. 16.12   Left panel: Differential conductance curve for D_2 contacted by Pt leads. The dI/dV curve (top) was recorded over 1 min, using a standard lock-in technique with a voltage bias modulation of 1 meV at a frequency of 700 Hz. The lower curve shows the numerically obtained derivative. The spectrum for H_2 in the inset shows two phonon energies, at 48 and 62 meV. Right panel: Distribution of vibrational energies observed for H_2, HD, and D_2 between Pt electrodes, with a bin size of 2 meV. The peaks in the distribution for H_2 are marked by arrows and their widths by error margins. These positions and widths were scaled by the expected isotope shifts, $\sqrt{2/3}$ for HD and $\sqrt{1/2}$ for D_2, from which the arrows and margins in the upper two panels have been obtained. Reprinted with permission from [567]. Copyright 2005 by the American Physical Society.

amplitude is due to the softening of the phonon modes with tension.

The first application of point-contact spectroscopy to the characterization of a molecular junction was carried out by Ruitenbeek's group in their study of the transport through hydrogen molecules that we discussed in section 14.1.3. The original study of Ref. [127] was extended in Ref. [567] with a thorough analysis of the stretching behavior of point-contact spectra as well as DFT calculations of the vibrational modes of the Pt-H_2 junctions.

The left panel of Fig. 16.12 shows examples for Pt-H_2 and Pt-D_2 junctions at a conductance near $1\,G_0$. The conductance is seen to drop by about 1 or 2%, symmetrically at positive and negative bias, very muck like in the Au atomic chains just described. The energies of the conductance drops are in the range of 50-60 meV, well above the Debye energy of $\sim$ 20 meV for Pt. A high energy for a vibrational mode implies that a light element is involved, since the frequency is given by $\omega = \sqrt{\kappa/M}$ with κ an effective spring constant and M the mass of the vibrating object. The proof that the spectral features are indeed associated with hydrogen vibrational modes came from further experiments where H_2 was substituted by the heavier isotopes D_2 and HD. The positions of the peaks in the spectra of $d^2 I / dV^2$ vary within some range between measurements on different junctions, which can be attributed to variations in the atomic geometry of the leads to which the molecules bind. Fig. 16.12 (right panel) shows histograms for the vibrational modes observed in a large number of spectra for each of the three isotopes. Two pronounced peaks are observed in each of the distributions, that scale approximately as the square root of the mass of the molecules, as expected. The two modes can often be observed together, as in the inset of the left panel of Fig. 16.12. For D_2 an additional mode appears near 90 meV. This mode cannot easily be observed for the other two isotopes, since the lighter HD and H_2 mass shifts the mode above 100 meV where the junctions become very unstable. For a given junction with spectra as in Fig. 16.12 (left panel), it is often possible to stretch the contact and follow the evolution of the vibrational modes. The frequencies for the two lower modes were seen to increase with stretching, while the high mode for D_2 is seen to shift downwards. This unambiguously identifies the lower two modes as transverse modes and the higher one as a longitudinal mode for the molecule. This interpretation agrees well with DFT calculations for a configuration of a Pt-H-H-Pt bridge in between Pt pyramidally shaped leads [567, 571]. The fact that the vibrational modes observed for HD that are intermediate between those for H_2 and D_2 confirms that the junction is formed by a molecule, not an atom.

A drop in the conductance as a fingerprint of the presence of a molecule in highly conductive junctions has also been reported, for instance, in Ref. [385]. In this work, PCS was used to identify the presence of oxygen intercalated in Au atomic chains. More recently, similar vibration-induced steps down in the conductance have been also observed in various small molecules directly bonded to Pt electrodes [474].

16.4 Crossover between PCS and IETS

As we have seen in the previous two subsections, electron-phonon interaction leads to an increase in the conductance for junctions in the tunnel regime (e.g. IETS done in STM); however, it decreases the conductance for junctions in the contact regime (e.g. PCS across a Pt-H_2 junction). In spite of this difference, all these physical systems have in common that the traversal time, $\tau = \hbar / \sqrt{\Delta E^2 + \Gamma^2}$, is much smaller than the time that it takes to interact with a vibrational mode, $\hbar / \lambda$. In IETS this is due to the fact that the tunneling is typically off-resonant, and therefore the injection energy ΔE is rather large. In the PCS case, however, this occurs because the molecule is strongly coupled to the leads and thus Γ is very large. In view of this similarity, one may wonder whether there is any fundamental difference between IETS and PCS in molecular junctions. As we shall discuss in the next chapter, recent theoretical work has shown that IETS and PCS are indeed two sides of the same coin and they can be described in a unified manner. In other words, these two techniques are based on the same underlying physics and they simply refer to two different limiting cases depending on the junction transparency.

In recent years, different experiments on highly conductive single-molecule junctions, but with conductances not to close to $1\,G_0$, have clearly suggested the idea that there is a smooth crossover between IETS and PCS. Thus for instance, experiments on Pt-H_2 junctions [734], Ag atomic wires decorated with oxygen [605] and Pt-benzene junctions [473] with conductances between 0.1 and $0.4\,G_0$ have shown that the signature of vibrational modes is a step up in the conductance at the vibrational energies, i.e. exactly like in the standard IETS case. The experiment that has finally clarified this issue was reported recently by Tal *et al.* [735] and we now proceed to describe it in certain detail.

In Ref. [735] the authors presented PCS and shot noise measurements across a single-molecule junction formed by Pt electrodes and H_2O molecules. The Pt/H_2O molecular junctions were formed using a MCBJ setup at about 5 K. The formation of a clean Pt contact was verified by conductance histograms, which exhibited a single peak around $1.4\,G_0$, providing so a fingerprint of a clean Pt contact [127]. Water molecules were then introduced to the junction through a heated capillary, while the Pt junction was broken and formed repeatedly. Following the introduction of water, the typical Pt peak in the conductance histogram was suppressed and contributions from a wide conductance range were detected with minor peaks around 0.2, 0.6, and $1.0\,G_0$. The continuum in the conductance counts implies a variety of stable junction configurations that the authors exploited for spectroscopy measurements on junctions with different conductance.

In Fig. 16.13 we reproduce the results for the differential conductance as a function of the voltage across the Pt/H_2O junction at two different linear conductance values: $1.02 \pm 0.01\,G_0$ (a) and $0.23 \pm 0.01\,G_0$ (b). Junctions with different zero-bias conductance were formed by altering the distance between the Pt contacts or by re-adjusting a new contact. The steps in

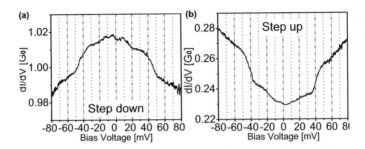

Fig. 16.13 Differential conductance (dI/dV) as a function of the bias voltage for two different Pt-H_2O-Pt junctions with linear conductance of $1.02 \pm 0.01\,G_0$ (a) and $0.23 \pm 0.01\,G_0$ (b). Reprinted with permission from [735]. Copyright 2008 by the American Physical Society.

the conductance that appear at 46 mV in Fig. 16.13(a), and 42 mV in Fig. 16.13(b) indicate the onset of a vibrational excitation at these voltages. Notice that while in (a) the differential conductance is decreased ("step down"), the curve (b) taken at lower linear conductance shows an increase in the differential conductance ("step up"). These two examples demonstrate that both conductance suppression and enhancement can be observed at a relatively high conductance (much higher than the typical tunneling conductance).

As we shall show in the next chapter, the theory predicts that in the regime of weak electron-phonon coupling, the transition from a step down to a step up in the conductance occurs at a transmission equal to 0.5 for a single channel model [736, 737]. In order to confirm these predictions Tal and coworkers collected many dI/dV spectra at different zero-voltage conductance values. They found that curves with steps up appear below $0.57 \pm 0.03\,G_0$ and curves with steps down were detected only above $0.72 \pm 0.03\,G_0$. Thus, they demonstrated that the crossover between conductance enhancement and conductance reduction by the electron-vibration interaction occurs between these two values. Since more than one conduction channel can contribute to the conductance in these junctions, the authors carried out shot noise measurements to determine the number of channels and their transmission probabilities. They concluded that there were typically two conduction channels. More importantly, they showed that the dominant channel had a transmission 0.51 ± 0.01 at the crossover conductance, which is nicely consistent with the predictions of single-channel models (see Ref. [735] for more details).

16.5   Resonant inelastic electron tunneling spectroscopy (RIETS)

In this section we shall discuss the signatures of the electron-phonon interaction in the case of resonant situations, when the traversal time is not small in comparison with the time $\hbar/\lambda$, where λ is the electron-phonon coupling constant. This occurs when the injection energy, ΔE, is rather small and the molecular orbital width, Γ, is not too large. In this regime, strong vibronic effects are expected and the transport characteristics provide in this case what is sometimes referred to as *resonant inelastic electron tunneling spectroscopy* (RIETS) [695]. This physical situation is realized, in particular, in the single-molecule transistors (SMT) discussed in the previous chapter. In these junctions the electronic states can be brought close to the chemical potentials of the reservoirs by means of a gate voltage. The additional flexibility provided by the third electrode together with the strong electron-phonon coupling give rise to a rich phenomenology that we now want to describe.

The experiment performed by Park *et al.* [22] that we described at the beginning of section 15.7 was also the first SMT experiment that revealed vibronic effects. Let us remind that in this work, the transport through a single C_{60} molecule was studied in a three-terminal device. In this case the fingerprint of the vibrational modes can be seen directly in the I-V characteristics, see Fig. 15.16. In that figure one can see that (for positive bias) the I-V curves exhibit a first step that corresponds to the crossing of an electronic resonance. There is a second step that is separated from the first one by a distance of around 5-10 mV. As discussed in section 15.7, there are good reasons to attribute that signature to the excitation of a vibrational mode that corresponds to the center-of-mass oscillation of C_{60}. Notice that, contrary to the cases discussed so far, the signature of a vibrational mode is now a step in the I-V curves (or a peak in the differential conductance). Moreover, this feature does not appear at a voltage equal to $\hbar\omega/e$, where ω is the vibration frequency, but rather at a bias voltage that is equal to the bias that is necessary to cross the electronic resonance (at a given gate voltage) plus a voltage of the order of $\hbar\omega/e$.[6]

What is the origin of this peculiar signature? This will be explained in detail in the next chapter, but let us briefly say that this feature originates from the same inelastic tunneling process (phonon emission) discussed

---

[6]The exact distance between the Coulomb peak and the first sideband depends on the voltage profile, i.e. on how the resonant level is shifted by the bias voltage.

above, see Fig. 16.2. The difference is now that this process is much more probable when the energy of the electron surpasses the energy of the electronic level by an amount that is equal to $\hbar\omega$ (corrected by a factor that depends on how the voltage drops across the junction). The reason is that an electron with that energy may lose an energy equal to $\hbar\omega$ (by emitting a vibrational mode) and then it crosses the molecule exactly at resonance with the molecular level. The enhanced probability of this inelastic process gives rise to a peak in the differential conductance at a bias voltage of the order of $\hbar\omega/e$ away from the Coulomb blockade peak. From this argument, it is also easy to understand that in order to observe a pronounced current step, the width of the electronic resonance, Γ, must be smaller than $\hbar\omega$. In any case, the bias voltage must be larger than $\hbar\omega$ for this inelastic process to take place.

As it was also explained in section 15.7, the signature of a vibrational mode can be also seen in the stability diagrams, see Fig. 15.17. In these plots, the peaks in the conductance, which correspond to the step-like features in Fig. 15.16, show up as lines. In particular, the vibration mode with energy 5 meV appears there as running lines that intersect the main diamonds or conductance gap regions. The energy of this excitation is too small to correspond to an electronic excitation. Moreover, some of these lines are observed for both charge states, which would be very unlikely for an electronic excitation. Even more convincing is the fact that multiple excitations with the same spacing are observed, see Fig. 15.17(d). This corresponds most likely to the excitation of several vibrational quanta of the same mode, i.e. multi-phonon processes. Let us also say to conclude this discussion that signatures of intrinsic vibrational modes of the C_{60} molecules were also observed in the stability diagram of some devices, see in particular Fig. 3 in Ref. [22].

The experiment just described was followed by other experiments with weakly coupled molecules where signatures of the vibrational modes in the transport characteristics were also observed. For instance, Zhitenev *et al.* [738] reported transport measurements through a small self-assembled monolayer of thiolated organic molecules in which the conductance exhibited a series of equally spaced peaks, the position of which could be controlled by a gate voltage. These peaks were attributed to the lowest molecular vibrations of the molecules. The most surprising thing in this experiment was the observation of a large number of conductance peaks with slowly decreasing amplitudes. This would mean that phonon processes of very high-order were taking place in these junctions. On the other hand,

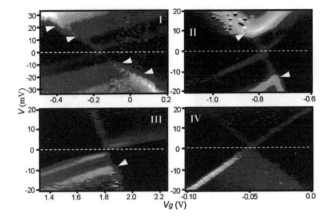

Fig. 16.14 Stability diagrams (dI/dV vs. V and V_g) for four C_{140} SMTs fabricated with the electromigration technique. White arrows indicate excited levels at 11 and 22 meV. dI/dV is represented by a color scale from black (zero) to white (maximum), with maximum values 200 nS (device I), 600 nS (II), 15 nS (III), and 100 nS (IV). Measurements were done at 1.5 K for I-III and 100 mK for IV. Reprinted with permission from [685]. Copyright 2005 American Chemical Society.

Park *et al.* [23] observed low-lying excitations in the stability diagrams of SMTs based on coordination complexes with Co ions. These excitations appeared in the Coulomb blockade regime in the two charge states observed in the diagrams, which clearly suggested that they might correspond to vibrational modes.

In the previous experiments it was difficult to determine the precise nature of the vibrational modes. In the transistors made from C_{60} [22] the mode observed was not intrinsic to the molecule itself. In this sense, experiments like the one of Pasupathy *et al.* [685] were important. In this case, the authors reported the study of single-molecule transistors made using a C_{140} molecule, in which it was possible to clearly identify low-energy internal vibrational modes. Such modes were clearly visible in the stability diagrams, see Fig. 16.14, and an excitation at 11 ± 1 meV was seen in most devices. By means of a detailed molecular modeling, it was possible to identify this mode as an internal stretching mode of the molecule. The modeling also explained the strong coupling of this mode to tunneling electrons, relative to other molecular modes.

An impressive example of resonant inelastic electron tunneling has been reported by Osorio *et al.* [678]. These authors performed transport measurements in electromigrated single-molecule junctions based on

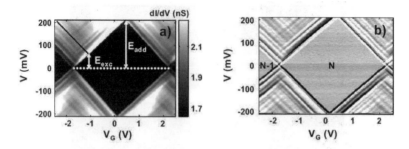

Fig. 16.15   Stability diagrams of a three-terminal junction with OPV-5, measured at 1.6 K. Plotted in (a) is dI/dV as measured with a lock-in technique (modulation amplitude 0.4 mV) and in (b) the numerically calculated second derivative, which serves to highlight the fine structure of the excitations. The current levels are the same near both degeneracy points, which is a strong indication that they belong to the same molecule. Three different charge states are probed. The $N+1$ state is not indicated; for low bias voltages it starts at gate voltages larger than 2.2 V. The data yield an addition energy of 210 meV and a gate coupling of 0.05. Reproduced with permission from [678]. Copyright Wiley-VCH Verlag GmbH & Co. KGaA.

an oligophenylenevinylene derivative (OPV-5). An example of the stability diagrams obtained in these experiments is shown in Fig. 16.15. This diagram clearly shows the presence of sets of excitation lines for all three charge states accessible in the experiment. A close inspection reveals that the point of intersection between the lines and the diamond edge are symmetric with respect to the bias polarity, and that their position is almost independent of the charge state. This observation makes it unlikely that the excitations are a result of electronic states, because these are expected to depend strongly on the charging of the molecule. Moreover, the 17 excitations present in the experimental data are unlikely to reflect precisely 17 available electronic states that differ by only 5-10 meV in energy. Therefore, the excitations were attributed to the vibrational modes of the single OPV-5 molecule trapped between the electrodes. The authors compared the vibrational modes probed in the transport experiment with those probed with light by using Raman and IR spectroscopy and found a good agreement for the ones with the highest energies (see Ref. [678] for further details).

Molecular junctions offer the possibility to examine transport regimes that are difficult to access in other systems. In particular, single-molecule transistors are ideal systems where to study the interplay between vibronic effects and Kondo physics. As we explained in section 15.7, the Kondo effect have been observed in SMTs by several groups. Already in the first observations of this effect there were clear hints of the coexistence of the

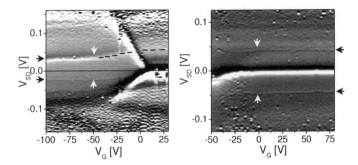

Fig. 16.16   Maps of d^2I/dV_{SD}^2 as a function of V_{SD} and V_G at 5 K for two Co-ion-based SMTs fabricated with the electromigration technique. Brightness scales are -8×10^{-5} A/V^2 (black) to 3×10^{-5} A/V^2 (white), and -2×10^{-5} A/V^2 (black) to 2×10^{-5} A/V^2, respectively. The zero-bias features correspond to Kondo peaks in $\partial I/\partial V_{SD}$. Prominent inelastic features are indicated by black arrows. In both devices, when the inelastic features approach the boundaries of the Coulomb blockade region, these levels shift and alter the line shape (white arrows). Black dashed line in left map traces an inelastic feature across the Coulomb blockade region boundary and into the Kondo regime. Reprinted with permission from [617]. Copyright 2004 by the American Physical Society.

Kondo resonance and vibrational sidebands [23, 662]. The first work in which this coexistence was studied in detail was reported by Natelson's group [617]. Using electromigration-based SMT junctions they analyzed the transport through a molecule comprising a single Co ion coordinated by conjugated ligands. In many devices they observed the Kondo effect and a Kondo temperature of ~ 40 K was deduced from the temperature dependence of the zero-bias conductance. Moreover, in some cases, the conductance in the classically blockaded region and/or outside the Kondo resonance was large enough to allow clean measurements of $\partial^2 I/\partial V_{SD}^2$. In Fig. 16.16 we reproduce results from Ref. [617] where maps of this quantity are shown as a function of V_{SD} (source-drain voltage) and V_G (gate voltage) in two different devices at 5 K. The left panel shows mainly a diamond-like region corresponding to a charge state exhibiting standard Coulomb blockade, while the right one focuses on the next diamond where the Kondo resonance is visible at zero bias. Two prominent features within the blockaded (Kondo) regime are indicated with black arrows. Features in $\partial^2 I/\partial V_{SD}^2$ of opposite sign are symmetrically located around zero source-drain bias, consistent with inelastic tunneling expectations.

The $\partial^2 I/\partial V_{SD}^2$ features in the blockaded region occur at essentially constant values of V_{SD} until V_G is varied such that the feature approaches the edge of the blockaded region. This independence of V_G resembles the signature of inelastic cotunneling (see section 15.6.1) which has been observed, for instance, in semiconductor single-electron devices [650]. The inelastic modes occur at energies low compared to the expected level spacing (> 100 meV), implying that the modes being excited are unlikely to be electronic. Indeed, the authors compared the energies of these features with Raman and IR data and found a nice correlation supporting the idea that they correspond to vibration modes.

Another work in which the interplay between the Kondo physics and vibration-assisted tunneling was investigated is that of Parks *et al.* [682]. In this experiment the Kondo resonance in a C_{60} junction was tuned mechanically using the MCBJ technique. They also observed pronounced peaks in the differential conductance at symmetric values of the bias voltage (see Fig. 4 in Ref. [682]). The main observed feature appeared at $\pm$ 33 mV, which was attributed to the lowest intracage vibrational mode of a C_{60} molecule. Additionally, as the electrodes were pulled apart, the energies of the modes were observed to shift due to the change in the strength of the metal-molecule coupling.

Let us mention that a Kondo resonance accompanied by vibrational sidebands has also been observed in STM experiments on the transport through a single molecular layer of a purely organic charge-transfer salt grown on a metal surface [739].

To conclude this section, we now want to briefly mention two experiments of special relevance. In the first one, Dekker and coworkers studied the current through a suspended single-wall carbon nanotube injected from a STM tip [740, 741]. They showed that the current exhibited usual features of the Coulomb blockade, i.e. a series of peaks in the differential conductance. Moreover, they found that these peak were accompanied not only by the usual RIETS satellite peaks on the right hand side of the Coulomb peaks (for positive bias), but also by peaks on the left hand side. The satellite peaks on the right are a signature of phonon emission (the mode excited in this experiment was believed to be the radial breathing modes of the tube). The peaks on the left were interpreted as a fingerprint of phonon absorption. Since the bath temperature of the experiment was much smaller than the energy of the modes (and therefore they could not be excited thermally), it was concluded that these anomalous peaks were the signature of nonequilibrium phonons that are created by the electrical current. This experiment illustrates that "hot" (or nonequilibrium) phonons can play an important role in the transport through a molecular structure.

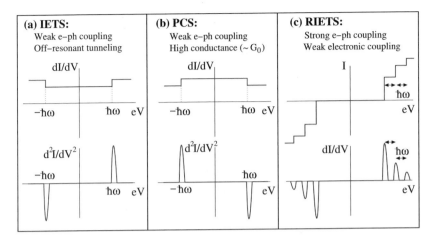

Fig. 16.17   Summary of the main vibrational signatures in the transport characteristics of molecular junctions in various transport regimes.

The other experiment that we want to briefly comment was reported by Ho's group [742]. The experiments described so far in this section were mainly performed with the electromigration technique and with the aid of a third terminal. However, this is not the only way to reach the transport regime that we are discussing. For instance, in Ref. [742] a STM was used to define a double-barrier junction by positioning the STM tip over an individual copper phthalocyanine molecule adsorbed on a thin (approximately 0.5 nm) insulating Al_2O_3 film grown on the NiAl(110) surface. The two tunnel barriers in the junction were the vacuum gap between the STM tip and the molecule, and the oxide film between the molecule and NiAl. The current through this double junction was found to exhibit clear signatures of molecular vibronic states that were observed to change dramatically by varying the tip-molecule separation, which in turn controls the ratio of electron tunneling rates through the two tunnel barriers.

16.6   Summary of vibrational signatures

Let us now briefly summarize the main vibrational signatures that we have shown to appear in the different transport regimes (see Fig. 16.17):

- In off-resonant situations (low transmissive junctions), in which the electron-phonon interaction is weak, the typical signature of vibrational modes is a small increase of the differential conductance at the mode energies. These features are usually better seen in the second derivative of the current (d^2I/dV^2), where they appear as peaks (for positive bias), see Fig. 16.17(a). This regime is realized in STM junctions in the tunneling limit and in strongly coupled contacts where the conductance is smaller than approximately $0.5\,G_0$. Let us stress that in the case of STM tunnel junctions, the observation of dips in d^2I/dV^2 is also possible.

- In the case of strongly coupled metal-molecule-metal junctions with conductances close to G_0 and weak electron-phonon interaction, the vibrations are manifested in the conductance as small drops at the mode energies and therefore, as dips in d^2I/dV^2, see Fig. 16.17(b).[7] This regime is realized, for instance, in atomic gold chains and hydrogen-based junctions.

- In the case of resonant transport, if the metal-molecule coupling is not very strong, like it is usually the case in SMTs, the excitation of vibrational modes leads to steps in the current versus the bias voltage, or a series of peaks in the dI/dV, see Fig. 16.17(c). If a gate electrode is available, the vibronic excitations can be seen in the Coulomb blockade regime as running lines in the stability diagrams. In the Kondo regime the modes induce sidebands at the phonon energies. These sidebands are seen in the stability diagrams as horizontal lines (i.e. independent of the gate voltage), parallel to the zero-bias Kondo resonance.

The reader should bear in mind that this summary is slightly oversimplified and more complex signatures are also possible. Thus for instance, Thijssen *et al.* [743] have reported the observation of anomalous spikes in the differential conductance of a variety of junctions, which were attributed to vibrationally induced two-level systems. On the other hand, vibronic effects can also be responsible for other strong non-linearities in the I-V characteristics (see section 8 of Ref. [695] for a detailed discussion of this issue).

---

[7]Here, we are assuming that the conductance is dominated by highly transmissive conduction channels. However, one can have situations in which several channels combine to give a conductance close to G_0. In this case, the signature of the vibrational modes can be a step down in the conductance, depending on the precise value of transmission coefficients.

Chapter 17

Vibrationally-induced inelastic current II: Theory

This chapter is devoted to the theoretical description of the vibrational effects detailed in the previous one. In particular, our main goal is to explain the origin of the different signatures summarized in section 16.6 (see also Fig. 16.17).

At the moment, there is no unified theory covering all the different regimes explored experimentally. However, a lot of progress has been made in several important limiting cases that will the main subject of our discussion here. The first one corresponds to the limit of weak electron-phonon coupling, in the sense explained in the introduction of the previous chapter. In this case, a perturbative approach has been quite successful in explaining the basic experimental observations. In the opposite limit of strong electron-phonon interaction, when the electronic metal-molecule coupling is weak, it is possible to describe the physics in terms of the rates equations that take into account the vibronic effects in a non-perturbative manner. In between these two extreme limits there is a loosely-defined crossover regime of intermediate electron-phonon coupling. The next three sections are devoted to the analysis of these three different regimes, and we shall finish this chapter with some comments and a brief discussion of the basic open problems for both theory and experiment.

17.1 Weak electron-phonon coupling regime

In this section we shall address the limit in which the traversal time is much smaller than the time needed for an electron to feel the molecular vibrations. In this case, the usual approach is to treat the electron-electron interaction at a mean field level and to make a perturbative expansion in the electron-phonon interaction. Our discussion of this regime will be divided into two subsections. In the first one, we shall discuss in detail the results obtained from the resonant tunneling model including in addition the coupling to a single phonon mode. This model will help us to understand the origin of the different vibrational signatures in this regime. Then, the next subsection will be devoted to a description of the ab initio methods that have been developed so far to elucidate the propensity rules in this regime and to establish a quantitative comparison with experimental results.

17.1.1 *Single-phonon model*

In this subsection we shall discuss the predictions of a toy model for the regime of weak electron-phonon interaction. As we shall see, this simple model explains the origin of the different experimental signatures described in sections 16.2-16.4 and provides a deep insight into the tunneling processes responsible for these signatures. We find this subsection particularly important and in order to make it accessible to everybody, we have avoided very technical discussions.[1]

The simplest model to study the electron-phonon interaction in a molecular junction is a natural extension of the resonant tunneling model, which includes the interaction with a single vibrational mode. Let us recall that in the resonant tunneling model an electronic level with energy ϵ_0 is coupled to two metallic reservoirs. The strength of this coupling is described by the scattering rates Γ_L and Γ_R, where L and R denote the left and right leads, respectively. For the sake of simplicity, we shall assume here that these rates are energy-independent. In order to describe the role of the electron-phonon interaction, we now assume that this resonant level is also coupled to a single vibrational mode of energy $\hbar\omega$ (see Fig. 8.2). The Hamiltonian describing this system has the following form

$$\mathbf{H} = \mathbf{H}_e + \hbar\omega \left(\mathbf{b}^\dagger \mathbf{b} + 1/2\right) + \lambda \mathbf{d}^\dagger \mathbf{d} \left(\mathbf{b}^\dagger + \mathbf{b}\right). \qquad (17.1)$$

Here, $\mathbf{H}_e$ describes the electronic part of this problem as it is given by Eq. (7.93). The second term corresponds to the vibrational (or phonon) mode, which is described here as a simple harmonic oscillator. The operators $\mathbf{b}^\dagger$ and $\mathbf{b}$ are the creation and annihilation operators related to the phonon mode, and they satisfy the bosonic commutation relations. Finally, the last term describes the electron-phonon interaction in the molecule, where λ is the electron-vibration coupling constant and $\mathbf{d}^\dagger$ and $\mathbf{d}$ are the

---

[1]The technical details of the calculations reported in this subsection can be found in section 8.2.1.

fermionic operators related to the electronic level.[2] Let us remark that we ignore here the electron-electron interaction in the molecule. This model is sometimes referred to as the (single-level) Holstein model.

This model has been analyzed in the last years by numerous authors to study different aspects of the problem that we are addressing here [736, 744–751]. In spite of its apparent simplicity, there is no known exact solution for this model and approximations have to be made. To keep our discussion as simple as possible, in what follows, unless we state otherwise, we shall make use of the following two approximations: (i) the electron-phonon interaction is treated perturbatively and we include only the lowest-order corrections (second order in λ) and (ii) we assume that the phonons are in thermal equilibrium at the bath temperature. The first approximation is referred to as the lowest-order expansion (LOE). The second one means that the phonon mode is occupied according to the Bose function and it requires the existence of a mechanism that equilibrates the local vibrations (e.g. coupling to bulk phonons). Later, we shall discuss the consequences of relaxing these two approximations.

Our goal is to compute the I-V characteristics when a constant bias voltage is applied. The details of the calculation can be found in section 8.2.1 and we concentrate here on the analysis of the results. In the absence of electron-phonon interaction, the transport characteristics of this model have been discussed in sections 7.4.3 and 13.2. Within the LOE approximation, i.e. collecting all the contributions up to order λ^2, the current can be written as (see section 8.2.1)

$$I = I_{el}^0 + \delta I_{el} + I_{inel}. \qquad (17.2)$$

Here, I_{el}^0 is the elastic current in the absence of electron-vibration interaction (see section 13.2). The other two terms constitute the correction to the current due to the electron-vibration interaction and we now proceed to explain their physical meaning.

The term I_{inel} is the inelastic contribution coming from the emission and absorption of a single vibrational mode. At temperatures much lower than $\hbar\omega/k_B$, the emission process dominates. This latter process is exactly the one considered in the standard IETS (see section 16.2) and we show it again schematically in Fig. 17.1(a). At zero temperature the emission process has a threshold voltage equal to $\hbar\omega/e$ below which it cannot occur. Above this voltage this term gives always a positive contribution to the

---

[2]The spin does not play any role in this problem and we have dropped it in the previous expression.

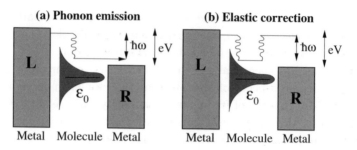

Fig. 17.1   Second-order inelastic processes contributing to the low-temperature current in a molecular junction due to electron-phonon interaction: (a) phonon emission and (b) elastic correction.

current, which means that it contributes to a step up in the conductance at a voltage equal to $\hbar\omega/e$.

The term δI_{el} corresponds to the contribution of a process which involves the emission and re-absorption of a virtual vibrational mode, see Fig. 17.1(b). In this process, which was first discussed by Davis back in 1970 [752], there is a net conservation of the energy of the electrons and for this reason we shall refer to its contribution as *elastic correction*. This process has in general no threshold voltage and it gives a contribution to the current that can be positive or negative depending on the voltage, transmission and other factors, as we shall show below. This process has traditionally been ignored in the context of IETS and also in many publications related to vibronic effects in molecular junctions, which has led to some confusion.

The additional elastic contribution δI_{el} can be interpreted as arising from the interference between the zero order elastic amplitude and the second order amplitude of the process in which a phonon is created and destroyed [752]. The idea goes as follow. The total quantum-mechanical amplitude of an electron tunneling event in the presence of the electron-phonon interaction can be written as a series: $A = A^{(0)} + A^{(1)} + A^{(2)} + \cdots$, where the superindex indicates the order of the contribution in the electron-phonon coupling constant, λ. The corresponding probability is obtained by taking the modulus square. Thus, collecting all the terms up to second order one gets[3] $|A|^2 \approx |A^{(0)}|^2 + |A^{(1)}|^2 + 2\mathrm{Re}\{A^{(0)}A^{(2)}\}$. The term $|A^{(1)}|^2$ corresponds to the processes involving the emission or absorption of a single phonon, while the last one arises from the interference mentioned above

---

[3]The term proportional to λ in this series vanishes because it does not conserve the number of phonons in the system.

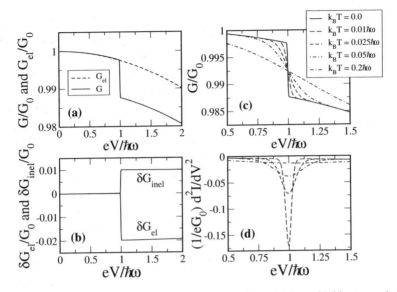

Fig. 17.2 Results of the single-phonon (or Holstein) model for a highly transmissive contact: $\epsilon_0 - E_F = 0$, $\lambda = 2\hbar\omega$ and $\Gamma_L = \Gamma_R = 10\hbar\omega$. (a) Zero-temperature total conductance ($G = dI/dV$) and elastic conductance ($G_{el} = dI_{el}^0/dV$) as a function of the bias voltage. (b) Elastic ($\delta G_{el} = d\delta I_{el}/dV$) and inelastic ($\delta G_{inel} = dI_{inel}/dV$) conductance corrections vs. voltage for the parameters of (a). (c) Temperature dependence of the total conductance. (d) The corresponding d^2I/dV^2 vs. voltage for the temperatures considered in (c).

and it is the origin of the elastic correction. Depending on whether this interference is constructive (enhancing the forward scattering probability) or destructive (enhancing the backscattering probability), this process can give a positive or a negative contribution to the conductance, respectively. So in short, the actual signature of the vibration modes observed in an experiment in the weak electron-phonon regime is a result of the competition between the emission term and the elastic correction.

Now we turn to the analysis of the results of this model. These results have been calculated numerically using the formulas detailed in section 8.2.1. Let us start by discussing the case of a highly conductive junction in the spirit of PCS, see section 16.3. In Fig. 17.2(a) we present the results for the differential conductance for an on-resonant situation where $\epsilon_0 = 0$ (measured with respect to the Fermi energy) and $\Gamma_L = \Gamma_R = 10\hbar\omega$. With these values the conductance is equal to G_0 at zero bias and it shows a very weak voltage dependence. As one can see in Fig. 17.2(a), the zero-temperature conductance (sum of the elastic and inelastic contributions) exhibits an abrupt step down (of about 1%) at $eV = \hbar\omega$. This result reproduces the typical signature observed in the gold atomic chains or in the Pt-H_2 junctions discussed in section 16.3. As one can see in Fig. 17.2(b), the step down in the conductance is due to the dominant negative contribution coming from the elastic correction ($\delta G_{el} = d\delta I_{el}/dV$). In other words, the elastic correction gives rise in this limit to a finite backscattering that reduces the conductance of the junction [745]. After all, this is natural because the (elastic) transmission is already close to one and thus, an incoming electron can only be backscattered.

In Fig. 17.2(c) and (d) we show for this high transmission case the temperature dependence of the differential conductance and the corresponding d^2I/dV^2, respectively. First, notice that the signature of the inelastic current in d^2I/dV^2 is a dip and second, notice also that for temperatures of the order of $0.2\hbar\omega/k_B$ the signature is no longer visible.

We now consider a low-transmissive situation by simply shifting the level away from the Fermi energy ($\epsilon_0 - E_F = 80\hbar\omega$), but keeping the values of the scattering rates of the previous example unchanged. Thus, the (elastic) zero-bias conductance is equal to $0.059\,G_0$. In Fig. 17.3(a) we show the contributions δG_{el} and δG_{inel} versus the voltage, as well as the sum of the two (δG). In this case, we have assumed that the temperature is $k_B T = 0.05\hbar\omega$. Notice that there several basic differences with respect to the previous example. First, the change in the conductance at $eV = \hbar\omega$ is dominated this time by the phonon emission process giving rise to a step up. Second, the contribution δG_{el} is positive for every voltage, but it decreases slightly at the phonon energy. Third, the emission term has no abrupt onset because of the finite temperature.

As one can see in Fig. 17.3(b), the signature of the vibrational mode is barely visible in the differential conductance and one has to resort to its derivative to see it clearly, see Fig. 17.3(c). Of course, the order of magnitude of the inelastic current depends primarily on the electron-phonon coupling constant, λ, which we have chosen small in comparison with the scattering rates to ensure the validity of the perturbative approach. On the other hand, notice that d^2I/dV^2 exhibits a linear background, typically seen in the experiments, which is due to the contribution of the elastic current, which contains a tiny cubic term ($\propto V^3$).

The model also describes the crossover between the two situations just described, as we illustrate in Fig. 17.4. In this example, we have kept constant the values of the scattering $\Gamma_L = \Gamma_R = 10\hbar\omega$ (symmetric junction) and changed the level position. As one can see in this figure, the vibrational

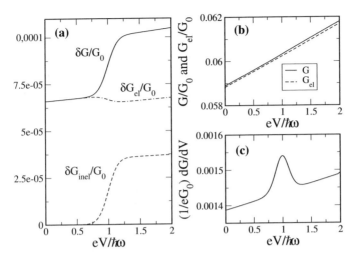

Fig. 17.3   Results of the single-phonon model for a low transmissive contact: $\epsilon_0 - E_F = 80\hbar\omega$, $\lambda = 2\hbar\omega$, $\Gamma_L = \Gamma_R = 10\hbar\omega$ and $k_BT = 0.05\hbar\omega$. (a) Elastic (δG_{el}), inelastic (δG_{inel}) and total ($\delta G = \delta G_{el} + \delta G_{inel}$) conductance corrections versus voltage. (b) Corresponding total conductance and elastic conductance versus voltage. (c) The corresponding d^2I/dV^2.

signature in d^2I/dV^2 evolves from dip in an on-resonant situation to a peak in an off-resonant case. It is also worth stressing that the crossing point of this transition occurs exactly at an (elastic) transmission equal to 0.5.

It is possible to get an analytical insight into the previous results by assuming that the elastic transmission is energy-independent. This is a good approximation in two cases: (i) when the coupling to the leads is so strong that the broadening of the resonant level ($\Gamma_L + \Gamma_R$) is much larger than $\hbar\omega, eV$ and $|E_F - \epsilon_0|$ or (ii) when the resonant level is far away from the Fermi energy, i.e. $|\epsilon_0 - E_F| \gg \Gamma_{L,R}, eV, \hbar\omega$. As we have shown in section 8.2.1, under this assumption, one can prove that the zero-temperature conductance for a symmetric contact ($\Gamma_L = \Gamma_R = \Gamma$) exhibits a jump at $eV = \pm\hbar\omega$ given by $(\lambda^2/\Gamma^2)\tau^2(1 - 2\tau)/4$, where τ is the transmission of the contact. This result suggests that in a symmetric situation the conductance shows a step down for $\tau > 1/2$ and a step down for $\tau < 1/2$, while the signature vanishes for $\tau = 1/2$. This result, which has been coined as the *1/2 rule*, was first derived by Paulsson *et al.* [736] using the model that we are discussing and by de la Vega *et al.* [737] using an alternative model. Both models differ in the exact transmission dependence of the conductance jump, but both of them predict a crossover at exactly

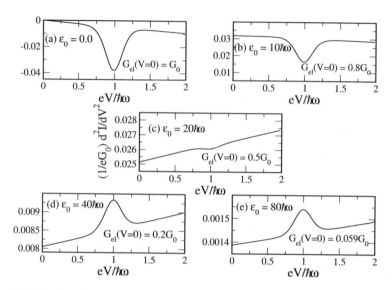

Fig. 17.4   Results of the single-phonon model for the crossover between PCS and IETS: $\lambda = 2\hbar\omega$, $\Gamma_L = \Gamma_R = 10\hbar\omega$ and $k_B T = 0.05\hbar\omega$. Second derivative of the current as a function of the voltage for different values of the level position (measured with respect to E_F) as indicated in the different panels. We also indicate in the panels the value of the zero-bias elastic conductance.

$\tau = 1/2$. Moreover, one can show that while the phonon emission term gives a contribution equal to $+(\lambda^2/\Gamma^2)\tau^2/4$ to the conductance jump, the elastic correction gives a negative contribution equal to $-(\lambda^2/\Gamma^2)\tau^3/2$. Notice that this result suggests that for very low transparencies, and if one is only interested in the signature at the phonon energy, the contribution of the elastic correction can be ignored, which is usually done in the IETS context.

With respect to the temperature dependence of the phonon signature, Paulsson *et al.* [736] have shown that for a symmetric contact, and ignoring the energy dependence of the elastic transmission, the full width at half maximum (FWHM) of the peak in $d^2 I / dV^2$ is approximately $5.4 k_B T$, i.e. like in the standard IETS case [705].

Let us address now the typical situation realized in the STM contacts, where there is a large asymmetry in the couplings between the molecule and the surface and the molecule and the STM tip. In Fig. 17.5 we show IET spectra for a junction in which $\Gamma_L = 10\hbar\omega$ and $\Gamma_R = 0.01\Gamma_L$. In this figure the level position has been varied from a resonant case in panel (a) to an off-resonant situation in panels (c) and (d). As one can see in

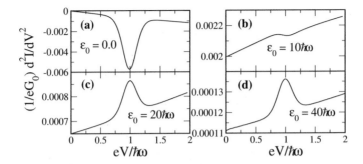

Fig. 17.5   Results of the single-phonon model for a very asymmetric contact simulating the situation typically realized in STM experiments: $\lambda = 2\hbar\omega$, $\Gamma_L = 10\hbar\omega$, $\Gamma_R = 0.01\Gamma_L$ and $k_B T = 0.05\hbar\omega$. Second derivative of the current as a function of the voltage for different values of the level position (measured wit respect to E_F) as indicated in the different panels.

the latter panels, the vibrational mode is manifested in the d^2I/dV^2 as a peak at the mode energy, as it is usually observed in most STM-IETS experiments. Notice, however, that when the level is brought close to the Fermi energy, the signature progressively changes into a dip, as it is shown in panel (a). As we explained in section 16.2, a dip is sometimes observed in STM experiments and these results nicely clarify the necessary conditions for the observation of dips. The origin of this crossover is the same as for symmetric junctions, i.e. the elastic correction gives a dominant negative contribution in the resonant case, while the phonon emission dominates in off-resonant situations leading to a peak in the spectra.

To our knowledge, Persson and Baratoff were the first to point out the possibility of a decrease in the conductance of a molecule in a STM experiment due to resonant tunneling [753]. The issue of peaks and dips observed, in particular, in the STM experiments has been revisited by Galperin and coworkers going beyond the LOE [748, 749]. More recently, Egger and Gogolin have reported analytic results for the zero-temperature inelastic current in a molecule within the LOE approximation [750]. They have established the criteria for the sign change of the step in the conductance. In particular, they have shown that this transition, in general, not only depends on the transmission of the junction, but it is governed by essentially all system parameters (scattering rates and level position), as we have shown here.

The single-phonon model is able to describe in a unified manner the ba-

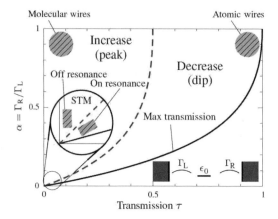

Fig. 17.6 Phase diagram for the single-phonon level model discussed in this section (inset) illustrating the sign of the conductance change at the onset of phonon emission. At a given asymmetry factor α the elastic transmission τ has an upper bound $\tau_{\max}$ (solid line), and the inelastic conductance change undergoes a sign change at $\tau_{\text{crossover}} = \tau_{\max}/2$ (dashed line). Reprinted with permission from [751]. Copyright 2008 by the American Physical Society.

sic vibrational signatures observed in the experiments in which the electron-phonon interaction is weak. We can summarize the results discussed so far following Paulsson *et al.* [751] with the phase diagram of Fig. 17.6. This diagram describes the parameter range in which an increase or decrease of the conductance is expected due to the phonon mode. This diagram has been constructed assuming that the transmission can be considered energy-independent (see discussion above). The diagram is plotted for the ratio of the coupling to the two leads $\alpha = \Gamma_R/\Gamma_L$ and the transmission τ at E_F. In this model the maximal transmission is $\tau_{\max} = 4\alpha/(1+\alpha)^2$ corresponding to the on-resonance case. Notice that the crossover from a decrease to an increase in the conductance is given by the *1/2 rule* [735–737], i.e. at $\tau_{\text{crossover}} = \tau_{\max}/2$.

So far, we have assumed that the phonon mode in this model is in thermal equilibrium at the bath temperature. In principle, the current flow can drive this mode out of the equilibrium creating a finite population even at zero temperature (as long as $eV > \hbar\omega$). From a technical point of view, the correct description of this nonequilibrium effect requires the evaluation of the "phonon self-energies" that contain the information about the phonon occupation and the phonon lifetime [748, 749]. This is a complicated task that it is not easy to carry out in a consistent manner. For this reason we will follow here Paulsson *et al.* [736] and describe this effect at a phenomenological level. The simplest way to include non-equilibrium heating is to write down a rate equation for the phonon occupation, n, including an external damping rate γ_d of the phonons [736]

$$\dot{n} = \frac{P}{\hbar\omega} + \gamma_d \left[n_B(\hbar\omega) - n \right], \tag{17.3}$$

where P is the power dissipated into the phonon mode and n_B is the Bose function. The external damping can be due to either the interaction with the phonons of the electrodes or the electron-phonon interaction in the molecule. From this equation, the steady state occupation n is easily found. To complete the calculation we need now an expression for both the power and the current in terms of the nonequilibrium phonon occupation. Assuming that the transmission is energy-independent and considering a symmetric junction ($\Gamma_L = \Gamma_R = \Gamma$), Paulsson *et al.* [736] showed that these quantities can be expressed within the LOE approximation as follows

$$P^{\mathrm{LOE}} = \gamma_{eh} \hbar\omega \left[n_B(\hbar\omega) - n \right] + \frac{\gamma_{eh}}{4} \frac{\pi\hbar}{\hbar\omega} P, \tag{17.4}$$

$$I^{\mathrm{LOE}} = \frac{2e^2}{h} \tau V + e\gamma_{eh} \frac{1 - 2\tau}{4} \frac{\pi\hbar}{e\hbar\omega} I^{\mathrm{Sym}}, \tag{17.5}$$

where $\gamma_{eh} = (\omega/\pi)\lambda^2\tau^2/\Gamma^2$ is the electron-hole damping rate.[4] Here, P and I^{Sym} are universal functions of the voltage, phonon frequency, temperature and phonon occupation given by

$$P = \frac{\hbar\omega}{\pi\hbar} \frac{\left[\cosh\left(\frac{eV}{k_B T}\right) - 1 \right] \coth\left(\frac{\hbar\omega}{2k_B T}\right) \hbar\omega - eV \sinh\left(\frac{eV}{k_B T}\right)}{\cosh\left(\frac{\hbar\omega}{k_B T}\right) - \cosh\left(\frac{eV}{k_B T}\right)} \tag{17.6}$$

$$I^{\mathrm{Sym}} = \frac{2e}{h} \left(2eVn + \frac{\hbar\omega - eV}{e^{\frac{\hbar\omega - eV}{k_B T}} - 1} - \frac{\hbar\omega + eV}{e^{\frac{\hbar\omega + eV}{k_B T}} - 1} \right). \tag{17.7}$$

Eq. (17.5) reproduces the zero-temperature transmission dependence of the conductance jump discussed above. However, there is a small discrepancy between these two results, namely the zero-temperature inelastic conductance in Eq. (17.5) vanishes for $eV < \hbar\omega$, while this is not the case in the results presented above. The origin of this little difference is unclear to us.

Eqs. (17.3)-(17.5) were used by Paulsson *et al.* in Ref. [736] to fit the experimental results of Pt-H$_2$ junctions [127, 567]. As one can see in

---

[4]There is difference of a factor 4 in the expression of γ_{eh} with respect to Ref. [736] because of the different definition of the scattering rates.

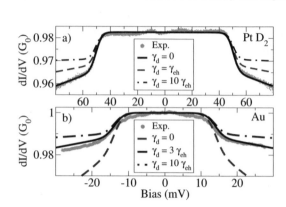

Fig. 17.7   Single level model [Eqs. (17.4) and (17.5)] fitted to the experimentally measured conductance through a deuterium molecule [567]. The parameters used for the fit are $\hbar\omega = 50$ meV, $\tau = 0.9825$, $\gamma_{eh} = 1.1 \times 10^{12}$ s^{-1}, and $T = 17$ K. (b) A simple model (see Ref. [736] for details) fitted to the measured conductance through an atomic gold wire (experimental data from Ref. [732]). The fit yields the following parameters: $\hbar\omega = 13.8$ meV, $T = 10$ K, $\gamma_{eh} = 12 \times 10^{10}$ s^{-1}, and $\gamma_d = 3\gamma_{eh}$. Reprinted with permission from [736]. Copyright 2005 by the American Physical Society.

Fig. 17.7(a), an excellent fit of the experimental data can be achieved by using γ_{eh} and γ_d as adjustable parameters. The best fit was obtained using a negligible external damping of the phonon mode ($\gamma_d \ll \gamma_{eh}$), which can be understood physically from the mass difference between the hydrogen molecule and the platinum atoms of the break-junction. The nonequilibrium occupation gives rise to the conductance slope that is seen in the experiments for $eV > \hbar\omega$. This feature in absent in the model with thermalized phonons.

Using a similar single-phonon model designed for the atomic gold chains, Paulsson *et al.* were also able to fit the experimental results of Ref. [732], as one can see in Fig. 17.7(b). In this case the external damping $\gamma_d = 3\gamma_{eh}$ is not negligible in contrast to the hydrogen case. This indicates that presumably there is a strong interaction between the modes of the gold chains and the phonons of the leads.

17.1.2   *Ab initio description of inelastic currents*

Although the simple model discussed in the previous section has proven to be very useful, there are still many basic questions that are out of its scope. Probably the most important one is related to the issue of the selection or propensity rules. An understanding of the factors that determine why certain vibrational modes show up in transport experiments, while others remain hidden, requires a microscopic modeling of the problem. In this section we shall briefly review the work done in this direction in recent years.

Probably the first ab initio calculations to investigate the inelastic current through molecules were carried out by Lorente and Persson [754, 755]. They used a combination of DFT and Green's function techniques to interpret the STM experiments of Ho and coworkers [709, 756, 757]. In particular, Ref. [755] represents a first attempt to formulate propensity rules for inelastic tunneling spectra. Since then, numerous authors have applied microscopic methods to the description of vibrationally-induced inelastic currents in molecular transport junctions with different levels of sophistication [202, 566, 751, 758–779].[5] In what follows, we shall first formulate the general problem of electron-phonon interaction in molecular junctions. This will serve us to appreciate the ingredients that are required to calculate vibrationally-induced inelastic currents. Then, we shall briefly comment on the different approximations that have been put forward to perform these calculations in realistic systems. Additionally, we shall describe the propensity rules that have been derived so far and we shall show some examples of the comparison between experiment and theory.[6]

17.1.2.1 *Formulation of the problem*

The general objective is to describe the effect of vibrations on the transport through molecular junctions, when a voltage is applied. The coupled system of electrons and vibrations in a molecular contact can be generically modeled by the following Hamiltonian: $\mathbf{H} = \mathbf{H}_e + \mathbf{H}_{vib} + \mathbf{H}_{e-vib}$, where

$$\mathbf{H}_e = \sum_{ij} \mathbf{d}_i^\dagger H_{ij} \mathbf{d}_j$$

$$\mathbf{H}_{vib} = \sum_\alpha \hbar\omega_\alpha \left(\mathbf{b}_\alpha^\dagger \mathbf{b}_\alpha + 1/2 \right)$$

$$\mathbf{H}_{e-vib} = \sum_{ij} \sum_\alpha \mathbf{d}_i^\dagger \lambda_{ij}^\alpha \mathbf{d}_j (\mathbf{b}_\alpha^\dagger + \mathbf{b}_\alpha). \tag{17.8}$$

Here ω_α are the vibrational frequencies, $H_{ij} = \langle i|H|j \rangle$ are the matrix elements of the single-particle electronic Hamiltonian H in the atomic-orbital

---

[5]This list is by no means complete, but it should be easy to trace back from it the whole relevant literature on this subject.

[6]What follows is more technical than usual and it is meant for the theoretical readership. The reader not interested in this theoretical discussion can jump directly to the description of the propensity rules in section 17.1.2.4.

basis $\{|i\rangle\}$, and λ_{ij}^{α} are the electron-vibration coupling constants. The index i denotes collectively the atomic sites and orbitals, and α runs from 1 to $3N_{vib}$, where N_{vib} is the number of atoms in the system, which are allowed to vibrate. The creation and annihilation operators for vibrational modes $\mathbf{b}_{\alpha}^{\dagger}$ and $\mathbf{b}_{\alpha}$ satisfy the bosonic commutation relation $[\mathbf{b}_{\alpha}, \mathbf{b}_{\beta}^{\dagger}] = \delta_{\alpha\beta}$. The electronic basis is in general non-orthogonal, with overlap matrix elements $S_{ij} = \langle i|j\rangle$. The calculation of the electronic structure of the junction, i.e. the determination of the Hamiltonian $\mathbf{H}_e$, is usually done within the DFT framework (see Chapter 10) or with sophisticated tight-binding parameterizations (see Chapter 9).

In practice, the calculation of the vibrational modes is restricted to a central region that includes the molecule and a small portion of the electrodes. In principle, one should also describe how these central (or primary) vibrations are coupled to the phonons in the electrodes. This is very difficult to do in a rigorous manner and such a coupling is usually taking into account by means of a phenomenological parameter that enters as a broadening in the density of states of the primary vibrations [695].

The solution of the inelastic transport problem involves a few rather separate sub-problems: (i) the optimization of the geometry and evaluation of the vibrational modes, (ii) computation of the electron-vibration coupling constants and (iii) the calculation of the transport. Let us now discuss these sub-problems in certain detail.

17.1.2.2   *Vibrational modes and electron-vibration coupling constants*

The calculation of the vibrational modes requires knowledge of the total ground-state energy of the system as a function $E(\vec{R}_k)$ of the ionic coordinates $\vec{R}_k$ with $k = 1, \ldots, N_{vib}$. This energy is usually determined in the framework of DFT. This energy needs to be minimized in order to find the equilibrium configuration $\{\vec{R}_k^{(0)}\}$. Now consider small displacements $\vec{Q}_k = \vec{R}_k - \vec{R}_k^{(0)}$ around the equilibrium positions. The Hamiltonian (in first quantization) describing the oscillations of the ions around $\vec{R}_k^{(0)}$ is given in the harmonic approximation by

$$H_{ion} = \frac{1}{2}\sum_{k\mu} M_k \dot{Q}_{k\mu}^2 + \frac{1}{2}\sum_{k\mu,l\nu} \mathcal{H}_{k\mu,l\nu} Q_{k\mu} Q_{l\nu}, \qquad (17.9)$$

where M_k are the ionic masses, $\mu, \nu = x, y, z$ denote the Cartesian components of vectors and $\mathcal{H}$ is the Hessian matrix: $\mathcal{H}_{k\mu,l\nu} =$

$\partial^2 E/\partial R_{k\mu}\partial R_{l\nu}$. This matrix can be diagonalized by the transformation $Q_{k\mu} = \sum_{\alpha=1}^{3N_{vib}} A_{k\mu,\alpha}q_\alpha$, where q_α are the normal coordinates. Thus, we obtain $H_{ion} = \frac{1}{2}\sum_\alpha(\dot{q}_\alpha^2 + \omega_\alpha^2 q_\alpha^2)$, where ω_α ($\alpha = 1, \ldots, 3N_{vib}$) are the vibrational frequencies. The transformation matrix A is normalized according to $A^T M A = 1$, M being the diagonal mass matrix $M_{ij} = M_i\delta_{ij}$. Using the canonical quantization prescription $q_\alpha = (\hbar/2\omega_\alpha)^{1/2}(b_\alpha^\dagger + b_\alpha)$ and $\dot{q}_\alpha = i(\hbar\omega_\alpha/2)^{1/2}(b_\alpha^\dagger - b_\alpha)$, one finally obtains $\mathbf{H}_{vib}$ in Eq. (17.8).

The electron-vibration interaction may be derived as follows [174, 202]. Assume that the electronic single-particle Hamiltonian H is a function of the ionic coordinates, denoted collectively as $\vec{R}$. Then, we may expand $H(\vec{R}^{(0)} + \vec{Q}) \approx H(\vec{R}^{(0)}) + \sum_k \vec{Q}_k \cdot \vec{\nabla}_k H|_{\vec{Q}=0}$. Defining $\mathbf{H}'_e = \sum_{ij} d_i^\dagger\langle i|H(\vec{R}^{(0)} + \vec{Q})|j\rangle d_j$, inserting the expansion, and using the canonical quantization for q_α again, one finds $\mathbf{H}'_e = \mathbf{H}_e + \mathbf{H}_{e-vib}$. Here $\mathbf{H}_e$ and $\mathbf{H}_{e-vib}$ are given by Eq. (17.8), with H being given by $H(\vec{R}^{(0)})$ and the electron-vibration coupling constants by

$$\lambda_{ij}^\alpha = \left(\frac{\hbar}{2\omega_\alpha}\right)^{1/2} \sum_{k\mu} M_{ij}^{k\mu} A_{k\mu,\alpha}, \qquad (17.10)$$

where $M_{ij}^{k\mu} = \langle i|\nabla_{k\mu}H|_{\vec{Q}=0}|j\rangle$. From Eq. (17.10) one can see that the calculation of the coupling constants requires to compute derivatives of the Hamiltonian matrix elements with respect to the atomic position. Indeed, since the employed basis sets are usually nonorthogonal, things are slightly more complicated and the coupling constants are often calculated using the ideas of Head-Gordon and Tully [780], see e.g. Ref. [775].

17.1.2.3 *Inelastic current*

The electric current is usually computed making use of the nonequilibrium Green's function (NEGF) techniques that we have described in Chapter 7. In section 8.2 we derived the general current expression for an interacting junction. As we explained there, there are indeed several possibilities for this formula. Following Caroli *et al.* [209], we write the current as the sum of two contributions, $I = I_{el} + I_{inel}$, where [see Eqs. (8.34)-(8.35)][7]

$$I_{el} = \frac{8e}{h}\int_{-\infty}^{\infty} dE\, \mathrm{Tr}\left[\mathbf{G}^r\mathbf{\Gamma}_R\mathbf{G}^a\mathbf{\Gamma}_L\right](f_L - f_R), \qquad (17.11)$$

---

[7]Here the current has been evaluated at the left interface. Let us recall that to compute the current one first divides the system into three parts: the leads (L and R) and a central region (C), which contains the molecule and part of the electrodes. The electron-phonon interaction is assumed to be restricted to this central part.

$$I_{inel} = \frac{4ie}{h} \int_{-\infty}^{\infty} dE \operatorname{Tr} \left\{ \mathbf{G}^a \mathbf{\Gamma}_L \mathbf{G}^r \left[(f_L - 1)\mathbf{\Sigma}^{+-}_{e-vib} - f_L \mathbf{\Sigma}^{-+}_{e-vib} \right] \right\},$$

where $f_{L,R}(E) = f(E - \mu_{L,R})$, $f(E) = [1 + \exp(\beta E)]^{-1}$ is the Fermi function and $\beta = 1/k_B T$ is the inverse temperature. Here, the full retarded and advanced Green functions $\mathbf{G}^{r,a}$ are given by $\mathbf{G}^r = [E\mathbf{S}_{CC} - \mathbf{H}_{CC} - \mathbf{\Sigma}^r_L - \mathbf{\Sigma}^r_R - \mathbf{\Sigma}^r_{e-vib}]^{-1}$ and $\mathbf{G}^a = [\mathbf{G}^r]^\dagger$. On the other hand, $\mathbf{\Sigma}^{r,a}_{L,R}$ are the electronic self-energies that describe the electronic coupling between the central region and the leads. The imaginary part of the advanced self-energies are the corresponding scattering rate matrices, $\mathbf{\Gamma}_{L,R}$. The self-energies $\mathbf{\Sigma}^r_{e-vib}$ and $\mathbf{\Sigma}^{\pm\mp}_{e-vib}$ are due to the electron-vibration interaction in the central region. Since they vanish in the absence of λ^α, we call the I_{inel} part an "inelastic" current, while I_{el} is the "elastic" part.

Up to now the expression of the current is exact, but we must now specify an approximation for the electron-vibration self-energies. Most of the realistic calculations done so far have been carried out within the lowest-order expansion (LOE), which we already used in the single-phonon model above [736, 202]. More accurate approximations, like the so-called self-consistent Born approximation (SCBA), have also been used [763], but this latter approximation is computationally very costly. In the LOE approximation the Green's functions are expanded to second order in λ^α, i.e. $\mathbf{G}^r = \tilde{\mathbf{G}}^r + \tilde{\mathbf{G}}^r \mathbf{\Sigma}^r_{e-vib} \tilde{\mathbf{G}}^r + \cdots$. In this way the elastic current is split into two parts as $I_{el} = I^0_{el} + \delta I_{el}$, where δI_{el} is an "elastic correction". We find

$$I^0_{el} = \frac{8e}{h} \int dE \operatorname{Tr}[\tilde{\mathbf{G}}^r \mathbf{\Gamma}_R \tilde{\mathbf{G}}^a \mathbf{\Gamma}_L](f_L - f_R) \tag{17.12}$$

$$\delta I_{el} = \frac{16e}{h} \int dE \operatorname{ReTr}[\mathbf{\Gamma}_L \tilde{\mathbf{G}}^r \mathbf{\Sigma}^r_{e-vib} \tilde{\mathbf{G}}^r \mathbf{\Gamma}_R \mathbf{G}^a](f_L - f_R)$$

$$I_{inel} = \frac{4ie}{h} \int dE \operatorname{Tr}\{\tilde{\mathbf{G}}^a \mathbf{\Gamma}_L \tilde{\mathbf{G}}^r [(f_{L,R} - 1)\mathbf{\Sigma}^{+-}_{e-vib} - f_{L,R}\mathbf{\Sigma}^{-+}_{e-vib}]\}.$$

Notice that this division was also made in the analysis of the single-phonon model. The expressions of the second-order self-energies can be found, for instance, in Appendix C of Ref. [202], and they are natural extension of those in Eq. (8.43). It is worth mentioning that within this approximation one can rigorously prove the conservation of the current.

Even in the LOE, the current formulas [see Eqs. (17.12)] involve double energy integrals which can be very cumbersome to evaluate. A further simplification is achieved by assuming that the elastic transmission has no pronounced energy dependence in the energy window where the vibrational modes show up in the current. This approximation is not valid in the case where sharp resonances are present, but it turns out to be quite reasonable in many situations of interest. With this assumption, the retarded and advanced Green's functions, as well as the scattering rates, can be evaluated at the Fermi energy and some of the integrals can be done analytically, which simplifies enormously the calculations. The detailed formulas for the current within this approximation can be found in Refs. [736, 202].

Another important issue is the expression of the phonon occupation that enters in the current formula via the electron-vibration self-energies. The simplest approximation, which is fully consistent with the LOE, is to assume that the phonons are in thermal equilibrium at the bath temperature. Heating effects, due to the nonequilibrium established at finite bias, can be described in a various ways. For instance, as we explained for the single-phonon model, the authors of Refs. [763, 736, 775] determine the phonon occupation in a self-consistent manner by imposing that the power transferred by electrons from the leads into to the device is balanced by the power transferred from the device electrons to the phonons. Another phenomenological way of introducing the nonequilibrium effects is discussed in Ref. [777].

17.1.2.4 *Propensity rules*

One of the most surprising aspects revealed by the experiments is the fact that only a small number out of the many possible vibrational modes gives a signal in the transport characteristics. Motivated by this fact, many researchers have employed the formalism detailed above, or variations of it, to establish the rules that govern the contribution of a mode to the inelastic signal. Let us emphasize that there are no strict selection rules like in optical spectroscopies, but rather *propensity rules*. It is also worth remarking that the contribution of a mode to the inelastic spectra does not only depend on the symmetry of the mode itself, but also on the nature of the orbitals that contribute to the current. In the search for these rules, some general trends have been identified. For instance, the most significant contributions typically come from modes with large longitudinal component, i.e. motion along the tunneling direction [760, 766, 768]. On the other hand, the calculations indicate a high sensitivity of the computed spectra to the structure of the molecular bridge [760, 767, 768].

One of the most systematic studies of the propensity rules has been carried out by Troisi and Ratner [765, 769, 770]. These authors have developed a simplified computational method that, although it does not allow them to compute the line shapes of the IET spectra, it provides a convenient way to determine the intensities of the peaks in off-resonant situations. More importantly, this appealing formulation has allowed the authors to get a deeper insight into the propensity rules. With this approach, Troisi and Ratner have again emphasized the importance of modes with large component in the tunneling direction. Thus for instance, they have shown that for a linear chain with one orbital per atom, only totally symmetric modes contribute to IETS signal. For molecules with side chains any normal mode dominated by side chain motion will contribute only weakly to IETS. The authors have also employed group theory to identify the main normal modes for planar conjugated molecules with C_{2h} symmetry.

Gagliardi *et al.* [776] have also presented a detailed study of the propensity rules in the case of low-transmissive junctions, extending the work of Troisi and Ratner. The approach of this work is based on the idea that both the elastic and inelastic current can be expressed as the sum of a small number of essentially noninteracting paths or conduction channels through the device.

More recently, Paulsson *et al.* [751] have reported a method to determine the propensity rules in junctions with arbitrary transparency (within the weak electron-phonon coupling regime). Similar to Ref. [751], the key idea in this work is to analyze the inelastic transport in terms of just a few selected electronic scattering states, namely those belonging to the most transmitting channels at the Fermi energy. These scattering states typically have the largest amplitude inside the junction and thus account for the majority of the electron-phonon scattering.

17.1.2.5 *Quantitative comparison with experiments*

The theory has been quite successful in reproducing the experimental inelastic spectra in the limit of weak electron-phonon interaction. By now, there are many examples of satisfactory agreement between experiment and theory. It is impossible review all these examples and here we shall just mention a few illustrative cases.

One of the first comparisons was reported by Pecchia *et al.* [762], who found reasonable agreement between their calculations on Au-octanethiolate-Au junctions and the IETS results of Ref. [713]. Frederiksen *et al.* [763] reported quantitative agreement between their calculation of the IETS signal for atomic gold wires and experimental results [732]. These authors found that the modes responsible for the inelastic signal are lon-

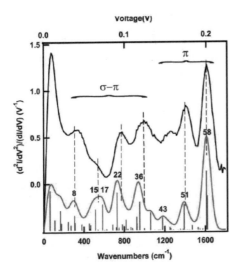

Fig. 17.8 IET spectra of an anthracene thiol junction. The upper curve corresponds to the experimental spectra and the lower curve to the computed one. Labels refer to the normal modes of the molecule computed in the absence of metal and numbered from the lowest energy vibration. Reprinted with permission from [718].

gitudinal ones with alternating bond length. Moreover, their calculations showed the decrease in conductance with increase in the inelastic signal and softening of the modes resulting from straining the wire. These calculations were extended by Viljas *et al.* [202] who studied systematically how the position and height of the conductance steps vary as a gold wire is stretched and more atoms are added to it, and found good agreement with the experiments.

Troisi and Ratner have applied their approach to several experimental examples and they have found consistently a good agreement in all these cases [765, 574, 718, 720]. In Fig. 17.8 we show a comparison between theory and experiment for an anthracene thiol junction that was reported in Ref. [718].

Another impressive example of agreement between theory and experiment has been reported by Paulsson and coworkers [751]. These authors studied very different model systems that range from atomic gold chains, as an example of highly conductive junction, to off-resonant situations typically realized in STM-IETS experiments, see Fig. 17.9. The satisfactory agreement with the experiment in these very different cases nicely illustrates the level of understanding achieved in the weak electron-phonon coupling

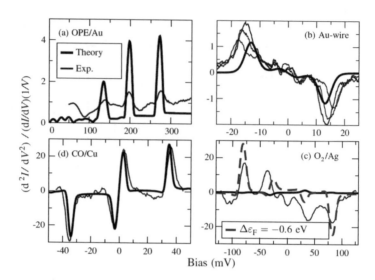

Fig. 17.9   Calculated (black lines) and experimental (blue lines) IETS. (a) OPE molecule with Au(111) leads, (b) Au chain connected to Au(100) leads, (c) O_2 molecule on Ag(110), and (d) CO molecule on Cu(111). In case (c) the Fermi energy has been shifted manually to match the experiment (dashed red line). The experimental data originates from Refs. [712, 732, 711, 781]. For the STM configurations (c) and (d), the calculated IETS is compared with a rescaled d^2I/dV^2. Reprinted with permission from [751]. Copyright 2008 by the American Physical Society.

regime. This agreement of the calculated and measured IET spectra makes this spectroscopy, in combination with theory, a very useful diagnostic tool.

17.2   Intermediate electron-phonon coupling regime

The perturbative methodology discussed above describes correctly off-resonant situations encountered in standard IETS experiments as well as the resonant tunneling regime in cases where weak vibronic coupling results from strong electronic coupling to the leads (large electronic width Γ) that ensures short electron lifetime on the bridge. The electronic transport through a junction with a strong electron-phonon interaction is very different from the weak coupling limit. Physically, in the course of the transmission process the electron occupies the bridge long enough to affect polarization of the bridge and its environment. In the ultimate limit of this situation, decoherence and thermal relaxation are sufficient to render the processes of bridge occupation and de-occupation, and often also transmission between different sites on the bridge, independent of each other. This makes it possible to treat the transmission process as a sequence of consecutive statistically decoupled kinetic events. In this section we want to discuss intermediate situations where effects of transient polaron formation on the bridge have to be accommodated, however dephasing is not fast enough to make a simple kinetic description possible.

In this intermediate regime, the LOE fails and new theoretical approaches are necessary. A strategy is to improve systematically the perturbation theory by including higher orders. Thus for instance, different authors have used the self-consistent Born approximation (SCBA) [746–749, 782]. In this case, the lowest-order Feynman diagrams taken into account in the LOE are "dressed" by using the full Green's functions. Additionally, the description of the phonons can be improved by including the corresponding phonon self-energies that describe the renormalization of the phonon energies and their finite lifetimes. This method provides a way to sum up certain diagrams up to infinite order, but it misses important contributions of other high-order diagrams (vertex corrections). In this sense, its validity is restricted to rather weak electron-phonon coupling.

In recent years, many other theoretical schemes have been introduced to describe this intermediate regime [783, 785–791]. It is important to emphasize that the application of all these methods has been restricted to model Hamiltonians, in particular, to the single-phonon mode discussed above. Some of these works are based on the NEGF methodology [783, 786–788, 790, 791] and others are based on an extension of the equation-of-motion (EOM) method described in section 5.4.3 to include the phonon dynamics [785, 789]. A central idea in most of these approaches is the application of the polaron (or Lang-Firsov) transformation [792, 174] to the single-phonon model. This transformation, which will be described below, replaces the additive electron-phonon coupling [last term in Eq. (17.1)] by a renormalization of the electronic coupling elements by phonon displacement operators. The renormalized electronic coupling contains then the effects of electron-phonon interaction to all orders and the transformed Hamiltonian provides a more adequate starting point for situations where the electron-phonon coupling is rather strong.

All the different approaches reported so far are approximate and the exact description of vibrational effects for arbitrary strength of the electron-phonon interaction, even ignoring electron-electron interaction, remains as an open problem, at least in nonequilibrium situations. However, the works mentioned above have been extremely useful to elucidate the essential

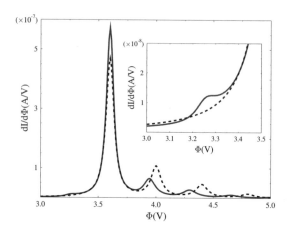

Fig. 17.10 Differential conductance versus source-drain voltage calculated with the EOM method applied to the single-phonon model of Eq. (17.1). The parameters have the following values: $\Gamma_L = \Gamma_R = 0.02$ eV, $T = 10$ K, $\epsilon_0 = 2$ eV, $\hbar\omega = 0.2$ eV, and $\lambda = 0.01$ eV. The solid line corresponds to the self-consistent result and the dashed line to the zero-order result (see Ref. [789] for details). The inset shows a blow-up of the phonon absorption peak that appears on the left of the main resonance. Reprinted with permission from [789]. Copyright 2006 by the American Physical Society.

physics in the intermediate regime. In particular, these approaches nicely describe the appearance of phonon sidebands, which is the main vibrational signature in resonant situations (when the metal-molecule coupling is not too strong). As an illustration, we reproduce in Fig. 17.10 results reported by Galperin *et al.* [789]. These authors applied the EOM method to the Holstein model of Eq. (17.1). In particular, Fig. 17.10 shows the differential conductance as a function of the bias voltage for a set of parameter values that corresponds to the case of a relatively narrow electronic resonance (see figure caption). The first thing to notice in this figure is the appearance of a main conductance peak at (at ~ 3.6 V). This is the usual elastic peak that appears when the resonant level crosses the chemical potential of one of the reservoirs. In the absence of electron-vibration coupling this peak would appear at 4 eV in this example because the voltage was applied symmetrically. As one can see in Fig. 17.10, the position of the level has been renormalized by the interaction with the vibrational mode. The most important consequence of phonon-assisted resonant tunneling is the

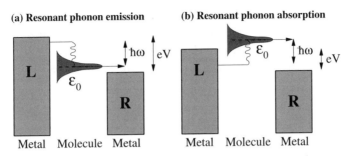

Fig. 17.11 Resonant inelastic tunneling processes in molecular junctions: (a) resonant phonon emission and (b) resonant phonon absorption. This latter process does not have a threshold voltage and it requires a finite occupation of the vibration modes due to a finite temperature or to nonequilibrium phonon generation.

appearance of satellite peaks on the right hand side of the elastic peak. Notice that these peaks are separated by a distance $\sim 2\hbar\omega/e$. The factor 2 is again due to the choice of the voltage profile.

As we explained earlier, the fact that vibrations are in this case manifested as peaks in the conductance, rather than peaks in d^2I/dV^2, is simply due to the fact that we are now dealing with a resonant situation. As shown schematically in Fig. 17.11(a), the probability of the phonon emission tunneling process is greatly enhanced when the energy of an incoming electron is such that by emitting a phonon it loses exactly the energy necessary to cross the molecular level on resonance. This implies the appearance of a peak in the differential conductance when the bias exceeds the voltage necessary to see the resonant level in a quantity equal to $\hbar\omega/e$ times a correction factor that accounts for the shift of the level due to the voltage (this factor equals 2 in Fig. 17.10). This argument applies for a single-phonon process. If the electron-phonon coupling is large enough, emission of several vibrational quanta becomes possible and it results in the appearance of additional peaks in the conductance separated in this example by a voltage equal to $\sim 2\hbar\omega/e$, see Fig. 17.10. It is important to emphasize that in order to resolve such satellite vibronic peaks, both the electronic coupling (Γ) and the thermal energy ($k_B T$) must be smaller than $\hbar\omega$, as in the example of Fig. 17.10. Of course, at very low temperatures the voltage must be larger than $\hbar\omega/e$ for the emission process to happen at all.

Another remarkable feature in Fig. 17.10 is the appearance of an additional peak at $\Phi \sim 3.25$ V (see inset). As we show schematically in Fig. 17.11(b), the resonant absorption of phonon could lead in this case to the appearance of a peak on the left side of the elastic one. However, at the temperature of the calculation, the probability to thermally excite phonons is negligible. Therefore, such feature must be a result of the heating of the phonon subsystem by electron flux. In other words, it probably originates from the absorption of nonequilibrium phonons generated by the current flow. As we explained at the end of section 16.5, such nonequilibrium absorption peaks were reported by LeRoy and coworkers in tunneling experiments with suspended carbon nanotubes [740, 741].

17.3 Strong electron-phonon coupling regime

The impact of the vibrations on the transport characteristics increases with both the strength of the electron-phonon interaction and the time that electrons reside in the molecule. The former factor is not easy to tune, while the second one can be controlled via the length of the molecule or the metal-molecule coupling. In particular, when the electronic coupling between the molecule and the electrodes is weak, as in the case of molecular transistors, the electrons in the molecule have sufficient time to interact strongly with vibrations leading to the polaron formation (a mixed state in which an electron is "dressed" by a phonon cloud [174]). What complicates the theoretical description of this strong coupling regime is the fact that the vibronic effects coexist with strong electronic correlations due to the Coulomb interaction. Thus, electron-phonon interaction and electron-electron interaction must be described in an equal footing.[8] In this section we shall review the present status of the theoretical understanding of this strong coupling regime, which is realized in molecular transistors. We have divided the discussion into two main parts. First, we shall consider vibronic effects in the Coulomb blockade regime and then, we shall focus in the interplay between Kondo physics and electron-phonon interaction.

17.3.1 *Coulomb blockade regime*

Let us now consider the case in which a molecule is weakly coupled to metallic electrodes so that the transport is dominated by the Coulomb blockade effect. In this regime the coupling to vibrational degrees of freedom leads to the emergence of sidebands in the I-V characteristics. This phenomenon, which we have already discussed in the previous section, was

---

[8]Notice that in the previous sections the electron-electron interaction was either ignored or just described at a mean-field level.

already described at the end of the 1980's [794, 795]. In the context of molecular transistors, Boese and Schoeller [796] were the first to analyze vibronic effects in the Coulomb blockade regime. Motivated by the experiments on C_{60} SMTs by Park *et al.* [22], these authors generalized the many-body master equations described in section 15.5.1 to include vibronic effects. Since then, many authors have used rate equations to study different aspects of this problem, see e.g. Refs. [747, 797–807]. In what follows, we shall first describe how this transport problem is formulated in terms of rate equations and then, we shall briefly discuss some of the main physical effects that have been predicted to occur in this regime.

The different formulations of rate (or master) equations differ only in minor details and we have chosen to follow Ref. [747]. The starting point is the single-phonon model (Holstein model) that we have extensively discussed in previous sections, but now with the inclusion of the electron-electron interaction in the molecule. In this model, often referred to as Anderson-Holstein model, the transport through the molecule is assumed to be dominated by a single level of degeneracy d_g with energy ε in the presence of one vibrational mode with frequency ω_0. This system is described by the Hamiltonian $\mathbf{H} = \mathbf{H}_{\text{mol}} + \mathbf{H}_{\text{leads}} + \mathbf{H}_{\text{t}}$, where[9]

$$\mathbf{H}_{\text{mol}} = \varepsilon \mathbf{n}_d + \frac{U}{2} \mathbf{n}_d(\mathbf{n}_d - 1) + \lambda \hbar \omega_0 (\mathbf{b}^\dagger + \mathbf{b}) \mathbf{n}_d + \hbar \omega_0 (\mathbf{b}^\dagger \mathbf{b} + 1/2),$$

$$\mathbf{H}_{\text{leads}} = \sum_{a=L,R} \sum_{p,\sigma} \epsilon_p \mathbf{c}_{ap\sigma}^\dagger \mathbf{c}_{ap\sigma},$$

$$\mathbf{H}_{\text{t}} = \sum_{a=L,R; \, i=1,d_g} \sum_{p,\sigma} \left(t_a \mathbf{c}_{ap\sigma}^\dagger \mathbf{d}_{i\sigma} + \text{h.c.} \right). \tag{17.13}$$

Here, $\mathbf{H}_{\text{mol}}$ describes the molecular degrees of freedom, $\mathbf{H}_{\text{leads}}$ the leads and $\mathbf{H}_{\text{t}}$ the tunneling between the leads and the molecule. The Coulomb blockade is taken into account via the charging energy U. We focus on the regime of strong Coulomb blockade, $U \to \infty$, appropriate when $eV, k_{\text{B}}T \ll U$. The operator $\mathbf{d}_{i\sigma}$ ($\mathbf{d}_{i\sigma}^\dagger$) annihilates (creates) an electron with spin projection σ on degenerate level i of the molecule and $\mathbf{n}_d = \sum_{i=1,d_g; \, \sigma} \mathbf{d}_{i\sigma}^\dagger \mathbf{d}_{i\sigma}$ denotes the corresponding occupation-number operator. Similarly, $\mathbf{c}_{ap\sigma}$ ($\mathbf{c}_{ap\sigma}^\dagger$) annihilates (creates) an electron in lead a ($a = L, R$) with momentum p and spin projection σ. Notice that now the strength of the electron-phonon interaction is measured in units of $\hbar \omega_0$ and it is characterized by the dimensionless constant λ.

---

[9]Here, we ignore the dependence of the hopping integrals (t_a) on the indexes i and p.

It is convenient to choose a representation which is diagonal in the molecule degrees of freedom. In the present model this is achieved via the polaron or Lang-Firsov canonical transformation [792, 174]. Defining $\mathbf{S} = \lambda(\sum_{i,\sigma} \mathbf{d}_{i,\sigma}^\dagger \mathbf{d}_{i,\sigma})(\mathbf{b}^\dagger - \mathbf{b})$ and transforming all operators $\mathbf{O}$ via $e^{\mathbf{S}} \mathbf{O} e^{-\mathbf{S}}$ leads to a transformed Hamiltonian $\mathbf{H}' = \mathbf{H}'_{\text{mol}} + \mathbf{H}_{\text{leads}} + \mathbf{H}'_t$ with

$$\mathbf{H}'_{\text{mol}} = \varepsilon' \mathbf{n}_d + \hbar\omega_0(\tilde{\mathbf{b}}^\dagger \tilde{\mathbf{b}} + 1/2) + \frac{\tilde{U}}{2} \mathbf{n}_d(\mathbf{n}_d - 1) \qquad (17.14)$$

$$\mathbf{H}'_t = \sum_{a=L,R;i} \sum_{p,\sigma} \left(t_a \mathbf{X} \mathbf{c}_{ap\sigma}^\dagger \mathbf{d}_{i\sigma} + h.c. \right), \qquad (17.15)$$

where the transformed phonon operator $\tilde{\mathbf{b}} = \mathbf{b} - \lambda \sum_{i,\sigma} \mathbf{d}_{i\sigma}^\dagger \mathbf{d}_{i\sigma}$, so that the phonon ground state depends on the dot occupancy. Moreover $\varepsilon' = \varepsilon - \lambda^2 \omega_0$ is the "polaron shift" in the energy for adding one electron to the molecule and the interaction parameter U is also renormalized: $\tilde{U} = U - 2\lambda^2 \hbar\omega_0$. This renormalization will not be important below, since we shall focus here on the limit $U \to \infty$. The crucial phonon renormalization of the electron-lead coupling is given by

$$\mathbf{X} = \exp\left[-\lambda \left(\tilde{\mathbf{b}}^\dagger - \tilde{\mathbf{b}} \right) \right]. \qquad (17.16)$$

We are now in a position to write rate (master) equations for the electron-phonon joint probabilities, which take the form[10]

$$\begin{aligned}
\dot{P}_q^n = \sum_{a,q'} \Big\{ &f_a\left((q-q')\hbar\omega_0 + U(n-1)\right) \Gamma_{q,q'}^a P_{q'}^{(n-1)} \\
&+ \left[1 - f_a\left((q'-q)\hbar\omega_0 + Un\right)\right] \Gamma_{q,q'}^a P_{q'}^{(n+1)} \\
&- \left[1 - f_a\left((q-q')\hbar\omega_0 + U(n-1)\right)\right] \Gamma_{q',q}^a P_q^n \\
&- f_a\left((q'-q)\hbar\omega_0 + Un\right) \Gamma_{q',q}^a P_q^n \Big\}.
\end{aligned} \qquad (17.17)$$

Here, P_q^n is the probability to find the molecule with n electrons n ($n = 0, \cdots, 2d_g$) and q phonons, while $f_a(x)$ is a short form for the Fermi function $f(x + \varepsilon' - \mu_a)$, μ_a being the chemical potential of lead a.

Thus, the rate for going from a state with n electrons and q phonons on the molecule to a state with $n - 1$ electrons and q' phonons is $W_{q \to q'}^{n \to n-1} = \sum_{a=L,R} f_a\left((q-q')\hbar\omega_0 + U(n-1)\right) \Gamma_{q,q'}^a$, where $\Gamma_{q',q}^a$ represents the transition rate involving hopping of an electron from the dot to lead a by changing the phonon occupancy from q (measured relative to the ground state of $\mathbf{H}'_{\text{mol}}$ with occupancy n) to q' (measured relative to the ground state

---

[10]These equations are a straightforward generalization of those discussed in sections 15.4.4 and 15.5.1 in the absence of the electron-phonon interaction.

of $\mathbf{H}'_{\text{mol}}$ with occupancy $n - 1$). This rate is equal to the transition rate involving hopping of an electron from the lead a to the dot by changing the phonon occupancy from q (measured relative to the ground state of $\mathbf{H}'_{\text{mol}}$ with occupancy $n - 1$) to q' (measured relative to the ground state of $\mathbf{H}'_{\text{mol}}$ with occupancy n). More explicitly

$$\Gamma^a_{q',q} = \Gamma_a \left| \langle q' | \mathbf{X} | q \rangle \right|^2. \tag{17.18}$$

The matrix elements $\langle q' | \mathbf{X} | q \rangle$ are known as the *Franck-Condon matrix elements* because they also govern the transitions between different vibrational states in molecular physics. They can be computed by standard methods [174] and their absolute value $|\langle q | X | q' \rangle|^2 \equiv X^2_{qq'}$, which are symmetric under interchange of q and q', are given by (see Exercise 17.1)

$$X^2_{q<q'} = \left| \sum_{k=0}^{q} \frac{(-\lambda^2)^k (q! q'!)^{1/2} \lambda^{|q-q'|} e^{-\lambda^2/2}}{(k)!(q-k)!(k+|q'-q|)!} \right|^2. \tag{17.19}$$

It is interesting to write down explicitly a few elements:

$$X_{0n} = e^{-\lambda^2/2} \frac{\lambda^n}{\sqrt{n!}} \; ; \; X_{11} = \left(1 - \lambda^2 \right) e^{-\lambda^2/2} \tag{17.20}$$

$$X_{21} = \sqrt{2}\lambda \left(1 - \frac{\lambda^2}{2} \right) e^{-\lambda^2/2} \; ; \; X_{22} = \left(1 - 2\lambda^2 + \frac{\lambda^4}{2} \right) e^{-\lambda^2/2}.$$

Notice that for certain values of λ some of the matrix elements vanish. This unusual behavior is an interference phenomenon. A state which has q phonons excited above the ground state of the system with $n = 0$ electrons is a superposition (with varying sign) of many multi-phonon states, when viewed in the basis which diagonalizes the $n = 1$ electron problem, and therefore the transition described by $X_{qq'}$ is really a superposition of many different transitions, which for some values of λ may destructively interfere.

The current through the lead a in terms of the joint probability distribution functions is given by

$$I_a = \sum_{n,q,q'} (2d_g - n) P^n_q f_a \left((q' - q)\hbar\omega_0 + Un \right) \Gamma^a_{q,q'} \tag{17.21}$$

$$- (n+1) P^{n+1}_q \left[1 - f_a \left((q - q')\hbar\omega_0 + Un \right) \right] \Gamma^a_{q',q},$$

where the sum on n is from 0 to $(2d_g - 1)$, $2d_g$ being the maximum occupation of the dot.

Eq. (17.17) describes the nonequilibrium dynamics of the molecular vibrations. We shall now discuss the opposite limit, of phonons equilibrated to an independent heat bath, assumed to be at the same temperature as the leads. To implement this, one forces the probability distributions on the right hand side of Eq. (17.17) to have the phonon-equilibrium form $P_q^n = P^n e^{-q\hbar\omega_0/k_B T}(1 - e^{-\hbar\omega_0/k_B T})$. In the $U \to \infty$ limit this Ansatz implies that the probability P^0 that the molecule is empty is given by

$$P^0 = \frac{\sum_{a,q,q'} \Gamma_{q,q'}^a e^{-q\hbar\omega_0/k_B T} \bar{f}_{a,q,q'}}{\sum_{a,q,q'} 2\Gamma_{q,q'}^a e^{-q'\hbar\omega_0/k_B T} f_{a,q,q'} + \Gamma_{q,q'}^a e^{-q\hbar\omega_0/k_B T} \bar{f}_{a,q,q'}}, \quad (17.22)$$

where $\bar{f}_{a,q,q'} = 1 - f_a\left((q - q')\hbar\omega_0\right)$, $f_{a,q,q'} = 1 - \bar{f}_{a,q,q'}$ and $P^1 = 1 - P^0$.

For both equilibrated and unequilibrated cases the rate equations may be written in the matrix form

$$\dot{P} = \hat{M}P. \quad (17.23)$$

Therefore under steady state conditions ($\dot{P}_n = 0$), the problem reduces to finding the eigenvector corresponding to the zero eigenvalue of the matrix $\hat{M}$, which is easy to do numerically.

Let us now turn to analysis of the results of this approach. In Fig. 17.12 we reproduce some results of Ref. [747] where the current is depicted as a function of the source-drain voltage ($V_{sd} = (\mu_L - \mu_R)/e$). In these examples the level position was assumed to be $\varepsilon' = 0$, i.e. the electronic level is at resonance at zero bias. The two panels in Fig. 17.12 correspond to two different values of the gate voltage defined as $V_g = (\mu_L + \mu_R)/e$. The upper panel corresponds to $V_g = 0$ ($\mu_L = -\mu_R$), while the lower one corresponds to $V_g = V_{sd}/2$ ($\mu_R = 0$). In both cases the results are shown for equilibrated and unequilibrated phonons.

As one can see in Fig. 17.12, steps (broadened by the temperature) in the current associated with "phonon sidebands" are observed when the source-drain voltage passes through an integer multiple of the phonon frequency. As we explained in the previous section, these steps originate from resonant phonon emission processes. Notice that these I-V characteristics reproduce the main features observed in the experiments in this regime, see e.g. Fig. 15.16. In the linear response limit $V_{sd} \to 0$ (not shown here), as V_g is varied one finds one main step in the I-V curves. This is natural since, as explained above, the appearance of phonon sidebands requires a bias voltage larger than $\hbar\omega_0$.

Fig. 17.12 also reveals that in some cases the current is larger for equilibrated phonons than for the unequilibrated case. This is surprising because one expects that in the unequilibrated case the phonons arrange themselves so as to maximize the current. The authors of Ref. [747] attributed this behavior to the special dependence of the Franck-Condon matrix elements on the coupling constant λ.

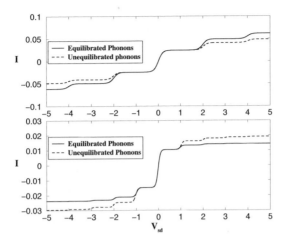

Fig. 17.12   Current (I) vs source-drain voltage V_{sd} for coupling constant $\lambda = 1.0$, $\hbar\omega_0 = 1$ and $k_B T = 0.05$. Upper panel is for $V_g = 0.0$, while lower panel is for $V_g = V_{sd}/2$, $\mu_R = 0$. I is in units of $ek_B T/\hbar$. Reprinted with permission from [747]. Copyright 2004 by the American Physical Society.

The steps in current may be conveniently parameterized by the height (or the area, as the width is simply proportional to T) of the corresponding peaks $G_{\max}$ in the differential conductance $G = dI/dV$. Ratios of peak heights (or areas) provide a convenient experimental measure of whether the phonons are in equilibrium. At low T, the equilibrium phonon distribution corresponds to occupancy only of the $n = 0$ phonon state, so the n-th sideband involves a transition from the 0 phonon to the n phonon state. Therefore the ratios of the peak heights or areas are controlled by ratios of $|X_{n0}|^2$. In particular, Eqs. (17.20) and (17.21) imply that if $\mu_L = -\mu_R$ and $k_B T \ll \hbar\omega_0$,

$$\left. \frac{G_{\max}^n}{G_{\max}^0} \right|_{\mathrm{eq}} = \frac{|X_{n0}|^2}{2|X_{00}|^2} = \frac{\lambda^{2n}}{2(n!)}. \qquad (17.24)$$

This equation also gives a simple rule of thumb to estimate how many phonon sidebands are expected for a given coupling constant λ. In particular, multiple steps arise only if λ is of the order of 1 or larger. Let us mention that Sapmaz *et al.* [793] reported I-V characteristics of suspended single-wall carbon nanotube quantum dots exhibiting a series of steps equally spaced in voltage. These features were attributed to the excitation of the stretching mode of the nanotubes. By comparing the I-V curves with the model above for equilibrated phonons, a reasonable agreement was found with coupling constants of order unity.

Now that we have described the basic formalism, let us briefly discuss some of the main physical effects that have been predicted in this regime:

Negative differential conductance (NDC). – It has been shown by several authors that the interaction with vibronic degrees of freedom can lead to negative differential conductance (NDC). This was first discussed by Boese and Schoeller [796] and then by others [797, 801, 803, 804]. See in particular Ref. [804] for a detailed discussion of the conditions for the appearance of NDC in this regime. This phenomenon has indeed been reported in tunneling experiments with suspended carbon nanotubes [793].

Franck-Condon blockade. – When the electron-phonon interaction is very strong, the Franck-Condon physics leads to a significant current suppression at low bias voltages, which has been termed as Franck-Condon (FC) blockade [801, 805]. This phenomenon is illustrated in Fig. 17.13 where we reproduce the results reported by Koch and von Oppen in Ref. [801]. This figure illustrates the strong dependence of the transport characteristics on the electron-phonon coupling strength λ. In particular, Fig. 17.13(a) shows the I-V curves for $\lambda = 1$ (intermediate coupling) and $\lambda = 4$ (strong coupling), as obtained from the rate-equation approach. These results correspond to $\varepsilon' = 0$, i.e. the molecular single-particle level and the lead Fermi energies are aligned at zero bias voltage. Notice that for $\lambda = 1$, the current increases sharply due to resonant tunneling when switching on a small bias voltage, and it exhibits the characteristic steps. In contrast, for $\lambda = 4$ the current is significantly suppressed at low bias voltages.

The current suppression originates from the behavior of the FC matrix elements determining the rates of phononic transitions $q_1 \rightarrow q_2$. For weak coupling, $\lambda \ll 1$, transitions mainly occur along the diagonal $q_1 \rightarrow q_1$. For intermediate coupling, $\lambda \approx 1$, the distribution of transition rates becomes wider, and transitions slightly off-diagonal are favored. For strong electron-phonon coupling, $\lambda \gg 1$, the distribution widens considerably and a gap of exponentially suppressed transitions between low-lying phonon states opens, see Fig. 17.13(b). Finally, let us mention that the observation of FC blockade has recently been reported in the context of suspended carbon nanotube quantum dots [808].

Pair-tunneling. – The coupling to molecular vibrations induces a polaron shift and can lead to a negative effective charging energy. In this case a ground state with even number of electrons is favored. Moreover, the charge transport through such molecules can be dominated by tunneling of electron pairs and the I-V characteristics can exhibit striking differences from the

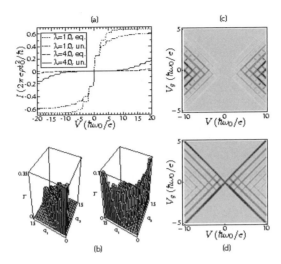

Fig. 17.13 (a) I-V characteristics for intermediate ($\lambda = 1$) and strong ($\lambda = 4$) electron-phonon coupling for $\varepsilon' = 0$ and $k_{\mathrm{B}}T = 0.05\hbar\omega_0$ for equilibrated and unequilibrated phonons. The strong electron-phonon coupling leads to a significant current suppression at low bias voltages. This Franck-Condon blockade arises from the behavior of the Franck-Condon rates for phonon transitions $q_1 \to q_2$ plotted in (b). The rates $\Gamma_{q_1 q_2}$, shown for $\lambda = 1$ (left) and $\lambda = 4$ (right), are given in units of the ordinary electronic coupling Γ (here a symmetric junction is considered). For strong electron-phonon coupling, transitions between low-lying phonon states are exponentially suppressed. The corresponding current suppression cannot be lifted by a gate voltage, which may serve as a fingerprint of FC blockade. This is depicted in the plot of dI/dV in the V–V_g plane for unequilibrated phonons with $\lambda = 4$ (c). The case of intermediate coupling with $\lambda = 1$ (d) is shown for comparison. Reprinted with permission from [801]. Copyright 2005 by the American Physical Society.

conventional Coulomb blockade. For a discussion of this phenomenology, see Ref. [809].

Absorption sidebands. – As we discussed in the previous section, the current flow can drive the vibrational modes far out of thermal equilibrium, which will, in turn, act back on the current. This can be reflected in the transport, in particular, with the appearance of vibrational sidebands in the differential conductance on the left side of the Coulomb peaks (for positive bias). This is due to resonant absorption of nonequilibrium phonons generated by the current, see Fig. 17.11(b). This phenomenon, observed in suspended nanotubes [740], has also been studied in the Coulomb blockade regime, see Refs. [810, 811].

Vibrational nonequilibrium effects with multiple electronic states. – The phenomena discussed above referred to a situation where the transport was assumed to be dominated by a single electronic molecular level. Härtle *et al.* [812] have shown recently that if multiple electronic states of the molecular bridge are involved in the transport, a number of additional vibronic processes take place and they may have a profound influence on the current-voltage characteristics.

To conclude this section, let us say that most of the theoretical investigations on vibrational effects in the Coulomb blockade regime have so far concentrated on model systems with only one vibrational mode. Only recently, several groups have started to combine ab initio methods with rate equations to investigate more realistic systems. Thus for instance, Chang *et al.* [806] reported the calculation of various phonon overlaps and their corresponding phonon emission probabilities for the problem of an electron tunneling onto and off of the fullerene-dimer molecular quantum dots C_{72} and C_{140}. In their approach, they do not assume that the vibrational modes are identical for different charge states, as it is usually done. Another example along this direction is the work of Seldenthuis *et al.* [807]. In this case, the authors have developed a method to calculate the vibrational spectrum of a sizable molecule in the sequential tunneling regime, based on DFT calculations to obtain the vibrational modes in a three-terminal setup. This method takes the charge state and contact geometry of the molecule into account and predicts the relative intensities of vibrational excitations. In addition, transitions from excited to excited vibrational state are accounted for by evaluating the Franck-Condon factors involving several vibrational quanta. Thus, this method can predict qualitatively different behavior compared to calculations that only include transitions from ground state to excited vibrational state.

17.3.2   *Interplay of Kondo physics and vibronic effects*

When the metal-molecule coupling is not too weak, high-order tunneling processes become possible and their interplay with the Coulomb repulsion in the molecules can lead to many-body phenomena like the Kondo effect (see section 15.6.2). As we discussed in section 16.5, different experiments have shown that the Kondo effect can coexist with vibronic effects. In this section we shall present a brief discussion of the theoretical work done to clarify the interplay between Kondo physics and electron-phonon interaction in molecular junctions.

Since the Kondo effect is a coherent many-body phenomenon, one may wonder under which circumstances this effect can survive in the presence of vibrationally-induced inelastic scattering. This question has been answered to a large extend by Cornaglia and coworkers in a series of papers [813–817]. These authors have applied the numerical renormalization group (NRG) to the Anderson-Holstein model in order to study the ground state and linear conductance for a broad range of parameters. They have found that at low temperatures and weak electron-phonon coupling $(2\lambda^2 \hbar\omega_0 \ll U)$ the properties of the conductance can be explained in terms of the standard Kondo model with renormalized parameters. In particular, the electron-phonon interaction leads surprisingly to an increase of the Kondo temperature in this regime. In the limit of strong electron-phonon interaction $(2\lambda^2 \hbar\omega_0 \gg U)$ the problem can be mapped onto an anisotropic Kondo model where the Kondo temperature decreases as λ increases [813].

Cornaglia *et al.* [817] also applied NRG to the Anderson-Holstein model to explain the anomalous gate voltage dependence of the Kondo temperature (T_K) found by Yu *et al.* [664] in SMTs based on transition metal complexes. They found that, as the frequency of the vibrational mode decreases, an anomalous gate dependence of T_K and of the transport properties emerges. This effect arises because soft vibrational modes in the molecular transistor drive the system into a new regime where the characteristic energy scales for spin and charge fluctuations are not related as in the conventional theory of the Kondo effect.

As shown in section 16.5, the clearest signature of the coexistence of Kondo physics and vibronic effects is the appearance of sidebands at the vibrational energies in the differential conductance versus the bias voltage. This is in clear contrast to what it is found the Coulomb blockade regime. What does the theory say about this nonequilibrium effect? This is an extremely challenging problem since, even in absence of electron-phonon interaction, there is no exact description of the finite-bias Kondo effect. In this context, we would like to mention the work of Paaske and Flensberg [818] where this problem has been addressed. These authors studied the Anderson-Holstein model for a very asymmetric contact $(\Gamma_L \gg \Gamma_R)$ and found that the nonlinear conductance exhibits Kondo sidebands located at bias voltages equal to multiples of the vibrational frequency. Moreover, due to selection rules, the side-peaks were found to have strong gate-voltage dependences. An example of the results is shown in Fig. 17.14. The left panel shows a gray-scale plot of $\partial^2 I/\partial V^2$ as a function of bias-voltage V and mean occupation number (gate voltage) $\mathcal{N} = C_g V_g/e$. The right panel shows three cuts revealing the side-band resonances on the flanks of the central zero-bias resonance.

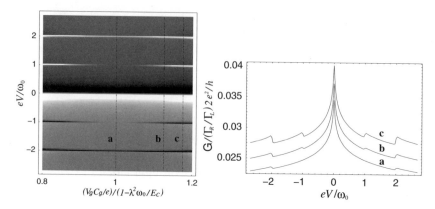

Fig. 17.14 Left panel: $\partial^2 I/\partial V^2$ vs. bias and gate voltage, for $\lambda^2 = 3$, $N(0)|t_L|^2 = 0.1\hbar\omega_0$, $U = 16\hbar\omega_0$, and $k_BT = 0.01\hbar\omega_0$. The junction is considered to be very asymmetric ($\Gamma_L \gg \Gamma_R$). Black/white indicates large negative/positive values. Right panel: Conductance vs. bias voltage for three values of V_g corresponding to the vertical black lines (**a,b,c**) in the upper panel. The lower curve (a) corresponds to the symmetric point $\mathcal{N} = 1 - \lambda^2\hbar\omega_0/E_C$. Reprinted with permission from [818]. Copyright 1999 by the American Physical Society.

We conclude here our brief discussion of this transport regime by recommending Refs. [819–826] for further details on this problem.

17.4   Concluding remarks and open problems

Although we have discussed many different vibrationally-induced transport phenomena in molecular junctions, it must be clear that our list is by no means complete. For instance, we have not touched at all the dramatic nonlinear effects that might appear due to a strong electron-phonon coupling. It has been predicted by Galperin *et al.* [509] that the charging of a molecular bridge (stabilized by the electron-phonon interaction) can lead to a modification of the molecular geometry and in turn to effects like negative differential conductance, multistability and hysteresis. Such issues have been addressed by several authors [827–829].

As we have seen in this chapter, the role of vibrations in the transport through molecular junctions is one of the most studied topics in molecular electronics. There several good reasons for that. On the one hand, as we have shown throughout this chapter, the understanding of the vibrational signatures in the transport characteristics is crucial for the detection of internal modes of the junctions. In turn, the observation of these modes provides a valuable in situ characterization of the contacts. The modes contain information not only about the presence of the molecule, which is a non-trivial issue with most experimental techniques, but also about the orientation of the molecule, its structure, the presence of defects and many other aspects. On the other hand, molecular junctions provide an ideal system to investigate new transport regimes where vibronic effects play an essential role. In this sense, molecular transport junctions are becoming an endless source of new physical phenomena.

The progress made in the last years in the understanding of vibronic effects in molecular electronics is certainly remarkable. However, there are still many challenges and basic open problems. Let us just mention a few of them. From the experimental side, it would highly desirable to have more IETS-like experiments with single-molecule junctions. Such experiments could be very important to obtain structural information which could help us to understand how the junctions are actually formed. In the context of three-terminal devices, it would be interesting to characterize in more detail the vibronic features in the Kondo regime. There are by now clear predictions, for instance, about how the Kondo temperature is affected by the electron-phonon interaction or about the gate dependence of the Kondo sidebands at finite bias. These predictions await for experimental confirmation.

The major open problem for the theory is the development of methods that are able to interpolate between the different transport regimes that we have discussed in this chapter. On the other hand, most theoretical models avoid considering the back action produced by the excitation of vibration modes. Such excitation, especially at high bias, may lead to structural changes that can affect dramatically the transport properties. With respect to the strong-coupling regime, little has done to describe microscopically how the energy and coupling of the modes depend on the charge state of the molecule. Of course, the description of the nonequilibrium "vibronic" Kondo effect needs further investigation. Finally, irrespective of the transport regime, more work is required to understand the role and signatures of anharmonicity.

17.5 Exercises

17.1 Franck-Condon matrix elements: Show that the Franck-Condon matrix elements in the Anderson-Holstein model are given by Eq. (17.19).

17.2 Phonon sidebands in the Coulomb blockade regime: Solve the rate equations both for equilibrated and unequilibrated phonons to reproduce the results of Fig. 17.12. Using the parameters of the upper panel of Fig. 17.12, compute the corresponding stability diagram.

17.3 Franck-Condon blockade: Use the rate-equation formalism described in section 17.3.1 to study the Franck-Condon blockade in the regime of strong electron-phonon coupling. In particular, solve the master equations for both for equilibrated and unequilibrated phonons to compute the different transport characteristics and reproduce the results of Fig. 17.13.

Chapter 18

The hopping regime and transport through DNA molecules

In Chapters 15-17 we have discussed how the electronic transport is modified when the quantum coherence is partially destroyed by either Coulomb correlations or the excitation of molecular vibrations. One of the central subjects of this chapter will be the analysis of the charge transport in situations in which this coherence is completely lost. As we explained in previous chapters, this incoherent regime is realized when the tunneling traversal time is considerably larger that the time scales associated to the inelastic interactions. Obviously, this becomes more likely as the length of a molecular bridge increases. In the extreme case in which the inelastic scattering time is much smaller than the tunneling time, the current is transported by electrons that hop sequentially from one segment of the molecule to another. For this reason this transport regime is also referred to as the *hopping regime.*

In long molecules, especially in biological ones, there are additional issues that should be considered when exploring the electronic transport through them. Thus for instance, the environment (solvent, atmosphere, etc.) in which the experiments are carried out plays a decisive role. In order to illustrate these issues, we shall also discuss in this chapter the transport through DNA molecules, which is one of the most emblematic and difficult topics in the field of molecular electronics.

The two main goals described above will be addressed in the following sections. First, we shall discuss in section 18.1 the characteristic signatures of the hopping transport regime. Then, in section 18.2 we shall describe some representative examples of experiments in which the hopping regime has been realized. Finally, section 18.3 is devoted to a brief review of the recent activities on the electronic transport through DNA-based molecular junctions.

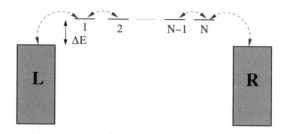

Fig. 18.1 Schematic representation of the model discussed in text to describe the incoherent tunneling through a molecular junction. Here, N sites with the same energy are connected via nearest-neighbor transfer rates $k_{j,j\pm1}$. The continua on the left and right correspond to the metallic states in the electrodes and ΔE is the activation energy.

18.1   Signatures of the hopping regime

The question that we want to address in this section is: How can we identify the occurrence of the hopping regime in an experiment? As we saw in Chapter 13, the coherent transport in off-resonant situations is manifested in the linear conductance as an exponential dependence on the length of the molecule and as an independence on the temperature. The hopping regime is however characterized by the following two main signatures:

- The conductance decays linearly with the length of the molecular wire.
- The conductance depends exponentially on the temperature as $\exp(-\Delta E/k_{\mathrm{B}}T)$, where ΔE is an activation energy that depends on the system under study.

Following the spirit of this monograph, we now proceed to discuss a simple model that illustrates how these two signatures come about. The model for a metal-molecule-metal junction, which is borrowed from the field of electron transfer [830, 831], is schematically represented in Fig. 18.1. Here, the molecular bridge has N sites (or states) and the incoherent tunneling between them is described by the transfer rates $k_{i,j}$ (from state j to state i).[1] For the sake of simplicity, we assume that all the states in the wire have the same energy, which differs by ΔE from the equilibrium Fermi energy of the leads. The quantity ΔE, which is nothing but the injection energy, plays here the role of an activation energy.

---

[1]We assume that only nearest-neighbor sites are directly connected, i.e. the only non-zero rates are $k_{j,j\pm1}$.

In this model the current between sites j and $j+1$ is determined by the occupations P_j and P_{j+1} in those two sites as follows[2]

$$I_j = e\left(k_{j+1,j}P_j - k_{j,j+1}P_{j+1}\right). \qquad (18.1)$$

Assuming that the tunneling between the different sites is incoherent, the occupations P_j fulfill then the following classical kinetic equations

$$\dot{P}_1 = -(k_{0,1} + k_{2,1})P_1 + k_{1,0}P_0 + k_{1,2}P_2$$

$$\vdots \qquad (18.2)$$

$$\dot{P}_j = -(k_{j-1,j} + k_{j+1,j})P_j + k_{j,j-1}P_{j-1} + k_{j,j+1}P_{j+1}$$

$$\vdots$$

$$\dot{P}_N = -(k_{N-1,N} + k_{N+1,N})P_N + k_{N,N-1}P_{N-1} + k_{N,N+1}P_{N+1},$$

where $\dot{P}_j$ stands for dP_j/dt, $P_0 = f_L$ and $P_{N+1} = f_R$, $f_{L,R}$ being the Fermi functions describing the electron occupations on the left and right electrodes. We are interested in a stationary situation where $\dot{P}_j = 0$. In this case, the previous kinetic equations reduce to the following algebraic equations

$$(k_{0,1} + k_{2,1})P_1 = k_{1,0}P_0 + k_{1,2}P_2 \qquad (18.3)$$

$$\vdots$$

$$(k_{j-1,j} + k_{j+1,j})P_j = k_{j,j-1}P_{j-1} + k_{j,j+1}P_{j+1}$$

$$\vdots$$

$$(k_{N-1,N} + k_{N+1,N})P_N = k_{N,N-1}P_{N-1} + k_{N,N+1}P_{N+1}.$$

As a further simplification, we assume that all the internal rates in the bridge are equal: $k_{j,j\pm1} = k$. Moreover, the detailed balance condition leads to the following relations for the rates involving the leads[3]

$$k_{1,0} = k_L e^{-(\Delta E - eV)/k_B T} \; ; \; k_{0,1} = k_L \qquad (18.4)$$

$$k_{N,N+1} = k_R e^{-\Delta E/k_B T} \; ; \; k_{N+1,N} = k_R. \qquad (18.5)$$

Here, we have taken into account the influence in the activation energy, ΔE, of the bias voltage, V, which we assume to be applied in the left electrode.

---

[2]The current is, of course, conserved and therefore, it is irrelevant where it is evaluated.

[3]In equilibrium the current must vanish and this leads to the relations: $k_{j+1,j}P_j^{eq} = k_{j,j+1}P_{j+1}^{eq}$, known as detailed balance conditions. Here, P_j^{eq} is the occupation probability of the site j in equilibrium. Therefore, $k_{j+1,j}/k_{j,j+1} = P_{j+1}^{eq}/P_j^{eq} = \exp[-(E_{j+1} - E_j)/k_B T]$, if $E_{j+1} > E_j$ and 0 otherwise.

It is straightforward to solve Eqs. (18.3) and to show that the charge current is given by (see Exercise 18.1)

$$I = e \frac{e^{-\Delta E/k_B T}}{[1/k_L + 1/k_R + (N-1)/k]} [e^{eV/k_B T} f_L - f_R]. \qquad (18.6)$$

Therefore, the corresponding linear conductance can be expressed as

$$G = \frac{e^2}{k_B T} \frac{e^{-\Delta E/k_B T}}{[1/k_L + 1/k_R + (N-1)/k]}. \qquad (18.7)$$

Here, for the sake of simplicity, we have neglected the temperature dependence coming from the Fermi functions of the leads.

From Eq. (18.7) one can deduce the two signatures described at the beginning of this section. First, notice that the conductance decays linearly with the number of sites (or incoherent segments) and therefore with the length of the molecular bridge. This is nothing else but the classical Ohm's law, which is a consequence of the loss of quantum coherence. Notice that if we ignore the activation process, the conductance simply adopts the standard expression of the conductance of a combination of resistors in series.[4] On the other hand, the conductance depends exponentially on the temperature, as in any thermally activated process. In our particular model, this process takes place at the metal-molecule interfaces, but in general it can occur at any point along the junction and there can even be several activation centers with their corresponding activation energies.

It is worth stressing that this model just provides a simple argument to understand the origin of the main signatures of the hopping regime, but one cannot expect quantitative predictions from it. An important issue that this model fails to describe is the transition from the coherent to the incoherent regime as function of the temperature and the length of the molecular bridge. Such transition, which is a key signature in the experiments (see next section), has been described by several authors using, for instance, the reduced density matrix formalism [830, 832–834]. In these models, the dephasing and relaxation is provided by a generic thermal bath. For a discussion on the unified description of coherent tunneling and the hopping mechanism, see Ref. [835].

The problem with these simple bath models is that they do not shed light on the microscopic origin of the loss of coherence. The main physical mechanism that makes the transport incoherent is believed to be the electron-phonon interaction inside the molecular bridge. In particular, when the

---

[4]Here, the resistors are the two metal-molecule interfaces, with resistances $k_B T/(e^2 k_{L,R})$, and the $N-1$ connections between the bridge sites, with a resistance $k_B T/(e^2 k)$ each.

coupling between different segments of a molecule is weak, the charge carriers can be localized over a single or a few segments. In this case, the molecule tends to change its conformation in order to lower its energy when charged. This process is known as *polaron formation*. As explained in the previous chapter, a polaron is a combination of a charge carrier and localized deformation. At room temperature, charge transport can be then dominated by incoherent hopping of polarons along the molecule. Such incoherent hopping gives rise to the exponential temperature dependence described above, with an activation energy which is the polaron binding energy.

The polaronic mechanism has been extensively studied in the context of conduction in solids [836]. This mechanism plays a fundamental role, for instance, in the conduction properties of organic materials used in organic electronics. In principle, the polaronic effects in molecular junctions can be described with an extension of the theoretical methods discussed in the previous chapter. However, such an extension is not straightforward. In practice, the polaron formation have mainly been analyzed with the help of single-level models [509, 827, 837], and polaron hopping in long molecules has typically been described using simple rate equations [838, 839]. For a discussion on the difficulties in describing polaron formation and hopping in molecular junctions and on recent advances in the treatment of this problem, see Ref. [840] and references therein.

18.2   Hopping transport in molecular junctions: Experimental examples

Experiments showing clear indications of the occurrence of hopping transport are rather scarce, especially at the level of single molecules. The main reason for this is the difficulty of measuring the temperature dependence of the transport characteristics. In this section we shall review a couple of representative examples in which the observation of hopping transport has been claimed.

The observation of thermally activated transport in single-molecule junctions was first claimed by Selzer *et al.* [841]. These authors studied the transport through individual 1-nitro-2,5-di(phenylethynyl-4'-mercapto)benzene molecules, see inset of Fig. 18.2(a), with gold electrodes using the electromigration technique. In this experiment, I-V measurements were taken at a temperature range of 13-296 K over a ± 1 V bias range.

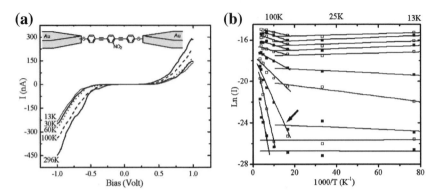

Fig. 18.2 (a) A set of I-V curves measured at different temperatures in Au-molecule-Au junctions fabricated with the electromigration technique. The molecule is shown in the upper inset, where a junction is schematically represented. (b) Arrhenius plots of Ln current (amperes) versus inverse T (K^{-1}) at different bias voltages showing a transition in conductance from T-independent tunneling behavior at low T to a thermally activated process at high T. The bias increment between curves is 0.1 V, and the bias of the lowest curve is 0.1 V. The transition temperatures between coherent and incoherent behavior are marked by the intersection between lines; see, for example, the arrow for 0.3 V. Reprinted with permission from [841]. Copyright 2004 American Chemical Society.

A representative set of I-V curves for different temperatures is shown in Fig. 18.2(a), where one can see that the current is quite sensitive to the bath temperature. An Arrhenius plot for a typical junction is shown in Fig. 18.2(b). Here, one can clearly see the transition from temperature-independent behavior at low T, where the conduction is dominated by coherent tunneling, to temperature-independent hopping behavior at high T, where presumably the transport is incoherent (hopping regime).

As one can see in Fig. 18.2(b), the transition from coherent to incoherent behavior is shifted to lower temperatures with increasing bias. As pointed out by the authors, there may be two complementary reasons for this behavior. First, the activation energy ΔE for hopping decreases as a function of bias. As the current in the hopping mechanism is proportional to $\exp(-\Delta E/k_B T)$, it is initiated at a lower bath temperature as ΔE decreases. Second, due to heat dissipation, the effective temperature of the molecule increases with bias, which can also induce a transition to incoherent tunneling at a lower bath temperature.

Let us also mention that the activation energy at zero bias was found to be 0.13 eV, which is probably too small to correspond to the injection energy (distance between the Fermi energy and the closest molecular level) for

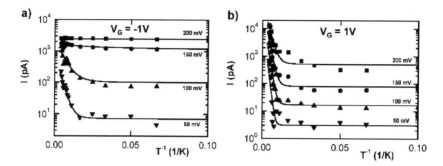

Fig. 18.3 Measurements of the temperature dependence of the current in three-terminal molecular junctions with sulfur end-functionalized tercyclohexylidenes. The different curves correspond to four values of the source-drain voltage as denoted in the figures. The panel (a) corresponds to a gate voltage of -1.0 V and the panel (b) to 1.0 V. The solid lines represent the best fits to the resonant tunneling model (see text). Reprinted with permission from [471]. Copyright 2006 American Chemical Society.

this molecule. This suggests that the rate-limiting process in the hopping mechanism is not thermal population of electrons/holes from the electrodes into the first hopping site, but rather an intramolecular hopping process (along the molecule).

Although the evidence presented in Ref. [841] is rather convincing, one cannot completely exclude an interpretation of the data of Fig. 18.2 in terms of coherent tunneling. As we explained in section 13.2, coherent tunneling can also lead to a pronounced temperature dependence of the I-V characteristics (see discussion below). This has been illustrated by Poot *et al.* [471], who reported data similar to those of Fig. 18.2(b) in a three-terminal device fabricated with the electromigration technique. In particular, these authors investigated the gate and temperature dependence of the current in molecular junctions containing sulfur end-functionalized tercyclohexylidenes. In Fig. 18.3 we reproduce some of the results of Ref. [471] in which one can see the current as a function of temperature for four different bias voltages at two gate voltages on a semilog scale. Notice that at low bias the curves of Fig. 18.3 show thermally activated transport at high temperature and temperature-independent transport at low temperature, i.e., very much like in Fig. 18.2(b). The crossover temperature is about 150 K in Fig. 18.3(a) and it decreases slightly as the bias is increased. The slope of the exponential increase above this crossover temperature yields and activation energy of 120 meV at low bias and this value decreases with increasing bias.

The authors of Ref. [471] use the simple resonant tunneling model described in detail in section 13.2 to analyze their data. Let us recall that in this model the temperature dependence comes from the Fermi distribution function in the leads and that the current becomes temperature-dependent when $k_B T$ is not too small in comparison with the injection energy (or level position measured with respect to the Fermi energy), which is the case in this experiment and in the previous one described above. As one can see in Fig. 18.3, the authors were able to fit the experimental data with this model using as adjustable parameters the level position, ϵ_0, and the scattering rates Γ_L and Γ_R. It is important to remark that the two gate voltages in Fig. 18.3 were far away from the degeneracy points, i.e. the transport is not completely at resonance. In the fits, the values found for ϵ_0 were very similar to the activation energy mentioned above of 120 meV, which shows the consistency of the fits. The total broadening of the level, $\Gamma = \Gamma_L + \Gamma_R$, was found to be in the range from 0.1 to 5 meV and it increased with increasing bias voltage.[5] In addition, the ratio Γ/ϵ_0 was found to range between 10^{-3} and 10^{-2}.[6]

The previous discussion shows that an unambiguous identification of the hopping regime requires additional information beyond the temperature dependence of the current. As we explained in the previous section, another key signature of the hopping regime is the linear decay of the conductance/current with the length of the molecule. To our knowledge, this signature, which is well-known in the context of electron transfer (see e.g. Ref. [842]), has not yet been reported in single-molecule junctions. However, Choi *et al.* [843] have reported recently the transition from coherent to hopping regime as a function of the molecular length in junctions based on monolayers of conjugated oligophenyleneimine (OPI) molecules ranging in length from 1.5 to 7.3 nm. The OPI wires were grown on a gold substrate and contacted by a metal-coated AFM as a second electrode. In Fig. 18.4(a) we reproduce the results of this experiment concerning the resistance (R) versus molecular length (L) for a series of OPI molecules with different numbers of phenyl units (n). As one can see, there is a clear transition of the length dependence near 4 nm (OPI 5). In short wires, the linear fit in Fig. 18.4(a) indicates that the data are well described with the standard formula of coherent non-resonant tunneling: $R = R_0 \exp(\beta L)$. The β value

---

[5]Here, Γ corresponds to the full width of the resonance at half maximum, while in section 13.2 it represents the half width at half maximum.

[6]The temperature dependence of the current within the resonant tunneling model is further discussed in Exercise 18.2.

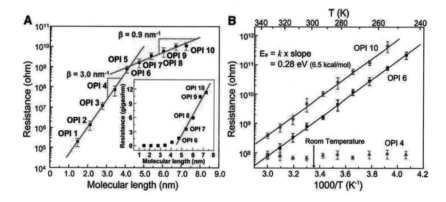

Fig. 18.4   Measurements of molecular wire resistance with a conducting probe AFM. A gold coated tip was brought into contact with an OPI monolayer on a gold substrate. (a) Semilog plot of R versus L for the gold/wire/gold junctions. Each data point is the average differential resistance obtained from 10 I-V traces in the range -0.3 to +0.3 V. Straight lines are linear fits to the data according to $R = R_0 \exp(\beta L)$. The inset shows a linear plot of R versus L, demonstrating linear scaling of resistance with length for the long OPI wires. (b) Arrhenius plot for OPI 4, OPI 6, and OPI 10. Each data point is the average differential resistance obtained at six different locations on samples in the range -0.2 to +0.2 V. Straight lines are linear fits to the data. From [843]. Reprinted with permission from AAAS.

was found to be 0.3 Å^{-1}, which is within the range of β values of typical conjugated molecules.

For long OPI wires, there is a much flatter resistance versus molecular length relation ($\beta \sim 0.09$ Å^{-1}). The extremely small β suggests that the principal transport mechanism is hopping. As one can see in the inset of Fig. 18.4(a), a plot of R versus L for long wires is linear, which is consistent with hopping. The change in transport mechanism was also verified by the temperature dependence. Fig. 18.4(b) shows that the resistance for OPI 4 is independent of temperature from 246 to 333 K, as expected for non-resonant coherent tunneling. However, both OPI 6 and OPI 10 display the strongly thermally activated transport that is characteristic of hopping. The activation energies determined from the slopes of the data are identical at 0.28 eV for both OPI 6 and OPI 10. Concerning the questions on the nature of the hopping sites and the origin of this activation energy, the authors suggested that three-repeat conjugated subunits are the charge-hopping sites in the long wires and that the hopping activation energy corresponds to the barrier for rotation of the aromatic rings, which transiently couples the conjugated subunits.

18.3   DNA-based molecular junctions

When the transport through a long molecule is investigated additional ingredients not discussed so far in this book, such as the presence of a solvent or the interaction with a substrate, may play a decisive role. This is specially clear in the case of biological molecules, where these factors can affect dramatically their conduction properties. In order to illustrate these ideas, we shall discuss in this section the transport through DNA molecules.

The great interest in the DNA molecule as a possible component of molecular electronic devices is due to its unique recognition and self-assembling properties. These properties offer in principle the possibility to build complex circuits with a bottom-up approach using this biological molecule as a building block. Obviously, the understanding of the electron transport in DNA molecules is a necessary prerequisite for the development of a DNA-based molecular electronics [844–847]. For this reason, a great effort has been devoted in the last 20 years to elucidate the transport properties of this molecule. However, due to the complexity of DNA, there are still many open questions in this subject. In this section, and taking into account the scope of this book, we shall briefly review the experiments in which the electronic transport through *single* DNA molecules has been investigated. For the vast literature on multi-molecule measurements we refer to the review articles by Porath, Cuniberti and Di Felice [848] and by Endres, Cox, and Singh [849].

Let us remind that natural DNA consists of two long polymers of simple units called nucleotides, with backbones made of sugars and phosphate groups joined by ester bonds. Each of these strands is called single-stranded DNA (ssDNA). These two strands form a double helix with the backbones pointing outwards, the so-called double-stranded DNA (dsDNA). Attached to each sugar is one of four types of molecules called bases: adenine (A), cytosine (C), guanine (G) and thymine (T). Each type of base on one strand forms a bond with just one type of base on the other strand. This is called complementary base pairing or Watson-Crick pairing. In particular, A binds only to T, while C binds only to G. This arrangement of two nucleotides binding together across the double helix is called a base pair.

Double-stranded DNA exists in several conformations, among which the B-conformation is the natural one, which is, however, only stable in aqueous environment. In the B-conformation CG and AT pairs are stacked above each other at a distance of 3.4 Å between each pair, see Fig. 18.5. Each strand is stabilized by the backbone keeping the bases at this distance. In

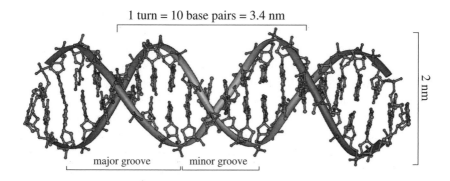

1 turn = 10 base pairs = 3.4 nm

2 nm

major groove   minor groove

Fig. 18.5   Double helical structure of DNA in B-conformation. Taken from Wikipedia Commons.

the B-conformation the long axis of neighboring base pairs are twisted with respect to each other by an average angle of 36° such that 10 base pairs make a full turn. This conformation is stabilized by water molecules and other counter ions. In dry conditions (less than 5 H_2O molecules per base pair) the stable conformation is the so-called A-conformation that differs from the B-conformation by an inclination of the base pairs with respect to the molecule's axis and a somewhat weaker twist [850].

With respect to the conduction mechanism in DNA, it is generally accepted that the electron transfer in DNA takes place via the overlap between the π-orbitals of neighboring base pairs. This is similar to what happens in certain stacked aromatic crystals, like the Beechgard salts, which are indeed metallic. This electron transfer mechanism in DNA suggests that the base sequence can be very important since the π-system of the individual bases may be different. Moreover, the conformation is important because it determines the overlap between the base pairs. Finally, in order to have a measurable electrical current in a DNA junction, it is crucial to make sure that the π-system hybridizes strongly with the metallic states of the electrodes.

Now, we turn to the discussion of the transport experiments in DNA-based junctions. Let us start by summarizing the main findings. Most of the transport measurements on single DNA molecules reported so far can be divided into three classes. First, there are experiments showing that DNA is an insulator for lengths larger than 40 nm at room temperature, with essentially no discernible conductance up to 10 V. This suggests that the electronic states of DNA are completely localized [851, 852]. Second, some experiments show that it is possible to transport charge through short DNA molecules of up to 20 nm with currents of the order of 1 nA at 1 V. This suggests that short DNA pieces behave as large-bandgap semiconductors [853, 854]. Third, some experiments show that if special care is taken to tailor the interaction with the substrate and to provide good contacts to the leads, semiconducting-like I-V characteristics with currents exceeding 200 nA at 1 V can be achieved even under ambient conditions [855].

It is worth remarking that there are several experiments that do not fall into any of these three categories. Thus for instance, Fink and Schönenberger [856] reported ohmic behavior in 16 μm-long DNA molecules in experiments performed with a field-emission microscope. On the other hand, Kasumov *et al.* [857] found the induction of proximity superconductivity in experiments where a peculiar contacting scheme was used (see Chapter 3). This observation can only be explained if DNA turns out to be a very good conductor.

Several experiments have been designed to elucidate the origin of the discrepancies mentioned above. Thus for instance, both Kasumov *et al.* [858] and Heim *et al.* [859] using a STM or a conducting AFM, respectively, showed that the interaction between DNA and the underlying substrate plays a fundamental role in the conduction properties of this molecule. Such interaction turns out to be in some cases strong enough to deform the DNA molecule and to induce conformational changes. These changes may be responsible for the blocking of the current along the molecule, which results in the insulating behavior observed in many experiments. The importance of the molecule-substrate interaction have also been emphasized by several groups in the case of different polymers [504, 860]. For DNA, it is known that external forces can stabilize several helical conformations [861].

As mentioned above, the transport through DNA is expected to depend on the exact base sequence. This has been nicely illustrated by Tao's group in experiments on short DNA pieces (eight GC base pairs plus a varying number of AT base pairs) performed with the STM break junction technique in liquid environment [862]. These authors found qualitative differences between the transport mechanism for GC base pairs and AT ones, see Fig. 18.6. The interpretation of this experiment is that coherent transport would be possible through CG-only DNA, while the AT base pairs act as tunneling barrier over which the transport takes place via incoherent hopping from site to site. The reported sequence dependence is in agreement with theoretical predictions [838, 863, 864].

It is interesting to mention at this point a related experiment by Giese

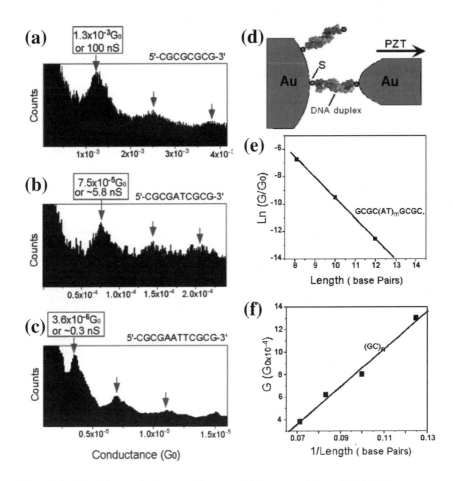

Fig. 18.6 The left panels show conductance histograms of three DNA duplexes measured with the STM break-junction technique: (a) 5′-CGCGCGCG-3′-thiol linker, (b) 5′-CGCGATCGCG-3′-thiol linker, and (c) 5′-CGCGAATTCGCG-3′-thiol linker. (d) Schematic illustration of a single DNA conductance measurement. (e) Natural logarithm of GCGC(AT)$_m$GCGC conductance vs. length (total number of base pairs). The solid line is a linear fit that reflects the exponential dependence of the conductance on length. (f) Conductance of (GC)$_n$ vs. 1/length (in total base pairs). Reprinted with permission from [862]. Copyright 2004 American Chemical Society.

et al. [842], where the charge transfer rate in DNA molecules was measured. Some of these results are shown in Fig. 18.7. As compared to the experiment just described, Giese *et al.* found a weaker length dependence of the transfer rates when several AT base pairs were inserted between CG base pairs. These experiments showed the existence of two different processes for

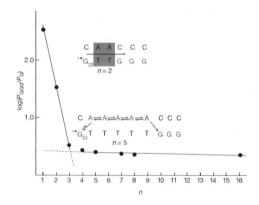

Fig. 18.7 Sequence dependence of the charge transfer in DNA. Plot of $\log(P_{GGG}/P_G)$ (P_{GGG}/P_G is proportional to the charge transfer rates) against the number n of the AT base pairs. Each experiment was performed three times, and their relative errors are within $\pm 10 - 20\%$. The steep line corresponds to the coherent superexchange charge transfer (tunneling). The flat line is drawn in order to make clear the weak distance dependence. The arrows in the depicted DNA strands indicate the superexchange charge transfer between the guanine radical cation G_{22}^+ and the GGG sequence for short distances ($n = 2$), or the hopping mechanism for long distances ($n = 5$), where - in addition adenines act as charge carriers. For clarity, only the double strands with $n = 2$ and $n = 5$ are shown. The nucleotides in grey indicate all charge carriers. Reprinted by permission from MacMillan Publishers Ltd: Nature [842], copyright 2001.

the *hole* transfer between guanines in DNA: (i) A coherent superexchange reaction (single-step tunneling), where the bridging adenines are indirectly affecting the transfer mechanism by mediating the electronic coupling between the guanines, and (ii) a thermally induced hopping process, where the guanines oxidize the intervening adenine bases and directly involve them in charge transport. The efficiency of the tunneling reaction decreases rapidly with the number of the intervening AT base pairs, whereas the hopping process is only slightly influenced by the number of the AT base pairs.

The discussion of these two experiments shows that one cannot talk about a single transport mechanism in DNA. From these experiments it is expected that CG-DNA could serve as a molecular wire. However, it has been shown that longer DNA species with a percentage of CG pairs above approximately 75% undergo a conformational change to the presumably conductive quartet geometry (nicknamed G4-wires) [865]. Recent experiments on G4 derivatives show an enhanced electrical polarizability, while dsDNA oligomers appeared electrically "silent" in an equivalent experiment [866]. This instability makes it difficult to investigate the sequence dependence of the transport mechanisms for longer molecules.

On the other hand, a lot of progress has been made in the last years concerning the relation between electronic properties and the sequence by using transverse scanning tunneling microscopy and spectroscopy [867, 868] or theoretical modeling [869]. Thus for instance, Shapir *et al.* [867] showed that GC and AT base pairs have very distinct I-V spectra. The authors of Ref. [868] used this difference for developing a method for determining the sequence of a DNA strand by measuring their electronic properties.

From the theory side, it has been predicted that a junction with a single DNA (in B-form) should exhibit a non-monotonic behavior of the electrical response as a function of its elongation [870]. This non-monotonic behavior originates from a competition between a stretching and a de-twisting process of the helical structure. To be precise, the elongation of a helix is predicted to reduce the angle between neighboring base pairs, which results in an enhancement of the overlap of the conducting orbitals. Simultaneously, the stretching enhances the distance and thus reduces the overlap again.

An important characteristic of the DNA molecule is its remarkable flexibility. It can be stretched in excess of 1.7 times its B-form length. Single-molecule stretching experiments have shown that DNA undergoes a pronounced and abrupt structural transformation to a yet unknown structure, which is elongated by more than 50% and called S-conformation [871]. This conformation is presumably associated with rotations of specific orbitals along the helix axis, in turn influencing the effective orbital overlap between neighboring base pairs. Theoretical calculations indicate that this pronounced conformational transition has a strong impact in the conduction properties of DNA molecules [872–874].

In summary, there are at least three important factors that influence the conduction properties of single DNA molecules:

- The environment: The presence (or the absence) of a solvent plays a key role. This is evident in the case of the most conductive form of DNA, namely the B-conformation, which only exists in solution. On the other hand, the interaction with an underlying substrate may give rise to a conformational change, which in turn can modify dramatically the transport properties of DNA-based junctions [858, 859].
- The contacting method: As usual, the metal-molecule interface plays a very important role. In this sense, the highest currents through DNA have been achieved with dithiolated molecules, i.e. with covalent bonds

to gold electrodes at both ends of the molecules [855, 862]. Also, relatively high currents have been measured through dsDNA molecules covalently attached to single-walled carbon nanotubes (SWNT) [875] and through duplex DNA coupled to SWNT electrodes via amide linkages [876].

- The sequence: The exact base sequence determines finally the level of the current that can flow through DNA molecules [842, 862, 864]. In particular, a sequence rich in CG pairs is expected to exhibit a higher current.

18.4   Exercises

18.1 Length and temperature dependence of the conductance in the hopping regime: Show that the conductance in the incoherent model of section 18.1 is given by Eq. (18.7). Hint: In order to learn how to solve Eq. (18.3), consider first the case in which the molecular bridge is composed of only two sites.

18.2 Activation-like temperature dependence in the resonant tunneling model: The goal of this exercise is to show that an exponential dependence of the current on temperature is also possible in the coherent regime. For this purpose, use the resonant tunneling model of section 13.2 and compute the current as a function of $\epsilon_0/k_\mathrm{B}T$ for $\Gamma_L = \Gamma_R = 0.005\epsilon_0$ for several bias voltages: $eV/\epsilon_0 = 0.1, 0.5, 1.0, 1.5, 2.0$. Hint: The solution can be found in Fig. 3(a) of Ref. [471].

Chapter 19

Beyond electrical conductance: Shot noise and thermal transport

In the previous chapters we have addressed the main transport regimes that are realized in molecular junctions. In our discussion so far, we have focused our attention on the analysis of the electrical conductance. However, there are many other transport properties that provide valuable information, which often is not contained in the conductance. A paradigmatic example is the current fluctuations or noise. Its investigation has contributed decisively to our understanding of the transport mechanisms in a great variety of mesoscopic and nanoscale devices [150]. On the other hand, the charge transport is not the only important aspect in the context of conduction in molecular junctions. Thermal transport is also a key issue in the field of molecular electronics from a fundamental as well as a from a practical viewpoint. Molecular-scale contacts provide a new territory to study heat conduction in regimes never explored before and, issues like heating will have to be faced and understood, if molecular electronics wants to become a viable technology. Obviously, the study of thermoelectric phenomena in molecular junctions, resulting from the interplay between electrical and thermal transport, can also give a new insight into the physics of these nanocircuits.

For these reasons, we shall put aside the electrical conductance for a while, and in this chapter we shall concentrate on the discussion of other transport properties. To be precise, in section 19.1 we shall discuss the basic physics of noise in molecular junctions and describe the first noise experiments in this field. Then, we shall turn our attention to thermal transport and in section 19.2 we shall present a detailed discussion of heating and heat conduction in molecular wires. Finally, section 19.3 is devoted to the analysis of the thermopower, which is becoming a vital source of novel information on molecular transport junctions.

As in previous chapters, we shall present here both a discussion of the basic concepts related to these transport properties as well as a review of the work reported on this subject in recent years. In any case, we shall concentrate on the analysis of aspects that have been already investigated experimentally or that are likely to be investigated in the near future. Let us also say that in some cases we have also included a brief description of the corresponding phenomena in metallic atomic contacts, since these systems often paved the way for a later analysis in the context of molecular junctions.

19.1   Shot noise in atomic and molecular junctions

The electrical current through any conductor exhibits temporal fluctuations (or noise). As it was already pointed out by Schottky in 1918 [877], when all sources of spurious noise are eliminated, there remain two types of noise in the electrical current, namely the *thermal noise* and the *shot noise*. The thermal noise, which is also known as Johnson-Nyquist noise (after the experimentalist [878] and the theorist [879] who investigated it), is due to the thermal motion of the electrons and occurs in any conductor. The nonequilibrium fluctuations known as shot noise are caused by the discreteness of the charge of the carriers of the electrical current. We have discussed this transport property within the scattering formalism in section 4.7 and for further details we recommend to the reader the excellent reviews of Refs. [150, 155].

Noise is characterized by its spectral density or power spectrum $P(\omega)$, which is the Fourier transform at frequency ω of the current-current correlation function,

$$P(\omega) = 2 \int_{-\infty}^{\infty} dt\, e^{i\omega t} \langle \Delta I(t + t_0) \Delta I(t_0) \rangle. \tag{19.1}$$

Here $\Delta I(t)$ denotes the time-dependent fluctuations in the current at a given voltage V and temperature T. The brackets $\langle \cdots \rangle$ indicate an ensemble average. Both thermal and shot noise have a white power spectrum, i.e. the noise power does not depend on ω over a very wide frequency range. Thermal noise ($V = 0$, $T \neq 0$) is directly related to the conductance G by the fluctuation-dissipation theorem [880],

$$P = 4k_B T G, \tag{19.2}$$

as long as $\hbar\omega \ll k_B T$. Therefore, the thermal noise of a conductor does not give any new information as compared to the conductance.

Shot noise ($V \neq 0$, $T = 0$) is more interesting, because it gives information on the temporal correlation of the electrons, which is not contained in the conductance. In devices such as tunnel junctions, Schottky barrier diodes, *p-n* junctions, and thermionic vacuum diodes, the electrons are transmitted randomly and independently of each other. The transfer of electrons can be described by the Poisson statistics, which is used to analyze events that are uncorrelated in time. For these devices the zero-frequency shot noise has its maximum value

$$P = 2eI \equiv P_{\text{Poisson}} , \qquad (19.3)$$

which is proportional to the time-averaged current I. Correlations suppress the low-frequency shot noise below P_{Poisson}. One source of correlations, operative even for non-interacting electrons, is the Pauli principle, which forbids multiple occupancy of the same single-particle state. A typical example is a ballistic point contact in a metal, where $P = 0$ because the stream of electrons is completely correlated by the Pauli principle in the absence of scattering. In single-channel quantum point contacts, and in the absence of inelastic scattering, shot noise is predicted to be suppressed by a factor proportional to $\tau(1 - \tau)$, where τ is the transmission probability of the conduction channel[1] [881–883]. This quantum suppression was first observed in point contact devices in a two-dimensional electron gas [884, 97]. For a general multichannel contact in the limit of very low temperatures the shot noise power is predicted to be [883]

$$P = 2eV G_0 \sum_n \tau_n(1 - \tau_n), \qquad (19.4)$$

where $G_0 = 2e^2/h$ is the quantum of conductance. For arbitrary temperature and voltage the noise is a mixture of thermal noise and shot noise and, assuming that the transmission coefficients do not depend on energy, it is given by Eq. (4.75), which we reproduce here[2]

$$P = 2G_0 \left[2k_B T \sum_n \tau_n^2 + eV \coth\left(\frac{eV}{2k_B T}\right) \sum_n \tau_n (1 - \tau_n) \right]. \qquad (19.5)$$

Since the shot noise depends on the sum over the second power of the transmission coefficients, this quantity is independent of the conductance, $G = G_0 \sum_n \tau_n$, and the simultaneous measurement of these two quantities

---

[1]Throughout this chapter we shall denote the transmission as τ in order to avoid confusions with the temperature.

[2]Here, we have taken into account the spin degeneracy that will be assumed in our discussion throughout this section.

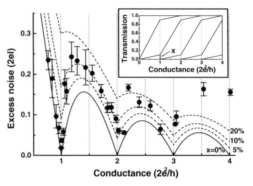

Fig. 19.1 Noise measurements on Au atomic contacts using the MCBJ technique. The symbols correspond to the measured excess noise values for 27 contacts at 4.2 K with a bias current of 0.9 μA. The different lines show the calculations using Eq. (19.5) in the case of one single partially transmitted channel (full curve) and for various amounts of contributions of other modes according to the model described in the inset (dashed curves). In the limit of zero conductance, these curves all converge to full shot noise, i.e. 2.9 10^{-25} A^2/Hz. *Inset:* transmission of modes in the case of x=10% contribution from neighboring modes. Reprinted with permission from [885]. Copyright 1999 by the American Physical Society.

should give information about the transmission coefficients of the contact. The relevant quantity is conveniently expressed in terms of the Fano factor F, which is the ratio of the shot noise to the noise that the same current would produce in the classical Schottky limit,

$$F = \frac{P}{2eI} = \frac{\sum_n \tau_n (1 - \tau_n)}{\sum_n \tau_n}. \tag{19.6}$$

Shot noise in atomic-scale contacts was first measured by van den Brom and J.M. van Ruitenbeek using the MCBJ technique [885]. The measurements were conducted at low temperatures to reduce the thermal noise. However, in these experiments the noise level of the pre-amplifiers in general exceeds the shot noise to be measured. Using two sets of pre-amplifiers in parallel and measuring the cross-correlation, this undesired noise is reduced. By subtracting the zero-bias thermal noise from the current-biased noise measurements, the pre-amplifier noise, present in both, is further eliminated. For currents up to 1 μA the shot noise level was found to have the expected linear dependence on current. For further details on the measurement technique, we refer to [885].

In Fig. 19.1 we show the results of Ref. [885] for the noise of gold atomic contacts as a function of the conductance of the junctions. The measured shot noise is given relative to the classical shot noise value $2eI$. All data are

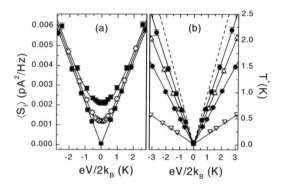

Fig. 19.2 Noise measurements on Al atomic contacts using microfabricated MCBJs. Symbols: measured average current noise power density $\langle S_I \rangle$ and noise temperature T^*, defined as $T^* = S_I/4k_BG$, as a function of reduced voltage, for a contact in the normal state at three different temperatures (from bottom to top: 20, 428, 765 mK). The solid lines are the predictions of Eq. (19.5) for the set of transmissions $\{0.21,0.20,0.20\}$ measured independently from the I-V in the superconducting state. (b) Symbols: measured effective noise temperature T^* versus reduced voltage for four different contacts in the normal state at $T = 20$ mK. The solid lines are predictions of Eq. (19.5) for the corresponding set of transmissions (from top to bottom: $\{0.21,0.20,0.20\}$, $\{0.40,0.27,0.03\}$, $\{0.68,0.25,0.22\}$, $\{0.996,0.26\}$. The dashed line is the Poisson limit. Reprinted with permission from [98]. Copyright 2001 by the American Physical Society.

strongly suppressed compared to the full shot noise value, with minima close to 1 and 2 times the conductance quantum. For contacts with conductance below $1\,G_0$ the data are consistent with a single conduction channel having a transmission probability $\tau = G/G_0$, as expected for this monovalent metal. For larger contacts there is a tendency for the channels to open one-by-one, but admixture of additional channels grows rapidly. There is a very strong suppression, down to $F = 0.02$, for $G = 1\,G_0$, which unambiguously shows that the current is carried dominantly by a single channel. It needs to be stressed that this holds for gold contacts. There is a fundamental distinction between this monovalent metal and the multivalent metal aluminum, which shows no systematic suppression of the shot noise at multiples of the conductance quantum, and the Fano factors lie between about 0.3 and 0.6 for G close to G_0 [886].

Shot noise measurements by Cron *et al.* [98] have provided a very stringent experimental test of the multichannel character of the electrical conduction in Al atomic contacts. In these experiments the set of transmissions τ_n were first determined independently by the technique of fitting the subgap structure in the superconducting state, discussed in section 11.4.

The knowledge of the transmission coefficients allows a direct quantitative comparison of the experimental results on the shot noise with the theoretical predictions of Eq. (19.5). The experiments were done using Al nanofabricated break-junctions which exhibit a large mechanical stability. The superconducting I-V curves for the smallest contacts were measured below 1 K and then a magnetic field of 50 mT was applied in order to switch into the normal state. The measured voltage dependence of the intrinsic current noise is shown in Fig. 19.2(a) for a typical contact in the normal state at three different temperatures, together with the predictions of Eq. (19.5), using the set of transmission coefficients measured independently. The noise measured at the lowest temperature for four contacts having different sets of transmission coefficients is shown in Fig. 19.2(b), together with the predictions of the theory. This excellent agreement between theory and experiments provides an unambiguous demonstration of the presence of several conduction channels in the smallest Al contacts and serves as a test of the accuracy that can be obtained in the determination of the τ's from the subgap structure in the superconducting I-V curve.

In the last years, van Ruitenbeek's group has performed shot noise measurements in highly conductive molecular junctions to determine the channel decomposition of the conductance. A first example was reported by Djukic and van Ruitenbeek for the hydrogen molecule bridge [568]. In this case, a Pt-H_2 junction was adjusted so as to have a clear vibrational mode signal, and the shot noise signal was measured for the same junction. An example of this measurement is shown in Fig. 19.3. Although shot noise generally does not allow determining the full set of transmission values, one can obtain information from the property that the noise increases the more channels are partially transmitted. The result of Fig. 19.3 for a junction with conductance $G = 1.021\,G_0$ was fitted using two channels with transmissions $\tau_1 = 1.000$ and $\tau_2 = 0.021$. In principle, the conductance in this example can be redistributed over more than just two channels. However, when the transmission $\tau_1 = 1.000$ is broken up into more channels this strongly increases the Fano factor. Thus, the only freedom is to redistribute the transmission $\tau_2 = 0.021$ over two or more channels, that will all have a very small contribution. Therefore, one can conclude that the conductance is largely dominated by a single channel with nearly perfect transmission, which was found to be a very robust result of these measurements. As we explained in section 14.1.3, this result was decisive to discriminate between the different possible geometries for the hydrogen bridge which had been proposed theoretically.

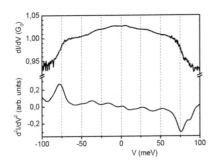 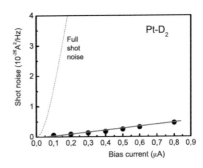

Fig. 19.3 Shot noise measurements in Pt-D2 break-junctions. The left panel shows the point-contact spectroscopy (PCS) signal for this junction, with a clear vibrational mode at 76 meV. The right panel shows the excess noise (the white noise in the current above the thermal noise) as a function of the current. The noise is strongly suppressed below the full Schottky noise for a tunnel junction. After each measurement of noise at a given current, the PCS was measured again to verify that the contact had not changed. The total conductance for this junction is $G = 1.021\, G_0$, and the shot noise can be fitted with two channels, $\tau_1 = 1.000$, and $\tau_2 = 0.021$, giving a Fano factor of $F = 0.020$. Adapted with permission from [568]. Copyright 2006 American Chemical Society.

More recently, shot noise measurements have also been used to characterize Pt-H_2O-Pt junctions [735] and Pt-benzene-Pt junctions [473]. In the former case, the noise results indicated that for conductance below $1\, G_0$ there are typically two conduction channels, although one clearly dominates the transport. These results were very important to understand the crossover between PCS and IETS and to test the so-called 1/2-rule (see section 16.4). In the case of Pt-benzene junctions (see section 14.1.4), the analysis of the shot noise results showed that for conductances around $1\, G_0$ (and also well below) several channels contribute significantly to the transport, while when the conductance is reduced to $0.2\, G_0$, the number of channels is eventually reduced to one. As opposed to Pt-H_2O-Pt junctions, in this case there is no dominant transmission channel when more than a single channel exists. It was shown theoretically in same work [473] that the number of channels is roughly determined by the number or carbon atoms directly coupled to the Pt electrodes.

So far the shot noise measurements in molecular junctions have been used to extract the channel transmissions in highly conductive junctions, where the transport is supposed to be coherent. Notice that this application is restricted to junctions with a high conductance, let us say above $0.1\, G_0$. Below that, the quadratic term in the transmission coefficients is negligible and the shot noise becomes proportional to the conductance (i.e.

linear in the transmission coefficients). Anyway, the shot noise can provide very important information also in other transport regimes. For instance, shot noise measurements in the Coulomb blockade regime [887, 888] or in the Kondo regime [889] have been only reported very recently in the contexts of carbon nanotubes and semiconductor quantum dots. In this sense, weakly coupled molecular junctions (molecular transistors) can be an ideal playground to further explore the noise in these transport regimes.

On the other hand, the vibrational effects discussed in previous chapters can be further investigated with the help of the noise. For instance, it has been predicted that the Franck-Condon blockade (see section 17.3.1) is characterized by remarkably large Fano factors (10^2-10^3 for realistic parameters), which arise due to avalanche-like transport of electrons [801].

The vibrationally-induced inelastic effects on noise properties of molecular junctions in different transport regimes have been studied using NEGF techniques by Zhu and Balatsky [784] and by Galperin and coworkers [890].[3] Very recently, several theoretical groups have discussed the noise induced by vibrations in the limit of weak electron-phonon coupling [891–893]. One of the central issues of these papers was the discussion of the sign of the inelastic noise as a function of the transmission, which is related to our discussion of the sign of the inelastic conductance in this regime (see section 16.4). The predictions of these papers could be in principle tested in the type of experiments discussed above.

19.2   Heating and heat conduction

As mentioned in the introduction, so far we have only discussed the transport of electrical charge in molecular junctions, but heat transport is also very important for several reasons. From a practical point of view, the understanding of heat generation in molecular contacts is crucial. When an electrical current flows through a junction, there is an energy transfer (Joule heating) from the electrons to the vibrations that might cause a large temperature increase that in turn can affect the stability and integrity of molecular junctions. From a more fundamental point of view, it is very exciting to investigate how the heat is conducted through the tiniest circuits ever built, namely atomic-scale junctions. These structures

---

[3]The approach used in these two references has been criticized in Ref. [893], where it is claimed that it misses vertex corrections even at the lowest order in the electron-phonon interaction.

have dimensions that are much smaller than the inelastic mean free path for phonons, even at room temperature, and thus they offer the possibility to study phonon transport (and their contribution to thermal conduction) in a very special regime.

Heat generation and heat conduction are intimately connected. Indeed, heat conduction is an essential ingredient in the balance of the processes that determines local heat generation. For this reason, we have chosen to organize this section in the following way. After some general comments about the problem of describing heating and heat conduction in molecular junctions, we shall briefly review the work done so far on thermal conductance. Then, we shall discuss the issue of heat generation in molecular junctions and, in particular, we shall describe the main experiments reported to date on this topic.

The subject of thermal properties of molecular junctions is presently dominated by the theory. The experiments are rather scarce due to the difficulties of measuring thermal transport at the nanoscale (for a review on this subject see Ref. [894]). Although it is a very interesting subject, we shall not discuss here in depth the theoretical techniques to describe heat transport in molecular junctions, and we shall merely point out the main ideas and challenges. For a more detailed discussion on the theory, see section 9 of Ref. [695].

19.2.1 *General considerations*

In general, both electrons and phonons contribute to the thermal transport properties. In insulators heat is conducted by atomic vibrations, while in metals electrons are the dominant carriers, at least at low temperatures. In molecular junctions, both types of carriers exist and mutually interact. Therefore, a complete description of the thermal transport in these systems requires to take into account the energy transport due to both electrons and phonons, as well as the energy exchange between them due to the electron-phonon interaction. This problem is quite complicated and so far no realistic calculations have been performed taking into account all the ingredients mentioned above.

Even though practical applications can be difficult, a unified description of both heat generation and heat transport is in principle possible within the framework of the nonequilibrium Green's function formalism (NEGF). This formalism was first applied to thermal transport by Datta and coworkers [895, 896] and it has been extended by several groups to treat different aspects of this problem [897–900].

The analysis of heat transport can be greatly simplified, in particular, when electrons and phonons can be considered separately (e.g. when the electron-phonon interaction is negligible). For instance, in highly conductive molecular junctions the low-temperature thermal conductance is expected to be dominated by electrons. Assuming that the transport is coherent, the contribution of electrons can be computed within the scattering formalism, as we have shown in section 4.8. In this case, the heat current is simply determined by the (electronic) transmission coefficient and it is given by Eq. (4.84). In this sense, it can be computed from the usual methods for coherent transport like DFT-based ones.

In the case of junctions with a low electrical conductance, the dominant contribution to thermal transport comes from phonons (or vibrations). Ignoring anharmonic effects and the electron-phonon interaction, the heat current can be expressed in terms of a Landauer-like formula [901–903], where the phonon transmission can be determined using Green's function techniques analogous to those of the corresponding electronic problem, see e.g. Ref. [899].

19.2.2   *Thermal conductance*

We consider now the heat conduction through a molecular wire suspended between two reservoirs characterized by different temperatures. In particular, we shall focus here on situations where the heat transfer is dominated by phonons. The theoretical analysis of the transport of phonons and the corresponding thermal transport goes back to Peierls' early work [904]. In the recent years, it has become clear that the thermal properties of nanowires can be very different from the corresponding bulk properties. For example, Rego and Kirczenow [901] have shown theoretically that in the low temperature ballistic regime, the phonon thermal conductance of one-dimensional (1D) quantum wires is quantized in units of $\pi^2 k_B^2 T/3h$, where T is the temperature. This prediction was confirmed experimentally by Schwab *et al.* [905] in a nanofabricated 1D structure, which behaves essentially like a phonon waveguide.

An aspect that has attracted a lot of attention in the last decades is the validity of the macroscopic Fourier law of heat conduction in 1D systems [906–912]. The Fourier law is a relationship between the heat current J per unit area A and the temperature gradient ∇T

$$J/A = -\tilde{K}\nabla T, \tag{19.7}$$

where A is the cross-section area normal to the direction of heat propagation and $\tilde{K}$ is the thermal conductivity (the thermal conductance K is defined as $K = J/\nabla T$). In spite of all the work on this subject, there is yet no convincing and conclusive result about the validity of this law in 1D systems.

Another aspect that has been the subject of recent discussions is the possibility of having an asymmetry in the directionality of heat transfer. Several authors have proposed both classical and quantum-mechanical models that exhibit heat rectification [913–918]. In these models, rectification is usually associated with a non-linear (anharmonic) response.

From the experimental point of view, remarkable progress has been made in the last decade in nanoscale thermometry, and measurements on the scale of the mean free path of phonons and electrons are now possible. Using scanning thermal microscopy methods one can obtain the spatial temperature distribution of the sample surface, study local thermal properties of materials, and perform calorimetry at nanometric scale [894, 919, 920]. These advances have allowed, for instance, studying the thermal transport in single carbon nanotubes (see e.g. [921] and references therein) and establishing a quantitative comparison with the theory (see e.g. [922] and references therein).

Experimental work on thermal transport in molecular junctions is however very limited. The first thermal conductance measurements that we are aware of were reported by Wang *et al.* [923]. These authors studied solid-solid junctions with an interfacial self-assembled monolayer (SAM). To be precise, Au-SAM-GaAs junctions were made using alkanedithiol SAMs and fabricated by nanotransfer printing. Measurements of thermal conductance were very robust and no thermal conductance dependence on alkane chain length was observed. The thermal conductances using octanedithiol, nonanedithiol, and decanedithiol SAMs at room temperature were found to be 27.6 ± 2.9, 28.2 ± 1.8, and 25.6 ± 2.4 MW m^{-2} K^{-1}, respectively.

The thermal conductance of an alkanedithiol SAM anchored to a gold substrate was studied by ultrafast heating of the gold with a femtosecond laser pulse in Ref. [924]. It was found that when the heat reached the methyl groups at the chain ends, a nonlinear coherent vibrational spectroscopy technique detected the resulting thermally induced disorder. The flow of heat into the chains was limited by the interface conductance. The leading

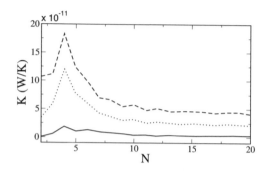

Fig. 19.4   Theoretical results for the heat transport coefficient (heat flux per unit T difference between hot and cold bath) displayed as a function of alkane bridge length, for a particular model of molecule-heat bath coupling at 50 K (full line), 300 K (dotted line) and 1000 K (dashed line). Reprinted with permission from [903]. Copyright 2003, American Institute of Physics.

edge of the heat burst traveled ballistically along the chains at a velocity of 1 kilometer per second. The molecular conductance per chain was 50 pW/K.

The thermal conductance of alkane-based junctions was indeed addressed theoretically by Segal *et al.* [903] a few years before the realization of the experiments mentioned above. These authors computed the phonon contribution to the heat current, which should be the dominant one in these low transmissive junctions. To be precise, they computed the heat flux for a harmonic molecule characterized by a set of normal modes and coupled through its end atoms to harmonic heat reservoirs. They have also performed classical mechanics simulations in order to assess the role played by anharmonicity. The general conclusions of this work are: (i) At room temperature and below, molecular anharmonicity is not an important factor in the heat transport properties of alkanes of length up to several tens of carbon atoms. (ii) At room temperature, the efficiency of heat transport by alkane chains decreases with chain size above 3-4 carbons, then saturates and becomes length independent for moderate sizes of up to a few tens of carbon atoms (this prediction agrees with the observations of Ref. [923] mentioned above). (iii) At low temperature, the heat transport efficiency increases with chain length. This is a quantum effect: at low temperatures only low frequency modes can be populated and contribute to phonon transport, however such modes are not supported by short molecules and become available only in longer ones. In Fig. 19.4 we reproduce the results for the thermal conductance of Segal *et al.* [903] that illustrate these conclusions.

To conclude this discussion, let us simply say that the investigation of heat transport in molecular junctions is still in its infancy both theoretically and experimentally and this is one of the issues in which a lot of progress is expected (and desired) in the next years.

19.2.3   *Heating and junction temperature*

In the context of thermal properties of molecular junctions, heating or heat generation is the most studied aspect from the experimental point of view. When an electrical current is driven through an atomic or molecular junction, there is a continuous energy transfer from the conduction electrons to the vibrational degrees of freedom that, loosely speaking, tends to increase the local temperature inside junction. This heating effect is partially alleviated by the conduction out of the junction via phonon thermal conduction. The balance between these two mechanisms determines the excess energy that is deposited in the phonon subsystem.

This energy transfer is usually described in terms of an effective local temperature. This is, of course, questionable since the system is out of equilibrium and the phonon distribution can differ significantly from that described by the Bose function. From a theoretical point of view, the effective temperature is sometimes defined by forcing the phonon occupation to adopt the form of the Bose function with an effective temperature, T_{eff}. Other definitions have been introduced and for a detailed discussion of this issue we refer to Refs. [894, 899].

From the experimental point of view, indirect information about the local temperature is obtained by measuring temperature-dependent properties like the switching rate between two different configurations in atomic sized contacts [925, 926], the fracture rate in atomic chains [927], the force required to break a molecule-electrode bond [928] or the distance over which molecular junctions can be stretched before breakdown [929]. The experimental data on the current-induced local heating are typically analyzed with the help of the theory of Todorov and coworkers [930, 267, 931]. This theory provides a simple estimate for the voltage dependence of the local effective temperature. In particular, it predicts that the temperature in the center of a general ballistic nanoscale junction of length L at a voltage V is given by

$$T_{eff} = (T_0^4 + T_V^4)^{1/4}, \tag{19.8}$$

where T_0 is the ambient temperature and $T_V = \gamma \sqrt{L|V|}$, where γ is a material-dependent parameter. For a typical metal contact, $\gamma = 60$ K $V^{-1/2}$ nm$^{-1/2}$. This means that with $L \approx 2$ nm and $V = 1.5$ V, $T_V \approx 100$ K. Three factors are taken into account in this estimate: (i) heating by the electrons due to creation of phonons in the junction, (ii) cooling by the electrons due to absorption of phonons, and (iii) cooling by the thermal transport of energy away from the contact into the metal reservoirs.

Eq. (19.8) was experimentally tested by Smit *et al.* [927]. These authors investigated the breaking mechanism of Au and Pt atomic chains as a function of the bias voltage. The chain breaking is a thermally activated process and the fracture rate contains information about the bias-dependent local temperature. An analysis of the data showed a reasonable agreement with the predictions based on Eq. (19.8). From this analysis, the authors could estimate the effective (lattice) temperature inside the atomic wire. It rises in proportion to the square root of the bias voltage for sufficiently high bias, and for a monatomic gold chain of length $L = 1$ nm at $V = 1$ V it reaches a temperature of 60 K, which is well above the bath temperature of 4.2 K in the experiments.

The derivation of Eq. (19.8) assumes a bulk T^3 law for heat capacity of the contact, which is an approximation only valid for temperatures well below the Debye temperature. Therefore, it should not be surprising to find deviations from the square-root dependence of Eq. (19.8) at elevated temperatures. In this respect, Tsutsui *et al.* [926] found the effective temperature of Zn atomic-sized contacts at 77 K rises more rapidly with bias than the $\sqrt{V}$ dependence of Eq. (19.8).

In the case of molecular junctions, Tao's group measured the local effective temperature. In this case, the authors studied the force required to break Au-octanedithiol-Au junctions under finite bias. The breakdown process is thermally activated, which can be used to extract the effective temperature. The data could be roughly fitted with Eq. (19.8). It was found that at a bias voltage of 1 V, the temperature of the junction is raised ~ 30 K above the ambient room temperature. Above this bias, the molecular junctions become increasingly unstable.

In another work of Tao's group, the effective temperature of single-molecule (n-alkanedithiol) junctions due to current-induced local heating was measured as a function of molecular length and applied bias voltage. In this case the method was based on analyzing the average stretching length over which a molecular junction can be stretched before breakdown, using the STM break-junction approach. By measuring the stretching length as a function of stretching rate and temperature, the authors showed that the breakdown of the molecular junctions is thermally activated and the dependence of the stretching length on temperature was used to extract the effective temperature of single-molecule junctions. They reported the following two notable findings. First, at a given bias, the local ionic heating increases with decreasing molecular length, in agreement with the theoretical predictions of Ref. [766] (see discussion below). Second, for a given molecule, the effective local temperature first increases with bias, and then decreases after reaching a maximum value at ~ 0.8 V. This is in agreement with a transport theory based on a hydrodynamic approach [933], which predicts that effective cooling of the temperature at high biases can occur due to electron-electron interaction with consequent local electron heating at the junction.

In attempts to go beyond the simple estimate of Eq. (19.8), Di Ventra and co-workers have reported quantitative calculations of the temperature rise in realistic models of atomic and molecular junctions [766, 932, 934]. These calculations are based on a microscopic description of the heat generation, while the heat conduction is estimated via a simplified approach. The following important observations based on these calculations have been made: (i) For the same voltage, the temperature rise in a benzenedithiol junction is considerably smaller than that of a gold wire of similar size because of the larger conduction (therefore higher current) in the latter. In absolute terms, the temperature rise is predicted to be about 15 and 130 K above ambient temperatures at a voltage bias of ~ 1 V [932]. (ii) In dithiolate alkane chains, the estimated temperature rise is a few tens degrees at 0.5 V and depends on the chain length, see Fig. 19.5. The temperature rise is smaller in longer chains characterized by smaller electrical conduction [766], which is in agreement with the experimental results of Ref. [929]. In this case, decreasing conduction with molecular lengths overshadows the less efficient heat dissipation in these systems. (iii) In contrast to alkanes, in Al wires the temperature rise in current carrying wires is more pronounced for longer chains [934]. In these good conductors the balance between the length effects on conduction and heat dissipation is tipped the opposite way from their molecular counterparts, because length dependence of conduction is relatively weak.

The main difficulty in the calculation of the effective temperature is the description of the energy transfer between the local vibrations and the phonons in the reservoirs. Some progress has been made in the last years, see e.g. Refs. [777, 935, 936], but the description of the phonon transport in molecular junctions has not yet reached the level of sophistication achieved

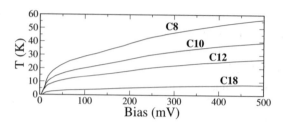

Fig. 19.5   Estimated junction temperature as a function of bias in alkane-dithiol junctions of various chain lengths. Reprinted with permission from [766]. Copyright 2005 American Chemical Society.

for the electron transport problem.

Recently, new experimental methods have been introduced to study the mechanisms of heating and heat dissipation induced by the flow of current across a single molecule. For instance, in the context of STM experiments, Schulze *et al.* [937] have used a method based on detecting the maximum power that one molecule can sustain. In particular, these authors used a low temperature STM to control the flow of electrons through a single C_{60} molecule at an increasing rate until the molecule decomposes. By comparing the power applied for decomposition of the molecule (P_{dec}) in the tunneling regime and in contact with the STM tip, they found that it depends significantly on two factors: (i) P_{dec} decreases when molecular resonances participate in the transport, evidencing that they enhance the heating; (ii) P_{dec} increases as the molecule is contacted to the source and drain electrodes, revealing the heat dissipation by phonon coupling to the leads. A good contact between the single-molecule device and the leads is hence an important requirement for its operation under large current densities.

Probably the most direct method to investigate local heating in molecular junctions has been reported by Ioffe *et al.* [938]. These authors have shown that the effective temperature of current-carrying junctions can be monitored with surface-enhanced Raman spectroscopy (SERS) that involves measuring both the Stokes and anti-Stokes components of the Raman scattering. The ratio of these two components for each Raman active vibrational mode gives direct information about its steady-state nonequilibrium population. This ratio can be translated into a mode-specific effective temperature [939]. In Ref. [938], Ag-SAMBPDT-Ag junctions were studied, where SAMBPDT stands for self-assembled monolayer of 4,4′-biphenyldithiol. In Fig. 19.6 we reproduce the results of this work for

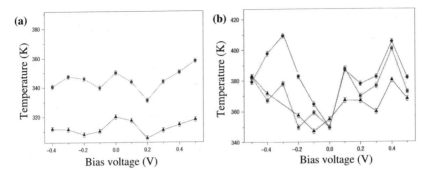

Fig. 19.6   Measurements using surface enhanced Raman spectroscopy of the mode-specific effective temperature $[T_{\text{eff}}(\nu)]$ as a function of bias for two representative Ag-SAMBPDT-Ag junctions. (a) Plot of $T_{\text{eff}}(\nu)$ as a function of bias voltage for a mode with 1,585 cm^{-1} (triangles) and 1,083 cm^{-1} (squares) modes (532 nm laser). (b) Plot of $T_{\text{eff}}(\nu)$ as a function of bias voltage for a mode with 1,585 cm^{-1} (triangles), 1,280 cm^{-1} (circles) and 1,083 cm^{-1} (squares) modes (671 nm laser). Reprinted by permission from Macmillan Publishers Ltd: Nature Nanotechnology [938], copyright 2008.

the effective temperature of several modes as a function of the voltage for two representative junctions. As one can see, the apparent dependence of the effective temperature on the applied bias reveals two types of behavior: (i) Between 0 and $\sim |0.2 \text{ V}|$ in both polarities an apparent cooling process is observed and (ii) at bias values higher than $|0.2 \text{ V}|$, heating of the vibrational modes takes place. As explained by the authors, these experimental results reveal a rich heating/cooling behavior that is inexplicable using existing models. The calculations of the bias dependence of heat dissipation in molecular junctions described above do not include the additional intermolecular dissipation channel prevailing in these monolayer junctions. The explanation of these experimental results constitutes at the moment an interesting open problem.

19.3   Thermoelectricity in molecular junctions

A property closely related to heat transport, namely the thermopower, has recently received considerable attention and it deserves a separate discussion. The thermopower (also called the Seebeck coefficient) of a material is a measure of the magnitude of an induced thermoelectric voltage in response to a temperature difference across that material.[4] Classically, an

---

[4] A brief discussion of thermoelectric phenomena in nanocontacts, including the Seebeck effect, can be found in section 4.8.

applied temperature difference causes charged carriers in the material to diffuse from the hot side to the cold side, similar to a gas that expands when heated.

Mobile charged carriers migrating to the cold side leave behind their oppositely charged and immobile nuclei at the hot side thus giving rise to a thermoelectric voltage.[5] Since a separation of charges also creates an electric field, the build-up of charged carriers on the cold side eventually ceases at some maximum value since there exists an equal amount of charge carriers drifting back to the hot side as a result of the electric field at equilibrium.

The thermopower of a (bulk) material, represented as S, depends on the material temperature and crystal structure. Typically, metals have small thermopower because most have half-filled bands. Electrons and holes both contribute to the induced thermoelectric voltage thus canceling each other's contribution to that voltage and making it small. In contrast, semiconductors can be doped with an excess amount of electrons or holes and thus can have large negative (for n-type materials) or positive values (for p-type materials) of the thermopower depending on the charge of the excess carriers. The sign of the thermopower can thus determine which charge carriers dominate the electric transport in both metals and semiconductors. This is one of the key ideas that makes the thermopower interesting for molecular electronics.

If the temperature difference ΔT between the terminals of a junction (or the two ends of a material) is small, then the thermopower of a material is conventionally defined as[6]

$$S = -\frac{\Delta V}{\Delta T}, \tag{19.9}$$

where ΔV is the thermoelectric voltage seen at the terminals. In general, there are two main contributions to the thermopower, namely an electronic one and the contribution of phonons, the so-called phonon drag[7] [940]. It has been argued that for point contacts (and in general for nanoconstric-

---

[5]Thermoelectric refers to the fact that the voltage is created by a temperature difference.

[6]Strictly speaking, this expression is only approximate. The numerator should be the difference in electrochemical potential divided by $-e$, not the electric potential, see Eq. (4.79). However, the chemical potential is often relatively constant as a function of temperature, so using electric potential alone is in these cases a very good approximation.

[7]Any thermal gradient gives rise to the transport of heat by the phonons, while an electric current, though carried by the electrons, cannot fail to transfer some of its momentum to the lattice vibration, and drag them along with it.

tions), the phonon drag contribution to the thermopower becomes negligible [941], which simplifies enormously the theoretical analysis of this property. As we have shown in section 4.8, if the transport is assumed to be phase-coherent (no inelastic scattering), the electronic contribution to the thermopower can be expressed in terms of the zero-bias transmission function $\tau(E)$ as [160]

$$S = \frac{1}{eT} \frac{\int_{-\infty}^{\infty} (E - \mu)\,\tau(E)\,[\partial f(E,T)/\partial E]\,dE}{\int_{-\infty}^{\infty} \tau(E)\,[\partial f(E,T)/\partial E]\,dE}, \qquad (19.10)$$

where $f(E,T) = \{\exp[(E - \mu)/k_B T] + 1\}^{-1}$ is the Fermi function and μ the chemical potential with $\mu \approx E_F$. From the numerator of Eq. (19.10), it is evident that a non-vanishing thermopower requires a certain electron-hole asymmetry in the transmission function. This asymmetry also determines the sign of this transport property.

At low temperatures, the leading-order term in the Sommerfeld expansion for the thermopower yields

$$S = -\frac{\pi^2 k_B^2 T}{3e} \frac{\tau'(E_F)}{\tau(E_F)}, \qquad (19.11)$$

where the prime denotes derivative with respect to energy. Let us remind that the linear conductance in this limit is given by $G = G_0 \tau(E_F)$.

As the in the case of shot noise, experiments in metallic atomic-sized contacts paved the way for the analysis of thermopower in molecular junctions. In 1999 Ludoph and van Ruitenbeek reported the first thermopower measurements in gold atomic contacts [942]. The principle of the measurement is illustrated in the left panel of Fig. 19.7. By applying a constant temperature difference over the contacts, the thermally induced potential could be measured simultaneously with the conductance. In this experiment, large thermopower values were obtained, which jump to new values simultaneously with the jumps in the conductance. The values are randomly distributed around zero with a roughly bell-shaped distribution, as one can see in the right panel of Fig. 19.7. Negative values of the thermopower are not expected in simple adiabatic models for point contacts [161, 163, 943]. Ludoph and van Ruitenbeek proposed a convincing interpretation in terms of coherent backscattering of the electrons with impurities near the contact [942]. As a result of the interference of waves with different path length, the transmission of the contact shows fluctuations as a function of energy, which according to Eq. (19.10) lead to a finite thermopower with a sign that can be either positive or negative. So in other

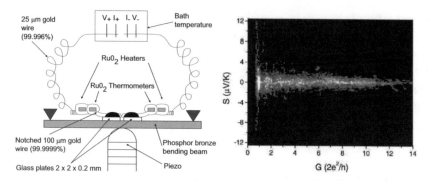

Fig. 19.7   Left panel: Schematic diagram of the modified MCBJ configuration, used for the simultaneous measurement of conductance and thermopower of metallic atomic contacts. Right panel: Density plot of thermopower of gold atomic contacts against conductance constructed from 220 breaking curves for for two samples. Black represents no data points and white more than 100. Reprinted with permission from [942]. Copyright 1999 by the American Physical Society.

words, the thermopower signal was shown to be dominantly of the same origin as the conductance fluctuations discussed in Chapter 11.

In the context of molecular junctions, Paulsson and Data stressed in a theoretical study the importance of measuring the thermopower [944]. They showed that in molecular contacts this transport property is large enough to be measured, it is rather insensitive to the detailed coupling to the contacts and it provides valuable information about the position of the Fermi energy relative to the molecular levels. Let us illustrate these ideas with a simple model. Following Ref. [944], let us assume that the transmission function exhibits a double-peak structure described by two independent Lorentzian:

$$\tau(E) = \sum_{i=1}^{2} \frac{4\Gamma_L \Gamma_R}{(E - \epsilon_i)^2 + (\Gamma_L + \Gamma_R)^2}, \tag{19.12}$$

were ϵ_i is the energy of the two levels, and Γ_L and Γ_R the broadenings by contacts L and R. For simplicity, we assume here that the broadenings are the same for both levels. Eq. (19.12) describes a typical situation that is realized in many organic molecules where the two levels typically correspond to the HOMO and LUMO of the molecule and the Fermi energy lies somewhere between them. The transmission function of Eq. (19.12) is illustrated in Fig. 19.8(a) for a case in which $\epsilon_1 = -7$ eV, $\epsilon_2 = -3$ eV and two values of the broadenings, 100 and 30 meV, which are assumed to be equal for both contacts. The exact value of the linear conductance depends

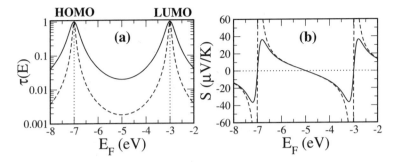

Fig. 19.8   (a) Transmission as a function of the Fermi energy computed from Eq. (19.12). The values of the parameter are: $\epsilon_1 = -7$ eV, $\epsilon_2 = -3$ eV and $\Gamma_L = \Gamma_R = 100$ meV (solid line) and $\Gamma_L = \Gamma_R = 30$ meV (dashed line). (b) Corresponding thermopower as a function of the Fermi energy calculated from Eqs. (19.12) and (19.11).

on the position of the Fermi level, which we leave as a free parameter. Using Eq. (19.11) one can compute the thermopower and the result is shown in Fig. 19.8(b). Notice that depending on the position of the Fermi energy with respect to molecular levels, the thermopower can be either positive or negative. If E_F is closer to the HOMO, the sign is positive and one talks about hole-dominated transport. If at the contrary, the LUMO is closer to E_F, then the thermopower is negative and one has electron-dominated transport. Notice also that the thermopower is in the range of $\mu V/K$ (or larger), like in the case of atomic contacts, and therefore it should be measurable. Finally, when E_F is not too close to one of the frontier orbitals, the thermopower is very similar for the two cases shown in Fig. 19.8(b), although the broadenings differ by a factor of three. Indeed, if the Fermi energy is located between the HOMO and LUMO and far away from them, it is easy to show from Eqs. (19.12) and (19.11) that, to first order, the thermopower is independent of the metal-molecule coupling [944].

The first experiment measuring the thermopower in single-molecule junctions was reported by Reddy *et al.* [101]. These authors used STM break-junctions to trap molecules between two gold electrodes with a temperature difference across them. In this way they were able to measure the thermopower (or Seebeck coefficient) of 1,4-benzenedithiol (BDT), 4,4'-dibenzenedithiol (DBDT), and 4,4''-tribenzenedithiol (TBDT) in contact with gold at room temperature and found the values $+8.7 \pm 2.1$ $\mu V/K$, $+12.9 \pm 2.2$ $\mu V/K$, and $+14.2 \pm 3.2$ $\mu V/K$, respectively. As explained above, the positive sign indicates p-type (hole) conduction in these heterojunctions, i.e. the transport is dominated by the HOMO of the molecules.

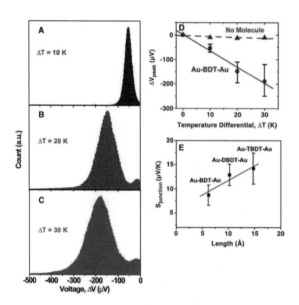

Fig. 19.9   Histograms obtained by analyzing approximately 1000 consecutive thermo-electric voltage curves obtained in measurements of Au-BDT-Au junctions with tip-substrate temperature differential (A) $\Delta T = 10$ K, (B) $\Delta T = 20$ K, and (C) $\Delta T = 30$ K. a.u., arbitrary units. (D) Plot of the peak values of the thermoelectric voltage in histograms as a function of the temperature differential. The error bars represent FWHM of the corresponding histograms. It can be seen that the measured voltage varies linearly with the temperature differential, as expected. (E) Plot of measured junction Seebeck coefficient as a function of molecular length for BDT, DBDT, and TBDT. From [101]. Reprinted with permission from AAAS.

It was also observed that S grows roughly linearly with the number N of the phenyl rings in the molecule. These results are illustrated in Fig. 19.9.

This pioneering experiment motivated new theoretical work on this subject. Thus for instance, Pauly *et al.* [546] presented an ab initio (DFT-based) study of the thermopower in metal-molecule-metal junctions made up of dithiolated oligophenylenes contacted to gold electrodes. It was found that, in agreement with the experiment, the transport is dominated by the HOMO of these molecules. Moreover, it was shown that while the conductance decays exponentially with increasing molecular length, the thermopower increases linearly as in the experiments of Ref. [101]. This is illustrated in Fig. 19.10, where the conductance and thermopower for oligophenylenes with up to 4 phenyl rings are shown in panel (c) and (d), respectively. Notice that the transmission functions for these molecules, see Fig. 19.10(a), resemble those obtained with the simple model of Eq. (19.12).

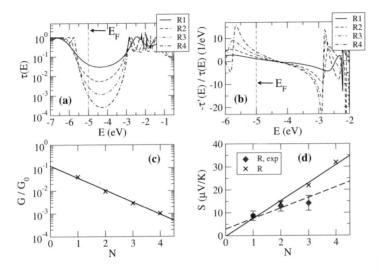

Fig. 19.10 Ab initio calculations for the conductance and thermopower of dithiolated oligophenylenes contacted to gold electrodes. N is the number of phenyl rings in the molecules. (a,b) Transmission function and the negative of its logarithmic derivative. (c,d) The corresponding conductance (G) and thermopower (S). The experimental data in (d) are form Ref. [101]. The straight lines are the best fits to the numerical results. Adapted with permission from [546]. Copyright 2008 by the American Physical Society.

As one can see in Fig. 19.10(b), there is a qualitative agreement with the experimental results which, taking into account the usual theory-experiment disagreement for the conductance, is certainly encouraging.

Pauly *et al.* also explained in simple terms the origin of the linear increase of the thermopower with the length of the molecules (see also Ref. [945]). As mentioned above, the transport in oligophenylenes proceeds through the tail of the HOMO and the off-resonant tunneling is reflected in the typical exponential decay of the linear conductance: $G/G_0 \sim e^{-\beta N}$, where N is the number of phenyl rings. This off-resonant transport is the origin of the linear increase in the thermopower. The idea goes as follows. Assuming that the transmission around $E = E_F$ is of the form $\tau(E) = \alpha(E)e^{-\beta(E)N}$, then Eq. (19.11) yields $S = S_C + \beta_S N$, where

$$S_C = -\frac{\pi^2 k_B^2 T}{3e}[\ln \alpha(E_F)]' \quad \text{and} \quad \beta_S = \frac{\pi^2 k_B^2 T}{3e}\beta'(E_F). \qquad (19.13)$$

It is important to notice that, while S_C depends on the prefactor $\alpha(E)$, β_S does not. Since $\alpha(E)$ contains the most significant uncertainties related to the contact geometries, one expects β_S to be described at a higher level of confidence than S_C. This linear dependence of the thermopower on molecular length has also been obtained in model calculations [222].

Another interesting suggestion of the work of Pauly *et al.* [546] is the idea that the thermopower can be tuned to a large extend by modifying the molecules with the inclusion of appropriate side-groups. In particular, in that work the introduction of methyl groups in the oligophenylenes molecules was shown to have a two-fold effect: (i) the substituents push the energies of the π electrons up as a result of their electron-donating behavior and (ii) they increase the tilt angles between the phenyl rings through steric repulsion. The latter effect tends to decrease both G and S due to a reduction of the degree of the π-electron delocalization, while the former opposes this tendency by bringing the HOMO closer to E_F, see Ref. [546] for further details.

Indeed, thermopower measurements were used by Majumdar's group [102] to elucidate the role of side-groups on the electronic structure and charge transport in molecular junctions. Again, this group used a STM break-junction technique to study the thermopower of several benzene derivatives. To be precise, 1,4-benzenedithiol (BDT) was modified by the addition of electron-withdrawing or -donating groups such as fluorine, chlorine, and methyl on the benzene ring. Moreover, the thiol end groups on BDT were replaced by cyanide end groups. It was observed that the thermopower of the molecular junction decreases for electron-withdrawing substituents (fluorine and chlorine) and increases for electron-donating substituents (methyl). The authors interpreted these results as follows. Electron-withdrawing groups remove electron density from the σ-orbital of the benzene ring allowing the rings high energy π-system to stabilize. Because the HOMO has a largely π-character, its energy is therefore decreased, shifting it further away from E_F. According to the simple model discussed above, such a shift results obviously in a decrease of the thermopower. Alternatively, the addition of electron-donating groups increases the σ-orbital electron density in the benzene ring, leading to an increase in the energy of the π-system and thereby shifting the HOMO closer to E_F. This shift causes in this case the enhancement of the thermopower. Finally, let us say that cyanide end groups were found to radically change transport relative to BDT. The thermopower in this case was found to be negative, which indicates that transport in 1,4-benzenedicyanide is dominated by the LUMO. For a recent theoretical study of thermopower of some of these molecules, see Ref. [946].

In yet another experiment of Majumdar's group, the alignment and coupling of the molecular orbitals with the states in the metal contacts were

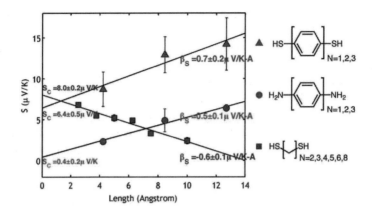

Fig. 19.11   Measurements of the Seebeck coefficient vs. molecular length for N-unit phenylenedithiols ($N = 1, 2, 3$), phenylenediamines ($N = 1, 2, 3$), and alkanedithiols ($N = 2, 3, 4, 5, 6, 8$). Fit lines to the data indicate that thermopower increases with length at a similar rate (β_S) for phenylenediamines and phenylenedithiols but decreases with length for alkanedithiols. Reprinted with permission from [947]. Copyright 2009 American Chemical Society.

investigated [947]. For this purpose, thermopower measurements were conducted for a series of phenylenes and alkanes with varying binding groups. As shown in Fig. 19.11, the thermopower increases linearly with length for phenylenediames and phenylenedithiols while it decreases linearly in alkanedithiols. The comparison between the two phenylenes series suggests that the molecular backbone determines the length dependence of S, while the binding group determines the zero length or contact S. Notice that for both thiol and amine end groups, the transport in phenylenes is dominated by the HOMO. Analyzing the data in terms of the model of Eq. (19.12), the authors concluded that for phenylenes the HOMO aligns closer to the Fermi energy of the contacts as L^{-1}, but becomes more decoupled from them as e^{-L}. Notice that this approximate behavior is reproduced by the ab initio results for the phenylenedithiols shown in Fig. 19.10(a). There, one can see that the HOMO shifts progressively towards the Fermi energy, while the corresponding resonance becomes narrower. The shift of the level can be traced back to the electronic structure of the isolated molecules. As shown in Fig. 2 of Ref. [546], the HOMO of the molecules moves closer to the gold Fermi energy as the number of phenyl rings increases.

The case of the alkanedithiols is more complicated to understand. As shown in Eq. (19.13), the linear coefficient β_S is determined by $\beta'(E_F)$, i.e. the derivative with respect to energy of the attenuation factor β at the Fermi energy. As shown in section 13.4, one can use simple bridge models to obtain an expression of the energy dependence of the β-factor. Assuming off-resonant tunneling, we obtained in section 13.4 the expression of Eq. (13.15) for β. Such a simple model predicts that a positive thermopower is accompanied by an increase with length. This explains the trend for the phenylenes, but not for the alkanes.[8] The authors of Ref. [947] suggested an explanation of this peculiar behavior in terms of gold-sulfur metal induced gap states residing between the HOMO and the LUMO. As we mentioned in section 14.1.2, several theoretical groups have concluded that the transport in alkanes can be influenced by states that originate from the hybridization of gold and sulfur orbitals with localized orbitals of the alkane chain [948–950]. Indeed, evidence of the existence of these states has been reported in STM experiments [951]. These hybrid states are localized at the interfaces and therefore, they are expected to have a major impact on the conductance for short molecules, while for long ones the transport is expected to be dominated by the HOMO of the alkane chains. The different length dependence of the metal induced gap states and the HOMO of the chains could be the origin of the decreasing thermopower [947].

As we already discussed in section 13.6, the transmission function of a molecular contact can exhibit lineshapes that completely differ from the double-peak structure shown in Fig. 19.8(a). One way to increase the thermopower is by "engineering" a much more pronounced energy dependence of the transmission function. As shown recently by Finch *et al.* [487], some molecules can exhibit sharp Fano resonances very close to the Fermi energy that in turn can lead to a huge thermopower in molecular junctions.

In the discussion so far we have focused on the thermopower in the coherent transport regime. However, this transport property can also provide very valuable information in many other transport regimes. For instance, Koch *et al.* [800] have shown theoretically that the thermopower of weakly coupled molecular junctions can give access to the electronic and vibrational excitation spectrum of the molecule even in a linear-response measurement.

To summarize, we have shown in this section that thermopower measurements in molecular junctions provide very important information not contained in the conductance. In this sense, we believe that measurements of this thermoelectric property will play a crucial role in the immediate future of molecular electronics.

---

[8]The simple bridge model of section 13.4 suggests that a decreasing S with length can only be obtained when the transport is dominated by the LUMO and therefore, the thermopower is negative.

Chapter 20

Optical properties of current-carrying molecular junctions

We have discussed so far different ways of controlling the current through a molecular junction such as gating or appropriate chemical synthesis. Another possibility is the use of an external electromagnetic field, which has been widely explored in larger mesoscopic structures [214]. In addition to controlling transport with external radiation, many other issues related to the optical properties of molecular junctions are of interest and some of them have been recently studied [211, 952]. In this sense, the goal of this chapter is to discuss the physical phenomena that emerge as a result of the interplay between current-carrying molecular junctions and an electromagnetic field.

The optical properties of molecular transport junctions involve many different aspects and it is certainly impossible to address all of them. Here, we shall focus our attention on the topics related to the following fundamental questions:

(1) Is it possible to use conventional optical spectroscopies to characterize molecular transport junctions?
(2) What is the effect of an electromagnetic radiation on molecular conduction?
(3) Can we use an external electromagnetic field to control the current or to learn something about the electronic structure of molecular contacts?
(4) What are the new transport phenomena than can be expected in ac driven molecular junctions?
(5) How does a molecular transport junction radiate?
(6) Different molecules have very peculiar and interesting optical properties. Can those molecules be used to design novel optoelectronic devices?

With those questions in mind, we have organized the rest of this chapter as follows. First, we shall describe recent experiments in which molecular junctions have been characterized using surface-enhanced Raman spectroscopy. Then, we shall discuss the physical mechanisms that are expected to play a major role in irradiated atomic and molecular junctions. In particular, we shall pay special attention to the so-called photon-assisted tunneling or current rectification.[1] Section 20.4 presents a description of some recent experimental results on the electronic transport through irradiated atomic and molecular contacts. In section 20.5 we shall briefly discuss the phenomenon of current amplification and other novel transport phenomena that have been predicted to appear in ac driven molecular junctions. Section 20.6 is devoted to the analysis of fluorescence of current carrying junctions. Finally, in section 20.7 we shall review some of the experiments in which the optical properties of certain molecules have been exploited to design primitive molecular optoelectronic devices.

20.1   Surface-enhanced Raman spectroscopy of molecular junctions

With respect to the first question posed above, it is obvious that combined optical and transport experiments on molecular transport junctions could reveal a wealth of additional information beyond that available from purely electronic measurements. It is, however, very challenging to use conventional optical spectroscopies to obtain local information about molecular junctions. First of all, it is not easy to inject light into slits of molecular size between two metal leads and second, the molecular emission may be strongly damped because of the proximity to a metal surface. Fortunately, recent work has shown that surface-enhanced Raman spectroscopy (SERS) can offer a way out of these problems [953–956]. The idea is based on the fact that metallic nanostructures, similar to those used to form the electrodes of molecular junctions, can act as effective plasmonic antennas, leading to a dramatic enhancement of the electric field locally at the junction region (see e.g. Ref. [957]). This enhanced field can then be used to perform Raman spectroscopy of objects placed in these nanogaps (for a review on SERS, see e.g. Ref. [958]). This idea has been explored recently in the context of molecular electronics, in particular, by Natelson's group

---

[1] These two terms are sometimes believed to refer to two different physical mechanisms. However, we shall show in section 20.3 that they are indeed identical.

in two works that we now proceed to describe [955, 956].[2]

This group has performed a series of optical experiments in Au nanogap structures prepared with the electromigration technique. The measurements were made with a confocal Raman microscope with a 785 nm diode laser at room temperature in air. The initial experiments of this group examined nanogaps as a potential SERS substrate [955], with paramercaptoaniline (pMA) as the molecule of interest. Following electromigration, the authors observed a SERS response strongly localized to the resulting gaps. Successive spectra measured directly over the SERS hot-spot revealed "blinking" and spectral diffusion, phenomena often associated with single-or few-molecule Raman sensitivity. Blinking was observed to occur when the Raman spectrum rapidly changed on the second timescale with the amplitudes of different modes changing independently of one another. Spectral shifts as large as ± 20 cm^{-1} were observed, making it difficult to directly compare SERS spectra with other published results. Blinking and spectral shifts are attributed to movement or rearrangement of the molecule relative to the metallic substrate. It is unlikely that an ensemble of molecules would experience the same rearrangements synchronously and thus blinking and wandering are expected to be observed only in situations where a few molecules are probed.

In a second experiment, the same group performed simultaneous SERS and transport measurements [956], including Raman microscope observations over the center of nanogap devices during electromigration. The molecules of interest, pMA or a fluorinated oligomer (FOPE), were assembled on the Au surface prior to electromigration. It was observed that once the resistance exceeds approximately 1 kΩ, SERS can be seen. This indicates that localized plasmon modes responsible for the large SERS enhancements may now be excited. As the gap further migrates the SERS response was seen to scale logarithmically with the device resistance until the resistance reaches approximately 1 MΩ. In most samples the Raman response and conduction of the nanogap became decoupled at this point with the conduction typically changing little while uncorrelated Raman blinking occurred.

In some devices, however, the Raman response and conduction showed very strong temporal correlations. A typical correlated SERS time spectrum and conductance measurement for a FOPE device are presented in Fig. 20.1. The temporal correlations between SERS and conduction are

---

[2]These experiments have been reviewed by the authors in Ref. [607].

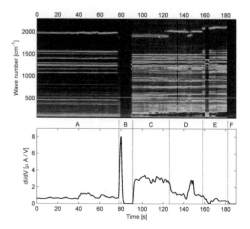

Fig. 20.1   The upper panel shows the Raman spectrum (1 s integration) for a single FOPE molecule in a gold junction formed by electromigration. The lower panel shows the correlated measurement of the conductance of this junction. The Raman mode observed between 1950 and 2122 cm^{-1} is believed to be for the same 2122 cm^{-1} mode associated with the C≡C stretch of the FOPE molecule. The large spectral shifts observed for this mode are attributed to interactions between the molecule and its nanogap environment. Clear correlations between the Raman structure and conductance can be seen. In particular in region B and for part of region E the Raman spectrum is observed to disappear while the conductance drops to zero. Reprinted with permission from [607]. Copyright 2008 IOP Publishing Ltd.

evident. Since the conduction in nanogaps is dominated by approximately a single molecular volume, the observed correlations between conductance and Raman measurements strongly indicate that the nanogaps have single-molecule Raman sensitivity. It is then possible to confirm that electronic transport is taking place through the molecule of interest, via the characteristic Raman spectrum. Data sets such as those of Fig. 20.1 contain implicitly an enormous amount of information about the configuration of the molecule in the junction.

Let us mention that Tian *et al.* [954] have used the MCBJ technique to study the intensity of the surface-enhanced Raman signal of molecular junctions. They showed that this signal depends critically on the separation of the electrodes and the incident light polarization. In particular, it was shown that when the incident laser polarization is along the two electrodes, the field in the nanogap is the strongest because of the coupling to the localized surface plasmon resonance of two gold electrodes [957].

It would be highly desirable to perform simultaneous measurements of

Raman and IET spectra, which, unfortunately, was not yet possible in the experiments just described. As we explained in the previous chapter, IETS requires cryogenic temperatures, at which it is not easy to operate a Raman microscope. However, in the experiment of Ioffe *et al.* [938], which was discussed in section 19.2.3, a comparison between vibrational modes revealed by Raman scattering and IETS could be established, although these two type of measurements were not performed at the same time. Let us remind that these authors reported SERS measurements of junctions based on biphenyldithiol SAMs[3]. In this experiment the Raman spectra were acquired at room temperature, while the IET spectra were obtained in transport measurements at 4 K. Interestingly, all Raman active vibrational modes were revealed in the IETS measurements, in spite of the fact that the selection/propensity rules are different in these two types of spectroscopies. Let us also recall that the main goal of this work was to measure the voltage dependence of the effective temperature of these current-carrying junctions. This was achieved by measuring both the Stokes and anti-Stokes components of the Raman scattering. Then, the effective temperature $T_{eff}(\nu)$ for each mode was calculated at each bias using the following expression[4]

$$\frac{I_{AS}}{I_S} = \frac{(\nu_L + \nu_\nu)^4}{(\nu_L - \nu_\nu)^4} \exp\left(-h\nu_\nu/k_B T_{eff}(\nu)\right) \frac{\sigma_{AS}}{\sigma_S} \frac{A_{AS}^2}{A_S^2}, \qquad (20.1)$$

where $I_{AS(S)}$ is the intensity of the anti-Stokes (Stokes) Raman mode, $\nu_{L(\nu)}$ is the frequency of the laser (Raman mode), $\sigma_{AS(S)}$ is the anti-Stokes (Stokes) scattering cross-section of the adsorbed molecules and $A_{AS(S)}$ is the average local field enhancement at the molecules at the anti-Stokes (Stokes) frequency. Strictly speaking, this expression is only valid in thermodynamical equilibrium and one may wonder whether this relation still holds at a finite bias voltage. For a discussion of this issue we refer the reader to Refs. [959, 960], where a detailed theoretical study of Raman scattering in current-carrying molecular junctions is presented.

20.2 Transport mechanisms in irradiated molecular junctions

A prerequisite to answer questions 2-4 in the list presented in the introduction, i.e. to understand how an electromagnetic field alters the electrical current of a molecular junction, is to identify the physical mechanisms that

---

[3]The use of SAMs facilitates the acquisition of Raman spectra.

[4]It was assumed that $\sigma_{AS}A_{AS}^2 = \sigma_S A_S^2$.

can play a role in this problem, which is by no means a trivial task. The theoretical and experimental work reported so far on this subject suggests that the main "suspects" are the following:

- *Current rectification or photon-assisted tunneling:* When an electrical contact is irradiated, the ac field may induce an alternating voltage in the junction with a frequency equal to that of the field. This ac bias in turn gives rise not only to an ac current, but also modifies the dc component. This phenomenon, in which an ac signal is converted to a dc current, is known as current rectification. The phenomenon is also known as photon-assisted tunneling (PAT) due to the nature of the inelastic tunneling processes that govern the electrical conduction in the presence of an ac bias (see discussion below).[5]

- *Internal molecular transitions:* While electronic transitions in the previous mechanism take place at the electrode-molecule interfaces, the radiation can also induce the standard optical transitions inside the molecules. This requires radiation frequencies comparable to the energy of the electronic excitations of the molecules, which are typically in the optical range. The induction of such transitions can in principle lead to phenomena like the resonant current amplification that will be discussed in section 20.5.

- *Hot electrons:* If the radiation frequency is close to the plasma frequency of the metal electrodes, the field can penetrate in the leads and excite the conduction electrons to high energy (hot electrons) . If these electrons are sufficiently close to the junction (closer than the inelastic mean free path at the corresponding energy), they can contribute significantly to the transport characteristics [961, 962].

- *Heating:* At optical frequencies a metal does not completely reflect the radiation and part of it can be absorbed. This absorption is usually accompanied by heating, which has several important consequences. First of all, heating can result in thermal expansion of the samples, which can be reflected in the junction current. This is well documented in the STM context [215], where a change in tip-sample distance due to thermal excitation has a dramatic effect on the tunneling current. On the other hand, heating can also create a temperature gradient

---

[5]Photon-assisted tunneling is maybe not a good name for this phenomenon since no real photons are emitted or absorbed in the tunneling processes. However, it is commonly used in the mesoscopic physics community, in which current rectification is rarely used. We shall use here both terms, but it must be clear that they refer to the same physical mechanism.

or a temperature difference between the electrodes. In both cases, thermoelectrical currents can appear in the junctions [215].

This list is not complete, and other effects can also play an important role. In particular, some readers might miss surface plasmons in this list. In this respect, we would like to say that for optical frequencies there is no doubt that plasmons play a key role. Surface plasmons are responsible for the local field distribution at the junction and, in particular, for its enhancement with respect to the incident field. In this sense, we can consider that plasmons determine the effective amplitude (and frequency dependence) of the ac potential induced in the junction, but the transport mechanism is still PAT or current rectification. In other words, we rather prefer to say that plasmons play an important role in the PAT mechanism than to say they constitute a different mechanism. After all, an ac field would also appear in the absence of plasmons and their role is only to modify the field distribution.

As we have already mentioned above, the importance of the different mechanisms depends primarily on the radiation frequency. For instance, in the microwave range PAT (or current rectification) largely dominates the transport. This has been firmly established in a great variety of mesoscopic structures [214] and more recently in atomic and molecular junctions (see discussion below). In the optical range, however, the other three mechanisms can also be very important.

20.3   Theory of photon-assisted tunneling

In the this section we shall present a description of the PAT theory for several reasons. First, this mechanism is likely to operate in almost any situation since when a junction is illuminated most of the radiation indeed impinges on the electrodes. Second, it is believed to be the dominant one at low frequencies and finally, recent experiments in atomic and molecular contacts seem to suggest that this mechanism is the dominant one even at optical frequencies. In what follows, we shall first present the basic theory of PAT and then, we shall discuss the basic predictions of this theory for atomic and molecular junctions.

20.3.1　*Basic theory*

In order to explain the steps in the current-voltage characteristics of microwave-irradiated superconductor-insulator-superconductor junctions [963], Tien and Gordon [213] proposed a heuristic theoretical treatment of electron tunneling in the presence of an ac field, which is of appealing simplicity. The central idea goes as follows.[6] First of all, the presence of the ac field is represented by a time-dependent voltage applied across the junction in addition to the dc bias[7]

$$V(t) = V + V_{ac}\cos(\omega t). \tag{20.2}$$

Indeed, Tien and Gordon assumed that this ac voltage is applied to one of the electrodes, while the other remains grounded. This applied voltage is assumed to modulate adiabatically the potential energy for each quasiparticle level on the ungrounded side of the barrier. This assumption is expected to be valid below the plasma frequencies of the two electrodes, typically well into the ultraviolet. The time dependence of the wave function for every single-electron state in the ungrounded electrode is therefore modified according to

$$\psi_i(x,t) = \psi_i(x)\exp\left[-\frac{i}{\hbar}\int^t dt'\,[E_i + eV(t')]\right] \tag{20.3}$$

$$= \psi_i(x)\exp\left[-i(E_i + eV)t/\hbar\right]\sum_{n=-\infty}^{\infty} J_n(\alpha)e^{-in\omega t},$$

where E_i is the unperturbed energy of the single-electron state, J_n is the Bessel function of the first kind (of order n) and $\alpha \equiv eV_{ac}/\hbar\omega$. The adiabatic modulation of the Fermi sea on this side of the junction can be thus viewed in terms of a probability amplitude $J_n(\alpha)$ for each quasiparticle level to be displaced in energy by $n\hbar\omega$. This interpretation is illustrated schematically in Fig. 20.2. Since all electron states are modulated together, these displacements in energy are equivalent to dc voltages $(V + n\hbar\omega/e)$ applied across the junction with a probability $J_n^2(\alpha)$ that depends upon the ac signal amplitude. The resulting dc tunneling current is, therefore, given by the expression

$$I(V;\alpha,\omega) = \sum_{n=-\infty}^{\infty} J_n^2(\alpha)I_0(V + n\hbar\omega/e), \tag{20.4}$$

---

[6]We present here an extension of the original Tien-Gordon argument due to Tucker [212].

[7]The dc part of the voltage will be simply denoted as V to follow the notation used so far. The total time-dependent voltage will always be denoted as $V(t)$, i.e. including explicitly the time argument.

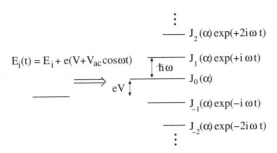

Fig. 20.2 Virtual energy levels generated according to Eq. (20.3) by adiabatic modulation of the energy $E_i(t)$ for each quasiparticle state within the ungrounded electrode of a junction in the presence of an ac field.

where $I_0(V)$ represents the current in the absence of radiation. This is a remarkable result that tell us that the I-V characteristics in the presence of radiation can be understood in terms of the I-V curves in the absence of external ac driving.

Notice that in this Tien-Gordon picture electrons undergo virtual transitions in the ungrounded electrode by "absorbing" or "emitting" an integer number of electromagnetic energy quanta $(n\hbar\omega)$ and then they tunnel elastically through the junction. In this sense, one can interpret the tunneling processes as photon-assisted events and this is the reason why the associated phenomenon is known as photon-assisted tunneling (PAT). Notice, however, that no "real" photons are involved in these processes and, in particular, the description of this phenomenon does not involve the quantization of the electromagnetic field. In this sense, the name PAT is maybe not very accurate, but since it is so commonly used, we shall also employ it here.

On the other hand, notice that Eq. (20.4) expresses how the dc current is modified by the radiation, i.e. it tells us how an ac signal results in a dc current. As we already mentioned above, this conversion process is known as current rectification and in this context, PAT and current rectification will be considered as synonyms.

A rigorous treatment of the electronic transport through an arbitrary junction subjected to an ac field can be done in the framework of different approaches such as the scattering formalism [165], the Floquet theory [211, 214] and the nonequilibrium Green's function formalism (NEGF) [214, 216–221]. This latter approach has been explained in detail in Chapter 8. In particular, we have shown there that the current through a junction in the presence of an ac voltage can be written in the form of the Tien-

Gordon formula of Eq. (20.4) under the following assumptions: (i) The energy dependence of the lead density of states is negligible and (ii) the ac potential does not vary spatially along the central part of the junction (i.e. across the molecule in the case of molecular junctions). While the first assumption is often justified, the second one may seem rather restrictive. However, Viljas *et al.* [221, 222] have shown that if the amplitude of the ac voltage is not too large, the precise shape of the profile does not play a crucial role.

In the derivation of Eq. (20.4) it was assumed that the ac voltage is applied to one of the electrodes, while the other is grounded. If we consider that the ac voltage is applied symmetrically, i.e. it drops equally at both interfaces, the current can be then written as

$$I(V; \alpha, \omega) = \sum_{n=-\infty}^{\infty} \left[J_n \left(\frac{\alpha}{2} \right) \right]^2 I_0(V + 2n\hbar\omega/e). \qquad (20.5)$$

Here, assuming that the transport in the absence of ac drive is elastic, the current I_0 is given by the standard Landauer formula. At low temperatures and in the linear response regime (vanishing dc bias), the conductance takes the particularly simple form

$$G(V = 0; \alpha, \omega) = G_0 \sum_{n=-\infty}^{\infty} \left[J_n \left(\frac{\alpha}{2} \right) \right]^2 \tau(E_F + n\hbar\omega), \qquad (20.6)$$

where $\tau(E)$ is the zero-bias equilibrium transmission and E_F is the Fermi energy. This quantity will be of special interest in our discussion below, and it will be referred to as *photoconductance*. Note that if the transmission does not depend on energy in the range probed by the inelastic processes, the conductance reduces to the conductance in the absence of drive, i.e. $G(V = 0; \alpha, \omega) = G_0\tau(E_F)$.[8]

It is interesting to consider the limit of small ac amplitudes. Using $J_0(x) \approx x^2/4$ and $J_{\pm n}(x) \approx (\pm x/2)^n/n!$ in Eq. (20.5) and retaining only the lowest-order terms in the ac potential V_{ac}, the correction to the current in this limit can be expressed as

$$\Delta I_{dc}(V; \alpha, \omega) \equiv I(V; \alpha, \omega) - I_0(V) \qquad (20.7)$$
$$= \frac{1}{4} V_{ac}^2 \left[\frac{I_0(V + 2\hbar\omega/e) - 2I_0(V) + I_0(V - 2\hbar\omega/e)}{(2\hbar\omega/e)^2} \right].$$

The quantity in large parentheses is a finite second difference of the I-V characteristics in the absence of radiation that reflects the emission or

---

[8]This can be easily shown using the relation $\sum_n J_n^2(x) = 1$.

absorption of a single quantum during the tunneling. All higher-order processes $n = 2, 3, \ldots$, contributing to the dc current may be neglected in the limit of small ac amplitude. When the photon energy $\hbar\omega/e$ is smaller than the voltage scale of the dc nonlinearity, the finite difference can be replaced by the second derivative of the current, i.e.

$$\Delta I_{dc}(V; \alpha, \omega) \approx \frac{1}{4} V_{ac}^2 \left(\frac{d^2 I_0}{dV^2} \right). \tag{20.8}$$

This expression reproduces the classical result for rectification [212] and illustrates the connection between PAT and current rectification. This is a very important relation since it provides a direct way to test whether the dominant transport mechanism is indeed PAT/current rectification. Such test requires to measure independently the induced dc current and the second derivative of the current with respect to the bias in the absence of radiation. Moreover, according to Eq. (20.8), the ratio of these two quantities gives the amplitude of the ac bias, which is typically unknown. This amplitude gives information about the field enhancement locally in the junction region.

Notice that Eq. (20.8) suggests that if the I-V characteristics in the absence of radiation exhibit an asymmetry at vanishingly bias voltage due to material and/or geometrical asymmetries (i.e. if $d^2 I_0/dV^2 \neq 0$ at $V = 0$), a radiation-induced current can flow in the system even in the absence of any dc bias voltage. This phenomenon of rectification at zero dc bias voltage was predicted by Cutler *et al.* in 1987 [964] and it was first reported by Walther's group in 1991 in laser-driven STM experiments on graphite surfaces [965] (for a detailed discussion of this phenomenon, see the review of Ref. [215]). The current generated by the ac field in the absence of dc bias is often referred to as *photocurrent*.[9]

The classical expression of Eq. (20.8) is probably valid in a wide range of molecular junctions for microwave frequencies, while in the optical range significant deviations from this expression are likely to appear and it has to be replaced by its quantum version of Eq. (20.7) [see Exercise 20.2(ii)]. On the other hand, it is interesting to derive similar expressions for the linear conductance. Defining the induced linear conductance correction as $\Delta G_{dc}(\alpha, \omega) \equiv G(V = 0; \alpha, \omega) - G(\omega = 0)$, where $G(\omega = 0) = G = G_0 \tau(E_F)$,

---

[9]Later in this chapter we shall discuss the so-called ratchet effect in molecular junctions, which is just another name for rectification at zero bias.

the relative correction for small ac amplitudes becomes

$$\frac{\Delta G_{dc}(\alpha, \omega)}{G} = \frac{(eV_{ac})^2}{16\tau(E_{\mathrm{F}})} \left[\frac{\tau(E_{\mathrm{F}} + \hbar\omega) - 2\tau(E_{\mathrm{F}}) + \tau(E_{\mathrm{F}} - \hbar\omega)}{(\hbar\omega)^2} \right]$$

$$\approx \frac{(eV_{ac})^2}{16} \frac{\tau''(E_{\mathrm{F}})}{\tau(E_{\mathrm{F}})}, \tag{20.9}$$

where the last expression has been obtained assuming that $\hbar\omega/e$ is smaller than the energy scale over which the transmission varies significantly. We thus see that in this limit ΔG_{dc} gives experimental access to the second derivative of the transmission function at the Fermi energy.

In the next section we shall review recent transport experiments performed with irradiated atomic and molecular junctions. In this sense, it is interesting to know the basic predictions of PAT theory for these systems. This is addressed in the next two subsections.

20.3.2   *Theory of PAT in atomic contacts*

In metallic atomic-sized contacts, the I-V curves in the absence of radiation are rather linear up to voltages of the order of 0.5-1.0 V. This is a consequence of the fact that the transmission does not change significantly around the Fermi level in an energy window of a few tenths of eV. According to Eqs. (20.8) and (20.9), this suggests that no significant changes in the transport characteristics are expected under irradiation up to frequencies close to the optical range.

As a side remark, let us say that in the superconducting state, atomic contacts are very sensitive to microwave frequencies. The reason is the presence of a gap in the spectrum, which ranges from 0.1 to 1 meV depending on the material. Recently, the subgap transport in superconducting atomic contacts under microwave irradiation was studied experimentally [966]. It was found that the subharmonic gap structure in the dc current is strongly modified in quantitative agreement with the theory of photon-assisted multiple Andreev reflections [967]. The importance of these results for our discussion here is that they provide firm support for the PAT mechanism in the microwave range.

A detailed theoretical study of PAT in atomic contacts has been reported by Viljas and one of the authors [221]. In this work the NEGF formalism of section 8.3 was used to compute the photoconductance as a function of frequency in one-atom thick contacts of several metals (Au, Pt and Al). In Fig. 20.3 we show an example of the results for a dimer con-

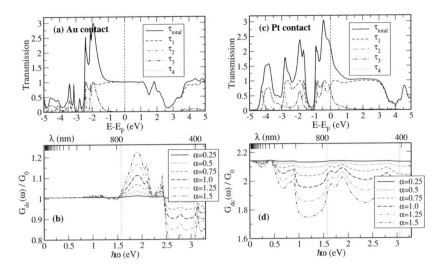

Fig. 20.3   Theoretical results for the photoconductance of Au and Pt atomic contacts. (a) Equilibrium transmission τ_{total} and its decomposition into conduction channels $\tau_{1,2,3,4}$ for an Au dimer contact. (b) Zero-temperature photoconductance for several values of α as a function of frequency ω computed using Eq. (20.6). In (b) the wavelengths λ with a tick spacing of 400 nm are shown. The range of visible light is indicated by vertical dotted lines. (c-d) The same as in panels (a-b) but for a Pt contact. Adapted with permission from [221]. Copyright 2007 by the American Physical Society.

tact[10] of Au and Pt. Here, we just reproduce the results obtained with the simple approximation of Eq. (20.6), which was usually found to reproduce qualitatively the more rigorous results obtained with the NEGF formalism [221]. In the case of Au, as can be seen in Fig. 20.3(a), the conductance for $\omega = 0$ is equal to $1\,G_0$ with a single open channel arising from the contribution of the $6s$ orbitals. Moreover, notice that the transmission around E_F is very flat. Due to this flatness, for frequencies up to $\hbar\omega \approx 1.5$ eV ($\lambda \approx 827$ nm) the effect of radiation is practically negligible. In the red part of the visible range ($\hbar\omega \lesssim 2$ eV) $\Delta G_{dc} > 0$ and it can reach up to 20% depending on the value of α.[11] This increase in the conductance is due to the contribution of the $5d$ bands located 2 eV below E_F, where the number of open transmission channels is higher than at E_F.

---

[10]The exact geometry of this dimer contact can be seen in Fig. 1 of Ref. [221]. This type of geometry is typically responsible for the last conductance plateau in the breaking process of an atomic contact.

[11]It is important to remark that for the case of Au it was found that the results were quite sensitive to the exact profile of the ac voltage. For the other metals the profile did not play a major role.

Fig 20.3(c-d) show the corresponding results for Pt. In the absence of radiation the conductance is close to $2.1\,G_0$ due to the contributions of mainly three conduction channels, which originate from the $6s$ and $5d$ orbitals. In this case, and in general in the contact regime for Pt, the effect of the radiation is always a significant reduction in conductance. This is understandable, since E_F lies at the edge of the d band, and photon absorption leads to an energy region where less open transmission channels are available and τ_{total} is smaller.

To conclude, the key message is that the photoconductance simply reflects the energy dependence of the transmission of the contacts. The sign of the induced correction can be both positive (like for Au) and negative (like for Pt) depending on the material. With respect to the order of magnitude of the correction, it can reach up to 50%-100% in some special cases depending on the geometry, frequency and power of the radiation, but it is usually below those values.

20.3.3 *Theory of PAT in molecular junctions*

From the discussion above, it is obvious that irradiation can lead to more dramatic effects in the case of molecular junctions. These junctions typically exhibit a much more pronounced energy dependence of the transmission function, which can lead to much larger modifications of the transport characteristics than in atomic contacts. In section 8.3.1 we have used the resonant tunneling model to illustrate some of the effects that may be expected from PAT in molecular junctions. The most prominent one is the resonant enhancement of the photoconductance. The idea is the following. The transmission function of most molecular junctions exhibit a deep pseudo-gap in the energy region between the HOMO and LUMO of the molecule, while the Fermi level lies somewhere in between. Then, if the photon energy is equal to the distance between the Fermi energy of the closest frontier orbital, the low-bias conductance can be greatly enhanced. This fact together with other interesting predictions are further illustrated in Exercise 20.2 at the end of the chapter with the use of the double-Lorentzian transmission function that we employed in section 19.3 to understand the thermopower of molecular junctions.

More realistic models of PAT in molecular junctions confirm these conclusions [968, 969]. Thus for instance, Viljas *et al.* [970] have reported a study of the photoconductance in organic single-molecule contacts. This study is based on Eq. (20.6), whereas the equilibrium transmission was computed using a DFT-based method. It was found that the radiation can indeed lead to large enhancements of the conductance of such contacts by bringing off-resonant levels into resonance through photon-assisted processes. The conductance enhancement was demonstrated for oligophenylene molecules between gold electrodes. It was shown that the exponential decay of the conductance with the length of the molecule can be replaced by a length-independent value in the presence of radiation. In other words, the photon-assisted processes turn the off-resonant tunneling into on-resonance transport. Results of this work are reproduced in Fig. 20.4. Notice first that the transmission in the absence of radiation exhibits a pseudo-gap in the region between the HOMO and LUMO (see panel (a) in the right figure). The HOMO lies closer to the Fermi energy in this case ($\sim$1 eV away). Second, in all of cases the low-bias conductance is greatly enhanced (for still reasonable values of α) and the onset of the enhancements is well inside the infrared region of the electromagnetic spectrum. Finally, panel (f) of the right figure shows how the typical exponential decay of the conductance with length is replaced by constant conductance in the presence of the radiation.

The fact that the conductance enhancement takes place in this case in the infrared region has important consequences. First, at these frequencies no internal transitions inside the molecules are possible. Second, metals do not absorb in the infrared and thus the associated heating effects are minimal. Therefore, this mechanism for enhanced photoconductance should be quite robust and, in principle, it could take place for a great variety of molecules since the only requirement is the existence of a pronounced gap in the transmission function. In this sense, molecular junctions can behave similarly to superconductor-insulator-superconductor systems, which have been used as microwave detectors in a variety of applications [212].

To conclude this section, let us also say that Viljas *et al.* [222] have studied the photon-assisted tunneling in more detail using simplified models and they have made further predictions that can be used as fingerprints of the PAT mechanism. First, in off-resonant situations, where the conductance in the absence of radiation decays exponentially with the length of the molecule, the correction to the dc linear conductance grows as the length square, i.e. the conductance enhancement is more pronounced for longer molecules. Second, at low frequencies additional steps can appear in I-V characteristics. Their separation, in the case of a symmetric junction, is roughly $2\hbar\omega$. For a discussion of the origin of these steps see section 8.3.1 or Exercise 20.2(iv) at the end of the chapter.

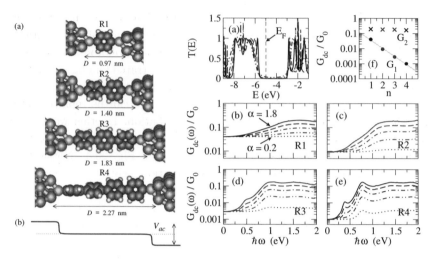

Fig. 20.4 Calculations of the photoconductance of oligophenylene-based single-molecule junctions. *Left figure:* (a) The four studied molecular contacts R1–R4, containing oligophenylenes with one to four phenyl rings and coupled to Au [111] pyramids through sulfur atoms. (b) In the calculation it was assumed the induced ac voltage V_{ac} to drop in a double-step manner. *Right figure:* (a) Transmission versus energy $[T(E)]$ for the contacts R1–R4 (dash-dot-dotted, dash-dotted, dashed, and solid lines, respectively). (b)–(e) The photoconductance versus external frequency ω for the contacts R1–R4, respectively. For each case the results for the following values of α are shown: 0.2, 0.6, 1.0., 1.4, and 1.8, in order of increasing conductance. (f) The dc conductances in the absence (G_1, dots) and presence (G_2, crosses) of radiation with $\hbar\omega = 1.5$ eV and $\alpha = 1.8$ for an increasing number n of phenyl rings. The gray line is a fit of the G_1 results to an exponential law. Reprinted with permission from [970]. Copyright 2007 by the American Physical Society.

20.4 Experiments on radiation-induced transport in atomic and molecular junctions

The experimental study of the electronic transport in irradiated atomic-scale junctions started around 20 years ago in the context of STM. In particular, in the early 1990's there was an intense activity related to the study of rectification (for a review see Ref. [215]). Although the observation of radiation-induced dc currents in STM experiments has been reported by many groups, it has always been difficult to show unambiguously that those currents were due to rectification and not to other mechanisms like, for instance, the generation of thermocurrents.

A convincing evidence of atomic-scale rectification was reported by Ho's group in 2006 [971]. This group presented STM experiments in which microwave-induced dc currents were measured for Mn atoms and MnCO molecules adsorbed on NiAl(110). The frequency of the microwave signal was 800 MHz. In Fig. 20.5(a) one can see results from this work for the differential conductance measured through a single Mn atom as a function of the bias voltage. In this figure V_B denotes the dc voltage, V_J is the local amplitude of the ac voltage and V_{IN} denotes the ac signal far away from the contact. In the absence of microwave ($V_{IN} = 0$), the differential conductance has a narrow peak at 2.0 V and a broad peak at 1.3 V, associated with the spin splitting of Mn sp states from magnetic interaction with its d electrons [971]. As V_{IN} is increased, the peak at 2.0 V becomes broader and eventually splits into two. In order to find out whether this modification of the dc current was due to rectification, the authors fitted the results with a classical rectification formula where the dc current is given by [971]

$$I(V_B, V_J) = \frac{\omega}{2\pi} \int_0^{2\pi/\omega} I\left[V_B + \sqrt{2}V_J \cos(\omega t)\right] dt. \qquad (20.10)$$

Here, the function $I(V)$ in the integrand has the same form as the static I-V characteristics measured in the absence of microwave, but with a time dependent argument. An example of the fits is shown in Fig. 20.5(b) for $V_{IN} = 5$ mV. Notice the high accuracy of the fit, which provides a strong support for the interpretation of the results in terms of rectification. Moreover, from the fits the value of V_J could be extracted. A plot of V_J versus V_{IN} is shown in Fig. 20.5(c), yielding a slope 45.5 from the best linear fit. This slope gives a direct information about the field enhancement at the contact.

On the other hand, it was also shown that the induced dc current as a function of voltage for a single Mn atom follows closely the d^2I/dV^2 spectrum in the absence of microwaves (see Fig. 2 of Ref. [971])[12]. This can be understood from Eq. (20.8), which tells us that the correction to the dc current is proportional to the d^2I/dV^2 spectrum without radiation. Notice that such relation can also be derived from Eq. (20.10) in the limit of small V_J by expanding the integrand up to second order in V_J. The close relation between the induced dc current and the d^2I/dV^2 spectra was also found in the case of transport through individual MnCO molecules, which constitutes a convincing proof of the fact that the rectification mechanism dominates the transport in irradiated atomic-scale junctions at microwave frequencies.

---

[12]The main difference between this experiment and previously reported ones was the use of low temperatures ($\sim$ 18 K) that made possible to measure directly d^2I/dV^2 spectra.

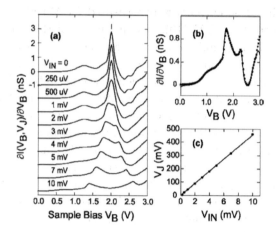

Fig. 20.5   Differential conductance spectra of a single Mn atom adsorbed on NiAl(110) surface with constant microwave input (no amplitude modulation). (b) The spectrum with $V_{IN} = 5$ mV was fitted numerically (line) to extract V_J, the microwave amplitude across the STM junction. (c) Plot of extracted V_J vs V_{IN}. The line is a linear fit with the constraint of zero intercept, which yields $V_J = 45.5 \times V_{IN}$. Reprinted with permission from [971]. Copyright 2006, American Institute of Physics.

Let us emphasize that the occurrence of PAT (or rectification) in the microwave regime has been firmly established in a variety of nanostructures [214], including carbon nanotube quantum dots [972], which are very closely related to our systems of interest.

Let us turn now to experiments in the optical regime. Recently, two different experiments have explored the influence of laser light on the transport through gold atomic contacts. The first one has been performed by the group of one of the authors [973]. In this case, the microfabricated version of the MCBJ technique was employed to fabricate gold atomic-sized contacts at room temperature. As a light source an argon-krypton cw laser was used, which allows to select a wavelength in the range between 480 nm and 650 nm. Moreover, pulsed light was used (with pulse durations of ~ 700 μs) to avoid irreversible deformations of the atomic junctions.[13] The conductance with and without light was measured simultaneously during the opening and closing of the atomic bridges. Fig. 20.6 shows an example of the results obtained for green light with $\lambda = 515$ nm. Notice that the

---

[13]It was found that continuous irradiation of the devices with $\lambda = 488$ nm for several seconds with a power of a few mW results in irreversible conductance changes.

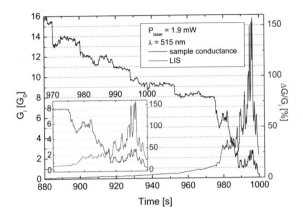

Fig. 20.6 Conductance and light-induced relative conductance change $\Delta G/G_i$ versus time when opening the break-junction continuously. Inset: close-up of the few-atom region with $G_i < 9\,G_0$. Reprinted with permission from [973]. Copyright 2007 by the American Physical Society.

relative change[14] in the conductance induced by the light is positive and it can reach up to more than 100% for $G \approx 2\,G_0$.

A statistical analysis revealed the following important findings: (i) Illumination always results in an enhancement of the conductance ($\Delta G > 0$), (ii) ΔG is usually smallest in the tunnel regime (when the contacts are broken) and (iii) ΔG depends very much on the wavelength of the light, the size and geometry of the contact and even on the exact spot in which laser light is focused on. In particular, the largest enhancements were found for $\lambda = 488$ nm with ΔG reaching up to 200%. However, for longer wavelengths the enhancements were typically below 20-30%. Findings (i) and (ii) rule out thermal expansion as a dominant mechanism in these experiments. On the other hand, the fact that $\Delta G > 0$ and the order of magnitude of the light-induced correction are compatible with the PAT mechanism explained in the previous section [221]. However, a quantitative comparison was not possible because the theoretical analysis of Ref. [221] focused on the case of single-atom contacts, which are not easy to stabilize at room temperature with the MCBJ technique.

Such a quantitative comparison with the theory has been done by Ittah *et al.* [974]. This group has developed a new method to form atomic contacts

---

[14]This relative change is defined as $\Delta G \equiv (G_f - G_i)/G_i$, where G_f is the conductance under illumination and G_i the conductance without light.

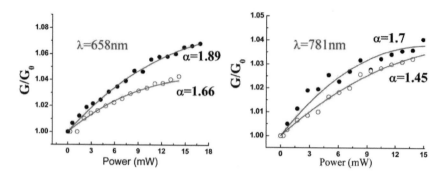

Fig. 20.7   Conductance measurements in $1\,G_0$ gold contacts under laser irradiation. A comparison between the conductance-enhancement as a function of laser power for two contacts (marked by dark and open circles) under irradiation with two wavelengths. The solid lines correspond to a fit with Eq. (20.6) using α as an adjustable parameter (see text). Adapted with permission from [974]. Copyright 2009 American Chemical Society.

in which the gold wires are fully anchored onto Si/SiO_2 substrates [975]. As a result the atomic junctions are mechanically highly stable even at room temperature, and under irradiation their heat dissipation characteristics are far more efficient than those of suspended MCBJs, resulting in only residual heating. Thanks to this method the authors were able to carry out a detailed study of the influence of laser light in the conductance of gold atomic contacts with conductances equal to $1\,G_0$.

In this work, the junctions were irradiated with three different lasers with wavelengths of 532 nm (2.33 eV), 658 nm (1.88 eV), 781 nm (1.58 eV). The maximum used power of the lasers was $\sim$ 20 mW, all measurements were performed under ambient conditions at room temperature and the junctions were placed with their long axis parallel to the laser polarization. Fig. 20.7 shows representative results of the conductance as a function of laser intensity for two different contacts and two different wavelengths. Notice that in the absence of light the conductance (measured at 30 mV) is $\sim 1\,G_0$. In all cases the conductance is enhanced by laser irradiation and the relative changes, which increase with decreasing wavelength, are below 10%. In order to establish a comparison with the results of PAT theory, Eq. (20.6) was used with the transmission curve of Fig. 20.3(a). The results for different values of $\alpha = eV_{ac}/\hbar\omega$ are shown in Fig. 20.3(b). Using α as an adjustable parameter the authors were able to fit the experimental results with a reasonable accuracy, see solid lines in Fig. 20.7. Notice in particular the nonlinear behavior, which is a remnant of the Bessel functions of Eq. (20.6).

These results suggest that the irradiated transport at the wavelengths in Fig. 20.7 is dominated by the PAT mechanism. However, for the 532 nm (2.33 eV) laser, deviations from the PAT theory were found. This is understandable since at this wavelength the absorption of Au is not negligible anymore (reflectance ~ 0.64 to be compared with 0.97 and 0.95 for Au at 781 and 658 nm, respectively). In this case, the enhancement of the conductance could additionally be due to the generation of hot electrons (see discussion in section 20.2). Such mechanism was termed photoinduced transport (PIT) in Ref. [974], where the authors presented a theoretical estimate of the contribution of photo-excited electrons (hot electrons) to the conductance and they showed that it could account for the discrepancy with PAT theory. Further support for the important role of PIT under the 532 nm laser was the finding of a linear dependence of the conductance on the laser power, which is expected from this mechanism.

In addition to the SERS experiments described in section 20.1, Natelson's group reported transport experiments in which significant dc currents in electromigrated molecular junctions under illumination were observed using different molecules such as para-mercaptoaniline (pMA) and fluorinated oligomers (FOPE) [607]. According to the authors, their observations are consistent with rectification at optical frequencies. Let us briefly repeat the arguments of Ref. [607]. In the presence of an oscillating potential $V(t) = V + V_{ac}\cos(\omega t)$ the current at small ac amplitudes can be written via a Taylor expansion as

$$I(t) = I(V_0) + \left(\frac{\partial I}{\partial V}\right)_V V_{ac}\cos(\omega t) + \frac{1}{2}\left(\frac{\partial^2 I}{\partial V^2}\right)_V V_{ac}^2\cos^2(\omega t) + \cdots .$$
(20.11)

Applying the trigonometric identity $2\cos^2(\omega t) = 1 + \cos(2\omega t)$, we see that the current nonlinearities lead to a second-harmonic ac signal as well as an additional dc current, both linearly proportional to $\partial^2 I/\partial V^2$. Notice that the expression of the additional dc current was already obtained in Eq. (20.8). This latter relation suggests that a comparison between measurements of the ac current at 2ω and the correction of the dc current could be used to test the occurrence of the rectification mechanism. In the case of ideal rectification, Eq. (20.11) tell us that the ratio of those two currents should be equal to one. Obviously, the current at 2ω can only be measured at frequencies much lower than the optical laser frequencies (e.g. 200 Hz), which weakens the usefulness of this test.

The measurements showed that the dc current under illumination changes proportionally to the low frequency ac current at 2ω, which is an indication of the occurrence of rectification, although it does not exactly follows the dependence expressed in Eq. (20.11). On the other hand, the induced dc currents at optical frequencies were found to depend linearly on the incident intensity, which is again consistent with the optical rectification mechanism. However, that this linear dependence (at small ac amplitudes) is also expected from other mechanisms like the PIT mentioned above. Therefore, the interpretation of these experimental results in terms of rectification is not unambiguous.

The same group repeated similar experiments but this time at lower temperature (80 K) and with radio frequencies (10 MHz). In this case it was possible to measure directly the second derivative of the current in the absence of radiation and it was shown that the rectified current followed exactly Eq. (20.8) or Eq. (20.11). This is again a demonstration that at low frequencies the rectification mechanism dominates the transport.

The last experiment that we shall mention in this section has been performed by Ho's group [976]. In this case the laser-assisted transport was studied in a single-molecule double-barrier junction that was defined by positioning a STM tip over an individual molecule adsorbed on a thin ($\sim$ 0.5 nm) insulating alumina film grown on a NiAl(110) surface. The two tunnel barriers in the junction are the vacuum gap between the STM tip and the molecule, and the oxide film between the molecule and NiAl. In this case the target molecule was a magnesium porphine (MgP), a simple metalloporphyrin molecule that is involved in photosynthesis, and the experiments were conducted at low temperatures ($\sim$ 10 K). In the absence of laser illumination, the differential tunneling conductance (dI/dV) spectra were shown to exhibit stepwise changes (with well-defined threshold voltages) and hysteresis. Upon illumination with three different lasers (532, 633 and 800 nm) the threshold voltage was shown to decrease linearly with the photon energy, suggesting a resonant mechanism. The authors argued that transport mechanism responsible for this behavior is a two-step process involving excited states of the tip (i.e. photo-induced resonant tunneling), as opposed to photon-assisted tunneling resonant tunneling from the tip directly to the molecule. In our opinion, without a quantitative analysis it is not easy to discriminate between these two different mechanism since both of them may lead to similar features in the current-voltage characteristics.

20.5 Resonant current amplification and other transport phenomena in ac driven molecular junctions

As it is clear from the previous section, photon-assisted tunneling is a crucial transport mechanism in atomic-scale junctions even at optical frequencies. However, it must be emphasized again that the PAT theory in the spirit of Tien-Gordon original work has clear limitations and it is only valid when the spatial dependence of the field-matter interaction along the junction can be ignored. If such spatial dependence becomes important, which must be always the case for large ac field amplitudes, the problem has to be addressed with other formalisms such as the so-called Floquet theory (see Ref. [211] for a review) or the NEGF approach detailed in section 8.3.

In most of the theoretical studies of the transport properties of irradiated molecular junctions the field-matter interaction is assumed to be restricted to the molecular bridge. Moreover, the molecular wire is usually described with a simple tight-binding Hamiltonian. This is represented schematically in Fig. 20.8. In this type of models, the time-dependent Hamiltonian that describes the molecular wire adopts a form like the following one [211]

$$\mathbf{H}_{\text{wire}}(t) = \Delta \sum_{n=1,\sigma}^{N-1} \left(\mathbf{c}_{n+1\sigma}^{\dagger} \mathbf{c}_{n\sigma} + h.c. \right) + \sum_{n,\sigma} [\epsilon_n + x_n a(t)] \mathbf{c}_{n\sigma}^{\dagger} \mathbf{c}_{n\sigma}, \quad (20.12)$$

where Δ is the hopping matrix elements that describes the coupling of each orbital to its nearest neighbors and ϵ_n stands for the one-site energies. The time-dependent part in the second term of this Hamiltonian describes the coupling to an oscillating dipole field that causes time-dependent level shifts $x_n a(t)$, where $x_n = (N + 1 - 2n)/2$ denotes the scaled position of site n. The energy $a(t)$, which is periodic in time, is determined by the electrical field strength multiplied by the electron charge and the distance between two neighboring sites.

At a first glance, one might have the impression that models based on the Hamiltonian of Eq. (20.12) describe very different physics from the Tien-Gordon PAT theory detailed above. However, one can show that if the spatial dependence of the field-matter interaction in Eq. (20.12) is neglected, i.e. if the driving shifts all the wire levels simultaneously, it is possible to map the driving field by a gauge transformation to oscillating chemical potentials. In other words, models based on Hamiltonians like the one in Eq. (20.12) reproduce the simple Tien-Gordon-like results in the limiting case of spatially homogeneous field-matter interaction.

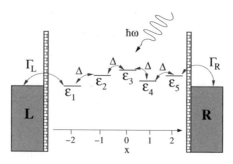

Fig. 20.8 Schematic representation of the level structure of a molecular bridge with $N = 5$ sites coupled to metallic electrodes.

The transport properties in models based on Eq. (20.12) are usually computed within the Floquet theory. It is not our intention to describe here this theory or other similar theoretical tools. Instead, the rest of the section is devoted to a brief description of the main physical effects that have been predicted with the help of models like the one in Eq. (20.12). Most of these effects have not yet been confirmed experimentally, but there are good reasons to believe that they will be observed in the near feature.

Resonant current amplification.– The application of laser fields in molecular junctions can lead to resonant excitation of the molecular bridge states, which in turn can be manifested as an enhancement of the dc current when the driving field is in resonance. This phenomenon, sometimes referred to as resonant current amplification, was already discussed in the context of PAT in section 20.3.3. It was first predicted by several authors using models similar to that of Eq. (20.12). Thus for instance, treating the driving as a perturbation, Keller *et al.* [978, 979] demonstrated that resonant electron excitations result in peaks of the current as a function of the driving frequency. Kohler *et al.* [980] studied the same problem including the driving exactly within a Floquet master equation approach and later derived an analytical expression [981]. In related work, Tikhonov *et al.* [968, 969] have studied this problem with more realistic models for the molecular bridge based on the extended Hückel approach. The central result of these studies is that such resonant excitations enhance the current significantly. In particular, Kohler *et al.* [981, 211] have shown that at the resonant frequencies the dc current decays linearly with the length of the molecule, in contrast to exponential decay of the current in the absence of ac driving.

Ratchet effect and light-induced currents.– A widely studied phenomenon in driven transport is the so-termed ratchet effect: the conversion of ac forces without any net bias into directed motion. In the context of molecular junctions the question is whether it is possible to induce a dc current with an ac field in the absence of a dc bias voltage. As we commented in section 20.3.1, in the context of STM it is well-known that this is indeed possible if there are left-right asymmetries in the junctions [964, 965, 215, 216]. There, the ratchet effect is referred to as rectification at zero dc bias. In the context of molecular junctions, this issue has been extensively studied by Lehmann *et al.* [982–984] and the main results have been reviewed in Ref. [211]. These authors have shown that it possible to generate a dc current with a pure ac driving (i.e. a photocurrent) by introducing certain asymmetries in the problem.[15] For instance, using a conductor with an asymmetric level structure, one can generate a dc current even with a purely harmonic dipole driving. Another possibility is to use a driving field in which several frequencies are mixed. In this case, a dc current is generated even in spatially symmetric molecules bridges.

Related to the ratchet effect, Galperin and Nitzan [977] have predicted that light-induced current in unbiased junctions (i.e. photocurrents) can flow when the bridging molecule is characterized by a strong charge-transfer transition. Such a current reaches its maximum when the light frequency matches the internal transition frequencies of the molecule. Using realistic estimates of molecule-lead coupling and molecule-radiation field interaction, these authors showed that such an effect should be observable.

Coherent destruction of tunneling. – As we saw in our discussion of PAT in section 20.3, the current as a function of the amplitude of the ac driving is modulated according to the behavior of the Bessel functions, see Eqs. (20.4) and (20.5). If the parameter α is such that J_0 vanishes, there is a pronounced reduction of the current [see Exercise 20.2(iv)]. This phenomenon appears in many different ac driven systems and it is known as coherent destruction of tunneling [985]. For a detailed discussion of this phenomenon in the context of molecular wires, see section 7 in Ref. [211] and references therein.

Role of electron excitation in the leads. – As we discussed in section 20.2, apart from modulating the electronic levels in the leads, the electromagnetic field can produce hot electrons in the leads by direct photon absorption. These electronic excitations can in turn contribute to the transport. Simple estimates of the contribution of these inelastic processes to the total current have been put forward long ago in the context of the STM [961].

---

[15]To be precise, the generation of a photocurrent requires the breaking of the so-called generalized parity [211].

More recently, Galperin and coworkers have studied in detail the influence of electronic excitations in the leads on the current through molecular transport junctions [962]. These authors have concluded that in certain situations such excitations can give a significant contribution to the current and moreover, this contribution can be distinguished from the direct current because it scales differently with the distance between the molecule and the leads.

Let us conclude this section by saying that the different physical effects discussed in the previous paragraphs can also have a big impact in other transport properties like shot noise. The theoretical activities along these lines have been reviewed by Kohler *et al.* in Ref. [211].

20.6 Fluorescence from current-carrying molecular junctions

In this section we shall face the question number 5 of our list in the introduction of this chapter. When a sufficiently high bias voltage is applied to a molecular junction, two molecular orbitals can be partially populated and then optical transitions between them become, in principle, possible with the subsequent light emission (or fluorescence). Is the light emission from a single molecule measurable? If so, what can we learn about the junctions from this local optical spectroscopy? The goal of this section is to briefly describe the recent experimental and theoretical efforts devoted to answer these and other basic questions related to the current-induced light emission from single molecules in transport junctions.

The fact that electron tunneling can lead to emission of light was first discovered by Lambe and McCarthy in 1976 in the context of metal-oxide-metal tunnel junctions [986]. In the context of atomic-scale junctions, light emission has been frequently observed in STM experiments. Thus for instance, it has been reported in clean metal [987, 988] and semiconductor surfaces [989], as well as for atomic and molecular adsorbates on metal substrates [990–993]. However, often the reported photon emission spectra do not show identifiable molecule-related features [992]. On a metal surface, the electronic levels of a molecule are considerably broadened whereas light emission is strongly quenched, making it difficult to detect and identify any molecule-specific emission.

In recent years, it has been demonstrated by means of STM experiments that electric-current flow through a molecule may indeed cause the molecule to luminesce due to electronic transitions [994, 995]. Photon emission from single-molecule contacts had already been discussed by Buker and Kirczenow in 2002 [996], and the appearance of those experiments has motivated additional theoretical work [977, 952, 997]. However, a theoretical understanding of single-molecule electroluminescence is in the earliest stages and contact between theory and specific experiments is just beginning to be made [998].

The basic idea of molecular electroluminescence as observed in STM experiments is as follows: By positioning a STM tip above a single molecule on a substrate and applying a bias voltage between the tip and the substrate, electron transmission through the molecule may occur, mediated by the molecule's electronic orbitals, and the molecule may be found to luminesce. In a simplified picture, when a bias voltage is applied, the molecule moves out of equilibrium with a flux of electrons passing through it. If two molecular orbitals are located in the energy window between the electrochemical potentials of the STM tip and substrate, they will both be partially occupied and if optical transitions between them are not forbidden, transitions from the higher-energy orbital to the lower-energy orbital will occur resulting in photon emission [996]. Such optical transitions will most likely involve vibrational levels of both electronic states. This is schematically represented in Fig. 20.9 as process B.

Molecular fluorescence always competes with the light emission channel known as inelastic electron tunneling that takes place even in the absence of molecules [961, 999, 1000]. This latter mechanism involves inelastic tunneling from the tip electronic states into the lower-lying states of the sample with a simultaneous release of the excess energy in the form of a plasmon.[16] The excited plasmon then decays into a far-field photon. The spectrum of this emission is typically quite broad and has a characteristic energy cut-off determined by the sample bias. This process is described schematically in Fig. 20.9 (process A).[17]

In order to avoid the quenching of molecular fluorescence, the metal-molecule coupling strength has to be reduced [996]. This was achieved by Ho's group [994] by adsorbing porphyrin molecules on an ultrathin alumina film grown on a NiAl(110) surface and using the STM as a second weakly

---

[16]By plasmon we mean here an electromagnetic mode of the tip-substrate system.

[17]Other light-emitting processes involving the injection of hot electrons or the creation of electron-hole pairs are also possible [961]. For photon energies $\hbar\omega < eV$, where V is the bias voltage, the single-electron process described above dominates the light emission. However, some of the additional processes have no threshold voltage and therefore, they can be responsible for the light emission with photon energies $\hbar\omega > eV$ [386].

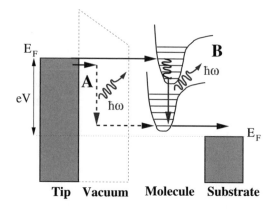

Fig. 20.9   Diagram showing the two major processes contributing to STM-excited light emission from a molecule adsorbed on a surface. In process A, the inelastic electron tunneling channel, an electron tunnels from the Fermi level of the STM tip into an unoccupied molecular orbital with simultaneous excitation of a plasmon. In process B, the fluorescence channel, an electron tunnels into the higher unoccupied orbital of the molecule. The charged molecule typically relaxes to a lower vibrational level of the same electronic level, with subsequent radiative (excitation of a plasmon) transition to the lower electronic level. The final step involves tunneling of this extra electron into the substrate.

coupled electrode. Some characteristic photon emission spectra obtained in this work are reproduced in Fig. 20.10. The results of panels A and B were obtained with the STM positioned directly above a molecule. As one can see, the spectra exhibit sharp features as compared with those related to light emission from NiAl and oxide film (also shown in panels A and B). Furthermore, the light-emission on top of the molecules was found to be very sensitive to the tip position inside the molecule, which indicates that the emission has submolecular resolution.

The series of emission spectra for different sample biases shown in panels C and D in Fig. 20.10 further clarifies the nature of the observed spectral features. In panel C, the spectral peaks do not shift when the bias voltage is varied. This result indicates that these peaks did not originate from transitions between the electronic states of the tip and those of the substrate. They were attributed to transitions inside the molecule. The existence of a cut-off voltage (approximately 2 V in panel C) for excitation of the sharp features is expected if an excited electronic level of the molecule participates in the emission. The difference between this cut-off voltage and the photon energy of the shortest-wavelength feature in the spectra ($\sim$ 1.57 eV), lies in the range of the low-energy dI/dV peak for this molecule, which further

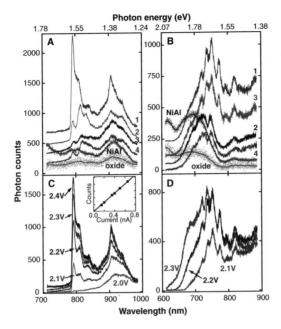

Fig. 20.10   (A-B) Light-emission spectra acquired on porphyrin molecules on an ultra-thin alumina film grown on a NiAl(110) surface using the STM. The spectra acquired on bare NiAl and Al_2O_3/NiAl(110) surfaces are also shown for comparison. The spectra are offset vertically for clarity. Series A and B were taken with two different Ag tips. In (A), the spectra were acquired at a voltage bias $V_{bias} = 2.35$ V and a current $I = 0.5$ nA, with an exposure time of 100 s; the NiAl and oxide spectra have been multiplied by factors of 4 and 15, respectively. In (B), $V_{bias} = 2.2$ V, $I = 0.5$ nA, and exposure time = 300 s; the oxide spectrum has been multiplied by a factor of 3. (C) Variation of curve 1 in (A) as a function of V_{bias} [the same tip was used as in (A)]. The inset shows the dependence of the 800-nm peak intensity on current ($V_{bias} = 2.35$ V). Linear dependence was found for all wavelengths in the measured spectral region. (D) Variation of curve 1 in (B) as a function of V_{bias} [the same tip was used as in (B)]. From [994]. Reprinted with permission from AAAS.

supports this interpretation.

The authors concluded that the total light emission was a result of the contribution of the two main processes discussed above (see Fig. 20.9). A careful analysis of the emission spectra revealed that in the optical transitions between two electronic states of the molecule, most likely the molecule relaxed to the vibrational ground state of the excited electronic state before exciting a plasmon. Moreover, the vibrational features observed in the light-emission spectra were found to depend sensitively on the different molecular conformations and the corresponding electronic states.

In another remarkable experiment Dong *et al.* [995] observed intrinsic molecular fluorescence from porphyrin molecules on Au(100) by using a nanoscale multimonolayer decoupling approach with nanoprobe excitation in the tunneling regime. They observed well-defined vibrationally resolved fluorescence excited by STM that matched nearly perfectly with the standard photoluminescence data of the molecule. The linewidths of spectral peaks were found to narrow down with increased thickness, i.e. by making the junctions more symmetric. On the other hand, a quantum efficiency of $\sim 10^{-5}$ photons per tunneling electron was obtained for the molecular fluorescence at both polarities. Interestingly, emission of photons with energies exceeding the energy of tunneling electrons was reported. The authors attributed tentatively this phenomenon to an excitation mechanism via hot electron injection from either tip or substrate.

From the theory side, simple bridge models have been used to elucidate the basic facts related to the fluorescence of current-carrying molecular junctions. Buker and Kirczenow [996] predicted that the photon emission rate is more sensitive than the electric current to coupling asymmetries between the molecule and contacts. They also showed that electroluminescence may be used to measure the HOMO-LUMO gap and the location of the Fermi level of the contacts relative to the HOMO and LUMO. This has already become clear from our description of the experiments above.

On the other hand, Galperin and Nitzan [977, 952] have used the NEGF formalism in combination with a simple two-level model to compute the dependence of the emission rate on essential parameters such as bias voltage, metal-molecule coupling strength and level separation. In particular, they have derived a very transparent and intuitive relation in which the emission rate is expressed in terms of the level occupations.

On the way to more quantitative descriptions of current-induced single-molecule light emission, Harbola *et al.* [997] have developed a nonequilibrium superoperator Green's function theory which can be combined with DFT. More recently, Buker and Kirczenow [998] have presented a detailed analysis of the experiments of Ref. [994] where the electronic structure calculations were done using the extended Hückel approach.

20.7   Molecular optoelectronic devices

Time has come to address the last question posed in the introduction. One of the dreams in molecular electronics is to use the amazing optical properties of molecules to develop novel optoelectronic devices. One idea is to use compounds in which the excited states have very different character as compared with the ground state. Thus, upon optical excitation one could in principle populate those states and in turn change the conduction properties of the junctions in which they are embedded. The problem with this idea and similar ones is the need to continuously inject light into the molecules to induce the transitions. As commented above, at present this is challenging and the fundamental limitations, if any, are not known.

A class of molecules known as *photochromic molecules* offers a way out for this problem. Photochromism is defined as a reversible photo-transformation of a chemical species between two forms having different absorption spectra. During the photoisomerization, not only the absorption spectra but also various physicochemical properties change, such as the refractive index, dielectric constant, oxidation/reduction potential, and geometrical structure. These molecular property changes can be exploited in various photonic devices, such as erasable optical memory media and photo-optical switch components, and they could also lead to applications in the context of molecular electronics. A key idea is that photoisomerization does not require to continuously photo-excite a molecule, and the absorption of a single photon is enough to trigger its transformation.

There are many photochromic compounds, sometimes referred to as photochromic molecular switches, but two classes of them have attracted special attention in the field of molecular electronics. The first one is formed by azobenzene and its derivatives. Azobenzene is composed of two phenyl rings linked by a N=N double bond. One of the most intriguing properties of azobenzene is the photoisomerization of trans and cis isomers. The two isomers can be switched with particular wavelengths of light: ultraviolet light for trans-to-cis conversion and blue light for cis-to-trans isomerization. This is schematically represented in Fig. 20.11(a). The cis isomer is less stable than the trans one and thus, cis-azobenzene thermally relaxes back to the trans via cis-to-trans isomerization.

A second promising group of switches is formed by diarylethene molecules, which were pioneered by Irie [1001, 1002]. These molecules can be converted from a conjugated ("on" or closed state) to a cross-conjugated ("off" or open) state upon illumination in the visible region, see Fig. 20.11(b). The reverse process is possible with ultraviolet (UV) light. Diarylethenes have additional attractive properties. First and foremost, they are fatigue resistant. Furthermore, their length change upon isomerization is negligible. This allows for minimal mechanical stress when

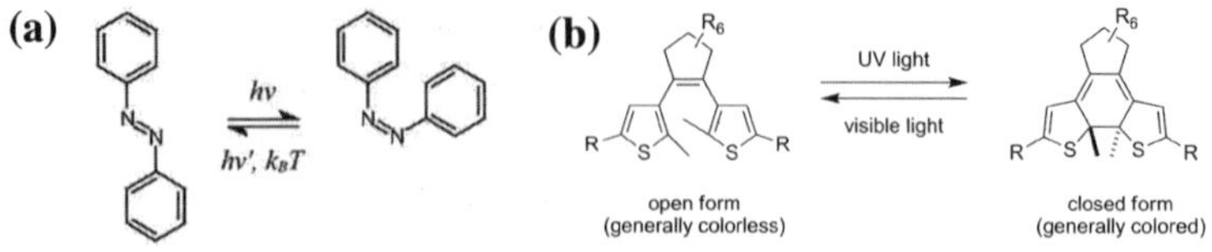

Fig. 20.11 Photochromic molecular switches. (a) Azobenzene photoisomerization. Dithienylethene (a member of the diarylethene class) photochemistry.

a molecule between two electrodes changes conformation.

Light-induced switching of conductance of a molecular junction based on a photochromic molecule was first reported by Dulić *et al.* [587]. In this work the MCBJ technique was used to investigate the transport of single thiophene-substituted diarylethenes. The switching of the molecules was observed from the conducting (closed) state to the insulating (open) state upon illumination with visible light (546 nm). This switching results in a significant resistance increase of over two orders of magnitude. However, the reverse process, which should occur upon illumination with UV light (313 nm), was not observed. This one-way switching was further confirmed by means of UV/Vis spectroscopy to measure absorption of these molecules self-assembled on gold. The authors attributed this observation to quenching of the excited state of the molecule in the open form by the presence of gold. Additional support for these conclusions was obtained by the same group in STM measurements on monothiol thiophene-substituted diarylethene switches in a dodecanethiol matrix [1003].

Theoretical studies have pointed out that the possibility to switch reversibly depends critically on the linker used [1004–1006]. In this sense, it is believed that the one-way switching of the experiments above might be due to the strong electronic hybridization between molecule and metal. Indeed, He *et al.* [1007] reported photoisomerization in both directions for diarylethenes with a methyl spacer and a phenyl linker in the para position. In this case the transport data were obtained with a break-junction method and the authors reported single-molecule resistances of 526 ± 90 MΩ in the open form and 4 ± 1 MΩ in the closed form. It is important to emphasize that the resistances of the two isomers were measured independently, i.e. no conductance switching was observed in situ in the same junction. In this experiment, the photoisomerization was demonstrated with optical spectroscopy of self-assembled monolayers of these molecules on gold surfaces. The crucial role of the linker was further illustrated by Katsonis *et al.* [1008]. These authors demonstrated in STM experiments light-controlled reversible conductance switching for meta-phenyl-linked diarylethenes on gold. This is drastically different from the behavior of the switch mentioned above [587], which had a thiophene linker. Interestingly, the meta-phenyl spacer forms a cross-conjugated system, whereas the thiophene is fully conjugated with the switching unit. This suggests that the reversibility of switching is directly related to the conjugation between the switching unit and the substrate.

In a remarkable experiment, Whalley *et al.* [1009] used single-walled carbon nanotubes (SWNTs) to contact single (or a few) diarylethene molecules. These authors showed that the thiophene-based devices can be switched from the insulating open form to the conductive closed form but not back again, as in the experiments of Dulić *et al.* [587]. However, pyrrole-based devices were shown to cycle between the open and closed states. In particular, in a device with semiconducting nanotubes and pyrrole-based diarylethenes initially in the open state, it was found that with UV irradiation the bridge transforms to the closed state and the current increases by more than 5 orders of magnitude. Irradiation with visible light did not restore the initial, low-conductance state; however, the low-conductance state reappeared when the device aged at room temperature overnight. The on/off cycle can be toggled many times.

One of the most impressive examples of reversible conductance switching in molecular devices has been reported by Kronemeijer *et al.* [1010]. In this work junctions with self-assembled monolayers of photochromic diarylethene-based switches were studied. Large-area molecular junctions were processed in vertical interconnects in an insulating photoresist matrix, see left panel of Fig. 20.12. The diarylethene monolayer is self-assembled in the individual interconnects and topped off with a highly conductive organic top electrode. This organic top electrode is used to prevent the formation of short-circuits from top to bottom electrode. Then, upon irradiation with a specific wavelength range, the conductance of these devices can be optically switched. The major advantage of this approach is that the two distinct isomers of the diarylethene can be individually synthesized and, therefore, separately assembled in a device. Consequently, the ON and OFF state can be independently measured in the devices, without any involvement of a switching event. Optically induced switching of the conductance of the devices in between these two states then provides a direct proof of the molecular origin of the switching events.

In Fig. 20.12 (right panel) we reproduce the results of this experiment for the current density versus voltage (J-V) for devices with molecules ex-

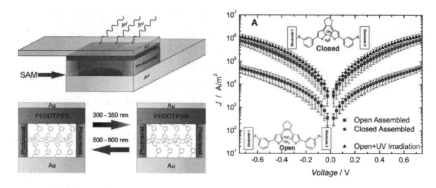

Fig. 20.12   *Left panel:* Schematic cross section of the device layout of a large-area molecular junction in which the diarylethene is sandwiched between Au and poly(3,4-ethylenedioxythiophene): poly(4-styrenesulphonic acid) (PEDOT:PSS)/Au. Using UV (312 nm) illumination the open, nonconjugated isomer can be converted to the closed, conjugated isomer. Visible irradiation of 532nm reverses the photoisomerization process. *Right panel:* Current density (J) versus voltage (V) of the closed (circles) and open (squares) isomers as self-assembled in the molecular junctions, and J-V characteristics of the junctions with the open isomer self-assembled and subsequently photoisomerized to the closed isomer with UV irradiation (triangles). Averaged data (at least 35 devices) from devices with diameters of 10-100 mm. Error bars by standard deviation. Reproduced with permission from [1010]. Copyright Wiley-VCH Verlag GmbH & Co. KGaA.

clusively in the open or closed states. The conductance through the closed, more-conducting state of the switch is shown to be 16 times higher at 0.75 V bias. On the other hand, devices with the open isomer were illuminated for 15 min with 312 nm UV irradiation to convert the molecular switches in the devices to the closed isomer. The J-V characteristics of the converted open-state isomer after UV irradiation show an increase of the conductance through the monolayer, as expected from the devices with closed isomers present, see Fig. 20.12 (right panel). Following UV irradiation and consecutive measurements, these devices were illuminated with 532 nm irradiation (visible light) to achieve ring opening of the switches in the SAM. The observed J-V characteristics show a significant decrease (by a factor 3) of the conductance upon visible light irradiation, but the conductance of the devices with the open isomer is not fully recovered. The origin of this behavior was not fully understood (see Ref. [1010] for more details).

Let us mention that light-controlled conductance switching of molecular devices based on photochromic diarylethene molecules has also been demonstrated by van der Molen *et al.* [135]. In this case, the devices consisted of ordered, two-dimensional lattices of gold nanoparticles, in which neighboring particles are bridged by the switchable molecules. In this work, it was independently confirmed by means of optical spectroscopy that reversible isomerization of the diarylethenes employed is at the heart of the room-temperature conductance switching.

As mentioned above, azobenzene and its derivatives are also possible candidates for molecular switches [1011]. These molecules perform a cis-trans isomerization upon illumination, accompanied by a significant change in molecular length and dipole moment. Precisely, this significant change in the length of the molecule can be a problem to fabricate photo-switchable metal-molecule-metal junctions based on these compounds. Indeed, the experiments reporting isomerization and reversible photo-switching of azobenzene molecules have been performed with these molecules bound to metallic surfaces [1012–1015], i.e. without a second electrode. In these experiments a STM (or a conductive AFM) tip was just used to probe the conformational changes of the molecules after isomerization.

20.8 Final remarks

As we have seen in this chapter, the interplay between electromagnetic fields and molecular transport junctions gives rise to fascinating possibilities and new physical phenomena. We have shown that thanks to the phenomenon of surface-enhanced Raman scattering the spectroscopy of single molecules in transport junctions is indeed possible. The challenge is now to combine SERS and transport measurements at low temperatures, which could provide an unprecedented characterization of molecular junctions.

On the other hand, we have seen that different transport mechanisms can play a role in the transport properties of irradiated junctions. The existent experiments in atomic and molecular contacts suggest that the photon-assisted tunneling mechanism (or rectification) is at work even at optical frequencies, although more experiments are certainly needed to confirm it.

From the theory side, many novel transport phenomena have been predicted and await for experimental confirmation. The problems to observe them are related both to the difficulty of injecting light in such tiny objects surrounded by metallic electrodes and to the fact that other effects can mask their appearance.

Current-induced light emission is an extraordinary physical phenomenon that can provide the most direct spectroscopy of the electronic states in a molecular junction. So far, there have been very few experiments where molecular signatures have been unambiguously identified in the fluorescence spectra and more experiments are needed. For the theory the challenge is now to provide quantitative predictions that can be directly compared with the experiments.

The last topic that we have discussed in this chapter is very important from the technological point of view. Molecular electronics is often seen as a field that aims at reproducing the standard microelectronic components and devices, but at a smaller scale. However, the future of molecular electronics depends crucially on our ability to provide devices with new functionalities out of the scope of more traditional technologies. The optical properties of many molecules may offer a down-to-earth possibility for the future. In principle, they can be used in transport junctions to control the current at will and many researchers believe by now that molecular optoelectronics will soon grow as a field of its own. However, so far it has been difficult to take advantage of those optical properties and the only successful implementations have made use of photochromic molecular switches. At present, it is not clear whether the difficulties encountered so far are just of technical nature or there are true fundamental limitations. In any case, the next years will be certainly exciting for the scientists working on this subject.

20.9   Exercises

20.1 Photon-assisted tunneling in atomic gold chains: In the Exercise 7.5 a tight-binding model was used to show that the conductance of atomic chains can exhibit parity oscillations, i.e. that it depends on whether the number of atoms N in the chain is even or odd. Use that model to show that the sign of correction to the linear conductance due to irradiation, ΔG_{dc}, can also exhibit an even-odd effect. In particular, show that for low frequencies $\Delta G_{dc} < 0$, if N is odd; while $\Delta G_{dc} > 0$, if N is even. Hint: For the last task use the low-frequency formula of Eq. (20.9).

20.2 Photon-assisted tunneling in molecular junctions: The goal of this exercise is to gain some insight into PAT in molecular junctions. For this purpose, let us assume that the transmission function of a molecular junction is given by the following double Lorentzian[18]

$$\tau(E) = \sum_{i=1}^{2} \frac{4\Gamma_L \Gamma_R}{(E - \epsilon_i)^2 + (\Gamma_L + \Gamma_R)^2},$$

---

[18]This model was extensively used in section 19.3 to describe the thermopower in molecular junctions.

were ϵ_1 and ϵ_2 is the energy of HOMO and LUMO, respectively, and Γ_L and Γ_R the broadenings by contacts L and R. For the sake of concreteness, we shall assume throughout this exercise that $\epsilon_1 = -7.0$ eV, $\epsilon_2 = -3.0$ eV and $\Gamma_L = \Gamma_R = 30$ meV.

(i) Use the second equality in Eq. (20.9), i.e. the classical rectification formula, to analyze the low-frequency radiation-induced correction to the conductance (ΔG_{dc}) as a function of the Fermi level position (E_F). In particular, show that ΔG_{dc} is negative only when E_F is very close to one of the frontier orbitals.

(ii) Let us now assume that the Fermi level lies in the middle of the HOMO-LUMO gap, i.e. $E_F = -5.0$ eV and consider the limit of small ac amplitudes ($\alpha \ll 1$). Compute ΔG_{dc} as a function of photon energy and show that for $\hbar\omega \gtrsim 0.5$ eV, the classical rectification formula fails to describe the correction to the conductance given by the first equality in Eq. (20.9).

(iii) Now assume that $E_F = -6.0$ eV and use Eq. (20.6) to compute ΔG_{dc} as a function of the photon energy in the interval $\hbar\omega \in [0$ eV$, 4$ eV$]$ for different values of α. For $\alpha \ll 1$ you will find the appearance of two peaks at 1 and 3 eV. What is the origin of these peaks? Finally, compare the results obtained with the exact formula of Eq. (20.6) and with the approximation of Eq. (20.9) to establish in which range of α this approximation is valid.

(iv) One of the key signatures of PAT is the appearance of additional steps in the I-V characteristics. Assume that $E_F = -6.0$ eV and $\hbar\omega = 0.5$ eV and use Eq. (20.5) or Eq. (8.74) to compute the I-V curves and the corresponding differential conductance for $\alpha = 0, 1, 2, 4$. Discuss the origin of current steps (or the corresponding peaks in the differential conductance) induced by the radiation. Finally, analyze the phenomenon of coherent destruction of tunneling by computing the I-V curves and differential conductance for values of α for which J_0 in Eq. (20.5) vanish, i.e. $\alpha/2 = 2.405, 5.520, \ldots$.

Chapter 21

What is missing in this book?

At this stage in the development of molecular electronics it is already impossible to cover all the aspects of this multidisciplinary field in a single monograph. Our selection of topics has been biased, as it could not be otherwise, by our own backgrounds and research interests and we are aware of the fact that important issues have been left out. Therefore, we would like to close this manuscript by pointing out some of those topics and suggesting some references where the reader can find information about them.

Among the topics not covered in this monograph, we believe that the following ones are of special relevance:

- **Molecules for molecular electronics:** In section 3.2 we presented a brief discussion about the typical molecules considered in molecular electronics. We also mentioned briefly the electronic functions for which they are well suited. A detailed discussion of this issue can be found in several monographs and review articles, see e.g. Refs. [583, 1002, 1016, 33, 47, 1017].
- **Electron transfer:** Electron transfer, the process by which an electron moves from one atom or molecule to another atom or molecule, is one the simplest and most important reactions both in chemistry and biology. In particular, electron transfer in donor-bridge-acceptor complexes is in many respects very similar to the conduction of an electron through a molecular transport junction.[1] Thus, it is obvious that molecular electronics can profit a lot from the much more mature field of electron transfer in chemistry and biology. In the sense, researchers working in our field should at least know the basics of the

---

[1] Although the driving forces are different in these two types of experiments, the mechanisms by which the electron is transferred through the bridge/molecular wire are essentially the same.

standard theory of electron transfer, known as *Marcus theory*. This theory is nicely explained in the monograph of Kuznetsov and Ulstrup [582], but we specially recommend Chapters 16 and 17 of Nitzan's book on chemical dynamics in condensed phases [30]. For an introduction to the techniques used to measure electron transfer rates, see Ref. [583]. This reference also contains countless experimental results for many different chemical compounds. Finally, for a discussion of the connection between electron transfer rates and electrical conductance, see e.g. Refs. [38, 830, 1018, 831].

- **SAM-based molecular junctions:** Since we are interested in the basic conduction mechanisms in molecular systems, we have focused our attention on the study of transport through single-molecule junctions. However, the technological applications of molecular electronics will surely come from devices containing of a large number of molecules, like in the junctions based on self-assembled monolayers (SAM). In this sense, there are many basic questions to be addressed like for instance whether or not we can straightforwardly extrapolate the results for single-molecule junctions to those devices.[2] On the other hand, such SAM-based devices require fabrication techniques that differ from the ones described here. Some of these issues are discussed in the review of Ref. [41].

- **Scaling and integration of molecular devices:** Important topics for the future of molecular electronics, which are related to the previous issue, are the reliable mass production of molecular devices and the integration of molecular junctions into macroscopic circuits. The strategies and ideas explored so far in this respect have been reviewed by Lu and Lieber in Ref. [1019].

- **Carbon nanotubes:** These molecules are considered to be something in between a solid and a molecule and the study of their electrical and thermal properties constitutes a field in its own. For this reason, we have rarely talked about carbon nanotubes (CN) in this monograph. However, it is obvious that many of the concepts, ideas and techniques that we have discussed here are directly applicable to the problem of transport through CN-based junctions. For a recent review on the electronic and transport properties of carbon nanotubes, see Ref. [1020].

---

[2]It is not obvious what is the role of inter-molecular interactions in these systems and in some cases, the transport characteristics may differ significantly from the corresponding ones of a single-molecule device.

- **Strongly correlated methods for molecular electronics:** One of the main challenges for the theory in our field is the development of new methods that are able to describe properly the transport through systems that exhibit strong electronic correlations, as it is often the case in molecular junctions. Such methods should (and they will) replace DFT in the near future as the main theoretical tool for the description of transport in molecular junctions. We have not said much about this topic in this monograph because those methods are still under development and their performance has still to be established. For recent advances on this subject we recommend [634, 1021–1023] and references therein.

PART 5
Appendixes

Appendix A

Second Quantization

All the relevant systems in molecular electronics are composed of many identical particles such as electrons, protons, phonons (or vibrations), etc. As we all know, these particles obey their corresponding quantum statistics depending on whether there are fermions or bosons. This statistics is reflected in the symmetry of the many-particle wave functions. Thus for instance, a fermionic wave function is expressed in the form of Slater determinants to ensure its antisymmetry (Pauli's exclusion principle).

The algebra with many-particle wave functions is quite cumbersome, and it has been shown that the description of a many-body system can be greatly simplified by using the so-called *second quantization* formalism of quantum mechanics. This is just an alternative formalism (with no new physics) in which the symmetry of the wave functions is transferred to some convenient operators which fulfill simple commutation rules. Then, in this formalism all the standard calculations can be done using the algebraic properties of these operators, rather than using lengthy many-particle wave functions.

The second quantization approach will be used throughout this book and therefore, in order to make this manuscript more self-contained, we have included a brief review of this formalism in this appendix. For a more detailed discussion of the second quantization formalism, we recommend the many-body textbooks of Refs. [173–176].

A.1   Harmonic oscillator and phonons

A.1.1   *Review of simple harmonic oscillator quantization*

One convenient way to introduce the subject of creation and annihilation operators, which is the essence of the second quantization formalism, is to review the physics of a quantum-mechanical harmonic oscillator. Let us consider a particle of mass m that is subjected to a one-dimensional (1D) harmonic potential. The Hamiltonian describing this system can be written as

$$H = \frac{p^2}{2m} + \frac{K}{2}x^2, \qquad (A.1)$$

where x and $p = -i\hbar(\partial/\partial x)$ are the position and momentum operators, respectively, which satisfy $[x, p] = xp - px = i\hbar$.

To diagonalize this Hamiltonian, we introduce the frequency ω and the dimensionless coordinate ξ:

$$\omega^2 = \frac{K}{m} \;\; ; \;\; \xi = x\left(\frac{m\omega}{\hbar}\right)^{1/2}. \qquad (A.2)$$

In terms of these new parameters, we can write the Hamiltonian simply as

$$H = \frac{\hbar\omega}{2}\left(-\frac{\partial^2}{\partial\xi^2} + \xi^2\right). \qquad (A.3)$$

The harmonic oscillator has a solution in terms of Hermite polynomials. The corresponding eigenvalues are given by

$$H\psi_n = \hbar\omega\left(n + \frac{1}{2}\right)\psi_n, \qquad (A.4)$$

where n is an integer. From now on, we shall use the Dirac notation for the eigenstate: $|n\rangle = \psi_n$.

We now introduce two dimensionless operators as follows

$$a = \frac{1}{\sqrt{2}}\left(\xi + \frac{\partial}{\partial\xi}\right) = \left(\frac{m\omega}{2\hbar}\right)^{1/2}\left(x + \frac{ip}{m\omega}\right) \qquad (A.5)$$

$$a^\dagger = \frac{1}{\sqrt{2}}\left(\xi - \frac{\partial}{\partial\xi}\right) = \left(\frac{m\omega}{2\hbar}\right)^{1/2}\left(x - \frac{ip}{m\omega}\right).$$

They are Hermitian conjugates of each other. They are sometimes called raising and lowering operator (or ladder operators), but here we call them *creation* ($a^\dagger$) and *annihilation* (a) operators. In terms of these new operators, the Hamiltonian can now be written as

$$H = \frac{\hbar\omega}{2}\left[aa^\dagger + a^\dagger a\right]. \qquad (A.6)$$

It is easy to show from their definitions that these operators satisfy the following *commutation relations*

$$[a, a^\dagger] = 1 \;\; ; \;\; [a, a] = 0 \;\; ; \;\; [a^\dagger, a^\dagger] = 0. \tag{A.7}$$

With these relations, the Hamiltonian adopts the following form

$$H = \frac{\hbar\omega}{2} \left[aa^\dagger + a^\dagger a \right] = \hbar\omega \left[a^\dagger a + \frac{1}{2} \right]. \tag{A.8}$$

The three commutators above plus this Hamiltonian completely specify the harmonic oscillator problem in terms of operators. With these four relationships, one can show that the eigenvalue spectrum is indeed that of Eq. (A.4). The eigenstates are

$$|n\rangle = \frac{(a^\dagger)^n}{\sqrt{n!}} |0\rangle, \tag{A.9}$$

where $|0\rangle$ is the ground state which obeys

$$a|0\rangle = 0 \tag{A.10}$$

and where the $n!$ is for normalization. Operating on this state by a creation operator gives

$$a^\dagger |n\rangle = \frac{(a^\dagger)^{n+1}}{\sqrt{n!}} |0\rangle = (n+1)^{1/2} |n+1\rangle \tag{A.11}$$

the state with the next highest integer. In the same way, one can show that

$$a|n\rangle = (n)^{1/2} |n-1\rangle, \tag{A.12}$$

which shows that the annihilation operator a lowers the quantum number. Then operating by the sequence

$$a^\dagger a|n\rangle = a^\dagger (n)^{1/2} |n-1\rangle = n|n\rangle \tag{A.13}$$

gives an eigenvalue n, which verifies the eigenvalue relation A.4. Furthermore, using the original definition of Eq. (A.5) permits us to express x and p in terms of these operators as

$$x = \left(\frac{\hbar}{2m\omega} \right)^{1/2} (a + a^\dagger) \tag{A.14}$$

$$p = i \left(\frac{m\hbar\omega}{2} \right)^{1/2} (a^\dagger - a). \tag{A.15}$$

The description of the harmonic oscillator in terms of operators is equivalent to the conventional method of using wave functions $\psi_n(\xi)$ of position.

A.1.2 *1D harmonic chain*

With the material of the previous subsection, we are now in position to discuss briefly our first many-body problem, namely the physics of phonons. This will illustrate in an informal way the second quantization formalism for bosons.

In a solid there are many atoms, which mutually interact. The vibration modes are collective motions involving many atoms. A simple introduction to this problem is obtained by studying the normal modes of an infinite one-dimensional harmonic chain:

$$H = \sum_i \frac{p_i^2}{2m} + \frac{K}{2} \sum_i (x_i - x_{i+1})^2. \qquad (A.16)$$

Here, we have assumed that an atom is only coupled to its nearest neighbors and that all the atoms are identical.

The classical solution is obtained by solving the equation of motion:

$$-m\ddot{x}_j = m\omega^2 x_j = K(2x_j - x_{j+1} - x_{j-1}). \qquad (A.17)$$

A solution is assumed of the form $x_j = x_0 \cos(kaj)$, where a is the interatomic distance. Then, the normal modes have the solution

$$\omega_k^2 = \frac{2K}{m}[1 - \cos(ka)] = \frac{4K}{m}\sin^2(ka/2). \qquad (A.18)$$

To quantize the theory, let us impose canonical commutation relations on the position and momentum of the lth and jth atoms: $[x_l, p_j] = i\hbar\delta_{lj}$ and construct collective variables which describe the modes themselves (recall k is wave vector, l is position):

$$x_l = \frac{1}{\sqrt{N}} \sum_k e^{ikal} x_k \; ; \; x_k = \frac{1}{\sqrt{N}} \sum_l e^{-ikal} x_l$$

$$p_l = \frac{1}{\sqrt{N}} \sum_k e^{-ikal} p_k \; ; \; p_k = \frac{1}{\sqrt{N}} \sum_l e^{ikal} p_l, \qquad (A.19)$$

which leads to canonical commutation relations in wave vector space:

$$[x_k, p_{k'}] = \frac{1}{N} \sum_{l,m} e^{-ikal} e^{ik'am} [x_l, p_m]$$

$$= \frac{i\hbar}{N} \sum_l e^{-ial(k-k')} = i\hbar\delta_{k,k'}. \qquad (A.20)$$

Let us now express the Hamiltonian of Eq. (A.16) in terms of the new variables. We have, with a little algebra,

$$\sum_l p_l^2 = \sum_k p_k p_{-k} \qquad (A.21)$$

$$\frac{K}{2} \sum_l (x_l - x_{l+1})^2 = \frac{m}{2} \sum_k \omega_k^2 x_k x_{-k}.$$

Then the Hamiltonian may be written in wave vector space as

$$H = \frac{1}{2m} \sum_k p_k p_{-k} + \frac{m}{2} \sum_k \omega_k^2 x_k x_{-k}. \tag{A.22}$$

Note that the energy is now expressed as the sum of kinetic plus potential energy of each mode k, and there is no more explicit reference to the motion of the atomic constituents. To *second quantize* the system, we write down creation and annihilation operators for each mode k. We define

$$a_k = \left(\frac{m\omega_k}{2\hbar}\right)^{1/2} \left(x_k + \frac{i}{m\omega_k} p_{-k}\right) \tag{A.23}$$

$$a_k^\dagger = \left(\frac{m\omega_k}{2\hbar}\right)^{1/2} \left(x_{-k} - \frac{i}{m\omega_k} p_k\right), \tag{A.24}$$

which can be shown, just as in the single harmonic oscillator case, to obey commutation relations

$$\left[a_k, a_{k'}^\dagger\right] = \delta_{k,k'}, \quad [a_k, a_{k'}] = 0, \quad \left[a_k^\dagger, a_{k'}^\dagger\right] = 0 \tag{A.25}$$

and the Hamiltonian can be simply expressed as

$$H = \sum_k \hbar\omega_k \left(a_k^\dagger a_k + \frac{1}{2}\right). \tag{A.26}$$

These collective modes of vibrations are called *phonons*. They are the quantized version of the classical vibrational modes in a solid. Each wave vector state behaves independently, as a harmonic oscillator, with a possible set of quantum numbers $n_k = 0, 1, 2, \ldots$ The state of the system at any time is

$$\Psi = |n_1, n_2, \ldots, n_n\rangle = \prod_k |n_k\rangle = \prod_k \frac{(a_k^\dagger)^{n_k}}{\sqrt{n_k!}} |0\rangle \tag{A.27}$$

so that the expectation value of the Hamiltonian is

$$\langle H\rangle = \sum_k \hbar\omega_k \left(n_k + \frac{1}{2}\right). \tag{A.28}$$

In thermal equilibrium the states have an average value of n_k which is given in terms of the temperature $\beta = 1/k_B T$ by the Bose distribution function:

$$\langle n_k\rangle \equiv N_k = \frac{1}{e^{\beta\hbar\omega_k} - 1} \equiv n_B(\hbar\omega_k). \tag{A.29}$$

So in summary, we have shown that the physics of these collective modes can be described in terms of creation and annihilation operators that satisfy simple commutation relations. In the next section we shall show in a more formal manner that these basic ideas can be extended to any many-body system, focusing on the case of fermionic particles.

A.2   Second quantization for fermions

The systems that we will be dealing with are composed of many identical particles such as electrons or phonons. In quantum mechanics those identical particles are indistinguishable. Thus for instance, no electron can be distinguished from another electron, except by saying where it is, what quantum state it is in, etc. Internal quantum-mechanical consistency requires that when we write down a many-identical-particle state, we make that state noncommittal as to which particle is in which single-particle state. For example, we say that we have electron 1 and electron 2, and we put them in states a and b respectively, but exchange symmetry requires (since electrons are fermions) that a satisfactory wave-function has the form

$$\Phi(\mathbf{r}_1, \mathbf{r}_2) = A\left[\phi_a(\mathbf{r}_1)\phi_b(\mathbf{r}_2) - \phi_a(\mathbf{r}_2)\phi_b(\mathbf{r}_1)\right], \qquad (A.30)$$

i.e. the wave function is antisymmetric with respect to the exchange of the two electrons: $\Phi(\mathbf{r}_1, \mathbf{r}_2) = -\Phi(\mathbf{r}_2, \mathbf{r}_1)$. This is a consequence of Pauli's exclusion principle that states that there cannot be two fermions in the same quantum sate.

If we have N particles, the wave functions must be either symmetric or antisymmetric under exchange depending on the nature of the particles:

$$\Phi^{\mathrm{B}}(\mathbf{r}_1, ..., \mathbf{r}_i, ..., \mathbf{r}_j, ..., \mathbf{r}_N) = \Phi^{\mathrm{B}}(\mathbf{r}_1, ..., \mathbf{r}_j, ..., \mathbf{r}_i, ..., \mathbf{r}_N) \ \ \text{(Bosons)}$$

$$\Phi^{\mathrm{F}}(\mathbf{r}_1, ..., \mathbf{r}_i, ..., \mathbf{r}_j, ..., \mathbf{r}_N) = -\Phi^{\mathrm{F}}(\mathbf{r}_1, ..., \mathbf{r}_j, ..., \mathbf{r}_i, ..., \mathbf{r}_N) \ \ \text{(Fermions)}.$$

In particular, in the fermionic case the antisymmetry of the wave function can be ensured by using Slater determinants. But, can we satisfy the antisymmetry principle without using Slater determinants? Second quantization is a formalism in which the antisymmetry property of the wave function has been transferred onto the algebraic properties of certain operators. Second quantization introduces no new physics. It is just another, although very elegant, way of treating many-electron systems, which shifts the emphasis away from N-electron wave functions to the one- and two electron matrix elements of the different operators. This has been illustrated already in the case of bosons with the analysis of the phonons in a 1D chain in the previous section, and we shall now concentrate on the case of fermions.

A.2.1   *Many-body wave function in second quantization*

The total wave function for the ground state and excited states of noninteracting particles is the product of single-particle wave functions. However, because we are considering identical fermions, this product must be anti-symmetrized (Pauli's exclusion principle) and the proper wave function is the *Slater determinant*

$$
\Phi_{k_1,...,k_N}(\mathbf{r}_1, ..., \mathbf{r}_N) = \frac{1}{\sqrt{N!}}
\begin{vmatrix}
\phi_{k_1}(\mathbf{r}_1) & \phi_{k_1}(\mathbf{r}_2) & \cdots & \phi_{k_1}(\mathbf{r}_N) \\
\phi_{k_2}(\mathbf{r}_1) & \phi_{k_2}(\mathbf{r}_2) & \cdots & \phi_{k_2}(\mathbf{r}_N) \\
\vdots & \vdots & \vdots & \vdots \\
\phi_{k_N}(\mathbf{r}_1) & \phi_{k_N}(\mathbf{r}_2) & \cdots & \phi_{k_N}(\mathbf{r}_N)
\end{vmatrix}, \qquad \text{(A.31)}
$$

where N is the number of particles and the ϕ_k's are a set of single-particle states with k_i as quantum number (e.g. energy). If the particles are allowed to interact with each other, or with an external potential, then the exact wave functions of the system are no longer that of Eq. (A.31), but a linear combination of Φ's:

$$
\Psi(\mathbf{r}_1, ..., \mathbf{r}_N) = \sum_{k_1,...,k_N} A_{k_1,...,k_N} \Phi_{k_1,...,k_N}(\mathbf{r}_1, ..., \mathbf{r}_N). \qquad \text{(A.32)}
$$

That is, the $\Phi_{k_1,...,k_N}(\mathbf{r}_1, ..., \mathbf{r}_N)$ for the non-interacting system are the basis states used to describe the interacting system.

Now these are rather clumsy expressions to carry around, so it would be desirable to have a more compact way of writing them. This may be achieved by noting that since all particles are indistinguishable, the essential information in Eq. A.31 is just how many particles there are in each single-particle state. Therefore, we could equally well specify the state of the non-interacting system by writing Φ as

$$
\Phi_{k_1,...,k_N}(\mathbf{r}_1, ..., \mathbf{r}_N) = \Phi_{n_{p_1},n_{p_2},...,n_{p_i},...}(\mathbf{r}_1, ..., \mathbf{r}_N). \qquad \text{(A.33)}
$$

For short, we shall represent this as

$$
\Phi_{n_{p_1},n_{p_2},...,n_{p_i},...}(\mathbf{r}_1, ..., \mathbf{r}_N) \equiv |n_{p_1}, n_{p_2}, ..., n_{p_i}, ...\rangle \qquad \text{(A.34)}
$$

meaning: n_{p_1} particles in state ϕ_{p_1}, n_{p_2} in ϕ_{p_2}, etc., where $n_k = 0$ or 1 by the Pauli principle. This is called "occupation number notation". For brevity, from now on we shall drop the p's and just use the numerical subscripts. Then

$$
\Phi = |n_1, n_2, ..., n_i, ...\rangle. \qquad \text{(A.35)}
$$

It is important to remember that the $|n_1, n_2, ..., n_i, ...\rangle$ are orthonormal because the $\Phi_{k_1,...,k_N}$ are, and we may write this in the following way

$$
\langle n_1', n_2', ..., n_i', ... | n_1, n_2, ..., n_i, ...\rangle = \delta_{n_1',n_1} \delta_{n_2',n_2} ... \delta_{n_i',n_i} ... \qquad \text{(A.36)}
$$

Up to this point we have been dealing with systems containing a fixed number of particles. Now we take an important step, and, even though the particle number in a real system is fixed, allow N to be variable, running from 0 to ∞. This generates the following set of basis functions

$$
\begin{array}{lll}
0 & \Phi_0 & |000...\rangle \\
1 & \Phi_1, \Phi_2, \Phi_3, ... & |100...\rangle, |010...\rangle, |001...\rangle, ... \\
2 & \Phi_{12}, \Phi_{13}, \Phi_{23}, ... & |1100...\rangle, |1010...\rangle, |0110...\rangle, ... \\
\vdots & \vdots & \vdots \\
N & \Phi_{k_1, k_2, ..., k_N} & |n_1, n_2, ..., n_i, ...\rangle \\
\vdots & \vdots & \vdots
\end{array}
\tag{A.37}
$$

The state Φ_0 or $|000...\rangle$ with no particles at all in it is called the true vacuum. The set of all $|n_1, ..., n_i, ...\rangle$ in Eq. (A.37) is a complete orthogonal set of basis functions in an extended Hilbert space in which the number of particles is variable. This set is often called occupation number basis, and the whole formalism is sometimes referred to as occupation number representation.

States like the ones appearing in Eq. (A.37) can describe the state of a non-interacting fermionic systems. In the presence of interactions the correct eigenstates of the system can be obtained as linear combination of the states $|n_1, ..., n_i, ...\rangle$, i.e.

$$
\Psi = \sum_{n_1, ..., n_i, ...} A_{n_1, ..., n_i, ...} |n_1, ..., n_i, ...\rangle. \tag{A.38}
$$

A.2.2   Creation and annihilation operators

We shall go on constructing the formalism of second quantization by showing how the properties of determinants can be transferred onto the algebraic properties of operators. For this purpose, we begin by associating a *creation operator* $c_i^\dagger$ and an *annihilation operator* c_i with each single-particle state ϕ_i. We define $c_i^\dagger$ and c_i by their action on an arbitrary Slater determinant $|n_1, ..., n_i, ...\rangle$ as follows

$$
c_i^\dagger |n_1, ..., n_i, ...\rangle = (-1)^{\Sigma_i}(1 - n_i)|n_1, ..., n_i + 1, ...\rangle \tag{A.39}
$$

$$
c_i |n_1, ..., n_i, ...\rangle = (-1)^{\Sigma_i} n_i |n_1, ..., n_i - 1, ...\rangle, \tag{A.40}
$$

where

$$
\Sigma_i = n_1 + n_2 + ... + n_{i-1}. \tag{A.41}
$$

That is, we get a factor of (-1) for each particle (i.e., each occupied state) standing to the left of the state i in the wave function. For example

$$c_i|..., 0_i, ...\rangle = 0 \quad , \quad c_i^\dagger|..., 1_i, ...\rangle = 0$$
$$c_3|11111000...\rangle = +|11011000...\rangle$$
$$c_4^\dagger|1110100...\rangle = -|11111000...\rangle$$
$$c_2^\dagger c_3 c_1^\dagger c_2 c_3^\dagger c_1|1100...\rangle = -|1100...\rangle. \tag{A.42}$$

One of the nice properties of the $c_i^\dagger$ operators is that by applying them repeatedly to the true vacuum state (state with no particles in it), it is possible to generate all other states, thus:

$$|n_1, n_2, ...\rangle = (c_1^\dagger)^{n_1}(c_2^\dagger)^{n_2}...|0000...\rangle. \tag{A.43}$$

For example

$$|011000...\rangle = c_2^\dagger c_3^\dagger|0000...\rangle. \tag{A.44}$$

Another important property of the $c_i^\dagger$, c_i operators is that they are hermitian adjoint of each other, i.e. $c_i^\dagger = (c_i)^\dagger$. The demonstration is left to the reader. This property shows that $c_i^\dagger$, c_i are non-hermitian and are therefore not observables. It is, however, easy to construct a hermitian operator from $c_i^\dagger$ and c_i as follows. The combination

$$\hat{n}_i = c_i^\dagger c_i \quad (\hat{N} = \sum_i c_i^\dagger c_i) \tag{A.45}$$

is obviously hermitian and is an extremely important observable called number operator ($\hat{N}$ = total number operator). To understand its properties, let it operate on some typical state vectors:

$$\begin{aligned}
c_i^\dagger c_i|n_1, n_2, ..., 1_i, ...\rangle &= (-1)^{\Sigma_i} c_i^\dagger|n_1, n_2, ..., 0_i, ...\rangle \\
&= (-1)^{\Sigma_i + \Sigma_i}|n_1, n_2, ..., 1_i, ...\rangle \\
&= (+1)|n_1, n_2, ..., 1_i, ...\rangle.
\end{aligned}$$

Similarly

$$c_i^\dagger c_i|n_1, n_2, ..., 0_i, ...\rangle = 0|n_1, n_2, ..., 0_i, ...\rangle,$$

so that in general

$$c_i^\dagger c_i|n_1, n_2, ..., n_i, ...\rangle = n_i|n_1, n_2, ..., n_i, ...\rangle. \tag{A.46}$$

Thus, the eigenvalue of the number operator for the state ϕ_i is just the occupation number for that state. Hence, in the occupation number basis, all number operators are diagonal and the total system wave function

$|n_1, ..., n_i, ...\rangle$ are just simultaneous eigenfunctions of the number operators $\hat{n}_1, ..., \hat{n}_i, ...$

The $c_i^\dagger$, c_i operators obey the following important *fermion commutation rules*:

$$\{c_l, c_k^\dagger\} = c_l c_k^\dagger + c_k^\dagger c_l = \delta_{lk} \; ; \; \{c_l, c_k\} = \{c_l^\dagger, c_k^\dagger\} = 0. \qquad (A.47)$$

These can be easily proved from the definitions of Eqs. (A.39) and (A.40). Thus for instance, the second relation can be shown as follows:

$$c_l c_k |n_1, ..., n_l, ..., n_k, ...\rangle = (-1)^{\Sigma_k} n_k c_l |n_1, ..., n_l, ..., n_k - 1, ...\rangle \qquad (A.48)$$
$$= (-1)^{\Sigma_k + \Sigma_l} n_k n_l |n_1, ..., n_l - 1, ..., n_k - 1, ...\rangle$$
$$c_k c_l |n_1, ..., n_l, ..., n_k, ...\rangle = (-1)^{\Sigma_l} n_l c_k |n_1, ..., n_l - 1, ..., n_k, ...\rangle$$
$$= (-1)(-1)^{\Sigma_k + \Sigma_l} n_k n_l |n_1, ..., n_l - 1, ..., n_k - 1, ...\rangle,$$

where the extra (-1) on line four comes from the fact that there is one less particle to the left of state k. Adding the two equations yields the second rule in Eq. (A.47). The other rules may be established in a similar fashion.

The importance of the above set of anti-commutation relations lies in the fact that all the antisymmetry properties are built into them. Therefore, by using them in the right places, we do not have to worry either about the symmetry of the wave functions themselves, or even about the awkward $(-1)^\Sigma$ factors.

A.2.3 *Operators in second quantization*

We have seen that we can represent determinants by using creation and annihilation operators, which obey a set of anti-commutation relations, and a vacuum state. To be able to develop the entire theory of many-electron systems without using determinants, we must express the many-body operators in terms of the creation and annihilation operators. This is the goal of this subsection.

All the operators that we shall encounter, in particular for electronic systems, can be written in first quantization as the sum of two types of operators. The first type is a sum of one-electron operators

$$\mathcal{O}_1 = \sum_{i=1}^{N} h(i), \qquad (A.49)$$

where $h(i)$ is any operator involving only the ith electron. These operators represent dynamic variables that depend only on the position or momentum of the electron in question, independent of the position or momentum of other electrons. Examples are operators for the kinetic energy, attraction of an electron to a nucleus, dipole moment, and most of the other operators that one encounters. The second type of operator is a sum of two-electron operators

$$\mathcal{O}_2 = \frac{1}{2} \sum_{i \neq j}^{N} V(i,j), \tag{A.50}$$

where $V(i,j)$ is an operator that depends on the position (or momentum) of both the ith and jth electron. The Coulomb interaction between two electrons

$$V(i,j) = \frac{e^2}{|\mathbf{r}_i - \mathbf{r}_j|} \tag{A.51}$$

is a two-electron operator.

Obviously, the expression for an operator $\mathcal{O}$ in second quantization must be such that the value of the matrix element $\langle K|\mathcal{O}|L \rangle$, $|K\rangle$ and $|L\rangle$ being two arbitrary Slater determinants, is the same irrespective of whether we obtain it using the properties of determinants or using the algebra of creation and annihilation operators. The appropriate expressions for $\mathcal{O}_1$ (our sum of one-electron operators) and $\mathcal{O}_2$ (the two-electron operator) in second quantization are

$$\mathcal{O}_1 = \sum_{ij} h_{ij} c_i^\dagger c_j \tag{A.52}$$

$$\mathcal{O}_2 = \frac{1}{2} \sum_{ijkl} V_{ijkl} c_i^\dagger c_j^\dagger c_l c_k, \tag{A.53}$$

where the sums run over the set $\{\psi_i\}$. Here, the different matrix elements are defined as follows

$$h_{ij} \equiv \int d\mathbf{r}_1 \, \psi_i^*(\mathbf{r}_1) h(\mathbf{r}_1) \psi_j(\mathbf{r}_1) \tag{A.54}$$

$$V_{ijkl} \equiv \int d\mathbf{r}_1 \int d\mathbf{r}_2 \, \psi_i^*(\mathbf{r}_1) \psi_j^*(\mathbf{r}_2) V(\mathbf{r}_1, \mathbf{r}_2) \psi_k(\mathbf{r}_1) \psi_l(\mathbf{r}_2). \tag{A.55}$$

Let us now sketch the demonstration of this result. Consider the following N-electron Slater determinant $|\Psi\rangle = |\psi_1, ..., \psi_a, \psi_b, ..., \psi_N\rangle = |1_1, ..., 1_a, 1_b, ..., 1_N\rangle$. From the first quantization formalism (this is a simple exercise), we know that the expectation value of the one-electron operator $\mathcal{O}_1$ is equal to

$$\langle \Psi|\mathcal{O}_1|\Psi \rangle = \sum_i h_{ii}. \tag{A.56}$$

Let us now demonstrate that we can also recover this result using the second quantization expression for the operator $\mathcal{O}_1$. In this case,

$$\langle\Psi|\mathcal{O}_1|\Psi\rangle = \sum_{ij} h_{ij}\langle\Psi|c_i^\dagger c_j|\Psi\rangle. \tag{A.57}$$

Since both c_j and $c_i^\dagger$ are trying to destroy an electron (c_j to the right and $c_i^\dagger$ to the left), the indices i and j must belong to the set $\{a, b, ...\}$ and thus

$$\langle\Psi|\mathcal{O}_1|\Psi\rangle = \sum_{ab} h_{ab}\langle\Psi|c_a^\dagger c_b|\Psi\rangle. \tag{A.58}$$

Using

$$c_a^\dagger c_b = \delta_{ab} - c_b c_a^\dagger \tag{A.59}$$

to move $c_a^\dagger$ to the right, we have

$$\langle\Psi|c_a^\dagger c_b|\Psi\rangle = \delta_{ab}\langle\Psi|\Psi\rangle - \langle\Psi|c_b c_a^\dagger|\Psi\rangle. \tag{A.60}$$

The second term on the right is zero since $c_a^\dagger$ is trying to create an electron in ψ_a, which is already occupied in $|\Psi\rangle$. Since $\langle\Psi|\Psi\rangle = 1$, we finally have

$$\langle\Psi|\mathcal{O}_1|\Psi\rangle = \sum_{ab} h_{ab}\delta_{ab} = \sum_a h_{aa}. \tag{A.61}$$

in agreement with the first quantization result above. We can proceed in a similar way to demonstrate the result for two-electron operators (see Exercise A.4 at the end of this appendix).

Note that the form of the operators above is independent of the number of electrons. One of the advantages of second quantization is that it treats systems with different numbers of particles on an equal footing. This is particularly convenient when one is dealing with infinite systems such as solids or molecular junctions.

A.2.4   *Some special Hamiltonians*

Our description of the electronic structure of any electronic system will start always by presenting the corresponding Hamiltonian. In this sense, it is important to get familiar with the form that some basic Hamiltonians adopt in second quantization.

The Hamiltonian of an electron system has the following generic form in first quantization

$$H = \sum_i \left[\frac{p_i^2}{2m} + U(\mathbf{r}_i)\right] + \frac{1}{2}\sum_{i\neq j} V(\mathbf{r}_i - \mathbf{r}_j) \tag{A.62}$$

with the electrons interacting with a potential $U(\mathbf{r})$, such as the lattice potential in a solid, and with each other through particle-particle interactions $V(\mathbf{r}_i - \mathbf{r}_j)$, typically the Coulomb interaction. As we have learned above, this Hamiltonian can be written in terms of the fermionic creation and annihilation operators as

$$H = \sum_{ij} h_{ij} c_i^\dagger c_j + \frac{1}{2} \sum_{ijkl} V_{ijkl} c_i^\dagger c_j^\dagger c_l c_k, \tag{A.63}$$

where

$$h_{ij} = \int d\mathbf{r} \, \psi_i^*(\mathbf{r}) \left[-\frac{\hbar^2 \nabla^2}{2m} + U(\mathbf{r}) \right] \psi_j(\mathbf{r})$$

$$V_{ijkl} = \int d\mathbf{r}_1 \int d\mathbf{r}_2 \, \psi_i^*(\mathbf{r}_1) \psi_j^*(\mathbf{r}_2) V(\mathbf{r}_1 - \mathbf{r}_2) \psi_k(\mathbf{r}_1) \psi_l(\mathbf{r}_2).$$

The precise form of this Hamiltonian depends primarily on the single-particle basis $\{\psi_i\}$ used. For the study of electron in solids, a popular basis set is plane waves: $\psi_i(\mathbf{r}) = \psi_{\mathbf{k},\sigma}(\mathbf{r}) = L^{-3/2} e^{i\mathbf{k}\cdot\mathbf{r}} u_\sigma$, where $\mathbf{k}$ is the electron momentum, σ is the spin index and u_σ is a spinor. The Hamiltonian then has the form

$$H = \sum_{\mathbf{k}\sigma} \epsilon_\mathbf{k} c_{\mathbf{k}\sigma}^\dagger c_{\mathbf{k}\sigma} + \sum_q U(\mathbf{q}) \rho_\mathbf{q} + \frac{1}{2V} \sum_{\mathbf{k}\mathbf{k}'\mathbf{q}\sigma\sigma'} v_\mathbf{q} c_{\mathbf{k}+\mathbf{q}\sigma}^\dagger c_{\mathbf{k}'-\mathbf{q}\sigma'}^\dagger c_{\mathbf{k}'\sigma'} c_{\mathbf{k}\sigma}, \tag{A.64}$$

where V is the total volume of the system and $\epsilon_\mathbf{k} = \hbar^2 k^2/2m$. The second term represents the interaction between the electrons and the atoms or ions of the solid, where $\rho_\mathbf{q}$ is the electron density operator given by

$$\rho_\mathbf{q} = \sum_{\mathbf{k}\sigma} c_{\mathbf{k}+\mathbf{q}\sigma}^\dagger c_{\mathbf{k}\sigma}. \tag{A.65}$$

Finally, $v_\mathbf{q}$ is the Fourier transform of the Coulomb potential e^2/r and it is given by $v_\mathbf{q} = 4\pi e^2/q^2$.

The full electron gas Hamiltonian of Eq. (A.64) is too complicated and it is often approximated by a model Hamiltonian which has a simpler form. Some of these popular models are discussed next.

The *homogeneous electron gas* is a model which is studied frequently to learn about correlation effects. It has the Hamiltonian

$$H = \sum_{\mathbf{k}\sigma} \epsilon_\mathbf{k} c_{\mathbf{k}\sigma}^\dagger c_{\mathbf{k}\sigma} + \frac{1}{2V} \sum_{\mathbf{k}\mathbf{k}',\mathbf{q}\neq 0, \sigma\sigma'} v_\mathbf{q} c_{\mathbf{k}+\mathbf{q}\sigma}^\dagger c_{\mathbf{k}'-\mathbf{q}\sigma'}^\dagger c_{\mathbf{k}'\sigma'} c_{\mathbf{k}\sigma}. \tag{A.66}$$

The basic premise is to get rid of the atoms and to replace them with a uniform positive background charge of density n_0. The homogeneous electron gas is also called *jellium* model. One can think of taking the positive charge of the ions and spreading it uniformly about the unit cell of the crystal. Of course, the homogeneous electron gas has no crystal structure. To preserve charge neutrality, the average particle density of the electron gas must also be n_0. This model, although a bit academic, has played a key role to understand basic issues about the Coulomb interaction of a many-particle system. A detailed discussion of this model can be found in Ref. [174].

The plane-wave model is often a poor approximation of electron behavior in solids where the electrons are localized on atomic sites and only occasionally hop to neighboring sites. This behavior is described by the *tight-binding model*, where the basis is formed by localized atomic-like orbitals.[3] One simple form of this model is bilinear in the operators:

$$H = \sum_{ij\sigma} t_{ij} c_{i\sigma}^{\dagger} c_{j\sigma}. \qquad (A.67)$$

The index j denotes a site at point $\mathbf{R}_j$, while i represents the nearest neighbor atoms. The matrix elements t_{ij} are given by

$$t_{ij} = \int d\mathbf{r}\, \phi^*(\mathbf{r} - \mathbf{R}_i) \left[-\frac{\hbar^2}{2m} \nabla^2 + U(\mathbf{r}) \right] \phi(\mathbf{r} - \mathbf{R}_j), \qquad (A.68)$$

where the orbitals $\phi(\mathbf{r})$ are localized in the sites $\mathbf{R}_i$ and $\mathbf{R}_j$. Thus, the element t_{ij} (for $j \neq i$) represents processes where the electron jumps from site j to i, while t_{ii} is the site energy. Simple versions of the model usually have a single orbital state for each atomic site. More realistic versions of the tight-binding model allow for multiple orbitals characteristic of p- or d-electrons (see Chapter 9).

The tight-binding Hamiltonian may also contain the Coulomb interaction between electrons. In its most general form, the interaction term is

$$\frac{1}{2} \sum_{ijkl} V_{ijkl} c_i^{\dagger} c_j^{\dagger} c_l c_k \qquad (A.69)$$

where

$$V_{ijkl} = \int d\mathbf{r}_1 \int d\mathbf{r}_2\, \phi^*(\mathbf{r}_1 - \mathbf{R}_i) \psi^*(\mathbf{r}_2 - \mathbf{R}_j) \frac{e^2}{|\mathbf{r}_1 - \mathbf{r}_2|} \psi(\mathbf{r}_1 - \mathbf{R}_k) \psi(\mathbf{r}_2 - \mathbf{R}_l).$$

The four orbitals could be centered on four different sites. These are called four-center integrals. They are usually small and often neglected in many-body calculations.

---

[3]Tight-binding models and their used in molecular electronics are the subject of Chapter 9.

The *Hubbard model* [177] retains only the Coulomb integral which is the largest, namely that in which all four orbitals $\phi(\mathbf{r})$ are centered on the same site. This term describes the interaction between two electrons which are on the same atom. Since two electrons cannot be in the same state, the two on the same atom must be in different atomic states. In the simplest model, which considers only a single orbital state on each atom, the two electrons must have different spin configurations. One has spin up, while the other has spin down. The Hubbard model considers the following Hamiltonian

$$H = \epsilon_0 \sum_i c_{i\sigma}^{\dagger} c_{i\sigma} + t \sum_{ij} c_{i\sigma}^{\dagger} c_{j\sigma} + U \sum_i n_{i\uparrow} n_{i\downarrow} \qquad (A.70)$$

$$U = V_{iiii} = \int d\mathbf{r}_1 \int d\mathbf{r}_2 \, |\phi^*(\mathbf{r}_1)|^2 \frac{e^2}{|\mathbf{r}_1 - \mathbf{r}_2|} |\phi^*(\mathbf{r}_1)|^2. \qquad (A.71)$$

The hopping term is usually limited to nearest neighbors. The Hamiltonian was also introduced by Gutzwiller [178], who studied the properties of electrons in d-bands in ferromagnets. It is thought to be a good model for electron conduction in narrow band materials, for example, in transition metal oxides. The Hubbard model has been investigated thoroughly over the past forty years, and its properties are starting to be understood [179].

A simplified version of the Hubbard model is the so-called *Anderson model* [180]. In this model the Hubbard-like interaction is considered to be only present in a single site. This model was introduced to study the interaction of localized magnetic impurities with the conduction electrons of a metal. In recent years, it has been widely used to study the electronic and transport properties of quantum dots and molecular transistors (see Chapters 15 and 17).

A.3 Second quantization for bosons

The second quantization formalism for bosons was already outlined when we discussed the physics of phonons in section A.1.2. Anyway, for the sake of completeness, we summarize here the main results of this formalism for the case of bosons:

(1) The many-body wave functions for a bosonic system has to be symmetric with respect to the particle exchange. In this sense, the Slater

determinant of the fermionic case is replaced now by

$$\Phi_{k_1,k_2,\ldots,k_N}(\mathbf{r}_1,\ldots,\mathbf{r}_N) = \left(\frac{n_1!n_2!\ldots}{N!}\right)^{1/2}\sum_{P}(+1)P[\phi_{k_1}(\mathbf{r}_1)\ldots\phi_{k_N}(\mathbf{r}_N)]$$

$$= \Phi_{n_1,\ldots,n_i,\ldots}(\mathbf{r}_1,\ldots,\mathbf{r}_N) = |n_1,\ldots,n_i,\ldots\rangle, \tag{A.72}$$

where P is the permutation operator which interchanges the $\mathbf{r}_i$'s in all possible ways. Moreover, the occupation number can take any integer value: $n_i = 0, 1, 2, 3, \ldots$.

(2) The $c_i^\dagger$, c_i operators are now defined by

$$c_i^\dagger|n_1,\ldots,n_i,\ldots\rangle = \sqrt{n_i+1}|n_1,\ldots,n_i+1,\ldots\rangle \tag{A.73}$$

$$c_i|n_1,\ldots,n_i,\ldots\rangle = \sqrt{n_i}|n_1,\ldots,n_i-1,\ldots\rangle. \tag{A.74}$$

(3) The commutation relations now read:

$$\left[c_l, c_k^\dagger\right] = c_l c_k^\dagger - c_k^\dagger c_l = \delta_{lk} \; ; \; [c_l, c_k] = \left[c_l^\dagger, c_k^\dagger\right] = 0. \tag{A.75}$$

(4) The one- and two-body operators are expressed in terms of the creation and annihilation operators in the same way as in the fermion case.

A.4   Exercises

A.1 Find $c_1 c_5^\dagger c_2|111000\ldots\rangle$.

A.2 Find $\langle\Psi|\hat{N}|\Psi\rangle$, where $|\Psi\rangle = A|100\ldots\rangle + B|111000\ldots\rangle$, and $\hat{N}$ is the total number operator.

A.3 Demonstrate the first and third fermion commutation rules of Eq. (A.47).

A.4 Verify that for a two-particle system, the matrix elements of the two-body operator $\mathcal{O}_2$ in Eq. A.53 between two-particle states $\langle 0\ldots 1_p\ldots 1_q\ldots|$ and $|0\ldots 1_r\ldots 1_s\ldots\rangle$ are the same as the matrix elements of the first quantization version of $\mathcal{O}_2$ taken between the corresponding two-particle Slater determinants.

A.5 Prove that the components of the total spin operator, S, in second quantized form are:

$$S_x = \frac{1}{2}\sum_k \left(c_{k\uparrow}^\dagger c_{k\downarrow} + c_{k\downarrow}^\dagger c_{k\uparrow}\right)$$

$$S_y = -\frac{i}{2}\sum_k \left(c_{k\uparrow}^\dagger c_{k\downarrow} - c_{k\downarrow}^\dagger c_{k\uparrow}\right)$$

$$S_z = \frac{1}{2}\sum_k \left(c_{k\uparrow}^\dagger c_{k\uparrow} - c_{k\downarrow}^\dagger c_{k\downarrow}\right).$$

Bibliography

[1] R. Chau, B. Doyle, S. Datta, J. Kavaliero, K. Zhang, *Integrated nanoelectronics for the future*, Nature Mat. **6**, 810 (2007).

[2] J.R. Heath and M.A. Ratner, *Molecular electronics*, Physics Today **56**, 43 (2003).

[3] H. Choi and C.C.M. Mody, *The long history of molecular electronics: Microelectronics origins of nanotechnology*, Social Studies of Science **39**, 11 (2009).

[4] A.R. von Hippel, *Molecular engineering*, Science **123**, 315 (1956).

[5] A.R. von Hippel (ed.), *Molecular Science and Molecular Engineering*, (Technology Press of MIT, Cambridge, USA and Wiley, New York, USA, 1959).

[6] B. Mann and H. Kuhn, *Tunneling through fatty acid salt monolayers*, J. Appl. Phys. **42**, 4398 (1971).

[7] E.E. Polymeropoulos, D. Möbius, H. Kuhn, *Monolayer assemblies with functional units of sensitizing and conducting molecular components: Photovoltage, dark conduction and photoconduction in systems with aluminium and barium electrodes*, Thin Solid Films **68**, 173 (1980).

[8] A. Aviram and M. Ratner, *Molecular rectifiers*, Chem. Phys. Lett. **29**, 277 (1974).

[9] R.P. Feynman, *There's plenty of room at the bottom*, Engineering and Science **43**, 22 (1960).

[10] G. Binnig, H. Rohrer, Ch. Gerber, E. Weibel, *Vacuum tunneling*, Physica B **109-110**, 2075 (1982).

[11] G. Binnig, H. Rohrer, Ch. Gerber, E. Weibel, *Surface studies by scanning tunneling microscopy*, Phys. Rev. Lett. **49**, 57 (1982).

[12] G. Binnig and H. Rohrer, *In touch with atoms*, Rev. Mod. Phys. **71**, S324 (1999).

[13] M. Fujihira, K. Nishiyama, H. Yamada, *Photoelectrochemical responses of optically transparent electrodes modified with Langmuir-Blodgett films consisting of surfactant derivatives of electron donor, acceptor, and sensitizer molecules*, Thin Solid Films **132**, 77 (1985).

[14] R.M. Metzger, B. Chen, U. Höpfner, M.V. Lakshmikantham, D. Vuil-

laume, T. Kawai, X. Wu, H. Tachibana, T.V. Hughes, H. Sakurai, J.W. Baldwin, C. Hosch, M.P. Cava, L. Brehmer, G.J. Ashwell, *Unimolecular electrical rectification in hexadecylquinolinium tricyanoquinodimethanide,* J. Am. Chem. Soc. **119**, 10455 (1997).

[15] N. Agraït, A. Levy Yeyati, J.M. van Ruitenbeek, *Quantum properties of atomic-sized conductors,* Phys. Rep. **377**, 81 (2003).

[16] M.A. Reed, C. Zhou, C.J. Muller, T.P. Burgin, and J.M. Tour, *Conductance of a molecular junction,* Science **278**, 252 (1997).

[17] C. Zhou, M.R. Deshpande, M.A. Reed, and J.M. Tour, *Fabrication of nanoscale metal/self-assembled monolayer/metal heterostructures,* Appl. Phys. Lett. **71**, 611 (1997).

[18] J. Chen, M.A. Reed, A.M. Rawlett, and J.M. Tour, *Large on-off ratios and negative differential resistance in a molecular electronic device,* Science **286**, 1550 (1999).

[19] C.P. Collier, E. Wong, M. Belohradsky, F. Raymo, J.F. Stoddart, P.J. Kuekes, R.S. Williams, J.R. Heath, *Electronically configurable molecular-based logic gates,* Science **285**, 391 (1999).

[20] C.P. Collier, G. Mattersteig, Y. Li, E.W. Wong, K. Beverly, J. Sampaio, F. Raymo, J.F. Stoddart, J.R. Heath, *A [2]-catenane based solid-state electronically reconfigurable switch,* Science **289**, 1172 (2000).

[21] H. Park, A.K.L. Lim, A.P. Alivisatos, J. Park, P.L. McEuen, *Fabrication of metallic electrodes with nanometer separation by electromigration,* Appl. Phys. Lett. **75**, 301 (1999).

[22] H. Park, J. Park, A.K.L. Lim, E.H. Anderson, A.P. Alivisatos, P.L. McEuen, *Nanomechanical oscillations in a single-C_{60} transistor,* Nature **407**, 57 (2000).

[23] J. Park, A.N. Pasupathy, J.I. Goldsmith, C. Chang, Y. Yaish, J.R. Petta, M. Rinkoski, J.P. Sethna, H.D. Abruña, P.L. McEuen, D.C. Ralph, *Coulomb blockade and the Kondo effect in single-atom transistors,* Nature **417**, 722 (2002).

[24] W. Liang, M.P. Shores, M. Bockrath, J.R. Long, H. Park, *Kondo resonance in a single-molecule transistor,* Nature **417**, 725 (2002).

[25] J.R. Heath, P.J. Kuekes, G. Snider, R.S. Williams, *A defect tolerant computer architecture: Opportunities for nanotechnology,* Science **280**, 1716 (1998).

[26] Y. Chen, G.-Y. Jung, D.A.A. Ohlberg, X. Li, D.R. Stewart, J.O. Jeppesen, K.A. Nielsen, J.F. Stoddart, R.S. Williams, *Nanoscale molecular-switch crossbar circuits,* Nanotechnology **14**, 462 (2003).

[27] D.R. Stewart, D.A.A. Ohlberg, P.A. Beck, Y. Chen, R.S. Williams, J.O. Jeppesen, K.A. Nielsen, J.F. Stoddart, *Molecule-independent electrical switching in Pt/organic monolayer/Ti devices,* Nano Lett. **4**, 133 (2004).

[28] R.F. Service, *Molecular electronics: Next-generation technology hits an early midlife crisis,* Science **302**, 556 (2003).

[29] J.E. Green, J.W. Choi, A. Boukai, Y. Bunimovich, E. Johnston-Halperin, E. DeIonno, Y. Luo, B.A. Sheriff, K. Xu, Y.S. Shin, H.-R. Tseng, J.F. Stoddart J.R. Heath, *A 160-kilobit molecular electronic memory patterned*

at 10^{11} *bits per square centimetre*, Nature **445**, 414 (2007).

[30] A. Nitzan, *Chemical Dynamics in Condensed Phases: Relaxation, Transfer, and Reactions in Condensed Molecular Systems*, (Oxford University Press, Oxford, UK, 2006).

[31] R. Waser (ed.), *Nanoelectronics and Information Technology*, (Wiley-VCH, Weinheim, D, 2003).

[32] M.A. Reed and J.M. Tour, *Computing with molecules*, Scientific American **282**, 86, (2000).

[33] C. Joachim, J.K. Gimzewski, A. Aviram, *Electronics using hybrid-molecular and mono-molecular devices*, Nature **408**, 541 (2000).

[34] A. Nitzan and M.A. Ratner *Electron transport in molecular wire junctions*, Science **300**, 1384 (2003).

[35] A.W. Ghost, P.S. Damle, S. Datta, A. Nitzan, *Molecular electronics: Theory and devie prospects*, MRS Bulletin June issue, 391 (2004).

[36] N.J. Tao, *Electron transport in molecular junctions*, Nature Nanotech. **1**, 173 (2006).

[37] K. Moth-Poulsen and T. Bjørnholm, *Molecular electronics with single molecules in solid-state devices*, Nature Nanotech. **4**, 551 (2009).

[38] A. Nitzan, *Electron transmission through molecules and molecular interfaces*, Ann. Rev. Phys. Chem. **52**, 681 (2001).

[39] Y. Selzer and D.L. Allara, *Single-molecule electrical junctions*, Annu. Rev. Phys. Chem. **57**, 593 (2006).

[40] F. Chen, J. Hihath, Z. Huang, X. Li, N.J. Tao, *Measurement of single-molecule conductance*, Annu. Rev. Phys. Chem. **58**, 535 (2007).

[41] H.B. Akkerman and B. de Boer, *Electrical conduction through single molecules and self-assembled monolayers*, J. Phys.: Condens. Matter **20**, 013001 (2008).

[42] J. Jortner and M.A. Ratner (eds.), *Molecular Electronics*, (Blackwell Science, USA, 1997).

[43] A. Aviram and M.A. Ratner (eds.), *Molecular Electronics I*, Vol. 852 of Annals of the New York Academy of Sciences (New York Academy of Sciences, New York, USA, 1998).

[44] A. Aviram, M.A. Ratner, V. Mujica (eds.), *Molecular Electronics II*, Vol. 960 of Annals of the New York Academy of Sciences (New York Academy of Sciences, New York, USA, 2002).

[45] J.R. Reimers, C.A. Picconatto, J.C. Ellenbogen, R. Shashidar (eds.), *Molecular Electronics III*, Vol. 1006 of Annals of the New York Academy of Sciences (New York Academy of Sciences, New York, USA, 2003).

[46] M.A. Reed and T. Lee (eds.), *Molecular Nanoelectronics*, (American Scientific Publishers, Los Angeles, USA, 2003).

[47] J.M. Tour, *Molecular Electronics: Commercial Insights, Chemistry, Devices, Architecture and Programming*, (World Scientific Publishers, New Jersey, USA, 2003).

[48] G. Cuniberti, G. Fagas, K. Richter (eds.), *Introducing Molecular Electronics*, Series: Lecture Notes in Physics, Vol. **680**, (Springer-Verlag, Heidel-

berg, Berlin, D, 2005).

[49] M.C. Petty, *Molecular Electronics: From Principles to Practice*, Wiley Series in Materials for Electronic and Optoelectronic Applications, (Wiley-VCH, Weinheim, D, 2008).

[50] S. Datta, *Electronic Transport in Mesoscopic Systems*, (Cambridge University Press, Cambridge, UK, 1995).

[51] S. Datta, *Quantum Transport: Atom to Transistor*, (Cambridge University Press, Cambridge, UK, 2005).

[52] M. Di Ventra, *Electrical Transport in Nanoscale Systems*, (Cambridge University Press, Cambridge, UK, 2008).

[53] Y.V. Nazarov and Y.M. Blater, *Quantum Transport: Introduction to Nanoscience*, (Cambridge University Press, Cambridge, UK, 2009).

[54] R. Wiesendanger, *Scanning Probe Microscopy and Spectroscopy*, (Cambridge University Press, Cambridge, UK, 1994).

[55] J.K. Gimzewski and R. Möller, *Transition from the tunneling regime to point contact studied using scanning tunneling microscopy*, Physica B **36**, 1284 (1987).

[56] L. Limot, J. Kröger, R. Berndt, A. Garcia-Lekue, W.A. Hofer, *Atom transfer and single-adatom contacts*, Phys. Rev. Lett. **94**, 126102 (2005).

[57] K. Hansen, S.K. Nielsen, M. Brandbyge, E. Lægsgaard, I. Stensgaard, F. Besenbacher, *Current-voltage curves of gold quantum point contacts revisited*, Appl. Phys. Lett. **77**, 708 (2000).

[58] G. Rubio, N. Agraït, S. Vieira, *Atomic-sized metallic contacts: mechanical properties and electronic transport*, Phys. Rev. Lett. **76**, 2302 (1996).

[59] T. Trenkler, T. Hantschel, R. Stephenson, P. De Wolf, W. Vandervorst, L. Hellemans, A. Malavé, D. Büchel, E. Oesterschulze, W. Kulisch, P. Niedermann, T. Sulzbach, and O. Ohlsson, *Evaluating probes for electrical atomic force microscopy*, J. Vac. Sci. Technol. B **18**, 418 (2000).

[60] J.L. Costa-Krämer, N. García, P. García-Mochales, P.A. Serena, *Nanowire formation in macroscopic metallic contacts: quantum mechanical conductance tapping a table top*, Surf. Sci. **342**, L1144 (1995).

[61] J.L. Costa-Krämer, N. García, P. García-Mochales, P.A. Serena, M.I. Marqués, and A. Correia, *Conductance quantization in nanowires formed between micro and macroscopic metallic electrodes*, Phys. Rev. B **55**, 5416 (1997).

[62] V. Rodrigues and D. Ugarte, *Real-time imaging of atomistic process in one-atom-thick metal junctions*, Phys. Rev. B **63**, 073405 (2001).

[63] H. Ohnishi, Y. Kondo, K. Takayanagi. *Quantized conductance through individual rows of suspended gold atoms*, Nature **395**, 780 (1998).

[64] T. Kizuka, *Atomic configuration and mechanical and electrical properties of stable gold wires of single-atom width*, Phys. Rev. B **77**, 155401 (2008).

[65] V. Rodrigues, J. Bettini, A.R. Rocha, L.G.C. Rego, D. Ugarte, *Quantum conductance in silver nanowires: correlation between atomic structure and transport properties*, Phys. Rev. B **65**, 153402 (2002).

[66] Y.G. Naidyuk and I.K. Yanson, *Point-contact spectroscopy*, (Springer-Verlag, Heidelberg, Berlin, D, 2005).

[67] J. Moreland and P.K. Hansma, *Electromagnetic squeezer for compressing squeezable electron tunneling junctions*, Rev. Sci. Instrum. **55**, 399 (1984).

[68] J. Moreland and J.W. Eki, *Electron tunneling experiments using Nb-Sn "break" junctions*, J. Appl. Phys. **58**, 3888 (1985).

[69] C.J. Muller, J.M. van Ruitenbeek, L.J. de Jongh, *Experimental observation of the transition from weak link to tunnel junction*, Physica C **191**, 485 (1992).

[70] J.M. Krans, C.J. Muller, I.K. Yanson, Th.C.M. Govaert, R. Hesper, J.M. van Ruitenbeek, *One-atom point contacts*, Phys. Rev. B **48**, 14721 (1993).

[71] J.M. van Ruitenbeek, A. Alvarez, I. Piñeyro, C. Grahmann, P. Joyez, M.H. Devoret, D. Esteve, C. Urbina, *Adjustable nanofabricated atomic size contacts*, Rev. Sci. Instrum. **67**, 108 (1996).

[72] A.I. Yanson, I.K. Yanson, J.M. van Ruitenbeek, *Observation of shell structure in sodium nanowires*, Nature **400**, 144 (1999).

[73] N. van der Post, *Superconductivity and magnetism in nano-scale junctions*, PhD thesis, Leiden, The Netherlands (1997).

[74] J.M. Krans, *Size effects in atomic-scale point contacts*, PhD thesis, Leiden, The Netherlands (1996).

[75] A.M.C. Valkering, A.I. Mares, C. Untiedt, K.B. Gavan, T.H. Oosterkamp, J.M. van Ruitenbeek, *A force sensor for atomic point contacts*, Rev. Sci. Instrum. **76**, 103903 (2005).

[76] S.A.G. Vrouwe, E. van der Giessen, S.J. van der Molen, D. Dulic, M.L. Trouwborst, B.J. van Wees, *Mechanics of lithographically defined break junctions*, Phys. Rev. B **71**, 35313 (2005).

[77] E. Scheer, P. Joyez, D. Esteve, C. Urbina, M.H. Devoret, *Conduction channel transmissions of atomic-size aluminum contacts*, Phys. Rev. Lett. **78**, 3535 (1997).

[78] L. Grüter, M.T. González, R. Huber, M. Calame, C. Schönenberger, *Electrical conductance of atomic contacts in liquid environments*, Small **1**, 1067 (2005).

[79] H.S.J. van der Zant, Y.-V, Kervennic, M. Poot, K. O'Neill, Z. de Groot, J.M. Thijssen, H.B. Heersche, N. Stuhr-Hansen, T. Bjørnholm, D. Vanmaekelbergh, C.A. van Walree, L.W. Jenneskens, *Molecular three-terminal devices: fabrication and experiment*, Faraday Discuss. **131**, 347 (2006).

[80] Z.M. Wu, M. Steinacher, R. Huber, M. Calame, S.J. van der Molen, C. Schönenberger, *Feedback controlled electromigration in four-terminal nano-junctions*, Appl. Phys. Lett. **91**, 053118 (2007).

[81] M.L. Trouwborst, S.J. van der Molen, B.J. van Wees, *The role of Joule heating in the formation of nanogaps by electromigration*, J. Appl. Phys. **99**, 114316 (2006).

[82] H.B. Heersche, G. Lientschnig, K. O'Neill, H.S.J. van der Zant, H.W. Zandbergen, *In situ imaging of electromigration-induced nanogap formation by transmission electron microscopy*, Appl. Phys. Lett. **91**, 072107 (2007).

[83] R. Hoffmann, D. Weissenberger, J. Hawecker, D. Stöffler, *Conductance of gold nanojunctions thinned by electromigration*, Appl. Phys. Lett. **93**,

043118 (2008).

[84] A. Lodder, *Calculations of the screening of the charge of a proton migrating in a metal*, Phys. Rev. B **74**, 045111 (2006).

[85] K.-N. Tu, J.W. Mayer, L.C. Feldman, *Electronic Thin Film Science for Electrical Engineers and Materials Scientists*, (McMillan, New York, USA, 1992).

[86] A.R. Champagne, A.N. Pasupathy, D.C. Ralph, *Mechanically adjustable and electrically gated single-molecule transistors*, Nano Lett. **5**, 305 (2005).

[87] C. Schirm, H.-F. Pernau, E. Scheer, *Switchable wiring for high-resolution electronic measurements at very low temperatures*, Rev. Sci. Instrum. **80**, 024704 (2009).

[88] J. Heinze, *Cyclic voltammetry-electrochemical spectroscopy*, Angew. Chem. Int. Ed. **23**, 831 (1984).

[89] F.-Q. Xie, L. Nittler, Ch. Obermair, Th. Schimmel, *Gate-controlled atomic quantum switch*, Phys. Rev. Lett. **93**, 128303 (2004).

[90] K. Terabe, T. Hasegawa, T. Nakayama, M. Aono, *Quantized-conductance atomic switch*, Nature **433**, 47 (2005).

[91] A.F. Morpurgo, C.M. Marcus, D.B. Robinson, *Controlled fabrication of metallic electrodes with atomic separation*, Appl. Phys. Lett. **74**, 2084 (1999).

[92] M. Reyes-Calvo, A.I. Mares, V. Climent, J.M. van Ruitenbeek, C. Untiedt, *Formation of atomic-sized contacts controlled by electrochemical methods*, phys. stat. sol. (a) **204**, 1677 (2007).

[93] C.Z. Li and N.J. Tao, *Quantum transport in metallic nanowires fabricated by electrochemical deposition/dissolution*, Appl. Phys. Lett. **72**, 23 (1998).

[94] R. Waitz, O. Schecker, E. Scheer, *Nanofabricated adjustable multicontact devices on membranes*, Rev. Sci. Instrum. **79**, 093901 (2008).

[95] R. Waitz, R. Hoffmann, E. Scheer, unpublished.

[96] C.A. Martin, R.H.M. Smit, H.S.J. van der Zant, J.M. van Ruitenbeek, *A nanoelectromechanical single-atom switch*, Nano Lett. **9**, 2940 (2009).

[97] A. Kumar, L. Saminadayar, D.C. Glattli, Y. Jin, B. Etienne, *Experimental test of the quantum shot noise reduction theory*, Phys. Rev. Lett. **76**, 2778 (1996).

[98] R. Cron, M.F. Goffman, D. Esteve, C. Urbina, *Multiple-charge-quanta shot noise in superconducting atomic contacts*, Phys. Rev. Lett. **86**, 4104 (2001).

[99] R. Aguado and L.P. Kouwenhoven, *Double quantum dots as detectors of high-frequency quantum noise in mesoscopic conductors*, Phys. Rev. Lett. **84**, 1986 (2000).

[100] R. Deblock, E. Onac, L. Gurevich, L.P. Kouwenhoven, *Detection of quantum noise from an electrically driven two-level system*, Science **301**, 203 (2003).

[101] P. Reddy, S.-Y. Jang, R. Segalman, A. Majumdar, *Thermoelectricity in molecular junctions*, Science **315**, 1568 (2007).

[102] K. Baheti, J.A. Malen P. Doak, P. Reddy, S.-Y. Jang, T.D. Tilley, A. Majumdar, R. A. Segalman, *Probing the chemistry of molecular hetero-*

junctions using thermoelectricity, Nano Lett. **8**, 715 (2008).

[103] P.J.F. Harris, *Carbon Nanotube Science - Synthesis, Properties and Applications*, (Cambridge University Press, Cambridge, UK, 2009).

[104] D. Krüger, H. Fuchs, R. Rousseau, D. Marx, M. Parinello, *Pulling monatomic gold wires with single molecules: An ab initio simulation*, Phys. Rev. Lett. **89**, 186402 (2002).

[105] J.Ch. Love, L.A. Estroff, J.K. Kriebel, R.G. Nuzzo, G.M. Whitesides, *Self-assembled monolayers of thiolates on metals as a form of nanotechnology*. Chem. Rev. **105**, 1103 (2005).

[106] A. Ulman, *An Introduction to Ultrathin Organic Films From Langmuir-Blodgett to Self-Assembly*, (Academic Press, San Diego, USA, 1991).

[107] I.R. Peterson, *Langmuir-Blodgett films*, J. Phys. D: Appl. Phys. **23**, 379 (1990).

[108] O.V. Salata, *Tools of nanotechnology: Electrospray*, Current Nanoscience **1**, 25 (2005).

[109] A. de Picciotto, J.E. Klare, C. Nuckolls, K. Baldwin, A. Erbe, and R. Willett, *Prevalence of Coulomb blockade in electro-migrated junctions with conjugated and non-conjugated molecules*, Nanotechnology **16**, 3110 (2005).

[110] H. van der Zant, Y-V. Kervennic, M. Poot, K. O'Neill, Z. de Groot, J.M. Thijssen, H.B. Heersche, N. Stuhr-Hansen, T. Bjørnholm, D. Vanmaekelbergh, C.A. van Walree, and L.W. Jenneskens, *Molecular three-terminal devices: fabrication and measurements*, Farad. Discuss. **131**, 347 (2006).

[111] A. Erbe, W. Jiang, Z. Bao, D. Abusch-Magder, D.M. Tennant, E. Garfunkel, N. Zhitenev, *Nanoscale patterning in application to materials and device structures*, J. Vac. Sci. Technol. B **23**, 3132 (2005).

[112] A.Yu. Kasumov, M. Kociak, M. Ferrier, R. Deblock, S. Guéron, B. Reulet, I. Khodos, O. Stéphan, and H. Bouchiat, *Quantum transport through carbon nanotubes: Proximity-induced and intrinsic superconductivity*, Phys. Rev. B **68**, 214521 (2003).

[113] A.Yu. Kasumov, K. Tsukagoshi, M. Kawamura, T. Kobayashi, Y. Aoyagi, K. Senba, T. Kodama, H. Nishikawa, I. Ikemoto, K. Kikuchi, V.T. Volkov, Yu.A. Kasumov, R. Deblock, S. Guéron, and H. Bouchiat, *Proximity effect in a superconductor-metallofullerene-superconductor molecular junction*, Phys. Rev. B **72**, 033414 (2005).

[114] T. Dadosh, Y. Gordin, R. Krahne, I. Khivrich, D. Mahalu, V. Frydman, J. Sperling, A. Yacoby, and I. Bar-Joseph, *Measurement of the conductance of single conjugated molecules*, Nature **436**, 677 (2005).

[115] C. Joachim, J.K. Gimzewski, R.R. Schlittler, C. Chavy, *Electronic transparence of a single C_{60} molecule*, Phys. Rev. Lett. **74**, 2102 (1995).

[116] N. Néel, J. Kröger, L. Limot, T. Frederiksen, M. Brandbyge, and R. Berndt, *Controlled contact to a C_{60} molecule*, Phys. Rev. Lett. **98**, 065501 (2007).

[117] J. Repp, G. Meyer, S. Paavilainen, F.E. Olsson, M. Persson, *Imaging bond formation between a gold atom and pentacene on an insulating surface*, Science **312**, 1196 (2006).

[118] F. Pump, R. Temirov, O. Neucheva, S. Soubatch, F.S. Tautz, M. Rohlfing, G. Cuniberti, *Quantum transport through STM-lifted single PTCDA molecules*, Appl. Phys. A **93**, 335 (2008).

[119] R. Temirov, A. Lassise, F.B. Anders, F.S. Tautz, *Kondo effect by controlled cleavage of a single-molecule contact*, Nanotechnology **19**, 065401 (2008).

[120] L. Gross, F. Mohn, P. Liljeroth, J. Repp, F.J. Giessibl, G. Meyer, *Measuring the charge state of an adatom with noncontact atomic force microscopy*, Science **324**, 1428 (2009).

[121] L. Gross, F. Mohn, N. Moll, P. Liljeroth, G. Meyer, *The chemical structure of a molecule resolved by atomic force microscopy*, Science **325**, 1110 (2009).

[122] R. Temirov, S. Soubatch, O. Neucheva, A.C. Lassise, F.S. Tautz, *A novel method achieving ultra-high geometrical resolution in scanning tunnelling microscopy*, New J. Phys. **10**, 053012 (2008).

[123] L.A. Bumm, J.J. Arnold, M.T. Cygan, T.D. Dunbar, T.P. Burgin, L. Jones, D.L. Allara, J. M. Tour, P. S. Weiss, *Are single molecular wires conducting?*, Science **271**, 1705 (1996).

[124] R.P. Andres, T. Bein, M. Dorogi, S. Feng, J.I. Henderson, C.P. Kubiak, W. Mahoney, R.G. Osifchin, R. Reifenberger, *"Coulomb staircase" at room temperature in a self-assembled molecular nanostructure.* Science **272**, 1323 (1996).

[125] X.D. Cui, A. Primak, X. Zarate, J. Tomfohr, O.F. Sankey, A.L. Moore, T.A. Moore, D. Gust, G. Harris, S.M. Lindsay, *Reproducible measurement of single-molecule conductivity*, Science **294**, 571 (2001).

[126] L. Venkataraman, J.E. Klare, I.W. Tam, C. Nuckolls, M.S. Hybertsen, M.L. Steigerwald, *Single-molecule circuits with well-defined molecular conductance*, Nano Lett. **6**, 458 (2006).

[127] R.H.M. Smit, Y. Noat, C. Untiedt, N.D. Lang, M.C. van Hemert, J.M. van Ruitenbeek, *Measurement of the conductance of a hydrogen molecule*, Nature **419**, 906 (2002).

[128] T. Böhler, A. Edtbauer, E. Scheer, *Conductance of individual C_{60} molecules measured with controllable gold electrodes*, Phys. Rev. B **76**, 125432 (2007).

[129] K.S. Ralls, R.A. Buhrman, R.C. Tiberio, *Fabrication of thin-film nanobridges*, Appl. Phys. Lett. **55**, 2459 (1989).

[130] W. Wang, T. Lee, M.A. Reed, *Mechanism of electron conduction in self-assembled alkanethiol monolayer devices*, Phys. Rev. B **68**, 035416 (2003).

[131] H.B. Akkerman, P.W.M. Blom, D.M. de Leeuw, B. de Boer, *Towards molecular electronics with large-area molecular junctions*, Nature **441**, 69 (2006).

[132] C. Kreuter, S. Bächle, E. Scheer, A. Erbe, *Electrical characterization of alkane monolayers using micro transfer printing: tunneling and molecular transport*, New J. Phys. **10**, 075001 (2008).

[133] J. Liao, L. Bernard, M. Langer, C. Schönenberger, M. Calame, *Reversible formation of molecular junctions in two-dimensional nanoparticle arrays*, Adv. Mater. **18**, 2444 (2006).

[134] L. Bernard, Y. Kamdzhilov, M. Calame, S.J. van der Molen, J. Liao, C. Schönenberger, *Spectroscopy of molecular junction networks obtained by place exchange in 2D nanoparticle ararys*, J. Phys. Chem. C **111**, 18445 (2007).

[135] S.J. van der Molen, J. Liao, T. Kudernac, J.S. Agustsson, L. Bernard, M. Calame, B.J. van Wees, B.L. Feringa, C. Schönenberger, *Light-controlled conductance switching of ordered metal-molecule-metal devices*, Nano Lett. **9**, 76 (2009).

[136] S. Washburn, *Resistance fluctuations in small samples: Be careful when playing with Ohm's law*, in B. Kramer (Ed.), *Quantum Coherence in Mesoscopic Systems*, p. 341, (Plenum Press, New York, USA, 1991).

[137] R. Landauer, *Spatial variation of currents and fields due to localized scatterers in metallic conduction*, IBM J. Res. Dev. **1**, 223 (1957).

[138] J.C. Maxwell, *A treatise on electricity and magnetism*, volume 1, (Dover Publication, New York, USA, 1954).

[139] Yu.V. Sharvin, *A possible method for studying Fermi surfaces*, Sov. Phys.-JETP **21**, 655 (1965) [Zh. Eksp. Teor. Fiz. **48**, 984 (1965)].

[140] I.O. Kulik, A.N. Omel'yanchuk, R.I. Shekhter, *Electrical conductivity of point microbridges and phonon and impurity spectroscopy in normal metals*, Sov. J. Low. Temp. Phys. **3**, 740 (1977).

[141] H. Weyl, *Gruppentheorie in Quantenmechanik*, (Hirzel, Leipzig, D, 1931).

[142] V.I. Fal'ko and G.B. Lesovik, *Quantum conductance fluctuations in 3d ballistic adiabatic wires*, Solid State Comm. **84**, 835 (1992).

[143] J.A. Torres, J.I. Pascual, J.J. Saenz, *Theory of conduction through narrow constrictions in a three-dimensional electron gas*, Phys. Rev. B **49**, 16581 (1994).

[144] Y. Imry, *Physics of mesoscopic systems*, in G. Grinstein and G. Mazenko (Eds.), *Directions in Condensed Matter Physics*, p. 101, (World Scientific Publishers, Singapore, 1986).

[145] D. Bohm, *Quantum Theory*, (Prentice-Hall, Englewood Cliffs, USA, 1951).

[146] A. Messiah, *Quantum Mechanics*, (Dover Publication, New York, USA, 1999).

[147] J.G. Simmons, *Generalized formula for the electric tunnel effect between similar electrodes separated by a thin insulating film*, J. Appl. Phys. **34**, 1793 (1963).

[148] K.H. Grundlach, *Zur Berechnung des Tunnel- stroms durch eine trapezförmige Potentialstufe*, Solid-State Electron. **9**, 946 (1966).

[149] M. Büttiker, *Scattering theory of current and intensity noise correlations in conductors and wave guides*, Phys. Rev. B **46**, 12485 (1992).

[150] Ya. M. Blanter and M. Büttiker, *Shot noise in mesoscopic conductors*, Phys. Rep. **336**, 2 (2000).

[151] C.W.J. Beenakker, H. van Houten, *Quantum Transport in Semiconductor Nanostructures*, Solid State Physics **44**, 1 (1991).

[152] H. van Houten, C.W.J. Beenakker, *Quantum Point Contacts*, Physics Today, p. 22, July (1996).

[153] B.J. van Wees, H. van Houten, C.W.J. Beenakker, J.G. Williamson, L.P.

Kouwenhoven, D. van der Marel, C.T. Foxon, *Quantized conductance of point contacts in a two-dimensional electron gas*, Phys. Rev. Lett. **60**, 848 (1988).

[154] D.A. Wharam, T.J. Thornton, R. Newbury, M. Pepper, H. Ahmed, J.E.F. Frost, D.G. Hasko, D.C. Peacock, D.A. Ritchie, G.A.C. Jones, *One-dimensional transport and the quantisation of the ballistic resistance*, J. Phys. C **21**, L209 (1988).

[155] M.J.M. de Jong and C.W.J. Beenakker, *Shot noise in mesoscopic systems*, in L.L. Sohn, L.P. Kouwenhoven, G. Schön (Eds.), *Mesoscopic Electron Transport, NATO-ASI Series E*, Vol. 345, p. 225, (Kluwer Academic Publishers, Dordrecht, NL, 1997).

[156] L. Saminadayar, D.C. Glattli, Y. Jin, B. Etienne, *Observation of the e/3 fractionally charged Laughlin quasiparticle*, Phys. Rev. Lett. **79**, 2526 (1997).

[157] R. de Picciotto, M. Reznikov, M. Heiblum, V. Umansky, G. Bunin, D. Mahalu, *Direct observation of a fractional charge*, Nature **389**, 162 (1997).

[158] J.C. Cuevas, A. Martin-Rodero, A. Levy Yeyati, *Shot noise and coherent multiple charge transfer in superconducting quantum point contacts*, Phys. Rev. Lett. **82**, 4086 (1999).

[159] Y. Naveh and D.V. Averin, *Nonequilibrium current noise in mesoscopic disordered superconductor-normal-metal-superconductor junctions*, Phys. Rev. Lett. **82**, 4090 (1999).

[160] U. Sivan and Y. Imry, *Multichannel landauer formula for thermoelectric transport with application to thermopower near the mobility edge*, Phys. Rev. B **33**, 551 (1986).

[161] P. Streda, *Quantised thermopower of a channel in the ballistic regime*, J. Phys.: Condens. Matter **1**, 1025 (1989).

[162] P.N. Butcher, *Thermal and electrical transport formalism for electronic microstructures with many terminals*, J. Phys.: Condens. Matt. **2**, 4869 (1990).

[163] H. van Houten, L.W. Molenkamp, C.W.J. Beenakker, C.T. Foxon, *Thermo-electric properties of quantum point contacts*, Semicond. Sci. Technol. **7**, B215 (1992).

[164] L.D. Landau and E.M. Lifschitz, *Statistical Physics*, (Pergamon Press, New York, USA, 1959).

[165] M.H. Pedersen and M. Büttiker, *Scattering theory of photon-assisted electron transport*, Phys. Rev. B **58**, 12993 (1998).

[166] M. Büttiker and T. Christen, *Basic Elements of Electrical Conduction*, in B. Kramer (Ed.), *Quantum Transport in Semiconductor Submicron Structures*, NATO ASI Series E, Vol. 326, p. 63, (Kluwer Academic Publishers, Dordrecht, NL, 1996).

[167] M. Büttiker and T. Christen, *Admittance and nonlinear transport in quantum wires, point contacts, and resonant tunneling barriers*, in L.L. Sohn, L.P. Kouwenhoven, G. Schön (Eds.), *Mesoscopic Electron Transport*, NATO ASI Series E, Vol. 345. p. 259., (Kluwer Academic Publishers, Dordrecht, NL, 1997).

[168]   Y. Imry and R. Landauer, *Conductance viewed as transmission*, Rev. Mod. Phys. **71**, S306 (1999).

[169]   Y. Imry, *Introduction to Mesoscopic Physics*, (Oxford University Press, Oxford, UK, 1997).

[170]   C.W.J. Beenakker, *Random-matrix theory of quantum transport*, Rev. Mod. Phys. **69**, 731 (1997).

[171]   M. Büttiker, *Role of quantum coherence in series resistors*, Phys. Rev. B **33**, 3020 (1986).

[172]   M. Büttiker, *Quantized transmission of a saddle-point constriction*, Phys. Rev. B **41**, 7906 (1990).

[173]   A.L. Fetter and J.D. Walecka, *Quantum Theory of Many-Particle Systems*, (McGraw-Hill, New York, USA, 1971).

[174]   G.D. Mahan, *Many-Particle Physics*, (Plenum Press, New York, USA, 1990).

[175]   R.D. Mattuck, *A Guide to Feynman Diagrams in the Many-Body Problem*, (Dover Publications, New York, USA, 1992).

[176]   A. Szabo and N. Ostlund, *Modern Quantum Chemistry: Introduction to Advanced Electronic Structure Theory*, (Dover Publication, New York, USA, 1996).

[177]   J. Hubbard, *Electron correlations in narrow energy bands*, Proc. R. Soc. London Ser. A **276**, 238 (1963).

[178]   M.C. Gutzwiller, *Effect of correlation on the ferromagnetism of transition metals*, Phys. Rev. Lett. **10**, 159 (1963).

[179]   A. Montorsi (editor), *The Hubbard Model*, (World Scientific Publishers, Singapore, 1992).

[180]   P.W. Anderson, *Ground state of a magnetic impurity in a metal*, Phys. Rev. **164**, 352 (1967).

[181]   E.N. Economou, *Green's Functions in Quantum Physics*, (Springer-Verlag, Heidelberg, D, 1990).

[182]   A.A. Abrikosov, L.P. Gorkov, I.E. Dzyaloshinski, *Methods of Quantum Field Theory in Statistical Physics*, (Dover Publication, New York, USA, 1963).

[183]   G. Rickayzen, *Green's Functions and Condensed Matter*, (Academic Press, San Diego, USA, 1980).

[184]   S. Doniach and E.H. Sondheimer, *Green's Functions for Solid State Physicists*, (Imperial College Press, London, UK, 1998).

[185]   H. Bruus and K. Flensberg, *Many-body Quantum Theory in Condensed Matter Physics: An Introduction*, (Oxford University Press, Oxford, UK, 2004).

[186]   L.P. Kadanoff and G. Baym, *Quantum Statistical Mechanics*, (WA Benjamin, Menlo Park, USA, 1962).

[187]   L.V. Keldysh, *Diagram technique for nonequilibrium processes*, Zh. Ekps. Teor. Fiz. **47**, 1515 (1964) [Sov. Phys. JETP **20**, 1018 (1965)].

[188]   J. Rammer and H. Smith, *Quantum field-theoretical methods in transport theory of metals*, Rev. Mod. Phys. **58**, 323 (1986).

[189]   H. Huang and A.P. Jauho, *Quantum Kinetics in Transport and Optics of*

Semiconductors, Springer Series in Solid State Science, Vol. 123, (Springer-Verlag, Heidelberg, Berlin, D, 1993).

[190] D.F. Du Bois, *Nonequilibrium Quantum Statistical Mechanics of Plasma and Radiation*, in W.E. Brittin and A.D. Barut (Eds.), *Lectures in Theoretical Physics* Vol. IX C, p. 469, (Gordon and Breach, New York, USA, 1967).

[191] D.C. Langreth, in J.T. Devreese and V.E. van Doren (Eds.), *Linear and Nonlinear Electron Transport in Solids*, NATO-ASI Series B, Vol. 17, (Plenum Press, New York, USA, 1976).

[192] M. Wagner, *Expansions of nonequilibrium Green's functions*, Phys. Rev. B **44**, 6104 (1991).

[193] A.I. Larkin and Yu.N. Ovchinnikov, in D.N. Langenberg and A.I. Larkin (Eds.), *Nonequilibrium Superconductivity*, p. 493, (Elsevier, Amsterdam, NL, 1986).

[194] J. Ferrer, A. Martin-Rodero, F. Flores, *Contact resistance in the scanning tunneling microscope at very small distances*, Phys. Rev. B **38**, 10113 (1988).

[195] J.M. Blanco, F. Flores, R. Perez, *STM-theory: Image potential, chemistry and surface relaxation*, Prog. Surf. Sci. **81**, 403 (2006).

[196] N. Néel, J. Kröger, L. Limot, R. Berndt, *Conductance of single atoms and molecules studied with a scanning tunnelling microscope*, Nanotechnology **18**, 044027 (2007).

[197] J. Kröger, N. Néel, L. Limot, *Contact to single atoms and molecules with the tip of a scanning tunnelling microscope*, J. Phys.: Condens. Matter **20**, 223001 (2008).

[198] F. Guinea, C. Tejedor, F. Flores, E. Louis, *Effective two-dimensional Hamiltonian at surfaces*, Phys. Rev. B **28**, 4397 (1983).

[199] D.S. Fisher and P.A. Lee, *Relation between conductivity and transmission matrix*, Phys. Rev. B **23**, 6851 (1981).

[200] E. Emberly and G. Kirczenow, *State orthogonalization by building a Hilbert space: A new approach to electronic quantum transport in molecular wires*, Phys. Rev. Lett. **81**, 5205 (1998).

[201] M. Brandbyge, N. Kobayashi, M. Tsukada, *Conduction channels at finite bias in single-atom gold contacts*, Phys. Rev. B **60**, 17064 (1999).

[202] J.K. Viljas, J.C. Cuevas, F. Pauly, M. Häfner, *Electron-vibration interaction in transport through atomic gold wires*, Phys. Rev. B **72**, 245415 (2005).

[203] K.S. Thygesen, *Electron transport through an interacting region: The case of a nonorthogonal basis set*, Phys. Rev. B **73**, 035309 (2006).

[204] N.F. Mott, *The electrical conductivity of transition metals*, Proc. R. Soc. London, Ser. A **153**, 699 (1936).

[205] N.F. Mott, *The resistance and thermoelectric properties of transition metals*, Proc. R. Soc. London, Ser. A **156**, 368 (1936).

[206] C. Caroli, R. Combescot, P. Nozieres, D. Saint-James, *Direct calculation of tunneling current*, J. Phys. C **4**, 916 (1971).

[207] C. Caroli, R. Combescot, D. Lederer, P. Nozieres, D. Saint-James, *Direct*

calculation of tunneling current. II. Free electron description, J. Phys. C **4**, 2598 (1971).

[208] R. Combescot, *Direct calculation of tunneling current. III. Effect of localized impurity states in barrier*, J. Phys. C **4**, 2611 (1971).

[209] C. Caroli, R. Combescot, P. Nozieres, and D. Saint-James, *Direct calculation of tunneling current. IV. Electron-phonon interaction effects*, J. Phys. C **5**, 21 (1972).

[210] Y. Meir and N.S. Wingreen, *Landauer formula for the current through an interacting electron region*, Phys. Rev. Lett. **68**, 2512 (1992).

[211] S. Kohler, J. Lehmann, P. Hänggi, *Driven quantum transport on the nanoscale*, Phys. Rep. **406**, 379 (2005).

[212] J.C. Tucker and M.J. Feldman, *Quantum detection at millimeter wavelengths*, Rev. Mod. Phys. **57**, 1055 (1985).

[213] P.K. Tien and J.P. Gordon, *Multiphoton process observed in the interaction of microwave fields with the tunneling between superconductor films*, Phys. Rev. **129**, 647 (1963).

[214] G. Platero and R. Aguado, *Photon-assisted transport in semiconductor nanostructures*, Phys. Rep. **395**, 1 (2004).

[215] S. Grafström, *Photoassisted scanning tunneling microscopy*, J. Appl. Phys. **91**, 1717 (2002).

[216] A. Levy Yeyati and F. Flores, *Photocurrent effects in the scanning tunneling microscope*, Phys. Rev. B **44**, 9020 (1991).

[217] S. Datta and M.P. Anantram, *Steady-state transport in mesoscopic systems illuminated by alternating fields*, Phys. Rev. B **45**, 13761 (1992).

[218] A.-P. Jauho, N.S. Wingreen, Y. Meir, *Time-dependent transport in interacting and noninteracting resonant-tunneling systems*, Phys. Rev. B **50**, 5528 (1994).

[219] T. Brandes, *Truncation method for Green's functions in time-dependent fields*, Phys. Rev. B **56**, 1213 (1997).

[220] B. Wang, J. Wang, H. Guo, *Current partition: A nonequilibrium Green's function approach*, Phys. Rev. Lett. **82**, 398 (1999).

[221] J.K. Viljas and J.C. Cuevas, *Role of electronic structure in photoassisted transport through atomic-sized contacts*, Phys. Rev. B **75**, 075406 (2007).

[222] J.K. Viljas, F. Pauly and J.C. Cuevas, *Modeling elastic and photoassisted transport in organic molecular wires: Length-dependence and current-voltage characteristics*, Phys. Rev. B **77**, 155119 (2008).

[223] N.W. Ashcroft and N.D. Mermin, *Solid State Physics*, (Saunders, USA, 1976).

[224] W.A. Harrison, *Electronic Structure and the Properties of Solids*, (Dover Publication, New York, USA, 1989).

[225] W.A. Harrison, *Elementary Electronic Structure*, (World Scientific Publishers, Singapore, 1999).

[226] R.M. Martin, *Electronic Structure: Basic Theory and Practical Methods*, (Cambridge University Press, Cambridge, UK, 2004).

[227] P.W. Atkins and R.S. Friedman, *Molecular Quantum Mechanics* (3rd edition), (Oxford University Press, Oxford, UK, 1997).

[228] R. Hoffmann. *An Extended Hückel Theory. I. Hydrocarbons*, J. Chem. Phys. **39**, 1397 (1963).

[229] E. Hückel, *Quantentheoretische Beiträge zum Benzolproblem. I Die Elektronenkonfiguration des Benzols und verwandter Verbindungen*, Zeitschrift für Physik **70**, 204 (1931).

[230] E. Hückel, *Quantentheoretische Beiträge zum Benzolproblem. II Quantentheorie der induzierten Polaritäten*, Zeitschrift für Physik **72**, 310 (1931).

[231] E. Hückel, *Quantentheoretische Beiträge zum Benzolproblem. III Quantentheoretische Beiträge zum Problem der aromatischen und ungesättigten Verbindungen.*, Zeitschrift für Physik **76**, 628 (1932).

[232] C.A. Coulson, B. O'Leary, R.B. Mallion, *Hückel Theory for Organic Chemists*, Academic Press, San Diego, USA, 1978).

[233] M. Wolfsberg and L.J. Helmholz, *The spectra and electronic structure of the tetrahedral ions MnO_4^-, CrO_4^{--}, and ClO_4^-*, J. Chem. Phys. **20**, 837 (1952).

[234] J.C. Slater and G.F. Koster. *Simplified LCAO method for the periodic potential problem*, Phys. Rev. **94**, 1498 (1954).

[235] D.A. Papaconstantopoulos, *Handbook of Electronic Structure of Elemental Solids*, (Plenum Press, New York, USA, 1986).

[236] M.D. Stiles, *Generalized Slater-Koster method for fitting band structures*, Phys. Rev. B **55**, 4168 (1997).

[237] P.R. Wallace, *The band theory of graphite*, Phys. Rev. **71**, 622 (1947).

[238] K.S. Novoselov, A.K. Geim, S.V. Morozov, D. Jiang, Y. Zhang, S.V. Dubonos, I.V. Grigorieva, A.A. Firsov, *Electric field effect in atomically thin carbon films*, Science **306**, 666 (2004).

[239] A.K. Geim and K.S. Novoselov, *The rise of graphene*, Nature Mat. **6**, 183 (2007).

[240] A.K. Geim and A.H. MacDonald, *Graphene: Exploring carbon flatland*, Physics Today **60**, 35 (2007).

[241] A.K. Geim and P. Kim, *Carbon Wonderland*, Scientific American **290**, 90 (2008).

[242] R.E. Cohen, M.J. Mehl, D.A. Papaconstantopoulos, *Tight-binding total-energy method for transition and noble metals*, Phys. Rev. B **50**, 15694 (1994).

[243] M.J. Mehl and D.A. Papaconstantopoulos, *Applications of a new tight-binding total energy method for transition and noble metals: Elastic constants, vacancies, and surfaces of monatomic metals*, Phys. Rev. B **54**, 4519 (1996).

[244] M.J. Mehl and D.A. Papaconstantopoulos, *Tight-binding parametrization of first-principles results*, in *Topics in Computational Materials Science*, p. 169., C.Y. Fong (Ed.) (World Scientific Publishers, Singapore, 1998).

[245] D.A. Papaconstantopoulos and M.J. Mehl, *The Slater-Koster tight-binding method: a computationally efficient and accurate approach*, J. Phys.: Condens. Matter **15**, R513 (2003).

[246] P.L. Pernas, A. Martin-Rodero, F. Flores, *Electrochemical-potential variations across a constriction*, Phys. Rev. B **41**, 8553 (1990).

[247] F. Pauly, J.K. Viljas, U. Huniar, M. Häfner, S. Wohlthat, M. Bürkle, J.C. Cuevas, G. Schön, *Clustfer-based density-functional approach to quantum transport through molecular and atomic contacts*, New J. Phys. **10**, 125019 (2008).

[248] T.N. Todorov, *Tight-binding simulation of current-carrying nanostructures*, J. Phys.: Condens. Matter **14**, 3049 (2002).

[249] A. Martin-Rodero, F. Flores, N.H. March, *Tight-binding theory of tunneling current with chemisorbed species*, Phys. Rev. B **38**, 10047 (1988).

[250] P. Sautet and C. Joachim, *Calculation of benzene on rhodium STM images*, Chem. Phys. Lett. **185**, 23 (1991).

[251] N. Mingo, L. Jurczyszyn, F.J. Garcia-Vidal, R. Saiz-Pardo, P.L. de Andres, F. Flores, S.Y. Wu, W. More, *Theory of the scanning tunneling microscope: Xe on Ni and Al*, Phys. Rev. B **54**, 2225 (1996).

[252] J. Cerdá, M.A. Van Hove, P. Sautet, M. Salmeron, *Efficient method for the simulation of STM images. I. Generalized Green-function formalism*, Phys. Rev. B **56**, 15885 (1997).

[253] P. Sautet and C. Joachim, *Electronic interference effects induced by a benzene ring embedded in a polyacetylene chain*, Chem. Phys. Lett. **153**, 511 (1988).

[254] P. Sautet and C. Joachim, *Electronic transmission coefficient for the single-impurity problem in the scattering matrix approach*, Phys. Rev. B **38**, 12238 (1988).

[255] V. Mujica, M. Kemp, M.A. Ratner, *Electron conduction in molecular wires. I. A scattering formalism*, J. Chem. Phys. **101**, 6849 (1994).

[256] V. Mujica, M. Kemp, M.A. Ratner, *Electron conduction in molecular wires. II. Application to scanning tunneling microscopy*, J. Chem. Phys. **101**, 6856 (1994).

[257] M. Kemp, V. Mujica, M.A. Ratner, *Molecular electronics: Disordered molecular wires*, J. Chem. Phys. **101**, 5172 (1994).

[258] A. Cheong, A.E. Roitberg, V. Mujica, M.A. Ratner, *Resonances and interference effects on the effective electronic coupling in electron transfer*, J. Photochem. Photobiol. A **82**, 81 (1994).

[259] M. Kemp, A. Roitberg, V. Mujica, T. Wanta, M. Ratner, *Molecular wires: Extended coupling and disorder effects*, J. Phys. Chem. **100**, 8349 (1996).

[260] M.P. Samanta, W. Tian, S. Datta, J.I. Henderson, C.P. Kubiak, *Electronic conduction through organic molecules*, Phys. Rev. B **53**, 7626 (1996).

[261] S. Datta, W. Tian, S. Hong, R. Reifenberger, J.I. Henderson, C.P. Kubiak, *Current-Voltage characteristics of self-assembled monolayers by scanning tunneling microscopy*, Phys. Rev. Lett. **79**, 2530 (1997).

[262] T.N. Todorov and A.P. Sutton, *Jumps in electronic conductance due to mechanical instabilities*, Phys. Rev. Lett. **70**, 2138 (1993).

[263] J.C. Cuevas, A. Levy Yeyati, A. Martin-Rodero, *Microscopic origin of the conducting channels in metallic atomic-size contacts*, Phys. Rev. Lett. **80**, 1066 (1998).

[264] J.C. Cuevas, A. Levy Yeyati, A. Martin-Rodero, G. Rubio Bollinger, C. Untiedt, N. Agraït, *Evolution of conducting channels in metallic atomic*

contacts under elastic deformation, Phys. Rev. Lett. **81**, 2990 (1998).

[265] T.N. Todorov, J. Hoekstra, A.P. Sutton, *Current-induced forces in atomic-scale conductors*, Phil. Mag. B **80**, 421 (2000).

[266] J. Hoekstra, A.P. Sutton, T.N. Todorov, A.P. Horsfield, *Electromigration of vacancies in copper*, Phys. Rev. B **62**, 8568 (2000).

[267] T.N. Todorov, J. Hoekstra, A.P. Sutton, *Current-induced embrittlement of atomic wires*, Phys. Rev. Lett. **86**, 3606 (2001).

[268] P. Hohenberg and W. Kohn, *Inhomogeneous electron gas*, Phys. Rev. **136**, B864 (1964).

[269] W. Kohn and L.J. Sham, *Self-consistent equations including exchange and correlation effects*, Phys. Rev. **140**, A1133 (1965).

[270] W. Kohn, *Nobel Lecture: Electronic structure of matter–wave functions and density functionals*, Rev. Mod. Phys. **71**, 1253 (1998).

[271] R.O. Jones and O. Gunnarsson, *The density functional formalism, its applications and prospects*, Rev. Mod. Phys. **61**, 689 (1989).

[272] R.G. Parr and W. Yang, *Density-Functional Theory of Atoms and Molecules*, (Oxford University Press, New York, USA, 1989).

[273] W. Koch and M.C. Holthausen, *A Chemist's Guide to Density Functional Theory*. (Wiley-VCH, Weinheim, D, 2001).

[274] T.A. Koopman, *Über die Zuordnung von Wellenfunktionen und Eigenwerten zu den einzelnen Elektronen eines Atoms*, Physica **1**, 104 (1934).

[275] M. Levy, J.P. Perdew, and V. Sahni, *Exact differential equation for the density and ionization energy of a mny-particle system*, Phys. Rev. A **30**, 2745 (1984).

[276] D.M. Ceperly and B.J. Alder, *Ground state of the electron gas by a stochastic method*, Phys. Rev. Lett. **45**, 566 (1980).

[277] A.D. Becke, *Density functional exchange-energy approximation with correct asymptotic bahavior*, Phys. Rev. A **38**, 3098 (1986).

[278] J.P. Perdew, *Density-functional approximation for the correlation energy of the inhomogeneous electron gas*, Phy. Rev. B **33**, 8822 (1986).

[279] J.P. Perdew, K. Burke, and M. Ernzerhof, *Generalized gradient approximation made simple*, Phys. Rev. Lett. **77**, 3865 (1996); Erratum: Phys. Rev. Lett. **78**, 1396 (1997).

[280] J.P. Perdew and Y. Wang, *Accurate and simple analytic representation ofthe electron-gas correlation energy*, Phy. Rev. B **45**, 13244 (1992).

[281] C. Lee, W. Yang, R.G. Parr, *Development of the Colle-Salvetti correlation-energy formula into a functional of the electron density*, Phys. Rev. B **37**, 785 (1988).

[282] A.D. Becke, *A new mixing of Hartree-Fock and local density-functional theories*, J. Chem. Phys. **98**, 1372 (1993).

[283] A.D. Becke, *Density functional thermochemistry III. The ole of exact exchange*, J. Chem. Phys. **98**, 5648 (1993).

[284] P.J. Stephens, J.F. Devlin, C.F. Chabalowski, M.J. Frisch, *Ab initio calculations of vibrational absorption and circular dichroism spectra using SCF, MP2, and density functional theory force fields*, J. Phys. Chem. **98**, 11623 (1994).

[285] A.P. Scott and L. Radom, *Harmonic vibrational frequencies: An evalua-
       tion of Hartree-Fock, Moller-Plesset, quadratic configuration interaction,
       density functional theory, and semiempirical scale factors*, J. Phys. Chem.
       100, 16502 (1996).

[286] A. Pecchia and A. Di Carlo, *Atomistic theory of transport in organic and
       inorganic nanostructures*, Rep. Prog. Phys. **67**, 1497 (2004).

[287] M. Brandbyge, J.-L. Mozos, P. Ordejón, J. Taylor, K. Stokbro, *Density-
       functional method for nonequilibrium electron transport*, Phys. Rev. B **65**,
       165401 (2002).

[288] S.N. Yaliraki, A.E. Roitberg, C. Gonzalez, V. Mujica, M.A. Ratner, *The
       injecting energy at molecule/metal interfaces: Implications for conduc-
       tance of molecular junctions from an ab initio molecular description*, J.
       Chem. Phys. **111**, 6997 (1999).

[289] Y. Xue, S. Datta, and M.A. Ratner, *Charge transfer and "band lineup" in
       molecular electronic devices: A chemical and numerical interpretation*, J.
       Chem. Phys. **115**, 4292 (2001).

[290] J.J. Palacios, A.J. Pérez-Jiménez, E. Louis, E. SanFabián, J.A. Vergés,
       *First-principles approach to electrical transport in atomic-scale nanostruc-
       tures*, Phys. Rev. B **66**, 035322 (2002).

[291] J. Heurich, J.C. Cuevas, W. Wenzel, G. Schön, *Electrical transport through
       single-molecule junctions: from molecular orbitals to conduction channels*,
       Phys. Rev. Lett. **88**, 256803 (2002).

[292] Y. Xue and M.A. Ratner, *Microscopic study of electrical transport through
       individual molecules with metallic contacts. I. Band lineup, voltage drop,
       and high-field transport*, Phys. Rev. B **68**, 115406 (2003).

[293] P.S. Damle, A.W. Ghosh, S. Datta, *Unified description of molecular con-
       duction: From molecules to metallic wires*, Phys. Rev. B **64**, 201403
       (2001).

[294] P.S. Damle, A.W. Ghosh, S. Datta, *First-principles analysis of molecu-
       lar conduction using quantum chemistry software*, Chem. Phys. **281**, 171
       (2002).

[295] M.P. López Sancho, J.M. López Sancho, J. Rubio, *Quick iterative scheme
       for the calculation of transfer matrices: application to Mo (100)*, J. Phys.
       F: Met. Phys. **14**, 1205 (1984).

[296] M.P. López Sancho, J.M. López Sancho, J. Rubio, *Highly convergent
       schemes for the calculation of bulk and surface Green functions*, J. Phys.
       F: Met. Phys. **15**, 851 (1985).

[297] J. Taylor, H. Guo, J. Wang, *Ab initio modeling of quantum transport
       properties of molecular electronic devices*, Phys. Rev. B **63**, 245407 (2001).

[298] A. Di Carlo, M. Gheorghe, P. Lugli, M. Sternberg, G. Seifert, T. Frauen-
       heim, *Theoretical tools for transport in molecular nanostructures*, Physica
       B **314**, 86 (2002).

[299] J.K. Tomfohr and O.F. Sankey, *Complex band structure, decay lengths,
       and Fermi level alignment in simple molecular electronic systems*, Phys.
       Rev. B **65**, 245105 (2002).

[300] K.S. Thygesen and K.W. Jacobsen, *Molecular transport calculations with*

Wannier functions, Chem. Phys. **319**, 111 (2005).

[301] A.R. Rocha, V.M. García-Suárez, S. Bailey, C. Lambert, J. Ferrer, S. Sanvito, *Spin and molecular electronics in atomically generated orbital landscapes*, Phys. Rev. B **73**, 085414 (2006).

[302] J.M. van Ruitenbeek, *Conductance quantization in metallic point contacts*, in K.H. Meiwes-Broer (Ed.), *Metal Clusters on Surfaces: Structure, Quantum Properties, Physical Chemistry*, p. 175 (Springer-Verlag, Heidelberg, Berlin, D, 2000). Available in cond-mat/9910394.

[303] A.P. Sutton and J.B. Pethica, *Inelastic flow processes in nanometre volumes of solids*, J. Phys.: Condens. Matter **2**, 5317 (1990).

[304] U. Landman, W.D. Luedtke, N.A. Burnham, R.J. Colton, *Atomistic mechanisms and dynamics of adhesion, nanoindentation and fracture*, Science **248**, 454 (1990).

[305] L. Olesen, E. Lægsgaard, I. Stensgaard, F. Besenbacher, J. Schiøtz, P. Stoltze, K.W. Jacobsen, J.K. Nørskov, *Reply on comment on quantized conductance in an atom-sized point contact*, Phys. Rev. Lett. **74**, 2147 (1995).

[306] J.M. Krans, J.M. van Ruitenbeek, V.V. Fisun, I.K. Yanson, L.J. Jongh, *The signature of conductance quantisation in metallic point contacts*, Nature **375**, 767 (1995).

[307] N. Agraït, J.G. Rodrigo, S. Vieira, *Conductance steps and quantization in atomic-size contacts*, Phys. Rev. B **47**, 12345 (1993).

[308] J.I. Pascual, J. Méndez, J. Gómez-Herrero, A.M. Baró, N. García, V. Thien Binh, *Quantum contact in gold nanostructures by scanning tunneling microscopy*, Phys. Rev. Lett. **71**, 1852 (1993).

[309] L. Olesen, E. Lægsgaard, I. Stensgaard, F. Besenbacher, J. Schiøtz, P. Stoltze, K.W. Jacobsen, J.K. Nørskov, *Quantised conductance in an atom-sized point contact*, Phys. Rev. Lett. **72**, 2251 (1994).

[310] M. Brandbyge, J. Schiøtz, M.R. Sørensen, P. Stoltze, K.W. Jacobsen, J.K. Nørskov, L. Olesen, E. Lægsgaard, I. Stensgaard, and F. Besenbacher, *Quantized conductance in atom-sized wires between two metals*, Phys. Rev. B **52**, 8499 (1995).

[311] J.L. Costa-Krämer, *Conductance quantization at room temperature in magnetic and nonmagnetic metallic nanowires*, Phys. Rev. B **55**, 4875 (1997).

[312] B. Ludoph, M.H. Devoret, D. Esteve, C. Urbina, J.M. van Ruitenbeek, *Evidence for saturation of channel transmission from conductance fluctuations in atomic-size point contacts*, Phys. Rev. Lett. **82**, 1530 (1999).

[313] B. Ludoph and J.M. van Ruitenbeek, *Conductance fluctuations as a tool for investigating the quantum modes in atomic size metallic contacts*, Phys. Rev. B. **61**, 2273 (2000).

[314] B. Ludoph, *Quantum conductance properties of atomic-size contacts*, Ph.D. thesis, Universiteit Leiden, The Netherlands (1999).

[315] A.I. Yanson, *Atomic chains and electronic shells: quantum mechanisms for the formation of nanowires*, Ph.D. Thesis, Universiteit Leiden, The

Netherlands (2001).
Available at http://www.physics.leidenuniv.nl/sections/cm/amc/.

[316] M. Brandbyge, M.R. Sørensen, K.W. Jacobsen, *Conductance eigenchannels in nanocontacts*, Phys. Rev. B. **56**, 14956 (1997).

[317] E.N. Bogachek, A.N. Zagoskin, I.O. Kulik, *Conductance jumps and magnetic flux quantization in ballistic point contacts*, Sov. J. Low Temp. Phys. **16**, 796 (1990).

[318] A.I. Yanson and J.M. van Ruitenbeek, *Do histograms constitute a proof for conductance quantization?* Phys. Rev. Lett. **79**, 2157 (1997).

[319] M. Häfner, P. Konrad, F. Pauly, J.C. Cuevas, E. Scheer, *Conduction channels of one-atom zinc contacts*, Phys. Rev. B **70**, 241404R (2004).

[320] E. Scheer, P. Konrad, C. Bacca, A. Mayer-Gindner, H. von Löhneysen, M. Häfner, J.C. Cuevas, *Correlation between transport properties and atomic configuration of atomic contacts of zinc by low-temperature measurements*, Phys. Rev. B **74**, 205430 (2006).

[321] R.H.M. Smit, A.I. Mares, M. Häfner, P. Pou, J.C. Cuevas, J.M. van Ruitenbeek, *Metallic properties of magnesium point contacts*, New. J. Phys. **11**, 073043 (2009).

[322] J.R. Schrieffer and J.W. Wilkins, *Two-particle tunneling processes between superconductors*, Phys. Rev. Lett. **10**, 17 (1963).

[323] T.M. Klapwijk, G.E. Blonder, M. Tinkham, *Explanation of subharmonic energy gap structure in superconducting contacts*, Physica B **109–110**, 1657 (1982).

[324] G.B. Arnold, *Superconducting tunneling without the tunneling Hamiltonian. II. Subgap harmonic structure*, J. Low Temp. Phys. **68**, 1 (1987).

[325] D. Averin and D. Bardas, *ac Josephson effect in a single quantum channel*, Phys. Rev. Lett. **75**, 1831 (1995).

[326] J.C. Cuevas, A. Martin-Rodero, A. Levy Yeyati, *Hamiltonian approach to the transport properties of superconducting quantum point contacts*, Phys. Rev. B **54**, 7366 (1996).

[327] E.N. Bratus, V.S. Shumeiko, G. Wendin, *Theory of subharmonic gap structure in superconducting mesoscopic tunnel contacts*, Phys. Rev. Lett. **74**, 2110 (1995).

[328] E.N. Bratus, V.S. Shumeiko, E.V. Bezuglyi, G. Wendin, *dc-current transport and ac Josephson effect in quantum junctions at low voltage*, Phys. Rev. B **55**, 12666 (1997).

[329] N. van der Post, E.T. Peters, I.K. Yanson, J.M. van Ruitenbeek, *Subgap structure as function of the barrier in superconducting tunnel junctions*, Phys. Rev. Lett. **73**, 2611 (1994).

[330] J.G. Rodrigo, N. Agraït, C. Sirvent, S. Vieira, *Josephson effect in nanoscopic structures*, Phys. Rev. B **50**, 12788 (1994).

[331] M. Häfner, *Phase-coherent electron transport through non- and ferromagnetic atomic-sized metal contacts*, Ph.D. thesis, Universität Karlsruhe, Germany (2009).
Available in http://www-tfp.physik.uni-karlsruhe.de/Publications/.

[332] P. Jelínek, R. Pérez, J. Ortega, F. Flores, *First-principles simulations of*

the stretching and final breaking of Al nanowires: Mechanical properties and electrical conductance, Phys. Rev. B **68**, 085403 (2003).

[333] P. Jelínek, R. Pérez, J. Ortega, F. Flores, *Universal behaviour in the final stage of the breaking process for metal nanowires*, Nanotechnology **16**, 1023 (2005).

[334] E. Scheer, W. Belzig, Y. Naveh, M.H. Devoret, D. Esteve, C. Urbina, *Proximity effect and multiple Andreev reflections in gold point contacts*, Phys. Rev. Lett. **86**, 284 (2001).

[335] E. Scheer, N. Agraït, J.C. Cuevas, A. Levy Yeyati, B. Ludoph, A. Martin-Rodero, G. Rubio Bollinger, J.M. van Ruitenbeek, C. Urbina, *The signature of chemical valence in the electrical conduction through a single-atom contact*, Nature **394**, 154 (1998).

[336] Y. Takai, T. Kawasaki, Y. Kimura, T. Ikuta, R. Shimizu, *Dynamic observation of an atom-sized gold wire by phase electron microscopy*, Phys. Rev. Lett. **87**, 106105 (2001).

[337] S.K. Nielsen, M. Brandbyge, K. Hansen, K. Stokbro, J.M. van Ruitenbeek, F. Besenbacher, *Current-voltage curves of atomic-sized transition metal contacts: An explanation of why Au is Ohmic and Pt is not*, Phys. Rev. Lett. **89**, 066804 (2002).

[338] A. Bagrets, N. Papanikolaou, I. Mertig, *Ab initio approach to the ballistic transport through single atoms*, Phys. Rev. B **73**, 045428 (2006).

[339] A. Bagrets, N. Papanikolaou, I. Mertig, *Conduction eigenchannels of atomic-sized contacts: Ab initio KKR Green's function formalism*, Phys. Rev. B **75**, 235448 (2007).

[340] A.M. Bratkovsky, A.P. Sutton, T.N. Todorov, *Conditions for conductance quantization in realistic models of atomic-scale metallic contacts*, Phys. Rev. B **52**, 5036 (1995).

[341] A. Nakamura, M. Brandbyge, L.B. Hansen, K.W. Jacobsen, *Density functional simulation of a breaking nanowire*, Phys. Rev. Lett. **82**, 1538 (1999).

[342] M. Dreher, F. Pauly, J. Heurich, J.C. Cuevas, E. Scheer, P. Nielaba, *Structure and conductance histogram of atomic-sized Au contacts*, Phys. Rev. B **72**, 075435 (2005).

[343] F. Pauly, M. Dreher, J.K. Viljas, M. Häfner, J.C. Cuevas, P. Nielaba, *Theoretical analysis of the conductance histograms and structural properties of Ag, Pt, and Ni nanocontacts*, Phys. Rev. B **74**, 235106 (2006).

[344] A. Hasmy, A.J. Pérez-Jiménez, J.J. Palacios, P. García-Mochales, J.L. Costa-Krämer, M. Díaz, E. Medina, P.A. Serena, *Ballistic resistivity in aluminum nanocontacts*, Phys. Rev. B **72**, 245405 (2005).

[345] A.I. Yanson, I.K. Yanson, J.M. van Ruitenbeek, *Supershell structure in alkali metal nanowires*, Phys. Rev Lett. **84**, 5832 (2000).

[346] A.I. Yanson, I.K. Yanson, J.M. van Ruitenbeek, *Shell effects in heavy alkali-metal nanowires*, in A.S. Alexandrov et al. (Eds.), *Molecular Nanowires and Other Quantum Objects*, p. 243 (Kluwer Academic Publishers, Dordrecht, 2004). Available in cond-mat/0405112.

[347] S. Kirchner, J. Kroha, E. Scheer, *Generalized conductance sum rule in*

atomic break junctions, in V. Chandrasekhar, C. v. Haesendonck, A. Zawadowski (Eds.), *Kondo effect and dephasing in low-dimensional metallic systems*, NATO Science Series II 50, p. 215, (Kluwer Acadmeic Publishers, Dordrecht, NL, 2001).

[348] E. Medina, M. Díaz, N. León, C. Guerrero, A. Hasmy, P.A. Serena, J.L. Costa-Krämer, *Ionic shell and subshell structures in aluminum and gold nanocontacts*, Phys. Rev. Lett. **91**, 026802 (2003).

[349] A.I. Mares, A.F. Otte, L.G. Soukiassian, R.H.M. Smit, J.M. van Ruitenbeek, *Observation of electronic and atomic shell effects in gold nanowires*, Phys. Rev B **70**, 073401 (2004).

[350] A.I. Mares and J.M. van Ruitenbeek, *Observation of shell effects in nanowires for the noble metals Cu, Ag, and Au*, Phys. Rev. B **72**, 205402 (2005).

[351] A. Hasmy, E. Medina, and P.A. Serena, *From favorable atomic configurations to supershell structures: A new interpretation of conductance histograms*, Phys. Rev. Lett. **86**, 5574 (2001).

[352] P.A.M. Holweg, J.A. Kokkedee, J. Caro, A.H. Verbruggen, S. Radelaar, A.G.M. Jansen, and P. Wyder, *Conductance fluctuations in a ballistic metallic point contact*, Phys. Rev. Lett. **67**, 2549 (1991).

[353] P.A.M. Holweg, J. Caro, A.H. Verbruggen, S. Radelaar, *Correlation energy of conductance fluctuations in ballistic silver point contacts*, Phys. Rev. B **48**, 2479 (1993).

[354] D.C. Ralph, K.S. Ralls, R.A. Buhrman, *Ensemble studies of nonlinear conductance fluctuations in phase coherent samples*, Phys. Rev. Lett. **70**, 986 (1993).

[355] V.I. Kozub, J. Caro, P.A.M. Holweg, *Local-interference theory of conductance fluctuations in ballistic metallic point contacts: combination of near and remote backscattered trajectories*, Phys. Rev. B **50**, 15126 (1994).

[356] A.I. Yanson, G. Rubio Bollinger, H.E. van den Brom, N. Agraït, J.M. van Ruitenbeek, *Formation and manipulation of a metallic wire of single gold atoms*, Nature **395**, 783 (1998).

[357] V. Rodrigues, T. Fuhrer, D. Ugarte, *Signature of atomic structure in the quantum conductance of gold nanowires*, Phys. Rev. Lett. **85**, 4124 (2000).

[358] G.M. Finbow, R.M. Lynden-Bell, I.R. McDonald, *Atomistic simulation of the stretching of nanoscale metal wires*, Molecular Physics **92**, 705 (1997).

[359] M.R. Sorensen, M. Brandbyge, K.W. Sorensen, *Mechanical deformation of atomic-scale metallic contacts: structure and mechanisms*, Phys. Rev. B **57**, 3283 (1998).

[360] D. Sánchez-Portal, E. Artacho, J. Junquera, P. Ordejón, A. García, and J.M. Soler, *Stiff monatomic gold wires with a spinning zigzag geometry*, Phys. Rev. Lett. **83**, 3884 (1999).

[361] S.R. Bahn an K.W. Jacobsen, *Chain formation of metal atoms*, Phys. Rev. Lett. **87**, 266101 (2001).

[362] E.Z. da Silva, A.J.R. da Silva, A. Fazzio, *How do gold wires break?*, Phys. Rev. Lett. **87**, 256102 (2001).

[363] H. Häkkinen, R.N. Barnett, A.G. Scherbakov, U. Landman, *Nanowire gold*

chains: formation mechanisms and conductance, J. Phys. Chem. B **104**, 9063 (2000).

[364] D. Sánchez-Portal, E. Artacho, J. Junquera, A. García, J.M. Soler, *Zigzag equilibrium structure in monatomic wires*, Surf. Sci. **482**, 1261 (2001).

[365] R.H.M. Smit, C. Untiedt, A.I. Yanson, J.M. van Ruitenbeek, *Common origin for surface reconstruction and the formation of chains of metal atoms*, Phys. Rev. Lett. **87**, 266102 (2001).

[366] R.H.M. Smit, C. Untiedt, G. Rubio-Bollinger, R.C. Segers, J.M. van Ruitenbeek, *Observation of a parity oscillation in the conductance of atomic wires*, Phys. Rev. Lett. **91**, 076805 (2003).

[367] C. Untiedt, A.I. Yanson, R. Grande, G. Rubio-Bollinger, N. Agraït, S. Vieira, J.M. van Ruitenbeek, *Calibration of the length of a chain of single gold atoms*, Phys. Rev. B **66**, 085418 (2002).

[368] G. Rubio-Bollinger, S.R. Bahn, N. Agraït, K.W. Jacobsen, S. Vieira, *Mechanical properties and formation mechanisms of a wire of single gold atoms*, Phys. Rev. Lett. **87**, 026101 (2001).

[369] N.D. Lang, *Anomalous dependence of resistance on length in atomic wires*, Phys. Rev. Lett. **79**, 1357 (1997).

[370] H.S. Sim, H.W. Lee, K.J. Chang, *Even-odd behavior of conductance in monatomic sodium wires*, Phys. Rev. Lett. **87**, 096803 (2001).

[371] Z.Y. Zeng and F. Claro, *Delocalization and conductance quantization in one-dimensional systems attached to leads*, Phys. Rev. B **65**, 193405 (2002).

[372] T.S. Kim and S. Hershfield, *Even-odd parity effects in conductance and shot noise of metal-atomic-wire-metal (superconducting) junctions*, Phys. Rev. B **65**, 214526 (2002).

[373] P. Havu, T. Torsti, M.J. Puska, R.M. Nieminen, *Conductance oscillations in metallic nanocontacts*, Phys. Rev. B **66**, 075401 (2002).

[374] R. Gutiérrez, F. Grossmann, R. Schmidt, *Resistance of atomic sodium wires*, Acta Phys. Pol. B **32**, 443 (2001).

[375] R.H.M. Smit, *From quantum point contacts to monatomic chains: Fabrication and characterization of the ultimate nanowire*, Ph.D. Thesis, Universiteit Leiden, The Netherlands (2003). Available in http://www.physics.leidenuniv.nl/sections/cm/amc/.

[376] L. de la Vega, A. Martin-Rodero, A. Levy Yeyati, A. Saúl, *Different wavelength oscillations in the conductance of 5d metal atomic chains*, Phys. Rev. B **70**, 113107 (2004).

[377] Y.J. Lee, M. Brandbyge, M.J. Puska, J. Taylor, K. Stokbro, R.M. Nieminen, *Electron transport through monovalent atomic wires*, Phys. Rev. B **69**, 125409 (2004).

[378] J. Fernandez-Rossier, D. Jacob, C. Untiedt, J.J. Palacios, *Transport in magnetically ordered Pt nanocontacts*, Phys. Rev. B **72**, 224418 (2005).

[379] V.M. García-Suárez, A.R. Rocha, S.W. Bailey, C.J. Lambert, S. Sanvito, J. Ferrer, *Conductance oscillations in zigzag platinum chains*, Phys. Rev. Lett. **95**, 256804 (2005).

[380] A. Smogunov, A. Dal Corso, E. Tosatti, *Magnetic phenomena, spin-orbit*

effects, and Landauer conductance in Pt nanowire contacts: Density-functional theory calculations, Phys. Rev. B **78**, 014423 (2008).

[381] V.M. García-Suárez, D.Zs. Manrique, C.J. Lambert, J. Ferrer, *Anisotropic magnetoresistance in atomic chains of iridium and platinum from first principles*, Phys. Rev. B **79**, 060408 (2009).

[382] J.M. Krans and J.M. van Ruitenbeek, *Subquantum conductance steps in atom-sized contacts of the semimetal Sb*, Phys. Rev. B **50**, 17659 (1994).

[383] J.L. Costa-Krämer, N. García, H. Olin, *Conductance quantization in bismuth nanowires at 4 K*, Phys. Rev. Lett. **78**, 4990 (1997).

[384] J. G. Rodrigo, A. García-Martín, J.J. Sáenz, S. Vieira, *Quantum conductance in semimetallic bismuth nanocontacts*, Phys. Rev. Lett. **88**, 246801 (2002).

[385] W.H.A. Thijssen, M. Strange, J.M.J. aan de Brugh, J.M. van Ruitenbeek, *Formation and properties of metal-oxygen atomic chains*, New J. Phys. **10**, 033005 (2008).

[386] G. Schull, N. Néel, P. Johansson, R. Berndt, *Electron-plasmon and electron-electron interactions at a single atom contact*, Phys. Rev. Lett. **102**, 057401 (2009).

[387] S.A. Wolf, D.D. Awschalom, R.A. Buhrman, J.M. Daughton, S. von Molnár, M.L. Roukes, A.Y. Chtchelkanova, D.M. Treger, *Spintronics: A spin-based electronics vision for the future*, Science **294**, 1488 (2001).

[388] I. Žutić, J. Fabian, S. Das Sarma, *Spintronics: Fundamentals and applications*, Rev. Mod. Phys. **76**, 323 (2004).

[389] R.C. O'Handley, *Modern Magnetic Materials: Principles and Applications* (Wiley, New York, USA, 2000).

[390] C. Sirvent, J.G. Rodrigo, S. Vieira, L. Jurczyszyn, N. Mingo, F. Flores, *Conductance step for a single-atom contact in the scanning tunneling microscope: Noble and transition metals*, Phys. Rev. B **53**, 16086 (1996).

[391] K. Hansen, E. Lægsgaard, I. Stensgaard, F. Besenbacher, *Quantized conductance in relays*, Phys. Rev. B **56**, 2208 (1997).

[392] F. Ott, S. Barberan, J.G. Lunney, J.M.D. Coey, P. Berthet, A.M. de Leon-Guevara, A. Revcolevschi, *Quantized conductance in a contact between metallic oxide crystals*, Phys. Rev. B **58**, 4656 (1998).

[393] H. Oshima and K. Miyano, *Spin-dependent conductance quantization in nickel point contacts*, Appl. Phys. Lett. **73**, 2203 (1998).

[394] T. Ono, Y. Ooka, H. Miyajima, Y. Otani, $2e^2/h$ *to* e^2/h *switching of quantum conductance associated with a change in nanoscale ferromagnetic domain structure*, Appl. Phys. Lett. **75**, 1622 (1999).

[395] F. Komori and K. Nakatsuji, *Quantized conductance through atomic-sized iron contacts at 4.2 K*, J. Phys. Soc. Jap. **68**, 3786 (1999).

[396] M. Viret, S. Berger, M. Gabureac, F. Ott, D. Olligs, I. Petej, J.F. Gregg, C. Fermon, G. Francinet, G. Le Goff, *Magnetoresistance through a single nickel atom*, Phys. Rev. B **66**, 220401 (2002).

[397] F. Elhoussine, S. Mátéfi-Tempfli, A. Encinas, L. Piraux, *Conductance quantization in magnetic nanowires electrodeposited in nanopores*, Appl. Phys. Lett. **81**, 1681 (2002).

[398] M. Shimizu, E. Saitoh, H. Miyajima, Y. Otani, *Conductance quantization in ferromagnetic Ni nano-constriction*, J. Magn. Magn. Mat. **239**, 243 (2002).

[399] D. Gillingham, I. Linington, J. Bland, *e^2/h quantization of the conduction in Cu nanowires*, J. Phys.: Condens. Matter **14**, L567 (2002).

[400] V. Rodrigues, J. Bettini, P.C. Silva, D. Ugarte, *Evidence for spontaneous spin-polarized transport in magnetic nanowires*, Phys. Rev. Lett. **91**, 96801 (2003).

[401] D. Gillingham, C. Müller, J. Bland, *Spin-dependent quantum transport effects in Cu nanowires*, J. Phys.: Condens. Matter **15**, L291 (2003).

[402] D. Gillingham, I. Linington, C. Müller, J. Bland, *e^2/h quantization of the conduction in Cu nanowires*, J. Appl. Phys. **93**, 7388 (2003).

[403] C. Untiedt, D.M.T Dekker, D. Djukic, J.M. van Ruitenbeek, *Absence of magnetically induced fractional quantization in atomic contacts*, Phys. Rev. B **69**, 081401 (2004).

[404] M. Gabureac, M. Viret, F. Ott, C. Fermon, *Magnetoresistance in nanocontacts induced by magnetostrictive effects*, Phys. Rev. B **69**, 100401 (2004).

[405] C.-S. Yang, C. Zhang, J. Redepenning, B. Doudin, *In situ magnetoresistance of Ni nanocontacts*, Appl. Phys. Lett. **84**, 2865 (2004).

[406] J.L. Costa-Krämer, M. Díaz, P.A. Serena, *Magnetic field effects on total and partial conductance histograms in Cu and Ni nanowires*, Appl. Phys. A **81**, 1539 (2005).

[407] A. Martín-Rodero, A. Levy Yeyati, J.C. Cuevas, *Transport properties of normal and ferromagnetic atomic-size constrictions with superconducting electrodes*, Physica C **352**, 67 (2001).

[408] P.S. Krstić, X.-G. Zhang, W.H. Butler, *Generalized conductance formula for the multiband tight-binding model*, Phys. Rev. B **66**, 205319 (2002).

[409] A. Smogunov, A. Dal Corso, E. Tosatti, *Selective d-state conduction blocking in nickel nanocontacts*, Surf. Sci. **507**, 609 (2002).

[410] A. Smogunov, A. Dal Corso, E. Tosatti, *Complex band structure with ultrasoft pseudopotentials: fcc Ni and Ni nanowire*, Surf. Sci. **532**, 549 (2003).

[411] A. Delin and E. Tosatti, *Magnetic phenomena in 5d transition metal nanowires*, Phys. Rev. B **68**, 144434 (2003).

[412] J. Velev and W.H. Butler, *Domain-wall resistance in metal nanocontacts*, Phys. Rev. B **69**, 094425 (2004).

[413] A.R. Rocha and S. Sanvito, *Asymmetric I-V characteristics and magnetoresistance in magnetic point contacts*, Phys. Rev. B **70**, 094406 (2004).

[414] A. Bagrets, N. Papanikolaou, I. Mertig, *Magnetoresistance of atomic-sized contacts: An ab initio study*, Phys. Rev. B **70**, 064410 (2004).

[415] D. Jacob, J. Fernández-Rossier, J.J. Palacios, *Magnetic and orbital blocking in Ni nanocontacts*, Phys. Rev. B **71**, 220403 (2005).

[416] M. Wierzbowska, A. Delin, E. Tosatti, *Effect of electron correlations in Pd, Ni, Co monowires*, Phys. Rev. B **72**, 035439 (2005).

[417] H. Dalgleish and G. Kirczenow, *Theoretical study of spin-dependent electron transport in atomic Fe nanocontacts*, Phys. Rev. B **72**, 155429 (2005).

[418] P.A. Khomyakov, G. Brocks, V. Karpan, M. Zwierzycki, P.J. Kelly, *Conductance calculations for quantum wires and interfaces: Mode matching and Green's functions*, Phys. Rev. B **72**, 35450 (2005).

[419] J. Fernández-Rossier, D. Jacob, C. Untiedt, J.J. Palacios, *Magnetic and orbital blocking in Ni nanocontacts*, Rev. B **72**, 224418 (2005).

[420] G. Autès, C. Barreteau, D. Spanjaard, M.C. Desjonquères, *Magnetism of iron: from the bulk to the monatomic wire*, J. Phys. Condens. Matter **18**, 6785 (2006).

[421] A. Smogunov, A. Dal Corso, E. Tosatti, *Ballistic conductance and magnetism in short tip suspended Ni nanowires*, Phys. Rev. B **73**, 75418 (2006).

[422] D. Jacob and J.J. Palacios, *Orbital eigenchannel analysis for ab initio quantum transport calculations*, Phys. Rev. B **73**, 075429 (2006).

[423] K. Xia, M. Zwierzycki, M. Talanana, P.J. Kelly, G.E.W. Bauer, *First-principles scattering matrices for spin transport*, Rev. B **73**, 64420 (2006).

[424] A.R. Rocha, T. Archer, S. Sanvito, *Search for magnetoresistance in excess of 1000% in Ni point contacts: Density functional calculations*, Phys. Rev. B **76**, 054435 (2007).

[425] J.C. Tung and G.Y. Guo, *Systematic ab initio study of the magnetic and electronic properties of all 3d transition metal linear and zigzag nanowires*, Phys. Rev. B **76**, 094413 (2007).

[426] A. Bagrets, N. Papanikolaou, I. Mertig, *Conduction eigenchannels of atomic-sized contacts: Ab initio KKR Green's function formalism*, Phys. Rev. B **75**, 235448 (2007).

[427] M. Häfner, J.K. Viljas, D. Frustaglia, F. Pauly, M. Dreher, P. Nielaba, J.C. Cuevas, *Theoretical study of the conductance of ferromagnetic atomic-sized contacts*, Phys. Rev. B **77**, 104409 (2008).

[428] M.J. Mehl, D.A. Papaconstantopoulos, I.I. Mazin, N.C. Bacalis, W.E. Pickett, *Applications of the NRL tight-binding method to magnetic systems*, Vol. 90, p. 6680, (American Institute of Physics, USA, 2001).

[429] D.R. Lide, *CRC Handbook of Chemistry and Physics, 79th Edition*, (CRC Press, Boca Raton, USA, 1998).

[430] A.D. Kent, J. Yu, U. Rüdiger and S.S.P. Parkin, *Domain wall resistivity in epitaxial thin film microstructures*, J. Phys.: Condens. Matter **13**, R461 (2001).

[431] P. Bruno, *Geometrically constrained magnetic wall*, Phys. Rev. Lett. **83**, 2425 (1999).

[432] P.M. Levy and S. Zhang, *Resistivity due to domain wall scattering*, Phys. Rev. Lett. **79**, 5110 (1997).

[433] S.S.P. Parkin, *Spin-polarized current in spin valves and magnetic tunnel junctions*, Annu. Rev. Mater. Sci. **25**, 357 (1995).

[434] N. García, M. Muñoz, Y.-W. Zhao, *Magnetoresistance in excess of 200% in ballistic Ni nanocontacts at room temperature and 100 Oe*, Phys. Rev. Lett. **82**, 2923 (1999).

[435] S.Z. Hua and H.D. Chopra, *100,000 % ballistic magnetoresistance in stable Ni nanocontacts at room temperature*, Phys. Rev. B **67**, 060401 (2003).

[436] W.F. Egelhoff Jr., L. Gan, H. Ettedgui, Y. Kadmon, C.J. Powell, P.J. Chen, A.J. Shapiro, R.D. McMichael, J.J. Mallet, T.P. Moffat, M.D. Stiles, E.B. Svedberg, *Artifacts that mimic ballistic magnetoresistance*, J. Magn. Magn. Mater. **287**, 496 (2005).

[437] E.Y. Tsymbal, O.N. Mryasov, P.R. LeClair, *Spin-dependent tunnelling in magnetic tunnel junctions*, J. Phys.: Condens. Matter **15**, R109 (2003).

[438] K.I. Bolotin, F. Kuemmeth, A.N. Pasupathy, D.C. Ralph, *From ballistic transport to tunneling in electromigrated ferromagnetic breakjunctions*, Nano Lett. **6**, 123 (2006).

[439] G. Tatara, Y.-W. Zhao, M. Muñoz, N. García, *Domain wall scattering explains 300% ballistic magnetoconductance of nanocontacts*, Phys. Rev. Lett. **83**, 2030 (1999).

[440] H. Imamura, N. Kobayashi, S. Takahashi, S. Maekawa, *Conductance quantization and magnetoresistance in magnetic point contacts*, Phys. Rev. Lett. **84**, 1003 (2000).

[441] J.B. van Hoof, K.M. Schep, A. Brataas, G.E. Bauer, P.J. Kelly, *Ballistic electron transport through magnetic domain walls*, Phys. Rev. B **59**, 138 (1999).

[442] W. Thomson, *On the electro-dynamic qualities of metals: Effects of magnetization on the electric conductivity of nickel and of iron*, Proc. R. Soc. London **8**, 546 (1857).

[443] T.R. Mcguire and R.I. Potter, *Anisotropic magnetoresistance in ferromagnetic 3d alloys*, IEEE Trans. Magn. **11**, 1018 (1975).

[444] J. Smit, *Magnetoresistance of ferromagnetic metals and alloys at low temperatures*, Physica (Amsterdam) **17**, 612 (1951).

[445] Z.K. Keane, L.H. Yu, D. Natelson, *Magnetoresistance of atomic-scale electromigrated nickel nanocontacts*, Appl. Phys. Lett. **88**, 062514 (2006).

[446] K.I. Bolotin, F. Kuemmeth, D.C. Ralph, *Anisotropic magnetoresistance and anisotropic tunneling magnetoresistance due to quantum interference in ferromagnetic metal break junctions*, Phys. Rev. Lett. **97**, 127202 (2006).

[447] S. Adam, M. Kindermann, S. Rahav, P.W. Brouwer, *Mesoscopic anisotropic magnetoconductance fluctuations in ferromagnets*, Phys. Rev. B **73**, 212408 (2006).

[448] M. Viret, M. Gabureac, F. Ott, C. Fermon, C. Barreteau, G. Autes, R. Guirardo-Lopez, *Giant anisotropic magneto-resistance in ferromagnetic atomic contacts*, Eur. Phys. J. B **51**, 1 (2006).

[449] A. Sokolov, E.Y. Tsymbal, J. Redepenning, B. Doudin, *Quantized magnetoresistance in atomic-size contacts*, Nature Nanotech. **2**, 171 (2007).

[450] J. Velev, R. F. Sabirianov, S. S. Jaswal, E. Y. Tsymbal, *Ballistic anisotropic magnetoresistance*, Phys. Rev. Lett. **94**, 127203 (2005).

[451] S.-F. Shi and D.C. Ralph, *Atomic motion in ferromagnetic break junctions*, Nature Nanotech. **2**, 522 (2007).

[452] S.F. Shi, K.I. Bolotin, F. Kuemmeth, D.C. Ralph, *Temperature dependence of anisotropic magnetoresistance and atomic rearrangements in ferromagnetic metal break junctions*, Phys. Rev. B **76**, 184438 (2007).

[453] J.D. Burton, R.F. Sabirianov, J.P. Velev, O.N. Mryasov, E.Y. Tsymbal, *Effect of tip resonances on tunneling anisotropic magnetoresistance in ferromagnetic metal break-junctions: A first-principles study*, Phys. Rev. B **76**, 144430 (2007).

[454] D. Jacob, J. Fernández-Rossier, J.J. Palacios, *Anisotropic magnetoresistance in nanocontacts*, Phys. Rev. B **77**, 165412 (2008).

[455] G. Autès, C. Barreteau, D. Spanjaard, M.-C. Desjonquères, *Electronic transport in iron atomic contacts: From the infinite wire to realistic geometries*, Phys. Rev. B **77**, 155437 (2008).

[456] M. Häfner, J.K. Viljas and J.C. Cuevas, *Theory of anisotropic magnetoresistance in atomic-sized ferromagnetic metal contacts*, Phys. Rev. B **79**, 140410 (2009).

[457] S.M. Sze, *Physics of Semiconductor Devices*, 2nd ed. (Wiley, New York, USA, 1981).

[458] R. Holmlin, R. Haag, M.L. Chabinyc, R.F. Ismagilov, A.E. Cohen, A. Terfort, M.A. Rampi, G.M. Whitesides, *Electron transport through thin organic films in metal-insulator-metal junctions based on self-assembled monolayers*, J. Am. Chem. Soc. **123**, 5075 (2001).

[459] M.A. Rampi and G.M. Whitesides, *A versatile experimental approach for understanding electron transport through organic materials*, Chem. Phys. **281**, 373 (2002).

[460] H.B. Akkerman, R.C.G. Naber, B. Jongbloed, P.A. van Hal, P.W.M. Blom, D.M. de Leeuw, B. de Boer, *Electron tunneling through alkanedithiol self-assembled monolayers in large-area molecular junctions*, Proc. Natl. Acad. Sci. USA **104**, 11161 (2007).

[461] J.M. Beebe, B.S. King, J.W. Gadzuk, C.D. Frisbie, J.G. Kushmerick, *Transition from direct tunneling to field emission in metal-molecule-metal junctions*, Phys. Rev. Lett. **97**, 026801 (2006).

[462] M. Mayor, C. von Hänisch, H. Weber, J. Reichert, D. Beckmann, *A trans-platinum(II) complex as single molecule insulator*, Angew. Chem. Int. Ed. Engl. **41**, 1183 (2002).

[463] G.J. Ashwell, J.R. Sambles, A.S. Martin, W.G. Parker, M. Szablewski, *Rectifying characteristics of Mg|($C_{16}H_{33}$-Q3CNQ LB film)|Pt structures*, J. Chem. Soc. Chem. Commun. 1374 (1990).

[464] A.S. Martin, J.R. Sambles, G.J. Ashwell, *Molecular rectifier*, Phys. Rev. Lett. **70**, 218 (1993).

[465] R.M. Metzger, T. Xu, I.R. Peterson, *Electrical rectification by a monolayer of hexadecylquinolinium tricyanoquinodimethanide measured between macroscopic gold electrodes*, J. Phys. Chem. B **105**, 7280 (2001).

[466] T. Xu, I.R. Peterson, M.V. Lakshmikantham, R.M. Metzger, *Rectification by a monolayer of hexadecylquinolinium tricyanoquinodimethanide between gold electrodes*, Angew. Chem. Int. Ed. **40**, 1749 (2001).

[467] R.M. Metzger, *Unimolecular rectifiers: Present status*, Chem. Phys. **326**, 176 (2006).

[468] M. Elbing, R. Ochs, M. Köntopp, M. Fischer, C. von Hänisch, F. Evers, H.B. Weber, Marcel Mayor, *A single-molecule diode*, Proc. Nat. Acad. Sci.

USA **102**, 8815 (2005).

[469] J. Reichert, R. Ochs, D. Beckmann, H.B. Weber, M. Mayor, H. von Löhneysen, *Driving current through single organic molecules*, Phys. Rev. Lett. **88**, 176804 (2002).

[470] L. Grüter, F. Cheng, T.T. Heikkilä, M.T. González, F. Diederich, C. Schönenberger, M. Calame, *Resonant tunneling through C60 molecular junction in liquid environment*, Nanotechnology **16**, 2143 (2005).

[471] M. Poot, E. Osorio, K. O'Neill, J.M. Thijssen, D. Vanmaekelbergh, C.A. van Walree, L.W. Jenneskens, H.S.J. van der Zant, *Temperature dependence of three-terminal molecular junctions with sulfur end-functionalized tercyclohexylidenes*, Nano Lett. **6**, 1031 (2006).

[472] L.A. Zotti, T. Kirchner, J.C. Cuevas, F. Pauly, T. Huhn, E. Scheer, A. Erbe, *Revealing the role of anchoring groups in the electrical conduction through single-molecule junctions*, to be published (2009).

[473] M. Kiguchi, O. Tal, S. Wohlthat, F. Pauly, M. Krieger, D. Djukic, J.C. Cuevas, J.M. van Ruitenbeek, *Highly conductive molecular junctions based on direct binding of benzene to platinum electrodes*, Phys. Rev. Lett. **101**, 046801 (2008).

[474] O. Tal, M. Kiguchi, W.H.A. Thijssen, D. Djukic, C. Untiedt, R.H.M. Smit, J.M. van Ruitenbeek, *Molecular signature of highly conductive metal-molecule-metal junctions*, Phys. Rev. B **80**, 085427 (2009).

[475] D.J. Wold, R. Haag, M.A. Rampi, C.D. Frisbie, *Distance dependence of electron tunneling through self-assembled monolayers measured by conducting probe atomic force microscopy: Unsaturated versus saturated molecular junctions*, J. Phys. Chem. B **106**, 2813 (2002).

[476] X.D. Cui, X. Zarate, J. Tomfohr, O.F. Sankey, A. Primak, A.L. Moore, T.A. Moore, D. Gust, G. Harrias, S.M. Lindsay, *Making electrical contacts to molecular monolayers*, Nanotechnology **13**, 5 (2002).

[477] L. Venkataraman, J.E. Klare, C. Nuckolls, M.S. Hybertsen, M.L. Steigerwald, *Dependence of single molecule junction conductance on molecular conformation*, Nature **442**, 904, (2006).

[478] F. Pauly, J.K. Viljas, J.C. Cuevas, G. Schön, *Density-functional study of tilt-angle and temperature-dependent conductance in biphenyl dithiol single-molecule junctions*, Phys. Rev. B **77**, 155312 (2008).

[479] E.G. Emberly and G. Kirczenow, *Theoretical study of electrical conduction through a molecule connected to metallic nanocontacts*, Phys. Rev. B **58**, 10911 (1998).

[480] R. Collepardo-Guevarra, D. Walter, D. Neuhauser, R. Baer, *A Hückel study of the effect of a molecular resonance cavity on the quantum conductance of an alkene wire*, Chem. Phys. Lett. **393**, 367 (2004).

[481] D.M. Cardamone, C.A. Stafford, S. Mazumdar, *Controlling Quantum transport through a single molecule*, Nano Lett. **6**, 2422 (2006).

[482] S.-H. Ke, W. Yang, H.U. Baranger, *Quantum interference controlled molecular electronics*, Nano Lett. **8**, 3257 (2008).

[483] A. Grigoriev, J. Skoldberg, G. Wendin, Z. Crljen, *Critical roles of metal-molecule contacts in electron transport through molecular-wire junctions*,

Phys. Rev. B **74**, 045401 (2006).

[484] T.A. Papadopoulos, I.M. Grace, C.J. Lambert, *Control of electron transport through Fano resonances in molecular wires*, Phys. Rev. B **74**, 193306 (2006).

[485] M. Ernzerhof, *A simple model of molecular electronic devices and its analytical solution*, J. Chem. Phys. **127**, 204709 (2007).

[486] X. Shi, Z. Dai, Z. Zeng, *Electron transport in self-assembled monolayers of thiolalkane: Symmetric I-V curves and Fano resonance*, Phys. Rev. B **76**, 235412 (2007).

[487] C.M. Finch, V.M. García-Suárez, C.J. Lambert, *Giant thermopower and figure of merit in single-molecule devices* Phys. Rev. B **79**, 033405 (2009).

[488] U. Fano, *Effects of configuration interaction on intensities and phase shifts*, Phys. Rev. **124**, 1866 (1961).

[489] N. Nilius, T.M. Wallis, W. Ho, *Localized molecular constraint on electron delocalization in a metallic chain*, Phys. Rev. Lett. **90**, 186102 (2003).

[490] A. Calzolari, C. Cavazzoni, M.B. Nardelli, *Electronic and transport properties of artificial gold chains*, Phys. Rev. Lett. **93**, 096404 (2004).

[491] Y. Xue, S. Datta, S. Hong, R. Reifenberger, J.I. Henderson, C.P. Kubiak, *Negative differential resistance in the scanning-tunneling spectroscopy of organic molecules*, Phys. Rev. B **59**, 7852 (1999).

[492] J. Chen, W. Wang, M.A. Reed, A.M. Rawlett, D.W. Price, J.M. Tour, *Room-temperature negative differential resistance in nanoscale molecular junctions*, Appl. Phys. Lett. **77**, 1224 (2000).

[493] F.F. Fan, J. Yang, L. Cai, D.W. Price Jr., S.M. Dirk, D.V. Kosynkin, Y. Yao, A.M. Rawlett, J.M. Tour, A.J. Bard, *Charge transport through self-assembled monolayers of compounds of interest in molecular electronics*, J. Am. Chem. Soc. **124**, 5550 (2002).

[494] I. Amlani, A.N. Rawlett, L.A. Nagahara, R.K. Tsui, *An approach to transport measurements of electronic molecules*, Appl. Phys. Lett. **80**, 2761 (2002).

[495] I. Kratochvilova, M. Kocirik, A. Zambova, J. Mbindyo, T.E. Mallouk, T.S. Mayer, *Room temperature negative differential resistance in molecular nanowires*, J. Mater. Chem. **12**, 2927 (2002).

[496] K. Walzer, E. Marx, N.C. Greenham, R.J. Less, P.R. Raithby, K. Stokbro, *Scanning tunneling microscopy of self-assembled phenylene ethynylene oligomers on Au(111) substrates*, J. Am. Chem. Soc. **126**, 1229 (2004).

[497] A.M. Rawlett, T.J. Hopson, L.A. Nagahara, R.K. Tsui, G.K. Ramachandran, S.M. Lindsay, *Electrical measurements of a dithiolated electronic molecule via conducting atomic force microscopy*, Appl. Phys. Lett. **81**, 3043 (2002).

[498] A.M. Rawlett, T.J. Hopson, I. Amlani, R. Zhang, J. Tresek, L.A. Nagahara, R.K. Tsui, H. Goronkin, *A molecular electronics toolbox*, Nanotechnology **14**, 377 (2003).

[499] J.D. Le, Y. He, T.R. Hoye, C.C. Mead, R.A. Kiehl, *Negative differential resistance in a bilayer molecular junction*, Appl. Phys. Lett. **83**, 5518 (2003).

[500] L.L. Chang, L. Esaki, R. Tsu, *Resonant tunneling in semiconductor double barriers*, Appl. Phys. Lett. **24**, 593 (1974).

[501] F. Capasso (Ed.), *Physics of Quantum Electron Devices*, (Springer-Verlag, Heidelberg, Berlin, D, 1990).

[502] L.L. Chang, E.E. Mendez, C. Tejedor (Eds.), *Resonant Tunneling in Semiconductors: Physics and Applications*, (Plenum Press, New York, USA, 1991).

[503] H. Mizuta and T. Tanoue, *The Physics and Applications of Resonant Tunneling Diodes*, (Cambridge University Press, Cambridge, UK, 1995).

[504] Z.J. Donhauser, B.A. Mantooth, K.F. Kelly, L.A. Bumm, J.D. Monnell, J.J. Stapleton, D.W. Price Jr., A.M. Rawlett, D.L. Allara, J.M. Tour, P.S. Weiss, *Conductance switching in single molecules through conformational changes*, Science **292**, 2303 (2001).

[505] J.M. Seminario, A.G. Zacarias, J.M. Tour, *Theoretical study of a molecular resonant tunneling diode*, J. Am. Chem. Soc. **122**, 3015 (2000).

[506] Y. Karzazi, J. Cornil, J.L. Brédas, *Theoretical investigation of the origin of negative differential resistance in substituted phenylene ethynylene oligomers*, Nanotechnology **14**, 165 (2003).

[507] Y. Karzazi, J. Cornil, J.L. Brédas, *Negative differential resistance behavior in conjugated molecular wires incorporating spacers: A quantum-chemical description*, J. Am. Chem. Soc. **123**, 10076 (2001).

[508] E.G. Emberly and G. Kirczenow, *Current-driven conformational changes, charging, negative differential resistance in molecular wires*, Phys. Rev. B **64**, 125318 (2001).

[509] M. Galperin, M.A. Ratner, A. Nitzan, *Hysteris, switching, negative differential resistance in molecular junctions: A polaron model*, Nano Lett. **5**, 125 (2005).

[510] S.Y. Quek, J.B. Neaton, M.S. Hybertsen, E. Kaxiras, S.G. Louie, *Negative differential resistance in transport through organic molecules on silicon*, Phys. Rev. Lett. **98**, 066807 (2007).

[511] K.H. Bevan, D. Kienle, H. Guo, S. Datta, *First-principles nonequilibrium analysis of STM-induced molecular negative-differential resistance on Si(100)*, Phys. Rev. B **78**, 035303 (2008).

[512] T. Rakshit, G.-C. Liang, A.W. Ghosh, S. Datta, *Silicon-based molecular electronics*, Nano Lett. **4**, 1803 (2004).

[513] T. Rakshit, G.-C. Liang, A.W. Ghosh, M.C. Hersam, S. Datta, *Molecules on silicon: Self-consistent first-principles theory and calibration to experiments*, Phys. Rev. B **72**, 125305 (2005).

[514] N.P. Guisinger, M.E. Greene, R. Basu, A.S. Baluch, M.C. Hersam, *Room temperature negative differential resistance through individual organic molecules on silicon surfaces*, Nano Lett. **4**, 55 (2004).

[515] N.P. Guisinger, R. Basu, M.E. Greene, A.S. Baluch, M.C. Hersam, *Observed suppression of room temperature negative differential resistance in organic monolayers on Si(100)*, Nanotechnology **15**, S452 (2004).

[516] N.P. Guisinger, N.L. Yoder, M.C. Hersam, *Probing charge transport at the single-molecule level on silicon by using cryogenic ultra-high vacuum*

scanning tunneling microscopy, Proc. Natl. Acad. Sci. U.S.A. **102**, 8838 (2005).

[517] W. Lu, V. Meunier, J. Bernholc, *Nonequilibrium quantum transport properties of organic molecules on silicon*, Phys. Rev. Lett. **95**, 206805 (2005).

[518] J.L. Pitters and R.A. Wolkow, *Detailed studies of molecular conductance using atomic resolution scanning tunneling microscopy*, Nano Lett. **6**, 390 (2006).

[519] A.-S. Hallbacka, B. Poelsemaa, H.J.W. Zandvliet, *Negative differential resistance of TEMPO molecules on Si(111)*, Appl. Surf. Sci. **253**, 4066 (2007).

[520] K.H. Bevan, F. Zahid, D. Kienle, H. Guo, *First-principles analysis of the STM image heights of styrene on Si(100)*, Phys. Rev. B **76**, 045325 (2007).

[521] G.C. Liang, A.W. Ghost, M. Paulsson, S. Datta, *Electrostatic potential profiles of molecular conductors*, Phys. Rev. B **69**, 115302 (2004).

[522] V. Mujica, A.E. Roitberg, M.A. Ratner, *Molecular wire conductance: Electrostatic potential spatial profile*, J. Chem. Phys. **112**, 6834 (2000).

[523] A. Nitzan, M. Galperin, G.-L. Ingold, H. Grabert, *On the electrostatic potential profile in biased molecular wires*, J. Chem. Phys. **117**, 10837 (2002).

[524] D.M. Adams, L. Brus, C.E.D. Chidsey, S. Creager, C. Creutz, C.R. Kagan, P.V. Kamat, M. Lieberman, S. Lindsay, R.A. Marcus, R.M. Metzger, M.E. Michel-Beyerle, J.R. Miller, M.D. Newton, D.R. Rolison, O. Sankey, K.S. Schanze, J. Yardley, X. Zhu, *Charge transfer on the nanoscale: Current status*, J. Phys. Chem. **107**, 6668 (2003).

[525] A. Salomon, D. Cahen, S. Lindsay, J. Tomfohr, V.B. Engelkes, C.D. Frisbie, *Comparison of electronic transport measurements on organic molecules*, Adv. Mater. **15**, 1881 (2003).

[526] J. Tomfohr, G. Ramachandran, O.F. Sankey, S.M. Lindsay, *Making contacts to single molecules: Are we there yet?*, in G. Cuniberti, G. Fagas, K. Richter (Eds.), *Introducing Molecular Electronics*, Series: Lecture Notes in Physics, Vol. 680, (Springer-Verlag, Berlin, Heidelberg, D, 2005).

[527] J. He, O. Sankey, M. Lee, X. Li, N.J. Tao, S.M. Lindsay, *Measuring single molecule conductance with break junctions*, Faraday Discuss. **131**, 145 (2006).

[528] S.M. Lindsay, *Molecular wires and devices: Advances and issues*, Faraday Discuss. **131**, 403 (2006).

[529] S.M. Lindsay and M.A. Ratner, *Molecular transport junctions: Clearing mists*, Adv. Mater. **19**, 23 (2007).

[530] X. Xiao, B. Xu, N. J. Tao, *Measurement of single molecule conductance: Benzenedithiol and benzenedimethanethiol*, Nano Lett. **4**, 267 (2004).

[531] S. Ghosh, H. Halimun, A.K. Mahapatro, J. Choi, S. Lodha, D. Janes, *Device structure for electronic transport through individual molecules using nanoelectrodes*, Appl. Phys. Lett. **87**, 233509 (2005).

[532] J. Ulrich, D. Esrail, W. Pontius, L. Venkataraman, D. Millar, L.H. Doerrer, *Device structure for electronic transport through individual molecules using nanoelectrodes*, J. Phys. Chem. B **110**, 2462 (2006).

[533] M. Tsutsui, Y. Teramae, S. Kurokawa, A. Sakai, *High-conductance states of single benzenedithiol molecules*, Appl. Phys. Lett. **89**, 163111 (2006).

[534] E. Lörtscher, H.B. Weber, H. Riel, *Statistical approach to investigating transport through single molecules*, Phys. Rev. Lett. **98**, 176807 (2007).

[535] C.A. Martin, D. Ding, H.S.J. van der Zant, J.M. van Ruitenbeek, *Lithographic mechanical break junctions for single-molecule measurements in vacuum: possibilities and limitations*, New J. Phys. **10**, 065008 (2008).

[536] C.A. Martin, D. Ding, J.K. Sørensen, T. Bjørnholm, J.M. van Ruitenbeek, H.S.J. van der Zant, *Fullerene-based anchoring groups for molecular electronics*, J. Am. Chem. Soc. **130**, 13198 (2008).

[537] M. Di Ventra, S. T. Pantelides, N. D. Lang, *First-principles calculation of transport properties of a molecular device*, Phys. Rev. Lett. **84**, 979 (2000).

[538] K. Stokbro, J. Taylor, M. Brandbyge, J.-L. Mozos, P. Ordejón, *Theoretical study of the nonlinear conductance of Di-thiol benzene coupled to Au(111) surfaces via thiol and thiolate bonds*, Comput. Mater. Sci. **27**, 151 (2003).

[539] E. G. Emberly and G. Kirczenow, *The smallest molecular switch*, Phys. Rev. Lett. **91**, 188301 (2003).

[540] J. Tomfohr and O. F. Sankey, *Theoretical analysis of electron transport through organic molecules*, J. Chem. Phys. **120**, 1542 (2004).

[541] F. Evers, F. Weigend, M. Koentopp, *Conductance of molecular wires and transport calculations based on density-functional theory*, Phys. Rev. B **69**, 235411 (2004).

[542] S. V. Faleev, F. Léonard, D. A. Stewart, M. van Schilfgaarde, *Ab initio tight-binding LMTO method for nonequilibrium electron transport in nanosystems*, Phys. Rev. B **71**, 195422 (2005).

[543] H. Kondo, H. Kino, J. Nara, T. Ozaki, T. Ohno, *Contact-structure dependence of transport properties of a single organic molecule between Au electrodes*, Phys. Rev. B **73**, 235323 (2006).

[544] K. Varga and S.T. Pantelides, *Quantum transport in molecules and nanotube devices*, Phys. Rev. Lett. **98**, 076804 (2007).

[545] M. Strange, I.S. Kristensen, K.S. Thygesen, K.W. Jacobsen, *Benchmark density functional theory calculations for nanoscale conductance*, J. Chem. Phys. **128**, 114714 (2008).

[546] F. Pauly, J.K. Viljas, J.C. Cuevas, *Length-dependent conductance and thermopower in single-molecule junctions of dithiolated oligophenylene derivatives: A density-functional study*, Phys. Rev. B **78**, 035315 (2008).

[547] P. Delaney and J. C. Greer, *Correlated electron transport in molecular electronics*, Phys. Rev. Lett. **93**, 036805 (2004).

[548] I. Bâldea and H. Köppel, *Electron transport through correlated molecules computed using the time-independent Wigner function: Two critical tests*, Phys. Rev. B **78**, 115315 (2008).

[549] B. Xu and N.J. Tao, *Measurement of single-molecule resistance by repeated formation of molecular junctions*, Science **301**, 1221 (2003).

[550] W. Haiss, R.J. Nichols, H. van Zalinge, S.J. Higgins, D. Bethell, D.J. Schiffrin, *Measurement of single molecule conductivity using the sponta-*

neous formation of molecular wires, Phys. Chem. Chem. Phys. **6**, 4330 (2004).

[551] X. Li, J. He, J. Hihath, B. Xu, S.M. Lindsay, N. Tao, *Conductance of single alkanedithiols: Conduction mechanism and effect of molecule-electrode contacts*, J. Am. Chem. Soc. **128**, 2135 (2006).

[552] M.T. González, R. Huber, S.J. van der Molen, C. Schönenberger, M. Calame, *Electrical conductance of molecular junctions by a robust statistical analysis*, Nano Lett. **6**, 2238 (2006).

[553] M. Fujihira, M. Suzuki, S. Fujii, A. Nishikawa, *Currents through single molecular junction of Au/hexanedithiolate/Au measured by repeated formation of break junction in STM under UHV: Effects of conformational change in an alkylene chain from gauche to trans and binding sites of thiolates on gold*, Phys. Chem. Chem. Phys. **8**, 3876 (2006).

[554] S.-Y. Jang, P. Reddy, A. Majumdar, R.A. Segalman, *Interpretation of stochastic events in single molecule conductance measurements*, Nano Lett. **6**, 2362 (2006).

[555] C. Li, I. Pobelov, T. Wandlowski, A. Bagrets, A. Arnold, F. Evers, *Charge transport in single Au-alkanedithiol-Au junctions: coordination geometries and conformational degrees of freedom*, J. Am. Chem. Soc. **130**, 318 (2008).

[556] M.T. González, J. Brunner, R. Huber, S. Wu, C. Schönenberger, M. Calame, *Conductance values of alkanedithiol molecular junctions*, New J. Phys. **10**, 065018 (2008).

[557] F. Chen, X. Li, J. Hihath, Z. Huang, N.J. Tao, *Effect of anchoring groups on single-molecule conductance: Comparative study of thiol-, amine-, carboxylic-acid-terminated molecules*, J. Am. Chem. Soc. **128**, 15874 (2006).

[558] L. Venkataraman, Y.S. Park, A.C. Whalley, C. Nuckolls, M.S. Hybertsen, M.L. Steigerwald, *Electronics and chemistry: Varying single molecule junction conductance with chemical substituent*, Nano Lett. **7**, 502 (2007).

[559] Y.S. Park, A.C. Whalley, M. Kamenetska, M.L. Steigerwald, M.S. Hybertsen, C. Nuckolls, L. Venkataraman, *Contact chemistry and single-molecule conductance: A comparison of phosphines, methyl sulfides, amines*, J. Am. Chem. Soc. **129**, 15768 (2007).

[560] P.E. Laibinis, G.M. Whitesides, D.L. Allara, Y.T. Tao, A.N. Parikh, R.G. Nuzzo, *Comparison of the structures and wetting properties of self-assembled monolayers of n-alkanethiols on the coinage metal surfaces, copper, silver, gold*, J. Am. Chem. Soc. **113**, 7152 (1991).

[561] A. Nishikawa, J. Tobita, Y. Kato, S. Fujii, M. Suzuki, M. Fujihira, *Accurate determination of multiple sets of single molecular conductance of Au/1,6-hexanedithiol/Au break junctions by ultra-high vacuum-scanning tunneling microscope and analyses of individual current separation curves*, Nanotechnology **18**, 424005 (2007).

[562] M.H. Lee, G. Speyer, O.F. Sankey, *Electron transport through single alkane molecules with different contact geometries on gold*, Phys. Stat. Sol. (b) **243**, 2021 (2006).

[563] K.H. Müller, *Effect of the atomic configuration of gold electrodes on the electrical conduction of alkanedithiol molecules*, Phys. Rev. B **73**, 045403 (2006).

[564] F. Picaud, A. Smogunov, A. Dal Corso, E.J. Tosatti, *Complex band structures and decay length in polyethylene chains*, J. Phys.: Condens. Matter **15**, 3731 (2003).

[565] G. Fagas, P. Delaney, J.C. Greer, *Independent particle descriptions of tunneling using the many-body quantum transport approach*, Phys. Rev. B **73**, 241314 (2006).

[566] M. Paulsson, C. Krag, T. Frederiksen, M. Brandbyge, *Conductance of alkanedithiol single-molecule junctions: A molecular dynamics study*, Nano Lett. **9**, 117 (2009).

[567] D. Djukic, K.S. Thygesen, C. Untiedt, R.H.M. Smit, K.W. Jacobsen, J.M. van Ruitenbeek, *Stretching dependence of the vibration modes of a single-molecule Pt-H2-Pt bridge*, Phys. Rev. B **71**, 161402 (2005).

[568] D. Djukic and J.M. Ruitenbeek, *Shot noise measurement on a single molecule*, Nano Lett. **6**, 789 (2006).

[569] J.C. Cuevas, J. Heurich, F. Pauly, W. Wenzel, G. Schön, *Theoretical description of the electrical conduction in atomic and molecular junctions*, Nanotechnology **14**, R29 (2003).

[570] Y. García, J.J. Palacios, E. SanFabián, J.A. Vergés, A.J. Pérez-Jiménez, E. Louis, *Electronic transport and vibrational modes in a small molecular bridge: H_2 in Pt nanocontacts*, Phys. Rev. B **69**, 041402 (2004).

[571] K.S. Thygesen, K.W. Jacobsen, *Conduction mechanism in a molecular hydrogen contact*, Phys. Rev. Lett. **94**, 036807 (2005).

[572] V.M. García-Suárez, A.R. Rocha, S.W. Bailey, C. Lambert, S. Sanvito, J. Ferrer, *Single-channel conductance of H_2 molecules attached to platinum or palladium electrodes*, Phys. Rev. B **72**, 045437 (2005).

[573] S. Csonka, A. Halbritter, G. Mihály, O.I. Shklyarevskii, S. Speller, H. van Kempen, *Conductance of Pd-H nanojunctions*, Phys. Rev. Lett. **93**, 016802 (2004).

[574] D.P. Long, J.L. Lazorcik, B.A. Mamtooth, M.H. Moore, M.A. Ratner, A. Troisi, Y. Yao, J.W. Ciszek, J.M. Tour, R. Shashidhar, *Effects of hydration on molecular junction transport*, Nature Mat. **5**, 901 (2006).

[575] R. Ahlrichs, M. Bär, M. Häser, H. Horn, C. Kölmel, *Electronic structure calculations on workstation computers: The program system turbomole*, Chem. Phys. Lett. **162**, 165 (1989).

[576] F. Pauly, J.K. Viljas, U. Huniar, M. Häfner, S. Wohlthat, M. Burkle, J.C. Cuevas, G. Schön, *Cluster-based density-functional approach to quantum transport through molecular and atomic contacts*, New J. Phys. **10**, 125019 (2008).

[577] A. Ulman, *Formation and structure of self-assembled monolayers*, Chem. Rev. **96**, 1533 (1996).

[578] S.Y. Quek, L. Venkataraman, H.J. Choi, S.G. Louie, M.S. Hybertsen, J.B. Neaton, *Amine-gold linked single-molecule circuits: Experiment and theory*, Nano Lett. **7**, 3477 (2007).

[579] M.S. Hybertsen, L. Venkataraman, J.E. Klare, A.C. Whalley, M.L. Steigerwald and C. Nuckolls, *Amine-linked single-molecule circuits: systematic trends across molecular families*, J. Phys.: Condens. Matter **20**, 374115 (2008).

[580] C. Rogero, J.I. Pascual, J. Gómez-Herrero, A.M. Baró, *Resolution of site-specific bonding properties of C_{60} adsorbed on Au(111)*, J. Chem. Phys. **116**, 832 (2002).

[581] M. Mayor, H.B. Weber, J. Reichert, M. Elbing, C. von Hänisch, D. Beckmann, M. Fischer, *Electric current through a molecular rod: Relevance of the position of the anchor groups*, Angew. Chem. Int. Ed. **42**, 5834 (2003).

[582] A.M. Kuznetsov, J. Ulstrup, *Electron Transfer in Chemistry and Biology: An Introduction to the Theory*, Wiley Series in Theoretical Chemistry, (Wiley, New York, USA 1999).

[583] V. Balzani (Ed.) *Electron Transfer in Chemistry* (Vols. 1-5), (Wiley-VCH, Weinheim, D, 2001).

[584] S. Woitellier, J.P. Launay, C. Joachim, *The possibility of molecular switching: Theoretical study of $[(NH_3)_5 Ru\text{-}4,4'\text{-}Bipy\text{-}Ru(NH_3)_5]^{5+}$*, Chem. Phys. **131**, 481 (1989).

[585] V. Mujica, A. Nitzan, Y. Mao, W. Davis, M. Kemp, A. Roitberg, M.A. Ratner, *Electron transfer in molecules and molecular wires: Geometry dependence, coherent transfer, control*, Adv. Chem. Phys. **107**, 403 (1999).

[586] F. Moresco, G. Meyer, K.H. Rieder, H. Tang, A. Gourdon, C. Joachim, *Conformational changes of single molecules induced by scanning tunneling microscopy manipulation: A route to molecular switching*, Phys. Rev. Lett. **86**, 672 (2001).

[587] D. Dulić, S.J. van der Molen, T. Kudernac, H.T. Jonkman, J.J.D. de Jong, T.N. Bowden, J. van Esch, B.L. Feringa, B.J. van Wees, *One-way optoelectronic switching of photochromic molecules on gold*, Phys. Rev. Lett. **91**, 207402 (2003).

[588] S. Yasuda, T. Nakamura, M. Matsumoto, H. Shigekawa, *Phase switching of a single isomeric molecule and associated characteristic rectification*, J. Am. Chem. Soc. **125**, 16430 (2003).

[589] C.M. Finch, S. Sirichantaropass, S.W. Bailey, I.M. Grace, V.M. García-Suárez, C.J. Lambert, *Conformation dependence of molecular conductance: Chemistry versus geometry*, J. Phys.: Condens. Matter **20**, 022203 (2008).

[590] E. Lörtscher, M. Elbing, M. Tschudy, C. von Hänisch, H.B. Weber, M. Mayor, H. Riel, *Charge transport through molecular rods with reduced π-conjugation*, Chem. Phys. Chem. **9**, 2252 (2008).

[591] J.G. Kushmerick, D.B. Holt, S.K. Pollack, M.A. Ratner, J.C. Yang, T.L. Schull, J. Naciri, M.H. Moore, R. Shashidhar, *Effect of bond-length alternation in molecular wires*, J. Am. Chem. Soc. **124**, 10654 (2002).

[592] L.T. Cai, H. Skulason, J.G. Kushmerick, S.K. Pollack, J. Naciri, R. Shashidhar, D.L. Allara, T.E. Mallouk, T.S. Mayer, *Nanowire-based molecular monolayer junctions: Synthesis, assembly, electrical characterization*, J. Phys. Chem. B **108**, 2827 (2004).

[593] R. Huber, M.T. González, S. Wu, M. Langer, S. Grunder, V. Horhoiu, M. Mayor, M.R. Bryce, C. Wang, R. Jitchati, C. Schönenberger, M. Calame, *Electrical conductance of conjugated oligomers at the single molecule level*, J. Am. Chem. Soc. **130** 1080 (2008).

[594] E. Leary, S.J. Higgins, H. van Zalinge, W. Haiss, R.J. Nichols, *Chemical control of double barrier tunnelling in* α, ω*-dithiaalkane molecular wires*, Chem. Commun., 3939 (2007).

[595] L.P. Hammett, *The Effect of structure upon the reactions of organic compounds. Benzene derivatives*, J. Am. Chem. Soc. **59**, 96 (1937).

[596] W. Haiss, H. van Zalinge, S.J. Higgins, D. Bethell, H. Hobenreich, D.J. Schiffrin, R.J. Nichols, *Redox state dependence of single molecule conductivity*, J. Am. Chem. Soc. **125**, 15294 (2003).

[597] W. Haiss, H. van Zalinge, H. Hobenreich, D. Bethell, D.J. Schiffrin, S.J. Higgins, R.J. Nichols, *Molecular wire formation from viologen assemblies*, Langmuir **20**, 7694 (2004).

[598] W. Haiss, H. van Zalinge, D. Bethell, J. Ulstrup, D.J. Schiffrin, R.J. Nichols, *Thermal gating of the single molecule conductance of alkanedithiol*, Faraday Discuss. **131**, 253 (2006).

[599] W. Haiss, C. Wang, A.S. Batsanov, D.J. Schiffrin, S.J. Higgins, M.R. Bryce, C.J. Lambert, R.J. Nichols, *Precision control of single-molecule electrical junctions*, Nature Mat. **5**, 995 (2006).

[600] L. Lafferentz, F. Ample, H. Yu, S. Hecht, C. Joachim, L. Grill, *Conductance of a single conjugated polymer as a continuous function of its length*, Science **323**, 1193 (2009).

[601] M. Büttiker and R. Landauer, *Traversal time for tunneling*, Phys. Rev. Lett. **49**, 1739 (1982).

[602] A. Nitzan, J. Jortner, J. Wilkie, A.L. Burin, M.A. Ratner, *Tunneling time for electron transfer reactions*, J. Phys. Chem. B **104**, 5661 (2000).

[603] C.H. Ahn, A. Bhattacharya, M. Di Ventra, J.N. Eckstein, C.D. Frisbie, M.E. Gershenson, A.M. Goldman, I.H. Inoue, J. Mannhart, A.J. Millis, A.F. Morpurgo, D. Natelson, J.-M. Triscone, *Electrostatic modification of novel materials*, Rev. Mod. Phys. **78**, 1185 (2006).

[604] D. Natelson, L.Y. Yu, J.W. Ciszek, Z.K. Keane, J.M. Tour, *Single-molecule transistors: Electron transfer in the solid state*, Chem. Phys. **324**, 267 (2006).

[605] J.M. Thijssen and H.S.J. van der Zant, *Charge transport and single-electron effects in nanoscale systems*, Phys. Stat. Sol. (b) **245**, 1455 (2008).

[606] E.A. Osorio, T. Bjørnholm, J.-M. Lehn, M. Ruben, H.S.J. van der Zant, *Single-molecule transport in three-terminal devices*, J. Phys.: Condens. Matter **20**, 374121 (2008).

[607] D.R. Ward, G.D. Scott, Z.K. Keane, N.J. Halas, D. Natelson, *Electronic and optical properties of electromigrated molecular junctions*, J. Phys.: Condens. Matter **20**, 374118 (2008).

[608] L.P. Kouwenhoven, C.M. Marcus, P.L. McEuen, S. Tarucha, R.M. Westervelt, N.D. Wingreen, *Electron transport in quantum dots*, in L.L. Sohn, L. P. Kouwenhoven, G. Schön (Eds.), *Mesoscopic Electron Transport*, NATO-

ASI Series E, Vol. 345, p. 105, (Kluwer Academic Publishers, Dordrecht, NL, 1997).

[609] D.V. Averin and K.K. Likharev, in B.L. Al'tshuler, P.A. Lee, R.A. Webb (Eds.), *Mesoscopic Phenomena in Solids*, (Elsevier, Amsterdam, NL, 1991).

[610] H. Grabert and M.H. Devoret (Eds.), *Single Charge Tunneling: Coulomb Blockade Phenomena in Nanostructures*, NATO ASI Series B, Vol. 294, (Plenum Press, New York, USA, 1992).

[611] J.P. Bird, *Electron Transport in Quantum Dots*, (Kluwer Academic Publishers, Dordrecht, NL, 2003).

[612] D.C. Ralph, C.T. Black, J.M. Hergenrother, J.G. Lu, M. Tinkham, in L. L. Sohn, L. P. Kouwenhoven, G. Schön (Eds.), *Mesoscopic Electron Transport*, NATO-ASI Series E, Vol. 345, p. 453, (Kluwer Academic Publishers, Dordrecht, NL, 1997).

[613] S.J. Tans, M.H. Devoret, H. Dai, A. Thess, R.E. Smalley, L.J. Geerligs, C. Dekker, *Individual single-wall carbon nanotubes as quantum wires*, Nature **386**, 474 (1997).

[614] P.L. McEuen, M. Fuhrer, H. Park, *Single-walled carbon nanotube electronics*, IEEE Trans. on Nanotechnol. **1**, 78 (2002).

[615] M.T. Björk, B.J. Ohlsson, T. Sass, A.I. Persson, C. Thelander, M.H. Magnusson, K. Deppert, L. R. Wallenberg, L. Samuelson, *One-dimensional steeplechase for electrons realized*, Nano Lett. **2**, 87 (2002).

[616] M.T. Björk, C. Thelander, A.E. Hansen, L.E. Jensen, M.W. Larsson, L.R. Wallenberg, L. Samuelson, *Few-electron quantum dots in nanowires*, Nano Lett. **4**, 1621 (2004).

[617] L.H. Yu, Z.K. Keane, J.W. Ciszek, L. Cheng, M.P. Stewart, J.M. Tour, D. Natelson, *Inelastic electron tunneling via molecular vibrations in single-molecule transistors*, Phys. Rev. Lett. **93**, 266802 (2004).

[618] E.A. Osorio, K. O'Neill, M. Wegewijs, N. Stuhr-Hansen, J. Paaske, T. Bjørnholm, H.S.J. van der Zant, *Electronic excitations of a single molecule contacted in a three-terminal configuration*, Nano Lett. **7**, 3336 (2007).

[619] H.B. Heersche, Z. de Groot, J.A. Folk, H.S.J. van der Zant, *Electron transport through single Mn_{12} molecular magnets*, Phys. Rev. Lett. **96**, 206801 (2006).

[620] M.H. Jo, J.E. Grose, K. Baheti, M.M. Deshmukh, J.J. Sokol, E.M. Rumberger, D.N. Hendrickson, J.R. Long, H. Park and D.C. Ralph, *Signatures of molecular magnetism in single-molecule transport spectroscopy*, Nano Lett. **6**, 2014 (2006).

[621] T. Taychatanapat, K.I. Bolotin, F. Kuemmeth, D.C. Ralph, *Imaging electromigration during the formation of break-junctions*, Nano Lett. **7**, 652 (2007).

[622] S. Kubatkin, A. Danilov, M. Hjort, J. Cornil, J.-L. Bredas, N. Stuhr-Hansen, P. Hedegard, T. Bjørnholm, *Single-electron transistor of a single organic molecule with access to several redox states*, Nature **425**, 698 (2003).

[623] S. Kubatkin, A. Danilov, M. Hjort, J. Cornil, J.-L. Bredas, N. Stuhr-

Hansen, P. Hedegard, T. Bjørnholm, *Single electron transistor with a single conjugated molecule*, Curr. Appl. Phys. **4**, 554 (2004).

[624] C.W.J. Beenakker, *Theory of Coulomb-blockade oscillations in the conductance of a quantum dot*, Phys. Rev. B **44**, 1646 (1991).

[625] H. van Houten, C.W.J. Beenakker, A.A.M. Staring, *Coulomb-blockade oscillations in semiconductor nanostructures*, in H. Grabert and M.H. Devoret (Eds.), *Single Charge Tunneling*, NATO ASI Series B, Vol. 294, (Plenum Press, New York, USA, 1992).

[626] S. Sapmaz, P. Jarillo-Herrero, J. Kong, C. Dekker, L.P. Kouwenhoven, H.S.J. van der Zant, *Electronic excitation spectrum of metallic carbon nanotubes*, Phys. Rev. B **71**, 153402 (2005).

[627] Y. Oreg, K. Byczuk, B.I. Halperin, *Spin configurations of a carbon nanotube in a nonuniform external potential*, Phys. Rev. Lett. **85**, 365 (2000).

[628] L.P. Kouwenhoven, D.G. Austing, S. Tarucha, *Few-electron quantum dots*, Rep. Prog. Phys. **64**, 701 (2001).

[629] D.V. Averin and A.N. Korotkov, *Correlated single-electron tunneling via mesoscopic metal-particle: Effects of the energy quantization*, Zh. Eksp. Teor. Fiz. **97**, 1661 (1990) [Sov. Phys. JETP **70**, 937 (1990)].

[630] A.N. Korotkov, D.V. Averin, K.K. Likharev, *Single-electron charging of the quantum-wells and dots*, Physica B **165** & **166**, 927 (1990).

[631] D.V. Averin, A.N. Korotkov, K.K. Likharev, *Theory of single-electron charging of quantum well and dots*, Phys. Rev. B **44**, 6199 (1991).

[632] Y. Meir, N.S. Wingreen, P.A. Lee, *Transport through a strongly interacting electron system: Theory of periodic conductance oscillations*, Phys. Rev. Lett. **66**, 3048 (1991).

[633] J. von Delft and D.C. Ralph, *Spectroscopy of discrete energy levels in ultrasmall metallic grains*, Phys. Rep. **345**, 61 (2001).

[634] M.H. Hettler, W. Wenzel, M.R. Wegewijs, H. Schoeller, *Current collapse in tunneling transport through benzene*, Phys. Rev. Lett. **90**, 076805 (2003).

[635] M.R. Wegewijs, M.H. Hettler, C. Romeike, A. Thielmann, K. Nowack, J. König, in G. Cuniberti, G. Fagas, K. Richter (Eds.), *Single Electron Tunneling in Small Molecules*, Series: Lecture Notes in Physics, Vol. **680**, p. 207, (Springer-Verlag, Berlin, Heidelberg, D, 2005).

[636] J.J. Palacios, *Coulomb blockade in electron transport through a C_{60} molecule from first principles*, Phys. Rev. B **72**, 125424 (2005).

[637] B. Muralidharan, A.W. Ghosh, S. Datta, *Probing electronic excitations in molecular conduction*, Phys. Rev. B **73**, 155410 (2006).

[638] C. Timm and F. Elste, *Spin amplification, reading, writing in transport through anisotropic magnetic molecules*, Phys. Rev. B **73**, 235304 (2006).

[639] F. Elste and C. Timm, *Transport through anisotropic magnetic molecules with partially ferromagnetic leads: Spin-charge conversion and negative differential conductance*, Phys. Rev. B **73**, 235305 (2006).

[640] C. Romeike, M.R. Wegewijs, H. Schoeller, *Spin quantum tunneling in single molecular magnets: Fingerprints in transport spectroscopy of current and noise*, Phys. Rev. Lett. **96**, 196805 (2006).

[641] H. Wang and G.K.-L. Chan, *Self-interaction and molecular Coulomb block-ade transport in ab initio Hartree-Fock theory*, Phys. Rev. B **76**, 193310 (2007).

[642] J. Lehmann and D. Loss, *Sequential tunneling through molecular spin rings*, Phys. Rev. Lett. **98**, 117203 (2007).

[643] R. Stadler, V. Geskin, J. Cornil, *Towards a theoretical description of molecular junctions in the Coulomb blockade regime based on density functional theory*, Phys. Rev. B **78**, 113402 (2008).

[644] R. Stadler, V. Geskin, J. Cornil, *Screening effects in a density functional theory based description of molecular junctions in the Coulomb blockade regime*, Phys. Rev. B **79**, 113408 (2009).

[645] L. Michalak, C.M. Canali, M.R. Pederson, V.G. Benza, M. Paulsson, *Theory of tunneling spectroscopy in a Mn_{12} single-electron transistor by DFT methods*, arXiv:0812.1058.

[646] K. Kaasbjerg and K. Flensberg, *Strong polarization-induced reduction of addition energies in single-molecule nanojunctions*, Nano Lett. **8**, 3809 (2008).

[647] J. König, H. Schoeller, G. Schön, *Zero-bias anomalies and boson-assisted tunneling through quantum dots*, Phys. Rev. Lett. **76**, 1715 (1996).

[648] J. König, J. Schmid, H. Schoeller, G. Schön, *Resonant tunneling through ultrasmall quantum dots: Zero-bias anomalies, magnetic-field dependence, boson-assisted transport*, Phys. Rev. B **54**, 16820 (1996).

[649] D.V. Averin and Y.V. Nazarov, *Macroscopic quantum tunneling of charge and co-tunneling* in *Single Charge Tunneling: Coulomb Blockade Phenomena in Nanostructures*, H. Grabert and M.H. Devoret (Eds.), NATO-ASI Series B, Vol. 294, p. 217 (Plenum Press, New York, USA, 1992).

[650] S. de Franceschi, S. Sasaki, J.M. Elzerman, W.G. van der Wiel, S. Tarucha, L.P. Kouwenhoven, *Electron cotunneling in a semiconductor quantum dot*, Phys. Rev. Lett. **86**, 878 (2001).

[651] A. C. Hewson, *The Kondo Problem to Heavy Fermions*, (Cambridge University Press, Cambridge, UK, 1993).

[652] J. Kondo, *Resistance minimum in dilute magnetic alloys*, Prog. Theor. Phys. **32**, 37 (1964).

[653] L. Kouwenhoven and L. Glazman, *Revival of the Kondo effect*, Physics World **14**, 33 (2001).

[654] D. Goldhaber-Gordon, H. Shtrikman, D. Mahalu, D. Abusch-Magder, U. Meirav, M.A. Kastner, *Kondo effect in a single-electron transistor*, Nature **391**, 156 (1998).

[655] D. Goldhaber-Gordon, J. Göres, M.A. Kastner, H. Shtrikman, D. Mahalu, U. Meirav, *From the Kondo regime to the mixed-valence regime in a single-electron transistor*, Phys. Rev. Lett. **81**, 5225 (1998).

[656] S.M. Cronenwett, T.H. Oosterkamp, L.P. Kouwenhoven, *A tunable Kondo effect in quantum dots*, Science **281**, 540 (1998).

[657] J. Li, W.-D Schneider, R. Berndt, B. Delley, *Kondo scattering observed at a single magnetic impurity*, Phys. Rev. Lett. **80**, 2893 (1998).

[658] V. Madhavan, W. Chen, T. Jamneala, M.F. Crommie, N.S. Wingreen,

Tunneling into a single magnetic atom: Spectroscopic evidence of the Kondo resonance, Science **280**, 567 (1998).

[659] H.C. Manoharan, C.P. Lutz and D.M. Eigler, *Quantum mirages: The coherent projection of electronic structure*, Nature **403**, 512 (2000).

[660] J. Nygard, D.H. Cobden and P.E. Lindelof, *Kondo physics in carbon nanotubes*, Nature **408**, 342 (2000).

[661] M.R. Buitelaar, A. Bachtold, T. Nussbaumer, M. Iqbal and C. Schönenberger, *Multiwall carbon nanotubes as quantum dots*, Phys. Rev. Lett. **88**, 156801 (2002).

[662] L.H. Yu and D. Natelson, *The Kondo effect in C_{60} single-molecule transistors*, Nano Lett. **4**, 79 (2004).

[663] A.N. Pasupathy, R.C. Bialczak, J. Martinek, J.E. Grose, L.A.K. Donev, P.L. McEuen, D.C. Ralph, *The Kondo effect in the presence of ferromagnetism*, Science **306**, 86 (2004).

[664] L.H. Yu, Z.K. Keane, J.W. Ciszek, L. Cheng, J.M. Tour, T. Baruah, M.R. Pederson, D. Natelson, *Kondo resonances and anomalous gate dependence in the electrical conductivity of single-molecule transistors*, Phys. Rev. Lett. **95**, 256803 (2005).

[665] M. Pustilnik, L.I. Glazman, D.H. Cobden, L.P. Kouwenhoven, *Magnetic field-induced Kondo effects in Coulomb blockade systems*, Lect. Notes Phys. **579**, 3 (2001); Preprint cond-mat/0010336.

[666] M. Pustilnik and L. Glazman, *Kondo effect in quantum dots*, J. Phys.: Condens. Matter **16**, R513 (2004).

[667] M. Grobis, I.G. Rau, R.M. Potok, D. Goldhaber-Gordon, *Kondo effect in mesoscopic quantum dots* in H. Kronmuller (Ed.), *Handbook of Magnetism and Advanced Magnetic Materials*, Vol. 5, (Wiley, New York, USA, 2007).

[668] F.D.M. Haldane, *Scaling theory of the asymmetric Anderson model*, Phys. Rev. Lett. **40**, 416 (1978).

[669] L.I. Glazman and M.E. Raikh, *Resonant Kondo transparency of a barrier with quasilocal impurity states*, JETP Lett. **47**, 452 (1988).

[670] T.A. Costi, A.C. Hewson and V. Zlatic, *Transport coefficients of the Anderson model via the numerical renormalization group*, J. Phys.: Condens. Matter **6**, 2519 (1994).

[671] W.G. van der Wiel, S. De Franceschi, T. Fujisawa, J.M. Elzerman, S. Tarucha, L.P. Kouwenhoven, *The Kondo effect in the unitary limit*, Science **289**, 2105 (2000).

[672] P. Jarillo-Herrero, J. Kong, H.S.J. van der Zant, C. Dekker, L.P. Kouwenhoven, S. De Franceschi, *Orbital Kondo effect in carbon nanotubes*, Nature **434**, 484 (2005).

[673] S. Sasaki, S. De Franceschi, J.M. Elzerman, W.G. van der Wiel, M. Eto, S. Tarucha and L.P. Kouwenhoven, *Kondo effect in an integer-spin quantum dot*, Nature **405**, 764 (2000).

[674] M. Pustilnik, Y. Avishai and K. Kikoin *Quantum dots with even number of electrons: Kondo effect in a finite magnetic field*, Phys. Rev. Lett. **84**, 1756 (2000).

[675] M. Eto and Y.V. Nazarov, *Enhancement of Kondo effect in quantum dots*

with an even number of electrons, Phys. Rev. Lett. **85**, 1306 (2000).

[676] C. Kergueris, J.-P. Bourgoin, S. Palacin, D. Esteve, C. Urbina, M. Magoga, C. Joachim, *Electron transport through a metal-molecule-metal junction*, Phys. Rev. B **59**, 12505 (1999).

[677] J. Heinze, J. Mortensen, K. Müllen, R. Schenk, *The charge storage mechanism of conducting polymers: A voltammetric study on defined soluble oligomers of the phenylene-vinylene type*, J. Chem. Soc. Chem. Commun. 701 (1987).

[678] E.A. Osorio, K. O'Neill, N. Stuhr-Hansen, O.F. Nielsen, T. Bjørnholm, H.S.J. van der Zant, *Addition energies and vibrational fine structure measured in electromigrated single-molecule junctions based on an oligophenylenevinylene derivative*, Adv. Mater. **19**, 281 (2007).

[679] A. Danilov, S. Kubatkin, S. Kafanov, P. Hedegard, N. Stuhr-Hansen, K. Moth-Poulsen, T. Bjørnholm, *Electronic transport in single molecule junctions: Control of the molecule-electrode coupling through intramolecular tunneling barriers*, Nano. Lett. **8**, 1 (2008).

[680] J. Martinek, Y. Utsumi, H. Imamura, J. Barnas, S. Maekawa, J. König, G. Schön, *Kondo effect in quantum dots coupled to ferromagnetic leads*, Phys. Rev. Lett. **91**, 127203 (2003).

[681] J. Martinek, M. Sindel, L. Borda, J. Barnas, J. König, G. Schön and J. von Delft, *NRG study of the Kondo effect in the presence of itinerant-electron ferromagnetism*, Phys. Rev. Lett. **91**, 247202 (2003).

[682] J.J. Parks, A.R. Champagne, G.R. Hutchison, S. Flores-Torres, H.D. Abruña, D.C. Ralph, *Tuning the Kondo effect with a mechanically controllable break juncton*, Phys. Rev. Lett. **99**, 026601 (2007).

[683] G.D. Scott, Z.K. Keane, J.W. Ciszek, J.M. Tour, D. Natelson, *Universal scaling of nonequilibrium transport in the Kondo regime of single molecule devices*, Phys. Rev. B **79**, 165413 (2009).

[684] N. Roch, S. Florens, V. Bouchiat, W. Wernsdorfer, F. Balestro, *Quantum phase transition in a single-molecule quantum dot*, Nature **453**, 633 (2008).

[685] A.N. Pasupathy, J. Park, C. Chang, A.V. Soldatov, S. Lebedkin, R.C. Bialczak, J.E. Grose, L.A.K. Donev, J.P. Sethna, D.C. Ralph, P.L. McEuen, *Vibration-assisted electron tunneling in C_{140} single-molecule transistors*, Nano Lett. **5**, 203 (2005).

[686] L. Bogani, W. Wernsdorfer, F. Balestro, *Molecular spintronics using single-molecule magnets*, Nature Mat. **7**, 179 (2008).

[687] G. Christou, D. Gatteschi, D.N. Hendrickson, R. Sessoli, *Single-molecule magnets*, Mater. Res. Soc. Bull. **25**, 66 (2000).

[688] D. Gatteschi, R. Sessoli, J. Villain, *Molecular Nanomagnets*, (Oxford University Press, New York, USA, 2007).

[689] D. Weinmann, W. Häusler, B. Kramer, *Spin blockades in linear and nonlinear transport through quantum dots*, Phys. Rev. Lett. **74**, 984 (1995).

[690] X. Waintal and P.W. Brouwer, *Tunable magnetic relaxation mechanism in magnetic nanoparticles*, Phys. Rev. Lett. **91**, 247201 (2003).

[691] J.E. Grose, E.S. Tam, C. Timm, M. Scheloske, B. Ulgut, J.J. Parks, H.D. Abrunña, W. Harneit, D.C. Ralph, *Tunneling spectra of individual mag-*

netic endofullerene molecules, Nature Mat. **7**, 884 (2008).

[692] F. Elste and C. Timm, *Theory for transport through a single magnetic molecule: Endohedral NC$_{60}$*, Phys. Rev. B **71**, 155403 (2005).

[693] M. Mannini, F. Pineider, P. Sainctavit, C. Danieli, E. Otero, C. Sciancalepore, A.M. Talarico, M.-A. Arrio, A. Cornia, D. Gatteschi, R. Sessoli, *Magnetic memory of a single-molecule quantum magnet wired to a gold surface*, Nature Mat. **8**, 194 (2009).

[694] E.B. Wilson, J.C. Decius, P.C. Cross, *Molecular vibrations*, (McGraw-Hill, New York, USA, 1955, reprinted by Dover Publication Inc., New York, USA, 1980).

[695] M. Galperin, M.A. Ratner and A. Nitzan, *Molecular transport junctions: vibrational effects*, J. Phys.: Condens. Matter **19**, 103201 (2007).

[696] M. Galperin, M.A. Ratner, A. Nitzan, A. Troisi, *Nuclear coupling and polarization in molecular transport junctions: Beyond tunneling to function*, Science **319**, 1056 (2008).

[697] R.C. Jaklevic and J. Lambe, *Molecular vibration spectra by electron tunneling*, Phys. Rev. Lett. **17**, 1139 (1966).

[698] P.K. Hansma, *Inelastic electron tunneling*, Phys. Rep. **30**, 145 (1977).

[699] W.H. Weinberg, *Inelastic electron tunneling spectroscopy: A probe of the vibrational structure of surface species*, Annu. Rev. Phys. Chem. **29**, 115 (1978).

[700] T. Wolfram, Ed., *Inelastic Electron Tunneling Spectroscopy*, (Springer-Verlag, Heidelberg, Berlin, D, 1978).

[701] P.K. Hansma (Ed.), *Tunneling Spectroscopy* (Plenum Press, New York, USA, 1982).

[702] E.L. Wolf, *Principles of Electron Tunnelling Spectroscopy*, (Oxford University Press, Oxford, UK, 1985).

[703] C.J. Adkins and W.A. Phillips, *Inelastic electron tunnelling spectroscopy*, J. Phys. C: Solid State Phys. **18**, 1313 (1985).

[704] K.W. Hipps and U. Mazur, *Inelastic electron-tunneling: An alternative molecular spectroscopy*, J. Phys. Chem. **97**, 7803 (1993).

[705] J. Lambe and R.C. Jaklevic, *Molecular vibration spectroscopy by inelastic electron tunneling*, Phys. Rev. **165**, 821 (1968).

[706] J. Klein, A. Léger, M. Belin, D. Défourneau, M.J. Sangster, *Inelastic-electron-tunneling spectroscopy of metal-insulator-metal junctions*, Phys. Rev. B **7**, 2336 (1973).

[707] J. Kirtley, D.J. Scalapino, P.K. Hansma, *Theory of vibrational mode intensities in inelastic electron tunneling spectroscopy*, Phys. Rev. B **14**, 3177 (1976).

[708] J. Kirtley and J.T. Hall, *Theory of intensities in inelastic-electron tunneling spectroscopy orientation of adsorbed molecules*, Phys. Rev. B **22**, 848 (1980).

[709] B.C. Stipe, M.A. Rezaei, W. Ho, *Single-molecule vibration spectroscopy and microscopy*, Science **280**, 1732 (1998).

[710] W. Ho, *Single-molecule chemistry*, J. Chem. Phys. **117**, 11033 (2002).

[711] J.R. Hahn, H.J. Lee, W. Ho, *Electronic resonance and symmetry in single-*

molecule inelastic electron tunneling, Phys. Rev. Lett. **85**, 1915 (2000).

[712] J.G. Kushmerick, J. Lazorcik, C.H. Patterson, R. Shashidhar, D.S. Seferos, G.C. Bazan, *Vibronic contributions to charge transport across molecular junctions*, Nano Lett. **4**, 639 (2004).

[713] W. Wang, T. Lee, I. Kretzschmar, M.A. Reed, *Inelastic electron tunneling spectroscopy of an alkanedithiol self-assembled monolayer*, Nano Lett. **4**, 643 (2004).

[714] L.H. Yu, C.D. Zangmeister, J.G. Kushmerick, *Origin of discrepancies in inelastic electron tunneling spectra of molecular junctions*, Phys. Rev. Lett. **98**, 206803 (2007).

[715] L.T. Cai, M.A. Cabassi, H. Yoon, O.M. Cabarcos, C.L. McGuiness, A.K. Flatt, D.L. Allara, J.M. Tour, A.S. Mayer, *Reversible bistable switching in nanoscale thiol-substituted oligoaniline molecular junctions*, Nano Lett. **5**, 2365 (2005).

[716] L.H. Yu, C.D. Zangmeister, J.G. Kushmerick, *Structural contributions to charge transport across Ni-octanedithiol multilayer junctions*, Nano Lett. **6**, 2515 (2006).

[717] W. Wang and C.A. Richter, *Spin-polarized inelastic electron tunneling spectroscopy of a molecular magnetic tunnel junction*, Appl. Phys. Lett. **89**, 153105 (2006).

[718] A. Troisi, J.M. Beebe, L.B. Picraux, R.D. van Zee, D.R. Stewart, M.A. Ratner, J.G. Kushmerick, *Tracing electronic pathways in molecules by using inelastic tunneling spectroscopy*, Proc. Natl. Acad. Sci. USA **104**, 14255 (2007).

[719] J.M. Beebe, H.J. Moore, T.R. Lee, J.G. Kushmerick, *Vibronic coupling in semifluorinated alkanethiol junctions: Implications for selection rules in inelastic electron tunneling spectroscopy*, Nano Lett. **7**, 1364 (2007).

[720] D.P. Long and A. Troisi, *Inelastic electron tunneling spectroscopy of alkane monolayers with dissimilar attachment chemistry to gold*, J. Am. Chem. Soc. **129**, 15303 (2007).

[721] W. Wang, A. Scott, N. Gergel-Hackett, C.A. Hacker, D.B. Janes, C.A. Richter, *Probing molecules in integrated silicon-molecule-metal junctions by inelastic tunneling spectroscopy*, Nano Lett. **8**, 478 (2008).

[722] J. Hihath, C.R. Arroyo, G. Rubio-Bollinger, N.J. Tao, N. Agraït, *Study of electron-phonon interactions in a single molecule covalently connected to two electrodes*, Nano Lett. **8**, 1673 (2008).

[723] H. Song, Y. Kim, J. Ku, Y.H. Jang, H. Jeong, T. Lee, *Vibrational spectra of metal-molecule-metal junctions in electromigrated nanogap electrodes by inelastic electron tunneling*, Appl. Phys. Lett. **94**, 103110 (2009).

[724] I.K. Yanson, O.I. Shklyarevskii, *Point-contact spectroscopy of metallic alloys and compounds*, Sov. J. Low Temp. Phys. **12**, 509 (1986).

[725] A.M. Duif, A.G.M. Jansen and P. Wyder, *Point-contact spectroscopy*, J. Phys.: Condens. Matter **1**, 3157 (1989).

[726] Yu.G. Naidyuk, I.K. Yanson, *Point-Contact Spectroscopy*, Springer Series in Solid-State Sciences, Vol. 145 (Springer-Verlag, Heidelberg, Berlin, D, 2004).

[727] I.K. Yanson, *Nonlinear effects in the electric conductivity of point junctions and electron-phonon interaction in metals*, Zh. Eksp. Teor. Fiz. **66**, 1035 (1974) [Sov. Phys. JETP **39**, 506 (1974)].

[728] A.G.M. Jansen, A.P. van Gelder, P. Wyder, *Point contact spectroscopy in metals*, J. Phys. C **13**, 6073 (1980).

[729] A.V. Khotkevich, I.K. Yanson, *Atlas of Point Contact Spectra of Electron-Phonon Interactions in Metals*, (Kluwer Academic Publishers, Dordrecht, NL, 1995).

[730] N.V. Zavaritskii, *Electron-phonon interaction and characteristics of metal electrons*, Sov. Phys. USP. **15**, 608 (1973) [Usp. Fiz. Nauk **108**, 241 (1972)].

[731] C. Untiedt, G. Rubio Bollinger, S. Vieira, N. Agraït, *Quantum interference in atomic-sized point-contacts*, Phys. Rev. B **62**, 9962 (2000).

[732] N. Agraït, C. Untiedt, G. Rubio-Bollinger, S. Vieira, *Onset of dissipation in ballistic atomic wires*, Phys. Rev. Lett. **88**, 216803 (2002).

[733] N. Agraït, C. Untiedt, G. Rubio-Bollinger, S. Vieira, *Electron transport and phonons in atomic wires*, Chem. Phys. **281**, 231 (2002).

[734] M. Kiguchi, R. Stadler, I.S. Kristensen, D. Djukic, J.M. van Ruitenbeek, *Evidence for a single hydrogen molecule connected by an atomic chain*, Phys. Rev. Lett. **98**, 146802 (2007).

[735] O. Tal, M. Krieger, B. Leerink, J. M. van Ruitenbeek *Electron-vibration interaction in single-molecule junctions: From contact to tunneling regimes*, Phys. Rev. Lett. **100**, 196804 (2008).

[736] M. Paulsson, T. Frederiksen, M. Brandbyge, *Modeling inelastic phonon scattering in atomic- and molecular-wire junctions*, Phys. Rev. B **72**, 201101 (2005).

[737] L. de la Vega, A. Martin-Rodero, N. Agraït, A. Levy Yeyati, *Universal features of electron-phonon interactions in atomic wires*, Phys. Rev. B **73**, 075428 (2006).

[738] N.B. Zhitenev, H. Meng, Z. Bao, *Conductance of small molecular junctions*, Phys. Rev. Lett. **88**, 226801 (2002).

[739] I. Fernández-Torrente, K.J. Franke, J.I. Pascual, *Vibrational Kondo effect in pure organic charge-transfer assemblies*, Phys. Rev. Lett. **101**, 217203 (2008).

[740] B.J. LeRoy, S.G. Lemay, J. Kong, C. Dekker, *Electrical generation and adsorption of phonons in carbon nanotubes*, Nature **432**, 371 (2004).

[741] B.J. LeRoy, J. Kong, V.K. Pahilwani, C. Dekker, S.G. Lemay, *Three-terminal scanning tunneling spectroscopy of suspended carbon nanotubes*, Phys. Rev. B **72**, 075413 (2005).

[742] S.W. Wu, G.V. Nazin, X. Chen, X.H. Qiu, W. Ho, *Control of relative tunneling rates in single molecule bipolar electron transport*, Phys. Rev. Lett. **93**, 236802 (2004).

[743] W.H.A. Thijssen, D. Djukic, A.F. Otte, R.H. Bremmer, J.M. van Ruitenbeek, *Vibrationally induced two-level systems in single-molecule junctions*, Phys. Rev. Lett. **97**, 226806 (2006).

[744] S. Tikhodeev, M. Natario, K. Makoshi, T. Mii, H. Ueba, *Contribution*

to a theory of vibrational scanning tunnelling spectroscopy of adsorbates: Nonequilibrium Green's function approach, Surf. Sci. **493**, 63 (2001).

[745] T. Mii, S. Tikhodeev and H. Ueba, *Theory of vibrational tunnelling spectroscopy of adsorbates on metal surfaces*, Surf. Sci. **502-503**, 26 (2002).

[746] T. Mii, S. Tikhodeev, H. Ueba, *Spectral features of inelastic electron transport via a localized state*, Phys. Rev. B **68**, 205406 (2003).

[747] A. Mitra, I. Aleiner, A.J. Millis, *Phonon effects in molecular transistors: quantal and classical treatment*, Phys. Rev. B **69**, 245302 (2004).

[748] M. Galperin, M.A. Ratner, A. Nitzan, *On the line widths of vibrational features in inelastic electron tunnelling spectroscopy*, Nano Lett. **4**, 1605 (2004).

[749] M. Galperin, M.A. Ratner, A. Nitzan, *Inelastic electron tunnelling spectroscopy in molecular junctions: Peaks and dips*, J. Chem. Phys. **121**, 11965 (2004).

[750] R. Egger and A.O. Gogolin, *Vibration-induced correction to the current through a single molecule*, Phys. Rev. B **77**, 113405 (2008).

[751] M. Paulsson, T. Frederiksen, H. Ueba, N. Lorente, M. Brandbyge, *Unified description of inelastic propensity rules for electron transport through nanoscale junctions*, Phys. Rev. Lett. **100**, 226604 (2008).

[752] L.C. Davis, *Impurity-assisted inelastic tunneling: Many-electron theory*, Phys. Rev. B **6**, 1714 (1970).

[753] B.N. Persson and A. Baratoff, *Inelastic electron tunneling from a metal tip: The contribution from resonant processes*, Phys. Rev. Lett. **59**, 339 (1987).

[754] N. Lorente and M. Persson, *Theory of single molecule vibrational spectroscopy and microscopy*, Phys. Rev. Lett. **85**, 2997 (2000).

[755] N. Lorente, M. Persson, L.J. Lauhon, W. Ho, *Symmetry selection rules for vibrationally inelastic tunneling*, Phys. Rev. Lett. **86**, 2593 (2001).

[756] B.C. Stipe, M.A. Rezaei, W. Ho, *Coupling of vibrational excitation to the rotational motion of a single adsorbed molecule*, Phys. Rev. Lett. **81**, 1263 (1998).

[757] B.C. Stipe, M.A. Rezaei, W. Ho, *Localization of inelastic tunneling and the determination of atomic-scale structure with chemical specificity*, Phys. Rev. Lett. **82**, 1724 (1999).

[758] M.J. Montgomery, J. Hoekstra, T.N. Todorov, A.P. Sutton *Inelastic current-voltage spectroscopy of atomic wires*, J. Phys.: Condens. Matter **15**, 731 (2003).

[759] A. Troisi, M.A. Ratner, A. Nitzan, *Vibronic effects in off-resonant molecular wire conduction*, J. Chem. Phys. **118**, 6072 (2003).

[760] Y. Chen, M. Zwolak, M. Di Ventra, *Inelastic current-voltage characteristics of atomic and molecular junctions*, Nano Lett. **4**, 1709 (2004).

[761] Y. Asai, *Theory of inelastic electric current through single molecules*, Phys. Rev. Lett. **93**, 246102 (2004).

[762] A. Pecchia, A. Di Carlo, A. Gagliardi, S. Sanna, T. Frauenheim, R. Gutierrez, *Incoherent electron-phonon scattering in octanethiols*, Nano Lett. **4**, 2109 (2004).

[763] T. Frederiksen, M. Brandbyge, N. Lorente, A.-P. Jauho, *Inelastic scattering and local heating in atomic gold wires*, Phys. Rev. Lett. **93**, 256601 (2004).

[764] N. Sergueev, D. Roubtsov, H. Guo, *Ab initio analysis of electron-phonon coupling in molecular devices*, Phys. Rev. Lett. **95**, 146803 (2005).

[765] A. Troisi and M.A. Ratner, *Modeling the inelastic electron tunneling spectra of molecular wire junctions*, Phys. Rev. B **72**, 033408 (2005).

[766] Y.-C. Chen, M. Zwolak, M. Di Ventra, *Inelastic effects on the transport properties of alkanethiols*, Nano Lett. **5**, 621 (2005).

[767] J. Jiang, M. Kula, W. Lu, Y. Luo, *First-principles simulations of inelastic electron tunneling spectroscopy of molecular electronic devices*, Nano Lett. **5**, 1551 (2005).

[768] J. Jiang, M. Kula, Y. Luo, *A generalized quantum chemical approach for elastic and inelastic electron transports in molecular electronics devices*, J. Chem. Phys. **124**, 034708 (2006).

[769] A. Troisi and M.A. Ratner, *Molecular transport junctions: propensity rules for inelastic electron tunneling spectra*, Nano Lett. **6**, 1784 (2006).

[770] A. Troisi and M.A. Ratner, *Propensity rules for inelastic electron tunneling spectroscopy of single-molecule transport junctions*, J. Chem. Phys. **125**, 214709 (2006).

[771] G.C. Solomon, A. Gagliardi, A. Pecchia, T. Frauenheim, A. Di Carlo, J.R. Reimers, N.S. Hush, *Understanding the inelastic electron-tunneling spectra of alkanedithiols on gold*, J. Chem. Phys. **124**, 094704 (2006).

[772] M. Paulsson, T. Frederiksen, M. Brandbyge, *Inelastic transport through molecules: Comparing first-principles calculations to experiments*, Nano Lett. **6**, 258 (2006).

[773] M. Kula, J. Jiang, Y. Luo, *Probing molecule-metal bonding in molecular junctions by inelastic electron tunneling spectroscopy*, Nano Lett. **6**, 1693 (2006).

[774] T. Frederiksen, N. Lorente, M. Paulsson, M. Brandbyge, *From tunneling to contact: Inelastic signals in an atomic gold junction from first principles*, Phys. Rev. B **75**, 235441 (2007).

[775] T. Frederiksen, M. Paulsson, M. Brandbyge, A.-P. Jauho, *Inelastic transport theory from first principles: Methodology and application to nanoscale devices*, Phys. Rev. B **75**, 205413 (2007).

[776] A. Gagliardi, G.C. Solomon, A. Pecchia, T. Frauenheim, A. Di Carlo, N.S. Hush, J.R. Reimers, *A priori method for propensity rules for inelastic electron tunneling spectroscopy of single-molecule conduction*, Phys. Rev. B **75**, 174306 (2007).

[777] A. Gagliardi, G. Romano, A. Pecchia, A. Di Carlo, T. Frauenheim, T.A. Niehaus, *Electron-phonon scattering in molecular electronics: from inelastic electron tunneling spectroscopy to heating effects*, New J. Phys. **10**, 065020 (2008).

[778] H. Nakamura, K. Yamashita, A.R. Rocha, S. Sanvito, *Efficient ab initio method for inelastic transport in nanoscale devices: Analysis of inelastic electron tunneling spectroscopy*, Phys. Rev. B **78**, 235420 (2008).

[779] I.S. Kristensen, M. Paulsson, K.S. Thygesen, K.W. Jacobsen, *Inelastic scattering in metal-H_2-metal junctions*, Phys. Rev. B **79**, 235411 (2009).

[780] M. Head-Gordon and J.C. Tully, *Vibrational relaxation on metal surfaces: Molecular-orbital theory and application to CO/Cu(100)*, J. Chem. Phys. **96**, 3939 (1992).

[781] A.J. Heinrich, C.P. Lutz, J.A. Gupta, D.M. Eigler, *Molecule cascades*, Science **298**, 1381 (2002).

[782] D.A. Ryndyk, M. Hartung, G. Cuniberti, *Nonequilibrium molecular vibrons: an approach based on the nonequilibrium Green function technique and the self-consistent Born approximation*, Phys. Rev. B **73**, 045420 (2006).

[783] U. Lunding and R.H. McKenzie, *Temperature dependence of polaronic transport through single molecules and quantum dots*, Phys. Rev. B **66**, 075303 (2002).

[784] J.-X. Zhu and A.V. Balatsky, *Theory of current and shot-noise spectroscopy in single-molecular quantum dots with a phonon mode*, Phys. Rev. B **67**, 165326 (2003).

[785] K. Flensberg, *Tunneling broadening of vibrational sidebands in molecular transistors*, Phys. Rev. B **68**, 205323 (2003).

[786] A. Alexandrov, A.M. Bratkovsky, R.S. Williams, *Bistable tunnelling current through a molecular quantum dot*, Phys. Rev. B **67**, 075301 (2003).

[787] A. Alexandrov and A.M. Bratkovsky, *Memory effect in a molecular quantum dot with strong electron- vibron interaction*, Phys. Rev. B **67**, 235312 (2003).

[788] D.A. Ryndyk and J. Keller, *Inelastic resonant tunneling through single molecules and quantum dots: Spectrum modification due to nonequilibrium effects*, Phys. Rev. B **71**, 073305 (2005).

[789] M. Galperin, A. Nitzan, M.A. Ratner, *Resonant inelastic tunneling in molecular junctions*, Phys. Rev. B **73**, 045314 (2006).

[790] D.A. Ryndyk and G. Cuniberti, *Nonequilibrium resonant spectroscopy of molecular vibrons*, Phys. Rev. B **76**, 155430 (2007).

[791] A. Zazunov and T. Martin, *Transport through a molecular quantum dot in the polaron crossover regime*, Phys. Rev. B **76**, 033417 (2007).

[792] I.G. Lang and Y.A. Firsov, *Kinetic theory of semiconductors with low mobility*, Zh. Eksp. Teor. Fiz. **43**, 1843 (1962) [Sov. Phys. JETP **16**, 1301 (1963)].

[793] S. Sapmaz, P. Jarillo-Herrero, Ya.M. Blanter, C. Dekker, H.S.J. van der Zant, *Tunneling in suspended carbon nanotubes assisted by longitudinal phonons*, Phys. Rev. Lett. **96**, 026801 (2006).

[794] L.I. Glazman and R.I. Shekther, *Inelastic resonant tunneling of electrons through a potential barrier*, Sov. Phys. JEPT **67**, 163 (1988).

[795] N.S. Wingreen, K.W. Jacobsen and J.W. Wilkins, *Resonant tunnelling with electron phonon interaction: an exactly solvable model*, Phys. Rev. Lett. **61**, 1396 (1988).

[796] D. Boese and H. Schoeller, *Influence of nano-mechanical properties on single electron tunneling: A vibrating single-electron transistor*, Europhys.

Lett. **54**, 668 (2001).

[797] K.D. McCarthy, N. Prokof'ev, M.T. Tuominen, *Incoherent dynamics of vibrating single-molecule transistors*, Phys. Rev. B **67**, 245415 (2003).

[798] S. Braig and K. Flensberg, *Vibrational sidebands and dissipative tunnelling in molecular transistors*, Phys. Rev. B **68**, 205324 (2003).

[799] S. Braig and K. Flensberg, *Dissipative tunneling and orthogonality catastrophe in molecular transistors*, Phys. Rev. B **70**, 085317 (2004).

[800] J. Koch, F. von Oppen, Y. Oreg, E. Sela, *Thermopower of single-molecule devices*, Phys. Rev. B **70**, 195107 (2004).

[801] J. Koch and F. von Oppen, *Franck-Condon blockade and giant Fano factors in transport through single molecules*, Phys. Rev. Lett. **94**, 206804 (2005).

[802] J. Koch and F. von Oppen, *Effects of charge-dependent vibrational frequencies and anharmonicities in transport through molecules*, Phys. Rev. B **72**, 113308 (2005).

[803] M.R. Wegewijs and K.C. Nowack, *Nuclear wavefunction interference in single-molecule electron transport*, New J. Phys. **7**, 239 (2005).

[804] A. Zazunov, D. Feinberg, T. Martin, *Phonon-mediated negative differential conductance in molecular quantum dots*, Phys. Rev. B **73**, 115405 (2006).

[805] J. Koch, F. von Oppen, A.V. Andreev, *Theory of the Franck-Condon blockade regime*, Phys. Rev. B **74**, 205438 (2006).

[806] C.T. Chang, J.P. Sethna, A.N. Pasupathy, J. Park, D.C. Ralph, P.L. McEuen, *Phonons and conduction in molecular quantum dots: Density functional calculations of Franck-Condon emission rates for bifullerenes in external fields*, Phys. Rev. B **76**, 045435 (2007).

[807] J.S. Seldenthuis, H.S.J. van der Zant, M.A. Ratner, J.M. Thijssen, *Vibrational excitations in weakly coupled single-molecule junctions: a computational analysis*, ACS Nano **2**, 1445 (2008).

[808] R. Leturcq, C. Stampfer, K. Inderbitzin, L. Durrer, C. Hierold, E. Mariani, M.G. Schultz, F. von Oppen, K. Ensslin, *Franck-Condon blockade in suspended carbon nanotube quantum dots*, Nature Phys. **5**, 327 (2009).

[809] J. Koch, M.E. Raikh, F. von Oppen, *Pair tunneling through single molecules*, Phys. Rev. Lett. **96**, 056803 (2006).

[810] J. Koch, M. Semmelhack, F. von Oppen, A. Nitzan, *Current-induced nonequilibrium vibrations in single-molecule devices*, Phys. Rev. B **73**, 155306 (2006).

[811] M.C. Lüffe, J. Koch, F. von Oppen, *Theory of vibrational absorption sidebands in the Coulomb-blockade regime of single-molecule transistors*, Phys. Rev. B **77**, 125306 (2008).

[812] R. Härtle, C. Benesch, M. Thoss, *Vibrational nonequilibrium effects in the conductance of single molecules with multiple electronic states*, Phys. Rev. Lett. **102**, 146801 (2009).

[813] P.S. Cornaglia, H. Ness, D.R. Grempel, *Many-body effects on the transport properties of single-molecule devices*, Phys. Rev. Lett. **93**, 147201 (2004).

[814] P.S. Cornaglia, D.R. Grempel, H. Ness, *Quantum transport through a deformable molecular transistor* Phys. Rev. B **71**, 075320 (2005).

[815] P.S. Cornaglia and D.R. Grempel, *Magnetoconductance through a vibrating molecule in the Kondo regime*, Phys. Rev. B **71**, 245326 (2005).

[816] C.A. Balseiro, P.S. Cornaglia, D.R. Grempel, *Electron-phonon correlation effects in molecular transistors*, Phys. Rev. B **74**, 235409 (2006).

[817] P.S. Cornaglia, G. Usaj, C.A. Balseiro, *Electronic transport through magnetic molecules with soft vibrating modes*, Phys. Rev. B **76**, 241403 (2007).

[818] J. Paaske and K. Flensberg, *Vibrational sidebands and Kondo-effect in molecular transistors*, Phys. Rev. Lett. **94**, 176801 (2005).

[819] L. Arrachea and M.J. Rozenberg, *Quantum Monte Carlo method for models of molecular nanodevices*, Phys. Rev. B **72**, 041301 (2005).

[820] J. Mravlje, A. Ramsak, T. Rejec, *Conductance of deformable molecules with interaction*, Phys. Rev. B **72**, 121403 (2005).

[821] Z.-Z. Chen, H. Lu, R. Lü, B. Zhu, *Phonon-assisted Kondo effect in a single-molecule transistor out of equilibrium*, J. Phys.: Condens. Matter **18**, 5435 (2006).

[822] K. Kikoin, M.N. Kiselev, M.R. Wegewijs, *Vibration-induced Kondo tunneling through metal-organic complexes with even electron occupation number*, Phys. Rev. Lett. **96**, 176801 (2006).

[823] M. Galperin, A. Nitzan, M.A. Ratner, *Inelastic effects in molecular junctions in the Coulomb and Kondo regimes: Nonequilibrium equation-of-motion approach*, Phys. Rev. B **76**, 035301 (2007).

[824] A. Martin-Rodero, A. Levy Yeyati, F. Flores, R.C. Monreal, *Interpolative approach for electron-electron and electron-phonon interactions: From the Kondo to the polaronic regime*, Phys. Rev. B **78**, 235112 (2008).

[825] R.C. Monreal and A. Martin-Rodero, *Equation of motion approach to the Anderson-Holstein Hamiltonian*, Phys. Rev. B **79**, 115140 (2009).

[826] L.G.G.V. Dias da Silva and E. Dagotto, *Phonon-assisted tunneling and two-channel Kondo physics in molecular junctions*, Phys. Rev. B **79**, 155302 (2009).

[827] A. Mitra, I. Aleiner and A.J. Millis *Semiclassical analysis of the nonequilibrium local polaron*, Phys. Rev. Lett. **94**, 076404 (2005).

[828] R.A. Kiehl, J.D. Le, P. Candra, R.C. Hoye, T.R. Hoye, *Charge storage model for hysteretic negative-differential resistance in metal-molecule-metal junctions*, Appl. Phys. Lett. **88**, 172102 (2006).

[829] D. Mozyrsky, M.B. Hastings, I. Martin, *Intermittent polaron dynamics: Born-Oppenheimer approximation out of equilibrium*, Phys. Rev. B **73**, 035104 (2006).

[830] D. Segal, A. Nitzan, W.B. Davis, M.R. Wasielewski, M.A. Ratner, *Electron transfer rates in bridged molecular systems 2. A steady state analysis of coherent tunneling and thermal transitions*, J. Phys. Chem. B **104**, 3817 (2000).

[831] A. Nitzan, *The relationship between electron transfer rate and molecular conduction. 2. The sequential hopping case*, Israel J. Chem. **42**, 163 (2002).

[832] D. Segal, A. Nitzan, M. Ratner, W.B. Davis, *Activated conduction in microscopic molecular junctions*, J. Phys. Chem. **104**, 2790 (2000).

[833] D. Segal and A. Nitzan, *Steady-state quantum mechanics of thermally*

relaxing systems, Chem. Phys. **268**, 315 (2001).

[834] D. Segal and A. Nitzan, *Heating in current carrying molecular junctions*, Chem. Phys **281**, 235 (2002).

[835] E.G. Petrov, Ye.V. Shevchenko, V.I. Teslenko, V. May, *Nonadiabatic donor-acceptor electron transfer mediated by a molecular bridge: A unified theoretical description of the superexchange and hopping mechanism*, J. Chem. Phys. **115**, 7107 (2001).

[836] H. Böttger and V.V. Bryksin, *Hopping conduction in Solids*, (Akademie Verlag, Berlin, D, 1985).

[837] A.S. Alexandrov and A.M. Bratkovsky, *Fast polaron switching in degenerate molecular quantum dots*, J. Phys.: Condens. Matter **19**, 255203 (2007).

[838] Y.A. Berlin, A.L. Burin, M.A. Ratner, *Charge hopping in DNA*, J. Am. Chem. Soc. **123**, 260 (2001).

[839] B.B. Schmidt, M.H. Hettler, G. Schön, *Nonequilibrium polaron hopping transport through DNA*, Phys. Rev. B **77**, 165337 (2008).

[840] B.B. Schmidt, M.H. Hettler, G. Schön, *Charge correlations in polaron hopping through molecules*, arXiv:0902.3183.

[841] Y. Selzer, M.A. Cabassi, T.S. Mayer, D.L. Allara, *Thermally activated conduction in molecular junctions*, J. Am. Chem. Soc. **126**, 4052 (2004).

[842] B. Giese, J. Amaudrut, A.K. Köhler, M.Spormann, S. Wessely, *Direct observation of hole transfer through DNA by hopping between adenine bases and by tunneling*, Nature **412**, 318 (2001).

[843] S.H.Choi, B. Kim, C.D. Frisbie, *Electrical resistance of long conjugated molecular wires*, Science **320**, 1482 (2008).

[844] E. Braun, Y. Eichen, U. Sivan, G. Ben-Yoseph, *DNA-templated assembly and electrode attachment of a conducting silver wire*, Nature **391**, 775 (1998).

[845] Z. Hermon, S. Caspi, E. Ben-Jacob, *Prediction of charge and dipole solitons in DNA molecules based on the behaviour of phosphate bridges as tunnel elements*, Europhys. Lett. **43**, 482 (1998).

[846] E. Ben-Jacob, Z. Hermon, S. Caspi, *DNA transistor and quantum bit element: Realization of nano-biomolecular logical devices*, Phys. Lett. A **263**, 199 (1999).

[847] N.C. Seeman, *DNA nicks and nodes and nanotechnology*, Nano Lett. **1**, 22 (2001).

[848] D. Porath, G. Cuniberti, R. Di Felice, *Charge transport in DNA-based devices*, Top. Curr. Chem. **237**, 183 (2004).

[849] R.G. Endres, D.L. Cox, R.R.P. Singh, *Colloquium: The quest for high-conductance DNA*, Rev. Mod. Phys. **76**, 195 (2004).

[850] V.A. Bloomfield, D.M. Crothers, I. Tinoco (Ed.), *Nucleic Acids: Structures, Properties, Functions*, p. 475 (University Science Books, New York, USA, 2000).

[851] P.J. de Pablo, F. Moreno-Herrero, J. Colchero, J. Gómez Herrero, P. Herrero, A.M. Baró, P. Ordejón, J.M. Soler, E. Artacho, *Absence of dc-conductivity in λ-DNA*, Phys. Rev. Lett. **85**, 4992 (2000).

[852] A.J. Storm, J. van Noort, S. de Vries, C. Dekker, *Insulating behavior for*

DNA molecules between nanoelectrodes at the 100 nm length scale, Appl. Phys. Lett. **79**, 3881 (2002).

[853] D. Porath, A. Bezryadin, S. de Vries and C. Dekker, *Direct measurement of electrical transport through DNA molecules*, Nature **403**, 635 (2000).

[854] K.-H. Yoo, D.H. Ha, J.-O. Lee, J.W. Park, J. Kim, J.J. Kim, H.-Y. Lee, T. Kawai, H.Y. Choi, *Electrical conduction through poly(dA)-poly(dT) and poly(dG)-poly(dC) DNA Molecules*, Phys. Rev. Lett. **87**, 198102 (2001).

[855] H. Cohen, C. Nogues, R. Naaman, D. Porath, *Direct measurement of electrical transport through single DNA molecules of complex sequence*, Proc. Nat. Acad. Sci. **102**, 11589 (2005).

[856] H.W. Fink and C. Schönenberger, *Electrical conduction through DNA molecules*, Nature **398**, 407 (1999).

[857] A.Yu. Kasumov, M. Kociak, S. Guéron, B. Reulet, V. Volkov, D. Klinov, H. Bouchiat, *Proximity-induced superconductivity in DNA*, Science **291**, 280 (2001).

[858] A.Yu. Kasumov, D.V. Klinov, P.-E. Roche, S. Guéron, H. Bouchiat, *Thickness and low-temperature conductivity of DNA molecules*, Appl. Phys. Lett. **84**, 1007 (2004).

[859] T. Heim, D. Deresmes, D. Vuillaume, *Conductivity of DNA probed by conducting-atomic force microscopy: Effects of contact electrode, DNA structure, surface interactions*, J. Appl. Phys. **96**, 2927 (2004).

[860] H.X. He, X.L. Li, N.J. Tao, *Discrete conductance switching in conducting polymer wires*, Phys. Rev. B **68**, 45302 (2003).

[861] J.F. Leger, G. Romano, A. Sarkar, J. Robert, L. Bourdieu, D. Chatenay, J.F. Marko, *Structural transitions of a twisted and stretched DNA molecule*, Phys. Rev. Lett. **83**, 1066 (1999).

[862] B. Xu, P. Zhang, X. Li, N. Tao, *Direct conductance measurement of single DNA molecules in aqueous solution*, Nano Lett. **4**, 1105 (2004).

[863] J. Jortner, M. Bixon, T. Langenbacher, M.E. Michel-Beyerle, *Charge transfer and transport in DNA*, Proc. Nat. Acad. Sci. USA **95**, 12759 (1998).

[864] R. Roche, *Sequence dependent DNA-mediated conduction*, Phys. Rev. Lett. **91**, 108101 (2003).

[865] D.J. Patel, S. Bouaziz, A. Kettani, Y. Wong, *Structures of guanine-rich and cytosine-rich quadruplexes formed in vitro by telomeric, centromeric, triplet repeat disease DNA sequences*, in S. Neidle (Ed.), *Oxford Handbook of Nucleic Acid Structure*, p. 389, (Oxford University Press, Oxford, UK, 1999).

[866] H. Cohen, T. Sapir, N. Borovok, T. Molotsky, R. di Felice, A. Kotlyar, D. Porath, *Polarizability of G4-DNA observed by electrostatic force microscopy measurements*, Nano Lett. **7**, 981 (2007).

[867] E. Shapir, H. Cohen, A. Calzolari, C. Cavazzoni, D.A. Ryndyk, G. Cuniberti, A. Kotlyar, R. Di Felice, D. Porath, *Electronic structure of single DNA molecules resolved by transverse scanning tunnelling spectroscopy*, Nature Mat. **7**, 68 (2008).

[868] H. Tanaka and T. Kawai, *Partial sequencing of a single DNA molecule*

with a scanning tunnelling microscope, Nature Nanotech. **4**, 518 (2009).

[869] S.S. Mallajosyula, J.C. Lin, D.L. Cox, S.K. Pati, R.R.P. Singh, *Sequence dependent electron transport in wet DNA: Ab initio and molecular dynamics*, Phys. Rev. Lett. **101**, 176805 (2008).

[870] B. Song, M. Elstner, G. Cuniberti, *Anomalous conductance response of DNA wires under stretching*, Nano Lett. **8**, 3217 (2008).

[871] M.L. Bennink, O.D. Schärer, R. Kanaar, K. Sakata-Sogawa, J.M. Schins, J.S. Kanger, B.G. de Grooth, J. Greve, *Single-molecule manipulation of double-stranded DNA using optical tweezers: Interaction studies of DNA with RecA and YOYO-1*, Cytometry **36**, 200 (1999).

[872] P. Cluzel, A. Lebrun, C. Heller, R. Lavery, J.-L. Viovy, D. Chatenay, F. Caron, *DNA: An extensible molecule*, Science **271**, 792 (1996).

[873] S.B. Smith, Y. Cui, C. Bustamante, *Overstretching B-DNA: The elastic response of individual double-stranded and single-stranded DNA molecules*, Science **271**, 795 (1996).

[874] P. Maragakis, R. L. Barnett, E. Kaxiras, M. Elstner, T. Frauenheim, *Electronic structure of overstretched DNA*, Phys. Rev. B **66**, 241104 (2002).

[875] S. Roy, H. Vedala, A. Roy, D.-H. Kim, M. Doud, K. Mathee, H.K Shin, N. Shimamoto, V. Prasad, W. Choi, *Direct observation of hole transfer through DNA by hopping between adenine bases and by tunnelling*, Nano Lett. **8**, 26 (2008).

[876] X.F. Guo, A.A. Gorodetsky, J. Hone, J.K. Barton, C. Nuckolls, *Conductivity of a single DNA duplex bridging a carbon nanotube gap*, Nature Nanotech. **3**, 167 (2008).

[877] W. Schottky, *Spontaneous current fluctuations in electron streams*, Ann. Phys. (Leipzig) **57**, 541 (1918).

[878] M. B. Johnson, *Thermal agitation of electricity in conductors*, Phys. Rev. **32**, 97 (1928).

[879] H. Nyquist, *Thermal agitation of electric charge in conductors*, Phys. Rev. **32**, 110 (1928).

[880] H. B. Callen and T. W. Welton, *Irreversibility and generalized noise*, Phys. Rev. **83**, 34 (1951).

[881] V.A. Khlus, *Current and voltage fluctuations in microjunctions between normal metals and superconductors*, Sov. Phys. JETP **66**, 1243 (1987).

[882] G.B. Lesovik, *Excess quantum noise in 2D ballistic point contacts*, JETP Lett. **49**, 592 (1989).

[883] M. Büttiker, *Scattering theory of thermal and excess noise in open conductors*, Phys. Rev. Lett. **65**, 2901 (1990).

[884] M. Reznikov, M. Heiblum, H. Shtrikman, D. Mahalu, *Temporal correlation of electrons: Suppression of shot noise in a ballistic quantum point contact*, Phys. Rev. Lett. **75**, 3340 (1995).

[885] H.E. van den Brom and J.M. van Ruitenbeek, *Quantum suppression of shot noise in metallic atomic size contacts*, Phys. Rev. Lett. **82**, 1526 (1999).

[886] H.E. van den Brom and J.M. van Ruitenbeek, *Shot noise suppression in metallic quantum point contacts*, in *Statistical and Dynamical Aspects*

of Mesoscopic Systems, D. Reguera, G. Platero, L.L. Bonilla, J.M. Rubí (Eds.), p.114, (Springer-Verlag, Berlin, Heidelberg, 2000).

[887] E. Onac, F. Balestro, B. Trauzettel, C.F.J. Lodewijk, L.P. Kouwenhoven, *Shot-noise detection in a carbon nanotube quantum dot*, Phys. Rev. Lett. **96**, 026803 (2006).

[888] Y. Zhang, L. Di Carlo, D.T. McClure, M. Yamamoto, S. Tarucha, C.M. Marcus, M.P. Hanson, A.C. Gossard, *Noise correlations in a Coulomb-blockaded quantum dot*, Phys. Rev. Lett. **99**, 036603 (2007).

[889] T. Delattre, C. Feuillet-Palma, L.G. Herrmann, P. Morfin, J.-M. Berroir, G. Fève, B. Plaçais, D.C. Glattli, M.-S. Choi, C. Mora and T. Kontos, *Noisy Kondo impurities*, Nature Phys. **5**, 208 (2009).

[890] M. Galperin, A. Nitzan, M.A. Ratner, *Inelastic tunneling effects on noise properties of molecular junctions*, Phys. Rev. B **74**, 075326 (2006).

[891] T.L. Schmidt and A. Komnik, *Charge transfer statistics of a molecular quantum dot with a vibrational degree of freedom*, Phys. Rev. B **80**, 041307 (2009).

[892] R. Avriller and A. Levy Yeyati, *Electron-phonon interaction and full counting statistics in molecular junctions*, Phys. Rev. B **80**, 041309 (2009).

[893] F. Haupt, T. Novotný, W. Belzig, *Phonon-assisted current noise in molecular junctions*, Phys. Rev. Lett. **103**, 136601 (2009).

[894] D. Cahill, W.K. Ford, K.E. Goodson, G.D. Mahan, A. Majumdar, H.J. Maris, R. Merlin, S.R. Phillpot, *Nanoscale thermal transport*, J. Appl. Phys. **93**, 793 (2003).

[895] R.K. Lake and S. Datta, *Nonequilibrium Green's-function method applied to double-barrier resonant-tunneling diodes*, Phys. Rev. B **45**, 6670 (1992).

[896] R.K. Lake and S. Datta, *Energy balance and heat exchange in mesoscopic systems*, Phys. Rev. B **46**, 4757 (1992).

[897] J.-S. Wang, J. Wang, N. Zeng, *Nonequilibrium Green's function approach to mesoscopic thermal transport*, Phys. Rev. B **74**, 033408 (2006).

[898] N. Mingo, *Anharmonic phonon flow through molecular-sized junctions*, Phys. Rev. B **74**, 125402 (2006).

[899] M. Galperin, A. Nitzan, M.A. Ratner, *Heat conduction in molecular transport junctions*, Phys. Rev. B **75**, 155312 (2007).

[900] J.T. Lü and J.-S. Wang, *Coupled electron and phonon transport in one-dimensional atomic junctions*, Phys. Rev. B **76**, 165418 (2007).

[901] L.G.C. Rego and G. Kirczenow, *Quantized thermal conductance of dielectric quantum wires*, Phys. Rev. Lett. **81**, 232 (1998).

[902] A. Ozpineci and S. Ciraci, *Quantum effects of thermal conductance through atomic chains*, Phys. Rev. B **63**, 125415 (2001).

[903] D. Segal, A. Nitzan and P. Hänggi, *Thermal conductance through molecular wires*, J. Chem. Phys. **119**, 6840 (2003).

[904] R. Peierls, *Zur kinetischen Theorie der Wärmeleitung in Kristallen*, Ann. Phys. **3**, 1055 (1929).

[905] K. Schwab, E.A. Henriksen, J.M. Worlock, M.L. Roukes, *Measurement of the quantum of thermal conductance*, Nature **404**, 974 (2000).

[906] Z. Rieder, J.L. Lebowitz, E. Lieb, *Properties of a harmonic crystal in a*

stationary non-equilibrium state, J. Chem. Phys. **8**, 1073 (1967).

[907] U. Zürcher and P. Talkner, *Quantum mechanical harmonic chain attached to heat baths II. Nonequilibrium properties*, Phys. Rev. A **42**, 3278 (1990).

[908] S. Lepri, R. Livi, A. Politi, *Heat conduction in chains of nonlinear oscillators*, Phys. Rev. Lett. **78**, 1896 (1997).

[909] B. Hu, B. Li, H. Zhao, *Heat conduction in one-dimensional chains*, Phys. Rev. E **57**, 2992 (1998).

[910] D.M. Leitner and P.G. Wolynes, *Heat flow through an insulating nanocrystal*, Phys. Rev. E **61**, 2902 (2000).

[911] P. Grassberger, W. Nadler, L. Yang, *Heat conduction and entropy production in a one-dimensional hard-particle gas*, Phys. Rev. Lett. **89**, 180601 (2002).

[912] G. Casati and T. Prosen, *Anomalous heat conduction in a one-dimensional ideal gas*, Phys. Rev. E **67**, 015203 (2003).

[913] M. Terraneo, M. Peyrard, G. Casati, *Controlling the energy flow in nonlinear lattices: a model for a thermal rectifier*, Phys. Rev. Lett. **88**, 094302 (2002).

[914] B. Li, L. Wang, G. Casati, *Thermal diode: rectification of heat flux*, Phys. Rev. Lett. **93**, 184301 (2004).

[915] D. Segal and A. Nitzan, *Spin-boson thermal rectifier*, Phys. Rev. Lett. **94**, 034301 (2005).

[916] D. Segal and A. Nitzan, *Heat rectification in molecular junctions*, J. Chem. Phys. **122**, 194704 (2005).

[917] B. Li, J. Lan, L. Wang, *Interface thermal resistance between dissimilar anharmonic lattices*, Phys. Rev. Lett. **95**, 104302 (2005).

[918] K. Saito, *Asymmetric heat flow in mesoscopic magnetic system*, J. Phys. Soc. Japan **75**, 034603 (2006).

[919] D. Cahill, K. Goodson, A. Majumdar, *Thermometry and thermal transport in micro/nanoscale solid-state devices and structures*, J. Heat Transfer **124**, 223 (2002).

[920] L. Shi and A. Majumdar, *Thermal transport mechanisms at nanoscale point contacts*, J. Heat Transfer **124**, 329 (2002).

[921] C. Yu, L. Shi, Z. Yao, D. Li, A. Majumdar, *Thermal conductance and thermopower of an individual single-wall carbon nanotube*, Nano Lett. **5**, 1842 (2005).

[922] N. Mingo and D.A. Broido, *Carbon nanotube ballistic thermal conductance and its limits*, Phys. Rev. Lett. **95**, 096105 (2005).

[923] R.Y. Wang, R.A. Segalman, A. Majumdar, *Room temperature thermal conductance of alkanedithiol self-assembled monolayers*, Appl. Phys. Lett. **89**, 173113 (2006).

[924] Z. Wang, J.A. Carter, A. Lagutchev, Y.K. Koh, N.-H. Seong, D.G. Cahill, D.D. Dlott, *Ultrafast flash thermal conductance of molecular chains*, Science **317**, 787 (2007).

[925] H.E. van den Brom, A.I. Yanson, J.M. van Ruitenbeek, *Characterization of individual conductance steps in metallic quantum point contacts*, Physica B **252**, 69 (1998).

[926] M. Tsutsui, S. Kurokawa, A. Sakai, *Bias-induced local heating in atom-sized metal contacts at 77 K*, Appl. Phys. Lett. **90**, 133121 (2007).

[927] R.H.M. Smit, C. Untiedt, J.M. van Ruitenbeek, *The high-bias stability of monatomic chains*, Nanotechnology **15**, S472 (2004).

[928] Z. Huang, B.Q. Xu, Y.C. Chen, M. Di Ventra, N.J. Tao, *Measurement of current-induced local heating in a single molecule junction*, Nano Lett. **6**, 1240 (2006).

[929] Z. Huang, F. Chen, R. D'Agosta, P.A. Bennett, M. Di Ventra, N.J. Tao, *Local ionic and electron heating in single-molecule junctions*, Nature Nanotech. **2**, 698 (2007).

[930] T.N. Todorov, *Local heating in ballistic atomic-scale contracts*, Phil. Mag. B **77**, 965 (1998).

[931] M.J. Montgomery, T.N. Todorov, A.P. Sutton, *Power dissipation in nanoscale conductors*, J. Phys.: Condens. Matter **14**, 5377 (2002).

[932] Y.C. Chen, M. Zwolak, M. Di Ventra, *Local heating in nanoscale conductors*, Nano Lett. **3**, 1691 (2003).

[933] R. D'Agosta, N. Sai, M. Di Ventra, *Local electron heating in nanoscale conductors*, Nano Lett. **6**, 2935 (2006).

[934] Z. Yang, M. Chshiev, M. Zwolak, Y.C. Chen, M. Di Ventra, *Role of heating and current-induced forces in the stability of atomic wires*, Phys. Rev. B **71**, 041402 (2005).

[935] A. Pecchia, G. Romano, A. Di Carlo, *Theory of heat dissipation in molecular electronics*, Phys. Rev. B **75**, 035401 (2007).

[936] G. Romano, A. Pecchia, A. Di Carlo, *Coupling of molecular vibrons with contact phonon reservoirs*, J. Phys.: Condens. Matter **19**, 215207 (2007).

[937] G. Schulze, K.J. Franke, A. Gagliardi, G. Romano, C.S. Lin, A.L. Rosa, T.A. Niehaus, Th. Frauenheim, A. Di Carlo, A. Pecchia, J.I. Pascual, *Resonant electron heating and molecular phonon cooling in single C_{60} junctions*, Phys. Rev. Lett. **100**, 136801 (2008).

[938] Z. Ioffe, T. Shamai, A. Ophir, G. Noy, I. Yutsis, K. Kfir, O. Cheshnovsky, Y. Selzer, *Detection of heating in current-carrying molecular junctions by Raman scattering*, Nature Nanotech. **3**, 727 (2008).

[939] M. Oron-Carl and R. Krupke, *Raman Spectroscopic evidence for hot-phonon generation in electrically biased carbon nanotubes*, Phys. Rev. Lett. **100**, 127401 (2008).

[940] J.M. Ziman, *Electrons and Phonons: The Theory of Transport Phenomena in Solids*, (Oxford Classic Texts in the Physical Sciences), (Oxford University Press, Oxford, UK, 2001).

[941] O.I. Shklyarevskii, A.G. Jansen, J.G. Hermsen, P. Wyder, *Thermoelectric voltage between identical metals in a point-contact configuration*, Phys. Rev. Lett. **57**, 1374 (1986).

[942] B. Ludoph and J.M. van Ruitenbeek, *Thermopower of atomic-size metallic contacts*, Phys. Rev. B **59**, 12290 (1999).

[943] E.N. Bogachek, A.G. Scherbakov, U. Landman, *Thermopower of quantum nanowires in a magnetic field*, Phys. Rev. B **54**, 11094 (1996).

[944] M. Paulsson and S. Datta, *Thermoelectric effect in molecular electronics*,

Phys. Rev. B **67**, 241403 (2003).

[945] D. Segal, *Thermoelectric effect in molecular junctions: A tool for revealing transport mechanisms*, Phys. Rev. B **72**, 165426 (2005).

[946] S.-H. Ke, W. Yang, S. Curtarolo, H.U. Baranger, *Thermopower of molecular junctions: An ab initio study*, Nano Lett. **9**, 1011 (2009).

[947] J.A. Malen, P. Doak, K. Baheti, T.D. Tilley, R.A. Segalman, A. Majumdar, *Identifying the length dependence of orbital alignment and contact coupling in molecular heterojunctions*, Nano Lett. **9**, 1164 (2009).

[948] Z. Crljen, A. Grigoriev, G. Wendin, K. Stokbro, *Nonlinear conductance in molecular devices: Molecular length dependence*, Phys. Rev. B **71**, 165316 (2005).

[949] C.C. Kaun and H. Guo, *Resistance of alkanethiol molecular wires*, Nano Lett. **3**, 1521 (2003).

[950] Y.X. Zhou, F. Jiang, H. Chen, R. Note, H. Mizuseki, Y. Kawazoe, *First-principles study of length dependence of conductance in alkanedithiols*, J. Chem. Phys. **128**, 044704 (2008).

[951] C. G. Zeng, B. Li, B. Wang, H.Q. Wang, K.D. Wang, J.L. Yang, J.G. Hou, Q.S. Zhu, *What can a scanning tunneling microscope image do for the insulating alkanethiol molecules on Au(111) substrates?*, J. Chem. Phys. **117**, 851 (2002).

[952] M. Galperin and A. Nitzan, *Optical properties of current carrying molecular wires*, J. Chem. Phys. **124**, 234709 (2006).

[953] A.M. Nowak and R.L. McCreery, *In situ Raman spectroscopy of bias-induced structural changes in nitroazobenzene molecular electronic junctions*, J. Am. Chem. Soc. **126**, 16621 (2004).

[954] J. Tian, B. Liu, X.L. Li, Z.L. Yang, B. Ren, S.T. Wu, N.J. Tao and Z.Q. Tian, *Study of molecular junctions with combined surface-enhanced Raman and mechanically controllable break junction method*, J. Am. Chem. Soc. **128**, 14748 (2006).

[955] D.R. Ward, N.K. Grady, C.S. Levin, N.J. Halas, Y. Wu, P. Nordlander, D. Natelson, *Electromigrated nanoscale gaps for surface-enhanced Raman spectroscopy*, Nano Lett. **7**, 1396 (2007).

[956] D.R. Ward, N.J. Halas, J.W. Ciszek, J.M. Tour, Y. Wu, P. Nordlander, D. Natelson, *Simultaneous measurements of electronic conduction and Raman response in molecular junctions*, Nano Lett. **8**, 919 (2008).

[957] P. Mühlschlegel, H.-J. Eisler, O.J.F. Martin, B. Hecht, D.W. Pohl, *Resonant optical antennas*, Science **308**, 1607 (2005).

[958] M. Moskovits, *Surface-enhanced Raman spectroscopy*, J. Raman Spectrosc. **36**, 485 (2005).

[959] M. Galperin, M.A. Ratner, A. Nitzan, *Raman scattering from nonequilibrium molecular conduction junctions*, Nano Lett. **9**, 758 (2009).

[960] M. Galperin, M.A. Ratner, A. Nitzan, *Raman scattering in current-carrying molecular junctions*, J. Chem. Phys. **130**, 144109 (2009).

[961] B.N. Persson and A. Baratoff, *Theory of photon emission in electron tunneling to metallic particles*, Phys. Rev. Lett. **68**, 3224 (1992).

[962] M. Galperin, A. Nitzan, M.A. Ratner, *Molecular transport junctions: Cur-*

rent from electronic excitations in the leads, Phys. Rev. Lett. **96**, 166803 (2006).

[963] A.H. Dayem and R.J. Martin, *Quantum interaction of microwave radiation with tunneling between superconductors*, Phys. Rev. Lett. **8**, 246 (1962).

[964] P.H. Cutler, T.E. Feuchtwang, T.T. Tsong, H. Nguyen, A.A. Lucas, *Proposed use of a scanning-tunneling-microscope tunnel junction for the measurement of a tunneling time*, Phys. Rev. B **35**, 7774 (1987).

[965] M. Völcker, W. Krieger, H. Walther, *Laser-driven scanning tunneling microscope*, Phys. Rev. Lett. **66**, 1717 (1991).

[966] M. Chauvin, P. vom Stein, H. Pothier, P. Joyez, M.E. Huber, D. Esteve, C. Urbina, *Superconducting atomic contacts under microwave irradiation*, Phys. Rev. Lett. **97**, 067006 (2006).

[967] J.C. Cuevas, J. Heurich, A. Martin-Rodero, A. Levy Yeyati, G. Schön, *Subharmonic Shapiro steps and assisted tunneling in superconducting point contacts*, Phys. Rev. Lett. **88**, 157001 (2002).

[968] A. Tikhonov, R.D. Coalson, Y. Dahnovsky, *Calculating electron transport in a tight binding model of a field-driven molecular wire: Floquet theory approach*, J. Chem. Phys. **116**, 10909 (2002).

[969] A. Tikhonov, R.D. Coalson, Y. Dahnovsky, *Calculating electron current in a tight-binding model of a field-driven molecular wire: Application to xylyl-dithiol*, J. Chem. Phys. **117**, 567 (2002).

[970] J.K. Viljas, F. Pauly and J.C. Cuevas, *Photoconductace of organic single-molecule contacts*, Phys. Rev. B **76**, 033408 (2007).

[971] X.W. Tu, J.H. Lee, W. Ho, *Atomic-scale rectification at microwave frequency*, J. Chem. Phys. **124**, 021105 (2006).

[972] C. Meyer, J.M. Elzerman, L.P. Kouwenhoven, *Photon-assisted tunneling in a carbon nanotube quantum dot*, Nano Lett. **7**, 295 (2007).

[973] D.C. Guhr, D. Rettinger, J. Boneberg, A. Erbe, P. Leiderer, E. Scheer, *Influence of laser light on electronic transport through atomic-size contacts*, Phys. Rev. Lett. **99**, 086801 (2007).

[974] N. Ittah, G. Noy, I. Yutsis, Y. Selzer, *Measurement of electronic transport through $1G_0$ gold contacts under laser irradiation*, Nano Lett. **9**, 1615 (2009).

[975] N. Ittah, I. Yutsis, Y. Selzer, *Fabrication of highly stable configurable metal quantum point contacts*, Nano Lett. **8**, 3922 (2008).

[976] S.W. Wu, N. Ogawa, W. Ho, *Atomic-scale coupling of photons to single-molecule junctions*, Science **312**, 1362 (2006).

[977] M. Galperin and A. Nitzan, *Current-induced light emission and light-induced current in molecular-tunneling junctions*, Phys. Rev. Lett. **95**, 206802 (2005).

[978] A. Keller, O. Atabek, M. Ratner, V. Mujica, *Laser-assisted conductance of molecular wires*, J. Phys. B **35**, 4981 (2002).

[979] I. Urdaneta, A. Keller, O. Atabek, V. Mujica, *Laser-assisted conductance of molecular wires: two-photon contributions*, Int. J. Quantum Chem. **99**, 460 (2004).

[980] S. Kohler, J. Lehmann, S. Camalet, P. Hänggi, *Resonant laser excitation of molecular wires*, Israel J. Chem. **42**, 135 (2002).

[981] S. Kohler, J. Lehmann, M. Strass, P. Hänggi, *Molecular wires in electromagnetic fields*, Adv. Solid State Phys. **44**, 157 (2004).

[982] J. Lehmann, S. Kohler, P. Hänggi, A. Nitzan, *Molecular wires acting as coherent quantum ratchets*, Phys. Rev. Lett. **88**, 228305 (2002).

[983] J. Lehmann, S. Kohler, P. Hänggi, A. Nitzan, *Rectification of laser-induced electronic transport through molecules*, J. Chem. Phys. **118**, 3283 (2003).

[984] S. Kohler, J. Lehmann, P. Hänggi, *Controlling currents through molecular wires*, Superlattice Microstruct. **34**, 419 (2004).

[985] F. Grossmann, T. Dittrich, P. Jung, P. Hänggi, *Coherent destruction of tunneling*, Phys. Rev. Lett. **67**, 516 (1991).

[986] J. Lambe and S.L. McCarthy, *Light emission from inelastic electron tunneling*, Phys. Rev. Lett. **37**, 923 (1976).

[987] R. Berndt, R. Gaisch, W.D. Schneider, J.K. Gimzewski, B. Reihl, R.R. Schlittler, M. Tschudy, *Atomic resolution in photon emission induced by a scanning tunneling microscope*, Phys. Rev. Lett. **74**, 102 (1995).

[988] Y. Uehara, T. Fujita, S. Ushioda, *Scanning tunneling microscope light emission spectra of Au(110)-(2×1) with atomic spatial resolution*, Phys. Rev. Lett. **83**, 2445 (1999).

[989] A. Downes and M.E. Welland, *Photon emission from Si(111)-(7×7) induced by scanning tunneling microscopy: Atomic scale and material contrast*, Phys. Rev. Lett. **81**, 1857 (1998).

[990] R. Berndt and J.K. Gimzewski, *Photon emission in scanning tunneling microscopy: Interpretation of photon maps of metallic systems*, Phys. Rev. B **48**, 4746 (1993).

[991] R. Berndt, R. Gaisch, J.K. Gimzewski, B. Reihl, R.R. Schlittler, W.D. Schneider, M. Tschudy, *Photon emission at molecular resolution induced by a scanning tunneling microscope*, Science **262**, 1425 (1993).

[992] G. Hoffmann, L. Libioulle, R. Berndt, *Tunneling-induced luminescence from adsorbed organic molecules with submolecular lateral resolution*, Phys. Rev. B **65**, 212107 (2002).

[993] Y. Uehara, T. Matsumoto, S. Ushioda, *Identification of O atoms on a Cu(110) surface by scanning tunneling microscope light emission spectra*, Phys. Rev. B **66**, 075413 (2002).

[994] X.H. Qiu, G.V. Nazin, W. Ho, *Vibrationally resolved fluorescence excited with submolecular precision*, Science **299**, 542 (2003).

[995] Z.-C. Dong, X.-L. Guo, A.S. Trifonov, P.S. Dorozhkin, K. Miki, K. Kimura, S. Yokoyama, S. Mashiko, *Vibrationally resolved fluorescence from organic molecules near metal surfaces in a scanning tunneling microscope*, Phys. Rev. Lett. **92**, 086801 (2004).

[996] J. Buker and G. Kirczenow, *Theoretical study of photon emission from molecular wires*, Phys. Rev. B **66**, 245306 (2002).

[997] U. Harbola, J.B. Maddox, S. Mukamel, *Many-body theory of current-induced fluorescence in molecular junctions*, Phys. Rev. B **73**, 075211 (2006).

[998] J. Buker and G. Kirczenow, *Understanding the electroluminescence emitted by single molecules in scanning tunneling microscopy experiments*, Phys. Rev. B **78**, 125107 (2008).

[999] P. Johansson, R. Monreal, P. Apell, *Theory for light emission from a scanning tunneling microscope*, Phys. Rev. B **42**, 9210 (1990).

[1000] Y. Uehara, Y. Kimura, S. Ushioda, K. Takeuchi, *Theory of visible light emission from scanning tunneling microscope*, Jpn. J. Appl. Phys. **31**, 2465 (1992).

[1001] M. Irie, *Photochromism: Memories and Switches*, Chem. Rev. **100**, 1685 (2000).

[1002] B.L. Feringa (Ed.), *Molecular Switches*, (Wiley-VCH, Weinheim, D, 2001).

[1003] S.J. van der Molen, H. van der Vegte, T. Kudernac, I. Amin, B.L. Feringa, B.J. van Wees, *Stochastic and photochromic switching of diarylethenes studied by scanning tunnelling microscopy*, Nanotechnology **17**, 310 (2006).

[1004] J. Li, G. Speyer, O.F. Sankey, *Conduction switching of photochromic molecules*, Phys. Rev. Lett. **93**, 248302 (2004).

[1005] M. Zhuang, M. Ernzerhof, *Mechanism of a molecular electronic photoswitch*, Phys. Rev. B **72**, 073104 (2005).

[1006] M. Kondo, T. Tada and K. Yoshizawa, *A theoretical measurement of the quantum transport through an optical molecular switch*, Chem. Phys. Lett. **412**, 55 (2005).

[1007] J. He, F. Chen, P.A. Liddell, J. Andreasson, S.D. Straight, D. Gust, T.A. Moore, A.L. Moore, J. Li, O.F. Sankey and S.M. Lindsay, *Switching of a photochromic molecule on gold electrodes: Single-molecule measurements*, Nanotechnology **16**, 695 (2005).

[1008] N. Katsonis, T. Kudernac, M. Walko, S.J. van der Molen, B.J. van Wees, B.L. Feringa, *Reversible switching of photochromic molecules self-assembled on gold*, Adv. Mater. **18**, 1397 (2006).

[1009] A.C. Whalley, M.L. Steigerwald, X. Guo, C. Nuckolls, *Reversible switching in molecular electronic devices*, J. Am. Chem. Soc. **129**, 12590 (2007).

[1010] A.J. Kronemeijer, H.B. Akkerman, T. Kudernac, B.J. van Wees, B.L. Feringa, P.W.M. Blom, B. de Boer, *Reversible conductance switching in molecular devices*, Adv. Mater. **20**, 1467 (2008).

[1011] M. Del Valle, R. Gutiérrez, C. Tejedor, G. Cuniberti, *Tuning the conductance of a molecular switch*, Nat. Nanotechnol. **2**, 176 (2007).

[1012] M. Alemani, M.V. Peters, S. Hecht, K.-H. Rieder, F. Moresco, L. Grill, *Electric field-induced isomerization of azobenzene by STM*, J. Am. Chem. Soc. **128**, 14446 (2006).

[1013] M.J. Comstock, N. Levy, A. Kirakosian, J. Cho, F. Lauterwasser, J.H. Harvey, D.A. Strubbe, J.M.J. Fréchet, D. Trauner, S.G. Louie, M.F. Crommie, *Reversible photochemical switching of individual engineered molecules at a metallic surface*, Phys. Rev. Lett. **99**, 038301 (2007).

[1014] A.S. Kumar, T. Ye, T. Takami, B.-C. Yu, A.K. Flatt, J.M. Tour, P.S. Weiss, *Reversible photo-switching of single azobenzene molecules in controlled nanoscale environments*, Nano Lett. **6**, 1644 (2008).

[1015] J.M. Mativetsky, G. Pace, M. Elbing, M.A. Rampi, M. Mayor, P. Samori, *Azobenzes as light-controlled molecular electronic switches in nanoscale metal-molecule-metal junctions*, J. Am. Chem. Soc. **130**, 9192 (2008).

[1016] J.M. Tour, *Molecular electronics. Synthesis and testing of components*, Acc. Chem. Res. **33**, 791 (2000).

[1017] N. Weibel, S. Grunder, M. Mayor, *Functional molecules in electronic circuits*, Org. Biomol. Chem. **5**, 2343 (2007).

[1018] A. Nitzan, *A relationship between electron transfer rates and molecular conduction*, J. Phys. Chem. A **105**, 2677 (2001).

[1019] W. Lu and C.M. Lieber, *Nanoelectronics from the bottom up*, Nature Mat. **6**, 841 (2007).

[1020] J.-C. Charlier, X. Blase, S. Roche, *Electronic and transport properties of nanotubes*, Rev. Mod. Phys. **79**, 677 (2007).

[1021] P. Darancet, A. Ferretti, D. Mayou, V. Olevano, *Ab initio GW electron-electron interaction effects in quantum transport*, Phys. Rev. B **75**, 075102 (2007).

[1022] K.S. Thygesen and A. Rubio, *Conserving GW scheme for nonequilibrium quantum transport in molecular contacts*, Phys. Rev. B **77**, 115333 (2008).

[1023] K.S. Thygesen, *Impact of exchange-correlation effects on the I-V characteristics of a molecular junction*, Phys. Rev. Lett. **100**, 166804 (2008).

Index